CLINICAL APPLICATIONS OF NONCODING RNAs IN CANCER

CLINICAL APPLICATIONS OF NONCODING RNAs IN CANCER

Edited by

SUBASH CHANDRA GUPTA

Department of Biochemistry, Institute of Science, Banaras Hindu University, Varanasi, India; and Department of Biochemistry, All India Institute of Medical Sciences, Guwahati, India

KISHORE B. CHALLAGUNDLA

Department of Biochemistry and Molecular Biology & The Fred and Pamela Buffett Cancer Center, University of Nebraska Medical Center, Omaha, NE, United States

Academic Press is an imprint of Elsevier
125 London Wall, London EC2Y 5AS, United Kingdom
525 B Street, Suite 1650, San Diego, CA 92101, United States
50 Hampshire Street, 5th Floor, Cambridge, MA 02139, United States
The Boulevard, Langford Lane, Kidlington, Oxford OX5 1GB, United Kingdom

Copyright © 2022 Elsevier Inc. All rights reserved.

No part of this publication may be reproduced or transmitted in any form or by any means, electronic or mechanical, including photocopying, recording, or any information storage and retrieval system, without permission in writing from the publisher. Details on how to seek permission, further information about the Publisher's permissions policies and our arrangements with organizations such as the Copyright Clearance Center and the Copyright Licensing Agency, can be found at our website: www.elsevier.com/permissions.

This book and the individual contributions contained in it are protected under copyright by the Publisher (other than as may be noted herein).

Notices
Knowledge and best practice in this field are constantly changing. As new research and experience broaden our understanding, changes in research methods, professional practices, or medical treatment may become necessary.

Practitioners and researchers must always rely on their own experience and knowledge in evaluating and using any information, methods, compounds, or experiments described herein. In using such information or methods they should be mindful of their own safety and the safety of others, including parties for whom they have a professional responsibility.

To the fullest extent of the law, neither the Publisher nor the authors, contributors, or editors, assume any liability for any injury and/or damage to persons or property as a matter of products liability, negligence or otherwise, or from any use or operation of any methods, products, instructions, or ideas contained in the material herein.

British Library Cataloguing-in-Publication Data
A catalogue record for this book is available from the British Library

Library of Congress Cataloging-in-Publication Data
A catalog record for this book is available from the Library of Congress

ISBN: 978-0-12-824550-7

For Information on all Academic Press publications
visit our website at https://www.elsevier.com/books-and-journals

Publisher: Stacy Masucci
Acquisitions Editor: Rafael E. Teixeira
Editorial Project Manager: Susan Ikeda
Production Project Manager: Maria Bernard
Cover Designer: Mark Rogers
Cover Credit: Created with BioRender.com

Typeset by MPS Limited, Chennai, India

Contents

List of contributors xiii
Preface xvii

Chapter 1 Noncoding ribonucleic acid for pancreatic cancer therapy 1
Lusine Demirkhanyan and Christopher S. Gondi

1.1 Introduction 1
1.2 Experimental methods and tools for analyzing noncoding RNAs in pancreatic cancer patients 2
1.3 Bioinformatics for analyzing noncoding RNAs in pancreatic cancer patients 3
1.4 Noncoding RNA 3
1.5 Pancreatic cancer–specific microRNAs 4
1.6 Pancreatic cancer–associated lncRNAs 4
1.7 Pancreatic cancer–associated circular RNAs 6
1.8 Diagnostic microRNA markers of PDAC 7
1.9 Diagnostic lncRNA markers of PDAC 7
1.10 Diagnostic circular RNA markers of PDAC 9
1.11 Summary and conclusion 9
Acknowledgment 9
References 9

Chapter 2 Applications of noncoding RNAs in brain cancer patients 17
Małgorzata Grabowska, Julia O. Misiorek, Żaneta Zarębska and Katarzyna Rolle

2.1 Introduction 17
    2.1.1 Types of brain cancer 17
    2.1.2 Types of ncRNAs and mechanism of action 19
    2.1.3 Circular RNA 20
2.2 Data sets for noncoding RNAs analysis 21
    2.2.1 RNA-seq deposits and data sets 24
2.3 Expression of noncoding RNAs in brain cancer patients 24
    2.3.1 Sample types used for analyzing noncoding RNAs in brain 24
    2.3.2 Models to study ncRNA expression 24
    2.3.3 NcRNA expression profiles in cancer stem cells and their consequences on tumor development 26
2.4 Experimental methods and tools for analyzing noncoding RNAs in brain cancer patients 30
    2.4.1 Studying ncRNA interactions with targets by luciferase assays and ncRNA–protein interactions by immunoprecipitation 30
    2.4.2 Validations of RNA-seq results by qRT-PCR and fluorescence in situ hybridization methods 31
2.5 Noncoding RNAs as predictive marker for brain cancer patients 35
    2.5.1 Diagnostic potential of noncoding RNAs in brain cancer patients 35
    2.5.2 Prognostic potential of noncoding RNAs in brain cancer patients 37
2.6 Potential of noncoding RNAs in predicting chemoresistance and radioresistance in brain cancer patients 40
2.7 Therapeutic potential and targeting of ncRNAs in brain cancer patients—challenges and perspectives 45
2.8 Summary and conclusions 47
References 49

Chapter 3 Noncoding RNAs in patients with colorectal cancer 65
Mohammad Amin Kerachian

3.1 Introduction 65
3.2 Experimental methods and tools for analyzing noncoding RNAs in colorectal cancer patients 66
3.3 Microarray 67

3.4 Serial analysis of gene expression 68
3.5 Cap analysis gene expression 68
3.6 RNA sequencing 69
3.7 Dataset and bioinformatics for analyzing noncoding RNAs in colorectal cancer patients 70
3.8 Expression of noncoding RNAs in colorectal cancer patients 72
3.9 Sample types used for analyzing noncoding RNAs 76
3.10 Cell signaling pathways modulated by noncoding RNAs in colorectal cancer patients 77
    3.10.1 Wnt/β-catenin signaling pathway 77
    3.10.2 JAK (Janus Kinase)/STAT (Signal Transducer and Activator of Transcription) signaling pathway 79
    3.10.3 PI3K (Phosphoinositide 3-Kinase)/PTEN (Phosphatase and Tensin Homolog)/AKT (AK Mouse Plus Transforming or Thymoma)/mTOR (Mechanistic Target of Rapamycin) signaling pathway 79
    3.10.4 Ras/MAPK (Mitogen-activated protein kinase)-signaling pathway 80
    3.10.5 p53 signaling pathway 80
    3.10.6 Notch signaling pathway 81
    3.10.7 NF-κB (Nuclear Factor Kappa B) signaling pathway 81
3.11 Several other mechanisms 82
    3.11.1 Caspase cleavage cascade 82
    3.11.2 Chemokine signaling 82
    3.11.3 Interleukin pathway 82
3.12 Clinical applications of noncoding RNAs as biomarkers in patients with colorectal cancer 82
    3.12.1 Noncoding RNAs as predictive markers for colorectal cancer patients 82
3.13 Diagnostic potential of noncoding RNAs in colorectal cancer patients 83
3.14 Prognostic potential of noncoding RNAs in colorectal cancer patients 84
3.15 Therapeutic potential of noncoding RNAs in colorectal cancer patients 85
3.16 Potential of noncoding RNAs in predicting chemoresistance and radioresistance in colorectal cancer patients 85
3.17 Conclusion 86
References 87

Chapter 4 Applications of noncoding ribonucleic acids in multiple myeloma patients 97

Simone Zocchi, Antoine David, Michele Goodhardt and David Garrick

4.1 Introduction 97
4.2 Samples and experimental methods for the analysis of noncoding RNAs in multiple myeloma patients 99
4.3 Datasets analyzing noncoding RNAs in multiple myeloma patients 100
    4.3.1 Datasets profiling expression of short noncoding RNAs in multiple myeloma 100
    4.3.2 Datasets profiling expression of long noncoding RNAs in multiple myeloma 107
4.4 Noncoding RNAs implicated in the etiology of multiple myeloma 108
    4.4.1 MicroRNAs implicated in the etiology of multiple myeloma 108
    4.4.2 Long noncoding RNAs implicated in multiple myeloma 113
    4.4.3 Other ncRNAs implicated in multiple myeloma 118
    4.4.4 Interactions between ncRNAs in multiple myeloma 119
4.5 Cell signaling pathways modulated by noncoding RNAs in multiple myeloma 119
    4.5.1 IL6/JAK/STAT signaling and noncoding RNAs in multiple myeloma 120
    4.5.2 Noncoding RNAs and the p53 pathway in multiple myeloma 121
    4.5.3 The influence of noncoding RNAs on PI3K/AKT/mTOR, MAPK, and NFκB signaling in multiple myeloma 123
4.6 Noncoding RNAs affecting interactions with the bone marrow niche 124
4.7 Noncoding RNAs as diagnostic and prognostic biomarkers in multiple myeloma 125
4.8 Therapeutic potential of noncoding RNAs in multiple myeloma patients 130
4.9 Summary and conclusion 131
Acknowledgments 132
References 132

Chapter 5 Clinical applications of noncoding RNAs in lung cancer patients 141

Santosh Kumar, Naveen Kumar Vishvakarma and Ajay Kumar

Abbreviations 141
5.1 Introduction 142
5.2 Experimental methods and tools for analyzing ncRNAs in lung cancer patients 143
5.3 Datasets and informatics for analyzing ncRNAs in lung cancer patients 143
5.4 Expression of ncRNAs in lung cancer patients 144
   5.4.1 Micro-RNA 145
   5.4.2 Long noncoding RNA 146
   5.4.3 tRNA-derived small RNA 146
   5.4.4 Circular RNA 146
   5.4.5 Small nucleolar RNA 147
   5.4.6 PIWI-interacting RNA 147
   5.4.7 Natural antisense transcripts 147
   5.4.8 Transcribed ultraconserved region 148
5.5 Sample types used for analyzing ncRNAs 148
   5.5.1 Blood 148
   5.5.2 Bronchoalveolar lavage fluid 150
   5.5.3 Sputum 151
   5.5.4 Tumor biopsies 152
5.6 Cell signaling pathways modulated by ncRNAs in lung cancer patients 153
5.7 NcRNAs as predictive markers for lung cancer patients 155
5.8 Diagnostic potential of ncRNAs in lung cancer patients 158
   5.8.1 LncRNAs as diagnostic markers 158
   5.8.2 miRNAs as diagnostic markers 159
5.9 Prognostic potential of ncRNAs in lung cancer patients 159
   5.9.1 LncRNAs as prognostic biomarkers 159
   5.9.2 miRNAs as prognostic biomarkers 160
5.10 Therapeutic potential of ncRNAs in lung cancer patients 161
   5.10.1 Therapeutic potential of lncRNAs 161
   5.10.2 Therapeutic potential of miRNAs 161
5.11 Potential of ncRNAs in predicting chemoresistance and radioresistance in lung cancer patients 162
   5.11.1 NcRNAs and their role in chemoresistance 162
   5.11.2 NcRNAs and their role in radioresistance 164
5.12 Summary and conclusion 165
Acknowledgments 166
References 166

Chapter 6 Noncoding RNAs in intraocular tumor patients 177

Daniel Fernandez-Diaz, Beatriz Fernandez-Marta, Nerea Lago-Baameiro, Paula Silva-Rodríguez, Laura Paniagua, María José Blanco-Teijeiro, María Pardo, Antonio Piñeiro and Manuel F. Bande

6.1 Introduction 177
6.2 Retinoblastoma 178
   6.2.1 Introduction 178
   6.2.2 Long noncoding RNAs and circular RNAs in retinoblastoma 178
   6.2.3 MicroRNAs in retinoblastoma 184
   6.2.4 Therapeutic potential of noncoding RNAs in retinoblastoma 189
   6.2.5 Conclusion 190
6.3 Uveal melanoma 190
   6.3.1 Introduction 190
   6.3.2 Long noncoding RNAs in uveal melanoma 191
   6.3.3 MicroRNAs in uveal melanoma 194
   6.3.4 Conclusion 199
6.4 Conclusion 200
References 201

Chapter 7 Applications of noncoding RNAs in renal cancer patients 211

Eman A. Toraih, Jessica A. Sedhom, Muhib Haidari and Manal S. Fawzy

7.1 Introduction 211
7.2 Datasets and informatics for analyzing noncoding RNAs in renal cancer patients 213
   7.2.1 Datasets and informatics for analyzing microRNAs in renal cancers 213
   7.2.2 Datasets and informatics for analyzing long noncoding RNAs in renal cancers 214

- 7.2.3 Datasets and informatics for analyzing circular RNAs in renal cancers 214
- 7.3 Expression of noncoding RNAs in renal cancer patients 228
  - 7.3.1 Expression of microRNAs in renal cancer patients 228
  - 7.3.2 Expression of long noncoding RNAs in renal cancer patients 228
  - 7.3.3 Expression of circular RNAs in renal cancer patients 233
- 7.4 Cell signaling pathways modulated by noncoding RNAs in renal cancer patients 240
  - 7.4.1 Cell signaling pathways modulated by microRNAs in renal cancer patients 241
  - 7.4.2 Cell signaling pathways modulated by long noncoding RNAs in renal cancer patients 242
  - 7.4.3 Cell signaling pathways modulated by circular RNAs in renal cancer patients 243
- 7.5 Diagnostic potential of noncoding RNAs in renal cancer patients 243
  - 7.5.1 Diagnostic potential of microRNAs in renal cancer patients 243
  - 7.5.2 Diagnostic potential of long noncoding RNAs in renal cancer patients 245
  - 7.5.3 Diagnostic potential of circular RNAs in renal cancer patients 245
- 7.6 Prognostic potential of noncoding RNAs in renal cancer patients 246
  - 7.6.1 Prognostic potential of miRNAs in RCC 246
  - 7.6.2 Prognostic potential of long noncoding RNAs in RCC 248
  - 7.6.3 Prognostic potential of circular RNAs in RCC 252
- 7.7 Therapeutic potential of noncoding RNAs in renal cancer patients 254
  - 7.7.1 Therapeutic potential of microRNAs in renal cancer patients 254
  - 7.7.2 Therapeutic potential of long noncoding RNAs in renal cancer patients 254
  - 7.7.3 Therapeutic potential of circular RNAs in renal cancer patients 255
- 7.8 Potential of noncoding RNAs in predicting chemoresistance and radioresistance in renal cancer patients 256
  - 7.8.1 Potential of microRNAs in predicting chemoresistance/radioresistance in renal cancer patients 256
  - 7.8.2 Potential of long noncoding RNAs in predicting chemoresistance/radioresistance in renal cancer patients 257
  - 7.8.3 Potential of ciRNAs in predicting chemoresistance/radioresistance in renal cancer patients 257
- 7.9 Summary and conclusion 258
- References 259

Chapter 8 Clinical significance of long noncoding RNAs in breast cancer patients 285

Nikee Awasthee, Anusmita Shekher, Vipin Rai, Pranjal K. Baruah, Anurag Sharma, Kishore B. Challagundla and Subash C. Gupta

- Abbreviations 285
- 8.1 Introduction 286
- 8.2 Potential of lncRNAs in the diagnosis of breast cancer 287
- 8.3 Potential of lncRNAs in the prognosis of breast cancer 288
- 8.4 Potential of lncRNAs in breast cancer therapy 289
- 8.5 Potential of lncRNAs in predicting breast cancer patient's response to therapeutics 290
- 8.6 Potential of lncRNAs in predicting chemoresistance and radioresistance in breast cancer patients 290
- 8.7 Experimental methods and tools for analyzing noncoding RNAs in cancer patients 291
- 8.8 Summary and conclusion 292
- Acknowledgment 292
- References 292

Chapter 9 Noncoding ribonucleic acids in gastric cancer patients 297

Rachel Sexton, Najeeb Al-Hallak, Bayan Al-Share, Anteneh Tesfaye and Asfar S. Azmi

- 9.1 Introduction 297
  - 9.1.1 MicroRNAs 298
  - 9.1.2 PiwiRNAs 298
  - 9.1.3 Circular RNAs 299

- 9.1.4 Long noncoding RNAs 299
- 9.2 Experimental methods and tools for analyzing noncoding RNAs in gastric cancer patients 300
- 9.3 Expression of noncoding RNAs in gastric cancer patients 302
- 9.4 Sample types used for analyzing noncoding RNAs (tumor biopsies, liquid biopsies, etc.) 303
- 9.5 Cell signaling pathways modulated by noncoding RNAs in gastric cancer patients 304
- 9.6 Noncoding RNAs as prognostic and predictive marker for gastric cancer patients 306
- 9.7 Diagnostic value of small noncoding RNAs in gastric cancer 307
- 9.8 Potential of noncoding RNAs in predicting chemotherapy resistance and radiotherapy resistance in gastric cancer patients 308
- 9.9 Summary and conclusion 309
- References 310

Chapter 10 Noncoding RNAs in prostate cancer patients 315

Atiyeh Al-e-Ahmad, Nahid Neamati, Emadoddin Moudi, Simin Younesi and Hadi Parsian

- 10.1 Introduction 315
- 10.2 Experimental methods and tools for analyzing ncRNAs in prostate cancer patients 317
  - 10.2.1 NcRNA profiling in prostate cancer 317
  - 10.2.2 Microarray 317
  - 10.2.3 RNA sequencing 317
  - 10.2.4 ncRNA validation in prostate cancer 317
  - 10.2.5 Investigation of ncRNA interactions 318
  - 10.2.6 Secondary structures of ncRNAs 319
- 10.3 Datasets and informatics for analyzing noncoding RNAs 319
  - 10.3.1 Homology-based methods 319
  - 10.3.2 De novo methods for ncRNA predictions 320
  - 10.3.3 Special miRNA, circularRNA, and lncRNA databases 321
- 10.4 Sample types used for analyzing ncRNAs (tumor biopsies, liquid biopsies, etc.) 321
  - 10.4.1 New aspects of RNA-based biomarkers discovery in prostate cancer 322
  - 10.4.2 Prostate cancer biospecimen repositories 322
- 10.5 Cell signaling pathways modulated by ncRNAs in prostate cancer 322
  - 10.5.1 Phosphatase and tensin homolog/phosphoinositide 3-kinase/AkT/mammalian target of rapamycin pathway 323
  - 10.5.2 Mitogen-activated protein kinase pathway 324
  - 10.5.3 c-Myc pathway 325
  - 10.5.4 AR signaling pathway 325
- 10.6 NcRNAs as biomarkers for prostate cancer 326
  - 10.6.1 Diagnostic value 327
  - 10.6.2 Prognostic value 327
  - 10.6.3 Predictive value 328
- 10.7 Therapeutic potential of ncRNAs in prostate cancer patients 330
- 10.8 Potential of ncRNAs in predicting chemo-resistance and radioresistance in prostate cancer patients 331
  - 10.8.1 Chemo-resistance 331
  - 10.8.2 Radioresistance 332
- 10.9 Conclusion 332
- References 333

Chapter 11 Noncoding RNAs in liver cancer patients 343

Julie Sanceau and Angélique Gougelet

- Abbreviations 343
- 11.1 Introduction 345
  - 11.1.1 Liver functions 345
  - 11.1.2 Liver cancers 346
- 11.2 NcRNA roles in liver development and functions 348
  - 11.2.1 MicroRNAs 348
  - 11.2.2 Long noncoding RNAs 350
- 11.3 Noncoding RNA detection 351
  - 11.3.1 Tissular versus circulating ncRNAs 351

    11.3.2 Methods of ncRNA analyses 352
11.4 Expression of ncRNAs in liver cancers 355
    11.4.1 MicroRNAs 355
    11.4.2 Long noncoding RNAs 358
11.5 Noncoding RNA relevance in liver cancer diagnosis and prognosis 360
    11.5.1 Noncoding RNA as potential diagnostic tools 360
    11.5.2 Prognostic potential of noncoding RNAs in liver cancer patients 363
    11.5.3 Noncoding RNAs as predictive markers for liver cancer patients 364
    11.5.4 Roles of microRNA in drug resistance 365
    11.5.5 Therapeutic potential of noncoding RNAs in liver cancer patients 366
11.6 Summary and conclusion 369
Acknowledgments 370
References 370

Chapter 12 Noncoding ribonucleic acids in gallbladder cancer patients 391

Bela Goyal, Tarunima Gupta, Sweety Gupta and Amit Gupta

12.1 Introduction 391
12.2 MiRNAs in gallbladder carcinoma 393
    12.2.1 Biogenesis and biological functions of miRNA 393
    12.2.2 MicroRNAs in pathogenesis and as a therapeutic target in gallbladder carcinoma 393
    12.2.3 MicroRNAs as biomarkers in gallbladder carcinoma 396
12.3 LncRNAs in gallbladder carcinoma 397
    12.3.1 Biogenesis and biological functions of LncRNA 398
    12.3.2 LncRNAs in pathogenesis and as a therapeutic target in gallbladder carcinoma 398
    12.3.3 LncRNA as a biomarker 401
12.4 PiRNAs in gallbladder carcinoma 402
12.5 Limitations of clinical utility of ncRNAs in gallbladder carcinoma 403
12.6 Conclusion 403
References 404

Chapter 13 Clinical implications of noncoding ribonucleic acids in neuroblastoma patients 409

Anup S. Pathania, Oghenetejiri V. Smith, Philip Prathipati, Subash C. Gupta and Kishore B. Challagundla

13.1 Introduction 409
13.2 Types of noncoding RNAs 410
    13.2.1 MicroRNAs 410
    13.2.2 Long noncoding RNAs 410
    13.2.3 P-element-induced wimpy testis (Piwi)-interacting RNAs 411
    13.2.4 Circular RNAs 411
13.3 Role of noncoding RNAs in neuroblastoma growth and development 412
    13.3.1 MicroRNAs 412
    13.3.2 Long noncoding RNAs 413
    13.3.3 P-element-induced wimpy testis (Piwi)-interacting RNAs 415
    13.3.4 Circular RNAs 415
13.4 Clinical significance of noncoding RNAs in neuroblastoma 416
    13.4.1 MicroRNAs 416
    13.4.2 Long noncoding RNAs 417
    13.4.3 P-element-induced wimpy testis (Piwi)-interacting RNAs 418
    13.4.4 Circular RNAs 419
13.5 Therapeutic implications and targeting strategies for noncoding RNAs in neuroblastoma 419
    13.5.1 Therapeutic potential of noncoding RNAs in neuroblastoma 419
    13.5.2 Targeting strategies for noncoding RNAs 420
13.6 Conclusion 422
Acknowledgments 423
Conflict of interest 423
References 424

Chapter 14 Potential clinical application of lncRNAs in pediatric cancer 433

Ravindresh Chhabra, Priyasha Neyol, Sonali Bazala, Ipsa Singh, Masang Murmu, Uttam Sharma, Tushar Singh Barwal and Aklank Jain

14.1 Introduction 433
    14.1.1 Pediatric cancer 433
    14.1.2 Long noncoding RNA 434

14.2 Experimental and bioinformatics tools for studying lncRNAs 435
14.3 LncRNAs in pediatric cancer 437
    14.3.1 Leukemia 437
    14.3.2 Neuroblastoma 441
    14.3.3 Osteosarcoma 442
    14.3.4 Retinoblastoma 443
    14.3.5 Wilms tumor 443
14.4 Conclusion and perspectives 444
Acknowledgment 444
References 445

Index 449

List of contributors

Atiyeh Al-e-Ahmad Student Research Committee, Babol University of Medical Sciences, Babol, Iran; Cellular and Molecular Biology Research Center, Health Research Institute, Babol University of Medical Sciences, Babol, Iran; Department of Clinical Biochemistry, Babol University of Medical Sciences, Babol, Iran

Najeeb Al-Hallak Department of Oncology, Wayne State University School of Medicine, Detroit, MI, United States

Bayan Al-Share Department of Oncology, Wayne State University School of Medicine, Detroit, MI, United States

Nikee Awasthee Department of Biochemistry, Institute of Science, Banaras Hindu University, Varanasi, India

Asfar S. Azmi Department of Oncology, Wayne State University School of Medicine, Detroit, MI, United States

Manuel F. Bande Department of Ophthalmology, University Hospital of Santiago de Compostela, Santiago de Compostela, Spain; Intraocular Tumors in Adults, Health Research Institute of Santiago de Compostela (IDIS), Santiago de Compostela, Spain

Pranjal K. Baruah Department of Applied Sciences, GUIST, Gauhati University, Guwahati, India

Tushar Singh Barwal Department of Zoology, Central University of Punjab, Bathinda, India

Sonali Bazala Department of Zoology, Central University of Punjab, Bathinda, India

María José Blanco-Teijeiro Department of Ophthalmology, University Hospital of Santiago de Compostela, Santiago de Compostela, Spain; Intraocular Tumors in Adults, Health Research Institute of Santiago de Compostela (IDIS), Santiago de Compostela, Spain

Kishore B. Challagundla Department of Biochemistry and Molecular Biology & The Fred and Pamela Buffett Cancer Center, University of Nebraska Medical Center, Omaha, NE, United States; The Children's Health Research Institute, University of Nebraska Medical Center, Omaha, NE, United States

Ravindresh Chhabra Department of Biochemistry, Central University of Punjab, Bathinda, India

Antoine David INSERM U976, Saint-Louis Institute for Research, University of Paris, Paris, France

Lusine Demirkhanyan Department of Internal Medicine, University of Illinois College of Medicine Peoria, Peoria, IL, United States

Manal S. Fawzy Department of Medical Biochemistry and Molecular Biology, Faculty of Medicine, Suez Canal University, Ismailia, Egypt; Department of Biochemistry, Faculty of Medicine, Northern Border University, Arar, Saudi Arabia

Daniel Fernandez-Diaz Department of Ophthalmology, University Hospital of Santiago de Compostela, Santiago de Compostela, Spain; Intraocular Tumors in Adults, Health Research Institute of Santiago de Compostela (IDIS), Santiago de Compostela, Spain

Beatriz Fernandez-Marta Department of Ophthalmology, University Hospital of Santiago de Compostela, Santiago de Compostela, Spain

David Garrick INSERM U976, Saint-Louis Institute for Research, University of Paris, Paris, France

Christopher S. Gondi Department of Internal Medicine, University of Illinois College of Medicine Peoria, Peoria, IL, United States; Department of Surgery, University of Illinois College of Medicine Peoria, Peoria, IL, United States; Department of Health Science Education and Pathology, University of Illinois College of Medicine Peoria, Peoria, IL, United States

Michele Goodhardt INSERM U976, Saint-Louis Institute for Research, University of Paris, Paris, France

Angélique Gougelet Centre de Recherche des Cordeliers, Sorbonne Université, Inserm, Université de Paris, Team "Oncogenic Functions of Beta-Catenin Signaling in the Liver", Paris, France

Bela Goyal Department of Biochemistry, All India Institute of Medical Sciences, Rishikesh, India

Małgorzata Grabowska Department of Molecular Neurooncology, Institute of Bioorganic Chemistry, Polish Academy of Sciences, Poznań, Poland

Amit Gupta Department of General Surgery, All India Institute of Medical Sciences, Rishikesh, India

Subash C. Gupta Department of Biochemistry, Institute of Science, Banaras Hindu University, Varanasi, India; Department of Biochemistry, All India Institute of Medical Sciences, Guwahati, India

Sweety Gupta Department of Radiation Oncology, All India Institute of Medical Sciences, Rishikesh, India

Tarunima Gupta Department of Biochemistry, All India Institute of Medical Sciences, Rishikesh, India

Muhib Haidari Department of Surgery, Tulane University School of Medicine, New Orleans, LA, United States

Aklank Jain Department of Zoology, Central University of Punjab, Bathinda, India

Mohammad Amin Kerachian Department of Medical Genetics, Faculty of Medicine, Mashhad University of Medical Sciences, Mashhad, Iran

Ajay Kumar Department of Zoology, Institute of Science, Banaras Hindu University, Varanasi, India

Santosh Kumar Department of Life Science, National Institute of Technology, Rourkela, India

Nerea Lago-Baameiro Obesidomics Group, Health Research Institute of Santiago de Compostela (IDIS), Santiago de Compostela, Spain

Julia O. Misiorek Department of Molecular Neurooncology, Institute of Bioorganic Chemistry, Polish Academy of Sciences, Poznań, Poland

Emadoddin Moudi Department of Urology, Babol University of Medical Sciences, Babol, Iran

Masang Murmu Department of Zoology, Central University of Punjab, Bathinda, India

Nahid Neamati Cellular and Molecular Biology Research Center, Health Research Institute, Babol University of Medical Sciences, Babol, Iran; Department of Clinical Biochemistry, Babol University of Medical Sciences, Babol, Iran

Priyasha Neyol Department of Biochemistry, Central University of Punjab, Bathinda, India

Laura Paniagua Department of Ophthalmology, University Hospital of Coruña, La Coruña, Spain

María Pardo Intraocular Tumors in Adults, Health Research Institute of Santiago de Compostela (IDIS), Santiago de Compostela, Spain; Obesidomics Group, Health Research Institute of Santiago de Compostela (IDIS), Santiago de Compostela, Spain

Hadi Parsian Cellular and Molecular Biology Research Center, Health Research Institute, Babol University of Medical Sciences, Babol, Iran; Department of Clinical Biochemistry, Babol University of Medical Sciences, Babol, Iran

Anup S. Pathania Department of Biochemistry and Molecular Biology & The Fred and Pamela Buffett Cancer Center, University of Nebraska Medical Center, Omaha, NE, United States

Antonio Piñeiro Department of Ophthalmology, University Hospital of Santiago de Compostela, Santiago de Compostela, Spain; Intraocular Tumors in Adults, Health Research Institute of Santiago de Compostela (IDIS), Santiago de Compostela, Spain

Philip Prathipati Laboratory of Bioinformatics, National Institutes of Biomedical Innovation, Health and Nutrition, Osaka, Japan

Vipin Rai Department of Biochemistry, Institute of Science, Banaras Hindu University, Varanasi, India

Katarzyna Rolle Department of Molecular Neurooncology, Institute of Bioorganic Chemistry, Polish Academy of Sciences, Poznań, Poland

Julie Sanceau Centre de Recherche des Cordeliers, Sorbonne Université, Inserm, Université de Paris, Team "Oncogenic Functions of Beta-Catenin Signaling in the Liver", Paris, France

Jessica A. Sedhom Department of Surgery, Tulane University School of Medicine, New Orleans, LA, United States

Rachel Sexton Department of Oncology, Wayne State University School of Medicine, Detroit, MI, United States

Anurag Sharma Division of Environmental Health and Toxicology, Nitte (Deemed to Be University), Nitte University Centre for Science Education and Research (NUCSER), Mangalore, India

Uttam Sharma Department of Zoology, Central University of Punjab, Bathinda, India

Anusmita Shekher Department of Biochemistry, Institute of Science, Banaras Hindu University, Varanasi, India

Paula Silva-Rodríguez Intraocular Tumors in Adults, Health Research Institute of Santiago de Compostela (IDIS), Santiago de Compostela, Spain; Galician Public Foundation of Genomic Medicine, University Hospital of Santiago de Compostela, Santiago de Compostela, Spain

Ipsa Singh Department of Zoology, Central University of Punjab, Bathinda, India

Oghenetejiri V. Smith Department of Biochemistry and Molecular Biology & The Fred and Pamela Buffett Cancer Center, University of Nebraska Medical Center, Omaha, NE, United States

Anteneh Tesfaye Department of Oncology, Wayne State University School of Medicine, Detroit, MI, United States

Eman A. Toraih Department of Surgery, Tulane University School of Medicine, New Orleans, LA, United States; Genetics Unit, Department of Histology and Cell Biology, Faculty of Medicine, Suez Canal University, Ismailia, Egypt

Naveen Kumar Vishvakarma Department of Biotechnology, Guru Ghasidas Vishwavidyalaya, Bilaspur, India

Simin Younesi School of Health and Biomedical Sciences, RMIT University, Melbourne, VIC, Australia

Żaneta Zarębska Department of Molecular Neurooncology, Institute of Bioorganic Chemistry, Polish Academy of Sciences, Poznań, Poland

Simone Zocchi INSERM U976, Saint-Louis Institute for Research, University of Paris, Paris, France

Preface

Despite significant advances in understanding the pathogenesis, cancer continues to be the second leading cause of death worldwide. This is partly due to delayed diagnosis, poor prognosis, recurrence, and resistance mechanisms by cancer cells. Almost 98% of the human genome consists of noncoding sequences. However, most of the currently available cancer biomarkers and therapeutic targets are based on coding genes. Of the noncoding sequences, up to 90% are transcribed, producing a vast number of noncoding RNAs (ncRNAs). Two major types of ncRNAs are microRNAs (miRNAs) (18–22 nucleotides) and long noncoding RNAs (lncRNAs, ≥ 200 nucleotides). Although much is known about miRNAs, very little is known about lncRNAs. This is evident from the PubMed database where keywords "microRNA+cancer" produced 59,353 articles on July 6, 2021. However, a steep rise in the field of lncRNAs has been witnessed in the recent past. The advent of sensitive, high-throughput genomic technologies such as microarrays and next-generation sequencing has enabled to detection of novel transcripts, the vast majority of which are derived from noncoding sequences of human genome. As of July 6, 2021, more than 17,900 entries were listed in the PubMed database with the keywords "long noncoding RNA+cancer." Some lncRNAs function as tumor suppressors, some as an oncogene, some as both oncogene and tumor suppressors, while the functions of several others remain unknown. Recent studies suggest that miRNAs and lncRNAs can be used as a biomarker. The ncRNAs also offer the potential to be a novel class of therapeutic cancer targets. The strategies could be either to suppress oncogenic function or to activate the tumor-suppressive activity of specific ncRNAs.

This book is an effort to provide in-depth information on the potential of ncRNAs in cancer diagnosis, prognosis, and therapy. The focus is on two major types of ncRNAs, that is, miRNAs and lncRNAs. The chapters cover the clinical applications of ncRNAs in diverse cancer types such as pancreatic, brain, colorectal, lung, bladder, renal, breast, gastric, prostate, liver, gallbladder, pediatric, and in patients with multiple myeloma, intraocular tumor, and neuroblastoma. The experimental methods and tools for analyzing ncRNAs in these cancer types are covered. In addition, the experts have discussed several other emerging topics with reference to the clinical utility of ncRNAs, such as datasets and informatics for analysis, sample types being used, cell signaling pathways being modulated, and potential of ncRNAs in predicting chemoresistance and radioresistance.

We hope that the book will be useful to the readers. We thank the authors for their outstanding contribution to this book and apologize to those whose contributions could not be solicited due to space limitations.

Subash Chandra Gupta and
Kishore B. Challagundla

Noncoding ribonucleic acid for pancreatic cancer therapy

Lusine Demirkhanyan[1] *and Christopher S. Gondi*[1,2,3]

[1]Department of Internal Medicine, University of Illinois College of Medicine Peoria, Peoria, IL, United States [2]Department of Surgery, University of Illinois College of Medicine Peoria, Peoria, IL, United States [3]Department of Health Science Education and Pathology, University of Illinois College of Medicine Peoria, Peoria, IL, United States

1.1 Introduction

Various types of noncoding RNAs (ncRNA) have been identified and the classification of these RNAs continues to evolve with steady progress. Historically RNAs are considered templates for protein synthesis, and RNAs that did not code for protein synthesis were considered by-products of transcription. The synergism of large-scale sequence with powerful computing tools has helped unravel the cryptic role of RNAs. This large-scale genome analysis revealed that most of the so-called junk DNA is transcribed to what is now called ncRNA (Eddy, 2001). It is now known that ncRNAs are found in almost all biological processes, including pathologies such as cancer (Pavet, Portal, Moulin, Herbrecht, & Gronemeyer, 2011). In this review, we will focus on the roles of some of these ncRNAs involved in pancreatic cancers (PCs) and the potential these ncRNAs can have in PC therapy. Pancreatic ductal adenocarcinoma (PDAC) is the third most common cause of cancer deaths in the United States and accounts for over 95% of all PCs. The combined 1- and 5-year survival rates for PDAC are very poor, at 25% and 9%, respectively. A major hallmark of PC is tumor recurrence and extremely poor response to chemotherapy and is able to form metastatic tumors indistinguishable from the parental tumors and contribute to chemo resistance (Clarke, Fuller, 2006; Dean, Fojo, & Bates, 2005; Dingli, Michor, 2006; Reya, Morrison, Clarke, & Weissman, 2001; Hermann, Huber et al., 2007; Ailles, Weissman, 2007; Moltzahn, Volkmer, Rottke, & Ackermann, 2008; Moriyama, Ohuchida et al., 2010; Subramaniam, Ramalingam, Houchen, & Anant, 2010; Asuthkar, Stepanova et al., 2013; Du, Qin et al., 2010; Du, Qin et al., 2011; Hamada, Shimosegawa, 2012;

Ischenko, Seeliger et al., 2010; Rossi, Rehman, & Gondi, 2014; Schober, Jesenofsky et al., 2014; Wang et al., 2011; Wei, Yin et al., 2011). PDAC tumors often reoccur after surgery and chemotherapy. These new tumors tend to be chemoresistant, which leads to poor survival and the survival rates have not increased in the last 50 years (American Cancer Society, 2013). PC is extensive local tumor invasion, early systemic dissemination, and extremely poor response to chemotherapy (Buell-Gutbrod, Cavallo, Lee, Montag, & Gwin, 2015; Harvey et al., 2003; Hasegawa et al., 2014; Hu, Scott et al., 2013; Iacopino, Angelucci et al., 2014; Ilmer, Boiles et al., 2015; Pandian, Ramraj, Khan, Azim, & Aravindan, 2015; Pearton, Smith et al., 2014; Shekhani, Jayanthy, Maddodi, & Setaluri, 2013; Gibbs, Schlieman et al., 2009; Watabe, Yoshida et al., 1998). The current status of PDAC therapy is still poor and a lot more needs to be done to improve therapeutic outcomes (Froeling, Casolino, Pea, Biankin, & Chang, 2021; Parrasia, Zoratti, Szabò, & Biasutto, 2021; Santofimia-Castaño, Iovanna, 2021; Sun, Russell, Scarlett, & McCluskey, 2020; Wu et al., 2021). Most PCs are associated with mutations in Kirsten rat sarcoma virus (KRAS), TP53, SMAD4, CDKN2A genes and associated pathways such as Wnt/Notch, Hippo, Hedgehog, and the Pi3k-Akt pathways (Waddell et al., 2015); recently a genome-wide metaanalysis has identifies new susceptibility loci for PC indicating that there may be multiple subtypes of PCs. All gene expressions are in some way controlled or modulated by ncRNAs and the involvement of numerous gene mutations in the ncRNA network can also add to the complexity of PC molecular profile (Obazee et al., 2018; Walsh et al., 2019; Zhang et al., 2016).

1.2 Experimental methods and tools for analyzing noncoding RNAs in pancreatic cancer patients

Considering the importance of early detection and diagnosis of PC at a stage potentially curable, still limited number of experimental methods and approaches are available. Imaging techniques and biomarkers are two major methods widely used for this purpose. Among imaging methods, the most efficient one is an endoscopic ultrasound (EUS) (Gheorghe, Bungau et al., 2020; Iordache, Albulescu, & Săftoiu, 2017) with high sensitivity and specificity to PC detection. It allows detecting deeply localized tumors, as well as obtaining a biological sample by fine needle aspiration (FNA) biopsy following histological and molecular examination. In complement to this, recent development of molecular diagnosis field brought up some new epigenetic biomarkers, circulating tumor deoxyribonucleic acids, microRNAs (miRNAs), and other ncRNAs with high potential for early PC detection. There is growing interest to regulatory ncRNAs, such as miRNAs, long noncoding RNA (lncRNAs), and circular RNAs (circRNAs), that are involved in many cellular processes, including tumorigenesis (Gheorghe, Bungau et al., 2020; Słotwiński, Lech, & Słotwińska, 2018; Vila-Navarro et al., 2017; Vila-Navarro, Duran-Sanchon et al., 2019). Evidences of dysregulation of these RNAs in PC tissue samples and pancreatic precursor lesions obtained by EUS—FNA or after a surgical biopsy, in plasma, saliva, stool samples of patients are accumulating at express speed. Most of the studies are based on messenger RNA (mRNA)/miRNA sequencing, lncRNA, circRNA, or Piwi-interacting RNA (piRNA) microarrays, following excessive bioinformatic analysis. Using genome-wide miRNA profiling by next-generation sequencing (NGS), Vila-Navarro et al. (2017) were able to identify

607 deregulated miRNAs in PDAC and 396 miRNAs in intraductal papillary mucinous neoplasm (IPMN), compared with healthy individuals. Total 30 overexpressed miRNAs were validated by quantitative reverse transcriptase polymerase chain reaction. The circRNA and miRNA microarray data from other group of investigators reveal differentially expressed 256 circRNAs and 20 miRNAs from PDAC tissues compared to normal. Using Kyoto Encyclopedia of Genes and Genomes pathway analysis demonstrated that 41 circRNAs out of 256 were enriched in 17 pathways (Zhang, Wang, Zhou, Yang, & Zhong, 2019). Raulefs et al. demonstrated altered miRNA regulation of coding gene expression in PDAC compared to normal pancreatic tissues using high-throughput NGS-based technologies, namely, Massive Analysis of complementary DNA Ends and small RNA sequencing. Differential expressed other ncRNAs (piRNA, several seed region RNA, long intergenic noncoding RNAs [lincRNAs], and natural antisense transcripts) were studied in PDAC, too (Muller et al., 2015). Although relevance and validity of ncRNA as diagnostics is well known, no ncRNAs are currently in the clinical diagnostic phase (Sharma, Okada, Von Hoff, & Goel, 2020).

1.3 Bioinformatics for analyzing noncoding RNAs in pancreatic cancer patients

Numerous approaches have been developed to characterize ncRNA expression profiles from published sequencing data. Recently researchers were able to demonstrate the use of a resource allocation technique in ncRNA—target—disease tripartite network and correlate with disease prediction (Mori, Ngouv, Hayashida, Akutsu, & Nacher, 2018). In another study specific to PDAC, researches used miRNA—mRNA interactions followed by high-throughput sequencing (CLIP-Seq) data from StarBas, identified differentially expressed 500 miRNAs, and 21 lncRNAs, and were further able to predict a dual transcription factor—miRNA—mRNA, lncRNA—miRNA—mRNA regulatory mechanisms in PC (Ye, Yang et al., 2014). Researchers have also used multiomics data (Mishra, Southekal et al., 2019), penalized algorithms (Lee, Lee et al., 2021), and integrating transcriptome analysis (Yang & Zeng, 2015) for identifying novel PDAC-specific ncRNA markers from available databases. It is worth mentioning that with the development of deep learning machine algorithms, the identification of novel and clinically relevant ncRNAs will increase significantly (Shaw, Chen, Xie, & Jiang, 2021; Yang et al., 2020; Zhang, Long, & Kwoh, 2020).

1.4 Noncoding RNA

Noncoding RNAs (ncRNAs) can be classified into two types, regular and regulatory (Fig. 1.1). Regular ncRNA consists of RNA molecules involved in the normal biological functioning of cells such as ribosomal RNA, transfer RNA (tRNA), small nuclear RNA, small nucleolar RNAs, telomerase RNA component, transfer RNA-derived RNA fragments, and tRNA-derived stress-induced RNAs, and regulatory ncRNAs consist of molecules such as miRNA (Ambros, 2001; Toscano-Garibay, Aquino-Jarquin, 2014), small interfering RNA (siRNA) (Chen et al., 2018; Shi, Jin, Song, & Chen, 2019), piRNA (Muller et al., 2015; Saito, Sakaguchi et al., 2007), enhancer RNA (Amirnasr, Sleijfer, & Wiemer, 2020; Parrasia, Zoratti et al., 2021; Shimamura, Nakagami, Sanada,

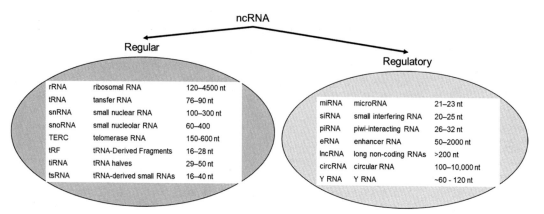

FIGURE 1.1 Classification of noncoding RNA based on function.

& Morishita, 2020), lncRNA (Ding, Li et al., 2020; Guo et al., 2020), circRNA (Arnaiz, Sole et al., 2018; Shang, Yang, Jia, & Ge, 2019; Wang et al., 2019), and small noncoding RNA (Zhang, Wu, Chen, & Chen, 2019). More and more types of ncRNAS continue to be identified on a regular basis, for example, tsRNAs (Zong et al., 2021). Significant amount of progress has been made in the identification and characterization of siRNA, miRNA, circRNA, and lncRNA in relation to PDAC; however, more amount of research is required for other types of ncRNA. We will focus on miRNA, circRNA, and lncRNA.

1.5 Pancreatic cancer–specific microRNAs

miRNAs regulate most biological processes. More than a decade ago, it was shown that miRNAs were small RNA molecules of about 22 nucleotides (∼22 nt) in length that probably function as antisense regulators of other RNAs (Ambros, 2001). The mechanisms of miRNA action are now much clearer though not complete (Toscano-Garibay, Aquino-Jarquin, 2014). Their capacity of regulating mRNAs and even other miRNAs is being reported with increasing frequency, making it a powerful tool for gene regulation. In more than 90% of PDACs show activating mutations of KRAS (Almoguera, Shibata et al., 1988). KRAS targeting miRs such as miRNA-96, 126, and 217 were observed to be downregulated in most PDACs. Diab, Muqbil, Mohammad, Azmi, and Philip (2016) recently published a detailed review of up- and downregulated miRs associated with PDAC. miR21, miR155, miR196s, and 196b are considered to be the most reliable preclinical PDAC markers (Wang et al., 2009) (Table 1.1).

1.6 Pancreatic cancer–associated lncRNAs

lncRNAs have multifunctional roles such as signaling regulation and cancer metastasis (Ming, Li et al., 2021). However, only a few lincRNAs have been identified that show specific association with PDACs (Table 1.2). SNHG16 and SOX2OT are known to be upregulated in PC tissues and function by sponging away miR-200a-3p and miR-200a/141 (Guo et al., 2020). GAS5 on the other hand is shown to be downregulated in PC cells and targets

TABLE 1.1 MicroRNAs involved in pancreatic cancer.

miRNA	Up- or downregulated in PC	Targets	PDAC/cell lines	References
miR-155	Up	SOCS3; p53 (TP53INP1); SOCS1	PDAC; cell lines	Calatayud, Dehlendorff et al. (2017), Hong, Park (2014), Wang, Guo, and Fan (2019)
miR-221	Up	SOCS3; HDAC6	PDAC; cell lines	Hong, Park (2014), Passadouro, Faneca (2016), Wu et al. (2020), Xie, Wen et al. (2018), Yang et al. (2018)
miR-21	Up	Spry2; VHL; PTEN; MMP-2; MMP-9; Erk1/2; VHL; MAPK/ERK; PI3K/AKT; PDGF; p27Kip1; p57kip2; PUMA; TIMP-2	PDAC; PC tissue, cell lines	Calatayud, Dehlendorff et al. (2017), Vychytilova-Faltejskova et al. (2015), Wang et al. (2009), Zhao, Chen et al. (2018), Ma, Wu et al. (2020), Sun et al. (2019)
miR-196a	Up	NFKBIA; Ing5	PDAC	Hong, Park (2014), Choi, Lee et al. (2014), Hernandez, Lucas (2016), Passadouro, Faneca (2016)
miR-200a	Up	Nrf2/ Keap1	PDAC	Hong, Park (2014)
miR-27a	Up	Spry2	PDAC	Hong, Park (2014)
miR-222	Up	p27; PPP2R2A/Akt PDGF; p27Kip1; p57kip2; PUMA; TIMP-2	PDAC	Hernandez, Lucas (2016), Passadouro, Faneca (2016)
miR-210	Up	E2F3; HOXA9	PDAC	Calatayud, Dehlendorff et al. (2017), Hernandez, Lucas (2016), Ni, Zhou, Yuan, Cen, and Yan (2019), Sun et al. (2018)
miR-100	Up	IGFR1	PDAC, cell lines	Passadouro, Faneca (2016)
miR-34a	Down	Bcl-2; Snail1; Notch 1; Notch 2 E2F3; c-MYC; Cyclin D1	PDAC, cell lines	Chen et al. (2019), Tang, Tang, and Cheng (2017), Vychytilova-Faltejskova et al. (2015), Zuo, Tao et al. (2020)
miR-300	Down	CUL4B	PDAC, cell lines	Zhang et al. (2018)
miR-203a-3p	Down	SLUG	PDAC, cell lines	An, Zheng (2020)
miR-145	Down	MUC13; ZEB1; Lin28; Nanog; Sox2; Oct4	PDAC, cell lines	Gao, Zhang et al. (2019), Setua, Khan et al. (2017)
miR-96	Down	FOXO3; AMOTL2; NUAK1	PDAC	Chang, Yu, Li, Yu, and Chen (2017), Hong, Park (2014)

(Continued)

TABLE 1.1 (Continued)

miRNA	Up- or downregulated in PC	Targets	PDAC/cell lines	References
miR-200	Down	BTBD1	PDAC	Hong, Park (2014)
miR-217	Down	E2F3; PRSS3; KRAS; SIRT1	PDAC; PC tissue	Hernandez, Lucas (2016), Hong, Park (2014), Vychytilova-Faltejskova et al. (2015), Yang et al. (2017)
miR-146	Down		PDAC	Papaconstantinou, Manta et al. (2013)
miR-148a	Down	CDC25B; CCKBR; Bcl-2; DNMT1	PDAC	Papaconstantinou, Manta et al. (2013), Passadouro, Faneca (2016)
miR-122	Down	CCNG1	PDAC	Dai, Zhang, Xu, and Jin (2020), Papaconstantinou, Manta et al. (2013)
miR-31	Down		PDAC	Papaconstantinou, Manta et al. (2013)
miR-205	Down		PC tissue, cell lines	Chaudhary, Mondal, Kumar, Kattel, and Mahato (2017)
miR-195	Down	DCLK1	PC tissue; cell lines	Zhou et al. (2017)
miR-34b	Down	Smad3	Cell lines	Liu, Cheng et al. (2013)
miR-10a	Down	HOX B1; HOX B2; HOX A1	cell lines	Ohuchida et al. (2012), Weiss, Marques et al. (2009)
miR-10b	Down	TIP30	PDAC	Passadouro, Faneca (2016), Schlick et al. (2020)
miR-375	Down	PDK1	PC tissue; cell lines	Zhou et al. (2014)

miRNA, MicroRNA; *PC*, pancreatic cancer; *PDAC*, pancreatic ductal adenocarcinoma.

miR-221/SOCS3 that are known contributors of chemoresistance and metastasis (Liu, Wu et al., 2018) (Table 1.2).

1.7 Pancreatic cancer—associated circular RNAs

One of the roles of this new class of RNAs is to function as sponges that tightly regulate the availability of miRNAs (Yang, Fu, & Zhou, 2018). Further, another regulatory layer is being classified at the RNA level, which includes various ncRNAs (Garajova, Balsano, Tommasi, & Giovannetti, 2019; Wong, Sorensen, Joglekar, Hardikar, & Dalgaard, 2018;

TABLE 1.2 Long noncoding RNAs (lncRNA) involved in pancreatic cancer.

lncRNA	Up- or downregulated in PC	Targets	PDAC/cell lines	References
SNHG16	Up	miR-200a-3p	Tissue	Guo et al. (2020)
SOX2OT	Up	miR-200a/141	Tissues; cell lines	Liu, Zhou et al. (2020)
UCA1	Up	miR-96; FOXO3; AMOTL2	Tissues; cell lines	Guo et al. (2020), Zhou et al. (2018)
DIO3OS	Up	miR-122	Tissue, cell lines	Cui, Jin et al. (2019)
GAS5	Down	miR-221; SOCS3	Cell lines	Liu, Wu et al. (2018)

PC, pancreatic cancer; *PDAC*, pancreatic ductal adenocarcinoma.

Zang, Lu, & Xu, 2018). circRNAs that were thought to belong to this group of ncRNAs are now slowly being recognized as belonging to a significantly importantly regulatory layer which can also code for proteins (Arnaiz, Sole et al., 2018; Shang, Yang et al., 2019; Wang et al., 2019). It is now recognized that most protein-coding genes not only produce linear mRNA but also produce circRNAs and this output ratio from linear to circular is dependent on the efficiency of pre-mRNA processing (Liang, Tatomer et al., 2017; Qu, Zhong et al., 2016) (Table 1.3).

1.8 Diagnostic microRNA markers of PDAC

Lately, liquid biopsy based on miRNAs continues to attract interest as early diagnostics of PCs. It has been demonstrated that miRNAs are highly conserved and are stable in biological fluids (Lagos-Quintana, Rauhut et al., 2001) making them attractive biomarkers for various pathologies. Numerous upregulated micro-RNAs have been identified in PDAC patients when compared to controls [for detailed list please see references (Diab, Muqbil et al., 2016; Wang et al., 2009)]. Among all the miRs up- or downregulated in PDACs, miR217 stands out (Hernandez, Lucas, 2016; Hong, Park, 2014; Vychytilova-Faltejskova et al., 2015; Yang, Zhang, & Qin, 2017). In PDACs, it is known that miR217 is downregulated. MiR217 is known to target KRAS that is significant in PDACs. It was recently shown that miR217 is a diagnostic biomarker and is involved in human podocyte cells apoptosis via targeting TNFSF11 (Li, Liu et al., 2017) and was shown to regulate the Rat sarcoma virus-mitogen-activated protein kinases signaling pathway in colorectal cancer (Zhang, Lu, & Chen, 2016) (Table 1.1).

1.9 Diagnostic lncRNA markers of PDAC

Numerous lncRNAs have been identified both experimentally and by computational analysis. A recent study has identified key lncRNAs involved in the progression of

TABLE 1.3 Circular RNAs (circRNAs) involved in pancreatic cancer.

Circular RNA	Up- or downregulated in PC	Targets	PDAC/PC tissue; cell lines	References
circ-LDLRAD3 (hsa_circ_0006988)	Up	miR-137–3p; pleiotrophin (PTN)	PDAC; cell lines	Yao, Zhang, Chen, and Gao (2019)
circ-IARS	Up	miR-122; ZO-1; RhoA; RhoA-GTP; ZO-1; F-actin	PDAC; cell lines; exosomes	Li, Li et al. (2018)
circ-PDE8A	Up	miR-338; MACC1; MET	PDAC; cell lines	Li, Yanfang et al. (2018)
circ_0007534	Up	miR-625; miR-892b	PDAC; cell lines	Hao, Rong et al. (2019)
circRNA_100782	Up	miR-124, IL6R, STAT3	PDAC; BxPC3	Chen, Shi, Zhang, and Sun (2017)
circZMYM2 (hsa_circ_0099999)	Up	miR-335–5p; JMJD2C	PDAC	Limb, Liu et al. (2020)
hsa_circ_0006215	Up	miR-378a-3p; SERPINA4	PDAC	Limb, Liu et al. (2020)
circ-RHOT1 (hsa_circ_0005397)	Up	miR-26b; miR-125a, miR-330; miR-382	PDAC; cell lines	Qu, Hao et al. (2019)
hsa_circ_0030235	Up	miR-1253; miR-1294	PDAC; cell lines	Xu, Yao, Gao, and Cui (2019)
ciRS-7 (Cdr1as)	Up	miR-7; EGFR; STAT3	PDAC	Liu, Liu et al. (2019)
hsa_circ_0007334	Up	miR-144–3p; miR-577; MMP7	PDAC	Yang et al. (2019)
circASH2L	Up	miR-34a; Notch 1	PDAC; cell lines	Chen et al. (2019)
circADAM9	Up	miR-217; PRSS3	PDAC; cell lines	Xing, Ye et al. (2019)
hsa_circ_001653	Up	miR-377; miR217; KRAS; HOXC6	PDAC; cell lines	Shi et al. (2020)
circHIPK3	Up	miR-330–5p; RASSF1	PDAC; cell lines	Liu, Xia et al. (2020)
circFOXK2	Up	miR-942; YBX1; hnRNPK; NUF2; PDXK	PDAC; cell lines	Wong et al. (2020)
circBFAR (hsa_circ_0009065)	Up	miR-34b-5p; MET	PDAC; cell lines	Guo et al. (2020)
hsa_circ_0013912	Up	miR-7–5p	PDAC; cell lines	Guo et al. (2020), Zhang et al. (2019)
hsa_circ_001587	Down	miR-223; SLC4A4	PDAC; PANC1	Zhang, Tan, Zhuang, and Du (2020)
hsa_circ_0001649	Down	Cas-3, Cas-9	PDAC; BxPC3; PANC1; Capan-2	Jiang, Wang et al. (2018)
circNFIB1 (hsa_circ_0086375)	Down	miR-486–5p; PIK3R1; VEGF-C	PDAC; cell lines	Kong, Li et al. (2020)

KRAS, Kirsten rat sarcoma virus; *PC*, pancreatic cancer; *PDAC*, pancreatic ductal adenocarcinoma.

intraductal IPMN to PDAC (Ding, Li et al., 2020). These newer findings have not yet been translated to a clinical setting (Table 1.2).

1.10 Diagnostic circular RNA markers of PDAC

After miRNAs, circRNAs are the most studied ncRNAs. circRNAs are circular in shape and lack a 5′ cap or 3′ poly-A tail and are more stable than linear mRNAs to traditional RNA degradation (Chen & Yang, 2015). Numerous studies have documented the expressions of circRNAs in PDAC using techniques such as microarray analysis and computational analysis from both cell lines and tissue samples (Table 1.3). However, no circRNA is currently being used to actively validate clinical outcome. It was recently shown that hsa_circ_001653 is significantly involved in the development of PDAC (Shi et al., 2020). It is known that hsa_circ_001653 targets miR217, which targets E2F3, PRSS3, KRAS, and SIRT1 (Table 1.3). As stated earlier in more than 90% of PDACs show activating mutations of KRAS (Almoguera, Shibata et al., 1988), therefore suggesting that hsa_circ_001653 expression activates KRAS expression and can be a significant diagnostic marker for PDAC.

1.11 Summary and conclusion

Recent studies have shown that lncRNA HOST2 (An, Cheng, 2020), has-miR125a (Yong, Yabin et al., 2017), and miR21 (Diab, Muqbil et al., 2016; Giovannetti et al., 2010; Sun et al., 2019; Wang et al., 2009; Zhao, Chen et al., 2018) correlate with Gemcitabine resistance and miR-320a modulates 5-FU response (Yong, Yabin et al., 2017). In another study, researchers found a core miRNA–mRNA regulatory network involving hsa-miR-643, hsa-miR-4644, hsa-miR-4650–5p, hsa-miR-4455, hsa-miR-1261, and hsa-miR-3676 as predictors of gemcitabine-resistant (Shen, Pan et al., 2015).

Acknowledgment

Support for this study by the McElroy Foundation and the Theresa Tracy Trott Foundation is acknowledged.

References

Ailles, L. E., & Weissman, I. L. (2007). Cancer stem cells in solid tumors. *Current Opinion in Biotechnology, 5,* 460–466.

Almoguera, C., Shibata, D., Forrester, K., Martin, J., Arnheim, N., & Perucho, M. (1988). Most human carcinomas of the exocrine pancreas contain mutant c-K-ras genes. *Cell, 4,* 549–554.

Ambros, V. (2001). microRNAs: Tiny regulators with great potential. *Cell, 7,* 823–826.

American Cancer Society. (2013). *Cancer facts & figures.* Atlanta, GA: American Cancer Society.

Amirnasr, A., Sleijfer, S., & Wiemer, E. A. C. (2020). Non-coding RNAs, a novel paradigm for the management of gastrointestinal stromal tumors. *International Journal of Molecular Sciences, 18,* E6975. Available from https://doi.org/10.3390/ijms21186975.

An, N., & Cheng, D. (2020). The long noncoding RNA HOST2 promotes gemcitabine resistance in human pancreatic cancer cells. *Pathology & Oncology Research, 1,* 425–431.

An, N., & Zheng, B. (2020). MiR-203a-3p inhibits pancreatic cancer cell proliferation, EMT, and apoptosis by regulating SLUG. *Technology in Cancer Research & Treatment, 19*, 1533033819898729.

Arnaiz, E., Sole, C., Manterola, L., Iparraguirre, L., Otaegui, D., & Lawrie, C. H. (2018). CircRNAs and cancer: Biomarkers and master regulators. *Seminars in Cancer Biology, 58*.

Asuthkar, S., Stepanova, V., Lebedeva, T., Holterman, A. L., Estes, N., Cines, D. B., ... Gondi, C. S. (2013). Multifunctional roles of urokinase plasminogen activator (uPA) in cancer stemness and chemoresistance of pancreatic cancer. *Molecular Biology of the Cell, 17*, 2620–2632.

Buell-Gutbrod, R., Cavallo, A., Lee, N., Montag, A., & Gwin, K. (2015). Heart and neural crest derivatives expressed transcript 2 (HAND2): A novel biomarker for the identification of atypical hyperplasia and type I endometrial carcinoma. *International Journal of Gynecological Pathology, 1*, 65–73.

Calatayud, D., Dehlendorff, C., Boisen, M. K., Hasselby, J. P., Schultz, N. A., Werner, J., ... Johansen, J. S. (2017). Tissue microRNA profiles as diagnostic and prognostic biomarkers in patients with resectable pancreatic ductal adenocarcinoma and periampullary cancers. *Biomarker Research, 5*, 8–017-0087-6. eCollection 2017.

Chang, X., Yu, C., Li, J., Yu, S., & Chen, J. (2017). hsa-miR-96 and hsa-miR-217 expression down-regulates with increasing dysplasia in pancreatic intraepithelial neoplasias and intraductal papillary mucinous neoplasms. *International Journal of Medical Sciences, 5*, 412–418.

Chaudhary, A. K., Mondal, G., Kumar, V., Kattel, K., & Mahato, R. I. (2017). Chemosensitization and inhibition of pancreatic cancer stem cell proliferation by overexpression of microRNA-205. *Cancer Letters, 402*, 1–8.

Chen, G., Shi, Y., Zhang, Y., & Sun, J. (2017). CircRNA_100782 regulates pancreatic carcinoma proliferation through the IL6-STAT3 pathway. *OncoTargets and Therapy, 10*, 5783–5794.

Chen, L. L., & Yang, L. (2015). Regulation of circRNA biogenesis. *RNA Biology, 4*, 381–388.

Chen, X., Mangala, L. S., Rodriguez-Aguayo, C., Kong, X., Lopez-Berestein, G., & Sood, A. K. (2018). RNA interference-based therapy and its delivery systems. *Cancer and Metastasis Reviews, 1*, 107–124.

Chen, Y., et al. (2019). Circ-ASH2L promotes tumor progression by sponging miR-34a to regulate Notch1 in pancreatic ductal adenocarcinoma. *Journal of Experimental & Clinical Cancer Research, 1*, 466–019-1436-0.

Choi, C. I., Lee, S. H., Hwang, S. H., Kim, D. H., Jeon, T. Y., Kim, D. H., ... Park, D. Y. (2014). Single-incision intragastric resection for upper and mid gastric submucosal tumors: A case-series study. *Annals of Surgical Treatment and Research, 6*, 304–310.

Clarke, M. F., & Fuller, M. (2006). Stem cells and cancer: Two faces of eve. *Cell, 6*, 1111–1115.

Cui, K., Jin, S., Du, Y., Yu, J., Feng, H., Fan, Q., & Ma, W. (2019). Long noncoding RNA DIO3OS interacts with miR-122 to promote proliferation and invasion of pancreatic cancer cells through upregulating ALDOA. *Cancer Cell International, 19*, 202-019-0922-y. eCollection 2019.

Dai, C., Zhang, Y., Xu, Z., & Jin, M. (2020). MicroRNA-122-5p inhibits cell proliferation, migration and invasion by targeting CCNG1 in pancreatic ductal adenocarcinoma. *Cancer Cell International, 20*, 98-020-01185-z. eCollection 2020.

Dean, M., Fojo, T., & Bates, S. (2005). Tumour stem cells and drug resistance. *Nature Reviews Cancer, 4*, 275–284.

Diab, M., Muqbil, I., Mohammad, R. M., Azmi, A. S., & Philip, P. A. (2016). The role of microRNAs in the diagnosis and treatment of pancreatic adenocarcinoma. *Journal of Clinical Medicine, 6*, 59. Available from https://doi.org/10.3390/jcm5060059.

Ding, J., Li, Y., Zhang, Y., Fan, B., Li, Q., Zhang, J., & Zhang, J. (2020). Identification of key lncRNAs in the tumorigenesis of intraductal pancreatic mucinous neoplasm by coexpression network analysis. *Cancer Medicine, 11*, 3840–3851.

Dingli, D., & Michor, F. (2006). Successful therapy must eradicate cancer stem cells. *Stem Cells, 12*, 2603–2610.

Du, Z., Qin, R., Wei, C., Wang, M., Shi, C., Tian, R., & Peng, C. (2011). Pancreatic cancer cells resistant to chemoradiotherapy rich in "stem-cell-like" tumor cells. *Digestive Diseases and Sciences, 3*, 741–750.

Du, Z., Qin, R., Wei, C., Wang, M., Shi, C., Tian, R., & Peng, C. (2010). Pancreatic cancer cells resistant to chemoradiotherapy rich in "Stem-Cell-Like" tumor cells. *Digestive Diseases and Sciences, 56*.

Eddy, S. R. (2001). Non-coding RNA genes and the modern RNA world. *Nature Reviews Genetics, 12*, 919–929.

Froeling, F. E. M., Casolino, R., Pea, A., Biankin, A. V., & Chang, D. K. (2021). Molecular subtyping and precision medicine for pancreatic cancer. *Journal of Clinical Medicine, 1*, 149. Available from https://doi.org/10.3390/jcm10010149.

Gao, Y., Zhang, Z., Li, K., Gong, L., Yang, Q., Huang, X., ... Yang, H. (2019). Author correction: Linc-DYNC2H1-4 promotes EMT and CSC phenotypes by acting as a sponge of miR-145 in pancreatic cancer cells. *Cell Death & Disease, 8*, 604–019-1852-2.

References

Garajova, I., Balsano, R., Tommasi, C., & Giovannetti, E. (2019). Noncoding RNAs emerging as novel biomarkers in pancreatic cancer. *Current Pharmaceutical Design, 24*.

Gheorghe, G., Bungau, S., Ilie, M., Behl, T., Vesa, C. M., Brisc, C., ... Diaconu, C. C. (2020). Early diagnosis of pancreatic cancer: The key for survival. *Diagnostics (Basel), 11*, 869. Available from https://doi.org/10.3390/diagnostics10110869.

Gibbs, J. F., Schlieman, M., Singh, P., Saxena, R., Martinick, M., Hutson, A. D., & Corasanti, J. (2009). A pilot study of urokinase-type plasminogen activator (uPA) overexpression in the brush cytology of patients with malignant pancreatic or biliary strictures. *HPB Surgery, 2009*, Article ID 805971.

Giovannetti, E., et al. (2010). MicroRNA-21 in pancreatic cancer: Correlation with clinical outcome and pharmacologic aspects underlying its role in the modulation of gemcitabine activity. *Cancer Research, 11*, 4528–4538.

Guo, J. Q., Yang, Z. J., Wang, S., Wu, Z. Z., Yin, L. L., & Wang, D. C. (2020). LncRNA SNHG16 functions as an oncogene by sponging miR-200a-3p in pancreatic cancer. *European Review for Medical and Pharmacological Sciences, 4*, 1718–1724.

Guo, W., Zhao, L., Wei, G., Liu, P., Zhang, Y., & Fu, L. (2020). Blocking circ_0013912 suppressed cell growth, migration and invasion of pancreatic ductal adenocarcinoma cells in vitro and in vivo partially through sponging miR-7-5p. *Cancer Management and Research, 12*, 7291–7303.

Guo, X., et al. (2020). Circular RNA circBFAR promotes the progression of pancreatic ductal adenocarcinoma via the miR-34b-5p/MET/Akt axis. *Molecular Cancer, 1*, 83-020-01196-4.

Guo, Z., Wang, X., Yang, Y., Chen, W., Zhang, K., Teng, B., ... Qiu, Z. (2020). Hypoxic tumor-derived exosomal long noncoding RNA UCA1 promotes angiogenesis via miR-96-5p/AMOTL2 in pancreatic cancer. *Molecular Therapy – Nucleic Acids, 22*, 179–195.

Hamada, S., & Shimosegawa, T. (2012). Pancreatic cancer stem cell and mesenchymal stem cell. In P. J. Grippo, & H. G. Munshi (Eds.), *Pancreatic cancer and tumor microenvironment*. Trivandrum: Transworld Research Network.

Hao, L., Rong, W., Bai, L., Cui, H., Zhang, S., Li, Y., ... Meng, X. (2019). Upregulated circular RNA circ_0007534 indicates an unfavorable prognosis in pancreatic ductal adenocarcinoma and regulates cell proliferation, apoptosis, and invasion by sponging miR-625 and miR-892b. *Journal of Cellular Biochemistry, 3*, 3780–3789.

Harvey, S. R., et al. (2003). Evaluation of urinary plasminogen activator, its receptor, matrix metalloproteinase-9, and von Willebrand factor in pancreatic cancer. *Clinical Cancer Research, 13*, 4935–4943.

Hasegawa, S., et al. (2014). MicroRNA-1246 expression associated with CCNG2-mediated chemoresistance and stemness in pancreatic cancer. *British Journal of Cancer, 8*, 1572–1580.

Hermann, P. C., Huber, S. L., Herrler, T., Aicher, A., Ellwart, J. W., Guba, M., ... Heeschen, C. (2007). Distinct populations of cancer stem cells determine tumor growth and metastatic activity in human pancreatic cancer. *Cell Stem Cell, 3*, 313–323.

Hernandez, Y. G., & Lucas, A. L. (2016). MicroRNA in pancreatic ductal adenocarcinoma and its precursor lesions. *World Journal of Gastrointestinal Oncology, 1*, 18–29.

Hong, T. H., & Park, I. Y. (2014). MicroRNA expression profiling of diagnostic needle aspirates from surgical pancreatic cancer specimens. *Annals of Surgical Treatment and Research, 6*, 290–297.

Hu, D., Scott, I. C., Snider, F., Geary-Joo, C., Zhao, X., Simmons, D. G., & Cross, J. C. (2013). The basic helix-loop-helix transcription factor Hand1 regulates mouse development as a homodimer. *Developmental Biology, 2*, 470–481.

Iacopino, F., Angelucci, C., Piacentini, R., Biamonte, F., Mangiola, A., Maira, G., ... Sica, G. (2014). Isolation of cancer stem cells from three human glioblastoma cell lines: Characterization of two selected clones. *PLoS One, 8*, e105166.

Ilmer, M., Boiles, A. R., Regel, I., Yokoi, K., Michalski, C. W., Wistuba, I. I., ... Vykoukal, J. (2015). RSPO2 enhances canonical wnt signaling to confer stemness-associated traits to susceptible pancreatic cancer cells. *Cancer Research, 9*, 1883–1896.

Iordache, S., Albulescu, D. M., & Săftoiu, A. (2017). The borderline resectable/locally advanced pancreatic ductal adenocarcinoma: EUS oriented. *Endosc Ultrasound, 6*, S83–S86.

Ischenko, I., Seeliger, H., Kleespies, A., Angele, M. K., Eichhorn, M. E., Jauch, K. W., & Bruns, C. J. (2010). Pancreatic cancer stem cells: New understanding of tumorigenesis, clinical implications. *Langenbeck's Archives of Surgery, 1*, 1–10.

Jiang, Y., Wang, T., Yan, L., & Qu, L. (2018). A novel prognostic biomarker for pancreatic ductal adenocarcinoma: hsa_circ_0001649. *Gene, 675*, 88–93. Available from https://doi.org/10.1016/j.gene.2018.06.099.

Kong, Y., Li, Y., Luo, Y., Zhu, J., Zheng, H., Gao, B., Guo, X., Li, Z., Chen, R., & Chen, C. (2020). circNFIB1 inhibits lymphangiogenesis and lymphatic metastasis via the miR-486-5p/PIK3R1/VEGF-C axis in pancreatic cancer. *Mol Cancer*, 19(1), 82-020-01205-6. PMCID: PMC7197141.

Lagos-Quintana, M., Rauhut, R., Lendeckel, W., & Tuschl, T. (2001). Identification of novel genes coding for small expressed RNAs. *Science*, 294(5543), 853–858.

Lee, J., Lee, H. S., Park, S. B., Kim, C., Kim, K., Jung, D. E., et al. (2021). Identification of Circulating Serum miRNAs as Novel Biomarkers in Pancreatic Cancer Using a Penalized Algorithm. *Int J Mol Sci*, 22(3), 1007. Available from https://doi.org/10.3390/ijms22031007.

Li, J., Li, Z., Jiang, P., Peng, M., Zhang, X., Chen, K., Liu, H., Bi, H., Liu, X., & Li, X. (2018). Circular RNA IARS (circ-IARS) secreted by pancreatic cancer cells and located within exosomes regulates endothelial monolayer permeability to promote tumor metastasis. *J Exp Clin Cancer Res*, 37(1), 177-018-0822-3. PMCID: PMC6069563.

Li, J., Liu, B., Xue, H., Zhou, Q. Q., & Peng, L. (2017). miR-217 is a useful diagnostic biomarker and regulates human podocyte cells apoptosis via targeting TNFSF11 in membranous nephropathy. *Biomed Res Int*, 2017, 2168767, PMCID PMC5682891.

Li, Z., Yanfang, W., Li, J., Jiang, P., Peng, T., Chen, K., Zhao, X., Zhang, Y., Zhen, P., Zhu, J., & Li, X. (2018). Tumor-released exosomal circular RNA PDE8A promotes invasive growth via the miR-338/MACC1/MET pathway in pancreatic cancer. *Cancer Lett*, 432, 237–250.

Liang, D., Tatomer, D. C., Luo, Z., Wu, H., Yang, L., Chen, L. L., Cherry, S., & Wilusz, J. E. (2017). The output of protein-coding genes shifts to circular RNAs when the pre-mRNA processing machinery is limiting. *Mol Cell*, 68(5), 940–954, e3. PMCID: PMC5728686.

Limb, C., Liu, D. S. K., Veno, M. T., Rees, E., Krell, J., Bagwan, I. N., Giovannetti, E., Pandha, H., Strobel, O., Rockall, T. A., & Frampton, A. E. (2020). The role of circular RNAs in pancreatic ductal adenocarcinoma and biliary-tract cancers. *Cancers (Basel)*, 12(11), 3250. Available from https://doi.org/10.3390/cancers12113250, PMCID: PMC7694172.

Liu, B., Wu, S., Ma, J., Yan, S., Xiao, Z., Wan, L., Zhang, F., Shang, M., & Mao, A. (2018). lncRNA GAS5 reverses EMT and tumor stem cell-mediated gemcitabine resistance and metastasis by targeting miR-221/SOCS3 in pancreatic cancer. *Mol Ther Nucleic Acids*, 13, 472–482. Available from https://doi.org/10.1016/j.omtn.2018.09.026. Epub 2018 Oct 6. PMID: 30388621; PMCID: PMC6205337.

Liu, C., Cheng, H., Shi, S., Cui, X., Yang, J., Chen, L., Cen, P., Cai, X., Lu, Y., Wu, C., Yao, W., Qin, Y., Liu, L., Long, J., Xu, J., Li, M., & Yu, X. (2013). MicroRNA-34b inhibits pancreatic cancer metastasis through repressing Smad3. *Curr Mol Med*, 13(4), 467–478.

Liu, C. S., Zhou, Q., Zhang, Y. D., & Fu, Y. (2020). Long noncoding RNA SOX2OT maintains the stemness of pancreatic cancer cells by regulating DEK via interacting with miR-200a/141. *Eur Rev Med Pharmacol Sci*, 24(5), 2368–2379.

Liu, L., Liu, F. B., Huang, M., Xie, K., Xie, Q. S., Liu, C. H., Shen, M. J., & Huang, Q. (2019). Circular RNA ciRS-7 promotes the proliferation and metastasis of pancreatic cancer by regulating miR-7-mediated EGFR/STAT3 signaling pathway. *Hepatobiliary Pancreat Dis Int*, 18(6), 580–586.

Liu, Y., Xia, L., Dong, L., Wang, J., Xiao, Q., Yu, X., & Zhu, H. (2020). CircHIPK3 promotes gemcitabine (GEM) resistance in pancreatic cancer cells by sponging miR-330-5p and targets RASSF1. *Cancer Manag Res*, 12, 921–929, PMCID: PMC7023912.

Ma, Q., Wu, H., Xiao, Y., Liang, Z., & Liu, T. (2020). Upregulation of exosomal microRNA-21 in pancreatic stellate cells promotes pancreatic cancer cell migration and enhances Ras/ERK pathway activity. *Int J Oncol*, 56(4), 1025–1033. Available from https://doi.org/10.3892/ijo.2020.4986, Epub 2020 Feb 14. PMID: 32319558.

Mishra, N. K., Southekal, S., & Guda, C. (2019). Survival Analysis of Multi-Omics Data Identifies Potential Prognostic Markers of Pancreatic Ductal Adenocarcinoma. *Front Genet*, 10, 624.

Moltzahn, F. R., Volkmer, J. P., Rottke, D., & Ackermann, R. (2008). "Cancer stem cells"-lessons from Hercules to fight the Hydra. *Urologic Oncology*, 6, 581–589.

Mori, T., Ngouv, H., Hayashida, M., Akutsu, T., & Nacher, J. C. (2018). ncRNA-disease association prediction based on sequence information and tripartite network. *BMC Systems Biology*, 37, 37–018-0527-4.

Moriyama, T., Ohuchida, K., Mizumoto, K., Cui, L., Ikenaga, N., Sato, N., & Tanaka, M. (2010). Enhanced cell migration and invasion of CD133 + pancreatic cancer cells cocultured with pancreatic stromal cells. *Cancer*, 14, 3357–3368.

Muller, S., et al. (2015). Next-generation sequencing reveals novel differentially regulated mRNAs, lncRNAs, miRNAs, sdRNAs and a piRNA in pancreatic cancer. *Molecular Cancer*, 14, 94-015-0358-5.

Ni, J., Zhou, S., Yuan, W., Cen, F., & Yan, Q. (2019). Mechanism of miR-210 involved in epithelial-mesenchymal transition of pancreatic cancer cells under hypoxia. *Journal of Receptor and Signal Transduction*, 5–6, 399–406.

References

Obazee, O., et al. (2018). Common genetic variants associated with pancreatic adenocarcinoma may also modify risk of pancreatic neuroendocrine neoplasms. *Carcinogenesis, 3*, 360–367.

Ohuchida, K., et al. (2012). MicroRNA-10a is overexpressed in human pancreatic cancer and involved in its invasiveness partially via suppression of the HOXA1 gene. *Annals of Surgical Oncology, 7*, 2394–2402.

Pandian, V., Ramraj, S., Khan, F. H., Azim, T., & Aravindan, N. (2015). Metastatic neuroblastoma cancer stem cells exhibit flexible plasticity and adaptive stemness signaling. *Stem Cell Research & Therapy, 1*, 2–015-0002–8.

Papaconstantinou, I. G., Manta, A., Gazouli, M., Lyberopoulou, A., Lykoudis, P. M., Polymeneas, G., & Voros, D. (2013). Expression of microRNAs in patients with pancreatic cancer and its prognostic significance. *Pancreas, 1*, 67–71.

Parrasia, S., Zoratti, M., Szabò, I., & Biasutto, L. (2021). Targeting pancreatic ductal adenocarcinoma (PDAC). *Cellular Physiology and Biochemistry, 1*, 61–90.

Passadouro, M., & Faneca, H. (2016). Managing pancreatic adenocarcinoma: A special focus in MicroRNA gene therapy. *International Journal of Molecular Sciences, 5*, 718. Available from https://doi.org/10.3390/ijms17050718.

Pavet, V., Portal, M. M., Moulin, J. C., Herbrecht, R., & Gronemeyer, H. (2011). Towards novel paradigms for cancer therapy. *Oncogene, 1*, 1–20.

Pearton, D. J., Smith, C. S., Redgate, E., van Leeuwen, J., Donnison, M., & Pfeffer, P. L. (2014). Elf5 counteracts precocious trophoblast differentiation by maintaining Sox2 and 3 and inhibiting Hand1 expression. *Developmental Biology, 2*, 344–357.

Qu, S., Hao, X., Song, W., Niu, K., Yang, X., Zhang, X., ... Liu, Z. (2019). Circular RNA circRHOT1 is upregulated and promotes cell proliferation and invasion in pancreatic cancer. *Epigenomics, 1*, 53–63.

Qu, S., Zhong, Y., Shang, R., Zhang, X., Song, W., Kjems, J., & Li, H. (2016). The emerging landscape of circular RNA in life processes. *RNA Biology, 14*, 1–8.

Reya, T., Morrison, S. J., Clarke, M. F., & Weissman, I. L. (2001). Stem cells, cancer, and cancer stem cells. *Nature, 6859*, 105–111.

Rossi, M. L., Rehman, A. A., & Gondi, C. S. (2014). Therapeutic options for the management of pancreatic cancer. *World Journal of Gastroenterology, 32*, 11142–11159.

Saito, K., Sakaguchi, Y., Suzuki, T., Suzuki, T., Siomi, H., & Siomi, M. C. (2007). Pimet, the Drosophila homolog of HEN1, mediates 2′-O-methylation of Piwi-interacting RNAs at their 3′ ends. *Genes & Development, 13*, 1603–1608.

Santofimia-Castaño, P., & Iovanna, J. (2021). Combating pancreatic cancer chemoresistance by triggering multiple cell death pathways. *Pancreatology, 21*.

Schlick, K., et al. (2020). Overcoming negative predictions of microRNA expressions to gemcitabine response with FOLFIRINOX in advanced pancreatic cancer patients. *Future Science OA, 2*, FSO644–2020-0128.

Schober, M., Jesenofsky, R., Faissner, R., Weidenauer, C., Hagmann, W., Michl, P., ... Lohr, J. M. (2014). Desmoplasia and chemoresistance in pancreatic cancer, . *Cancers (Basel)* (4, pp. 2137–2154).

Setua, S., Khan, S., Doxtater, K., Yallapu, M. M., Jaggi, M., & Chauhan, S. C. (2017). miR-145: Revival of a dragon in pancreatic cancer. *Journal of Natural Sciences, 3*, e332.

Shang, Q., Yang, Z., Jia, R., & Ge, S. (2019). The novel roles of circRNAs in human cancer. *Molecular Cancer, 1*, 6–018-0934-6.

Sharma, G. G., Okada, Y., Von Hoff, D., & Goel, A. (2020). Non-coding RNA biomarkers in pancreatic ductal adenocarcinoma. *Seminars in Cancer Biology*, S1044-579X.

Shaw, D., Chen, H., Xie, M., & Jiang, T. (2021). DeepLPI: A multimodal deep learning method for predicting the interactions between lncRNAs and protein isoforms. *BMC Bioinformatics, 1*, 24-020-03914-7.

Shekhani, M. T., Jayanthy, A. S., Maddodi, N., & Setaluri, V. (2013). Cancer stem cells and tumor transdifferentiation: Implications for novel therapeutic strategies. *American Journal of Stem Cells, 1*, 52–61.

Shen, Y., Pan, Y., Xu, L., Chen, L., Liu, L., Chen, H., ... Meng, Z. (2015). Identifying microRNA-mRNA regulatory network in gemcitabine-resistant cells derived from human pancreatic cancer cells. *Tumor Biology, 6*, 4525–4534.

Shi, H., Li, H., Zhen, T., Dong, Y., Pei, X., & Zhang, X. (2020). hsa_circ_001653 implicates in the development of pancreatic ductal adenocarcinoma by regulating microRNA-377-mediated HOXC6 axis. *Molecular Therapy – Nucleic Acids, 20*, 252–264.

Shi, S., Jin, Y., Song, H., & Chen, X. (2019). MicroRNA-34a attenuates VEGF-mediated retinal angiogenesis via targeting Notch1. *Biochemistry and Cell Biology, 4*, 423–430.

Shimamura, M., Nakagami, H., Sanada, F., & Morishita, R. (2020). Progress of gene therapy in cardiovascular disease. *Hypertension, 4*, 1038–1044.

Słotwiński, R., Lech, G., & Słotwińska, S. M. (2018). MicroRNAs in pancreatic cancer diagnosis and therapy. *Central European Journal of Immunology, 3*, 314–324.

Subramaniam, D., Ramalingam, S., Houchen, C. W., & Anant, S. (2010). Cancer stem cells: A novel paradigm for cancer prevention and treatment. *Mini-Reviews in Medicinal Chemistry, 5*, 359–371.

Sun, F. B., Lin, Y., Li, S. J., Gao, J., Han, B., & Zhang, C. S. (2018). MiR-210 knockdown promotes the development of pancreatic cancer via upregulating E2F3 expression. *European Review for Medical and Pharmacological Sciences, 24*, 8640–8648.

Sun, J., Jiang, Z., Li, Y., Wang, K., Chen, X., & Liu, G. (2019). Downregulation of miR-21 inhibits the malignant phenotype of pancreatic cancer cells by targeting VHL. *OncoTargets and Therapy, 12*, 7215–7226.

Sun, J., Russell, C. C., Scarlett, C. J., & McCluskey, A. (2020). *Small molecule inhibitors in pancreatic cancer,* . RSC *Medicinal Chemistry* (2, pp. 164–183).

Tang, Y., Tang, Y., & Cheng, Y. S. (2017). miR-34a inhibits pancreatic cancer progression through Snail1-mediated epithelial-mesenchymal transition and the Notch signaling pathway. *Scientific Reports, 7*, 38232.

Toscano-Garibay, J. D., & Aquino-Jarquin, G. (2014). Transcriptional regulation mechanism mediated by miRNA-DNA*DNA triplex structure stabilized by Argonaute. *Biochimica et Biophysica Acta, 11*, 1079–1083.

Vila-Navarro, E., et al. (2017). MicroRNAs for detection of pancreatic neoplasia: Biomarker discovery by next-generation sequencing and validation in 2 independent cohorts. *Annals of Surgery, 6*, 1226–1234.

Vila-Navarro, E., Duran-Sanchon, S., Vila-Casadesús, M., Moreira, L., Ginès, À., Cuatrecasas, M., ... Gironella, M. (2019). Novel circulating miRNA signatures for early detection of pancreatic neoplasia. *Clinical and Translational Gastroenterology, 4*, e00029.

Vychytilova-Faltejskova, P., et al. (2015). MiR-21, miR-34a, miR-198 and miR-217 as diagnostic and prognostic biomarkers for chronic pancreatitis and pancreatic ductal adenocarcinoma. *Diagnostic Pathology, 10*, 38-015-0272-6.

Waddell, N., et al. (2015). Whole genomes redefine the mutational landscape of pancreatic cancer. *Nature, 7540*, 495–501.

Walsh, N., et al. (2019). Agnostic pathway/gene set analysis of genome-wide association data identifies associations for pancreatic cancer. *Journal of the National Cancer Institute, 6*, 557–567.

Wang, J., Chen, J., Chang, P., LeBlanc, A., Li, D., Abbruzzesse, J. L., ... Sen, S. (2009). MicroRNAs in plasma of pancreatic ductal adenocarcinoma patients as novel blood-based biomarkers of disease. *Cancer Prevention Research (Phila), 9*, 807–813.

Wang, J., Guo, J., & Fan, H. (2019). MiR-155 regulates the proliferation and apoptosis of pancreatic cancer cells through targeting SOCS3. *European Review for Medical and Pharmacological Sciences, 12*, 5168–5175.

Wang, J., Zhu, M., Pan, J., Chen, C., Xia, S., & Song, Y. (2019). Circular RNAs: A rising star in respiratory diseases. *Respiratory Research, 1*, 3–018-0962-1.

Wang, Z., Li, Y., Ahmad, A., Banerjee, S., Azmi, A. S., Kong, D., & Sarkar, F. H. (2011). Pancreatic cancer: Understanding and overcoming chemoresistance. *Nature Reviews Gastroenterology & Hepatology, 1*, 27–33.

Watabe, T., Yoshida, K., Shindoh, M., Kaya, M., Fujikawa, K., Sato, H., ... Fujinaga, K. (1998). The Ets-1 and Ets-2 transcription factors activate the promoters for invasion-associated urokinase and collagenase genes in response to epidermal growth factor. *International Journal of Cancer, 1*, 128–137.

Wei, H. J., Yin, T., Zhu, Z., Shi, P. F., Tian, Y., & Wang, C. Y. (2011). Expression of CD44, CD24 and ESA in pancreatic adenocarcinoma cell lines varies with local microenvironment. *Hepatobiliary & Pancreatic Diseases International, 4*, 428–434.

Weiss, F. U., Marques, I. J., Woltering, J. M., Vlecken, D. H., Aghdassi, A., Partecke, L. I., ... Bagowski, C. P. (2009). Retinoic acid receptor antagonists inhibit miR-10a expression and block metastatic behavior of pancreatic cancer. *Gastroenterology, 6*, 2136-45.e1–7.

Wong, C. H., Lou, U. K., Li, Y., Chan, S. L., Tong, J. H., To, K. F., & Chen, Y. (2020). CircFOXK2 promotes growth and metastasis of pancreatic ductal adenocarcinoma by complexing with RNA-binding proteins and sponging MiR-942. *Cancer Research, 11*, 2138–2149.

Wong, W. K. M., Sorensen, A. E., Joglekar, M. V., Hardikar, A. A., & Dalgaard, L. T. (2018). Non-coding RNA in pancreas and beta-cell development. *Noncoding RNA, 4*. Available from https://doi.org/10.3390/ncrna4040041.

References

Wu, X., Huang, J., Yang, Z., Zhu, Y., Zhang, Y., Wang, J., & Yao, W. (2020). MicroRNA-221-3p is related to survival and promotes tumour progression in pancreatic cancer: A comprehensive study on functions and clinicopathological value. *Cancer Cell International, 20*, 443-020-01529-9. eCollection 2020.

Wu, Y., Zhang, C., Jiang, K., Werner, J., Bazhin, A. V., & D'Haese, J. G. (2021). The role of stellate cells in pancreatic ductal adenocarcinoma: Targeting perspectives. *Frontiers in Oncology, 10*, 621937.

Xie, J., Wen, J. T., Xue, X. J., Zhang, K. P., Wang, X. Z., & Cheng, H. H. (2018). MiR-221 inhibits proliferation of pancreatic cancer cells via down regulation of SOCS3. *European Review for Medical and Pharmacological Sciences, 7*, 1914–1921.

Xing, C., Ye, H., Wang, W., Sun, M., Zhang, J., Zhao, Z., & Jiang, G. (2019). Circular RNA ADAM9 facilitates the malignant behaviours of pancreatic cancer by sponging miR-217 and upregulating PRSS3 expression. *Artificial Cells, Nanomedicine, and Biotechnology, 1*, 3920–3928.

Xu, Y., Yao, Y., Gao, P., & Cui, Y. (2019). Upregulated circular RNA circ_0030235 predicts unfavorable prognosis in pancreatic ductal adenocarcinoma and facilitates cell progression by sponging miR-1253 and miR-1294. *Biochemical and Biophysical Research Communications, 1*, 138–142.

Yang, J., & Zeng, Y. (2015). Identification of miRNA-mRNA crosstalk in pancreatic cancer by integrating transcriptome analysis. *European Review for Medical and Pharmacological Sciences, 5*, 825–834.

Yang, J., Cong, X., Ren, M., Sun, H., Liu, T., Chen, G., . . . Yang, Q. (2019). Circular RNA hsa_circRNA_0007334 is predicted to promote MMP7 and COL1A1 expression by functioning as a miRNA sponge in pancreatic ductal adenocarcinoma. *Journal of Oncology, 2019*, 7630894.

Yang, J., Zhang, H. F., & Qin, C. F. (2017). MicroRNA-217 functions as a prognosis predictor and inhibits pancreatic cancer cell proliferation and invasion via targeting E2F3. *European Review for Medical and Pharmacological Sciences, 18*, 4050–4057.

Yang, L., Fu, J., & Zhou, Y. (2018). Circular RNAs and their emerging roles in immune regulation. *Frontiers in Immunology, 9*, 2977.

Yang, S., Wang, Y., Lin, Y., Shao, D., He, K., & Huang, L. (2020). LncMirNet: Predicting LncRNA-miRNA interaction based on deep learning of ribonucleic acid sequences. *Molecules, 19*, 4372. Available from https://doi.org/10.3390/molecules25194372.

Yang, Y., et al. (2018). MicroRNA-221 induces autophagy through suppressing HDAC6 expression and promoting apoptosis in pancreatic cancer. *Oncology Letters, 6*, 7295–7301.

Yao, J., Zhang, C., Chen, Y., & Gao, S. (2019). Downregulation of circular RNA circ-LDLRAD3 suppresses pancreatic cancer progression through miR-137-3p/PTN axis. *Life Sciences, 239*, 116871.

Ye, S., Yang, L., Zhao, X., Song, W., Wang, W., & Zheng, S. (2014). Bioinformatics method to predict two regulation mechanism: TF-miRNA-mRNA and lncRNA-miRNA-mRNA in pancreatic cancer. *Cell Biochemistry and Biophysics, 3*, 1849–1858.

Yong, S., Yabin, Y., Bing, Z., Chuanrong, Z., Dianhua, G., Jianhuai, Z., . . . Ling, L. (2017). Reciprocal regulation of DGCR5 and miR-320a affects the cellular malignant phenotype and 5-FU response in pancreatic ductal adenocarcinoma. *Oncotarget, 53*, 90868–90878.

Zang, J., Lu, D., & Xu, A. (2018). The interaction of circRNAs and RNA binding proteins: An important part of circRNA maintenance and function. *Journal of Neuroscience Research, 98*.

Zhang, J. Q., Chen, S., Gu, J. N., Zhu, Y., Zhan, Q., Cheng, D. F., . . . Peng, C. H. (2018). MicroRNA-300 promotes apoptosis and inhibits proliferation, migration, invasion and epithelial-mesenchymal transition via the Wnt/β-catenin signaling pathway by targeting CUL4B in pancreatic cancer cells. *Journal of Cellular Biochemistry, 1*, 1027–1040.

Zhang, M., et al. (2016). Three new pancreatic cancer susceptibility signals identified on chromosomes 1q32.1, 5p15.33 and 8q24.21. *Oncotarget, 41*, 66328–66343.

Zhang, N., Lu, C., & Chen, L. (2016). miR-217 regulates tumor growth and apoptosis by targeting the MAPK signaling pathway in colorectal cancer. *Oncology Letters, 6*, 4589–4597.

Zhang, P., Wu, W., Chen, Q., & Chen, M. (2019). Non-coding RNAs and their integrated networks. *Journal of Integrative Bioinformatics, 3*. Available from https://doi.org/10.1515/jib-2019-0027, 20190027.

Zhang, Q., Wang, J. Y., Zhou, S. Y., Yang, S. J., & Zhong, S. L. (2019). Circular RNA expression in pancreatic ductal adenocarcinoma. *Oncology Letters, 3*, 2923–2930.

Zhang, X., Tan, P., Zhuang, Y., & Du, L. (2020). hsa_circRNA_001587 upregulates SLC4A4 expression to inhibit migration, invasion, and angiogenesis of pancreatic cancer cells via binding to microRNA-223. *American Journal of Physiology-Gastrointestinal and Liver Physiology, 6*, G703–G717.

Zhang, Y., Long, Y., & Kwoh, C. K. (2020). Deep learning based DNA:RNA triplex forming potential prediction. *BMC Bioinformatics, 1*, 522-020-03864-0.

Zhao, Q., Chen, S., Zhu, Z., Yu, L., Ren, Y., Jiang, M., ... Li, B. (2018). miR-21 promotes EGF-induced pancreatic cancer cell proliferation by targeting Spry2. *Cell Death & Disease, 12*, 1157-018-1182-9.

Zhou, B., Sun, C., Hu, X., Zhan, H., Zou, H., Feng, Y., ... Zhang, B. (2017). MicroRNA-195 suppresses the progression of pancreatic cancer by targeting DCLK1. *Cellular Physiology and Biochemistry, 5*, 1867–1881.

Zhou, J., Song, S., He, S., Zhu, X., Zhang, Y., Yi, B., ... Li, D. (2014). MicroRNA-375 targets PDK1 in pancreatic carcinoma and suppresses cell growth through the Akt signaling pathway. *International Journal of Molecular Medicine, 4*, 950–956.

Zhou, Y., Chen, Y., Ding, W., Hua, Z., Wang, L., Zhu, Y., ... Dai, T. (2018). LncRNA UCA1 impacts cell proliferation, invasion, and migration of pancreatic cancer through regulating miR-96/FOXO3. *IUBMB Life, 4*, 276–290.

Zong, T., et al. (2021). tsRNAs: Novel small molecules from cell function and regulatory mechanism to therapeutic targets. *Cell Proliferation, 54*, e12977.

Zuo, L., Tao, H., Xu, H., Li, C., Qiao, G., Guo, M., ... In, X. (2020). Exosomes-coated miR-34a displays potent antitumor activity in pancreatic cancer both in vitro and in vivo. *Drug Design, Development and Therapy, 14*, 3495–3507.

CHAPTER 2

Applications of noncoding RNAs in brain cancer patients

Małgorzata Grabowska, Julia O. Misiorek, Żaneta Zarębska and Katarzyna Rolle

Department of Molecular Neurooncology, Institute of Bioorganic Chemistry, Polish Academy of Sciences, Poznań, Poland

2.1 Introduction

The number of patients suffering from cancer is constantly growing and, according to the data from the International Agency for Research on Cancer, it reached 18 million new cases worldwide in 2018. Cancer is also a leading cause of death in the world. Over 9 million of cancer patients were reported to die in the year 2018, although the incidence of brain and central nervous system cancers accounted only for 1.6% of all aforementioned cancer cases; high tumor aggressiveness, poor prognosis, and high mortality rate prompt researches to focus especially on this type of cancer (Ferlay et al., 2019). Brain tumors are either developed de novo or less commonly as a result of metastasis. The primary cancers that most frequently metastasize to brain are lung cancer (43.2%), breast cancer (15.7%), melanoma (16.4%), colorectal cancer (9.3%), and renal cell carcinoma (9.1%) (Berghoff et al., 2016).

2.1.1 Types of brain cancer

Brain tumors are generally classified based on the World Health Organization (WHO) system which is primarily focused on tumors' pathological features. Grade I and II tumors are considered benign ones, while grade III and IV tumors belong to malignant ones. Here, we aim to focus on glioblastoma multiforme (GBM) and medulloblastoma (MB), the most malignant brain tumors of grade IV. They are further subclassified based on the tumor location, the predominating type of the neoplastic cells which are either oligodendrocytes or astrocytes, and the major molecular events which contribute to the tumor formation (Louis et al., 2016).

2.1.1.1 Medulloblastoma

MB is defined as an embryonal tumor of the posterior fossa (space located in the bottom of the skull, close to the brainstem and cerebellum) which mainly affects children (Millard & De Braganca, 2016; Udaka & Packer, 2018). MB is classified, according to the WHO guidelines, into five histological subtypes: (1) classic, (2) desmoplastic, (3) anaplastic, (4) large cell, and (5) MB with extensive nodularity (MBEN) (Kijima & Kanemura, 2016; Louis et al., 2016). When considering the molecular events responsible for tumorigenesis, MB is classified into (1) wingless (WNT)-activated, (2) sonic hedgehog (SHH)-activated TP53-mutant, (3) SHH-activated/TP53-wildtype, and (4) non-WNT/non-SHH with further sub-classification into group 3 (G3) and group 4 (G4) of poorly defined pathogenesis and undetermined signaling pathways (Louis et al., 2016).

The most common WNT-MBs display the most promising prognosis among MBs, with the survival rate of 95% (Northcott et al., 2012). WNT-MBs harbor activating somatic mutations in gene-encoding β-catenin (*CTNNB1*) which leads to constitutively active WNT signaling through the stabilization of β-catenin (Northcott et al., 2019). Approximately 30% of all MB cases are SHH-MBs with survival rate of 75% (Northcott et al., 2012). SHH-active tumors are characterized with the dysregulation of SHH signaling. Germline mutations occur in the genes of negative pathway regulators such as *PTCH1* and *SUFU*. Moreover, aberrations in the copy number of target genes such as *MYCN* and *GLI2* are present (Cavalli et al., 2017). Additional mutations of *TP53* in SHH-MBs are associated with extremely poor outcomes (Northcott et al., 2012). The most aggressive MBs, with 5-year overall survival (OS) of <60%, are within G3 subtype (Northcott et al., 2019). To this group belong 25% of all MBs, nearly half metastatic at the time of diagnosis (Northcott et al., 2012). These tumors are characterized with aberrant expression of *MYC* (**my**elocyto**m**atosis) proto-oncogene (Northcott et al., 2019). G4, the most common and least defined subtype, accounts for 35% of all MB tumors and almost 50% of adolescent patients (Northcott et al., 2019). Interestingly, this subtype effects three times more frequently men than women (Northcott et al., 2012).

2.1.1.2 Glioblastoma multiforme

GBM is the most lethal among infiltrating gliomas. The WHO classification of brain tumors from 2016 assigns glioblastoma into a group of diffuse astrocytic and oligodendroglial tumors (Louis et al., 2016). GBM is characterized by genetic instability and alterations in chromosome structure and copy number. To the most significant, somatically mutated genes in GBM belong *IDH (isocitrate dehydrogenase), TP53, PTEN (phosphatase and tensin homolog), NF1, EGFR, RB1,* and *PIK3R1*. Based on mutations in *IDH1* and *IDH2* genes encoding isocitrate dehydrogenase, GBM is classified into IDH-mutant and IDH wild-type subtypes. The most common GBM is of IDH wild-type since it occurs in 90% of all cases (Louis et al., 2016). On the other hand, tumors with mutations in *IDH1* are less aggressive and have better prognosis (Waitkus et al., 2018). *IDH* mutations together with methylation status of the *O*-methylguanine-DNA methyltransferase (*MGMT*) gene promoter are the main biomarkers used by clinicians for treatment selection so far.

2.1.2 Types of ncRNAs and mechanism of action

One of the most novel, promising biomarkers of brain cancers can be noncoding RNAs (ncRNAs) (International Human Genome Sequencing Consortium, 2004), expression of which is generally dependent on the status of cell development, various pathway activation, response to environmental stimuli, as well as pathological conditions (Hombach & Kretz, 2016; Latowska et al., 2020). What is more important in that context is their cell- and tissue-specific expression pattern. In herein chapter, we focus on three types of ncRNAs: microRNA (miRNA), long noncoding RNA (lncRNA), and circular RNA (circRNA), and describe their applications in MB and GBM patients.

2.1.2.1 MicroRNA

MiRNAs are short [19−25 nucleotide (nt)] ncRNAs. They regulate gene expression of more than 30% mammalian genes at the posttranscriptional level (Lewis et al., 2005). Recent analysis based on combined in silico high- and experimental low-throughput validations indicated the existence of 2300 mature miRNAs (Alles et al., 2019).

The canonical pathway of miRNA biogenesis in mammals begins with the transcription of pri-miRNA, which is then cleaved by a ribonuclease III Drosha to the 60−70 nt long pre-miRNA and exported then to the cytoplasm. RNase III Dicer catalyzes the formation of 19−35 nt double-stranded miRNA duplex, which is subsequently loaded into AGO (Argonaute) protein and proceed in a process named strand selection, when generally a strand with the lower thermodynamic stability at the 5′ end becomes degraded (Denli et al., 2004; Han et al., 2004; Meijer et al., 2014; Schwarz et al., 2003). MiRNAs processing can be also through noncanonical pathways, for example, Dicer-independent ones (Stavast & Erkeland, 2019).

In most cases, miRNA interact with the 3′ UTR (untranslated region) of target messenger RNAs (mRNAs) and suppress gene expression. Nonetheless, it is also possible for miRNAs to interact with the 5′ UTR and the coding sequence (Spengler et al., 2016; Wongfieng et al., 2017). Interaction between miRNA and mRNA occurs based on sequence complementarity. The mRNA response elements (MRE) present on transcript pair with a seed region located at position 2−8 nt on the 5′ end of the miRNA (Kiriakidou et al., 2004; Wang et al., 2006). Many sites that match the seed region, particularly those in the 3′ UTR, are preferentially conserved (Friedman et al., 2009). One miRNA can target many different mRNAs and one transcript can be regulated by multiple miRNAs (Balachandran et al., 2020). Interestingly, almost half of the miRNA target genes are located near genomic regions which are associated with the processes responsible for neoplasm formation (Kozomara & Griffiths-Jones, 2011). Therefore the expression of miRNAs in tumors differs from healthy cells and is usually associated with the characteristics of malignancy. Suppressor miRNAs inhibit cancer cells proliferation (Godlewski et al., 2008; Lee et al., 2014; Ruan et al., 2015) and invasion (Lee et al., 2014; Lv et al., 2017; Que et al., 2015; Ruan et al., 2015) and also make chemotherapy more effective (Shi et al., 2014; Tian et al., 2016; Xiao et al., 2016), while oncomiRs promote metastasis (Esquela-Kerscher & Slack, 2006; Huang et al., 2017; Kaid et al., 2015).

2.1.2.2 Long noncoding RNA

LncRNAs are linear transcripts generally longer than 200 nt. They are mostly transcribed by RNA polymerase II, have cap at their 5′ end, and undergo splicing similar to mRNAs. Some of lncRNAs also have 3′ polyadenylated ends, include promoter regions, and display a coding potential for short peptides. Based on their genomic location, lncRNAs are classified as intergenic (located between protein-coding genes), bidirectional (transcribed form the same promoter in the opposite directions), antisense (transcribed form noncoding strand), and sense-overlapping (located on one or more introns/exons in the coding strand direction) lncRNAs (Balas & Johnson, 2018). The fundamental role of lncRNAs is the regulation of transcription; thus their presence may serve as an indicator of transcription activity of many genes (Quinn & Chang, 2016). Moreover, lncRNAs can also interact, and thus, remodel chromatin structure and be involved in posttranslational regulation. The regulatory role of lncRNA is performed through various mechanisms which still need to be uncovered. LncRNAs have the ability to form secondary structures and are assigned as guides, decoys, precursors of small ncRNAs or miRNA sponges, which all strongly affect gene expression in cells (Jarroux et al., 2017). Not only lncRNA products can be regulative but also the process of lncRNA transcription itself may have a regulatory character. This emphasizes the importance of studies on the regulation of lncRNAs transcription. LncRNAs can be localized and, thus, act in either nucleus or cytoplasm or both localizations (Carlevaro-Fita & Johnson, 2019).

A strong tissue-specific expression pattern of lncRNAs has been observed, especially in testis and brain (Derrien et al., 2012). The importance of lncRNAs in different biological processes has also been demonstrated like development, stress, and tumorigenesis (Carlevaro-Fita et al., 2020; Perry & Ulitsky, 2016; Valadkhan & Valencia-Hipolito, 2016). The pathogenic role of lncRNAs has been reported in diseases often affecting nervous system like Alzheimer's disease or heritable Huntington's disease (Johnson, 2012; Zhou & Xu, 2015). Some data also point to the role of lncRNAs in the pathogenesis and progression of brain tumors, including GBM (Han et al., 2012; Li et al., 2016), in which a strong lncRNA deregulation is observed. LncRNAs are found to act either as promoters or suppressors of GBM by serving as miRNA sponges, modifying chromatin or methylating DNA (Mazor et al., 2019).

2.1.3 Circular RNA

CircRNAs are a class of noncoding, single-stranded RNAs with a characteristic structure of covalently closed loop. Compared to mRNA, they lack typical modifications: a cap at 5′ end and a poly(A) tail at 3′ end (Chen & Yang, 2015). CircRNAs are evolutionary conserved among the species, abundantly expressed in human brain, as well as exhibit cell- and tissue-specific expression pattern (Jeck et al., 2013; Memczak et al., 2013; Rybak-Wolf et al., 2015; Salzman et al., 2013). The basic functions of circRNA include the regulation of gene expression through miRNA sponging, interactions with RNA-binding proteins (RBPs) and their sponging. Moreover, circRNAs act as scaffolds for protein complex assembly, participate in transcription regulation, or serve as templates for protein translation (Bose & Ain, 2018; Huang et al., 2020; Mo et al., 2019; Pamudurti et al., 2017; Panda, 2018; Schneider & Bindereif, 2017; Zang et al., 2020; Zlotorynski, 2015). However, the knowledge regarding their basic mechanisms of action and comprehensive interaction network is still missing. CircRNAs can arise from both exons and introns and, thus, are

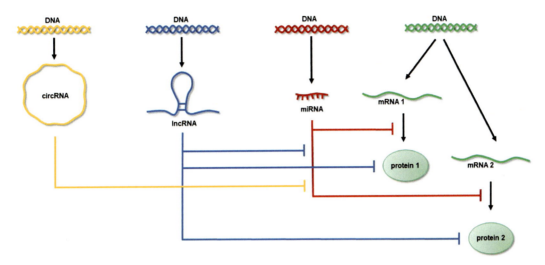

FIGURE 2.1 Competitive endogenous RNA hypothesis. In general, miRNA can inhibit the translation of multiple proteins, circRNAs together with lncRNAs by binding to miRNAs, prevent them from influencing the translation processes. *circRNA*, circular RNA; *lncRNA*, long noncoding RNA; *miRNA*, microRNA.

classified as exonic circRNAs, intronic circRNAs, and exon—intron circRNAs. The biogenesis of exon—intron circRNAs is similar to exon circRNA, except that the intron sequence is preserved and the fusion of circRNA occurs via chromosomal translocations (Greene et al., 2017; Guarnerio et al., 2016; Liu et al., 2019). Most circRNAs are of exonic type, derived from protein-coding genes (Guo et al., 2014). Majority of circRNAs are generated via back-splicing process (Barrett et al., 2015).

CircRNAs are involved in cancer initiation and progression by acting either as tumor suppressors or tumor promoters (also known as oncogenic circRNAs) (Bach et al., 2019). Comprehensive studies on circRNA implicated their role in processes like proliferation, angiogenesis, invasion, metastasis, and senescence (Kristensen et al., 2018). Recent investigations provide evidence for circRNA involvement in human diseases by disrupting circRNA—miRNA—mRNA axis and deregulating signaling of various pathways. NcRNAs like lncRNA or circRNA have a common functionality in their ability to modify gene expression by miRNA titration in a phenomenon known as the competitive endogenous RNA (ceRNA) (Salmena et al., 2011) (Fig. 2.1).

2.2 Data sets for noncoding RNAs analysis

Progressive advancement in the field of bioinformatics enables to deposit and analyze huge amounts of genomic data. Moreover, novel integration methods and developed algorithms facilitate the translation of obtained data into research findings of potential clinical application. In herein section, we aim to present some of commonly used RNA-seq deposit platforms as well as databases that are useful in the analysis of ncRNAs in the studies on brain cancer. Some of commonly used tools for miRNA, lncRNA, and circRNA analyses are listed in Table 2.1.

TABLE 2.1 Deposits and databases for microRNA (miRNA), long noncoding RNA (lncRNA), and circular RNAs (circRNAs) used in brain cancer research.

Databases and tools	Description	Web source	Reference
miRDB	miRNA targets prediction with gene ontology enrichment analysis	http://www.mirdb.org	Chen and Wang (2020)
DIANA	miRNA target prediction, databases of experimentally validated targets and identifying of altered molecular pathways	http://diana.imis.athena-innovation.gr/DianaTools	Paraskevopoulou et al. (2013)
PITA	miRNA targets prediction	https://genie.weizmann.ac.il/pubs/mir07/mir07_prediction.html	Kertesz et al. (2007)
TargetScan		http://www.targetscan.org/vert_72	Agarwal et al. (2015)
RNA22		https://cm.jefferson.edu/rna22	Miranda et al. (2006)
RNAhybrid	Searching for minimal free energy needed for hybridization between long and short RNA	https://bibiserv.cebitec.uni-bielefeld.de/rnahybrid	Rehmsmeier et al. (2004)
LncLocator	lncRNA subcellular localization prediction	http://www.csbio.sjtu.edu.cn/bioinf/lncLocator	Cao et al. (2018)
RNAInter	Gathers experimentally and computationally predicted RNA interactor from over 30 databases	https://www.rna-society.org/rnainter	Lin et al. (2020)
SFPEL-LPI	Prediction of lncRNA-protein interactions	http://www.bioinfotech.cn/SFPEL-LPI	Zhang et al. (2018)
CPPred	Determining lncRNA probability to encode proteins	http://www.rnabinding.com/CPPred	Tong and Liu (2019)
CNIT		http://cnit.noncode.org/CNIT	Guo et al. (2019)
circBase	Unified data sets of circRNAs and scripts to identify known and novel circRNAs	http://www.circbase.org	Glazar et al. (2014)
CIRCpedia v2	Comprehensive circRNA annotation across six different species	http://www.picb.ac.cn/rnomics/circpedia	Dong et al. (2018)
Circbank	Predicting of miRNA binding, coding potential, conservation, circRNA mutation, and methylation	http://www.circbank.cn	Liu et al. (2019)
CircAtlas 2.0	Contains circRNAs expression pattern, conservation, and functional annotation	http://circatlas.biols.ac.cn	Wu et al. (2020)
TSCD	Tissue-specific circRNAs in human and mouse	http://gb.whu.edu.cn/TSCD	Xia et al. (2017)

(Continued)

TABLE 2.1 (Continued)

Databases and tools	Description	Web source	Reference
exoRBase	circRNA, lncRNA, and mRNA in human blood exosomes	http://www.exoRBase.org	Li et al. (2018)
CircFunBase	Experimentally validated circRNAs and predicted functions and interactions	http://bis.zju.edu.cn/CircFunBase	Meng et al. (2019)
deepBase v2.0	The decoding evolution, expression patterns, and functions of ncRNAs	http://biocenter.sysu.edu.cn/deepBase	Zheng et al. (2016)
TRCirc	Transcriptional regulation of circRNAs, expression, and methylation levels	http://www.licpathway.net/TRCirc	Tang et al. (2019)
CSCD	Cancer-related circRNAs	http://gb.whu.edu.cn/CSCD	Xia et al. (2018)
MiOncoCirc		https://mioncocirc.github.io	Vo et al. (2019)
circ2GO		https://circ2go.dkfz.de	Lyu et al. (2020)
CircRiC		https://hanlab.uth.edu/cRic	Ruan et al. (2019)
Circ2Traits	Disease-related circRNAs	http://gyanxet-beta.com/circdb	Ghosal et al. (2013)
Circad		http://clingen.igib.res.in/circad	Rophina et al. (2020)
Circ2Disease		http://bioinformatics.zju.edu.cn/Circ2Disease/index.html	Yao et al. (2018)
CircR2Disease		http://bioinfo.snnu.edu.cn/CircR2Disease	Fan et al. (2018)
circRNA disease		http://cgga.org.cn:9091/circRNADisease	Zhao et al. (2018)
LncRNADisease 2.0		http://www.rnanut.net/lncrnadisease	Bao et al. (2019)
CircInteractome	RBP- and miRNA-binding sites mapping, primers design, and siRNAs, with potential IRES identification	http://circinteractome.nia.nih.gov	Dudekula et al. (2016)
circRNAprofiler	circRNAs detected by multiple annotation-based detection tools combining	https://github.com/Aufiero/circRNAprofiler	Aufiero et al. (2020)
Cerina	Inferring the circRNA function based on the competing endogenous RNA model	https://github.com/jcardenas14/CERINA	Cardenas et al. (2020)
CircCode	Investigation of the coding ability of circRNAs	https://github.com/PSSUN/CircCode	Sun and Li (2019)

mRNA, Messenger RNA; *RBP*, RNA-binding protein; *siRNA*, short interfering RNA.

2.2.1 RNA-seq deposits and data sets

Up-to-date next-generation sequencing (NGS) technologies enable to sequence a noncoding transcriptome, including miRNAs, lncRNAs, and circRNAs, even at a single-cell resolution. The raw data obtained from RNA-seq can be deposited in numerous data repositories and easily accessed by other users. For example, The Cancer Genome Atlas (TCGA), supervised by the National Cancer Institute's (NCI), Center for Cancer Genomics, and the National Human Genome Research Institute, has gathered over 2.5 petabytes of genomic, epigenomic, transcriptomic, and proteomic data using processed sequencing data of 33 cancer types with matched samples from healthy individuals so far. Since processing such a large amount of data remains challenging, a platform to search and download cancer data for analysis has been established. Genomic Data Commons is a research program of the NCI, which serves as a unified data repository and enables sharing cancer genomic data. It provides an access to high-quality data sets from NCI programs, like among others TCGA and Therapeutically Applicable Research to Generate Effective Therapies (Grossman et al., 2016).

2.3 Expression of noncoding RNAs in brain cancer patients

In the last few years, new technologies, such as next generation sequencing (NGS), have collected a significant amount of data, making it easier to compare different specimens. Here, we describe the sample types collected from brain cancer patients which are used to determine ncRNA expression for diagnostic and prognostic purposes. We also present different models to evaluate ncRNA expression for research purposes, with a special emphasis on cancer stem cells (CSCs).

2.3.1 Sample types used for analyzing noncoding RNAs in brain

The samples used to determine the ncRNA expression profile in brain cancer patients may come from dissected tumor tissue (both ante- and postmortem) or biopsy, including liquid biopsy, such as cerebrospinal fluid (CSF) and blood (Fontanilles et al., 2018).

The components of blood-derived biomarkers have been described in brain tumor patients and include circulating tumor cells, nucleic acids, and exosomes. CSF, collected during lumbar puncture or surgical operation, contains circulating tumor DNA, exosomes, and proteins derived from the brain tumor (Fontanilles et al., 2018). MiRNA, lncRNA, and circRNA can be detected in both blood plasma (Chen et al., 2020; Li et al., 2019; Liao et al., 2019; Roth et al., 2011; Wang et al., 2019) and in CSF (Akers et al., 2015; He et al., 2019; Kopková et al., 2019; Wang et al., 2018).

2.3.2 Models to study ncRNA expression

Basic research in a cancer field is mainly focused on studying the involvement of individual molecules, as a part of complex pathways, which are involved in tumorigenesis and cancer progression. In vitro studies are usually conducted on primarily established or commercially available cell lines and organoids. These in vitro models are relatively cheap and easy to cultivate and conducted in a wide range of experiments. Currently, there are 46 human GBM and 4 human MB cell lines banked and available from the American Type Culture Collection (https://www.lgcstandards-atcc.org/). The disadvantage of using GBM established lines, which needs to be considered, is their astrocytic differentiation appearing upon medium supplementation with serum (Robertson et al., 2019). Another drawback is that DNA profile of long-term grown cells may differ from that of the newly purchased cells, which can lead to misidentification of widely used cell lines (Allen et al., 2016). Also, global gene expression revealed that established cell lines display many genomic changes and gene expression fluctuations not present in primary tissues (Li et al., 2008).

To maintain the physiological microenvironment, patient-derived tumor cells can be implanted into immunodeficient mice—an orthotopic patient-derived xenograft model that can arise either by direct inoculation of cell suspension right after surgery, or by inoculation of patient-derived cell cultures with enhanced tumor-initiating potential (Xu et al., 2018).

Another research model possible to be applied in brain cancer studies is genetically engineered mouse model (GEMM) in which tumor is developed de novo via the introduction of gene alterations previously identified in human cancer (Simeonova & Huillard, 2014). This model is obtained by genetic engineering techniques, via utero electroporation and viral gene transfer (Roussel & Stripay, 2020). Unfortunately most of GEMMs possess a *TP53* negative background, whereas *TP53* mutations are observed only in 10% of MB patients (Zhukova et al., 2013). Moreover GBM mouse models lack the intratumor heterogeneity observed in human tumors (Lenting et al., 2017), which indicates the need to adjust the model to the type of cancer under study.

As cell cultures grown in a two-dimensional monolayer do not fully reflect the complexity of tumor mass plus they lack tumor microenvironment which stimuli substantially contribute to tumor initiation and progression, thus there is strong need to introduce the models better reflecting the native tumor (Ivanov et al., 2016). Stem-cell biology has been extensively explored in the past decade, with a special emphasis on CSCs. Along the studies dedicated to CSCs, it has been discovered that these cells display a high proliferation rate (Hamburger & Salmon, 1977) and, thus, can self-renew, as well as are able to differentiate into neurons, astrocytes, and oligodendrocytes (Singh et al., 2004).

GBM stem cells (GSCs) and MB stem cells (MBSCs) could be isolated from primary brain tumor cells using a surface marker of neural stem cells—CD133 (prominin 1) (Ding et al., 2013). Additionally, the pool of stem-cell markers, which gene expression is also increased in CSCs, including NANOG, OCT4, SOX2, BMI1, NESTIN (neuroepithelial stem cell protein), OLIG2, MYC, IDI1, and surface markers CD44, CD15, L1CAM, A2B5 may be used to verify the presence of CSCs (Guo et al., 2011; Ruiz-Garcia et al., 2020). Under neural stem-cell culture conditions, GSC and MBSC grow as 3D culture models—floating multicellular spheroids, neurospheres (Ahmad et al., 2015; Jayakrishnan et al., 2019). Currently, neurospheres are the most commonly used 3D culture model for GBM (Andreatta et al., 2020; Azari et al., 2011; Badodi & Marino, 2019).

GSC have been already shown to be capable of differentiating into endothelial cells allowing for tumor blood vessel formation and may also be involved in the infiltration of GBM into surrounding tissue (Ricci-Vitiani et al., 2010). Furthermore, hypoxia niche for

GSC is hypostatized to promote quiescence, a GSC phenotype that could highly contribute to the enrichment of chemo- and radioresistant subpopulations (Seidel et al., 2010). These features of CSCs have been shown to induce tumor recurrence and progression (Singh et al., 2004). Treatment resistance is widely known phenomenon of glioma patients, as it significantly narrows the possibilities of choosing therapeutic strategy. CSC-mediated resistance response is the subject of wide and comprehensive studies, while this particular fraction of cells is considered to be responsible for glioma aggressive phenotype.

Organoids are 3D structures, commonly derived from patient stem cells, embedded in a matrix. They grow in medium supplemented with a cocktail of growth factors and begin to proliferate and differentiate, self-organizing into an organotypic structure (Andreatta et al., 2020). The first protocol for establishing a culture of cerebral organoids was generated by Lancaster et al. (2013). In a similar way, it is also possible to generate a tumor model directly from patients' biopsies. This method enables to capture the heterogeneity within patients' tumors (Jacob et al., 2020), what makes them a valuable tool for drug screening and precision medicine (Andreatta et al., 2020). Ogawa et al. developed a cancer model of gliomas in human cerebral organoids allowing the observation of tumor initiation. Cells with integrated oncogenic *RAS* (rat sarcoma) expression cassette in the suppressor *TP53* locus rapidly become invasive overwhelming the entire organoid (Ogawa et al., 2018). Another newly proposed model is a neoplastic cerebral organoid, where via transposon- and CRISPR/Cas9-mediated mutagenesis authors introduced into the organoids 18 single gene mutations or amplifications as well as 15 of the most common clinically relevant combinations observed in brain tumors (Bian et al., 2018). Recently, cerebral organoid glioma assembly (GLICO) model has been presented, in which hESC- or iPSC-derived cerebral organoids are cocultured with patient-derived glioma stem cells (Linkous et al., 2019).

2.3.3 NcRNA expression profiles in cancer stem cells and their consequences on tumor development

The presence of CSCs in brain tumors stimulates tumor aggressiveness; thus it is of special interest to identify factors, including ncRNAs, which could shape the specific properties of CSCs and at the same time have potential in therapies. MiRNA expression profile in human neural stem cells and glioma stem cells from patient revealed 116 upregulated and 62 downregulated miRNAs in GSCs. The upregulation of miR-198, miR-146b-5p, miR-152, and miR125a-3p, and downregulation of miR-137 were further confirmed by quantitative reverse transcription polymerase chain reaction (qRT-PCR) analysis (Liu et al., 2014). Studies applying combined microarray and deep sequencing analyses pointed out a set of 10 miRNAs with different expression between GSC and neural stem cells (NSC). Among them, 5 miRNAs were upregulated in GSC including: miR-10a, miR-10b, and miR-140-5p, while downregulated: miR-124 and miR-874 (Lang et al., 2012). Another miRNA microarray research notices 69 differentially expressed miRNAs between the CD133+ and CD133− GBM cells. Lower expression of miR-125b, let-7a, and let-7c, and higher expression of miR-638 and miR-149 were subsequently verified in GSCs. Overexpression of miR-125b inhibited the self-renewal ability of CD133+ cells (Wu et al., 2012).

Global miRNA expression analysis of 10 paired GSC and nonstem GBM cultures concluded with the presentation of 51 most deregulated miRNAs. Total 14 of them correlated to both SOX2 and NESTIN expression: miR-3195, miR-3141, miR-4656, miR-100-5p, miR-4739, miR-3180, miR-1260b, miR-1233-5p, miR4674, miR-328-5p, miR-378h, miR-4505, miR-5787, and miR-1207-5p (Sana et al., 2018). Using the established neurosphere cell lines with and without SOX2-knockdown, the influence of SOX2 on the miRNA profile of GSC was investigated. Forced expression of SOX2 increased levels of miR-128b and miR-425-5p. Additionally an increase in miR-425-5p expression in GBM tumors was proved. Further investigations indicated that miR-425-5p inhibition repressed neurosphere formation, activates apoptosis, and suppress survival pathways in GSCs (de la Rocha et al., 2020).

Studies on the influence of miRNAs on the ability of GSCs to create neurospheres are also a valuable source of information, as they indicate the association of short ncRNAs with the malignant nature of cancer. Overexpression of miR-128 (Godlewski et al., 2008), miR-218 (Tu et al., 2013), and miR-181c (Ruan et al., 2015) decreases the number and volume of neurospheres affecting self-renewal capacity of GSCs.

Expression profiling of MB primary specimens pointed out 21 upregulated and 12 downregulated miRNAs in relative to CD133+ neural stem cells. The most upregulated, with fold change >6, were miR-127-3p, miR-495, miR-409-3p, miR-376c, miR-144*, miR-143, miR-146a, miR-126, miR-126*, and miR-223, while the most downregulated, with the fold change <5, were miR-10a, miR-935, miR-219−2-3p, and miR-504 (Genovesi et al., 2011). Another small RNA sequencing study was performed on SHH MBSCs, and its results were compared to neural stem cells miRNome. In SHH MBSCs, 35 upregulated and 133 downregulated miRNAs were identified. After validation by real-time PCR, higher expression of miR-20a-5p and miR-193a-5p and lower expression of miR-222-5p, miR-34a-5p, miR-345-5p, miR-210-5p, and miR-200a-3p were confirmed (Po et al., 2018). An miRNA profile analysis of the MBSC before and after retinoic acid−induced differentiation revealed 22 miRNAs differentially expressed after 48 h of treatment. It has been confirmed that after differentiation, the levels of miR-195 and miR-145 decrease, while the level of miR-135b increases (Catanzaro et al., 2016).

Total 61 MB patients' analysis indicated higher expression of miR-199b-5p in nonmetastatic cases. Overexpression of that miRNA decreased the MBSC subpopulation in DAOY, medulloblastoma cell line (Garzia et al., 2009). Decrease of CD133+ in DAOY cells, and a reduction in NESTIN staining, was observed also upon miR-34a overexpression (de Antonellis et al., 2011). Studies carried out on various MB cell lines indicated that cells overexpressing miR-367 are more capable of generating expressing CD133 neurospheres than control cells (Kaid et al., 2015). Instead increasing the level of miR-584-5p reduces neurospheres formation together with the reduction of CD133 and MYC markers (Abdelfattah et al., 2018).

An atypical teratoid rhabdoid tumor (ATRT) is a rare, extremally malignant pediatric brain tumor. In CSC subset population isolated from ATRT, miR142-3p downregulation was observed. Overexpression of that miRNA inhibited migration, invasion, and self-renewal of ATRT CD133+ cells (Lee et al., 2014).

lncRNAs have been shown to play essential, regulatory roles in GSCs. Recent studies have revealed differences in splicing and expression profiles of lncRNA present between mesenchymal and proneural subtypes of GSCs, which renders these molecules as

GSC-type specific biomarkers. Moreover, these lncRNAs from differentially expressed pool were found to be associated with GBM prognosis and survival and included CTD-2589M5.5, MYOSLID, CRNDE (Colorectal Neoplasia Differentially Expressed), AC005264.2, SOX21-AS1, RP11–575F12.1 as well as SOX21-AS1 lncRNAs (Guardia et al., 2020). The aberrant lncRNA expression levels have also been identified earlier and showed to affect GSC self-renewal and proliferation processes which, in turn, influence tumor development and progression. Initial microarray studies indicated that uc.283-plus lncRNA was highly expressed in glioma tumors and could affect cell pluripotency (Galasso et al., 2014). Another example is a high expression of lncRNA-ROR (regulator of reprogramming) which has been found in GSCs to negatively correlate to an expression of stemness marker Klf4 and lead to the inhibition of self-renewal and proliferation of these cells (Feng et al., 2015). The expression of other stemness markers SOX2 and nestin was found to be regulated by highly conserved lncRNA MALAT. Its downregulation has been shown to suppress the expression of these two factors and, thus, GSC proliferation (Han et al., 2016). Short interfering RNA (siRNA) downregulation of MALAT1 has also lead to the increased susceptibility of GSC to temozolomide (TMZ) treatment (Kim et al., 2018). Fang et al. (2016) revealed that the downregulation of lncRNA HOTAIR (HOX antisense intergenic RNA) leads to the inhibition of GSC proliferation and a loss of tumor ability to invade adjacent tissues. LncRNAs may also facilitate the interactions between nascent transcripts, like lncRNA FOXM1-AS promoting ALKBH5 and FOXM1 interactions which maintain the pool of GSCs (Zhang et al., 2017). A recent study has indicated TP73-AS1 lncRNA to be clinically significant in GBM. Its high expression levels have been identified specifically in GSCs and linked with poor prognosis as well as resistance to TMZ (Mazor et al., 2019). Current studies related to GSCs biology also explore the function of lncRNAs as "miRNA sponges" since differential expression of lncRNAs has been shown to strongly affect miRNA expression levels. For example, the lncRNA CRNDE has been found highly expressed in GBM tissues, specifically in GSCs, in parallel to downregulated levels of miRNA-186 (Zheng et al., 2015). Further mechanistic studies have revealed that CRNDE binds to and negatively regulates miRNA-186 which, in turn, cannot inhibit the transcription of its target genes *XIAP* and *PAK7*. Thus the overexpression of CRNDE in both in vitro cell culture and in vivo xenograft models promotes GSC proliferation as well as increases cell invasion and migration properties. Described herein studies indicate a strong regulatory role of lncRNA in GSC maintenance and potential to serve as prognostic biomarkers and future targets in therapies against brain cancer.

The expression of circRNA is characterized by dynamic changes during oncogenesis. In cancers and other diseases characterized by increased cell proliferation, decreased circRNA levels are often observed. Presumably, this effect was noticed due to a specific "dilution" caused by high division rate (Bachmayr-Heyda et al., 2015; Moldovan et al., 2019). A growing number of evidence revealed that dysregulated expression of circRNAs is implicated in the onset and development of glioma. A breakthrough event in study of the variability of circRNAs expression was the development of computational pipeline named UROBORUS that allowed for the detection of thousands of circRNAs in RNA-seq data obtained from glioma tissue in comparison with normal tissue samples (Song et al., 2016). The 1411 circRNAs differentially expressed in GBM patients were identified, whereas 1205 were shown to be downregulated and much fewer—206 upregulated

(Zhu et al., 2017). Furthermore, Yuan et al. (2018) took a step further and identified altered expression of 501 lncRNAs, 1999 mRNAs, 2038 circRNAs, and 143 miRNAs between glioblastoma tissue and matched normal brain tissue, revealing their downstream targets in the ceRNA network suggesting that glutamate metabolism might be involved in glioma genesis, development, and infiltration.

circRNAs have been reported to have a role in glioma chemoresistance; however, the detailed underlying mechanisms still needs to be unraveled (Hua et al., 2020; Yin & Cui, 2020; Zhao et al., 2020). The role of exosomes as a new way of circRNAs transfer has been highlighted for cancer diagnosis, therapy, prognosis, and chemoresistance. CircNFIX has been described as upregulated in glioma patient tissues, which is associated with increased glioma cell proliferation and migration by sponging miR-34a-5p. It has been shown that exosomal circNFIX is upregulated in the serum of temozolomide resistant glioma patients and is associated with poor prognosis. Furthermore, the downregulation of circNFIX led to the increase of the TMZ sensitivity in resistant glioma cells in vivo and in vitro. The authors assumed that the regulation of the TMZ resistance in glioma by circNFIX/miR-132 axis might be achieved by the modulation of ABCG2 expression, which have been reported to be a functional target of miR-132 in clear cell renal cell carcinoma (Reustle et al., 2018). Moreover, ABCG2 protein expression is regulated by the circNFIX/miR-132 axis in glioma cells under TMZ exposure and was shown to play a role in the chemoresistance of glioblastoma stem cells (Ding et al., 2020).

The potential role of circASAP1 in glioma chemoresistance has been recently investigated, as the circASAP1 expression was highly upregulated in recurrent GBM patient tissues and TMZ-resistant cell lines. The authors reported increased GBM cell proliferation and the TMZ resistance reduction after circASAP1 knockdown. Moreover, the study showed restored sensitivity to TMZ treatment of TMZ-resistant xenografts obtained by the reduction of circASAP1. The significant regulatory network of circASAP1/miR-502-5p/neuroblastoma Ras (NRAS) has been identified, which shows that the circASAP1 might increase the expression of NRAS via sponging miR-502-5p (Wei et al., 2021).

Blood—brain barrier (BBB) states one of the major challenges of efficient drug delivery in brain tumors. Therefore surgical resection to the extent feasible following by adjuvant radiation therapy is widely used in the clinics for gliomas treatment (Stupp et al., 2005). Since the use of radiotherapy in glioblastoma treatment is constantly evolving and often used as adjuvant temozolomide therapy, there is a way for personalized treatment implementation taking into account patients' performance status (Barani & Larson, 2015). Since circRNAs are known as regulatory and competing endogenous RNAs, the activation of several pathways might have an impact on the chemo- and radioresistance (Jeyaraman et al., 2019). Su et al. revealed 57 upregulated and 17 downregulated circRNAs in human radioresistant esophageal cancer cell line KYSE-150R in comparison with parental cell line KYSE-150, which clearly shows the association of circRNAs with radioresistance in this type of cancer and might be associated with other types of cancer (Jin et al., 2020; Su et al., 2016).

Up to date much remains to be discovered in the field of circRNAs impact on glioblastoma patients' radioresistance. Zhao et al. identified the upregulation of 63 and downregulation of 48 circRNAs in extracellular vesicles (EVs) of radioresistant U251 cell line in comparison to nontreated U251 cells. CircATP8B4 have been discovered to be significantly

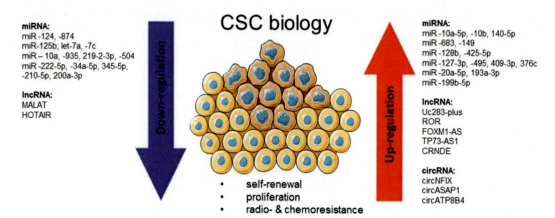

FIGURE 2.2 Aberrant expression profiles of ncRNAs in brain CSCs. The levels of indicated ncRNAs have been found either downregulated (left to the *blue arrow*) or upregulated (right to the *red arrow*) in CSCs isolated from patients' tissues and tissue-derived cell lines. These ncRNAs have been found to influence the biology of CSC, including their ability to self-renew and proliferate. Moreover, modulation in the expression of ncRNA indicated a strong impact of selected ncRNA on tumor resistance to radio- and chemotherapy. *CSC*, cancer stem cell; *ncRNA*, noncoding RNA.

elevated in radioresistant cells and acting as a sponge of miR-766-promoting cell radioresistance. However, future study of the circATP8B4/miR-766/mRNA axis mechanism of action is still needed; hence it is has been discovered that miR-766 acts as a tumor promoter or suppressor in multiple cancer types (Li et al., 2015; Zhao et al., 2019). Fig. 2.2 summarizes the level of different ncRNA expressions in CSC.

2.4 Experimental methods and tools for analyzing noncoding RNAs in brain cancer patients

2.4.1 Studying ncRNA interactions with targets by luciferase assays and ncRNA—protein interactions by immunoprecipitation

Firefly luciferase is commonly used as a reporter to evaluate in vitro transcriptional activity and it is considered the gold standard for the gene regulation studies (Chen et al., 2019; Clément et al., 2015; Jin et al., 2013; Sun et al., 2020; Tomasello et al., 2019; Zhang et al., 2019). The best characterized trait determining miRNA target recognition is pairing within the seed region. However, even 60% of seed interactions could be noncanonical, containing bulged or mismatched nucleotides (Helwak et al., 2013). Therefore target prediction methods based on finding complementarity between the MRE and the seed region may miss a large part of the miRNA—mRNA interaction. Later we present a description of a completely different approach, based on the detection of actually existing miRNA—mRNA interactions, in which the identification of individual elements is the final result, not the initial assumption of the experiment. The use of immunoprecipitation in glioma allowed for the recognition of interactions between lncRNA and miRNA (Fu et al., 2018), as well as circRNA and miRNA (Xu et al., 2018).

Cross-linking immunoprecipitation (CLIP) associated to high-throughput sequencing (CLIP-seq) is a technique used to identify RNA directly bound to RBPs. The application of CLIP-seq to AGO (AGO-CLIP) has been used to identify the miRNA-binding sites (Mato Prado et al., 2016). CLIP protocol begins with ultraviolet (UV) light irradiation of cells, during which covalent bonds between RNA−protein complexes are generated. This reaction only takes place between closely adjacent molecules so that only direct protein−RNA contacts can be cross-linked. Afterword RBPs are purified, mainly by the usage of antibodies. Simultaneously, to facilitate identification of binding sites, cross-linked RNA is partially RNase digested to ∼50 nt. Next proteins are removed with proteinase K leaving pure RNA, which could be used to complementary DNA (cDNA) synthesis and reverse transcription generating templates for sequencing (Jensen & Darnell, 2008). Main disadvantages of CLIP are mutations caused by UV light (Mato Prado et al., 2016). The efficiency of the protein−RNA cross-linking reaction is in the range of 1%−5% (Darnell, 2010). The CLIP technique is undergoing development and currently its new variants are available: the high-throughput sequencing of RNA isolated by CLIP (HITS-CLIP) (Licatalosi et al., 2008), the high-throughput sequencing of RNA isolated from photoactivable−ribonucleoside-enhanced-CLIP (PAR-CLIP) (Hafner et al., 2010), and the individual-nucleotide resolution CLIP (iCLIP) (König et al., 2010).

HITS-CLIP, supplementing the basic technique with bioinformatics approaches, reduces costs and increases the scope of received data by using NGS (Darnell, 2010). In PAR-CLIP a highly photoreactive ribonucleosides, such as 4-thiouridine (4SU) or 6-thioguanosine (6SG), is added into cell culture. As a result during reverse transcription, a characteristic mutation at the position of the cross-linked nucleotide, T−C for 4SU, G−A for 6SG, are introduced, allowing for the separation of cross-linked from noncross-linked input RNAs (Garzia et al., 2018; Spitzer et al., 2014). With HITS-CLIP and PAR-CLIP identification of only the cDNAs that have read through the cross-link site is possible. iCLIP overcoming that limitation by using a primer complementary to the 3′ adaptor, which also contains a 5′ adaptor sequence reducing a mispriming artifacts. Moreover, iCLIP protocol needs relatively low amount of starting material (Gillen et al., 2016; Huppertz et al., 2014).

2.4.2 Validations of RNA-seq results by qRT-PCR and fluorescence in situ hybridization methods

The disease-specific circRNAs have to be not only identified through RNA-seq technique, but they need to be also properly validated. Despite the variety of bioinformatic pipelines using different approaches to identify circular transcripts, there is still a lack of gold standard method for the validation of predicted circRNAs, and the combination of complementary methods is still advised (Holdt et al., 2018). The most widely used methods of circRNAs validation is combination of RNase R treatment and reverse transcription followed by qPCR (qRT-PCR) followed by Sanger sequencing or Northern blotting (Panda & Gorospe, 2018; Xiao & Wilusz, 2019). However, microscopic imaging such as in situ hybridization has become more and more popular method for establishing the abundance and localization of circular transcripts recently (Zaghlool et al., 2018; Hansen et al., 2013). Taking into account the specific closed structure of circRNAs, the basic methods of molecular biology have to be adjusted to differentiate between linear and circular gene transcripts. The presence of both linear and circular transcripts requires precision in experimental design so that the validation process includes only circular transcripts, excluding linear isoforms. The RNA quality plays key role in validation procedure, as well as RNA-seq library preparation since the characteristic head-to-tail junction might get partially degraded during those steps leading to misinterpreted results (Gallego Romero et al., 2014; Sarantopoulou et al., 2019).

qRT-PCR is considered one of the most powerful tools for a circRNAs validation. The fundamental step is divergent primer design, which spans the characteristic head-to-tail junction sequence characteristic only for circRNAs, allowing to differentiate them from linear counterparts. In some cases, low-abundant circRNAs have to be enriched using RNase R treatment prior to qRT-PCR analysis. RNase R exoribonuclease treatment allows for the hydrolysis of linear transcripts, while circular ones are almost fully resistant for the digestion due to its closed structure. Digital droplet PCR (ddPCR) is particularly useful, as it is based on determining the absolute concentration of circRNA by using the quantitative ratio of positive droplets, in which the qPCR product is present to negative ones, allowing for the elimination of errors related to the formation of exon concatamers (Chen, Zhang, Tan, & Jing, 2017). Therefore ddPCR allows to quantitatively analyze genome-wide data sets, providing highly accurate results, what is valuable in low-abundant RNAs research (Pandey et al., 2020). Despite the fact that this method is widely used and relatively simple, it offers some limitations as most of the reverse transcriptase are known to introduce mutations such as template-switching artifacts occurring while joining two distinct RNA molecules or concatamers generation, what might lead to the misinterpretation of the results (Barrett & Salzman, 2016; Szabo & Salzman, 2016). Additionally, it has been reported by many groups that RNase R might deplete also some circRNA or decay the noncircular RNAs not efficiently due to its complex secondary structure or length (Holdt et al., 2018; Vincent & Deutscher, 2006).

Detection and subcellular localization of circRNAs by microscopy is challenging due to the fact that the only way to distinguish circular transcripts from the linear ones is the presence of a head-to-tail junction. The circRNA molecules can be visualized and quantified using fluorescence in situ hybridization (FISH) coupled with high-resolution microscopy, applying highly sensitive junction-specific probes to avoid simultaneous detection of their linear counterparts (Zirkel & Papantonis, 2018). Variety of available protocols enable high specificity and sensitivity of circRNAs detection, allowing for single RNA molecule visualization. Short transcripts detection might be weak due to the insufficient fluorescence signal from a single bound probe; therefore it can be enhanced by the hybridization of multiple amplifier probes (Itzkovitz & Van Oudenaarden, 2011). However, this approach has been reported to produce less quantitative results (Kocks et al., 2018). FISH, combined with immunohistochemistry, may be applied to examine the colocalization of circRNAs with proteins (Bejugam et al., 2020; Huang et al., 2020).

Northern blotting can be applied to validate circRNAs presence, but it requires probes spanning the head-to-tail junction site. However, Northern blotting is still limited by the circRNAs length, to molecules ranging 0.2–1 kb (Pandey et al., 2020). Circular and linear transcripts vary in the case of migratory potential in polyacrylamide gels, where circRNAs tend to migrate slowly in comparison to the linear one what makes 2D gel methods useful in circRNAs validation process. CircRNAs can be either trapped in agarose gel by sulfide bridges (Gel trap) or reported to migrate slower in polyacrylamide gels. The methods presented in this section contain a number of advantages and disadvantages, which are summarized in Table 2.2.

TABLE 2.2 The advantages and disadvantages of the noncoding RNA (ncRNA) analysis and manipulation methods.

Methods	Advantages	Disadvantages
Approaches used to study and validate the ncRNAs		
NGS	High-throughput method, high sensitivity, and discovery power	High costs and complexity, RNA quality, and RNA-seq library preparation play key role to prevent misleading results
qRT-PCR	Low costs, simultaneous amplification, and detection during amplification	Usage of error-prone reverse transcriptases, limited primer designing in circRNA and miRNA, problems of specificity due to miRNA families similarity, circRNA concatemers formation, low-abundant circRNAs have to be enriched
ddPCR	Absolute quantification of a target with no standard curve requirements provides highly accurate results even in low-abundant RNA sample, lower sensitivity to PCR inhibitors, elimination of errors related to the formation of concatemers in circRNA study	High costs and complexity, limited primer design in circRNA and miRNA
FISH	High sensitivity and specificity in recognizing targeted sequences, visualization of hybridization signals at the single-cell level and ability to detect cell-to-cell variations	Background noises, limitation to cover abundant ncRNAs in cell, off-target hybridization, limited probe design in circRNA and miRNA
Northern blotting	Detection of RNA size and alternative splice products, the membranes can be reprobed long time after blotting	High complexity, requires a large amounts of total RNA, limited probe design in circRNA and miRNA, length limited to molecules ranging 0.2–1 kb
RNase R treatment	Low costs, easy circRNAs enrichment method	Possibility to deplete some circRNA or decay the noncircular RNAs not efficiently
Luciferase assay	Low cost, quickness, and precision of the experiment	Requirement for exogenous substrates, the delay from stimulus to response
CLIP	Determine the binding site with high accuracy, avoid protein–protein cross-links	Mutations caused by UV light
Pull-down	Relatively low input, detection of weak interactions, higher experimental flexibility	Background noises due to nonspecifically binding proteins may occur

(Continued)

TABLE 2.2 (Continued)

Methods		Advantages	Disadvantages
Approaches used to manipulate the expression level of ncRNAs			
miRNA mimics		No or easy available carrier system needed, no vector-based toxicity, easy to use	Relatively high costs, susceptible to ceRNAs and competitive endogenous RNA-binding protein interactions, off-target effects, risk of interferon response
Antisense oligonucleotides	Antagomirs	Nuclease resistant, can be delivered into cells directly without any delivery vehicles	High usage dose, possible off-target effects
	2'-O-methyl group–modified AMOs	Less sensitive to degradation in serum or by endogenous cellular exonucleases and endonucleases	2'-O-methyl modification decreases the binding of the AMO to its target miRNA
	LNAs	High efficiency of miRNA activity blocking, high stabilization of the LNA/RNA duplex, strongly resistant to nuclease degradation, applied in detection methods	Design requires careful examination, sequences without extensive self-complementary segments
	PMOs	Characterized by stronger steric blocking of nucleases and preventing degradation, highly efficient penetration in multiple tissues in vivo	Requires careful optimization regarding the PMOs length, lower binding affinity than equivalent PNAs
	PNAs	High target affinity, specificity, nucleases resistance, and cell penetration, applied in detection methods	Limitation of solubility and the structural flexibility of nucleic acid recognition
miRNA sponges		Transient and stable miRNA sponges depending on the needs, binding sequences for a family of miRNAs can be constructed together in one vector	Number of binding sequences affects sponge activity, different degrees of inhibition in different contexts, challenging optimization and validation
CRISPR/Cas9		Low costs, high simplicity and efficiency, new CRISPR-Cas13 approach represents a promising method for targeting a specific circRNA	Off-target effects, challenging and time-consuming optimization, single site of gene editing may not be sufficient to manipulate microRNA stem-loop structure and function, depleting the circRNAs without affecting the existing genes remains challenging
siRNAs		Low costs, easily introduced to the cells, predesigned reagents available, many chemical modifications available	Temporary gene silencing, incomplete silencing, off-target effects, transfection difficulties, stimulation of immune response

(Continued)

TABLE 2.2 (Continued)

Methods	Advantages	Disadvantages
shRNAs	Permanent gene silencing, high efficiency, reduced off-target effects, predesigned reagents available	Virus-mediated toxic effects, technologically challenging and time-consuming, BSL2 requirement
Small molecule chemical compounds	Typically low costs, easily introduced to the cells, various half-live, and bioavailability	Off-target effects, adverse reactions possible, various half-live, and bioavailability

AMOs, anti-miRNA oligonucleotides; *ceRNA*, competitive endogenous RNA; *circRNA*, circular RNA; *CLIP*, cross-linking immunoprecipitation; *ddPCR*, digital droplet PCR; *FISH*, fluorescence in situ hybridization; *LNA*, locked nucleic acid; *miRNA*, microRNA; *ncRNA*, noncoding RNA; *NGS*, next-generation sequencing; *PMO*, phosphorodiamidate morpholino oligonucleotide; *PNA*, peptide nucleic acid; *qRT-PCR*, quantitative reverse transcription polymerase chain reaction; *shRNA*, short hairpin RNA; *siRNA*, short interfering RNA; *UV*, ultra violet.

2.5 Noncoding RNAs as predictive marker for brain cancer patients

2.5.1 Diagnostic potential of noncoding RNAs in brain cancer patients

Diagnosis of brain cancer is based on neurological examination and confirmed by standard imaging technique—computed tomography and magnetic resonance imaging scans. Other techniques that are used to diagnose brain cancer include diffusion tensor imaging, single photon emission–computerized tomography, positron emission tomography, cerebral angiography, magnetic resonance spectroscopy, and biopsy (Shah & Kochar, 2018; Shahpar et al., 2016). Nowadays, molecular analysis detecting ncRNAs of diagnostic potential in tissue biopsies and body fluids collected from the patients is also possible. By reason of the extensive identification of disease-related ncRNAs and their cell- and tissue specificity, it is believed that they might serve as promising biomarkers and provide new targets for the treatment. Traditional biomarkers currently used in the clinics allow to establish more accurate diagnosis but have disadvantages in the form of lower organ specificity and low positive detection level (Meng et al., 2017). NcRNAs can support the traditionally used diagnostic biomarkers to increase the positive diagnosis rate, as they have been reported to be abundant in saliva, exosomes, blood samples, or CSF. Given in this chapter examples, summarized in the Table 2.3 show the diagnostic and stratification value for brain cancers' patients. Given examples, however, show that the level of ncRNA expression in the tissue does not necessarily correspond to the amount of molecules released into the serum, for example, miR-454-3p.

Determining the degree of malignancy of the neoplasm is important since it allows to decide on the further treatment method and also allows to determine the initial prognosis for the patient. Thus it has become an important aspect to look for biomarkers that help to distinguish low-grade gliomas (LGG) from high-grade gliomas (HGG)—Table 2.3. TCGA data analysis indicated miRNAs that may be considered cerebral LGG prognosis markers: miR-1287, miR-326, and miR-1275 (Liu et al., 2017). Furthermore, comparison of the miRNA profile with the occurrence of mutations in *IDH1/2* genes determined that 74% of 487 considered miRNAs were differentially expressed according to *IDH1/2* mutation status in LGGs (Cheng et al., 2017).

TABLE 2.3 MicroRNA, long noncoding RNA, and circular RNAs with diagnostic potential in brain cancers.

RNA	Tumor type	Expression level	Sample type	Reference
Diagnosis				
miR-129	GBM, MB, ATRT, ependymoma, pilocytic astrocytoma	Down	Fresh tumor tissue	Birks et al. (2011)
miR-142-5p, miR-25		Up		
miR-1259, miR-21, miR-222	Pediatric brain cancers	Up	Fresh tumor tissue	Yuan et al. (2018)
miR-128		Down		
miR-221, miR-222	GBM	Up	Blood	Swellam et al. (2019)
miR-222, miR-7, miR-137	GBM	Down	Formalin-fixed parafin-embedded tissue	Visani et al. (2014)
miR-454-3p	GBM	Down	Fresh tumor tissue and cell lines	Shao et al. (2019)
	GBM	Up	Preoperative exosomes	
	GBM	Down	Postoperative exosomes	
miR-15b	GBM	Up	Cerebrospinal fluid	Baraniskin et al. (2012)
miR210HG	GBM	Up	Fresh tumor tissue and serum	Min et al. (2016)
NFIX	GBM	Up	Fresh tumor tissue	Ding et al. (2020)
ITCH	GBM	Up	Fresh tumor tissue	Li et al. (2018)
SKA3 DTL	MB	Up	Fresh tumor tissue	Lv et al. (2018)
Stratification				
miR-7, miR-10b, miR-137, miR-223	GBM WHO grade I	Down	Fresh tumor tissue	Visani et al. (2014)
miR-21, miR-34a		Up		
miR-7, miR-10b, miR-137, miR-22	GBM WHO grade II	Down		
miR-34a		Up		
miR-7, miR-101, miR-137, miR-22	GBM WHO grade III	Down		
miR-4516	GBM WHO grade IV (compared to grade II and III)	Up	Fresh tumor tissue	Cui et al. (2019)

(Continued)

TABLE 2.3 (Continued)

RNA	Tumor type	Expression level	Sample type	Reference
miR-21	HGG and LGG	Up	Cerebrospinal fluid exosomes	Shi et al. (2015)
miR-21	GBM WHO grade IV (compared to WHO grades II–III)	Up	Fresh tumor tissue	Berthois et al. (2014)
miR-200a	GBM WHO grade IV (compared to WHO grades II–III)	Down		
miR-637	GBM	Down	Fresh tissue fresh tumor tissue	Que et al. (2015)
miR-1825	GBM	Down	Serum	Xing and Zeng (2017)
miR-1238	GBM recurrent (compared to primary tumor)	Up	Serum	Yin et al. (2019)
miR-204	MB WNT and 80% of MB G4	Up	Fresh tumor tissue	Bharambe et al. (2019)
	MB SHH and 60% of MB G3	Down		
miR-193a-3p, miR-224, miR-148a, miR-23b, miR-365, miR-10b,	MB WNT	Up	Fresh tumor tissue	Kunder et al. (2013)
miR-182, miR-135b, miR-204	MB SHH	Down		
miR-135b	MB G3 and MB G4	Up		
HOTAIR CRNDE	High-grade astrocytoma	Up	Fresh tumor tissue	Zhang et al. (2012)
LOC286002, C21orf131-A, C21orf131-B	High-grade oligodendrogliomas	Up		

ATRT, an atypical teratoid rhabdoid tumor; *GBM*, glioblastoma multiforme; *HGG*, high-grade gliomas; *LGG*, low-grade gliomas; *MB*, medulloblastoma; *SHH*, sonic hedgehog; *WNT*, wingless.

Noteworthy the diagnostic and prognostic value of miRNA was used to provide the stratification analysis. It allowed to distinguish five subgroups of GBM with different clinical properties and of different origins: (1) oligoneuronal, (2) radial glial, (3) neural, (4) neuromesenchymal, and (5) astrocytic (Kim et al., 2011). Undoubtedly, finding new biomarkers could help in the treatment of GBM at early onset and prevent from disease progression.

2.5.2 Prognostic potential of noncoding RNAs in brain cancer patients

There is a wide relationship between prognostic and predictive biomarkers, which allows to predict the response of the patient to a targeted therapy, as some of them can perform both functions and distinguishing them can be difficult. In the case of brain tumors, especially gliomas, classical prognostic evaluation is based on the analysis of phenotypical traits, including age, tumor location, histological grade, and performance status. Poor prognosis is associated with older age, where patients aged 70–79 years exhibited a median survival of 2.9 months and patients above 80 years only 1.9 months, poor preoperative performance status, and size and location of the tumor, with the worst outcome for GBM, which is characterized as extensively spreading tumor type (McDonald et al., 2012; Ohgaki et al., 2004). Prognostic markers can predict the course of the disease and help determine which patient should be treated and how. A prime example of a prognostic marker is the tumor histology: LGG have a better prognosis than HGG (Davis, 2018). New affordable methods are needed to help prognosis and to suggest the most appropriate treatment for patients' brain cancer. Recognition of the patient's prognosis could be based on the level of ncRNAs expression in a biopsy or liquid biopsy collected during standard treatment.

Comparing the medical history of patients suffering from brain cancers, with the results of miRNA profiling, allows for the identification of miRNA sets predicting possible outcomes of a disease. Expressions of miR-767-5p and miR-105 are positively correlated to OS, whereas miR-miR-584, miR-296-5p, and miR-196a negatively correlated to OS in anaplastic gliomas regardless of histology type (Yan et al., 2014). Study based on formalin-fixed paraffin-embedded GBM tissue sections developed a four-miRNA signature that could identify GBM patients: let-7a-5p, let-7b-5p, and miR-125a-5p have appositive effect on OS, hsa-miR-615-5p negative (Niyazi et al., 2016). Analysis of microarray data from GBM tissues elicited a set of deregulated miRNA. The expression of miR-124a, miR-129, miR-139, miR-218, and miR-7 was downregulated, and the expression of miR-15b and miR-21 was upregulated. The association with disease-free survival (DFS) demonstrates that the risk score established by those miRNAs was more effective than other criteria that are traditionally used, including Karnofsky Performance Status (KPS), tumor location, recurrence status, *MGMT* methylation, *IDH1* mutation, smoking, and family history of cancer (Chen et al., 2016). High abundance of mentioned well-known oncomiR, miR-21 (Pfeffer et al., 2015), defined a worse prognosis in primary glioma patients also in another study (Shi et al., 2015).

Relapsed GBM tissues were characterized by a lower expression level of miR-203 with simultaneous overexpression of its target, snail family transcriptional repressor 2. OS of patients with higher expression levels of miR-203 was longer (Liao et al., 2015). Better OS was observed as well in patients with higher miR-29c level, which increases chemotherapy efficacy (Xiao et al., 2016). Longer OS together with DFS was related to higher miR-29a/b/c levels. The opposite relation applied to the tumor necrosis factor receptor-associated factor-4 (TRAF4), which promotes apoptosis through the TRAF4/AKT/MDM2 pathway in a p53-dependent manner (Shi et al., 2018). Worse OS prognosis had patients with low expression levels of miRNA-320c, which impair migration and invasion of glioma cells via reducing the expression of matrix metalloproteinases 2 and 9 (MMP2, MMP9) *N*-cadherin and integrin $\beta1$ (Lv et al., 2018). The results of Kaplan–Meier analysis revealed that patients with low expression of miR-599 had not only significantly shorter OS but also poorer progression-free survival (PFS). MiR-599 downregulation was significantly correlated to KPS and WHO grading (Zhu et al., 2018). PFS was shorter likewise in patients with lower expression of miR-101, considering it sensitizes resistant GBM cells to chemotherapeutic drug (Tian et al., 2016).

Decreased exosomal miR-151 level in patients' CSF indicated worse prognosis, as induced expression of miR-151 in cell lines inhibited XRCC4 DNA repair mechanism (Zeng et al., 2018). miR-454-3p high expression in serum exosomes or on the other hand low expression in glioma tissue were correlated to a poor survival rate. Forced expression of miR-454-3p in cell lines autophagy as its direct target is mRNA coding autophagy-related 12 protein (ATG12) (Shao et al., 2019). Another prognostic marker was found in a glioma patients' serum—miR-1825. Its level was significantly decreased in patients compared with healthy controls and those with high miR-1825 expression had a longer survival rate (Xing & Zeng, 2017).

For brain tumors other than GBM, research on prognostic miRNAs is not that well developed; however, also here it is worth paying attention to a few examples. In the combined cohort of G3 and G4 MBs, tumors with miR-182 overexpression were found to correlate to worse survival rates, while those with miR-592 overexpression with better survival rates (Kunder et al., 2013). The 5-year OS rate of patients with high miR-495 expression in MB tissue was approximately 25 percentage points higher, compared with patients with low miR-495 expression (Wang et al., 2015). In the case of patients with metastatic inflammatory breast cancer higher serum miR-141 level was found, compared to patients with locally advanced breast cancer. High miR-141 levels were associated with shorter brain metastasis—free survival. Furthermore, knockdown of miR-141 inhibited metastatic colonization to brain in mice (Debeb et al., 2016).

The aberrant expression of lncRNAs has also been shown to correlate to tumor progression and patients' survival which indicates that lncRNA could serve as biomarkers of prognostic potential. Among many the most well-studied is HOTAIR that acts as miR-326 and miR-148b-3p sponge and displays high expression linked to poor prognosis and patient survival (Zhang et al., 2015). While CRNDE stimulates GBM proliferation and migration, serves as a sponge for miR-337-3p, and its elevated levels also correlate to poor prognosis (Gao et al., 2020). Among lncRNAs which act as tumor suppressors and their downregulation that leads to tumor progression and is linked with poor patient survival are MALAT1, TUSC7, and CASC2. MALTA1 executes its function via the deactivation of ERK/mitogen-activated protein kinase (MAPK) pathway, while TUSC7 is known as miR-23b and miR-10a sponge and CASC2 interacts with miR-181a and miR-193-5p affecting WNT signaling (Han et al., 2016; Jiang et al., 2018; Liao et al., 2017; Shang et al., 2016).

The clinicopathological and prognostic value of circRNA in onset and progression of glioma was recently reviewed. Metaanalysis included 24 eligible studies and a total of 1390 patients, with the analysis of two groups of circRNAs—tumor suppressors and carcinogenic tumor promoters. They revealed the correlation between high expression of tumor-promoter circRNAs and significantly poor clinicopathological features, and high expression of tumor-suppressor circRNAs and better clinicopathological features, indicating the circRNAs being related to the occurrence and development of glioma (Ding et al., 2020). Moreover, recent reports indicate that some circRNAs might undergo the translation process generating peptides and proteins. It has been shown that protein products encoded by ncRNA state a promising prognostic biomarker for cancer patients. SHPRH-146aa, encoded by circRNA SHPRH (SNF2 histone linker PHD RING helicase), has been identified as decreased in GBM patients, leading to tumorigenicity reduction through protecting full-length SHPRH, which ubiquitinates proliferating cell nuclear antigen (Zhu et al., 2018).

2.6 Potential of noncoding RNAs in predicting chemoresistance and radioresistance in brain cancer patients

Treatment strategies for brain cancer are either to excise the tumor or stop its propagation. Surgery, chemotherapy, and radiation therapy are paramount therapeutic approaches. Despite the advances in medicine, the treatment results of brain tumors are usually not spectacular since tumors display a strong resistance to radiotherapy and chemotherapy, often infiltrate and recur stimulated by stem cells. CD133+ stem cells in GBM are then considered to be associated with increased resistance to radiation and chemotherapy.

miRNA profiles comparison between two laboratory-derived GBM cell lines with acquired resistance to TMZ and their primary counterparts, together with the comparison between 12 recurrent TMZ-refractory GBM samples and primary tumors, revealed a group of miRNAs with altered expression. TMZ-resistant cell lines and specimens were distinctive for the upregulation of miR-9, miR-182, and downregulation of miR-29c, miR-93, and miR-101. Among them, miR-101 expression rendered glioma cells more sensitive to TMZ both in vitro and in vivo as it has been shown by increased cell apoptosis and decreased tumor growth. Direct target of miR-101 is a glycogen synthase kinase 3 beta (Pyko et al., 2013), which has been reported to affect MGMT expression via promoter methylation (Xiao et al., 2016). As MGMT is the DNA repair protein, it plays a critical role in TMZ resistance. MGMT expression has been shown to be also dependent on miR-29c, which suppresses its expression indirectly via targeting specificity protein 1 (Xiao et al., 2016).

The miRNA expression profiles from 82 primary GBM samples indicated that the TZM chemoresistant subtype is characterized by the overexpression of miR-1280, miR-1238, miR-938, and miR-423-5p and decreased expression of let-7i, miR-151-3p, and miR-93 (Yan et al., 2015). Further research also confirmed that miR-1238 (Yin et al., 2019) and miR-432-5p (Li et al., 2017) expression levels were significantly higher in secondary GBM samples than in primary ones, while miR-151a (Zeng et al., 2018) was significantly decreased in recurrent compared to paired primary GBM samples. Both established and primary human GBM cell line with induced TMZ chemoresistance displayed expressed higher miR-1238 levels than the respective initial cell lines. The proposed mechanism of action presents that loss of CAV1, the direct target of miR-1238, induced the activation of the EGFR-PI3K-Akt-mTOR pathway, leading to enhanced antiapoptosis of TMZ-treated GBM cells (Yin et al., 2019). Overexpression of miR-432-5p in glioma cell lines decreased TMZ chemosensitivity, which could be reversed by forced expression of Ing-4, its validated target (Li et al., 2017). In the case of miR-151a, its overexpression in xenografts decreases a tumor volume translating into better survival of animals (Zeng et al., 2018).

miR-302c was significantly downregulated in glioma brain tissues when compared with control tissue, furthermore its level was decreased in progressive disease samples compared with partial response glioma tissues. miR-302c was also reduced in cell line with induced TMZ resistance, and its overexpression resensitized cells to TMZ by direct targeting of P-glycoprotein known as multidrug resistance protein 1 (Wu et al., 2019). Another study found that tissue specimens are also characterized by the downregulation of miR-124 targeting—related Ras viral oncogene homolog (R-Ras) and NRAS viral oncogene homolog (N-Ras). Forced miR-124 expression in glioma cell lines significantly increased their chemosensitivity to TMZ treatment (Shi et al., 2014). GBM tissue samples and cell line have lower expression levels of miR-128-3p, which confirmed that target c-Met is responsible for chemoresistance. miR-128-3p enhanced the therapeutic effect of TMZ by the induction of apoptosis from 16% to 28% in vitro and reduction of tumor volume in vivo (Zhao et al., 2020). miR-224-3p induced enhanced chemosensitivity that was observed both in vitro and in vivo. At the molecular level, miR-224-3p downregulated hypoxia-inducible factor 1-alpha and autophagy-related gene 5 (Huang et al., 2019). MiRNA profile of TMZ-resistant glioma cell lines revealed the upregulation of miR-138. Overexpression of that miRNA or knockdown of its target Bcl-2-like protein 11 enhanced autophagy, which increases the chemoresistance of cells (Stojcheva et al., 2016). Significantly upregulated in tumor tissues compared with peritumoral brain edema miR-223 directly targets paired box protein (PAX6) and promotes the proliferation of GSCs after TMZ treatment by PI3K/Akt pathway (Huang et al., 2017).

With reference to CSC, it was observed that miR-29a was significantly downregulated in cisplatin-treated CD133+ GBM cells. Moreover, miR-29a significantly increased apoptosis of CD133+ after cisplatin treatment. Similarly, in xenograft mice, when combined with miRNA-29a overexpression, treatment with cisplatin significantly inhibited tumor growth than treatment with cisplatin alone (Yang et al., 2018). Profiling of glioma cell line with generated cisplatin resistance revealed a set of miRNAs with differential expression, among them let-7b greatly resensitized cells to cisplatin (Guo et al., 2013). miR-873, downregulated miRNA in glioma samples, has also lower expression level in cisplatin-resistant cell lines to the respective wild-type cells. miR-873 increased induced apoptosis in both the cisplatin-resistant and wild-type glioma cells. Its effect could be reversed by the reexpression of its targets, apoptosis regulator protein Bcl-2 (Chen et al., 2015).

miRNA microarray analyses of three pairwise radioresistant and parental GBM cell lines identified between these two groups 113 statistically significant deregulated miRNAs. The most upregulated miRNAs in radioresistant GBM cells were miR-145, miR-10b*, miR-204, miR-1231, miR-4721, miR-4697-3p, miR-10a-star, miR-4725-3p, and miR-4498, while the most downregulated were miR-1271, miR-29b, miR-3065-5p, miR-2467-3p, and miR-1290 (Ondracek et al., 2017).

Expression analysis of glioma tissue specimens from patients treated with radiotherapy for over 6 months showed that miR-320 is notably decreased in radioresistant glioma tumors compared with radiosensitive samples. Overexpression of miR-320 increased caspase-3 activity and DNA double-stranded breaks frequency in irradiated cells. Proposed model of action assumes a direct targeting of forkhead box protein M1 followed by the downregulation of sirtuin type 1 (Li et al., 2018). Another tissue miRNA expression research revealed that the expression of miR-183 was significantly increased in resistant specimens. Cell line transfected with miR-183 mimic and then radiated decreased cell apoptosis, while in vivo miR-183 decreased tumor volume (Fan et al., 2018). Irradiated glioma cell lines were characterized by increased level of miR-21. PI3K-AKT signaling pathway inactivation impairs DNA repair following γ-irradiation. Anti-miR-21 suppressed the phosphorylation of protein kinase B, thus sensitized cells to radiation (Gwak et al., 2012).

GSCs are more radioresistant than glioma cell lines, demonstrating no reactive oxygen species generation after 1 h of 8 Gy X-ray irradiation. miR-153 overexpression made GSCs more radiosensitive and slightly raised their oxygen enhancement ratio (Yang et al., 2015). CSC subset isolated from ATRT was defined by a low expression level of miR142-3p. After an ionizing radiation the proliferation of CD133+ cells with overexpressed miR142-3p was significantly lower than those with basal expression. Moreover, miR142-3p overexpression markedly reduced tumor growth, and with additional irradiation, reduced the tumor volume to a barely detectable size (Lee et al., 2014).

lncRNAs have also been reported to contribute to brain tumor radio- and chemoresistance. The most distinctive lncRNAs involved in cell resistance presents Table 2.4. The highest potential so far displays lncRNA H19. Its upregulated levels have been noted in glioma cells and its downregulation by shRNA (short hairpin RNA) technology leads to decrease in β-catenin expression and increased susceptibility to TMZ (Jia et al., 2018). Similar, oncogenic features of H19 were reported by Jiang et al. (2016) who silenced H19 with siRNA and also observed tumor decrease in TMZ chemoresistance. H19 has also been shown to activate a key protein of NF-κB (nuclear factor kappa-light-chain-enhancer of activated B cells) signaling which was revealed by both H19 silencing and overexpression experiments (Duan et al., 2018). H19 overexpression and simultaneous treatment with NF-κB inhibitor has led to the eradication of TMZ resistance in glioma cells. As mentioned earlier, lncRNAs often serve as miRNA sponges thus aberrations in lncRNA expression affects also miRNA and its target expression. The expression of lncRNA CASC2 was downregulated in glioma cell lines and tissues, but its overexpression leads to the inhibition of miR-181a expression which affected its target expression PTEN and p-AKT. As a result, the cell resistance to TMZ weakened (Liao et al., 2017). Two clinical studies on glioma tissues have revealed a high expression level of UCA1 lncRNA which ensures chemoresistance. In vitro studies have shown that its knockdown not only suppresses the activity of WNT/β-catenin signaling but can also affect levels of fibronectin, COL5 A1 and ZEB1, by sponging miR-204-5p (Liang et al., 2018; Zhang et al., 2019). The pathway, which involvement has also been noted in glioma chemoresistance, is PI3K/Akt. This pathway can be regulated by two lncRNAs DANCR and MSC-AS1. DANCR increased levels promote resistance to cisplatin treatment, while the increase in MSC-AS1 expression induces TMZ resistance. Less studies are dedicated to the involvement of lncRNAs in radioresistance especially in MB. Recent studies have shown that lncRNA TPTEP1 inhibits glioma radioresistance by competitive interaction with miR-106a-5p and activation of p38 MAPK signaling (Tang et al., 2020). Interesting studies have been performed by Zheng et al. (2016) who first established that lncRNA SNHG18 represses semaphorin 5S and thus promotes radioresistance in glioma cells and xenograft models. Further mechanistic studies revealed that SNHG18 inhibits nucleocytoplasmic transport of ENO1 protein that promotes glioma cell motility (Zheng et al., 2019). In the case of MB, increased chemosensitivity to cisplatin has been observed upon the inhibition of lncRNA CRNDE, which also serves as a sponge of miR-29c-3p (Sun et al., 2020).

CircASAP1 which biogenesis is induced by eukaryotic translation initiation factor 4A3 (EIF4A3) has been shown to promote GBM proliferation and TMZ resistance and to be associated with poor GBM patients' prognosis. Moreover, circASAP1 sponges miR-502-5p, what leads to the deregulation of its target—NRAS expression, enhancing tumor growth and TMZ resistance in vitro and in vivo. NRAS is a member of the RAS family, encoding membrane-bound protein known to function as regulatory element in the signal transduction events of molecules regulating cancer cells survival and proliferation (Wei et al., 2021).

Recently, circRNAs shown to play a relevant role in cancer have been also identified in EVs present in patients' blood. The growing number of evidence shows that

TABLE 2.4 The list of long noncoding RNA (lncRNAs) and circular RNAs (circRNAs) targeting microRNAs (miRNAs) and proteins as well as cross-talked with pathways assuring brain tumor chemo- and radioresistance.

	lncRNA/ circRNA	miRNA sponged	Target protein	Pathway involved	Reference
Chemoresistance in glioma	H19		β-Catenin, c-myc, survivin	Wnt/β-catenin	Jia et al. (2018)
	H19		MDR, MRP, ABCG		Jiang et al. (2016)
	H19		NF-κB	NF-κB	Duan et al. (2018)
	XIST	miR-29c	SP1, MGMT		Du et al. (2017)
	NCK1-AS1	miR-22-3p	IGF1R		Wang et al. (2020)
	CCAT2	miR-424	CHK1		Ding et al. (2020)
	CASC2	miR-181a	PTEN, p-AKT	PTEN	Liao et al. (2017)
	LINC00174	miR-138-5p	SOX9		Li et al. (2020)
	MALAT1	miR-203	TS		Chen et al. (2017)
	MALAT1	miR-101			Cai et al. (2018)
	HOXD-AS1	miR-130a-3p	ZEB1		Chi et al. (2018)
	KCNQ1OT1	miR-761	PIM1		Wang et al. (2020)
	UCA1			Wnt/β-catenin	Zhang et al. (2019)
	UCA1	miR-204-5p	Fibronectin, COL5 A1, ZEB1		Liang et al. (2018)
	CASC2	miR-193a-5p	mTOR		Jiang et al. (2018)
	BC200	miR-218-5p			Su et al. (2020)
	DANCR	miR-33a-5p, miR-33b-5p, miR-1-3p, miR-206, miR-613	AXL	PI3K/Akt	Ma et al. (2018)
	AC023115.3	miR-26a	GSK3β		Ma et al. (2017)
	GAS5				Huo and Chen (2019)
	MSC-AS1	miR-373-3p	CPEB4	PI3K/Akt	Li et al. (2020)
	LIFR-AS1	miR-4262	NF-κB		Ding et al. (2020)
	LINC01198		NEDD4-1, PTEN		Chen et al. (2019)
	SBF2-AS1	miR-151a-3p	ZEB1		Zhang et al. (2019)
	ASAP1	miR-502-5p	NRAS		Wei et al. (2021)
	NFIX	miR-132	ABCG2		Ding et al. (2020)

(Continued)

TABLE 2.4 (Continued)

	lncRNA/circRNA	miRNA sponged	Target protein	Pathway involved	Reference
Radioresistance in glioma	TPTEP1	miR-106a-5p	MAPK14	p38 MAPK	Tang et al. (2020)
	NCK1-AS1	miR-22-3p	IGF1R		Wang et al. (2020)
	TP53TG1	miR-524-5p	RAB5A		Gao et al. (2020)
	RA1				Zheng et al. (2020)
	SNHG18		Semaphorin 5A		Zheng et al. (2016)
	SNHG18		ENO1		Zheng et al. (2019)
	p21	miR-146b-5p	HuR		Yang et al. (2016)
	AHIF				Liao et al. (2019)
	PSMB8-AS1	miR-22-3p	DDIT4		Hu et al. (2020)
	AHIF				Dai et al. (2019)
	ATP8B4	miR-766			Zhao et al. (2019)
	AKT3			PI3K/Akt	Xia et al. (2019)
Chemoresistance in medulloblastoma	CRNDE	miR-29c-3p			Sun et al. (2020)

tumor-derived exosomes might have a role in a variety of cancer-related processes such as tumor microenvironment remodeling, metastasis, and drug resistance (Mashouri et al., 2019). CircNFIX has been shown to be upregulated in TMZ-resistant patients' serum and enhanced glioma cell migration, invasion, and reduced cancer cell apoptosis under TMZ exposure. Furthermore, authors proved that circNFIX enhances glioma TMZ resistance by sponging miR-132, which has been confirmed in vitro and in vivo. miR-132 is considered tumor-suppressor in glioma and circNFIX knockdown allowed for miR-132 overexpression leading to the inhibition of tumor cell migration, invasion, and the enhancement of cell apoptosis. Authors indicated that the circNFIX/miR-132 axis regulates the ABCG2 expression in glioma cells under TMZ exposure (Ding et al., 2020). Another target—miR-34a-5p has been identified as being sponged by circNFIX, thus directly influences NOTCH1 receptor and Notch signaling pathway, to promote glioma progression (Xu et al., 2018).

CircRNAs expression in EVs isolated both from U251 and radioresistant U251 was established by RNA-seq. Total 63 upregulated and 48 downregulated circRNAs in EVs isolated from radioresistant cell line compared with those untreated cells. CircATP8B4 expression level is significantly upregulated in EV isolated from radioresistant cell line and have been proved to act as a sponge of miR766 to promote cell radioresistance. The exact mechanism of circATP8B4/miR-766/mRNA still needs to be elucidated; however, miR-766 has been shown to impact the expression level of tumor-suppressor genes in colorectal cancer cell lines by the inhibition of DNMT3B gene (Zhao et al., 2019).

Not only circRNAs itself play a significant role in cancer events. CircRNAs that were considered mainly regulatory molecules have been recently shown to be capable of encoding peptides and proteins (Lei et al., 2020; Shang et al., 2019; Wawrzyniak et al., 2020). Turner et al. (2015) reported a significant role of AKT3—the dominant Akt isoform in promoting glioma progression and activating resistance to radiation and chemotherapy in GBM cases by activating DNA repair pathways of human glioblastoma cells. Interesting example is previously uncharacterized transcript variant of AKT gene—circ-AKT3, encoding a novel protein named AKT3–174aa. The authors showed that circ-AKT3 exhibits lower expression level in comparison to paired adjacent normal brain tissues, and its knockdown enhanced the malignant phenotypes of astrocytoma cells. Furthermore, the overexpression of AKT3–174aa decreased the proliferation, radiation resistance, and tumorigenicity of GBM cells and proved a negative regulatory role of AKT3–174aa in modulating the PI3K/AKT signal intensity via interaction with phosphorylated PDK1 (Xia et al., 2019).

The circRNAs expression profiling in MB tissue in comparison to human normal cerebellum tissues revealed a number of differentially expressed circRNAs. The downregulated circRNAs have been shown to be associated with detection of tumor cells, synaptic transmission, and activation of MAPK. Activation of the RAS/MAPK pathway is crucial in drug resistance, tumor growth, and enhancement of metastatic behavior in patients with Shh pathway-dependent MB (Lv et al., 2018; Zhao et al., 2015).

2.7 Therapeutic potential and targeting of ncRNAs in brain cancer patients—challenges and perspectives

Potential therapies based on the manipulation of ncRNA levels are based on tools which aim to increase the level of tumor suppressors or reduce the level of oncomiRs. Currently the most widely used approach involves delivering synthetic antisense oligonucleotides (ASOs) that imitate the native miRNA, or sequester the endogenous miRNA of interest (Anthiya et al., 2018). MiRNA mimics are synthetic double-stranded RNA molecules with identical sequence as native miRNAs that are able to integrate into the RISC acting as the missing miRNAλ. Antagomirs are single-stranded modified RNA molecules, perfectly complementary to mature miRNAs. They are forming duplexes with their miRNA target, what leads to miRNA degradation and the recycling of the antagomir (Krützfeldt et al., 2005). The discovery of RNA interference (RNAi) has found a new application for ASO—gene silencing, using synthetic, double-stranded, siRNA (Wittrup & Lieberman, 2015). The levels of an miRNA can be deregulated also indirectly, by targeting hypermethylation of miRNA promoter sites (Davalos et al., 2012), restoring a deleted genomic locus at the DNA level using CRISPR/Cas9 (Chang et al., 2016) or by inhibiting possible miRNA sponges like lncRNAs and circRNAs (Petrescu et al., 2019) which were proven to create complex coregulatory networks in brain (Kleaveland et al., 2018; Piwecka et al., 2017). MiRNA biogenesis, maturation, or function could be prevented by small molecule chemical compounds that bind precursor or mature miRNAs (Monroig et al., 2015).

When designing oligonucleotides with therapeutic potential, several rules should be followed. Molecules must not be toxic to cells, which is related to the use of safe concentrations. The influence of therapeutic oligo on cell proliferation should be individually established. It is accepted to use the concentration not exceeding 50 nM for cells cultured in six well dishes (Matsui & Corey, 2017). Toxicity is also associated with the "off-target" effect, the partial complementarity to unintended targets. Also, one cannot forget about possible interactions of RNA with proteins, as nucleic acids can induce interferon response by binding to proteins on the surface or inside cells (Matsui & Corey, 2017). To avoid confusing the expected effect with the results "off-target," when screening for ASOs or duplexes, it is worth using at least two or three potent compounds at a time that are capable of reducing gene expression and are complementary to the same target RNA by, thus, reducing the possibility that the phenotype is due to nonspecific binding of compounds to unintended targets. It is also crucial to design controls either "scrambled," with swapped groups of bases, or "mismatched," with introduced mismatched bases (Matsui & Corey, 2017).

The most important additional obstacles in ASO therapy are their poor extracellular and intracellular stability, and low efficacy of intracellular delivery. Unmodified ASOs encounter substantial stability and delivery barriers, and to overcome them, modifications protecting against nucleases and enhancing RNA-binding affinity are introduced (Adams et al., 2017). The most basic and widely used RNA modifications include 2′-O-methyl substitution in the sugar of RNA backbone (Meister et al., 2004), change in the 2′-O-methoxyethyl backbone (Khatsenko et al., 2000), locked nucleic acid bases (LNA), and oligomerization with phosphorothioate bonds instead of the canonical phosphodiester bond (Irie et al., 2020). When silencing the circRNA expression, not to disturb the linear version of the transcript, it should also be remembered to design the effector molecule in such a way that it targets the junction site, which is unique for the circular version (Karedath et al., 2019).

Clinical studies with successful lncRNA and circRNA downregulation or upregulation are still missing. However, in the basic research the loss- and gain-of-function studies are often performed, as indicated in examples in previous section. Generally, RNAi platform applies siRNA and shRNA to silence cytoplasmic lncRNAs, while ASOs aim to knockdown lncRNAs located in the nucleus. For example, siRNA knockdown of lncRNA MALAT1 in glioblastoma cell line U87 has been shown to lead to the apoptosis of tumor cells (Liang et al., 2019). On the other hand, shRNA knockdown of lncRNA SCAMP1 has been applied on patient-derived glioma cells and resulted in repression of their proliferation, migration, and invasion (Zong et al., 2019). Studies with xenografted model of glioma, which was intravenously injected with ASOs targeting lncRNA TUG1, showed a significant repression of tumor growth (Katsushima et al., 2016). A more recent approach used is CRISPRi by which the knock-outs and knock-ins of lncRNA are possible (Zhao et al., 2020). Moreover, this tool enables to screen and identify lncRNA of therapeutic potential like in the studies of Liu et al. (2020) CRISPRi screen found lncRNA called lncGRS-1 that sensitizes glioma cells to radiation. ASO-targeting of that lncRNA in brain organoids sensitized cultures to radiation therapy. However, further studies are needed to apply these tools in clinics. A summary of the advantages and disadvantages of ncRNA expression manipulation methods is provided in Table 2.2.

Expression of miR-10b is higher in GBM than in LGG (Visani et al., 2014) or healthy brains. Inhibition of that miRNA in GSC-derived xenograft using ASO inhibitors attenuated growth and progression tumor without observed systemic toxicity (Teplyuk et al., 2016). miR-10b loss-of-function by CRISPR/Cas9 on cell lines and GBM xenografts led to glioma cell apoptosis and impairs tumor growth (El Fatimy et al., 2017). A decrease in the survival of GBM cells as well as their arrest in G2/M phase upon TMZ treatment was also observed after the coinhibition of miR-21 and miR-10b with their antagomiRs (Ananta et al., 2016). Currently underway clinical trial tests the hypothesis that mir-10b expression patterns could serve as a prognostic and diagnostic marker in primary glioma samples. Tumor tissue, blood, and CSF will be analyzed in terms of the expression level, allowing to link patient survival, tumor grade, and genotypic variation with miR-10b. Moreover, in vitro studies of the sensitivity of individual primary tumors to anti-mir-10b treatment are ongoing (ClinicalTrials.gov Identifier: NCT01849952).

The main challenge that is associated with the use of therapeutics in the treatment of brain tumors is the way of potential drug delivery. One of the approaches enabling drug delivery through the BBB was a monthly repeated opening of the BBB with a pulsed ultrasound in combination with systemically injected microbubbles. Barrier was disrupted at acoustic pressure levels up to 1.1 megapascals with no detectable adverse effects in a clinical trial (Carpentier et al., 2016). This concept was successfully applied in delivery of short hairpin RNA targeting an apoptosis inhibitor Birc5 into rats with orthotopic glioma (Zhao et al., 2018). An effective and noninvasive system for the delivery of drugs to the brain is a transport via nanoparticles. The nanoparticles used to penetrate the BBB are mainly polymer particles composed of poly(butylcyanoacrylate), poly(lactic-*co*-glycolic acid), poly(lactic acid), carbon quantum dots, liposomes, and inorganic composites such as gold, silver, and zinc oxide (Zhou et al., 2018). Natural nanoparticles, which are EVs or exosomes derived from cells, are suitable for the transport of therapeutic nucleic acid molecules. Naturally, they contain endogenous small nucleic acids, although they can also be loaded with synthetic oligonucleotides of tumor suppressive potential (Haraszti et al., 2017; Yuan et al., 2017). Nevertheless, further studies are needed to apply described herein tools in clinical applications. So far, none of lncRNAs and circRNAs have been used in clinical studies.

2.8 Summary and conclusions

As indicated in herein chapter, ncRNAs display a strong diagnostic, prognostic, and therapeutic potential, thus, can be broadly applied in brain cancer field. The process of ncRNA discovery important in brain tumors is shown in Fig. 2.3. This is of a tremendous importance since brain tumors are mostly incurable, highly heterogeneous, and strongly resistant to variable therapies. Tumor heterogeneity constitutes nowadays a major challenge to cancer diagnosis and treatment. Moreover, different tumor stages as well as

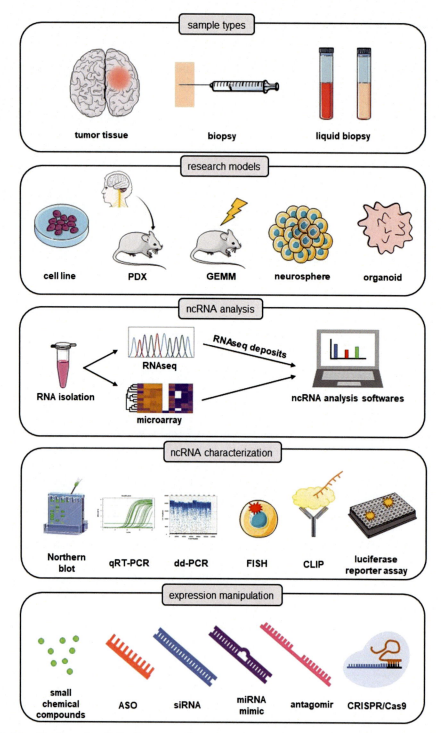

FIGURE 2.3 Pipeline of ncRNA discovery, characterization, and expression-level manipulation. *ASO*, antisense oligonucleotide; *CLIP*, cross-linking immunoprecipitation; *CRISPR/Cas9*, clustered regularly interspaced short palindromic repeats/CRISPR associated protein 9; *dd-PCR*, digital droplet polymerase chain reaction *FISH*, fluorescence in situ hybridization; *GEMM*, genetically modified mouse model; *ncRNA*, noncoding RNA; *PDX*, patient derived xenograft; *qRT-PCR*, quantitative reverse transcription polymerase chain reaction; *RNAseq*, RNA sequencing.

genetic and molecular events in brain tumors are associated with distinct outcomes of therapeutic responses. Specific pattern of ncRNA expression could help to classify these tumors as well as enable for precise treatment selection. Numerous publications focused on ncRNAs also reveal their molecular role in the biology of CSC which are crucial for the development of future successful therapies for brain cancer.

References

Abdelfattah, N., Rajamanickam, S., Panneerdoss, S., Timilsina, S., Yadav, P., Onyeagucha, B. C., ... Rao, M. K. (2018). MiR-584-5p potentiates vincristine and radiation response by inducing spindle defects and DNA damage in medulloblastoma. *Nature Communications, 9*(1), 4541.

Adams, B. D., Parsons, C., Walker, L., Zhang, W. C., & Slack, F. J. (2017). Targeting noncoding RNAs in disease. *Journal of Clinical Investigation, 127*(3), 761–771.

Agarwal, V., Bell, G. W., Nam, J. W., & Bartel, D. P. (2015). Predicting effective microRNA target sites in mammalian mRNAs. *Elife, 4*.

Ahmad, Z., Jasnos, L., Gil, V., Howell, L., Hallsworth, A., Petrie, K., ... Chesler, L. (2015). Molecular and in vivo characterization of cancer-propagating cells derived from MYCN-dependent medulloblastoma. *PLoS One, 10*(3), e0119834.

Akers, J. C., Ramakrishnan, V., Kim, R., Phillips, S., Kaimal, V., Mao, Y., ... Chen, C. C. (2015). miRNA contents of cerebrospinal fluid extracellular vesicles in glioblastoma patients. *Journal of Neuro-Oncology, 123*(2), 205–216.

Allen, M., Bjerke, M., Edlund, H., Nelander, S., & Westermark, B. (2016). Origin of the U87MG glioma cell line: Good news and bad news. *Science Translational Medicine, 8*(354).

Alles, J., Fehlmann, T., Fischer, U., Backes, C., Galata, V., Minet, M., ... Meese, E. (2019). An estimate of the total number of true human miRNAs. *Nucleic Acids Research, 47*(7), 3353–3364.

Ananta, J. S., Paulmurugan, R., & Massoud, T. F. (2016). Tailored nanoparticle codelivery of antimiR-21 and antimiR-10b augments glioblastoma cell kill by temozolomide: Toward a "Personalized" anti-microRNA therapy. *Molecular Pharmaceutics, 13*(9), 3164–3175.

Andreatta, F., Beccaceci, G., Fortuna, N., Celotti, M., De Felice, D., Lorenzoni, M., ... Alaimo, A. (2020). The organoid era permits the development of new applications to study glioblastoma. *Cancers (Basel), 12*(11), 1–16.

Anthiya, S., Griveau, A., Loussouarn, C., Baril, P., Garnett, M., Issartel, J. P., & Garcion, E. (2018). MicroRNA-based drugs for brain tumors. *Trends Cancer, 4*(3), 222–238.

Aufiero, S., Reckman, Y. J., Tijsen, A. J., Pinto, Y. M., & Creemers, E. E. (2020). circRNAprofiler: An R-based computational framework for the downstream analysis of circular RNAs. *BMC Bioinformatics, 21*(1), 164.

Azari, H., Millette, S., Ansari, S., Rahman, M., Deleyrolle, L. P., & Reynolds, B. A. (2011). Isolation and expansion of human glioblastoma multiforme tumor cells using the neurosphere assay. *Journal of Visualized Experiments* (56), e3633.

Bach, D. H., Lee, S. K., & Sood, A. K. (2019). Circular RNAs in cancer. *Molecular Therapy – Nucleic Acids, 16*, 118–129.

Bachmayr-Heyda, A., Reiner, A. T., Auer, K., Sukhbaatar, N., Aust, S., Bachleitner-Hofmann, T., ... Pils, D. (2015). Correlation of circular RNA abundance with proliferation – Exemplified with colorectal and ovarian cancer, idiopathic lung fibrosis, and normal human tissues. *Scientific Reports, 5*, 8057.

Badodi, S., Marino, S., & Guglielmi, L. (2019). Establishment and culture of patient-derived primary medulloblastoma cell lines. *Methods in Molecular Biology, 1869*, 23–36.

Balachandran, A. A., Larcher, L. M., Chen, S., & Veedu, R. N. (2020). Therapeutically significant microRNAs in primary and metastatic brain malignancies. *Cancers (Basel), 12*(9), 1–29.

Balas, M. M., & Johnson, A. M. (2018). Exploring the mechanisms behind long noncoding RNAs and cancer. *Non-coding RNA Research, 3*(3), 108–117.

Bao, Z., Yang, Z., Huang, Z., Zhou, Y., Cui, Q., & Dong, D. (2019). LncRNADisease 2.0: An updated database of long non-coding RNA-associated diseases. *Nucleic Acids Research, 47*(D1), D1034-D7.

Barani, I. J., & Larson, D. A. (2015). Radiation therapy of glioblastoma. *Cancer Treatment and Research, 163*, 49–73.

Baraniskin, A., Kuhnhenn, J., Schlegel, U., Maghnouj, A., Zöllner, H., Schmiegel, W., ... Schroers, R. (2012). Identification of microRNAs in the cerebrospinal fluid as biomarker for the diagnosis of glioma. *Neuro-Oncology*, *14*(1), 29–33.

Barrett, S. P., & Salzman, J. (2016). Circular RNAs: Analysis, expression and potential functions. *Development*, *143*(11), 1838–1847.

Barrett, S. P., Wang, P. L., & Salzman, J. (2015). Circular RNA biogenesis can proceed through an exon-containing lariat precursor. *Elife*, *4*, e07540.

Bejugam, P. R., Das, A., & Panda, A. C. (2020). Seeing is believing: Visualizing circular RNAs. *Noncoding RNA*, *6*(4), 45.

Berghoff, A. S., Schur, S., Füreder, L. M., Gatterbauer, B., Dieckmann, K., Widhalm, G., ... Preusser, M. (2016). Descriptive statistical analysis of a real life cohort of 2419 patients with brain metastases of solid cancers. *ESMO Open*, *1*(2), e000024.

Berthois, Y., Delfino, C., Metellus, P., Fina, F., Nanni-Metellus, I., Al Aswy, H., ... Boudouresque, F. (2014). Differential expression of miR200a-3p and miR21 in grade II-III and grade IV gliomas: Evidence that miR200a-3p is regulated by O6-methylguanine methyltransferase and promotes temozolomide responsiveness. *Cancer Biology and Therapy*, *15*(7), 938–950.

Bharambe, H. S., Paul, R., Panwalkar, P., Jalali, R., Sridhar, E., Gupta, T., ... Shirsat, N. V. (2019). Downregulation of miR-204 expression defines a highly aggressive subset of group 3/group 4 medulloblastomas. *Acta Neuropathologica Communications*, *7*(1), 52.

Bian, S., Repic, M., Guo, Z., Kavirayani, A., Burkard, T., Bagley, J. A., ... Knoblich, J. A. (2018). Genetically engineered cerebral organoids model brain tumor formation. *Nature Methods*, *15*(8), 631–639.

Birks, D. K., Barton, V. N., Donson, A. M., Handler, M. H., Vibhakar, R., & Foreman, N. K. (2011). Survey of microRNA expression in pediatric brain tumors. *Pediatric Blood and Cancer*, *56*(2), 211–216.

Bose, R., & Ain, R. (2018). Regulation of transcription by circular RNAs. *Advances in Experimental Medicine and Biology*, *1087*, 81–94.

Cai, T., Liu, Y., & Xiao, J. (2018). Long noncoding RNA MALAT1 knockdown reverses chemoresistance to temozolomide via promoting microRNA-101 in glioblastoma. *Cancer Medicine*, *7*(4), 1404–1415.

Cao, Z., Pan, X., Yang, Y., Huang, Y., & Shen, H. B. (2018). The lncLocator: A subcellular localization predictor for long non-coding RNAs based on a stacked ensemble classifier. *Bioinformatics*, *34*(13), 2185–2194.

Cardenas, J., Balaji, U., & Gu, J. (2020). Cerina: Systematic circRNA functional annotation based on integrative analysis of ceRNA interactions. *Scientific Reports*, *10*(1), 22165.

Carlevaro-Fita, J., & Johnson, R. (2019). Global positioning system: Understanding long noncoding RNAs through subcellular localization. *Molecular Cell*, *73*(5), 869–883.

Carlevaro-Fita, J., Lanzos, A., Feuerbach, L., Hong, C., Mas-Ponte, D., Pedersen, J. S., ... Johnson, R., & PCAWG Consortium. (2020). Cancer LncRNA Census reveals evidence for deep functional conservation of long non-coding RNAs in tumorigenesis. *Communications Biology*, *3*(1), 56.

Carpentier, A., Canney, M., Vignot, A., Reina, V., Beccaria, K., Horodyckid, C., ... Idbaih, A. (2016). Clinical trial of blood–brain barrier disruption by pulsed ultrasound. *Science Translational Medicine*, *8*(343), 343re2.

Catanzaro, G., Besharat, Z. M., Garg, N., Ronci, M., Pieroni, L., Miele, E., ... Ferretti, E. (2016). MicroRNAs-proteomic networks characterizing human medulloblastoma-SLCs. *Stem Cells International*, *2016*, 2683042.

Cavalli, F. M. G., Remke, M., Rampasek, L., Peacock, J., Shih, D. J. H., Luu, B., ... Taylor, M. D. (2017). Intertumoral heterogeneity within medulloblastoma subgroups. *Cancer Cell*, *31*(6), 737–754, e6.

Chang, H., Yi, B., Ma, R., Zhang, X., Zhao, H., & Xi, Y. (2016). CRISPR/cas9, a novel genomic tool to knock down microRNA in vitro and in vivo. *Scientific Reports*, *6*, 22312.

Chen, A., Zhong, L., Ju, K., Lu, T., Lv, J., & Cao, H. (2020). Plasmatic circrna predicting the occurrence of human glioblastoma. *Cancer Management and Research*, *12*, 2917–2923.

Chen, D. F., Zhang, L. J., Tan, K., & Jing, Q. (2017). Application of droplet digital PCR in quantitative detection of the cell-free circulating circRNAs. *Biotechnology & Biotechnological Equipment*, *32*(1), 116–123.

Chen, J., Chen, T., Zhu, Y., Li, Y., Zhang, Y., Wang, Y., ... Ke, Y. (2019). circPTN sponges miR-145-5p/miR-330-5p to promote proliferation and stemness in glioma. *Journal of Experimental & Clinical Cancer Research*, *38*(1), 398.

Chen, L. L., & Yang, L. (2015). Regulation of circRNA biogenesis. *RNA Biology*, *12*(4), 381–388.

Chen, W., Xu, X. K., Li, J. L., Kong, K. K., Li, H., Chen, C., ... Li, F. C. (2017). MALAT1 is a prognostic factor in glioblastoma multiforme and induces chemoresistance to temozolomide through suppressing miR-203 and promoting thymidylate synthase expression. *Oncotarget*, *8*(14), 22783–22799.

References

Chen, W., Yu, Q., Chen, B., Lu, X., & Li, Q. (2016). The prognostic value of a seven-microRNA classifier as a novel biomarker for the prediction and detection of recurrence in glioma patients. *Oncotarget, 7*(33), 53392–53413.

Chen, W. L., Chen, H. J., Hou, G. Q., Zhang, X. H., & Ge, J. W. (2019). LINC01198 promotes proliferation and temozolomide resistance in a NEDD4-1-dependent manner, repressing PTEN expression in glioma. *Aging (Albany NY), 11*(16), 6053–6068.

Chen, X., Zhang, Y., Shi, Y., Lian, H., Tu, H., Han, S., ... He, X. (2015). MiR-873 acts as a novel sensitizer of glioma cells to cisplatin by targeting Bcl-2. *International Journal of Oncology, 47*(4), 1603–1611.

Chen, Y., & Wang, X. (2020). MiRDB: An online database for prediction of functional microRNA targets. *Nucleic Acids Research, 48*(D1), D127–D131.

Cheng, W., Ren, X., Zhang, C., Han, S., & Wu, A. (2017). Expression and prognostic value of microRNAs in lower-grade glioma depends on IDH1/2 status. *Journal of Neuro-Oncology, 132*(2), 207–218.

Chi, C., Mao, M., Shen, Z., Chen, Y., Chen, J., & Hou, W. (2018). HOXD-AS1 exerts oncogenic functions and promotes chemoresistance in cisplatin-resistant cervical cancer cells. *Human Gene Therapy, 29*(12), 1438–1448.

Clément, T., Salone, V., & Rederstorff, M. (2015). Dual luciferase gene reporter assays to study miRNA Function. *Methods in Molecular Biology, 1296*, 187–198.

Cui, T., Bell, E. H., McElroy, J., Becker, A. P., Gulati, P. M., Geurts, M., ... Chakravarti, A. (2019). miR-4516 predicts poor prognosis and functions as a novel oncogene via targeting PTPN14 in human glioblastoma. *Oncogene, 38*(16), 2923–2936.

Dai, X., Liao, K., Zhuang, Z., Chen, B., Zhou, Z., Zhou, S., ... Lin, R. (2019). AHIF promotes glioblastoma progression and radioresistance via exosomes. *International Journal of Oncology, 54*(1), 261–270.

Darnell, R. B. (2010). HITS-CLIP: Panoramic views of protein-RNA regulation in living cells. *Wiley Interdisciplinary Reviews: RNA, 1*(2), 266–286.

Davalos, V., Moutinho, C., Villanueva, A., Boque, R., Silva, P., Carneiro, F., ... Esteller, M. (2012). Dynamic epigenetic regulation of the microRNA-200 family mediates epithelial and mesenchymal transitions in human tumorigenesis. *Oncogene, 31*(16), 2062–2074.

Davis, M. E. (2018). Epidemiology and overview of gliomas. *Seminars in Oncology Nursing, 34*(5), 420–429.

de Antonellis, P., Medaglia, C., Cusanelli, E., Andolfo, I., Liguori, L., De Vita, G., ... Zollo, M. (2011). MiR-34a targeting of Notch ligand delta-like 1 impairs CD15+/CD133+ tumor-propagating cells and supports neural differentiation in medulloblastoma. *PLoS One, 6*(9), e24584.

de la Rocha, A. M. A., Gonzalez-Huarriz, M., Guruceaga, E., Mihelson, N., Tejada-Solis, S., Diez-Valle, R., ... Lopez-Bertoni, H. (2020). miR-425-5p, a SOX2 target, regulates the expression of FOXJ3 and RAB31 and promotes the survival of GSCs. *Archives of Clinical and Biomedical Research, 4*(3), 221–238.

Debeb, B. G., Lacerda, L., Anfossi, S., Diagaradjane, P., Chu, K., Bambhroliya, A., ... Woodward, W. A. (2016). miR-141-mediated regulation of brain metastasis from breast cancer. *Journal of the National Cancer Institute, 108*(8), djw026.

Denli, A. M., Tops, B. B. J., Plasterk, R. H. A., Ketting, R. F., & Hannon, G. J. (2004). Processing of primary microRNAs by the microprocessor complex. *Nature, 432*(7014), 231–235.

Derrien, T., Johnson, R., Bussotti, G., Tanzer, A., Djebali, S., Tilgner, H., ... Guigo, R. (2012). The GENCODE v7 catalog of human long noncoding RNAs: Analysis of their gene structure, evolution, and expression. *Genome Research, 22*(9), 1775–1789.

Ding, B. S., James, D., Iyer, R., Falciatori, I., Hambardzumyan, D., Wang, S., ... Hormigo, A. (2013). Prominin 1/CD133 endothelium sustains growth of proneural glioma. *PLoS One, 8*(4), e62150.

Ding, C., Yi, X., Wu, X., Bu, X., Wang, D., Wu, Z., ... Kang, D. (2020). Exosome-mediated transfer of circRNA CircNFIX enhances temozolomide resistance in glioma. *Cancer Letters, 479*, 1–12.

Ding, H., Cui, L., & Wang, C. (2020). Long noncoding RNA LIFR-AS1 suppresses proliferation, migration and invasion and promotes apoptosis through modulating miR-4262/NF-kappaB pathway in glioma. *Neurological Research*, 1–10.

Ding, J., Zhang, L., Chen, S., Cao, H., Xu, C., & Wang, X. (2020). lncRNA CCAT2 enhanced resistance of glioma cells against chemodrugs by disturbing the normal function of miR-424. *OncoTargets and Therapy, 13*, 1431–1445.

Ding, X., Yang, L., Geng, X., Zou, Y., Wang, Z., Li, Y., ... Yu, H. (2020). CircRNAs as potential biomarkers for the clinicopathology and prognosis of glioma patients: A *meta*-analysis. *BMC Cancer, 20*(1), 1005.

Dong, R., Ma, X. K., Li, G. W., & Yang, L. (2018). CIRCpedia v2: An updated database for comprehensive circular RNA annotation and expression comparison. *Genomics Proteomics Bioinformatics, 16*(4), 226–233.

Du, P., Zhao, H., Peng, R., Liu, Q., Yuan, J., Peng, G., & Liao, Y. (2017). LncRNA-XIST interacts with miR-29c to modulate the chemoresistance of glioma cell to TMZ through DNA mismatch repair pathway. *Bioscience Reports, 37*(5), BSR20170696.

Duan, S., Li, M., Wang, Z., Wang, L., & Liu, Y. (2018). H19 induced by oxidative stress confers temozolomide resistance in human glioma cells via activating NF-kappaB signaling. *OncoTargets and Therapy, 11*, 6395–6404.

Dudekula, D. B., Panda, A. C., Grammatikakis, I., De, S., Abdelmohsen, K., & Gorospe, M. (2016). CircInteractome: A web tool for exploring circular RNAs and their interacting proteins and microRNAs. *RNA Biology, 13*(1), 34–42.

El Fatimy, R., Subramanian, S., Uhlmann, E. J., & Krichevsky, A. M. (2017). Genome editing reveals glioblastoma addiction to microRNA-10b. *Molecular Therapy, 25*(2), 368–378.

Esquela-Kerscher, A., & Slack, F. J. (2006). Oncomirs — MicroRNAs with a role in cancer. *Nature Reviews Cancer, 6*(4), 259–269.

Fan, C., Lei, X., Fang, Z., Jiang, Q., & Wu, F. X. (2018). CircR2 disease: A manually curated database for experimentally supported circular RNAs associated with various diseases. *Database (Oxford), 2018*, bay044.

Fan, H., Yuan, R., Cheng, S., Xiong, K., Zhu, X., & Zhang, Y. (2018). Overexpressed miR-183 promoted glioblastoma radioresistance via down-regulating LRIG1. *Biomedicine and Pharmacotherapy, 97*, 1554–1563.

Fang, K., Liu, P., Dong, S., Guo, Y., Cui, X., Zhu, X., ... Wu, Y. (2016). Magnetofection based on superparamagnetic iron oxide nanoparticle-mediated low lncRNA HOTAIR expression decreases the proliferation and invasion of glioma stem cells. *International Journal of Oncology, 49*(2), 509–518.

Feng, S., Yao, J., Chen, Y., Geng, P., Zhang, H., Ma, X., ... Yu, X. (2015). Expression and functional role of reprogramming-related long noncoding RNA (lincRNA-ROR) in glioma. *Journal of Molecular Neuroscience, 56*(3), 623–630.

Ferlay, J., Colombet, M., Soerjomataram, I., Mathers, C., Parkin, D. M., Pineros, M., ... Bray, F. (2019). Estimating the global cancer incidence and mortality in 2018: GLOBOCAN sources and methods. *International Journal of Cancer, 144*(8), 1941–1953.

Fontanilles, M., Duran-Peña, A., & Idbaih, A. (2018). Liquid biopsy in primary brain tumors: Looking for stardust!. *Current Neurology and Neuroscience Reports, 18*(3), 13-.

Friedman, R. C., Farh, K. K. H., Burge, C. B., & Bartel, D. P. (2009). Most mammalian mRNAs are conserved targets of microRNAs. *Genome Research, 19*(1), 92–105.

Fu, C., Li, D., Zhang, X., Liu, N., Chi, G., & Jin, X. (2018). LncRNA PVT1 facilitates tumorigenesis and progression of glioma via regulation of MiR-128-3p/GREM1 axis and BMP signaling pathway. *Neurotherapeutics, 15*(4), 1139–1157.

Galasso, M., Dama, P., Previati, M., Sandhu, S., Palatini, J., Coppola, V., ... Volinia, S. (2014). A large scale expression study associates uc.283-plus lncRNA with pluripotent stem cells and human glioma. *Genome Medicine, 6*(10), 76.

Gallego Romero, I., Pai, A. A., Tung, J., & Gilad, Y. (2014). RNA-seq: Impact of RNA degradation on transcript quantification. *BMC Biology, 12*, 42.

Gao, J., Chen, Q., Zhao, Y., & Hou, R. (2020). lncRNA CRNDE is upregulated in glioblastoma multiforme and facilitates cancer progression through targeting miR-337-3p and ELMOD2 axis. *OncoTargets and Therapy, 13*, 9225–9234.

Gao, W., Qiao, M., & Luo, K. (2020). Long noncoding RNA TP53TG1 contributes to radioresistance of glioma cells via miR-524-5p/RAB5A axis. *Cancer Biotherapy and Radiopharmaceuticals*.

Garzia, A., Morozov, P., Sajek, M., Meyer, C., & Tuschl, T. (2018). PAR-CLIP for discovering target sites of RNA-binding proteins. *Methods in Molecular Biology, 1720*, 55–75.

Garzia, L., Andolfo, I., Cusanelli, E., Marino, N., Petrosino, G., De Martino, D., ... Zollo, M. (2009). MicroRNA-199b-5p impairs cancer stem cells through negative regulation of HES1 in medulloblastoma. *PLoS One, 4*(3), e4998.

Genovesi, L. A., Carter, K. W., Gottardo, N. G., Giles, K. M., & Dallas, P. B. (2011). Integrated analysis of miRNA and mRNA expression in childhood medulloblastoma compared with neural stem cells. *PLoS One, 6*(9), e23935.

References

Ghosal, S., Das, S., Sen, R., Basak, P., & Chakrabarti, J. (2013). Circ2Traits: A comprehensive database for circular RNA potentially associated with disease and traits. *Frontiers in Genetics, 4*, 283.

Gillen, A. E., Yamamoto, T. M., Kline, E., Hesselberth, J. R., & Kabos, P. (2016). Improvements to the HITS-CLIP protocol eliminate widespread mispriming artifacts. *BMC Genomics, 17*(1), 338.

Glazar, P., Papavasileiou, P., & Rajewsky, N. (2014). circBase: A database for circular RNAs. *RNA, 20*(11), 1666−1670.

Godlewski, J., Nowicki, M. O., Bronisz, A., Williams, S., Otsuki, A., Nuovo, G., ... Lawler, S. (2008). Targeting of the Bmi-1 oncogene/stem cell renewal factor by microRNA-128 inhibits glioma proliferation and self-renewal. *Cancer Research, 68*(22), 9125−9130.

Greene, J., Baird, A. M., Brady, L., Lim, M., Gray, S. G., McDermott, R., & Finn, S. (2017). Circular RNAs: Biogenesis, function and role in human diseases. *Frontiers in Molecular Biosciences, 4*, 38.

Grossman, R. L., Heath, A. P., Ferretti, V., Varmus, H. E., Lowy, D. R., Kibbe, W. A., & Staudt, L. (2016). Toward a shared vision for cancer genomic data. *New England Journal of Medicine, 375*(12), 1109−1112.

Guardia, G. D. A., Correa, B. R., Araujo, P. R., Qiao, M., Burns, S., Penalva, L. O. F., & Galante, P. A. F (2020). Proneural and mesenchymal glioma stem cells display major differences in splicing and lncRNA profiles. *NPJ Genomic Medicine, 5*, 2.

Guarnerio, J., Bezzi, M., Jeong, J. C., Paffenholz, S. V., Berry, K., Naldini, M. M., ... Pandolfi, P. P. (2016). Oncogenic role of fusion-circRNAs derived from cancer-associated chromosomal translocations. *Cell, 165*(2), 289−302.

Guo, J. C., Fang, S. S., Wu, Y., Zhang, J. H., Chen, Y., Liu, J., ... Zhao, Y. (2019). CNIT: A fast and accurate web tool for identifying protein-coding and long non-coding transcripts based on intrinsic sequence composition. *Nucleic Acids Research, 47*(W1), W516−W522.

Guo, J. U., Agarwal, V., Guo, H., & Bartel, D. P. (2014). Expanded identification and characterization of mammalian circular RNAs. *Genome Biology, 15*(7), 409.

Guo, Y., Liu, S., Wang, P., Zhao, S., Wang, F., Bing, L., ... Hao, A. (2011). Expression profile of embryonic stem cell-associated genes Oct4, Sox2 and Nanog in human gliomas. *Histopathology, 59*(4), 763−775.

Guo, Y., Yan, K., Fang, J., Qu, Q., Zhou, M., & Chen, F. (2013). Let-7b expression determines response to chemotherapy through the regulation of cyclin D1 in Glioblastoma. *Journal of Experimental and Clinical Cancer Research, 32*(1), 41.

Gwak, H. S., Kim, T. H., Jo, G. H., Kim, Y. J., Kwak, H. J., Kim, J. H., ... Park, J. B. (2012). Silencing of microRNA-21 confers radio-sensitivity through Inhibition of the PI3K/AKT pathway and enhancing autophagy in malignant glioma cell lines. *PLoS ONE, 7*(10), e47449.

Hafner, M., Landthaler, M., Burger, L., Khorshid, M., Hausser, J., Berninger, P., ... Tuschl, T. (2010). Transcriptome-wide Identification of RNA-binding protein and microRNA target sites by PAR-CLIP. *Cell, 141*(1), 129−141.

Hamburger, A. W., & Salmon, S. E. (1977). Primary bioassay of human tumor stem cells. *Science, 197*(4302), 461−463.

Han, J., Lee, Y., Yeom, K. H., Kim, Y. K., Jin, H., & Kim, V. N. (2004). The Drosha-DGCR8 complex in primary microRNA processing. *Genes and Development, 18*(24), 3016−3027.

Han, L., Zhang, K., Shi, Z., Zhang, J., Zhu, J., Zhu, S., ... Kang, C. (2012). LncRNA pro fi le of glioblastoma reveals the potential role of lncRNAs in contributing to glioblastoma pathogenesis. *International Journal of Oncology, 40*(6), 2004−2012.

Han, Y., Wu, Z., Wu, T., Huang, Y., Cheng, Z., Li, X., ... Du, Z. (2016). Tumor-suppressive function of long non-coding RNA MALAT1 in glioma cells by downregulation of MMP2 and inactivation of ERK/MAPK signaling. *Cell Death & Disease, 7*, e2123.

Han, Y., Zhou, L., Wu, T., Huang, Y., Cheng, Z., Li, X., ... Du, Z. (2016). Downregulation of lncRNA-MALAT1 affects proliferation and the expression of stemness markers in glioma stem cell line SHG139S. *Cellular and Molecular Neurobiology, 36*(7), 1097−1107.

Hansen, T. B., Jensen, T. I., Clausen, B. H., Bramsen, J. B., Finsen, B., Damgaard, C. K., & Kjems, J. (2013). Natural RNA circles function as efficient microRNA sponges. *Nature, 495*(7441), 384−388.

Haraszti, R., Coles, A., Aronin, N., Khvorova, A., & Didiot, M.-C. (2017). Loading of extracellular vesicles with chemically stabilized hydrophobic siRNAs for the treatment of disease in the central nervous system. *Bio-Protocol, 7*(12), e2338.

He, J., Ren, M., Li, H., Yang, L., Wang, X., & Yang, Q. (2019). Exosomal circular RNA as a biomarker platform for the early diagnosis of immune-mediated demyelinating disease. *Frontiers in Genetics, 10*, 860.

Helwak, A., Kudla, G., Dudnakova, T., & Tollervey, D. (2013). Mapping the human miRNA interactome by CLASH reveals frequent noncanonical binding. *Cell, 153*(3), 654–665.

Holdt, L. M., Kohlmaier, A., & Teupser, D. (2018). Molecular functions and specific roles of circRNAs in the cardiovascular system. *Non-coding RNA Research, 3*(2), 75–98.

Hombach, S., & Kretz, M. (2016). Non-coding RNAs: Classification, biology and functioning. *Advances in Experimental Medicine and Biology, 937*, 3–17.

Hu, T., Wang, F., & Han, G. (2020). LncRNA PSMB8-AS1 acts as ceRNA of miR-22-3p to regulate DDIT4 expression in glioblastoma. *Neuroscience Letters, 728*, 134896.

Hua, L., Huang, L., Zhang, X., & Feng, H. (2020). Downregulation of hsa_circ_0000936 sensitizes resistant glioma cells to temozolomide by sponging miR-1294. *Journal of Biosciences, 45*, 101.

Huang, A., Zheng, H., Wu, Z., Chen, M., & Huang, Y. (2020). Circular RNA-protein interactions: Functions, mechanisms, and identification. *Theranostics, 10*(8), 3503–3517.

Huang, B. S., Luo, Q. Z., Han, Y., Huang, D., Tang, Q. P., & Wu, L. X. (2017). MiR-223/PAX6 axis regulates glioblastoma stem cell proliferation and the chemo resistance to TMZ via regulating PI3K/Akt pathway. *Journal of Cellular Biochemistry, 118*(10), 3452–3461.

Huang, S., Qi, P., Zhang, T., Li, F., & He, X. (2019). The HIF-1α/miR-224-3p/ATG5 axis affects cell mobility and chemosensitivity by regulating hypoxia-induced protective autophagy in glioblastoma and astrocytoma. *Oncology Reports, 41*(3), 1759–1768.

Huo, J. F., & Chen, X. B. (2019). Long noncoding RNA growth arrest-specific 5 facilitates glioma cell sensitivity to cisplatin by suppressing excessive autophagy in an mTOR-dependent manner. *Journal of Cellular Biochemistry, 120*(4), 6127–6136.

Huppertz, I., Attig, J., D'Ambrogio, A., Easton, L. E., Sibley, C. R., Sugimoto, Y., ... Ule, J. (2014). iCLIP: Protein-RNA interactions at nucleotide resolution. *Methods, 65*(3), 274–287.

International Human Genome Sequencing Consortium. (2004). Finishing the euchromatic sequence of the human genome. *Nature, 431*(7011), 931–945.

Irie, A., Sato, K., Hara, R. I., Wada, T., & Shibasaki, F. (2020). An artificial cationic oligosaccharide combined with phosphorothioate linkages strongly improves siRNA stability. *Scientific Reports, 10*(1), 14845.

Itzkovitz, S., & Van Oudenaarden, A. (2011). Validating transcripts with probes and imaging technology. *Nature Methods, 8*(4 Suppl), S12–S19.

Ivanov, D. P., Coyle, B., Walker, D. A., & Grabowska, A. M. (2016). In vitro models of medulloblastoma: Choosing the right tool for the job. *Journal of Biotechnology, 236*, 10–25.

Jacob, F., Salinas, R. D., Zhang, D. Y., Nguyen, P. T. T., Schnoll, J. G., Wong, S. Z. H., ... Song, H. (2020). A patient-derived glioblastoma organoid model and biobank recapitulates inter- and intra-tumoral heterogeneity. *Cell, 180*(1), 188–204, e22.

Jarroux, J., Morillon, A., & Pinskaya, M. (2017). History, discovery, and classification of lncRNAs. *Advances in Experimental Medicine and Biology, 1008*, 1–46.

Jayakrishnan, P., Venkat, E., Ramachandran, G., Kesavapisharady, K., Nair, S., Bharathan, B., ... Gopala, S. (2019). In vitro neurosphere formation correlates with poor survival in glioma. *IUBMB Life, 71*(2), 244–253.

Jeck, W. R., Sorrentino, J. A., Wang, K., Slevin, M. K., Burd, C. E., Liu, J., ... Sharpless, N. E., et al. (2013). Circular RNAs are abundant, conserved, and associated with ALU repeats. *RNA, 19*(2), 141–157.

Jensen, K. B., & Darnell, R. B. (2008). CLIP: Crosslinking and immunoprecipitation of in vivo RNA targets of RNA-binding proteins. *Methods in Molecular Biology, 488*, 85–98.

Jeyaraman, S., Hanif, E. A. M., Ab Mutalib, N. S., Jamal, R., & Abu, N. (2019). Circular RNAs: Potential regulators of treatment resistance in human cancers. *Frontiers in Genetics, 10*, 1369.

Jia, L., Tian, Y., Chen, Y., & Zhang, G. (2018). The silencing of LncRNA-H19 decreases chemoresistance of human glioma cells to temozolomide by suppressing epithelial-mesenchymal transition via the Wnt/beta-Catenin pathway. *OncoTargets and Therapy, 11*, 313–321.

Jiang, C., Shen, F., Du, J., Fang, X., Li, X., Su, J., ... Liu, Z. (2018). Upregulation of CASC2 sensitized glioma to temozolomide cytotoxicity through autophagy inhibition by sponging miR-193a-5p and regulating mTOR expression. *Biomedicine & Pharmacotherapy, 97*, 844–850.

Jiang, P., Wang, P., Sun, X., Yuan, Z., Zhan, R., Ma, X., & Li, W. (2016). Knockdown of long noncoding RNA H19 sensitizes human glioma cells to temozolomide therapy. *OncoTargets and Therapy, 9*, 3501–3509.

References

Jin, X., Yuan, L., Liu, B., Kuang, Y., Li, H., Li, L., ... Li, Q (2020). Integrated analysis of circRNA-miRNA-mRNA network reveals potential prognostic biomarkers for radiotherapies with X-rays and carbon ions in non-small cell lung cancer. *Annals of Translational Medicine, 8*(21), 1373.

Jin, Y., Chen, Z., Liu, X., & Zhou, X. (2013). Evaluating the MicroRNA targeting sites by luciferase reporter gene assay. *Methods in Molecular Biology, 936*, 117−127.

Johnson, R. (2012). Long non-coding RNAs in Huntington's disease neurodegeneration. *Neurobiology of Disease, 46*(2), 245−254.

Kaid, C., Silva, P. B. G., Cortez, B. A., Rodini, C. O., Semedo-Kuriki, P., & Okamoto, O. K. (2015). miR-367 promotes proliferation and stem-like traits in medulloblastoma cells. *Cancer Science, 106*(9), 1188−1195.

Karedath, T., Ahmed, I., Al Ameri, W., Al-Dasim, F. M., Andrews, S. S., Samuel, S., ... Malek, J. A. (2019). Silencing of ANKRD12 circRNA induces molecular and functional changes associated with invasive phenotypes. *BMC Cancer, 19*(1), 565.

Katsushima, K., Natsume, A., Ohka, F., Shinjo, K., Hatanaka, A., Ichimura, N., & Kondo, Y. (2016). Targeting the Notch-regulated non-coding RNA TUG1 for glioma treatment. *Nature Communications, 7*, 13616.

Kertesz, M., Iovino, N., Unnerstall, U., Gaul, U., & Segal, E. (2007). The role of site accessibility in microRNA target recognition. *Nature Genetics, 39*(10), 1278−1284.

Khatsenko, O., Morgan, R., Truong, L., York-Defalco, C., Sasmor, H., Conklin, B., & Geary, R. S. (2000). Absorption of antisense oligonucleotides in rat intestine: Effect of chemistry and length. *Antisense and Nucleic Acid Drug Development, 10*(1), 35−44.

Kijima, N., & Kanemura, Y. (2016). Molecular classification of medulloblastoma. *Neurologia Medico-Chirurgica (Tokyo), 56*(11), 687−697.

Kim, S. S., Harford, J. B., Moghe, M., Rait, A., Pirollo, K. F., & Chang, E. H. (2018). Targeted nanocomplex carrying siRNA against MALAT1 sensitizes glioblastoma to temozolomide. *Nucleic Acids Research, 46*(3), 1424−1440.

Kim, T. M., Huang, W., Park, R., Park, P. J., & Johnson, M. D. (2011). A developmental taxonomy of glioblastoma defined and maintained by microRNAs. *Cancer Research, 71*(9), 3387−3399.

Kiriakidou, M., Nelson, P. T., Kouranov, A., Fitziev, P., Bouyioukos, C., Mourelatos, Z., & Hatzigeorgiou, A. (2004). A combined computational-experimental approach predicts human microRNA targets. *Genes and Development, 18*(10), 1165−1178.

Kleaveland, B., Shi, C. Y., Stefano, J., & Bartel, D. P. (2018). A network of noncoding regulatory RNAs acts in the mammalian brain. *Cell, 174*(2), 350−362, e17.

Kocks, C., Boltengagen, A., Piwecka, M., Rybak-Wolf, A., & Rajewsky, N. (2018). Single-molecule fluorescence in situ hybridization (FISH) of circular RNA CDR1as. *Methods in Molecular Biology, 1724*, 77−96.

König, J., Zarnack, K., Rot, G., Curk, T., Kayikci, M., Zupan, B., & Ule, J. (2010). ICLIP reveals the function of hnRNP particles in splicing at individual nucleotide resolution. *Nature Structural and Molecular Biology, 17*(7), 909−915.

Kopková, A., Šána, J., Večeřa, M., Fadrus, P., Lipina, R., Smrčka, M., ... Slaby, O. (2019). MicroRNAs in cerebrospinal fluid as biomarkers in brain tumor patients. *Klinicka Onkologie, 32*(3), 181−186.

Kozomara, A., & Griffiths-Jones, S. (2011). miRBase: Integrating microRNA annotation and deep-sequencing data. *Nucleic Acids Research, 39*. (Database issue):D152-7.

Kristensen, L. S., Hansen, T. B., Veno, M. T., & Kjems, J. (2018). Circular RNAs in cancer: Opportunities and challenges in the field. *Oncogene, 37*(5), 555−565.

Krützfeldt, J., Rajewsky, N., Braich, R., Rajeev, K. G., Tuschl, T., Manoharan, M., ... Stoffel, M. (2005). Silencing of microRNAs in vivo with "antagomirs". *Nature, 438*(7068), 685−689.

Kunder, R., Jalali, R., Sridhar, E., Moiyadi, A., Goel, N., Goel, A., & Shirsat, N. V. (2013). Real-time PCR assay based on the differential expression of microRNAs and protein-coding genes for molecular classification of formalin-fixed paraffin embedded medulloblastomas. *Neuro-Oncology, 15*(12), 1644−1651.

Lancaster, M. A., Renner, M., Martin, C. A., Wenzel, D., Bicknell, L. S., Hurles, M. E., ... Knoblich, J. A. (2013). Cerebral organoids model human brain development and microcephaly. *Nature, 501*(7467), 373−379.

Lang, M. F., Yang, S., Zhao, C., Sun, G., Murai, K., Wu, X., ... Shi, Y. (2012). Genome-wide profiling identified a set of miRNAs that are differentially expressed in glioblastoma stem cells and normal neural stem cells. *PLoS One, 7*(4), e36248.

Latowska, J., Grabowska, A., Zarębska, Ż., Kuczyński, K., Kuczyńska, B., & Rolle, K. (2020). Non-coding RNAs in brain tumors, the contribution of lncRNAs, circRNAs, and snoRNAs to cancer development—Their diagnostic and therapeutic potential. *International Journal of Molecular Sciences, 21*(19), 1−31.

Lee, Y. Y., Yang, Y. P., Huang, M. C., Wang, M. L., Yen, S. H., Huang, P. I., ... Chen, M. T. (2014). MicroRNA142-3p promotes tumor-initiating and radioresistant properties in malignant pediatric brain tumors. *Cell Transplantation, 23*(4–5), 669–690.

Lei, M., Zheng, G., Ning, Q., Zheng, J., & Dong, D. (2020). Translation and functional roles of circular RNAs in human cancer. *Molecular Cancer, 19*(1), 30.

Lenting, K., Verhaak, R., ter Laan, M., Wesseling, P., & Leenders, W. (2017). Glioma: Experimental models and reality. *Acta Neuropathologica, 133*(2), 263–282.

Lewis, B. P., Burge, C. B., & Bartel, D. P. (2005). Conserved seed pairing, often flanked by adenosines, indicates that thousands of human genes are microRNA targets. *Cell, 120*(1), 15–20.

Li, A., Walling, J., Kotliarov, Y., Center, A., Steed, M. E., Ahn, S. J., ... Fine, H. A. (2008). Genomic changes and gene expression profiles reveal that established glioma cell lines are poorly representative of primary human gliomas. *Molecular Cancer Research, 6*(1), 21–30.

Li, B., Zhao, H., Song, J., Wang, F., & Chen, M. (2020). LINC00174 down-regulation decreases chemoresistance to temozolomide in human glioma cells by regulating miR-138-5p/SOX9 axis. *Human Cell, 33*(1), 159–174.

Li, C., Feng, S., & Chen, L. (2020). MSC-AS1 knockdown inhibits cell growth and temozolomide resistance by regulating miR-373-3p/CPEB4 axis in glioma through PI3K/Akt pathway. *Molecular and Cellular Biochemistry, 476*(2), 699–713.

Li, F., Ma, K., Sun, M., & Shi, S. (2018). Identification of the tumor-suppressive function of circular RNA ITCH in glioma cells through sponging miR-214 and promoting linear ITCH expression. *American Journal of Translational Research, 10*(5), 1373–1386.

Li, Q., Jia, H., Li, H., Dong, C., Wang, Y., & Zou, Z. (2016). LncRNA and mRNA expression profiles of glioblastoma multiforme (GBM) reveal the potential roles of lncRNAs in GBM pathogenesis. *Tumor Biology, 37*(11), 14537–14552.

Li, S., Li, Y., Chen, B., Zhao, J., Yu, S., Tang, Y., ... Huang, S. (2018). exoRBase: A database of circRNA, lncRNA and mRNA in human blood exosomes. *Nucleic Acids Research, 46*(D1), D106–D112.

Li, S., Zeng, A., Hu, Q., Yan, W., Liu, Y., & You, Y. (2017). miR-423-5p contributes to a malignant phenotype and temozolomide chemoresistance in glioblastomas. *Neuro-Oncology, 19*(1), 55–65.

Li, T., Ma, J., Han, X., Jia, Y., Yuan, H., Shui, S., & Guo, D. (2018). MicroRNA-320 enhances radiosensitivity of glioma through down-regulation of sirtuin type 1 by directly targeting forkhead box protein M1. *Translational Oncology, 11*(2), 205–M12.

Li, Y., Zhao, J., Yu, S., Wang, Z., He, X., Su, Y., ... Huang, S. (2019). Extracellular vesicles long RNA sequencing reveals abundant mRNA, circRNA, and lncRNA in human blood as potential biomarkers for cancer diagnosis. *Clinical Chemistry, 65*(6), 798–808.

Li, Y. C., Li, C. F., Chen, L. B., Li, D. D., Yang, L., Jin, J. P., & Zhang, B. (2015). MicroRNA-766 targeting regulation of SOX6 expression promoted cell proliferation of human colorectal cancer. *OncoTargets and Therapy, 8*, 2981–2988.

Liang, C., Yang, Y., Guan, J., Lv, T., Qu, S., Fu, Q., & Zhao, H. (2018). LncRNA UCA1 sponges miR-204-5p to promote migration, invasion and epithelial-mesenchymal transition of glioma cells via upregulation of ZEB1. *Pathology – Research and Practice, 214*(9), 1474–1481.

Liang, Z., Wang, Y., Li, H., Sun, Y., & Gong, Y. (2019). lncRNAs combine and crosstalk with NSPc1 in ATRA-induced differentiation of U87 glioma cells. *Oncology Letters, 17*(6), 5821–5829.

Liao, H., Bai, Y., Qiu, S., Zheng, L., Huang, L., Liu, T., ... Guo, H. (2015). MiR-203 downregulation is responsible for chemoresistance in human glioblastoma by promoting epithelial-mesenchymal transition via SNAI2. *Oncotarget, 6*(11), 8914–8928.

Liao, K., Ma, X., Chen, B., Lu, X., Hu, Y., Lin, Y., ... Qiu, Y. (2019). Upregulated AHIF-mediated radioresistance in glioblastoma. *Biochemical and Biophysical Research Communications, 509*(2), 617–623.

Liao, Y., Shen, L., Zhao, H., Liu, Q., Fu, J., Guo, Y., & Cheng, L. (2017). LncRNA CASC2 interacts with miR-181a to modulate glioma growth and resistance to TMZ through PTEN pathway. *Journal of Cellular Biochemistry, 118*(7), 1889–1899.

Liao, Y., Zhang, B., Zhang, T., Zhang, Y., & Wang, F. (2019). LncRNA GATA6-AS promotes cancer cell proliferation and inhibits apoptosis in glioma by downregulating lncRNA TUG1. *Cancer Biotherapy and Radiopharmaceuticals, 34*(10), 660–665.

Licatalosi, D. D., Mele, A., Fak, J. J., Ule, J., Kayikci, M., Chi, S. W., ... Darnell, R. B. (2008). HITS-CLIP yields genome-wide insights into brain alternative RNA processing. *Nature, 456*(7221), 464–469.

Lin, Y., Liu, T., Cui, T., Wang, Z., Zhang, Y., Tan, P., ... Wang, D. (2020). RNAInter in 2020: RNA interactome repository with increased coverage and annotation. *Nucleic Acids Research, 48*(D1), D189−D197.

Linkous, A., Balamatsias, D., Snuderl, M., Edwards, L., Miyaguchi, K., Milner, T., ... Fine, H. A. (2019). Modeling patient-derived glioblastoma with cerebral organoids. *Cell Reports, 26*(12), 3203−3211, e5.

Liu, J., Zhao, K., Huang, N., & Zhang, N. (2019). Circular RNAs and human glioma. *Cancer Biology & Medicine, 16*(1), 11−23.

Liu, M., Wang, Q., Shen, J., Yang, B. B., & Ding, X. (2019). Circbank: A comprehensive database for circRNA with standard nomenclature. *RNA Biology, 16*(7), 899−905.

Liu, S., Yin, F., Zhang, J., Wicha, M. S., Chang, A. E., Fan, W., & Li, Q. (2014). Regulatory roles of miRNA in the human neural stem cell transformation to glioma stem cells. *Journal of Cellular Biochemistry, 115*(8), 1368−1380.

Liu, S. J., Malatesta, M., Lien, B. V., Saha, P., Thombare, S. S., Hong, S. J., ... Lim, D. A. (2020). CRISPRi-based radiation modifier screen identifies long non-coding RNA therapeutic targets in glioma. *Genome Biology, 21*(1), 83.

Liu, X., Song, B., Li, S., Wang, N., & Yang, H. (2017). Identification and functional analysis of the risk microRNAs associated with cerebral low-grade glioma prognosis. *Molecular Medicine Reports, 16*(2), 1173−1179.

Louis, D. N., Perry, A., Reifenberger, G., Von Deimling, A., Figarella-Branger, D., Webster, K. C., ... Ellison, D. W. (2016). The 2016 World Health Organization classification of tumors of the central nervous system: A summary. *Acta Neuropathologica, 131*(6), 803−820.

Lv, Q. L., Du, H., Liu, Y. L., Huang, Y. T., Wang, G. H., Zhang, X., ... Zhou, H. H. (2017). Low expression of microRNA-320b correlates with tumorigenesis and unfavorable prognosis in glioma. *Oncology Reports, 38*(2), 959−966.

Lv, Q. L., Zhu, H. T., Li, H. M., Cheng, X. H., Zhou, H. H., & Chen, S. H. (2018). Down-regulation of miRNA-320c promotes tumor growth and metastasis and predicts poor prognosis in human glioma. *Brain Research Bulletin, 139*, 125−132.

Lv, T., Miao, Y. F., Jin, K., Han, S., Xu, T. Q., Qiu, Z. L., & Zhang, X. H. (2018). Dysregulated circular RNAs in medulloblastoma regulate proliferation and growth of tumor cells via host genes. *Cancer Medicine, 7*(12), 6147−6157.

Lyu, Y., Caudron-Herger, M., & Diederichs, S. (2020). circ2GO: A database linking circular RNAs to gene function. *Cancers (Basel), 12*(10).

Ma, B., Yuan, Z., Zhang, L., Lv, P., Yang, T., Gao, J., & Zhang, B. (2017). Long non-coding RNA AC023115.3 suppresses chemoresistance of glioblastoma by reducing autophagy. *Biochimica et Biophysica Acta − Molecular Cell Research, 1864*(8), 1393−1404.

Ma, Y., Zhou, G., Li, M., Hu, D., Zhang, L., Liu, P., & Lin, K. (2018). Long noncoding RNA DANCR mediates cisplatin resistance in glioma cells via activating AXL/PI3K/Akt/NF-kappaB signaling pathway. *Neurochemistry International, 118*, 233−241.

Mashouri, L., Yousefi, H., Aref, A. R., Ahadi, A. M., Molaei, F., & Alahari, S. K. (2019). Exosomes: Composition, biogenesis, and mechanisms in cancer metastasis and drug resistance. *Molecular Cancer, 18*(1), 75.

Mato Prado, M., Frampton, A. E., Giovannetti, E., Stebbing, J., Castellano, L., & Krell, J. (2016). Investigating miRNA-mRNA regulatory networks using crosslinking immunoprecipitation methods for biomarker and target discovery in cancer. *Expert Review of Molecular Diagnostics, 16*(11), 1155−1162.

Matsui, M., & Corey, D. R. (2017). *Non-coding RNAs as drug targets* (pp. 167−179). Nature Publishing Group.

Mazor, G., Levin, L., Picard, D., Ahmadov, U., Caren, H., Borkhardt, A., & Rotblat, B. (2019). The lncRNA TP73-AS1 is linked to aggressiveness in glioblastoma and promotes temozolomide resistance in glioblastoma cancer stem cells. *Cell Death & Disease, 10*(3), 246.

McDonald, K. L., Aw, G., & Kleihues, P. (2012). Role of biomarkers in the clinical management of glioblastomas: What are the barriers and how can we overcome them? *Frontiers in Neurology, 3*, 188.

Meijer, H. A., Smith, E. M., & Bushell, M. (2014). Regulation of miRNA strand selection: Follow the leader? *Biochemical Society Transactions, 42*(4), 1135−1140.

Meister, G., Landthaler, M., Dorsett, Y., & Tuschl, T. (2004). Sequence-specific inhibition of microRNA-and siRNA-induced RNA silencing. *RNA, 10*(3), 544−550.

Memczak, S., Jens, M., Elefsinioti, A., Torti, F., Krueger, J., Rybak, A., ... Rajewsky, N. (2013). Circular RNAs are a large class of animal RNAs with regulatory potency. *Nature, 495*(7441), 333−338.

Meng, S., Zhou, H., Feng, Z., Xu, Z., Tang, Y., Li, P., & Wu, M. (2017). CircRNA: Functions and properties of a novel potential biomarker for cancer. *Molecular Cancer, 16*(1), 94.

Meng, X., Hu, D., Zhang, P., Chen, Q., & Chen, M. (2019). CircFunBase: A database for functional circular RNAs. *Database (Oxford), 2019*, baz003.

Millard, N. E., & De Braganca, K. C. (2016). Medulloblastoma. *Journal of Child Neurology, 31*(12), 1341–1353.

Min, W., Dai, D., Wang, J., Zhang, D., Zhang, Y., Han, G., … Yue, Z. (2016). Long Noncoding RNA miR210HG as a potential biomarker for the diagnosis of glioma. *PLoS One, 11*(9), e0160451.

Miranda, K. C., Huynh, T., Tay, Y., Ang, Y. S., Tam, W. L., Thomson, A. M., … Rigoutsos, I. (2006). A pattern-based method for the identification of microRNA binding sites and their corresponding heteroduplexes. *Cell, 126*(6), 1203–1217.

Mo, D., Li, X., Raabe, C. A., Cui, D., Vollmar, J. F., Rozhdestvensky, T. S., … Brosius, J. (2019). A universal approach to investigate circRNA protein coding function. *Scientific Reports, 9*(1), 11684.

Moldovan, L. I., Hansen, T. B., Veno, M. T., Okholm, T. L. H., Andersen, T. L., Hager, H., … Kristensen, L. S. (2019). High-throughput RNA sequencing from paired lesional- and non-lesional skin reveals major alterations in the psoriasis circRNAome. *BMC Medical Genomics, 12*(1), 174.

Monroig, Pd. C., Chen, L., Zhang, S., & Calin, G. A. (2015). Small molecule compounds targeting miRNAs for cancer therapy. *Advanced Drug Delivery Reviews, 81*, 104–116.

Niyazi, M., Pitea, A., Mittelbronn, M., Steinbach, J., Sticht, C., Zehentmayr, F., … Unger, K. (2016). A 4-miRNA signature predicts the therapeutic outcome of glioblastoma. *Oncotarget, 7*(29), 45764–45775.

Northcott, P. A., Jones, D. T. W., Kool, M., Robinson, G. W., Gilbertson, R. J., Cho, Y. J., … Pfister, S. M. (2012). Medulloblastomics: The end of the beginning. *Nature Reviews Cancer, 12*, 818–834.

Northcott, P. A., Robinson, G. W., Kratz, C. P., Mabbott, D. J., Pomeroy, S. L., Clifford, S. C., & Pfister, S. M. (2019). Medulloblastoma. *Nature Reviews Disease Primers, 5*(1), 11.

Ogawa, J., Pao, G. M., Shokhirev, M. N., & Verma, I. M. (2018). Glioblastoma model using human cerebral organoids. *Cell Reports, 23*(4), 1220–1229.

Ohgaki, H., Dessen, P., Jourde, B., Horstmann, S., Nishikawa, T., Di Patre, P. L., … Kleihues, P. (2004). Genetic pathways to glioblastoma: A population-based study. *Cancer Research, 64*(19), 6892–6899.

Ondracek, J., Fadrus, P., Sana, J., Besse, A., Loja, T., Vecera, M., … Slaby, O. (2017). Global microRNA expression profiling identifies unique microRNA pattern of radioresistant glioblastoma cells. *Anticancer Research, 37*(3), 1099–1104.

Pamudurti, N. R., Bartok, O., Jens, M., Ashwal-Fluss, R., Stottmeister, C., Ruhe, L., … Kadener, S. (2017). Translation of circRNAs. *Molecular Cell, 66*(1), 9–21, e7.

Panda, A. C., & Gorospe, M. (2018). Detection and analysis of circular RNAs by RT-PCR. *Bio-protocol, 8*(6).

Panda, A. C. (2018). Circular RNAs act as miRNA sponges. *Advances in Experimental Medicine and Biology, 1087*, 67–79.

Pandey, P. R., Munk, R., Kundu, G., De, S., Abdelmohsen, K., & Gorospe, M. (2020). Methods for analysis of circular RNAs. *Wiley Interdisciplinary Reviews: RNA, 11*(1), e1566.

Paraskevopoulou, M. D., Georgakilas, G., Kostoulas, N., Vlachos, I. S., Vergoulis, T., Reczko, M., … Hatzigeorgiou, A. G. (2013). DIANA-microT web server v5.0: Service integration into miRNA functional analysis workflows. *Nucleic Acids Research, 41*(Web Server issue), W169-73.

Perry, R. B., & Ulitsky, I. (2016). The functions of long noncoding RNAs in development and stem cells. *Development, 143*(21), 3882–3894.

Petrescu, G. E. D., Sabo, A. A., Torsin, L. I., Calin, G. A., & Dragomir, M. P. (2019). MicroRNA based theranostics for brain cancer: Basic principles. *Journal of Experimental & Clinical Cancer Research, 38*(1), 231.

Pfeffer, S. R., Yang, C. H., & Pfeffer, L. M. (2015). The Role of MIR-21 in cancer. *Drug Development Research, 76*(6), 270–277.

Piwecka, M., Glazar, P., Hernandez-Miranda, L. R., Memczak, S., Wolf, S. A., Rybak-Wolf, A., … Rajewsky, N. (2017). Loss of a mammalian circular RNA locus causes miRNA deregulation and affects brain function. *Science, 357*(6357), eaam8526.

Po, A., Abballe, L., Sabato, C., Gianno, F., Chiacchiarini, M., Catanzaro, G., … Besharat, Z. M. (2018). Sonic hedgehog medulloblastoma cancer stem cells mirnome and transcriptome highlight novel functional networks. *International Journal of Molecular Sciences, 19*(8), 2326.

Pyko, I. V., Nakada, M., Sabit, H., Teng, L., Furuyama, N., Hayashi, Y., … Hamada, J. (2013). Glycogen synthase kinase 3β inhibition sensitizes human glioblastoma cells to temozolomide by affecting O6-methylguanine DNA methyltransferase promoter methylation via c-Myc signaling. *Carcinogenesis, 34*(10), 2206–2217.

References

Que, T., Song, Y., Liu, Z., Zheng, S., Long, H., Li, Z., & Qi, S. (2015). Decreased miRNA-637 is an unfavorable prognosis marker and promotes glioma cell growth, migration and invasion via direct targeting Akt1. *Oncogene, 34*(38), 4952–4963.

Quinn, J. J., & Chang, H. Y. (2016). Unique features of long non-coding RNA biogenesis and function. *Nature Reviews Genetics, 17*(1), 47–62.

Rehmsmeier, M., Steffen, P., Höchsmann, M., & Giegerich, R. (2004). Fast and effective prediction of microRNA/target duplexes. *RNA, 10*(10), 1507–1517.

Reustle, A., Fisel, P., Renner, O., Buttner, F., Winter, S., Rausch, S., ... Schaeffeler, E. (2018). Characterization of the breast cancer resistance protein (BCRP/ABCG2) in clear cell renal cell carcinoma. *International Journal of Cancer, 143*(12), 3181–3193.

Ricci-Vitiani, L., Pallini, R., Biffoni, M., Todaro, M., Invernici, G., Cenci, T., ... De Maria, R. (2010). Tumour vascularization via endothelial differentiation of glioblastoma stem-like cells. *Nature, 468*(7325), 824–828.

Robertson, F. L., Marqués-Torrejón, M. A., Morrison, G. M., & Pollard, S. M. (2019). Experimental models and tools to tackle glioblastoma. *Disease Models & Mechanisms, 12*(9).

Rophina, M., Sharma, D., Poojary, M., & Scaria, V. (2020). Circad: A comprehensive manually curated resource of circular RNA associated with diseases. *Database (Oxford), 2020*, baaa019.

Roth, P., Wischhusen, J., Happold, C., Chandran, P. A., Hofer, S., Eisele, G., & Keller, A. (2011). A specific miRNA signature in the peripheral blood of glioblastoma patients. *Journal of Neurochemistry, 118*(3), 449–457.

Roussel, M. F., & Stripay, J. L. (2020). Modeling pediatric medulloblastoma. *Brain Pathology, 30*(3), 703–712.

Ruan, H., Xiang, Y., Ko, J., Li, S., Jing, Y., Zhu, X., ... Han, L. (2019). Comprehensive characterization of circular RNAs in ~ 1000 human cancer cell lines. *Genome Medicine, 11*(1), 55.

Ruan, J., Lou, S., Dai, Q., Mao, D., Ji, J., & Sun, X. (2015). Tumor suppressor miR-181c attenuates proliferation, invasion, and self-renewal abilities in glioblastoma. *NeuroReport, 26*(2), 66–73.

Ruiz-Garcia, H., Alvarado-Estrada, K., Schiapparelli, P., Quinones-Hinojosa, A., & Trifiletti, D. M. (2020). Engineering three-dimensional tumor models to study glioma cancer stem cells and tumor microenvironment. *Front Cell Neurosci, 14*, 558381.

Rybak-Wolf, A., Stottmeister, C., Glazar, P., Jens, M., Pino, N., Giusti, S., & Rajewsky, N. (2015). Circular RNAs in the mammalian brain are highly abundant, conserved, and dynamically expressed. *Molecular Cell, 58*(5), 870–885.

Salmena, L., Poliseno, L., Tay, Y., Kats, L., & Pandolfi, P. P. (2011). A ceRNA hypothesis: The Rosetta Stone of a hidden RNA language? *Cell, 146*(3), 353–358.

Salzman, J., Chen, R. E., Olsen, M. N., Wang, P. L., & Brown, P. O. (2013). Cell-type specific features of circular RNA expression. *PLOS Genetics, 9*(9), e1003777.

Sana, J., Busek, P., Fadrus, P., Besse, A., Radova, L., Vecera, M., & Slaby, O. (2018). Identification of microRNAs differentially expressed in glioblastoma stem-like cells and their association with patient survival. *Scientific Reports, 8*(1), 2836.

Sarantopoulou, D., Tang, S. Y., Ricciotti, E., Lahens, N. F., Lekkas, D., Schug, J., ... Grant, G. R. (2019). Comparative evaluation of RNA-Seq library preparation methods for strand-specificity and low input. *Scientific Reports, 9*(1), 13477.

Schneider, T., & Bindereif, A. (2017). Circular RNAs: Coding or noncoding? *Cell Research, 27*(6), 724–725.

Schwarz, D. S., Hutvágner, G., Du, T., Xu, Z., Aronin, N., & Zamore, P. D. (2003). Asymmetry in the assembly of the RNAi enzyme complex. *Cell, 115*(2), 199–208.

Seidel, S., Garvalov, B. K., Wirta, V., von Stechow, L., Schanzer, A., Meletis, K., ... Acker, T. (2010). A hypoxic niche regulates glioblastoma stem cells through hypoxia inducible factor 2 alpha. *Brain, 133*(Pt 4), 983–995.

Shah, V., & Kochar, P. (2018). Brain cancer: Implication to disease, therapeutic strategies and tumor targeted drug delivery approaches. *Recent Patents on Anti-Cancer Drug Discovery, 13*(1), 70–85.

Shahpar, S., Mhatre, P. V., & Huang, M. E. (2016). *Update on brain tumors: New developments in neuro-oncologic diagnosis and treatment, and impact on rehabilitation strategies* (pp. 678–689). Elsevier Inc.

Shang, C., Guo, Y., Hong, Y., & Xue, Y. X. (2016). Long non-coding RNA TUSC7, a target of miR-23b, plays tumor-suppressing roles in human gliomas. *Frontiers in Cellular Neuroscience, 10*, 235.

Shang, Q., Yang, Z., Jia, R., & Ge, S. (2019). The novel roles of circRNAs in human cancer. *Molecular Cancer, 18*(1), 6.

Shao, N., Xue, L., Wang, R., Luo, K., Zhi, F., & Lan, Q. (2019). MiR-454-3p is an exosomal biomarker and functions as a tumor suppressor in glioma. *Molecular Cancer Therapeutics, 18*(2), 459−469.

Shi, C., Rao, C., Sun, C., Yu, L., Zhou, X., Hua, D., . . . Yu, S. (2018). miR-29s function as tumor suppressors in gliomas by targeting TRAF4 and predict patient prognosis. *Cell Death & Disease, 9*(11), 1078.

Shi, R., Wang, P. Y., Li, X. Y., Chen, J. X., Li, Y., Zhang, X. Z., . . . Cheng, S. J. (2015). Exosomal levels of miRNA-21 from cerebrospinal fluids associated with poor prognosis and tumor recurrence of glioma patients. *Oncotarget, 6*(29), 26971−26981.

Shi, Z., Chen, Q., Li, C., Wang, L., Qian, X., Jiang, C., . . . Jiang, B. H. (2014). MiR-124 governs glioma growth and angiogenesis and enhances chemosensitivity by targeting R-Ras and N-Ras. *Neuro-Oncology, 16*(10), 1341−1353.

Simeonova, I., & Huillard, E. (2014). In vivo models of brain tumors: Roles of genetically engineered mouse models in understanding tumor biology and use in preclinical studies. *Cellular and Molecular Life Sciences, 71*(20), 4007−4026.

Singh, S. K., Hawkins, C., Clarke, I. D., Squire, J. A., Bayani, J., Hide, T., . . . Dirks, P. B. (2004). Identification of human brain tumour initiating cells. *Nature, 432*(7015), 396−401.

Song, X., Zhang, N., Han, P., Moon, B. S., Lai, R. K., Wang, K., & Lu, W. (2016). Circular RNA profile in gliomas revealed by identification tool UROBORUS. *Nucleic Acids Research, 44*(9), e87.

Spengler, R. M., Zhang, X., Cheng, C., McLendon, J. M., Skeie, J. M., Johnson, F. L., . . . Boudreau, R. L. (2016). Elucidation of transcriptome-wide microRNA binding sites in human cardiac tissues by Ago2 HITS-CLIP. *Nucleic Acids Research, 44*(15), 7120−7131.

Spitzer, J., Hafner, M., Landthaler, M., Ascano, M., Farazi, T., Wardle, G., . . . Tuschl, T. (2014). PAR-CLIP (Photoactivatable Ribonucleoside-Enhanced Crosslinking and Immunoprecipitation): A step-by-step protocol to the transcriptome-wide identification of binding sites of RNA-binding proteins. *Methods in Enzymology, 539*, 113−161.

Stavast, C. J., & Erkeland, S. J. (2019). The non-canonical aspects of MicroRNAs: Many roads to gene regulation. *Cells, 8*(11).

Stojcheva, N., Schechtmann, G., Sass, S., Roth, P., Florea, A. M., Stefanski, A., . . . Happold, C. (2016). MicroRNA-138 promotes acquired alkylator resistance in glioblastoma by targeting the Bcl-2-interacting mediator BIM. *Oncotarget, 7*(11), 12937−12950.

Stupp, R., Mason, W. P., van den Bent, M. J., Weller, M., Fisher, B., Taphoorn, M. J., . . . Mirimanoff, R. O. European Organisation for Research and Treatment of Cancer Brain Tumor and Radiotherapy Groups. National Cancer Institute of Canada Clinical Trials Group. (2005). Radiotherapy plus concomitant and adjuvant temozolomide for glioblastoma. *The New England Journal of Medicine, 352*(10), 987−996.

Su, H., Lin, F., Deng, X., Shen, L., Fang, Y., Fei, Z., . . . Xie, C. (2016). Profiling and bioinformatics analyses reveal differential circular RNA expression in radioresistant esophageal cancer cells. *Journal of Translational Medicine, 14*(1), 225.

Su, Y. K., Lin, J. W., Shih, J. W., Chuang, H. Y., Fong, I. H., Yeh, C. T., & Lin, C. M. (2020). Targeting BC200/miR218-5p signaling axis for overcoming temozolomide resistance and suppressing glioma stemness. *Cells, 9*(8), 1859.

Sun, P., & Li, G. (2019). CircCode: A powerful tool for identifying circRNA coding ability. *Frontiers in Genetics, 10*, 981.

Sun, S. L., Shu, Y. G., & Tao, M. Y. (2020). LncRNA CCAT2 promotes angiogenesis in glioma through activation of VEGFA signalling by sponging miR-424. *Molecular and Cellular Biochemistry, 468*(1−2), 69−82.

Sun, X. H., Fan, W. J., An, Z. J., & Sun, Y. (2020). Inhibition of long noncoding RNA CRNDE increases chemosensitivity of medulloblastoma cells by targeting miR-29c-3p. *Oncology Research, 28*(1), 95−102.

Swellam, M., Ezz El Arab, L., Al-Posttany, A. S., & B. Said, S. (2019). Clinical impact of circulating oncogenic MiRNA-221 and MiRNA-222 in glioblastoma multiform. *Journal of Neuro-Oncology, 144*(3), 545−551.

Szabo, L., & Salzman, J. (2016). Detecting circular RNAs: Bioinformatic and experimental challenges. *Nature Reviews Genetics, 17*(11), 679−692.

Tang, T., Wang, L. X., Yang, M. L., & Zhang, R. M. (2020). lncRNA TPTEP1 inhibits stemness and radioresistance of glioma through miR106a5pmediated P38 MAPK signaling. *Molecular Medicine Reports, 22*(6), 4857−4867.

Tang, Z., Li, X., Zhao, J., Qian, F., Feng, C., Li, Y., . . . Li, C. (2019). TRCirc: A resource for transcriptional regulation information of circRNAs. *Briefings in Bioinformatics, 20*(6), 2327−2333.

Teplyuk, N. M., Uhlmann, E. J., Gabriely, G., Volfovsky, N., Wang, Y., Teng, J., ... Krichevsky, A. M. (2016). Therapeutic potential of targeting microRNA-10b in established intracranial glioblastoma: First steps toward the clinic. *EMBO Molecular Medicine*, 8(3), 268–287.

Tian, T., Mingyi, M., Qiu, X., & Qiu, Y. (2016). MicroRNA-101 reverses temozolomide resistance by inhibition of GSK3β in glioblastoma. *Oncotarget*, 7(48), 79584–79595.

Tomasello, L., Cluts, L., & Croce, C. M. (2019). Experimental validation of microRNA targets: Luciferase reporter assay. *Methods in Molecular Biology*, 1970, 315–330.

Tong, X., & Liu, S. (2019). CPPred: Coding potential prediction based on the global description of RNA sequence. *Nucleic Acids Research*, 47(8), e43.

Tu, Y., Gao, X., Li, G., Fu, H., Cui, D., Liu, H., ... Zhang, Y. (2013). MicroRNA-218 inhibits glioma invasion, migration, proliferation, and cancer stem-like cell self-renewal by targeting the polycomb group gene Bmi1. *Cancer Research*, 73(19), 6046–6055.

Turner, K. M., Sun, Y., Ji, P., Granberg, K. J., Bernard, B., Hu, L., ... Zhang, W. (2015). Genomically amplified Akt3 activates DNA repair pathway and promotes glioma progression. *Proceedings of the National Academy of Sciences of the United States of America*, 112(11), 3421–3426.

Udaka, Y. T., & Packer, R. J. (2018). Pediatric brain tumors. *Neurologic Clinics*, 36(3), 533–556.

Valadkhan, S., & Valencia-Hipolito, A. (2016). lncRNAs in stress response. *Current Topics in Microbiology and Immunology*, 394, 203–236.

Vincent, H. A., & Deutscher, M. P. (2006). Substrate recognition and catalysis by the exoribonuclease RNase R. *Journal of Biological Chemistry*, 281(40), 29769–29775.

Visani, M., de Biase, D., Marucci, G., Cerasoli, S., Nigrisoli, E., Bacchi Reggiani, M. L., ... Pession, A. PERNO study group. (2014). Expression of 19 microRNAs in glioblastoma and comparison with other brain neoplasia of grades I-III. *Molecular Oncology*, 8(2), 417-430.

Vo, J. N., Cieslik, M., Zhang, Y., Shukla, S., Xiao, L., Zhang, Y., ... Chinnaiyan, A. M. (2019). The landscape of circular RNA in. *Cancer Cell*, 176(4), 869–881, e13.

Waitkus, M. S., Diplas, B. H., & Yan, H. (2018). Biological role and therapeutic potential of IDH mutations in cancer. *Cancer Cell*, 34(2), 186–195.

Wang, B., Love, T. M., Call, M. E., Doench, J. G., & Novina, C. D. (2006). Recapitulation of short RNA-directed translational gene silencing in vitro. *Molecular Cell*, 22(4), 553–560.

Wang, B., Wang, K., Jin, T., Xu, Q., He, Y., Cui, B., & Wang, Y. (2020). NCK1-AS1 enhances glioma cell proliferation, radioresistance and chemoresistance via miR-22-3p/IGF1R ceRNA pathway. *Biomedicine & Pharmacotherapy*, 129, 110395.

Wang, C., Yun, Z., Zhao, T., Liu, X., & Ma, X. (2015). MIR-495 is a predictive biomarker that downregulates GFI1 expression in medulloblastoma. *Cellular Physiology and Biochemistry*, 36(4), 1430–1439.

Wang, J., Zhou, T., Wang, T., & Wang, B. (2018). Suppression of lncRNA-ATB prevents amyloid-β-induced neurotoxicity in PC12 cells via regulating miR-200/ZNF217 axis. *Biomedicine and Pharmacotherapy*, 108, 707–715.

Wang, L. Q., Sun, W., Wang, Y., Li, D., & Hu, A. M. (2019). Downregulation of plasma miR-124 expression is a predictive biomarker for prognosis of glioma. *European Review for Medical and Pharmacological Sciences*, 23(1), 271–276.

Wang, W., Han, S., Gao, W., Feng, Y., Li, K., & Wu, D. (2020). Long noncoding RNA KCNQ1OT1 confers gliomas resistance to temozolomide and enhances cell growth by retrieving PIM1 from miR-761. *Cellular and Molecular Neurobiology*.

Wawrzyniak, O., Zarebska, Z., Kuczynski, K., Gotz-Wieckowska, A., & Rolle, K. (2020). Protein-related circular RNAs in human pathologies. *Cells*, 9(8), 1841.

Wei, Y., Lu, C., Zhou, P., Zhao, L., Lyu, X., Yin, J., ... You, Y. (2021). EIF4A3-induced circular RNA ASAP1 promotes tumorigenesis and temozolomide resistance of glioblastoma via NRAS/MEK1/ERK1-2 signaling. *Neuro-Oncology*, 23(4), 611–624.

Wittrup, A., & Lieberman, J. (2015). Knocking down disease: A progress report on siRNA therapeutics. *Nature Reviews Genetics*, 16(9), 543–552.

Wongfieng, W., Jumnainsong, A., Chamgramol, Y., Sripa, B., & Leelayuwat, C. (2017). 5′-UTR and 3′-UTR regulation of MICB expression in human cancer cells by novel microRNAs. *Genes*, 8(9), 213.

Wu, N., Xiao, L., Zhao, X., Zhao, J., Wang, J., Wang, F., ... Lin, X. (2012). MiR-125b regulates the proliferation of glioblastoma stem cells by targeting E2F2. *FEBS Letters*, 586(21), 3831–3839.

Wu, W., Ji, P., & Zhao, F. (2020). CircAtlas: An integrated resource of one million highly accurate circular RNAs from 1070 vertebrate transcriptomes. *Genome Biology, 21*(1), 101.

Wu, Y., Yao, Y., Yun, Y., Wang, M., & Zhu, R. (2019). MicroRNA-302c enhances the chemosensitivity of human glioma cells to temozolomide by suppressing P-gp expression. *Bioscience Reports, 39*(9), BSR20190421.

Xia, S., Feng, J., Chen, K., Ma, Y., Gong, J., Cai, F., ... He, C. (2018). CSCD: A database for cancer-specific circular RNAs. *Nucleic Acids Research, 46*(D1), D925-D9.

Xia, S., Feng, J., Lei, L., Hu, J., Xia, L., Wang, J., ... He, C. (2017). Comprehensive characterization of tissue-specific circular RNAs in the human and mouse genomes. *Briefings in Bioinformatics, 18*(6), 984—992.

Xia, X., Li, X., Li, F., Wu, X., Zhang, M., Zhou, H., ... Zhang, N. (2019). A novel tumor suppressor protein encoded by circular AKT3 RNA inhibits glioblastoma tumorigenicity by competing with active phosphoinositide-dependent Kinase-1. *Molecular Cancer, 18*(1), 131.

Xiao, M. S., & Wilusz, J. E. (2019). An improved method for circular RNA purification using RNase R that efficiently removes linear RNAs containing G-quadruplexes or structured 3' ends. *Nucleic Acids Research, 47*(16), 8755—8769.

Xiao, S., Yang, Z., Qiu, X., Lv, R., Liu, J., Wu, M., ... Liu, Q. (2016). MiR-29c contribute to glioma cells temozolomide sensitivity by targeting O6-methylguanine-DNA methyltransferases indirectly. *Oncotarget, 7*(31), 50229—50238.

Xing, W., & Zeng, C. (2017). A novel serum microRNA-based identification and classification biomarker of human glioma. *Tumor Biology, 39*(5), 1010428317705339.

Xu, H., Zhang, Y., Qi, L., Ding, L., Jiang, H., & Yu, H. (2018). NFIX circular RNA promotes glioma progression by regulating miR-34a-5p via notch signaling pathway. *Frontiers in Molecular Neuroscience, 11*, 225.

Xu, Z., Kader, M., Sen, R., & Placantonakis, D. G. (2018). Orthotopic patient-derived glioblastoma xenografts in mice. *Methods in Molecular Biology, 1741*, 183—190.

Yan, W., Li, R., Liu, Y., Yang, P., Wang, Z., Zhang, C., ... Jiang, T. (2014). MicroRNA expression patterns in the malignant progression of gliomas and a 5-microRNA signature for prognosis. *Oncotarget, 5*(24), 12908—12915.

Yan, W., Liu, Y., Yang, P., Wang, Z., You, Y., & Jiang, T. (2015). MicroRNA profiling of Chinese primary glioblastoma reveals a temozolomide-chemoresistant subtype. *Oncotarget, 6*(13), 11676—11682.

Yang, L., Li, N., Yan, Z., Li, C., & Zhao, Z. (2018). MiR-29a-mediated CD133 expression contributes to cisplatin resistance in CD133 + glioblastoma stem cells. *Journal of Molecular Neuroscience, 66*(3), 369—377.

Yang, W., Shen, Y., Wei, J., & Liu, F. (2015). MicroRNA-153/Nrf-2/GPx1 pathway regulates radiosensitivity and stemness of glioma stem cells via reactive oxygen species. *Oncotarget, 6*(26), 22006—22027.

Yang, W., Yu, H., Shen, Y., Liu, Y., Yang, Z., & Sun, T. (2016). MiR-146b-5p overexpression attenuates stemness and radioresistance of glioma stem cells by targeting HuR/lincRNA-p21/beta-catenin pathway. *Oncotarget, 7*(27), 41505—41526.

Yao, D., Zhang, L., Zheng, M., Sun, X., Lu, Y., & Liu, P. (2018). Circ2Disease: A manually curated database of experimentally validated circRNAs in human disease. *Scientific Reports, 8*(1), 11018.

Yin, H., & Cui, X. (2020). Knockdown of circHIPK3 facilitates temozolomide sensitivity in glioma by regulating cellular behaviors through miR-524-5p/KIF2A-mediated PI3K/AKT pathway. *Cancer Biotherapy and Radiopharmaceuticals*.

Yin, J., Zeng, A., Zhang, Z., Shi, Z., Yan, W., & You, Y. (2019). Exosomal transfer of miR-1238 contributes to temozolomide-resistance in glioblastoma. *EBioMedicine, 42*, 238—251.

Yuan, D., Zhao, Y., Banks, W. A., Bullock, K. M., Haney, M., Batrakova, E., & Kabanov, A. V. (2017). Macrophage exosomes as natural nanocarriers for protein delivery to inflamed brain. *Biomaterials, 142*, 1—12.

Yuan, Y., Jiaoming, L., Xiang, W., Yanhui, L., Shu, J., Maling, G., & Qing, M. (2018). Analyzing the interactions of mRNAs, miRNAs, lncRNAs and circRNAs to predict competing endogenous RNA networks in glioblastoma. *Journal of Neuro-Oncology, 137*(3), 493—502.

Zaghlool, A., Ameur, A., Wu, C., Westholm, J. O., Niazi, A., Manivannan, M., ... Feuk, L. (2018). Expression profiling and in situ screening of circular RNAs in human tissues. *Scientific Reports, 8*(1), 16953.

Zang, J., Lu, D., & Xu, A. (2020). The interaction of circRNAs and RNA binding proteins: An important part of circRNA maintenance and function. *Journal of Neuroscience Research, 98*(1), 87—97.

Zeng, A., Wei, Z., Yan, W., Yin, J., Huang, X., Zhou, X., ... You, Y. (2018). Exosomal transfer of miR-151a enhances chemosensitivity to temozolomide in drug-resistant glioblastoma. *Cancer Letters, 436*, 10—21.

Zhang, B., Fang, S., Cheng, Y., Zhou, C., & Deng, F. (2019). The long non-coding RNA, urothelial carcinoma associated 1, promotes cell growth, invasion, migration, and chemo-resistance in glioma through Wnt/beta-catenin signaling pathway. *Aging (Albany NY)*, 11(19), 8239–8253.

Zhang, C., Chen, Q., Zhu, J. W., & Liu, Z. F. (2019). MicroRNA-199a-5p regulates glioma progression via targeting MARCH8. *European Review for Medical and Pharmacological Sciences*, 23(17), 7482–7487.

Zhang, K., Sun, X., Zhou, X., Han, L., Chen, L., Shi, Z., ... Kang, C. (2015). Long non-coding RNA HOTAIR promotes glioblastoma cell cycle progression in an EZH2 dependent manner. *Oncotarget*, 6(1), 537–546.

Zhang, S., Zhao, B. S., Zhou, A., Lin, K., Zheng, S., Lu, Z., ... Huang, S. (2017). m(6)A demethylase ALKBH5 maintains tumorigenicity of glioblastoma stem-like cells by sustaining FOXM1 expression and cell proliferation program. *Cancer Cell*, 31(4), 591–606, e6.

Zhang, W., Yue, X., Tang, G., Wu, W., Huang, F., & Zhang, X. (2018). SFPEL-LPI: Sequence-based feature projection ensemble learning for predicting LncRNA-protein interactions. *PLOS Computational Biology*, 14(12), e1006616.

Zhang, X., Sun, S., Pu, J. K., Tsang, A. C., Lee, D., Man, V. O., ... Leung, G. K. K. (2012). Long non-coding RNA expression profiles predict clinical phenotypes in glioma. *Neurobiology of Disease*, 48(1), 1–8.

Zhang, Z., Yin, J., Lu, C., Wei, Y., Zeng, A., & You, Y. (2019). Exosomal transfer of long non-coding RNA SBF2-AS1 enhances chemoresistance to temozolomide in glioblastoma. *Journal of Experimental & Clinical Cancer Research*, 38(1), 166.

Zhao, C., Gao, Y., Guo, R., Li, H., & Yang, B. (2020). Microarray expression profiles and bioinformatics analysis of mRNAs, lncRNAs, and circRNAs in the secondary temozolomide-resistant glioblastoma. *Investigational New Drugs*, 38(5), 1227–1235.

Zhao, C., Guo, R., Guan, F., Ma, S., Li, M., Wu, J., ... Yang, B. (2020). MicroRNA-128-3p enhances the chemosensitivity of temozolomide in glioblastoma by targeting c-Met and EMT. *Scientific Reports*, 10(1), 9471.

Zhao, G., Huang, Q., Wang, F., Zhang, X., Hu, J., Tan, Y., ... Cheng, Y. (2018). Targeted shRNA-loaded liposome complex combined with focused ultrasound for blood brain barrier disruption and suppressing glioma growth. *Cancer Letters*, 418, 147–158.

Zhao, M., Xu, J., Zhong, S., Liu, Y., Xiao, H., Geng, L., & Liu, H. (2019). Expression profiles and potential functions of circular RNAs in extracellular vesicles isolated from radioresistant glioma cells. *Oncology Reports*, 41(3), 1893–1900.

Zhao, X., Ponomaryov, T., Ornell, K. J., Zhou, P., Dabral, S. K., Pak, E., ... Segal, R. A. (2015). RAS/MAPK activation drives resistance to Smo inhibition, metastasis, and tumor evolution in SHH pathway-dependent tumors. *Cancer Research*, 75(17), 3623–3635.

Zhao, Y., Teng, H., Yao, F., Yap, S., Sun, Y., & Ma, L. (2020). Challenges and strategies in ascribing functions to long noncoding RNAs. *Cancers (Basel)*, 12(6), 1458.

Zhao, Z., Wang, K., Wu, F., Wang, W., Zhang, K., Hu, H., ... Jiang, T. (2018). circRNA disease: A manually curated database of experimentally supported circRNA-disease associations. *Cell Death & Disease*, 9(5), 475.

Zheng, J., Li, X. D., Wang, P., Liu, X. B., Xue, Y. X., Hu, Y., ... Liu, Y. (2015). CRNDE affects the malignant biological characteristics of human glioma stem cells by negatively regulating miR-186. *Oncotarget*, 6(28), 25339–25355.

Zheng, J., Wang, B., Zheng, R., Zhang, J., Huang, C., Zheng, R., & Yuan, Y. (2020). Linc-RA1 inhibits autophagy and promotes radioresistance by preventing H2Bub1/USP44 combination in glioma cells. *Cell Death & Disease*, 11(9), 758.

Zheng, L. L., Li, J. H., Wu, J., Sun, W. J., Liu, S., Wang, Z. L., ... Qu, L. (2016). deepBase v2.0: Identification, expression, evolution and function of small RNAs, LncRNAs and circular RNAs from deep-sequencing data. *Nucleic Acids Research*, 44(D1), D196-202.

Zheng, R., Yao, Q., Li, X., & Xu, B. (2019). Long noncoding ribonucleic acid SNHG18 promotes glioma cell motility via disruption of alpha-enolase nucleocytoplasmic transport. *Frontiers in Genetics*, 10, 1140.

Zheng, R., Yao, Q., Ren, C., Liu, Y., Yang, H., Xie, G., ... Yuan, Y. (2016). Upregulation of long noncoding RNA small nucleolar RNA host gene 18 promotes radioresistance of glioma by repressing semaphorin 5A. *International Journal of Radiation Oncology, Biology, Physics*, 96(4), 877–887.

Zhou, X., & Xu, J. (2015). Identification of Alzheimer's disease-associated long noncoding RNAs. *Neurobiology of Aging*, 36(11), 2925–2931.

Zhou, Y., Peng, Z., Seven, E. S., & Leblanc, R. M. (2018). Crossing the blood-brain barrier with nanoparticles. *Journal of Controlled Release*, 270, 290–303.

Zhu, J., Ye, J., Zhang, L., Xia, L., Hu, H., Jiang, H., ... Luo, C. (2017). Differential expression of circular RNAs in glioblastoma multiforme and its correlation with prognosis. *Translational Oncology, 10*(2), 271–279.

Zhu, S., Wang, J., He, Y., Meng, N., & Yan, G. R. (2018). Peptides/proteins encoded by non-coding RNA: A novel resource bank for drug targets and biomarkers. *Frontiers in Pharmacology, 9*, 1295.

Zhu, X. Y., Li, G. X., & Liu, Z. L. (2018). miR-599 as a potential biomarker for prognosis of glioma. *European Review for Medical and Pharmacological Sciences, 22*(2), 294–298.

Zhukova, N., Ramaswamy, V., Remke, M., Pfaff, E., Shih, D. J. H., Martin, D. C., ... Tabori, U. (2013). Subgroup-specific prognostic implications of TP53 mutation in medulloblastoma. *Journal of Clinical Oncology, 31*(23), 2927–2935.

Zirkel, A., & Papantonis, A. (2018). Detecting circular RNAs by RNA fluorescence in situ hybridization. *Methods in Molecular Biology, 1724*, 69–75.

Zlotorynski, E. (2015). Non-coding RNA: Circular RNAs promote transcription. *Nature Reviews Molecular Cell Biology, 16*(4), 206.

Zong, Z., Song, Y., Xue, Y., Ruan, X., Liu, X., Yang, C., ... Liu, Y. (2019). Knockdown of LncRNA SCAMP1 suppressed malignant biological behaviours of glioma cells via modulating miR-499a-5p/LMX1A/NLRC5 pathway. *Journal of Cellular and Molecular Medicine, 23*(8), 5048–5062.

CHAPTER 3

Noncoding RNAs in patients with colorectal cancer

Mohammad Amin Kerachian

Department of Medical Genetics, Faculty of Medicine, Mashhad University of Medical Sciences, Mashhad, Iran

3.1 Introduction

High-throughput technologies have displayed that most of the eukaryotic genome can be transcribed, in the region of 20,000–25,000 protein-encoding genes (Consortium, 2004), although the vast majority of untranslated fractions of the transcriptome are transcribed as noncoding RNAs (ncRNAs) by the HUGO Gene Nomenclature Committee (http://www.genenames.org) with a number of 8475 discovered until now. The number of identified ncRNAs has been rapidly increased in recent years and currently, the multiple types of ncRNAs have built up the noncoding transcriptome (Rodriguez-Casanova et al., 2021). Today, ncRNAs were recognized as a novel class of RNA, involved in the regulation of several biological processes, including gene expression, epigenetic processes, cell differentiation and proliferation, migration, invasion, apoptosis, transcriptional and posttranscriptional regulation, organ regeneration, and pathogenesis of human diseases (Huang, Peng, & Guo, 2015; Kaikkonen, Lam, & Glass, 2011; Li & Belmonte, 2015; Piccoli, Gupta, & Thum, 2015). They can be classified into two groups based on their size, small ncRNAs (sncRNAs) and long ncRNAs (lncRNAs), which are shorter and longer than 200 nucleotides (nt), respectively. It is estimated to have nearly 60,000 lncRNA genes in the human genome (Iyer et al., 2015). The sncRNAs can be subdivided further into microRNAs (miRNAs), small nucleolar RNAs, small interfering RNAs (siRNAs), small nuclear RNAs, piwi-interacting RNAs, promoter-associated small RNAs, termini-associated small RNAs, transcription initiation RNAs, and defective-interfering RNAs (Kapranov et al., 2007). Of note, some ncRNAs have variable lengths and could be attributed to both categories at the same time such as enhancer RNAs (eRNAs) (Zhang, Wu, Chen, & Chen, 2019). The classification of lncRNAs is based on four main features, that is genomic location and context, mechanism of action, effect conducted on DNA sequences, and their targeting mechanisms

(Ma, Bajic, & Zhang, 2013). Moreover, they can be also classified as follows: intergenic, intronic, bidirectional, sense, and antisense lncRNAs (Han et al., 2015).

Among sncRNAs, miRNAs have been the most widely investigated ncRNAs. MiRNAs are single-stranded molecules of 18–25 nt that bind to specific regions of target messenger RNA (mRNAs), mediating posttranscriptional gene silencing by two possible mechanisms: blocking transcription or triggering mRNA degradation (Lee & Calin, 2011). Therefore a single miRNA can control the expression of many genes, by regulating key pathways in cancer tumorigenesis (Lee & Dutta, 2009) and function as a tumor suppressor or oncogene or both as a dual regulator in cancer (Bayraktar, Pichler, & Kanlikilicer, 2017; Volinia, Calin, & Liu, 2006). Importantly, different types of cancers have shown specific miRNA signatures, defining the molecular characteristics of tumors (Lu et al., 2005).

Although miRNAs are the most described ncRNAs, lncRNAs have recently been shown to represent the vast majority of noncoding transcripts (Hon, Ramilowski, & Harshbarger, 2017). LncRNAs do not basically have the potential to encode proteins, although they may have some mRNA-like properties (Derrien, Johnson, & Bussotti, 2012; Guttman, Amit, & Garber, 2009). Besides, they have relevant regulatory functions in the process of gene expression, for example during transcriptional regulation and splicing (Kotake, Nakagawa, & Kitagawa, 2011; Léveillé, Melo, & Rooijers, 2015).

Because of the close link of ncRNAs especially lncRNAs and miRNAs to human diseases, including cancer, ncRNAs have gained great attention by researchers.

They might play different roles as oncogenes or tumor suppressors in various cancers such as colorectal cancer (CRC) and be offered as potential diagnostic or prognostic biomarkers with possible clinical applications (Xie, Tang, & Xiao, 2016).

According to GLOBOCAN 2020 data, more than 1.9 million new CRC (including anus) cases and 935,000 deaths were estimated to occur in 2020, representing about 1 in 10 cancer cases and deaths. Totally, CRC ranks third and second in terms of incidence and mortality (Sung, Ferlay, & Siegel, 2021), respectively.

In this chapter, we aim to describe the potential clinical relevance of ncRNAs in CRC patients and explain how the ncRNAs might be involved in shaping the future of diagnosis, prognosis, prediction, and treatment of colon cancer.

3.2 Experimental methods and tools for analyzing noncoding RNAs in colorectal cancer patients

Although miRNAs represent the most widely studied ncRNAs, lncRNAs are emerging as the cancer key regulators (Adriaens & Marine, 2017; Amodio et al., 2013, 2018; Caracciolo et al., 2019; Tang et al., 2019; Tokgun, Tokgun, Kurt, Tomatir, & Akca, 2019). Arrays and next-generation sequencing (NGS) as high-throughput methods could detect and quantify ncRNAs. However, several cautious should be considered. Typically, both miRNAs and lncRNA are expressed at lower abundance in comparison to mRNAs (~0.01% and ~0.1% of total RNA, respectively) (Pritchard, Cheng, & Tewari, 2012). Thus miRNA profiling requires RNA isolation procedures to retain the small RNA fraction. Moreover, miRNAs lack a common sequence, such as poly(A) tail that is typically observed in mRNAs. So, it is necessary to selectively detect this class of ncRNAs among different RNA species and select appropriate high-throughput assays. In addition, miRNAs of the same family may show high similarity or differ from the reference sequence due to posttranscriptional modifications. On the other hand, since lncRNAs have similarities with mRNAs such as size, RNA polymerase II transcription, 5′-capping, RNA splicing, a poly(A) tail reported in about 60% of cases, they could be profiled with mRNAs. However, the design of probes for many lncRNAs is challenging, since most lncRNAs are located at intergenic regions with high GC (Glucocorticoid) content or are antisense transcripts of known protein-coding genes (Wong, Huang, Islam, & Yip, 2018). Here, we describe different methodologies to study the pattern of gene expression to discover ncRNAs potentially involved in the pathogenesis of CRC.

3.3 Microarray

Microarray could profile both miRNAs and lncRNAs, although it was initially designed for protein-coding mRNAs. This technology is according to nucleic acid hybridization between labeled RNA targets and their specific and complementary probes. The advantages of microarrays are the high parallel analysis as well as its relatively low cost and the ability to detect low levels of RNAs without the need for PCR (Polymerase Chain Reaction)-enrichment steps. Various platforms for miRNA profiling include several direct miRNA labeling procedures without amplification. Importantly, lncRNA microarray platforms systematically profile lncRNAs together with mRNAs. LncRNA platforms usually include an in vitro transcription-based amplification step and are distinguished by less technical variations concerning miRNA platforms and mostly differ for the number of lncRNAs analyzed. For example, the *Agilent SurePrint G3 Gene Expression v3 microarray*, which targets 30,606 human lncRNA transcripts, covers all of LNCipedia 2.1; the *Arraystar LncRNA microarray* (release human v5.0) profiles 39,317 lncRNAs; *Clariom D human array* provided by Thermo Fisher Scientific covers more than 55,900 lncRNA NONCODE transcripts. Several microarray platforms have been adapted for miRNA quantification, including GeneChip (Affymetrix), miRCURY LNA (Exiqon), and SurePrint (Agilent). All platforms have designed probes specific for mature miRNA sequences, although they have major differences such as hybridization, washing procedures, and fluorescent dyes. Microarray technology can be used to measure multiple miRNAs simultaneously, but several drawbacks have limited their utility in clinic (Moody, He, Pan, & Chen, 2017). Limitations of microarrays for ncRNA analysis include (1) a restricted linear range of quantification, (2) limitation of probe design, (3) a continuous need to update annotations, (4) the relative quantification is limited to compare different statuses (e.g., healthy vs affected), (5) difficulty in data normalization that is also time-consuming, with no single method universally accepted to analyze microarray data (Meyer, Kaiser, Wagner, Thirion, & Pfaffl, 2012; Wu et al., 2013).

An example of microarray technology application has been reported by Zhou et al. (2018) who profiled 389 CRC patients and identified a signature of six lncRNAs (linc0184, AC105243.1, LOC101928168, ILF3-AS1, mir31HG, and AC006329.1) associated with the risk of cancer recurrence, or by Wang, Xiang, and Li (2015) who identified a panel of miR-409-3p, miR-7, and miR-93 that yielded high diagnostic performance in distinguishing CRC patients from healthy individuals (AUC (Area Under the Curve): 0.866 and 0.897 for training and validation dataset, respectively). Furthermore, Tang, Yan, and Bao (2019) also identified through microarray, lncRNA GLCC1, significantly upregulated under glucose starvation in CRC cells.

Another technique based on the microarray is the Tiling array that uses probes that may cover either specific chromosomal sequences and contiguous regions or even the whole genome (Bertone, Stolc, & Royce, 2004). Genome tiling arrays have demonstrated that the amount of noncoding sequences is at least four times larger than the amount of coding sequences, which indicates that only 1% of the human genome is composed of protein-coding genes and the remaining 4%–9% is transcribed into ncRNAs (MercerTR & Mattick, 2009; Xu, Qi, & Du, et al., 2014). This technique has also been used to discover ncRNAs. For instance, Rinn et al. focused on lncRNAs expressed in the region of the human *HOX* genes and compared skin fibroblasts isolated from different anatomical regions of the body (Xu et al., 2014). A similar *HOX* tiling array was used to identify lncRNAs specifically expressed in metastatic breast carcinoma (Gupta, Shah, & Wang, 2010). The lncRNA HOTAIRM1 was discovered intergenically, between the *HOXA1* and *HOXA2* genes with commercially available tiling arrays covering the human *HOXA* gene cluster (Zhang, Lian, & Padden, 2009). Taking advantage of this knowledge, Guttman et al. (2009) prepared DNA tiling arrays with 2.1 million oligonucleotide probes representing 350 K3-K36 domains and hybridized them with polyadenylated RNA to identify 1600 mouse lincRNAs (Long Intergenic Noncoding RNA). A similar tiling array was used to identify 300 lincRNAs in human cells (Khalil, Guttman, & Huarte, 2009).

3.4 Serial analysis of gene expression

Serial analysis of gene expression (SAGE), invented in the mid1990s, is the first high-throughput sequencing technology developed to analyze the transcriptome in terms of identification and quantification of transcripts, including ncRNAs (Velculescu, Zhang, Vogelstein, & Kinzler, 1995). It is based on the restriction enzymes-mediated generation of short stretches of unbiased cDNA (Complementary DNA) sequences (9-bp SAGE tags) followed by concatenation, cloning, and sequencing (Matsumura, Kruger, Kahl, & Terauchi, 2008). To date, numerous SAGE libraries representing a diversity of human and mouse, normal and malignant tissues, and cell lines have become publicly available (Barrett, Troup, & Wilhite, 2010). Of the 755 human SAGE libraries in the Gene Expression Omnibus (GEO) database, approximately 276 are SAGE libraries derived from human cancers or dysplasias (Strausberg, Buetow, Emmert-Buck, & Klausner, 2000). Gibb, Vucic, and Enfield (2011) reported lncRNA expression profiles across 26 normal and 19 tumoral tissues (breast, brain, and lung cancer) by analyzing 24 million SAGE tags. By this time, there is no investigation studying the ncRNA expression profile of CRC.

3.5 Cap analysis gene expression

Cap analysis gene expression (CAGE) is another high-throughput technology allowing the generation of a snapshot of the 5′ end of the mRNA. Similar to SAGE, sequencing is preceded by cDNA-tag generation, concatenation, and cloning. However, the main advantage is the ability of CAGE in identifying the exact location of the 5′ capped transcript. Compared to RNA sequencing (RNA-seq), CAGE could locate the transcriptionally active promoter regions and RNA polymerase II transcription start sites. This method could be used for identifying dysregulation of coding and noncoding genes in cancers, including CRC (Horie, Kaczkowski, & Ohshima, 2017).

3.6 RNA sequencing

This technique allows the detection and quantification of all sorts of ncRNAs through the construction of different cDNA libraries that are specific for each type of ncRNA. Following cDNA library preparation, massively parallel sequencing of transcripts of interest is performed. Small RNA-seq and total RNA-seq are suitable for the sequencing of sncRNAs and lncRNAs, respectively. The advantage of RNA-seq to microarray is to offer more comprehensive coverage of the whole transcriptomes. Furthermore, it is a design-free probe technique allowing the detection of novel transcripts and also sequences even with a single-nucleotide difference, like transcripts harboring mutations. Its main limitations are the difficulty of data analysis and the high deep reads needed to detect low amounts of ncRNAs (Yamada et al., 2018). The identification of transcripts from RNA-seq data is conducted using algorithms following different approaches: mapping-first algorithms like Cufflinks and Scripture and assembly first methods like Trinity, SOAPdenovo, and Oases (Arrigoni, Ranzani, & Rossetti, 2016). Then, some peculiar characteristics of noncoding transcripts are leveraged to isolate lncRNAs from protein-coding transcripts in datasets that are typically constituted by thousands of previously unidentified transcripts. The analysis of evolutionary patterns across different species [PhyloCSF (Lin, Jungreis, & Kellis, 2011)] and classifiers trained on linguistic features [iSeeRNA (Sun et al., 2013), CPAT (Coding-Potential Assessment Tool) (Wang et al., 2013), PLEK (Predictor of long noncoding RNAs and messenger RNAs based on an improved k-mer scheme) (Li, Zhang, & Zhou, 2014)] are used for this purpose. Furthermore, the use of chromatin immunoprecipitation (ChIP) coupled with NGS (ChIP-seq) provides important and complementary information, which assists the novel lncRNAs discovery (Barski, Cuddapah, & Cui, 2007).

Yamada et al. (2018) identified 27 and 22 up- and downregulated lncRNAs, respectively, as CRC biomarkers by integrating data from RNA-seq.

The most advanced application of RNA-seq is sequencing from very low amounts of target ncRNAs and/or a single-cell transcriptomic sequencing (Gawronski & Kim, 2017). For instance, the Designed Primer−based RNA-seq strategy allows the amplification of RNA from 50 pg of a sample or transcripts longer than 4 kb (Bhargava, Ko, Willems, Mercola, & Subramaniam, 2013; Sun & Chen, 2020). Similarly, a single-cell RNA-seq method called Quartz-Seq can reveal genetic changes between single cells into the same cell type and also into the same cell-cycle phase by reducing background noise (Kashi, Henderson, Bonetti, & Carninci, 2016; Sasagawa, Nikaido, & Hayashi, 2013).

Analysis of ncRNAs data may have different purposes such as gene discovery and annotation of novel ncRNAs, profiling of expression patterns, validation and structural reconstruction of known ncRNAs, and functional integrative analysis. Importantly, the application of various types of RNA-seq is becoming more popular to explore the patient-specific genetic background, covering intra- and interindividual variabilities, a key player in the era of personalized medicine (Arbitrio, Di Martino, & Barbieri, 2016; Di Martino, Scionti, & Sestito, 2016; Rossi, Teresa Di Martino, & Morelli, 2012; Scionti, Di Martino, & Sestito, 2017).

3.7 Dataset and bioinformatics for analyzing noncoding RNAs in colorectal cancer patients

Bioinformatics data analyses of high-throughput ncRNAs expression profiling are described next.

In the case of data generated by microarray technology, bioinformatics data analysis includes (1) the identification of differentially expressed genes between two groups such as normal vs tumor specimens, (2) clustering of genes in terms of expression level, (3) classification, and (4) analysis of pathways and interaction networks.

Several well-known algorithms of microarray data preprocessing include DFW (Distribution Free Weighted) (Chen, McGee, Liu, & Scheuermann, 2007), FARMS (Factor Analysis for Robust Microarray Summarization) (Hochreiter, Clevert, & Obermayer, 2006), MBEI (Model-Based Expression Indexes) (Li & Wong, 2001), VSN (Variance Stabilizing Normalization) (Huber, von Heydebreck, Sultmann, Poustka, & Vingron, 2002), MAS4.0, MAS5.0, RMA (Robust Multiarray Average), PLIER (Probe Logarithmic Error Intensity Estimate), and GCRMA (GeneChip RMA). Moreover, the array raw data processing involves several steps: (1) preprocessing, including background correction, normalization, and summarization. *Background adjustment* is essential to remove noise in the optical detection system due to nonspecific hybridization. *Normalization*, within and between arrays, is needed to remove systematic technical artifacts that could be due to different efficiency of reverse transcription, labeling or hybridization reactions, or other laboratory conditions. *Summarization* unifies signals generated from multiple probes, designed for the same transcripts, with multiple locations on the array; (2) annotation to enrich preprocessed data; data can be annotated by adding information such as gene symbols or functions; (3) statistical and/or data mining analysis is a process by which groups of samples are compared to find differentially expressed genes based on their expression values; and (4) visualization and biological interpretation. Many of the methods for visualization and interpretation of gene expression data include clustering analysis, gene set enrichment analysis, and pathway (Gene Ontology, KEGG (Kyoto Encyclopedia of Genes and Genomes), Ingenuity, Reactome, WikiPathways, NCI Pathways, and Pathway Commons) or network analysis (Mubeen et al., 2019; Olson, 2006; Pirim, Ekşioğlu, Perkins, & Yüceer, 2012). These methods could also be used for the interpretation of the RNA-seq experiment results.

With the development of microarray profiles and RNA-seq, a series of public gene expression analyses were published, including the Cancer Genome Atlas (TCGA; http://portal.gdc.cancer.gov/) and GEO (http://www.ncbi.nlm.nih.gov/gds) datasets. These databases provide powerful tools that can be used to identify cancer-associated mRNAs, miRNAs, and lncRNAs. For example, Poursheikhani, Abbaszadegan, Nokhandani, and Kerachian (2020) identified 2995 mRNA, 205 lncRNAs, and 345 miRNAs in CRC using the TCGA dataset.

In the case of RNA-seq, ncRNAs analysis workflow starts from raw NGS data and the first step is the *filtering* of low-quality reads from raw data. This process is generally conducted by using tools for preprocessing of files containing short reads encoded into FASTA-FASTQ files, which map the sequences to the reference databases. Examples of such programs are the FASTX-Toolkit, Blat, SHRiMP, and LastZ, MAQ (Liao, Li, Cui, & Zheng, 2018). Once filtered, the second step is to construct *transcript assembly* using, for instance, BowTie (Langmead & Salzberg, 2012) or TopHat (Brueffer & Saal, 2016). After the assembly, known genomic sequences or known coding genes are *filtered* by tools, like Bowtie. By this point, all the sequences may represent potential ncRNAs that have to be assessed and *mapped* in reference to existing ncRNA databases using assessment tools such as CPAT or Pfamscan. The NCBI (National Center for Biotechnology Information) nt (nucleotide) and nr (protein) databases are the preferred mapping databases in this step since they include sequences for all species. Finally, the function of ncRNAs has to be investigated by analyzing existing databases hosting a large number of ncRNA sequences and, when available, information about biological studies. In recent years, with advances in bioinformatics resources involving the development of a series of computational approaches and software tools for the analysis of extensive biological data (Table 3.1), many lncRNAs have been identified to be dysregulated in colon cancer. For instance, by a bioinformatic approach, Linc00659 was classified as a novel oncogenic lncRNA involved in the tumorigenesis of CRC by modulating the progression of the cell cycle. The downregulation of

TABLE 3.1 The long noncoding RNA (lncRNA) bioinformatics resources.

Bioinformatics resources	Web Link
Noncode	http://www.noncode.org
H-InvDB rel 8.3	http://www.h-invitational.jp
LncRNADisease	http://cmbi.bjmu.edu.cn/lncrnadisease
LincSNP	http://bioinfo.hrbmu.edu.cn/LincSNP
Rfam 11.0	http://rfam.janelia.org
Human Body Map lincRNAs	http://www.broadinstitute.org/genome_bio/human_lincrnas/
lncRNAdb	http://www.lncrnadb.org/
ncFANs	http://www.ebiomed.org/ncFANs/
Noncoder	http://noncoder.mpi-bn.mpg.de
NRED	http://nred.matticklab.com/cgi-bin/ncrnadb.pl/
ChIPBase	http://deepbase.sysu.edu.cn/chipbase/
LNCipedia 2.0	http://www.lncipedia.org/
DIANA-LncBase	http://www.microrna.gr/LncBase/
iSeeRNA	http://www.myogenesisdb.org/iSeeRNA/
GENCODE	https://www.gencodegenes.org/

Linc00659 expression resulted in severe cell-cycle arrest and enhancement of colon cancer cell apoptosis (Tsai, Lo, & Liu, 2018). Similarly, based on bioinformatics analysis of TCGA and/or the GEO datasets, as well as subsequent experimental validation, metastasis-associated lung adenocarcinoma transcript 1 (*MALAT1*) and small nuclear host gene 1 (*SNHG1*) have been identified to be oncogenic lncRNAs, serving as potential diagnostic and therapeutic targets in CRC (Wu, Meng, Jie, & Zhao, 2018; Xu, Zhang, & Hu, 2018; Xu, Chen, & Lin, 2018; Yang, Wang, & Kang, 2018).

3.8 Expression of noncoding RNAs in colorectal cancer patients

In the era of bioinformatic methods, interdisciplinary studies, and high technologies, such as microarray assays, several expressed miRNAs or lncRNAs were found to be associated with CRC (Wang, Song, & Ma, 2015). In the following sections, some of the lncRNAs and miRNAs most commonly expressed in CRC tissues are discussed to reveal their potential role in CRC tumorigenesis.

HOTAIR; During normal growth, HOX transcript antisense intergenic RNA (*HOTAIR*) is transcribed from the HOXC locus, which encodes nearly 2.2-kb lncRNA (Liu, Sun, & Nie, 2014). According to recent studies, *HOTAIR* acts in *trans* by adding polycomb repressive complex 2 (PRC2), LSD1, and CoREST/REST H3K4 demethylase complex to their target genes (Dou, Ni, & He, 2016; Rinn, Kertesz, & Wang, 2007), affecting cell epigenetics and gene expression (Xie et al., 2016). Overexpression of *HOTAIR* was discovered in cancer epithelial cells, resulting in histone methylation and cancer invasion (Svoboda, Slyskova, & Schneiderova, 2014). Furthermore, its expression is related to several proteins involved in CRC invasion, metastasis, tumor stage, and angiogenesis, including vimentin, matrix metalloproteinase, and E-cadherin (Luo et al., 2017). *HOTAIR* expression has been related to metastasis and poor clinical outcome, in a variety of cancers, including CRC, breast, pancreatic, and human epithelial ovarian cancers, as well as hepatocellular carcinoma (Deng et al., 2017). Svoboda et al. (2014) found that overexpression of *HOTAIR* is associated with a high mortality rate in CRC patients and can be used as a prognostic marker for sporadic CRC in blood and tissue samples. As a result, *HOTAIR* may be a promising cancer therapeutic target, but further research is still required.

MALAT1; *MALAT1* is an abundant ubiquitously expressed lncRNA that associates with pre-mRNA to localize transcriptionally active genes in chromatin and regulates alternative splicing by interacting with serine/arginine splicing factor (Schmitt & Chang, 2016). It is processed posttranscriptionally into a long nuclear-retained transcript of about 6.7 kb and a tRNA (Transfer Ribonucleic Acid)-like small RNA, which translocates into the cytoplasm (Wilusz, Freier, & Spector, 2008).

MALAT1 has several functions, including regulating pre-mRNA splicing and activating E2F target genes by relocating them from polycomb bodies to transcriptionally active nuclear sites in a serum-dependent manner in human cells (Tripathi, Ellis, & Shen, 2010; Tripathi, Shen, & Chakraborty, 2013) (Yang, Lin, & Liu, 2011). *MALAT1* is thought to play a critical role in controlling gene expression at both the transcriptional and posttranscriptional levels in different cancers by acting in a context-dependent manner (Arun, Diermeier, & Spector, 2018).

According to a study by Yang, Hu, and Xu (2015), *MALAT1* promotes cancer cell proliferation, invasion, migration, and lymph-node metastasis in CRC tissues and cells, by activating *AKAP-9*, a gene linked to the development and metastasis of several other cancers, including oral, melanoma, breast, thyroid, and lung cancers. *MALAT1* promotes the expression of *SRPK1* and phosphorylates *SRSF1* in CRC cells. This has been confirmed in a study by Hu et al. that found *MALAT1* improved *AKAP-9* expression by encouraging the expression of *SRPK1* and phosphorylating *SRSF1* in CRC cells (Wang, Li, Xie, Zhao, & Chen, 2008).

CCAT1; Colon cancer–associated transcript 1 (*CCAT1*), a lncRNA also known as *CARLo-5* (Kim, Cui, & Jeon, 2014) or onco-lncRNA-40 (Cabanski et al., 2015; Ye, Zhu, Qiu, Xu, & Wei, 2015), correlates with the cancer-associated variant rs6983267 that is found within the human 8q24 *MYC* (MYC Proto-Oncogene, BHLH Transcription Factor) superenhancer region, and related to increased cancer susceptibility (Kim et al., 2014). It was discovered to be upregulated in colorectal adenomas and adenocarcinomas in all stages and it is used as a target for real-time in vivo imaging (Yang et al., 2015). *CCAT1* in human CRC tissues specifically promotes cell proliferation and invasion by binding directly to the promoter region of *MYC*, which is the target of an oncogenic *miR-155* (Wang et al., 2015). Ma et al., on the other hand, discovered that overexpression of *CCAT1* in gallbladder cancer tissues and *CCAT1* knockdown is related to gallbladder cancer cell invasion and proliferation through miR218-5p (Smolle, Uranitsch, Gerger, Pichler, & Haybaeck, 2014). *CCAT1* can be used as a CRC biomarker since it can be identified at all stages of tumorigenesis, from premalignant lesions to metastasis, and it has been found in the peripheral blood of 40% of patients with CRC (Alaiyan, Ilyayev, & Stojadinovic, 2013; Nissan, Stojadinovic, & Mitrani-Rosenbaum, 2012). *CCAT1* has recently been identified as being highly sensitive to BET (bromodomain and extraterminal) protein inhibitors like JQ1 and was proposed to be a clinically relevant biomarker for patients who could benefit from BET inhibitors treatment in CRC (McCleland, Mesh, & Lorenzana, 2016).

CCAT2; *CCAT2* or Colon cancer–associated transcript 2 is located on chromosome 8q24.21 (Xu et al., 2014). *CCAT2* interacts directly with transcription factor 7–like 2, causing *MYC* overexpression that regulates *miR-17p* and *miR-20a*. This facilitates cancer metastasis by causing genomic instability (Le, Zhang, & Zhang, 2016). Furthermore, due to direct interactions between *CCAT2* and *TCF7L2*, this lncRNA has been shown to increase *MYC* expression by enhancing Wnt signaling via the transcription factor TCF7L2 (Ling, Spizzo, & Atlasi, 2013). Furthermore, *CCAT2* is a potential target for lncRNA-related therapy and can alter clinical outcomes (Cătană, Pichler, Giannelli, Mader, & Berindan-Neagoe, 2017). In addition, *CCAT2* acts as an activator of the *MYC* oncogene by forming a loop between the *MYC* promoter and the genomic locus rs6983267 (Xu et al., 2014).

UCA1; Human urothelial carcinoma-associated 1 (*UCA1*) is an oncofetal gene involved in embryonic development that was first identified in bladder cancer patients (Han, Yang, & Yuan, 2014). *UCA1* is also expressed in CRC tissues (Bian, Jin, & Zhang, 2016). It plays several roles in cancer cell biology, including promoting cell transformation, proliferation, invasion, and inducing drug resistance. *UCA1* is upregulated in CRC tissues, and thus silencing it could help slow the progression of the disease (Wang, Li, Xie, Zhao, & Chen, 2008). Furthermore, increased *UCA1* expression may result in larger tumor size and greater tumor depth (Han et al., 2014).

CASC11; Cancer susceptibility candidate 11 (*CASC11*) is located on chromosome 8q24. It is overexpressed in CRC cells and tissues, and its expression in CRC patients has been associated with tumor size, similar to *UCA1* (Zhang, Zhao, & Kim, 2015). Furthermore, inhibiting *CASC11* in CRC prevents tumor cell metastasis and proliferation by interacting with heterogeneous ribonucleoprotein, which protects catenin from degradation and transcription activation (Shen, Pichler, Chen, Calin, & Ling, 2017).

CRNDE; It has been shown that CRC tissues had 90 percent differentially overexpression of colorectal neoplasia differentially expressed (*CRNDE*). Insulin/IGFs (Insulin-Like Growth Factor) have been demonstrated to control *CRNDE* through metabolic changes in cancer cells, triggering the Warburg effect (Ye, Zhu, Qiu, Xu, & Wei, 2015). Its transcript variant 1 (Homo sapiens) (*CRNDE-h*) was found to be highly expressed in CRC tissues and its expression levels were linked to tumor size, distant metastasis, poor overall survival (OS) rate, and lymph-node metastasis (Liu, Zhang, Yang, Du, & Wang, 2016). *CRNDE* levels were also elevated in CRC tissues, and its upregulation was positively associated with tumor size and stage of cancer. Furthermore, *CRNDE* knockdown can increase CRC cell apoptosis in vitro and in vivo (Ding, Li, & Wang, 2017). *CRNDE* and miR181a-5p knockdown inhibited cell proliferation and decreased chemoresistance, according to Han, Li, and Zhang (2017) study, through inhibiting the Wnt/β-catenin signaling pathway.

PCAT-1; PCAT 1 or prostate cancer-associated ncRNA transcript 1 was first discovered in prostate cancer and later has been identified in a variety of other cancers (Zhao, Lu, Wang, Li, & He, 2016). For example, high levels of *PCAT-1* have been linked to distant metastasis and lower patient survival rates in CRC (Xie et al., 2016). By encouraging *PRC2* expression, upregulation of *PCAT-1* in cancer cells was related to an increase in vitro cell proliferation (Smolle, Uranitsch, Gerger, Pichler, & Haybaeck, 2014). *PCAT-1* has also been shown to function in the cytoplasm as an endogenous RNA (ceRNA), preventing miR-34-1 from binding to *MYC* transcripts, resulting in increased MYC protein levels in human prostate cancer cells (Prensner, Chen, & Han, 2014).

PCAT-1 was found to be overexpressed in non—small cell lung cancer cells and tissues promoting invasion, cell proliferation, and migration (Zhao, Hou, & Zhan, 2015). Furthermore, *PCAT-1* was found to be upregulated in CRC tissues, and its overexpression was associated with distant metastasis (Ge, Chen, & Liao, 2013). Surprisingly, the expression of *PCAT-1* in colon tissues was strongly related to CRC patients' OS since patients with high expression of *PCAT-1* had a lower survival rate than those with lower expression level (Ge et al., 2013). Moreover, downregulation of *PCAT-1* in CRC cells inhibited cell proliferation and blocked cell-cycle transition by suppressing the expression of cyclins and c-myc (Qiao et al., 2017).

CRCAL; LncRNA *RP11-138J23.1* (*CRCAL3*) could play a role in the development of CRC tumors (Yamada et al., 2018). In SW620 and HCT116 cell lines, transient *CRCAL3* knockdown suggested *CRCAL3* as a potential cell proliferation regulator (Yamada et al., 2018).

H19; A paternally imprinted gene, initially found to be expressed in embryonic tissues during mouse development and silenced in most tissues at birth, was one of the first lncRNAs described. It is overexpressed in a wide range of cancer types, including hepatocellular carcinoma, CRC, and breast cancer (Bartolomei, Zemel, & Tilghman, 1991; Davis, Weintraub, & Lassar, 1987). Imprinting loss triggers *H19* reexpression, which has been related to several stages of tumorigenesis in mouse models and human cell lines (Berteaux, Lottin, & Monté, 2005; Gabory, Ripoche, & Le Digarcher, 2009). The tumor suppressor and master cell-cycle regulator *p53*, as well as the ubiquitous oncogene *MYC*, play a significant role in *H19* transcription. On the other hand, increased *H19* expression is related to the loss of functional p53 or upregulation of *MYC* in various cancers (Barsyte-Lovejoy, Lau, & Boutros, 2006; Dugimont, Montpellier, & Adriaenssens, 1998). Hypoxia-induced factor 1 can also upregulate *H19* in the absence of wild-type p53 (Matouk, Mezan, & Mizrahi, 2010). Based on TCGA data, *H19* levels were found to be higher in colorectal and stomach cancers, but not in other types of cancer (Weidle, Birzele, Kollmorgen, & Rueger, 2017).

Overexpression of *H19* cDNA has been shown to minimize the tumorigenicity of human rhabdoid tumor cell lines in vivo (Hao, Crenshaw, Moulton, Newcomb, & Tycko, 1993). Furthermore, *H19* knockout mice developed more polyps and a faster onset of tumorigenesis than wild-type mice in an Apc mouse model of CRC, indicating a tumor-suppressive function (Gabory et al., 2009; Yoshimizu, Miroglio, & Ripoche, 2008). Such discrepancies in findings based on whether *H19* can function as an oncogene or a tumor suppressor may be explained by tumor genetic heterogeneity, the use of different model systems (i.e., transgenic mouse models vs human cancer cell lines), or alternatively, by a dual, context-dependent position of *H19* in tumorigenesis. However, this still remains to be confirmed.

Several hypotheses have been suggested to clarify how *H19* affects tumor progression. In CRC, *H19* is a precursor for miR-675, which has been shown to target the tumor suppressor protein retinoblastoma (RB) (Yoshimizu et al., 2008). Increased levels of *H19* in this model result in overexpression of miR-675 and low expression of RB, which promotes cell proliferation in CRC cell lines (Tsang, Ng, & Ng, 2010). *H19*, in addition to being a precursor for miR-675, appears to have the ability to function as a ceRNA for a variety of miRNAs (Poliseno et al., 2010). Several different miRNAs, including let-7 family members, *miR-200* family members (including *miR-141*), as well as *miR-138, miR-630, miR-138*, and *miR200a*, have been suggested to function as a "sponge" for *H19* (Kallen, Zhou, & Xu, 2013; Tsang et al., 2010). The primarily cytoplasmic localization of *H19* supports this model of action (Brannan, Dees, & Ingram, 1990).

It is assumed that enhanced H19/miR-675 levels in different cancers might lead to increased cancer cell proliferation and successful tumor metastasis. As a result, targeting *H19* could be a promising therapeutic strategy in a variety of cancers as it acts as an oncogene (Arun et al., 2018).

GAS5; Growth arrest-specific transcript 5 or *GAS5* was first discovered to be one of six genes that are expressed preferentially in growth-arrested mammalian cells (Schneider, King, & Philipson, 1988). It is one of the most widely expressed lncRNAs, found in all human tissues and active in embryogenesis similar to *H19* (Hudson, Pickard, & De Vera, 2014). Breast, prostate, bladder, gastric, colorectal, pancreatic, and cervical cancers all show a substantial reduction in its expression (Pickard & Williams, 2015). Its expression is also inversely linked to clinicopathological characteristics, including tumor size, staging, and metastasis (Pickard & Williams, 2015). Furthermore, overexpression of *GAS5* in xenografted breast cancer cell lines in nude mice has been reported to inhibit breast tumor growth in vivo by inducing cell-cycle arrest and apoptosis, confirming its function as a tumor suppressor (Pickard & Williams, 2015). *GAS5* serves as a decoy for the glucocorticoid receptor (GR), through suppressing GR-dependent gene regulation in HeLa and HepG2 cell lines at a molecular level (Kino, Hurt, Ichijo, Nader, & Chrousos, 2010). Other steroid hormone receptors, such as androgen and progesterone receptors, can bind to *GAS5* and play a crucial role in hormone-dependent cancers (Hudson et al., 2014). In CRC, *GAS5* is specifically downregulated and its lower expression is linked to an advanced stage and lymph-node metastasis. *GAS5* expression at high levels inhibits CRC cell proliferation and promotes apoptosis. Furthermore, *GAS5* suppresses the expression of *miR182-5p*, while the upregulation of *miR182-5p* reverses the effects of *GAS5* overexpression in CRC cells. In addition, *FOXO3a* is upregulated by *GAS5* in CRC cells (Cheng, Zhao, Wang, Wang, & Zhu, 2018). This lncRNA was found to inhibit cell division, triggering G0/G1 block and apoptosis, suggesting it as a possible therapeutic target in CRC (Yang, Shen, & Yan, 2017).

In conclusion, all these results indicate that *GAS5* plays a significant role in the development of CRC metastasis and could be used as a CRC prognostic marker.

3.9 Sample types used for analyzing noncoding RNAs

In recent years, ncRNAs have attracted increasing attention from researchers, as they could play crucial roles in regulating gene transcription. For example, miRNAs participate in the cellular process by targeting plenty of mRNAs and serve as oncogenes, tumor suppressor genes, or both to promote or inhibit the initiation and development of cancer (Iorio & Croce, 2012). Over the last few years, many types of research have been shifted from tissue biopsy to developing minimally invasive biomarkers in the form of liquid biopsy, which is the sampling and analysis of various types of cells or molecules collected from biological fluids. In this field, multiple ncRNAs have been detected in body fluids such as blood, urine, saliva, breast milk, pleural effusions, cerebrospinal, and bronchoalveolar lavage fluids (Shi, Gao, & Cao, 2016; Weber, Baxter, & Zhang, 2010). To date, the diverse cells and molecules analyzed in the liquid biopsy are circulating tumor cells (CTCs), circulating free DNA (cfDNA), circulating tumor DNA (ctDNA), circulating mRNAs, miRNAs, lncRNAs, circular RNAs (circRNAs), tumor-educated platelets, proteins, peptides, metabolites, microvesicles, and exosomes. The eclectic information that can be provided via liquid biopsy is the clinical potential of diagnostic, prognostic, and predictive biomarkers of different diseases, including cancer (Kerachian, Poudineh, & Thiery, 2019). Some evidence for the crucial role of liquid biopsies in patients' management has been growing across many cancers, especially in CRC. Several types of liquid biopsies have already demonstrated their potential role in CRC diagnosis, prognosis, and therapy prediction. For CRC diagnosis, a study has demonstrated that tumor heterogeneity could be detected in blood via ctDNA with 97% accuracy. However, due to tumor heterogeneity in some cases, it might be better than tissue biopsy (Palazzo & Lee, 2015). Furthermore, we can potentially diagnose patients with *KRAS* (Kirsten Rat Sarcoma Virus)-mutant CRC through a liquid biopsy, since *KRAS*-mutant fragments are detected in patients' ctDNA (Kerachian, Azghandi, Javadmanesh, Ghaffarzadegan, & Mozaffari-Jovin, 2020). There is an FDA-approved screening tool that detects the *SEPT9* promoter region methylation in plasma, which is offered as a specific biomarker for the early stages of CRC. It has been shown with high sensitivity and specificity in many studies as a CRC diagnostic tool (Jamialahmadi et al., 2021). Some prognostic biomarkers include total cfDNA levels that correlate with disease-free survival (DFS) and OS, irrespective of tumor stage, use of adjuvant chemotherapy, tumor marker, and sample type. Additionally, it has been demonstrated that ctDNA is a more accurate predictor of relapse than carcinoembryonic antigen (CEA). Another prognostic biomarker is cell-free miRNA (cf-miR), such as *cf-miR-21, 203, or 1290*, the high expression of which is associated with poor prognosis. In addition, the high expression of *cf-miR-200c* in serum is associated with lymph-node involvement and distant metastasis. Furthermore, CTCs positivity is associated with decreased DFS and poor OS, irrespective of staging. Liquid biopsies have also been able to identify mutations that cause resistance to EGFR (Epidermal Growth Factor Receptor) inhibitors (Hamzehzadeh, Khadangi, Ghayoor Karimiani, Pasdar, & Kerachian, 2018; Hamzehzadeh, Khadangi, Karimiani, Pasdar, & Kerachian, 2018). Furthermore, the ncRNAs could be detected in the form of exosomes introduced as different types of biomarkers. The abundance and stability of exosomes in circulation have made them a potentially suitable entity to pursue. Furthermore, several studies have demonstrated that aberrant expressions of exosomal ncRNAs, such as miRNAs, lncRNAs, and circRNAs, are involved in the regulation of various cellular mechanisms in tumorigenesis. They carry genetic information and other biomolecules, which could potentially serve as potential diagnostic, prognostic, and predictive biomarkers or even therapeutic targets for cancer (Team RC, 2013).

3.10 Cell signaling pathways modulated by noncoding RNAs in colorectal cancer patients

Metastasis and relapse are the leading causes of failure in the treatment of CRC. Therefore discovering the molecular mechanisms underlying CRC progression and metastasis is critical to better understand the failure of cancer therapy and relapse. Recent evidence supports the involvement of various signaling pathways in the development and progression of CRC as listed here.

3.10.1 Wnt/β-catenin signaling pathway

Activation of the Wnt/β-catenin cascade is a general phenomenon in CRC, and its downstream target genes can induce tumorigenesis by inducing cell-cycle progression and abnormal proliferation. In CRC, the Wnt/β-catenin signaling pathway regulates the epithelial—mesenchymal transition (EMT) as well as the cell proliferation, invasion, and migration. This pathway is also essential for the initiation of CRC because it controls epithelial stem cell self-renewal. Alterations in the Wnt/β-catenin signaling pathway have an impact on CRC, making it a possible therapeutic option. It is known that β-catenin acts as a transcription factor that works in tandem with *TCF1* and *LEF1* to activate downstream target genes. When Wnt ligands bind to the FZD (Frizzled receptor) or LRP (Receptor-Related Protein) receptors, β-catenin is activated from the β-catenin destructive complex, which consists of Axin, APC (Adenomatous Polyposis Coli), and GSK3β. Mutations in *APC* can cause the Wnt cascade to become activated. Then, through hcCF/LEF (Lymphoid Enhancer Binding Factor), β-catenin accumulates and translocates into the nucleus, activates downstream genes of the Wnt/β-catenin pathway, and likely helps to cause EMT by increasing the expression of main proteins. EMT has been implicated in cancer cell metastasis and invasion in a growing number of studies over the last few years. Some lncRNAs have been shown to influence EMT by or not by the Wnt/β-catenin pathway. Multiple ncRNAs affect the Wnt/β-catenin cascade to regulate the CRC. *CCAL* (Colorectal Cancer-Associated LncRNA), *CASC11*, and *CCAT2* are examples of lncRNAs that can bind to proteins or miRNAs and thus indirectly influence the Wnt/β-catenin pathway. *CCAL* is a lncRNA that can bind to activator protein 2 and form a complex with β-catenin and APC, compromising β-catenin/TCF4 interactions in CRC. The *lncRNA CASC11* may activate the Wnt/β-catenin cascade by binding to hnRNP-K, which can form a complex with GSK3β, TCF4, β-catenin, and Axin to inhibit the degradation of β-catenin. TCF7L2's physical interaction with the *lncRNA CCAT2* can enhance the function of the Wnt/β-catenin signaling pathway. *MALAT1* and *SNHG1* are two lncRNAs that facilitate the nuclear localization of β-catenin to upregulate the Wnt/β-catenin cascade. MiRNAs are also effective targets for lncRNAs because they help them exert their effects on signaling pathways by binding to a complex. The *lncRNAs CRNDE, ZEB1-AS1, H19, lincRNA-p21*, and XIST, for example, by sponging or binding to miRNAs, influence the Wnt/β-catenin signaling pathway. In addition, lncRNAs can influence the activation of the Wnt/β-catenin cascade by changing the expression of certain key proteins. The *lncRNA CTD903* may suppress the EMT mechanism, which is linked to the production of CRC, by downregulating Wnt/β-catenin signaling and decreasing the expression of *TCF, Snail, and Twist*. By upregulating the expression of *β-catenin, cyclin D1*, and *c-Myc*, the *lncRNA HNF1A-AS1* can induce CRC cell formation, invasion, and migration mediated by the Wnt/β-catenin cascade. Overexpression of *LncRNA-BCAT1* in CRC cells can also inhibit *β-catenin* expression (Baassiri et al., 2020). Dysregulation of *miR-552* has also been attributed to an increase in Wnt signaling in CRC cells (Cao, Yan, & Liu, 2017; Wang, Li, & Wang, 2016). Furthermore, by regulation of *GSK3* and *SFRP2*, Li, Lai, and Wang (2016) demonstrated mechanistic similarities between *miR-224* and Wnt/β-catenin in the pathogenesis of CRC. Alternatively, miRNAs can suppress pathway components and thereby modulate Wnt signaling. MiR-135a/b, for example, which is overexpressed in CRC, can specifically target *APC*, resulting in Wnt signaling upregulation (Nagel, le Sage, & Diosdado, 2008). *MiR-135a/b* is also thought to attack and inhibit secreted frizzled-related protein 4, which binds and represses Wnt proteins outside the cell (Kawano & Kypta, 2003). *TCF4/β-catenin* transcriptionally activates *miR-135b*, which is significantly enhanced in colonic tumors in mice with inactivated Apc and in sporadic human CRC in a potentially positive feedback loop (Valeri, Braconi, & Gasparini, 2014). The miR-34 family (*miR-34a/b/c*) operates upstream of Wnt and specifically targets and represses several Wnt signaling effectors, like *WNT1, WNT3, LRP6* (a Wnt ligand coreceptor), *β-catenin*, and *LEF1* (an HMG-box transcription factor that, which as *TCF4*, interacts with *β-catenin*). TP53 stimulates *miR-34* transcription directly, giving insight into how *TP53* suppresses Wnt signaling (Kim, Kim, & Kim, 2011). *MiR-29b, miR-29c*, and *miR-93*, which target *BCL9L* (a β-catenin coactivator and miR-29b target), are other inhibitors of Wnt ligands or β-catenin-mediated activity. In conclusion, certain miRNAs serve as Wnt pathway regulators at multiple levels of the signaling cascade, and some miRNAs, such as miR-34, can inhibit both Wnt and Notch signaling pathways. Given that the Wnt pathway is a key oncogenic regulator of CRC, Wnt-modulating miRNAs merit special attention as potential therapeutic targets. In CRC, however, additional mutations in genes like *KRAS* and *BRAF* (B-Raf and V-Raf murine sarcoma viral oncogene homolog) are inevitable, meaning that multiple pathways are therapeutically involved.

3.10.2 JAK (Janus Kinase)/STAT (Signal Transducer and Activator of Transcription) signaling pathway

JAKs and STATs are critical components of cytokine signaling pathways. Despite their long-recognized pivotal roles in immunological responses, JAKs and STATs have been related to cancer initiation and progression, reported in several studies. The JAK/STAT pathway, when activated, can prevent cells from dying and facilitate tumor cell proliferation and invasion. Some lncRNAs play a role in CRC by modulating the JAK/STAT pathway, functioning as ceRNAs, or modifying the expression of phosphorylated STAT3. STAT3 phosphorylation is suppressed by the *lncRNAs AB073614* and *SBDSP1*, while the *lncRNAs GACAT3* and *CASC2* serve as ceRNAs for particular miRNAs to indirectly upregulate *SP1* and *STAT3* expression (Baassiri et al., 2020).

MiRNAs have been related to the stage of CRC at diagnosis and survival in many studies (Slattery, Herrick, & Mullany, 2015; Slattery, Herrick, & Pellatt, 2016), since *miR-21*, *miR-29a, miR-29b-1,* and *miR-155* have been demonstrated to be induced by the JAK-STAT pathway (Witte & Muljo, 2014). MiRNAs, such as *miR-19a* (Collins, McCoy, Lloyd, O'Farrelly, & Stevenson, 2013) and *miR-9* (Zhuang, Wu, & Jiang, 2012), have been activated in the JAK-STAT pathway in different cell lines and model organisms by repressing mRNAs, including SOCS (Suppressor of Cytokine Signaling). Furthermore, *miR-199a5p, -205, -373,* and *-139-5p* have been found to function as oncogenes that upregulate STAT3 in osteosarcoma and CRC (Eyking, Reis, Frank, Gerken, & Schmid, 2016; Wang, Ba, & Guo, 2017; Zou, Mao, & Yang, 2016). STAT3 stimulates cell proliferation and migration while inhibiting cell apoptosis in CRC via upregulating *miR-181b* (Lei, Du, & Lin, 2016). Conversely, STAT3 downregulates *miR-34a* in colorectal (Li, Rokavec, & Hermeking, 2015; Rokavec, Oner, & Li, 2014) and breast cancer (Avtanski, Nagalingam, & Kuppusamy, 2015).

3.10.3 PI3K (Phosphoinositide 3-Kinase)/PTEN (Phosphatase and Tensin Homolog)/AKT (AK Mouse Plus Transforming or Thymoma)/mTOR (Mechanistic Target of Rapamycin) signaling pathway

The PI3K/PTEN/AKT/mTOR signaling pathway governs gene expression, metabolism, cell development, and metastasis among other biological processes in cells. PTEN's effects are often regulated by the PI3K/PTEN/AKT/mTOR signaling pathway's negative regulation, but it can also act as a suppressor of its own. The PI3K/PTEN/AKT/mTOR cascade, which plays a key role in CRC cell proliferation and EMT, has also been shown to control cancer progression through different lncRNAs. Approximately 40% of malignant tumors are linked to abnormalities in the PI3K/PTEN/AKT/mTOR cascade, suggesting that the PI3K/PTEN/AKT/mTOR pathway may be used as a therapeutic goal. LncRNAs, such as

AB073614, *DUXAP10*, *RP11−708H21.4*, and *lncRNA-422*, influence the expression of certain associated proteins to modulate signaling pathways. In this regard, *PlncRNA-1* regulates the phosphorylation of *PI3K* and *AKT* (Baassiri et al., 2020).

MiR-135b is supplemented by PI3K inhibition of FoxO transcription factors (FOXO1 and FOXO3A), which represses cell-cycle initiation in the PI3K pathway and is negatively regulated by PTEN. By suppressing negative regulators of the PI3K signaling pathway, *miR-221*, *miR-21*, and *miR-17/106* improve PI3K signaling activation (Strubberg & Madison, 2017). Many of the miRNAs have previously been reported to be associated with genes or biological responses that could influence CRC. In the PI3K/AKT pathway, *miR-590-5p*, *miR-106b*, and *miR-93* have all been related to PTEN (Li, Miao, & Shan, 2017; Soleimani et al., 2018).

3.10.4 Ras/MAPK (Mitogen-activated protein kinase)-signaling pathway

Several studies have elucidated that most human cancers are related to a biomolecular event, known as the activation of Ras/MAPK signaling. Activation of the Ras/MAPK pathway facilitates cell proliferation, migration, and differentiation. In contrast, inhibition of this pathway in CRC cells suppresses tumor development. *CRNDE* is an important lncRNA in the regulation of Ras/MAPK pathways. HnRNPUL2 can improve the stability of *CRNDE*, allowing it to serve as a key mediator and trigger of the Ras/MAPK cascade. *lncRNA NNT-AS1* regulates the function of the MAPK/Erk signaling pathway, and the activation of the MAPK/Erk signaling pathway could induce EMT in CRC cells, increasing CRC cell proliferation, invasion, and migration (Baassiri et al., 2020; Wang, Yang, & Hu, 2017).

MiR-101 has been revealed to target MAPK phosphatase-1 (MKP-1), which is involved in the inflammatory response. While *miR-101* has been linked to MKP-1, several other miRNAs have also been linked to immune response components, meaning that other miRNAs could be also associated with MKP-1 (Zhu, Liu, Chen, Lan, & Ge, 2010). Although studies have shown the role of MAPK signaling in the cause and progression of CRC, the interactions between miRNAs and genes in the MAPK-signaling pathway in CRC have not been fully investigated (Slattery et al., 2018).

In CRC cell lines, HCT116, the oncoprotein *KRAS* plays an important role in tumor growth by modulating various miRNAs such as *miR200c*, *miR-221*, and *miR222* (Tsunoda, Takashima, & Yoshida, 2011). Only in three-dimensional (3D) cultures in a colonic crypt model, DLD1 cells expressed *miR181a*, *miR200c*, and *miR210*. In CRC clinical samples, these miRNAs are also overexpressed (Ota, Doi, & Fujimoto, 2012).

3.10.5 p53 signaling pathway

P53 is a well-known tumor suppressor linked to CRC. Cell-cycle arrest, apoptosis, and tumor suppression are all induced by p53 overexpression. LncRNAs are regulators and mediators of the p53 cascade, and the effect of many lncRNAs exerts their impacts by modulating p53. For example, the *lncRNAs PURPL, ROR, SNHG1*, and *ZFAS1* promote tumorigenicity in CRC by suppressing basal p53 levels. *HNF1A-AS1* can also repress the miR-34a/SIRT1/p53 feedback loop, allowing p53 to be downregulated and promoting CRC metastasis (Baassiri et al., 2020).

MiRNAs have been identified as essential components of the p53 transcriptional network into the mechanisms by which p53 is triggered in tumors. *MiR-25* and *miR-30d* have been found to inhibit the *TP53* gene (Kumar, Lu, & Takwi, 2011). Several other miRNAs have been identified activated in the TP53 network, including *miR-161*, *miR-143*, *miR-145*, *miR-34*, *miR-194*, *miR-192*, *miR-215*, and *miR-29*, either through being specifically altered by *p53* or by their interactions with downstream genes targeted by *p53* (Slattery et al., 2019).

3.10.6 Notch signaling pathway

The Notch signaling pathway plays a role in the tumorigenesis of different cancers, including CRC. The cascade is used in the proliferation, invasion, and metastasis of CRC cells. In addition, Notch signaling regulates EMT. The *lncRNA FAM83HAS1* is overexpressed in CRC cell lines and tissues, which coincides with the upregulation of two Notch signaling molecules; *Notch1* and *Hes1*, and hence, controls CRC progression. In CRC, FOXD2-AS1 controls Notch signaling and acts as a tumor promoter (Baassiri et al., 2020).

MiR-34a, *miR-34b*, and *miR-34c* are members of the miR-34 family, which has been related to the modulation of the p53 and Notch signaling pathway (Prokopi, Kousparou, & Epenetos, 2014). The *miR-200* family has been involved in the control of cancer stem cells (CSCs) in recent research, with examples, including breast, colorectal, kidney, and brain CSCs. Low levels of *miR-200c* and *miR-141* are associated with high expression of *ZEB1*, an EMT activator, which in turn activates the Notch signaling pathway by targeting the Notch ligand Jagged-1 and Notch coactivators *Maml 2* and *3* (Brabletz, Bajdak, & Meidhof, 2011).

3.10.7 NF-κB (Nuclear Factor Kappa B) signaling pathway

In 1986 the transcription factor NF-κB was identified as a nuclear factor, and it was discovered to bind to enhancer components, causing tumorigenesis. Many proteins suppress NF-κB under physiological conditions. NF-κB is activated and translocated to the nucleus as these proteins are phosphorylated, promoting the initiation and development of CRC. The majority of lncRNAs control this cascade by directly or indirectly modulating NF-κB. For example, *lnc-GNAT1-1* exerts suppressive functions in CRC through the RKIP−NF-κB−Snail circuit, and *lnc-GNAT1-1* knockdown downregulates PKIP and increases NF-κB expression. The *lncRNA GAS5* can cause NF-κB signaling by decreasing NF-κB phosphorylation. Furthermore, the NF-κB cascade is associated with 5-fluoropyrimidine resistance caused by *lncRNA HOTAIR* in CRC (Baassiri et al., 2020).

NF-κB is a transcriptional target of many miRNAs, including *let-7*, *miR-9*, *miR-21*, *miR-143*, *miR-146*, and *miR-224* (Hoesel & Schmid, 2013). These miRNAs are involved in feedback mechanisms that influence the NF-κB signaling pathway by either targeting upstream signaling molecules or members of the NF-κB family (Iliopoulos, Hirsch, & Struhl, 2009). It has been suggested that *miR-520e* can affect cell proliferation by attacking NIK (NF-κB Inducing Kinase) (Zhang et al., 2012). *MiR-15a*, *miR-16*, and *miR-223* have been shown in other experiments to affect IKKα protein expression (Li et al., 2010).

3.11 Several other mechanisms

3.11.1 Caspase cleavage cascade

Caspase expression can be caused by apoptotic signaling, which can influence the development of CRC. The overexpression of *lncRNA loc554202* can promote G1 arrest and cause apoptosis to inhibit CRC cell proliferation, mainly via caspase cleavage cascade.

3.11.2 Chemokine signaling

Chemokines play an important role in promoting CRC invasion and metastasis. The *lncRNA MALAT1* could promote the progression of CRC through tumor-associated dendritic cells (TADCs). Furthermore, blocking chemokine ligand 5 with neutralizing antibodies or siRNA transfection could reduce the promotion of colon cancer caused by TADCs.

3.11.3 Interleukin pathway

Interleukins (ILs) are related to the development of cancer. The *lncRNA DILC* is negatively correlated with CRC proliferation and metastasis through the IL-6/STAT3 cascade. IL-10 is associated with CRC cell proliferation and colony formation in relation to *lncRNA GAS5* (Baassiri et al., 2020).

3.12 Clinical applications of noncoding RNAs as biomarkers in patients with colorectal cancer

3.12.1 Noncoding RNAs as predictive markers for colorectal cancer patients

The predictive biomarkers can be utilized to identify therapeutic targets, detect drug resistance, and monitor response to treatment (Kerachian et al., 2019). Several ncRNAs have been reported as predictive biomarkers in CRC patients. For example, Salendo et al. conducted a genome-wide miRNA profiling in multiple CRC lines and reported in vitro signatures of chemosensitivity. They analyzed the selected miRNAs in pretherapeutic biopsies from rectal cancer patients undergoing neoadjuvant (preoperative) chemoradiotherapy (nCRT) and reported a good prognosis with the overexpression of let-7 (Salendo, Spitzner, & Kramer, 2013). In another study, silencing *miR-21* restored the sensibility of CRC cells to 5-FU since the overexpression of *miR-21* significantly inhibited apoptosis and increased cell growth, invasion, and resistance of tumor cells to chemotherapeutic agent 5-FU and radiation in vivo. In contrast, overexpression of *miR-129* increased the cytotoxic effect of 5-FU both in vitro and in vivo, suggesting a potential of developing *miR-129*-based therapeutic strategies to enhance or overcome the resistance to 5-FU in CRC. *MiR-429* expression is also related to the enhancement of malignant potential, poor prognosis, and chemosensitivity in CRC. Another study observed that 5-FU-resistant CRC cells show elevated levels of *miR-577*, targeting heat shock protein 27 (HSP27). *MiR-1915* may play a key role in the production of multidrug resistance in CRC cells by modulating apoptosis via Bcl-2. In another study, it was found that miR-122 was differentially expressed between 5-FU-resistant and -sensitive CRC cells. Overexpression of miR-122 in 5-FU-resistant cells resensitizes 5-FU resistance through the inhibition of glycolytic enzyme pyruvate kinase type M2 both in vitro and in vivo. To et al. investigated the expression of an efflux transporter, ABCG2, in CRC and normal colonic mucosa. *MiR-203* was found to be downregulated in CRC specimens in comparison to normal colon mucosa. Downregulation of *miR-203* induced ABCG2 promoter methylation, through its target DNA methyltransferase *DNMT3b* activation, and a significant reduction in *ABCG2* expression. Lopes-Ramos et al. analyzed miRNA expression profile in rectal tumor biopsies prior to nCRT and found three miRNAs were overexpressed in complete responders, including *miR-21-5p, miR-1246*, and *miR-1290-3p*, and one, *miR205-5p*, was overexpressed in incomplete responders. Moreover, in rectal cancer patients who obtained the complete response, overexpression of *mi-R21-5p* was observed. Another study showed the possible correlation between miRNAs and oxaliplatin resistance in CRC patients. Overexpression of *miR-153* was correlated with the advanced CRC stage. Furthermore, upregulation of *miR-153* was observed to promote CRC invasiveness indirectly by inducing matrix metalloprotease enzyme 9 production and resistance to oxaliplatin and cisplatin directly by inhibiting the Forkhead transcription factor Forkhead box O3a. Kjersem et al. identified three circulating miRNAs in plasma of CRC patients (*miR-106a, miR-130b*, and *miR-484*), the upregulation of which was significantly correlated with resistance to the FOLFOX regimen. In another study, Zhang et al. performed a global analysis of miRNA expression in the serum of 250 CRC patients treated with chemotherapy and were able to identify and validate a 5-miRNA signature predictive of chemotherapy sensitivity. Among the lncRNAs involved in chemotherapy resistance, *TUG1* was proposed as a mediator of methotrexate resistance through its upregulation in methotrexate-resistant CRC and sponge activity on *miR-186* (Li, Gao, Li, & Ding, 2017). Li et al. investigated serum samples and tissue from patients with metastatic CRC, the role of the *lncRNA MALAT1* in oxaliplatin resistance. High expression of *MALAT1* in CRC patients was associated with poor prognosis and chemoresistance to FOLFOX treatment, through a mechanism related to E-cadherin downregulation. In addition, a study by Bian et al. described a regulatory network involving the *lncRNA UCA1, miR-204-5p*, and the transcription factor CREB1 in the 5-FU sensitivity of CRC cells. Moreover, LINE-1 (long interspersed nuclear element-1) has been proposed as a biomarker for colon cancer being its hypomethylation associated with a worse prognosis (Kerachian & Kerachian, 2019). A few studies had shown that patients, with a hypomethylated LINE-1 and treated with FOLFOX, had a greater risk of early postoperative recurrence, poor diagnosis, and a shorter period of DFS in stage III (Garajová, Ferracin, & Porcellini, 2017).

In conclusion, miRNAs and lncRNAs possess high potential as predictive markers for therapeutic response to chemotherapeutic drugs.

3.13 Diagnostic potential of noncoding RNAs in colorectal cancer patients

Diagnostic biomarkers could facilitate cancer screening and tumor heterogeneity detection. Several investigations have indicated the diagnostic biomarker values of ncRNAs especially lncRNAs and miRNAs in CRC detection. For example, it has been shown that *lncRNA CCAT1* and *HOTAIR* were significantly upregulated in serum and plasma of CRC patients in comparison with normal individuals. Overexpression of *CCAT1* accompanied by upregulated *HOTAIR* provided a more effective diagnostic value for *HOTAIR* or *CCAT1* alone. Serum exosomal *lncRNA CRNDE-h* has been upregulated in CRC patients, indicating a significant diagnostic performance. *CRNDE-h* overexpression was correlated to lower OS and lymph-node invasion and distant metastasis in CRC patients. The poor OS was also reported for overexpression of *lncRNA HOTTIP, PVT1,* and *UCA1* in advanced CRC patients. The diagnostic performance of lncRNAs has been compared with different types of cancer markers. For example, the diagnostic value of *lncRNA HOTAIRM1* was compared to CEA, CA19-9, and CA125. The results concluded that *HOTAIRM1* diagnostic performance was comparable to CEA, presenting better results in comparison to CA19-9, and CA125. Combining *HOTAIRM1* with CEA increased the sensitivity and specificity.

Not only lncRNA expression level but also methylation of lncRNA has been suggested as a diagnostic biomarker. For instance, hypermethylation of *lncRNA Colorectal Adenocarcinoma Hypermethylated* has been demonstrated to significantly increase plasma cfDNA in CRC patients, indicating a promising plasma diagnostic biomarker (Poursheikhani, Abbaszadegan, & Kerachian, 2021).

Furthermore, multiple studies have shown the utility of miRNAs as CRC diagnostic biomarkers (Ju, 2010). For example, Ng, Chong, and Jin (2009) revealed that the expression of *miR-92* raised significantly in the plasma of CRC patients and can be considered a potential noninvasive molecular marker for colon cancer. Huang et al. (2010) also reported that *miR-29a* and *miR-92a* have significant diagnostic values in advanced CRC and *miR-92a* expression levels discriminated CRC patients from normal individuals with a quite promising sensitivity and specificity of 89% and 70%, respectively.

3.14 Prognostic potential of noncoding RNAs in colorectal cancer patients

Prognostic biomarkers could be used to estimate the risk for progression versus relapse. Interestingly, several investigations have indicated that aberrant ncRNAs expression may be associated with poor prognosis and worsen clinicopathological characteristics in CRC patients. For example, serum exosomal *lncRNA H19, HULC,* and *HOTTIP* were downregulated in nonmetastatic CRC. *HOTTIP* expression level was significantly correlated to poor OS. Similarly, overexpression of *lncRNA MIR4435-2HG* and *MALAT1* was related to poor DFS and OS. Furthermore, more examples are upregulation of *lncRNA PVT1, PCAT1, CCAT2,* and *AK098783* and downregulation of *lncRNA RP11-462C24.1* and *GHRLOS* that were remarkably correlated to poor prognosis, potentially suggesting them as prognostic biomarkers for CRC (Poursheikhani et al., 2021).

From 37 identified miRNAs, differentially expressed in a microarray profiling study, overexpression of *miR-21* was reported as an independent predictor of OS (Schetter, Leung, & Sohn, 2008). Similar conclusions were drawn from a study in which *miR-21* expression was associated with poor survival and therapeutic outcome in stages II and III of CRC (Guo et al., 2008). Pdcd4, a 64-kDa protein that inhibits tumor progression by interacting with translation initiation factors, eIF4A and eIF4G, has shown downregulation by *miR-21*. This suggests a possible mechanism by which *miR-21* might promote tumor invasion and metastasis and potentially influence the clinical outcome (Asangani, Rasheed, & Nikolova, 2008).

3.15 Therapeutic potential of noncoding RNAs in colorectal cancer patients

Several ncRNAs have been considered novel cancer-targeted therapies since they contribute to a variety of prime mechanisms in cancer. lncRNA-based therapies have been demonstrated to have a high anticancer chemotherapeutic benefit that effectively could inhibit tumorigenesis of CRC. LncRNAs expression can provide clues for developing novel approaches in cancer therapy since some are up- or downregulated as oncogenes or tumor suppressor genes, respectively (Zhou et al., 2018). Ectopic expression of tumor-suppressive lncRNAs or knocking down the oncogenic lncRNAs with siRNA or other mechanisms could modulate lncRNA-based cancer therapy in CRC. Several investigations concluded that hyper- or hypomethylation of the genome can induce or repress lncRNAs expression, which might have clinical applications in cancer patients. It has been reported that overexpression of *LINC00152* by Yes-associated protein 1 (*YAP1*) (transcription coactivator) can promote Fascin actin–bundling protein 1 (*FSCN1*) expression via sponging *miR-632* and *miR-185-3p*, which consequently results in proliferation and metastasis in CRC. Targeting the YAP1/LINC00152/FSCN1 axis has been proposed as a new therapeutic approach in CRC. Overexpression of lncRNA *FLANC* has been related to poor prognosis in two independent CRC cohort studies. It has been shown that *FLANC* may enhance cellular growth, apoptosis, migration, and invasion. Targeting *FLANC* by 1,2-dioleoyl-sn-glycero-3-phosphatidylcholine nanoparticles loaded with a specific siRNA reduced metastasis without any significant toxicity, proposing a therapeutic new approach in CRC therapy (Poursheikhani et al., 2021).

Altogether, available molecular tools targeting ncRNAs are improving gradually in CRC therapy, but much more research is needed before they can be used in clinics.

3.16 Potential of noncoding RNAs in predicting chemoresistance and radioresistance in colorectal cancer patients

CRT is a combination of chemotherapy with radiotherapy, preferably used in rectal cancer. CRT can be used either prior to or after surgery. It exerts its action inducing DNA damage mostly through irradiation or the production of chemical radicals. The concomitant administration of chemotherapeutic agents may serve as a radiosensitizer. Preoperative radiotherapy (neoadjuvant) is used in rectal cancer patients to reduce the risk of cancer recurrence after surgery, as well as to shrink the tumor for favoring the complete removal of the mass. Since about 15% of all rectal cancer patients are diagnosed with an unrespectable disease and thus primary surgical resection is not fully practical, preoperative radiotherapy or neoadjuvant nCRT is recommended. Although CRT represents an effective treatment against CRC, not all patients experience the same response rate. In the era of personalized medicine, it is essential to choose the right patient for the appropriate treatment, thus the identification of a predictive biomarker for CRT could further improve survival for rectal cancer patients. Among several biomarker entities, ncRNAs, in particular, miRNAs are important determinants that may affect the response and/or resistance to radiotherapy. High-throughput gene expression profiling analysis has revealed that miRNA deregulation in CRC tissues influences the activity of signaling pathways that may be correlated to prognosis and response to CRT. For instance, *miR-622* and *miR-630* displayed a high efficacy in predicting pathological complete response to CRT. They probably act through regulating genes and signaling pathways important in cell repair. It has been demonstrated that *miR-630* reduces the ability of cells to repair DNA damage after cisplatin-based chemotherapy in non–small cell lung cancer thus providing a possible explanation for the benefit seen in a CRC cohort receiving oxaliplatin-based CRT, which may not be transferable to a more standard 5-FU-based neoadjuvant treatment. However, there is some conflicting data reported by Ma et al. in rectal cancer cell lines, reporting that *miR-622* is significantly upregulated in CRC cell lines exposed to ionizing radiations. Interestingly, this upregulation is maintained and persisted in surviving cells treated with continuous low-dose radiation, supporting the fact that *miR-622* induces radioresistance in vitro. They also reported that *miR-622* inhibits RB protein (Rb) by directly targeting the 3′ untranslated region of RB1, and *miR-622*-induced radioresistance may be reversed by overexpressing Rb. Thus *miR-622* may be a radioresistance biomarker. In contrast, *miR-630* is positively associated with radiosensitivity in CRC cell lines. The levels of *miR-630* are also significantly reduced after repeated ionic radiation confirming the possible role of *miR-630* in regulating pathways crucial for radiosensitivity. The main targets of *miR-630* are *BCL2L2* and *TP53RK*, both important in cell survival and apoptosis inhibition. Therefore *miR-630* may be potentially considered a radiosensitivity biomarker, because its upregulation negatively influences the expression of *BCL2L2* and *TP53RK*, causing the activation of apoptotic mechanisms. Another radioresistance biomarker is *miR100* that plays an important role in regulating the radiosensitivity of CRC, and it may act as a new clinical target for CRC radiotherapy (Fanale, Castiglia, Bazan, & Russo, 2016).

3.17 Conclusion

NcRNAs particularly miRNAs and lncRNAs have many functions related to a variety of cellular and biological processes as well as in tumorigenesis and metastasis. Numerous studies have indicated the aberrant expression of ncRNAs, modulated in CRC through different modes of actions, including cell proliferation, apoptosis, cell cycle, DNA repair response, epigenetics, and immune responses. Thus they could be served as promising diagnostic, predictive, or prognostic biomarkers for CRC. Moreover, they could be used for therapeutic targeting leading to better outcomes for patients with CRC especially in the era of precision medicine.

References

Adriaens, C., & Marine, J. C. (2017). NEAT1-containing paraspeckles: Central hubs in stress response and tumor formation. *Cell Cycle, 16*(2), 137.

Alaiyan, B., Ilyayev, N., Stojadinovic, A., Izadjoo, M., Roistacher, M., Pavlov, V., ... Nissan, A. (2013). Differential expression of colon cancer associated transcript1 (CCAT1) along the colonic adenoma-carcinoma sequence. *BMC Cancer, 13*, 196.

Amodio, N., Bellizzi, D., Leotta, M., Raimondi, L., Biamonte, L., D'Aquila, P., ... Tassone, P. (2013). miR-29b induces SOCS-1 expression by promoter demethylation and negatively regulates migration of multiple myeloma and endothelial cells. *Cell Cycle, 12*(23), 3650−3662.

Amodio, N., Stamato, M. A., Juli, G., Morelli, E., Fulciniti, M., Manzoni, M., ... Tassone, P. (2018). Drugging the lncRNA MALAT1 via LNA gapmeR ASO inhibits gene expression of proteasome subunits and triggers anti-multiple myeloma activity. *Leukemia, 32*(9), 1948−1957.

Arbitrio, M., Di Martino, M. T., Barbieri, V., Agapito, G., Guzzi, P. H., Botta, C., ... Tagliaferri, P. (2016). Identification of polymorphic variants associated with erlotinib-related skin toxicity in advanced non-small cell lung cancer patients by DMET microarray analysis. *Cancer Chemotherapy and Pharmacology, 77*(1), 205−209.

Arrigoni, A., Ranzani, V., Rossetti, G., Panzeri, I., Abrignani, S., Bonnal, R., J. P., & Pagani, M. (2016). *Analysis RNA-seq and Noncoding RNA. Polycomb group proteins* (pp. 125−135). Springer.

Arun, G., Diermeier, S. D., & Spector, D. L. (2018). Therapeutic targeting of long non-coding RNAs in cancer. *Trends in Molecular Medicine, 24*(3), 257−277.

Asangani, I. A., Rasheed, S. A., Nikolova, D. A., Leupold, J. H., Colburn, N. H., Post, S., & Allgayer, H. (2008). MicroRNA-21 (miR-21) post-transcriptionally downregulates tumor suppressor Pdcd4 and stimulates invasion, intravasation and metastasis in colorectal cancer. *Oncogene, 27*(15), 2128−2136.

Avtanski, D. B., Nagalingam, A., Kuppusamy, P., Bonner, M. Y., Arbiser, J. L., Saxena, N. K., & Sharma, D. (2015). Honokiol abrogates leptin-induced tumor progression by inhibiting Wnt1-MTA1-beta-catenin signaling axis in a microRNA-34a dependent manner. *Oncotarget, 6*(18), 16396−16410.

Baassiri, A., Nassar, F., Mukherji, D., Shamseddine, A., Nasr, R., & Temraz, S. (2020). Exosomal non coding RNA in LIQUID biopsies as a promising biomarker for colorectal cancer. *International Journal of Molecular Sciences, 21*(4), 1398.

Barrett, T., Troup, D. B., Wilhite, S. E., Ledoux, P., Evangelista, C., Kim, I. F., ... Soboleva, A. (2010). NCBI GEO: Archive for functional genomics data sets—10 years on. *Nucleic Acids Research, 39*(suppl_1), D1005−D1010.

Barski, A., Cuddapah, S., Cui, K., Roh, T., Schones, D. E., Wang, Z., ... Zhao, K. (2007). High-resolution profiling of histone methylations in the human genome. *Cell, 129*(4), 823−837.

Barsyte-Lovejoy, D., Lau, S. K., Boutros, P. C., Khosravi, K., Jurisica, I, Andrulis, I. L., ... Penn, L. Z. (2006). The c-Myc oncogene directly induces the H19 noncoding RNA by allele-specific binding to potentiate tumorigenesis. *Cancer Research, 66*(10), 5330−5337.

Bartolomei, M. S., Zemel, S., & Tilghman, S. M. (1991). Parental imprinting of the mouse H19 gene. *Nature, 351*(6322), 153−155.

Bayraktar, R., Pichler, M., Kanlikilicer, P., Ivan, C., Bayraktar, E., Kahraman, N., ... Ozpolat, B. (2017). MicroRNA 603 acts as a tumor suppressor and inhibits triple-negative breast cancer tumorigenesis by targeting elongation factor 2 kinase. *Oncotarget, 8*(7), 11641.

Berteaux, N., Lottin, S., Monté, D., Pinte, S., Quantannens, B., Coll, J., ... Adriaenssens, E. (2005). H19 mRNA-like noncoding RNA promotes breast cancer cell proliferation through positive control by E2F1. *Journal of Biological Chemistry, 280*(33), 29625−29636.

Bertone, P., Stolc, V., Royce, T. E., Rozowsky, J. S., Urban, A. E., Zhu, X., ... Snyder, M. (2004). Global identification of human transcribed sequences with genome tiling arrays. *Science, 306*(5705), 2242−2246.

Bhargava, V., Ko, P., Willems, E., Mercola, M., & Subramaniam, S. (2013). Quantitative transcriptomics using designed primer-based amplification. *Scientific Reports, 3*(1), 1−9.

Bian, Z., Jin, L., Zhang, J., Yin, Y., Quan, C., Hu, Y., ... Huang, Z. (2016). LncRNA—UCA1 enhances cell proliferation and 5-fluorouracil resistance in colorectal cancer by inhibiting miR-204-5p. *Scientific Reports, 6*(1), 1−12.

Brabletz, S., Bajdak, K., Meidhof, S., Burk, U., Niedermann, G., Firat, E., ... Brabletz, T. (2011). The ZEB1/miR-200 feedback loop controls notch signalling in cancer cells. *The EMBO Journal, 30*(4), 770−782.

Brannan, C. I., Dees, E. C., & Ingram, R. S. (1990). Tilghman SMJM, biology c. The product of the H19 gene may function as an RNA. *Molecular and Cellular Biology, 10*(1), 28−36.

Brueffer, C., & Saal, L. H. (2016). TopHat-recondition: A post-processor for TopHat unmapped reads. *BMC Bioinformatics, 17*(1), 1–3.

Cabanski, C. R., White, N. M., Dang, H. X., Silver-Fisher, J. M., Rauck, C. E., Cicka, D., & Maher, C. A. (2015). Pan-cancer transcriptome analysis reveals long noncoding RNAs with conserved function. *RNA Biology, 12*(6), 628–642.

Cao, J., Yan, X. R., Liu, T., Han, X., Yu, J., Liu, S., & Wang, L. (2017). MicroRNA-552 promotes tumor cell proliferation and migration by directly targeting DACH1 via the Wnt/beta-catenin signaling pathway in colorectal cancer. *Oncology Letters, 14*(3), 3795–3802.

Caracciolo, D., Di Martino, M. T., Amodio, N., Morelli, E., Montesano, M., Botta, C., … Tassone, P. (2019). miR-22 suppresses DNA ligase III addiction in multiple myeloma. *Leukemia, 33*(2), 487–498.

Cătană, C.-S., Pichler, M., Giannelli, G., Mader, R. M., & Berindan-Neagoe, I. (2017). Non-coding RNAs, the Trojan horse in two-way communication between tumor and stroma in colorectal and hepatocellular carcinoma. *Oncotarget, 8*(17), 29519.

Chen, Z., McGee, M., Liu, Q., & Scheuermann, R. H. (2007). A distribution free summarization method for Affymetrix GeneChip arrays. *Bioinformatics, 23*(3), 321–327.

Cheng, K., Zhao, Z., Wang, G., Wang, J., & Zhu, W. (2018). lncRNA GAS5 inhibits colorectal cancer cell proliferation via the miR1825p/FOXO3a axis. *Oncology Reports, 40*(4), 2371–2380.

Collins, A. S., McCoy, C. E., Lloyd, A. T., O'Farrelly, C., & Stevenson, N. J. (2013). miR-19a: An effective regulator of SOCS3 and enhancer of JAK-STAT signalling. *PLoS One, 8*(7), e69090.

Consortium, I. H. G. S. (2004). Finishing the euchromatic sequence of the human genome. *Nature, 431*(7011), 931.

Davis, R. L., Weintraub, H., & Lassar, A. B. (1987). Expression of a single transfected cDNA converts fibroblasts to myoblasts. *Cell, 51*(6), 987–1000.

Deng, J., Yang, M., Jiang, R., An, N., Wang, X., & Liu, B. (2017). Long non-coding RNA HOTAIR regulates the proliferation, self-renewal capacity, tumor formation and migration of the cancer stem-like cell (CSC) subpopulation enriched from breast cancer cells. *PLoS One, 12*(1), e0170860.

Derrien, T., Johnson, R., Bussotti, G., Tazer, A., Djebali, S., Tilgner, H., … Guigo, R. (2012). The GENCODE v7 catalog of human long noncoding RNAs: Analysis of their gene structure, evolution, and expression. *Genome Research, 22*(9), 1775–1789.

Di Martino, M. T., Scionti, F., Sestito, S., Nicoletti, A., Arbitrio, M., Guzzi, P. H., … Pensabene, L. (2016). Genetic variants associated with gastrointestinal symptoms in Fabry disease. *Oncotarget, 7*(52), 85895.

Ding, J., Li, J., Wang, H., Xie, M., He, X., Ji, H., … Guozhong, J. (2017). Long noncoding RNA CRNDE promotes colorectal cancer cell proliferation via epigenetically silencing DUSP5/CDKN1A expression. *Cell Death & Disease, 8*(8), e2997-e2997.

Dou, J., Ni, Y., He, X., Wu, D., Li, M., Wu, S., … Zhao, F. (2016). Decreasing lncRNA HOTAIR expression inhibits human colorectal cancer stem cells. *American Journal of Translational Research, 8*(1), 98.

Dugimont, T., Montpellier, C., Adriaenssens, E., Lottin, S., Dumont, L., Iotsova, V., … Curgy, J. J. (1998). The H19 TATA-less promoter is efficiently repressed by wild-type tumor suppressor gene product p53. *Oncogene, 16*(18), 2395–2401.

Eyking, A., Reis, H., Frank, M., Gerken, G., Schmid, K. W., & Cario, E. (2016). MiR-205 and MiR-373 are associated with aggressive human mucinous colorectal cancer. *PLoS One, 11*(6), e0156871.

Fanale, D., Castiglia, M., Bazan, V., & Russo, A. (2016). Involvement of non-coding RNAs in chemo-and radioresistance of colorectal cancer. *Advances in Experimental Medicine and Biology, 937*, 207–228.

Gabory, A., Ripoche, M.-A., Le Digarcher, A., Watrin, F., Ziyyat, A., Forne, T., … Dandolo, L. (2009). H19 acts as a trans regulator of the imprinted gene network controlling growth in mice. *Development, 136*(20), 3413–3421.

Garajová, I., Ferracin, M., Porcellini, E., Palloni, A., Abbati, F., Biasco, G., & Brandi, G. (2017). Non-coding RNAs as predictive biomarkers to current treatment in metastatic colorectal cancer. *International Journal of Molecular Sciences, 18*(7), 1547.

Gawronski, K. A., & Kim, J. (2017). Single cell transcriptomics of noncoding RNAs and their cell-specificity. *Wiley Interdisciplinary Reviews: RNA, 8*(6), e1433.

Ge, X., Chen, Y., Liao, X., Liu, D., Li, F., Ruan, H., & Jia, W. (2013). Overexpression of long noncoding RNA PCAT-1 is a novel biomarker of poor prognosis in patients with colorectal cancer. *Medical Oncology, 30*(2), 588.

Gibb, E. A., Vucic, E. A., Enfield, K. S., Stewart, G. L., Lonergan, K. M., Kennett, J. Y., … Lam, W. L. (2011). Human cancer long non-coding RNA transcriptomes. *PLoS One, 6*(10), e25915.

Guo, C., Sah, J. F., Beard, L., Willson, J. K., Markowitz, S. D., & Guda, K. (2008). The noncoding RNA, miR-126, suppresses the growth of neoplastic cells by targeting phosphatidylinositol 3-kinase signaling and is frequently lost in colon cancers. *Genes Chromosomes Cancer, 47*(11), 939–946.

Gupta, R. A., Shah, N., Wang, K. C., Kim, J., Horlings, H. M., Wong, D. J., ... Chang, H. Y. (2010). Long noncoding RNA HOTAIR reprograms chromatin state to promote cancer metastasis. *Nature, 464*(7291), 1071–1076.

Guttman, M., Amit, I., Garber, M., French, C., Lin, M. F., Feldser, D., ... Lander, E. S. (2009). Chromatin signature reveals over a thousand highly conserved large non-coding RNAs in mammals. *Nature, 458*(7235), 223–227.

Hamzehzadeh, L., Khadangi, F., Ghayoor Karimiani, E., Pasdar, A., & Kerachian, M. A. (2018). Common KRAS and NRAS gene mutations in sporadic colorectal cancer in Northeastern Iranian patients. *Current Problems in Cancer, 42*(6), 572–581.

Hamzehzadeh, L., Khadangi, F., Karimiani, E. G., Pasdar, A., & Kerachian, M. A. (2018). Common KRAS and NRAS gene mutations in sporadic colorectal cancer in Northeastern Iranian patients. *Current Problems in Cancer, 42*(6), 572–581.

Han, D., Wang, M., Ma, N., Xu, Y., Jiang, Y., & Gao, X. (2015). Long noncoding RNAs: Novel players in colorectal cancer. *Cancer Letters, 361*(1), 13–21.

Han, P., Li, J.-W., Zhang, B.-M., Lv, J., Li, Y., Gu, X., ... Cui, B. (2017). The lncRNA CRNDE promotes colorectal cancer cell proliferation and chemoresistance via miR-181a-5p-mediated regulation of Wnt/β-catenin signaling. *Molecular Cancer, 16*(1), 1–13.

Han, Y., Yang, Y. N., Yuan, H. H., Zhang, T. T., Sui, H., Wei, X. L., ... Bai, Y. X. (2014). UCA1, a long non-coding RNA up-regulated in colorectal cancer influences cell proliferation, apoptosis and cell cycle distribution. *Pathology, 46*(5), 396–401.

Hao, Y., Crenshaw, T., Moulton, T., Newcomb, E., & Tycko, B. (1993). Tumour-suppressor activity of H19 RNA. *Nature, 365*(6448), 764–767.

Hochreiter, S., Clevert, D. A., & Obermayer, K. (2006). A new summarization method for Affymetrix probe level data. *Bioinformatics, 22*(8), 943–949.

Hoesel, B., & Schmid, J. A. (2013). The complexity of NF-kappaB signaling in inflammation and cancer. *Molecular Cancer, 12*, 86.

Hon, C.-C., Ramilowski, J. A., Harshbarger, J., Bertin, N., Rackham, O. J. L., Gough, J., ... Forrest, A. R. R. (2017). An atlas of human long non-coding RNAs with accurate 5' ends. *Nature, 543*(7644), 199–204.

Horie, M., Kaczkowski, B., Ohshima, M., Matuzak, H., Noguchi, S., Mikam, Y., ... Lander, E. S. (2017). Integrative CAGE and DNA methylation profiling identify epigenetically regulated genes in NSCLC. *Molecular Cancer Research, 15*(10), 1354–1365.

Huang, J., Peng, J., & Guo, L. (2015). Non-coding RNA: A new tool for the diagnosis, prognosis, and therapy of small cell lung cancer. *Journal of Thoracic Oncology, 10*(1), 28–37.

Huang, Z., Huang, D., Ni, S., Peng, Z., Sheng, W., & Du, X. (2010). Plasma microRNAs are promising novel biomarkers for early detection of colorectal cancer. *International Journal of Cancer, 127*(1), 118–126.

Huber, W., von Heydebreck, A., Sultmann, H., Poustka, A., & Vingron, M. (2002). Variance stabilization applied to microarray data calibration and to the quantification of differential expression. *Bioinformatics, 18*(Suppl 1), S96–104.

Hudson, W. H., Pickard, M. R., De Vera, I. M. S., Kuiper, E. G., Mourtada-Maarabouni, M., Conn, G. L., ... Ortlund, E. A. (2014). Conserved sequence-specific lincRNA–steroid receptor interactions drive transcriptional repression and direct cell fate. *Nature Communications, 5*(1), 1–13.

Iliopoulos, D., Hirsch, H. A., & Struhl, K. (2009). An epigenetic switch involving NF-kappaB, Lin28, Let-7 MicroRNA, and IL6 links inflammation to cell transformation. *Cell, 139*(4), 693–706.

Iorio, M. V., & Croce, C. M. (2012). MicroRNA dysregulation in cancer: Diagnostics, monitoring and therapeutics. A comprehensive review. *EMBO Molecular Medicine, 4*(3), 143–159.

Iyer, M., Niknafs, Y., Malik, R., Singhal, U., Sahu, A., Hosono, Y., ... Chinnaiyan, A. M. (2015). The landscape of long noncoding RNAs in the human transcriptome. *Nature Genetics, 47*, 199–208.

Jamialahmadi, K., Azghandi, M., Javadmanesh, A., Zardadi, M., Shams Davodly, E., & Kerachian, M. A. (2021). DNA methylation panel for high performance detection of colorectal cancer. *Cancer Genet, 252–253*, 64–72.

Ju, J. (2010). miRNAs as biomarkers in colorectal cancer diagnosis and prognosis. *Bioanalysis, 2*(5), 901–906.

Kaikkonen, M. U., Lam, M. T., & Glass, C. K. (2011). Non-coding RNAs as regulators of gene expression and epigenetics. *Cardiovascular Research, 90*(3), 430–440.

Kallen, A. N., Zhou, X.-B., Xu, J., Qiao, C., Ma, J., Yan, L., ... Huang, Y. (2013). The imprinted H19 lncRNA antagonizes let-7 microRNAs. *Molecular Cell, 52*(1), 101–112.

Kapranov, P., Cheng, J., Dike, S., Nix, D., Duttagupta, R., & Willingham, A. (2007). RNA maps reveal new and a possible classes pervasive transcription RNA function. *Science, 316*(1486), 10.1126.

Kashi, K., Henderson, L., Bonetti, A., & Carninci, P. (2016). Discovery and functional analysis of lncRNAs: Methodologies to investigate an uncharacterized transcriptome. *Biochimica et Biophysica Acta, 1859*(1), 3–15.

Kawano, Y., & Kypta, R. (2003). Secreted antagonists of the Wnt signalling pathway. *Journal of Cell Science, 116*(Pt 13), 2627–2634.

Kerachian, M. A., Azghandi, M., Javadmanesh, A., Ghaffarzadegan, K., & Mozaffari-Jovin, S. (2020). Selective capture of plasma cell-free tumor DNA on magnetic beads: A sensitive and versatile tool for liquid biopsy. *Cellular Oncology (Dordr), 43*(5), 949–956.

Kerachian, M. A., & Kerachian, M. (2019). Long interspersed nucleotide element-1 (LINE-1) methylation in colorectal cancer. *Clinica Chimica Acta, 488*, 209–214.

Kerachian, M. A., Poudineh, A., & Thiery, J. P. (2019). Cell free circulating tumor nucleic acids, a revolution in personalized cancer medicine. *Critical Reviews in Oncology/Hematology, 144*, 102827.

Khalil, A. M., Guttman, M., Huarte, M., Garber, M., Raj, A., Morales, D. R., ... Rinn, J. L. (2009). Many human large intergenic noncoding RNAs associate with chromatin-modifying complexes and affect gene expression. *Proceedings of the National Academy of Sciences, 106*(28), 11667–11672.

Kim, N. H., Kim, H. S., Kim, N. G., Lee, I, Choi, H. S., Li, X. Y., ... Weiss, S. J. (2011). p53 and microRNA-34 are suppressors of canonical Wnt signaling. *Science Signaling, 4*(197), ra71.

Kim, T., Cui, R., Jeon, Y. J., Lee, J. H., Sim, H., Park, J. K., ... Croce, C. M. (2014). Long-range interaction and correlation between MYC enhancer and oncogenic long noncoding RNA CARLo-5. *Proceedings of the National Academy of Sciences of the United States of America, 111*(11), 4173–4178.

Kino, T., Hurt, D. E., Ichijo, T., Nader, N., & Chrousos, G. P. (2010). Noncoding RNA gas5 is a growth arrest–and starvation-associated repressor of the glucocorticoid receptor. *Science Signaling, 3*(107), ra8-ra8.

Kotake, Y., Nakagawa, T., Kitagawa, K., Suzuki, S., Liu, N., Kitagawa, M., & Xiong, Y. (2011). Long non-coding RNA ANRIL is required for the PRC2 recruitment to and silencing of p15 INK4B tumor suppressor gene. *Oncogene, 30*(16), 1956–1962.

Kumar, M., Lu, Z., Takwi, A. A., Chen, W., Callander, N. S., Ramos, K. S., ... Li, Y. (2011). Negative regulation of the tumor suppressor p53 gene by microRNAs. *Oncogene, 30*(7), 843–853.

Langmead, B., & Salzberg, S. L. (2012). Fast gapped-read alignment with Bowtie 2. *Nature Methods, 9*(4), 357.

Le, Wu. L. J., Zhang, W., & Zhang, L. J. (2016). Msmimjoe, research c. Roles of long non-coding RNA CCAT2 in cervical cancer cell growth and apoptosis. *Medical Science Monitor, 22*, 875.

Lee, S. K., & Calin, G. A. (2011). Non-coding RNAs and cancer: New paradigms in oncology. *Discovery Medicine, 11*(58), 245–254.

Lee, Y. S., & Dutta, A. (2009). MicroRNAs in cancer. *Signal Transduction and Targeted Therapy, 4*, 199–227.

Lei, K., Du, W., Lin, S., Yang, L., Xu, Y., Gao, Y., ... Liu, J. (2016). 3B, a novel photosensitizer, inhibits glycolysis and inflammation via miR-155-5p and breaks the JAK/STAT3/SOCS1 feedback loop in human breast cancer cells. *Biomed Pharmacother, 82*, 141–150.

Léveillé, N., Melo, C. A., Rooijers, K., Diaz-Lagares, A., Melo, S. A., Korkmaz, G., ... Agami, R. (2015). Genome-wide profiling of p53-regulated enhancer RNAs uncovers a subset of enhancers controlled by a lncRNA. *Nature Communications, 6*(1), 1–12.

Li, A., Zhang, J., & Zhou, Z. (2014). PLEK: A tool for predicting long non-coding RNAs and messenger RNAs based on an improved k-mer scheme. *BMC Bioinformatics, 15*, 311.

Li, C., Gao, Y., Li, Y., & Ding, D. (2017). TUG1 mediates methotrexate resistance in colorectal cancer via miR-186/CPEB2 axis. *Biochemical and Biophysical Research Communications, 491*(2), 552–557.

Li, C., & Wong, W. H. (2001). Model-based analysis of oligonucleotide arrays: Model validation, design issues and standard error application. *Genome Biology, 2*(8), 1–11.

Li, H., Rokavec, M., & Hermeking, H. (2015). Soluble IL6R represents a miR-34a target: Potential implications for the recently identified IL-6R/STAT3/miR-34a feed-back loop. *Oncotarget, 6*(16), 14026–14032.

Li, M., & Belmonte, J. C. I. (2015). Roles for noncoding RNAs in cell-fate determination and regeneration. *Nature Structural & Molecular Biology, 22*(1), 2–4.

References

Li, N., Miao, Y., Shan, Y., Liu, B., Li, Y., Zhao, L., & Jia, L. (2017). MiR-106b and miR-93 regulate cell progression by suppression of PTEN via PI3K/Akt pathway in breast cancer. *Cell Death & Disease, 8*(5), e2796.

Li, T., Lai, Q., Wang, S., Cai, J., Xiao, Z., Deng, D., ... Liao, W. (2016). MicroRNA-224 sustains Wnt/beta-catenin signaling and promotes aggressive phenotype of colorectal cancer. *Journal of Experimental & Clinical Cancer Research, 35*, 21.

Li, T., Morgan, M. J., Choksi, S., Zhang, Y., Kim, Y. S., & Liu, Z. G. (2010). MicroRNAs modulate the noncanonical transcription factor NF-kappaB pathway by regulating expression of the kinase IKKalpha during macrophage differentiation. *Nature Immunology, 11*(9), 799–805.

Liao, P., Li, S., Cui, X., & Zheng, Y. (2018). A comprehensive review of web-based resources of non-coding RNAs for plant science research. *International Journal of Biological Sciences, 14*(8), 819.

Lin, M. F., Jungreis, I., & Kellis, M. (2011). PhyloCSF: A comparative genomics method to distinguish protein coding and non-coding regions. *Bioinformatics, 27*(13), i275–i282.

Ling, H., Spizzo, R., Atlasi, Y., Nicoloso, M., Shimizu, M., Redis, R. S., ... Calin, G. A. (2013). CCAT2, a novel noncoding RNA mapping to 8q24, underlies metastatic progression and chromosomal instability in colon cancer. *Genome Research, 23*(9), 1446–1461.

Liu, T., Zhang, X., Yang, Y.-M., Du, L.-T., & Wang, C. X. (2016). Therapy. Increased expression of the long noncoding RNA CRNDE-h indicates a poor prognosis in colorectal cancer, and is positively correlated with IRX5 mRNA expression. *OncoTargets and Therapy, 9*, 1437.

Liu, X. H., Sun, M., Nie, F. Q., Ge, Y. B., Zhang, E. B., Yin, D. D., ... Wang, Z. X. (2014). Lnc RNA HOTAIR functions as a competing endogenous RNA to regulate HER2 expression by sponging miR-331-3p in gastric cancer. *Molecular Cancer, 13*(1), 1–14.

Lu, J., Getz, G., Miska, E. A., Alvarez-Saavedra, E., Lamb, J., Peck, D., & Golub, R. T. (2005). MicroRNA expression profiles classify human cancers. *Nature, 435*(7043), 834–838.

Luo, J., Qu, J., Wu, D.-K., Lu, Z.-L., Sun, Y.-S., & Qu, Q. (2017). Long non-coding RNAs: A rising biotarget in colorectal cancer. *Oncotarget, 8*(13), 22187.

Ma, L., Bajic, V. B., & Zhang, Z. (2013). On the classification of long non-coding RNAs. *RNA Biology, 10*(6), 924–933.

Matouk, I. J., Mezan, S., Mizrahi, A., Ohana, P., Abu-Lail, R., Fellig, Y., ... Hochberg, A. (2010). The oncofetal H19 RNA connection: Hypoxia, p53 and cancer. *Biochimica et Biophysica Acta, 1803*(4), 443–451.

Matsumura, H., Kruger, D. H., Kahl, G., & Terauchi, R. (2008). SuperSAGE: A modern platform for genome-wide quantitative transcript profiling. *Current Pharmaceutical Biotechnology, 9*(5), 368–374.

McCleland, M. L., Mesh, K., Lorenzana, E., Chopra, V. S., Segal, E., Watanabe, C., ... Firestein, R. (2016). CCAT1 is an enhancer-templated RNA that predicts BET sensitivity in colorectal cancer. *Journal of Clinical Investigation, 126*(2), 639–652.

MercerTR, D., & Mattick, J. S. (2009). Long non- coding RNAs: Insights into functions. *Nature Reviews Genetics, 10*(3), 155.

Meyer, S. U., Kaiser, S., Wagner, C., Thirion, C., & Pfaffl, M. W. (2012). Profound effect of profiling platform and normalization strategy on detection of differentially expressed microRNAs—a comparative study. *PLoS One, 7*(6), e38946.

Moody, L., He, H., Pan, Y.-X., & Chen, H. (2017). Methods and novel technology for microRNA quantification in colorectal cancer screening. *Clinical Epigenetics, 9*(1), 1–13.

Mubeen, S., Hoyt, C. T., Gemund, A., Hofmann-Apitius, M., Frohlich, H., & Domingo-Fernandez, D. (2019). The impact of pathway database choice on statistical enrichment analysis and predictive modeling. *Frontiers in Genetics, 10*, 1203.

Nagel, R., le Sage, C., Diosdado, B., van der Waal, M., Oude Vrielink, J. A., Bolijin, A., ... Agami, R. (2008). Regulation of the adenomatous polyposis coli gene by the miR-135 family in colorectal cancer. *Cancer Research, 68*(14), 5795–5802.

Ng, E. K., Chong, W. W., Jin, H., Lam, E. K. Y., Shin, V. Y., Yu, J., ... Sung, J. J. Y. (2009). Differential expression of microRNAs in plasma of patients with colorectal cancer: A potential marker for colorectal cancer screening. *Gut, 58*(10), 1375–1381.

Nissan, A., Stojadinovic, A., Mitrani-Rosenbaum, S., Halle, D., Grinbaum, R., Roistacher, M., ... Gure, A. O. (2012). Colon cancer associated transcript-1: A novel RNA expressed in malignant and pre-malignant human tissues. *International Journal of Cancer, 130*(7), 1598–1606.

Olson, N. E. (2006). The microarray data analysis process: From raw data to biological significance. *NeuroRx*, *3*(3), 373–383.

Ota, T., Doi, K., Fujimoto, T., Tanako, Y., Ogawa, M., Matsuzaki, H., ... Tsunoda, T. (2012). KRAS up-regulates the expression of miR-181a, miR-200c and miR-210 in a three-dimensional-specific manner in DLD-1 colorectal cancer cells. *Anticancer Research*, *32*(6), 2271–2275.

Palazzo, A. F., & Lee, E. S. (2015). Non-coding RNA: What is functional and what is junk? *Frontiers in Genetics*, *6*, 2.

Piccoli, M.-T., Gupta, S. K., & Thum, T. (2015). Noncoding RNAs as regulators of cardiomyocyte proliferation and death. *Journal of Molecular and Cellular Cardiology*, *89*, 59–67.

Pickard, M. R., & Williams, G. T. (2015). Molecular and cellular mechanisms of action of tumour suppressor GAS5 LncRNA. *Genes (Basel)*, *6*(3), 484–499.

Pirim, H., Ekşioğlu, B., Perkins, A. D., & Yüceer, Ç. (2012). Clustering of high throughput gene expression data. *Computers & Operations Research*, *39*(12), 3046–3061.

Poliseno, L., Salmena, L., Zhang, J., Carver, B., Haveman, W. J., & Pandolfi, P. P. (2010). A coding-independent function of gene and pseudogene mRNAs regulates tumour biology. *Nature*, *465*(7301), 1033–1038.

Poursheikhani, A., Abbaszadegan, M. R., & Kerachian, M. A. (2021). Mechanisms of long non-coding RNA function in colorectal cancer tumorigenesis. *Asia-Pacific Journal of Clinical Oncology*, *17*(1), 7–23.

Poursheikhani, A., Abbaszadegan, M. R., Nokhandani, N., & Kerachian, M. A. (2020). Integration analysis of long non-coding RNA (lncRNA) role in tumorigenesis of colon adenocarcinoma. *BMC Medical Genomics*, *13*(1), 1–16.

Prensner, J. R., Chen, W., Han, S., Iyer, M. K., Cao, Q., Kothari, V., ... Feng, F. Y. (2014). The long non-coding RNA PCAT-1 promotes prostate cancer cell proliferation through cMyc. *Neoplasia*, *16*(11), 900–908.

Pritchard, C. C., Cheng, H. H., & Tewari, M. (2012). MicroRNA profiling: Approaches and considerations. *Nature Reviews Genetics*, *13*(5), 358–369.

Prokopi, M., Kousparou, C. A., & Epenetos, A. A. (2014). The secret role of microRNAs in cancer stem cell development and potential therapy: A notch-pathway approach. *Frontiers in Oncology*, *4*, 389.

Qiao, L., Liu, X., Tang, Y., Zhao, Z., Zhang, J., & Feng, Y. (2017). Down regulation of the long non-coding RNA PCAT-1 induced growth arrest and apoptosis of colorectal cancer cells. *Life Sciences*, *188*, 37–44.

Rinn, J. L., Kertesz, M., Wang, J. K., Squazzo, S. L., Xu, X., Brugmann, S. A., ... Chang, H. Y. (2007). Functional demarcation of active and silent chromatin domains in human HOX loci by noncoding RNAs. *Cell*, *129*(7), 1311–1323.

Rodriguez-Casanova, A., Costa-Fraga, N., Bao-Caamano, A., López-López, R., Muinelo-Romay, L., & Diaz-Lagares, A. (2021). Epigenetic landscape of liquid biopsy in colorectal cancer. *Frontiers in Cell and Developmental Biology*, *9*.

Rokavec, M., Oner, M. G., Li, H., Jackstadt, R., Jiang, L., Lodygin, D., ... Hermeking, H. (2014). IL-6R/STAT3/miR-34a feedback loop promotes EMT-mediated colorectal cancer invasion and metastasis. *Journal of Clinical Investigation*, *124*(4), 1853–1867.

Rossi, M., Teresa Di Martino, M., Morelli, E., Leotta, M., Rizzo, A., Grimaldi, A., ... Caraglia, M. (2012). Molecular targets for the treatment of multiple myeloma. *Current Cancer Drug Targets*, *12*(7), 757–767.

Salendo, J., Spitzner, M., Kramer, F., Zhang, X., Jo, P., Wolff, H. A., ... Gaedcke, J. (2013). Identification of a microRNA expression signature for chemoradiosensitivity of colorectal cancer cells, involving miRNAs-320a, -224, -132 and let7g. *Radiotherapy and Oncology*, *108*(3), 451–457.

Sasagawa, Y., Nikaido, I., Hayashi, T., Danno, H., Uno, K. D., Imai, T., & Ueda, H. R. (2013). Quartz-Seq: A highly reproducible and sensitive single-cell RNA sequencing method, reveals non-genetic gene-expression heterogeneity. *Genome Biology*, *14*(4), 1–17.

Schetter, A. J., Leung, S. Y., Sohn, J. J., Zanetti, K. A., Bowman, E. D., Yanaihara, N., ... Harris, C. C. (2008). MicroRNA expression profiles associated with prognosis and therapeutic outcome in colon adenocarcinoma. *JAMA*, *299*(4), 425–436.

Schmitt, A. M., & Chang, H. Y. (2016). Long noncoding RNAs in cancer pathways. *Cancer Cell*, *29*(4), 452–463.

Schneider, C., King, R. M., & Philipson, L. (1988). Genes specifically expressed at growth arrest of mammalian cells. *Cell*, *54*(6), 787–793.

Scionti, F., Di Martino, M. T., Sestito, S., Nicoletti, A., Falvo, F., Roppa, K., ... Pensabene, L. (2017). Genetic variants associated with Fabry disease progression despite enzyme replacement therapy. *Oncotarget*, *8*(64), 107558.

Shen, P., Pichler, M., Chen, M., Calin, G. A., & Ling, H. (2017). To Wnt or lose: The missing non-coding linc in colorectal cancer. *International Journal of Molecular Sciences*, *18*(9), 2003.

Shi, T., Gao, G., & Cao, Y. (2016). Long Noncoding RNAs as novel biomarkers have a promising future in cancer diagnostics. *Disease Markers, 2016*, 9085195.

Slattery, M. L., Herrick, J. S., Mullany, L. E., Valeri, N., Stevens, J., Caan, B. J., ... Wolff, R. K. (2015). An evaluation and replication of miRNAs with disease stage and colorectal cancer-specific mortality. *International Journal of Cancer, 137*(2), 428–438.

Slattery, M. L., Herrick, J. S., Pellatt, D. F., Mullany, L. E., Stevens, J. R., Wolff, E., ... Samowitz, W. (2016). Site-specific associations between miRNA expression and survival in colorectal cancer cases. *Oncotarget, 7*(37), 60193–60205.

Slattery, M. L., Mullany, L. E., Sakoda, L. C., Wolff, R. K., Samowitz, W. S., & Herrick, J. S. (2018). The MAPK-signaling pathway in colorectal cancer: Dysregulated genes and their association with microRNAs. *Cancer Informatics, 17*, 1176935118766522.

Slattery, M. L., Mullany, L. E., Wolff, R. K., Sakoda, L. C., Samowitz, W. S., & Herrick, J. S. (2019). The p53-signaling pathway and colorectal cancer: Interactions between downstream p53 target genes and miRNAs. *Genomics, 111*(4), 762–771.

Smolle, M., Uranitsch, S., Gerger, A., Pichler, M., & Haybaeck, J. (2014). Current status of long non-coding RNAs in human cancer with specific focus on colorectal cancer. *International Journal of Molecular Sciences, 15*(8), 13993–14013.

Soleimani, A., Rahmani, F., Ferns, G. A., Ryzhikov, M., Avan, A., & Hassanian, S. M. (2018). Role of regulatory oncogenic or tumor suppressor miRNAs of PI3K/AKT signaling axis in the pathogenesis of colorectal cancer. *Current Pharmaceutical Design, 24*(39), 4605–4610.

Strausberg, R. L., Buetow, K. H., Emmert-Buck, M. R., & Klausner, R. D. (2000). The cancer genome anatomy project: Building an annotated gene index. *Trends in Genetics, 16*(3), 103–106.

Strubberg, A. M., & Madison, B. B. (2017). MicroRNAs in the etiology of colorectal cancer: Pathways and clinical implications. *Disease Models & Mechanisms, 10*(3), 197–214.

Sun, K., Chen, X., Jiang, P., Song, X., Wang, H., & Sun, H. (2013). iSeeRNA: Identification of long intergenic non-coding RNA transcripts from transcriptome sequencing data. *BMC Genomics, 14*(Suppl 2), S7.

Sun, Y. M., & Chen, Y. Q. (2020). Principles and innovative technologies for decrypting noncoding RNAs: From discovery and functional prediction to clinical application. *Journal of Hematology & Oncology, 13*(1), 109.

Sung, H., Ferlay, J., Siegel, R. L., Laversanne, M., Soerjomataram, I., Jemal, A., & Bray, F. (2021). Global cancer statistics 2020: GLOBOCAN estimates of incidence and mortality worldwide for 36 cancers in 185 countries. *CA: A Cancer Journal for Clinicians*.

Svoboda, M., Slyskova, J., Schneiderova, M., Makovicky, P., Bielik, L., Levy, M., ... Vodicka, P. (2014). HOTAIR long non-coding RNA is a negative prognostic factor not only in primary tumors, but also in the blood of colorectal cancer patients. *Carcinogenesis, 35*(7), 1510–1515.

Tang, J., Yan, T., Bao, Y., Shen, C., Yu, C., Zhu, X., ... Fang, J. Y. (2019). LncRNA GLCC1 promotes colorectal carcinogenesis and glucose metabolism by stabilizing c-Myc. *Nature Communications, 10*(1), 1–15.

Tang, Q., Zheng, F., Liu, Z., Wu, J. J., Chai, X. S., He, C. X., ... Hann, S. S. (2019). Novel reciprocal interaction of lncRNA HOTAIR and miR-214-3p contribute to the solamargine-inhibited PDPK1 gene expression in human lung cancer. *Journal of Cellular and Molecular Medicine, 23*(11), 7749–7761.

Team RC. (2013). *R: A language and environment for statistical computing*.

Tokgun, P. E., Tokgun, O., Kurt, S., Tomatir, A. G., & Akca, H. (2019). MYC-driven regulation of long non-coding RNA profiles in breast cancer cells. *Gene, 714*, 143955.

Tripathi, V., Ellis, J. D., Shen, Z., Song, D. Y., Pan, Q., Watt, A. T., ... Prasanth, K. V. (2010). The nuclear-retained noncoding RNA MALAT1 regulates alternative splicing by modulating SR splicing factor phosphorylation. *Molecular Cell, 39*(6), 925–938.

Tripathi, V., Shen, Z., Chakraborty, A., Giri, S., Freier, S. M., Wu, X., ... Prasanth, K. V. (2013). Long noncoding RNA MALAT1 controls cell cycle progression by regulating the expression of oncogenic transcription factor B-MYB. *PLOS Genetics, 9*(3), e1003368.

Tsai, K. W., Lo, Y. H., Liu, H., Yeh, C. Y., Chen, Y. Z., Hsu, C. W., ... Wang, J. H. (2018). Linc00659, a long non-coding RNA, acts as novel oncogene in regulating cancer cell growth in colorectal cancer. *Molecular Cancer, 17*(1), 72.

Tsang, W. P., Ng, E. K., Ng, S. S., Jin, H., Yu, J., Sung, J. J. Y., & Kwok, T. T. (2010). Oncofetal H19-derived miR-675 regulates tumor suppressor RB in human colorectal cancer. *Carcinogenesis, 31*(3), 350–358.

Tsunoda, T., Takashima, Y., Yoshida, Y., Doi, K., Tanaka, Y., Fujimoto, T., ... Shirasawa, S. (2011). Oncogenic KRAS regulates miR-200c and miR-221/222 in a 3D-specific manner in colorectal cancer cells. *Anticancer Research*, *31*(7), 2453–2459.

Valeri, N., Braconi, C., Gasparini, P., Murgia, C., Lampis, A., Paulus-Hock, V., ... Croce, C. M. (2014). MicroRNA-135b promotes cancer progression by acting as a downstream effector of oncogenic pathways in colon cancer. *Cancer Cell*, *25*(4), 469–483.

Velculescu, V. E., Zhang, L., Vogelstein, B., & Kinzler, K. W. (1995). Serial analysis of gene expression. *Science*, *270*(5235), 484–487.

Volinia, S., Calin, G. A., Liu, C. G., Ambs, S, Cimmino, A., Petrocca, F., ... Croce, C. M. (2006). A microRNA expression signature of human solid tumors defines cancer gene targets. *Proceedings of the National Academy of Sciences of the United States of America*, *103*(7), 2257–2261.

Wang, C., Ba, X., Guo, Y., Sun, D., Jiang, H., Li, W., ... Chen, J. (2017). MicroRNA-199a-5p promotes tumour growth by dual-targeting PIAS3 and p27 in human osteosarcoma. *Scientific Reports*, *7*, 41456.

Wang, F., Li, X., Xie, X., Zhao, L., & Chen, W. (2008). UCA1, a non-protein-coding RNA up-regulated in bladder carcinoma and embryo, influencing cell growth and promoting invasion. *FEBS Letters*, *582*(13), 1919–1927.

Wang, J., Li, H., Wang, Y., Wang, L., Yan, X., Zhang, D., ... Yang, Y. (2016). MicroRNA-552 enhances metastatic capacity of colorectal cancer cells by targeting a disintegrin and metalloprotease 28. *Oncotarget*, *7*(43), 70194–70210.

Wang, J., Song, Y. X., Ma, B., Wang, J. J., Sun, J. X., Chen, X. W., ... Wang, Z. N. (2015). Regulatory roles of non-coding RNAs in colorectal cancer. *International Journal of Molecular Sciences*, *16*(8), 19886–19919.

Wang, L., Park, H. J., Dasari, S., Wang, S., Kocher, J. P., & Li, W. (2013). CPAT: Coding-Potential Assessment Tool using an alignment-free logistic regression model. *Nucleic Acids Res*, *41*(6), e74.

Wang, Q., Yang, L., Hu, X., Jiang, Y., Hu, Y., Liu, Z., ... Feng, G. (2017). Upregulated NNT-AS1, a long noncoding RNA, contributes to proliferation and migration of colorectal cancer cells in vitro and in vivo. *Oncotarget*, *8*(2), 3441–3453.

Wang, S., Xiang, J., Li, Z., Lu, S., Hu, J., Gao, X., ... Zhu, H. (2015). A plasma microRNA panel for early detection of colorectal cancer. *International Journal of Cancer*, *136*(1), 152–161.

Weber, J. A., Baxter, D. H., Zhang, S., Huang, D. Y., Huang, K. H., Lee, M. J., ... Wang, K. (2010). The microRNA spectrum in 12 body fluids. *Clinical Chemistry*, *56*(11), 1733–1741.

Weidle, U. H., Birzele, F., Kollmorgen, G., & Rueger, R. (2017). Long non-coding RNAs and their role in metastasis. *Cancer Genomics Proteomics*, *14*(3), 143–160.

Wilusz, J. E., Freier, S. M., & Spector, D. L. (2008). 3′ end processing of a long nuclear-retained noncoding RNA yields a tRNA-like cytoplasmic RNA. *Cell*, *135*(5), 919–932.

Witte, S., & Muljo, S. A. (2014). Integrating non-coding RNAs in JAK-STAT regulatory networks. *JAK-STAT*, *3*(1), e28055.

Wong, N. K., Huang, C.-L., Islam, R., & Yip, S. P. (2018). Long non-coding RNAs in hematological malignancies: Translating basic techniques into diagnostic and therapeutic strategies. *Journal of Hematology & Oncology*, *11*(1), 1–22.

Wu, D., Hu, Y., Tong, S., Williams, B. R., Smyth, G. K., & Gantier, M. P. (2013). The use of miRNA microarrays for the analysis of cancer samples with global miRNA decrease. *RNA*, *19*(7), 876–888.

Wu, Q., Meng, W. Y., Jie, Y., & Zhao, H. (2018). LncRNA MALAT1 induces colon cancer development by regulating miR-129-5p/HMGB1 axis. *Journal of Cellular Physiology*, *233*(9), 6750–6757.

Xie, X., Tang, B., Xiao, Y. F., Xie, R., Li, B. S., Dong, H., ... Yang, S. M. (2016). Long non-coding RNAs in colorectal cancer. *Oncotarget*, *7*(5), 5226.

Xu, M., Chen, X., Lin, K., Zeng, K., Liu, X., Pan, B., ... Wang, S. (2018). The long noncoding RNA SNHG1 regulates colorectal cancer cell growth through interactions with EZH2 and miR-154-5p. *Molecular Cancer*, *17*(1), 141.

Xu, M.-d., Qi, P., & Du, X. (2014). Long non-coding RNAs in colorectal cancer: Implications for pathogenesis and clinical application. *Modern Pathology*, *27*(10), 1310–1320.

Xu, Y., Zhang, X., Hu, X., Zhou, W., Zhang, P., Zhang, J., ... Liu, Y. (2018). The effects of lncRNA MALAT1 on proliferation, invasion and migration in colorectal cancer through regulating SOX9. *Molecular Medicine*, *24*(1), 52.

Yamada, A., Yu, P., Lin, W., Okugawa, Y., Boland, C. R., & Goel, A. (2018). A RNA-Sequencing approach for the identification of novel long non-coding RNA biomarkers in colorectal cancer. *Scientific Reports*, *8*(1), 575.

Yang, H., Wang, S., Kang, Y. J., Wang, c., Xu, Y., Zhang, Y., & Jiang, Z. (2018). Long non-coding RNA SNHG1 predicts a poor prognosis and promotes colon cancer tumorigenesis. *Oncology Reports, 40*(1), 261–271.

Yang, L., Lin, C., Liu, W., Zhang, J., Ohgi, K. A., Grinstein, J. D., ... Rosenfeld, M. G. (2011). ncRNA- and Pc2 methylation-dependent gene relocation between nuclear structures mediates gene activation programs. *Cell, 147*(4), 773–788.

Yang, M. H., Hu, Z. Y., Xu, C., Xie, L. Y., Wang, X. Y., Chen, S. Y., & Li, Z. G. (2015). MALAT1 promotes colorectal cancer cell proliferation/migration/invasion via PRKA kinase anchor protein 9. *Biochimica et Biophysica Acta, 1852*(1), 166–174.

Yang, Y., Shen, Z., Yan, Y., Wang, B., Zhang, J., Shen, C., ... Wang, S. (2017). Long non-coding RNA GAS5 inhibits cell proliferation, induces G0/G1 arrest and apoptosis, and functions as a prognostic marker in colorectal cancer. *Oncology Letters, 13*(5), 3151–3158.

Ye, L. C., Zhu, D. X., Qiu, J. J., Xu, J., & Wei, Y. (2015). Involvement of long non-coding RNA in colorectal cancer: From benchtop to bedside. *Oncology Letters, 9*(3), 1039–1045.

Yoshimizu, T., Miroglio, A., Ripoche, M. A., Gabory, A., Vernucci, M., Riccio, A., ... Luisa, D. (2008). The H19 locus acts in vivo as a tumor suppressor. *Proceedings of the National Academy of Sciences of the United States of America, 105*(34), 12417–12422.

Zhang, A., Zhao, J. C., Kim, J., Fong, K. W., Yang, Y. A., Chakravarti, D., ... Yu, J. (2015). LncRNA HOTAIR enhances the androgen-receptor-mediated transcriptional program and drives castration-resistant prostate cancer. *Cell Reports, 13*(1), 209–221.

Zhang, P., Wu, W., Chen, Q., & Chen, M. (2019). Non-coding RNAs and their integrated networks. *Journal of Integrative Bioinformatics, 16*(3).

Zhang, S., Shan, C., Kong, G., Du, Y., Ye, L., & Zhang, X. (2012). MicroRNA-520e suppresses growth of hepatoma cells by targeting the NF-kappaB-inducing kinase (NIK). *Oncogene, 31*(31), 3607–3620.

Zhang, X., Lian, Z., Padden, C., Gerstein, M. B., Rozowsky, J., Snyder, M., ... Newburger, P. E. (2009). A myelopoiesis-associated regulatory intergenic noncoding RNA transcript within the human HOXA cluster. *Blood, The Journal of the American Society of Hematology, 113*(11), 2526–2534.

Zhao, B., Hou, X., & Zhan, H. (2015). Medicine e. Long non-coding RNA PCAT-1 over-expression promotes proliferation and metastasis in non-small cell lung cancer cells. *International Journal of Clinical and Experimental Medicine, 8*(10), 18482.

Zhao, B., Lu, M., Wang, D., Li, H., & He, X. (2016). Genome-wide identification of long noncoding RNAs in human intervertebral disc degeneration by RNA sequencing. *BioMed Research International*, 2016.

Zhou, M., Hu, L., Zhang, Z., Wu, N., Sun, J., & Su, J. (2018). Recurrence-associated long non-coding RNA signature for determining the risk of recurrence in patients with colon cancer. *Molecular Therapy-Nucleic Acids, 12*, 518–529.

Zhu, Q. Y., Liu, Q., Chen, J. X., Lan, K., & Ge, B. X. (2010). MicroRNA-101 targets MAPK phosphatase-1 to regulate the activation of MAPKs in macrophages. *Journal of Immunology, 185*(12), 7435–7442.

Zhuang, G., Wu, X., Jiang, Z., Kasman, I., Yao, J., Guan, Y., ... Ferrara, N. (2012). Tumour-secreted miR-9 promotes endothelial cell migration and angiogenesis by activating the JAK-STAT pathway. *The EMBO Journal, 31*(17), 3513–3523.

Zou, F., Mao, R., Yang, L., Lin, S., Lei, K., Zheng, Y., ... Liu, J. (2016). Targeted deletion of miR-139-5p activates MAPK, NF-kappaB and STAT3 signaling and promotes intestinal inflammation and colorectal cancer. *The FEBS Journal, 283*(8), 1438–1452.

CHAPTER 4

Applications of noncoding ribonucleic acids in multiple myeloma patients

Simone Zocchi, Antoine David, Michele Goodhardt and David Garrick

INSERM U976, Saint-Louis Institute for Research, University of Paris, Paris, France

4.1 Introduction

Multiple myeloma (MM) is a malignant disease caused by the transformation of plasma cells, the immunoglobulin-secreting terminally differentiated B lymphocytes. It is primarily a tumor of elderly individuals, with a median age at diagnosis of around 70 years and an incidence that increases strongly with advancing age (Haematological Malignancy Research Network, 2020). MM accounts for around 10% of all hematological malignancies, an incidence that increases to over 20% when it is considered together with its premalignant precursor stage disease, monoclonal gammopathy of undetermined significance (MGUS) (Haematological Malignancy Research Network, 2020). MM constitutes approximately 2% of all cancers, with an estimated incidence of 5–7 cases per 100 000 in Western populations (Sant, 2010; Haematological Malignancy Research Network, 2020; National Cancer Institute USA, 2020). Probably due to both population aging and improved surveillance and survival, the prevalence of MM has steadily increased over the last 40 years (Turesson, 2018; National Cancer Institute USA, 2020).

MM is a progressive disease, evolving from the precursor MGUS stage, through an indolent intermediate stage known as smoldering multiple myeloma (SMM), before developing into full symptomatic MM. The clinical manifestations of the disease arise primarily due to infiltration of the bone marrow by malignant plasma cells, as well as due to the accumulation of monoclonal immunoglobulins secreted by these cells in high quantities (Kumar et al., 2017; Rajkumar et al., 2014). Invasion of tumor plasma cells in the marrow induces bone demineralization and leads eventually to lesions, pain, and bone fragility. Cellular deficiencies include anemia, thrombo-, and neutropœnia with associated immune deficiency. Due to the excess of soluble immunoglobulin, free light chains accumulate in the kidney, frequently leading to renal insufficiency and eventual acute failure in some cases (Yadav, Cook, & Cockwell, 2016).

The last 20 years has seen the introduction of new classes of drugs, in particular proteasome inhibitors (PIs), immunomodulatory drugs, and monoclonal antibodies, which complemented the traditional combinations of high-dose chemotherapy and autologous hematopoietic stem cell transplantation (Landgren & Iskander, 2017). These modern combination therapies have resulted in a more sustained clinical response and improved survival times. Emerging immunotherapeutic approaches, including bispecific T-cell engager antibodies and Chimeric Antigen Receptor (CAR) T cells, have been recently approved for clinical use or are well advanced in development and promise to further improve patient outcome (Yang, Li, Gu, Dong, & Cai, 2020). However, despite this significant progress, disease relapse occurs frequently, and MM is still considered to be essentially incurable (Ravi et al., 2018).

MM is a biologically heterogeneous disease. Primary driver genetic events, which are detectable from the MGUS stage, can be broadly divided into two groups (van Nieuwenhuijzen, Spaan, Raymakers, & Peperzak, 2018). The first group consists of chromosomal aneuploidies, most frequently co-occurring trisomies of particular chromosomes or deletion of chromosome 13q. A second major group consists of recurrent chromosomal translocations that involve the immunoglobulin heavy chain (IgH) locus on chromosome 14 [most frequently t(4;14) and t(11;14), which are detected in around 15% of patients each] (Kumar et al., 2017). These different underlying founder events are associated with distinct prognostic outcomes. For example, while trisomy of chromosome 3 is associated with prolonged overall survival (OS), trisomy 21 in MM is associated with a more severe prognostic outcome (Chretien et al., 2015). Similarly, the t(4;14) translocation is associated with a more rapid disease progression than t(11;14) (Rajkumar et al., 2013). Following one of these founder events and the generation of a premalignant clone, progression to SMM and eventual symptomatic disease involves the acquisition of multiple secondary hits, finally resulting in a high level of clonal complexity (Boyle et al., 2021; Manier et al., 2017a; Walker et al., 2015). These secondary drivers of disease evolution include the acquisition of further genetic alterations, such as chromosomal translocations involving the MYC locus, gains, and losses of specific chromosome arms (particularly gain of 1q, loss of 1p or 17p), as well as recurrent mutations affecting *FAM46C*, genes of the Mitogen-activated protein kinase (MAPK) pathway (most frequently *KRAS*, *NRAS*, and *BRAF*), and DNA repair pathway (most frequently *TP53*, *ATRX*, *ATM*) (Bolli et al., 2014; Boyle et al., 2021; van Nieuwenhuijzen et al., 2018; Walker et al., 2015). Epigenetic alterations have also been linked to disease evolution, including changes in DNA methylation and covalent histone modifications, which are themselves frequently driven by mutations in critical epigenetic modifying proteins (Caprio, Sacco, Giustini, & Roccaro, 2020).

Within this biological complexity, accumulating evidence has demonstrated the important contribution of both short- (conventionally considered as < 200 nt) and long (> 200 nt) noncoding RNAs (ncRNAs) to clonal evolution, disease progression, therapeutic response, and eventual outcome in MM patients. In this chapter, we review the current state of knowledge of the noncoding transcriptome in MM, highlight important ncRNAs that have been studied in detail, discuss their mechanism(s) of action in the disease, and their potential use as

4.2 Samples and experimental methods for the analysis of noncoding RNAs in multiple myeloma patients

Most analyses of ncRNAs in MM patients have been performed in plasmocytes isolated from bone marrow aspirates. Typically, mononuclear cell fractions are purified from bone marrow samples by Ficoll density gradient centrifugation, and plasmocytes are then isolated by immunomagnetic selection for the expression of the transmembrane glycoprotein CD138 (Syndecan 1), a specific marker of plasmocytes among bone marrow populations (Chilosi et al., 1999). It is recommended that the purity of the CD138$^+$ population be subsequently validated by Fluorescence-activated Cell Sorting (FACS) (>80% purity is typically observed). It should be noted that since CD138 does not distinguish normal from neoplastic plasmocytes (Chilosi et al., 1999), one potential complication is the presence of normal plasma cells in the sample analyzed. This is particularly important at the earlier MGUS and SMM stages of the disease, where tumor plasma cells are less abundant. In addition to bone marrow plasma cells (BMPC), prognostically relevant information has also been obtained by analyzing circulating ncRNAs in patient serum, or in serum exosomal fractions (Isin et al., 2014; Pan et al., 2018; Sedlarikova et al., 2018). Unlike single-site bone marrow aspirates, serum samples reflect changes in ncRNAs occurring at all sites of the tumor. Expression of individual ncRNAs in total RNA from these MM samples is analyzed by standard quantitative reverse transcription Polymerase Chain Reaction (RT-PCR). High-throughput transcriptomic profiling of ncRNA has been carried out using standard microarray platforms, and more recently, by RNA-seq approaches (see Section 4.3).

Functional studies of ncRNAs deregulated in MM patients have primarily been carried out in a diverse range of MM cell lines, which collectively represent the different major cytogenetic groups of the disease. In these cell lines, the levels of microRNAs (miRNAs) can be altered with relative ease, using commercially available modified RNA molecules that inhibit or mimic the activity of the endogenous miRNA, or by transducing cells with lentiviral vectors to overexpress or sponge mature miRNAs (e.g., see Che, Chen, Wan, & Huang, 2020; Gowda et al., 2018). Functional studies of long noncoding RNA (lncRNA) have also been carried out using lentiviral-mediated overexpression and short hairpin RNA (shRNA) knockdown approaches (e.g., see Gao, Lv, Li, Han, & Zhang, 2017; Xu, Li, & Zhou, 2018). An important caveat of some of these experiments is that molecular mimics and lentiviral overexpression frequently elevate the levels of the ncRNA beyond those observed in primary MM cells, leading to an exaggerated interpretation of the true functional significance of the ncRNA in the disease. Further, classical knockdown approaches using small interfering RNA (siRNA) or shRNA can be challenging for lncRNAs, many of which exhibit a nuclear localization and are largely protected from the RNA interference machinery in the cytoplasm (Lennox & Behlke, 2016). Alternative approaches using locked nucleic acid antisense oligonucleotides that trigger endonucleolytic cleavage of the target lncRNA by endogenous RNAseH have also been successfully used in MM cells (Amodio et al., 2018a). More recently, CRISPR-mediated deletion and CRISPR inhibition-mediated silencing of lncRNA expression have also been employed to test lncRNA function in MM cell lines (David et al., 2020).

In preclinical models of MM, standard in vitro assays have been used to measure the effects of increasing or depletion of ncRNAs on key growth parameters, including cellular proliferation, cell cycle and apoptosis, adhesion and migration properties as well the effects on stromal cell layers in coculture assays (e.g., see Amodio et al., 2018a; David et al., 2020; Meng et al., 2017; Roccaro et al., 2009). The effects on tumorigenic growth in vivo are analyzed in standard xenograft experiments in immunodeficient mouse models, either by intramuscular injection followed by monitoring of tumor mass and size, or by intracaudal injection and subsequently quantitating human MM cells in the bone marrow by FACS or live-cell imaging (e.g., David et al., 2020; Gowda et al., 2018; Pichiorri et al., 2008). In addition to tumor growth, engraftment experiments can also provide information about the effects of ncRNAs on the ability of MM tumor cells to remodel the bone marrow niche and evade host immune responses (Gowda et al., 2018).

4.3 Datasets analyzing noncoding RNAs in multiple myeloma patients

MM cell lines and tissue samples are not included in the major public databases compiling expression datasets for short noncoding RNAs (sncRNAs) (including DASHR, deepBase, and miRbase) and lncRNAs (including LncExpDB, NONCODE, and FANTOM). However, several databases have now been published that contain expression profiles of ncRNAs [including miRNAs, lncRNAs, and circular RNAs (circRNAs)] in tumor plasma cells or serum of MM patients and healthy controls. Depending on the study, ncRNA expression profiles have been compared either between healthy control and patient samples, between different stages of disease progression, between patient groups bearing different underlying molecular driver events, and between circulating serum or exosomes of MM patients and healthy individuals (Tables 4.1 and 4.2).

4.3.1 Datasets profiling expression of short noncoding RNAs in multiple myeloma

As in other cancer types, the onset and progression of MM are characterized by dysregulated expression of many sncRNAs, and in particular miRNAs (Table 4.1). The first databases of miRNA expression in MM patients, published by Pichiorri et al. (2008), used a custom microarray to profile expression of 235 miRNAs in CD138$^+$ BMPC from normal individuals, MGUS, and MM patients. This pioneer study was followed by subsequent profiling of larger patient sets at diagnosis (Chi et al., 2011; Corthals et al., 2011; Gutierrez et al., 2010; Zhou et al., 2010), as well as a set of patients with relapsed/refractory MM (Roccaro et al., 2009). Several studies compare the miRNA expression profile between different molecular subtypes of MM, with supervised analyses identifying changes in miRNAs that are specific to the different translocations and chromosomal alterations (Chi et al., 2011; Corthals et al., 2011; Gutierrez et al., 2010; Lionetti et al., 2009; Wu et al., 2013). Many of these studies also profile expression of messenger RNA (mRNA) in the same

TABLE 4.1 Datasets of small noncoding RNA expression in multiple myeloma patients and cell lines.

Study	Experimental platform	Noncoding RNAs interrogated	Stage of disease	Sample type	Supervised comparisons/ analyses performed	Deregulated miRNAs (first group relative to second)	Accession
Pichiorri et al. (2008)	Custom microarray	235 human miRNAs	Various	BMPC and MM cell lines	MGUS ($n = 6$) V healthy control ($n = 6$)	41 miRNAs ↑; 7 miRNAs ↓	E-TABM-508[a]
					Symptomatic MM ($n = 16$) V healthy control ($n = 6$)	37 miRNAs ↑; 37 miRNAs ↓	
					Symptomatic MM ($n = 16$) and MM cell lines ($n = 49$) V healthy control ($n = 6$)	60 miRNAs ↑; 36 miRNAs ↓	
Roccaro et al. (2009)	Luminex microbead miRNA profiling	318 human miRNAs	Relapsed/ refractory	BMPC	rel./ref. MM ($n = 15$) and MM cell lines ($n = 3$) V healthy control ($n = 4$)	5 miRNAs ↑; 2 miRNAs ↓	Contact authors
Zhou et al. (2010)	Agilent Technologies Human miRNA Microarray V2	464 human miRNAs	Diagnosis	BMPC	Symptomatic MM ($n = 52$) V healthy control ($n = 2$)	39 miRNAs ↑; 1 miRNA ↓	GSE17306[b]
Gutierrez et al. (2010), Corthals et al. (2011)	Applied Biosystems TaqMan gene expression arrays	365 human miRNAs	Diagnosis	BMPC	t(4;14)$^+$ MM ($n = 17$) V healthy control ($n = 5$)	11 miRNAs ↓	GSE16558[b]
					t(11;14)$^+$ MM ($n = 11$) V healthy control ($n = 5$)	7 miRNAs ↓	
					t(14;16)$^+$ MM ($n = 4$) V healthy control ($n = 5$)	3 miRNAs ↑; 5 miRNAs ↓	
					MM with deletion of RB[c] ($n = 4$) V healthy control ($n = 5$)	14 miRNAs ↓	
					All MM ($n = 45$) V healthy control ($n = 4$)	1 miRNA ↑; 40 miRNAs ↓	

(Continued)

TABLE 4.1 (Continued)

Study	Experimental platform	Noncoding RNAs interrogated	Stage of disease	Sample type	Supervised comparisons/analyses performed	Deregulated miRNAs (first group relative to second)	Accession
Chi et al. (2011)	Custom microarray	655 human miRNAs	Various	BMPC	MGUS (n = 5) V healthy control (n = 9)	28 miRNAs ↑; 11 miRNAs ↓	GSE243371[b]
					Symptomatic MM (n = 33) V healthy control (n = 9)	109 miRNAs ↑; 20 miRNAs ↓	
					MM with IgH translocation (n = 14) V MM without IgH translocation (n = 12)	3 miRNAs ↑; 4 miRNAs ↓	
					t(11;14)$^+$ MM (n = 8) V other MM (n = 18)	6 miRNAs ↑; 11 miRNAs ↓	
					t(4;14)$^+$ MM (n = 3) V other MM (n = 23)	6 miRNAs ↑; 2 miRNAs ↓	
					MM with del 13q (n = 7) V other MM (n = 19)	27 miRNAs ↓	
Lionetti et al. (2009)	Agilent Technologies Human miRNA Microarray V2	723 human miRNAs	Diagnosis	BMPC	Healthy control (n = 3)		GSE17498[b]
					Group TC1[d]: t(11;14) or t(6;14) translocation (n = 9) V all others	3 miRNAs ↑	
					Group TC2[d]: moderate CCND1 without IgH translocation (n = 10) V all others	6 miRNAs ↑	
					Group TC3[d]: not in other groups (n = 9) V all others	None	
					Group TC4[d]: t(4;14) translocation (n = 7) V all others	7 miRNAs ↑	
						10 miRNAs ↑	

(Continued)

TABLE 4.1 (Continued)

Study	Experimental platform	Noncoding RNAs interrogated	Stage of disease	Sample type	Supervised comparisons/ analyses performed	Deregulated miRNAs (first group relative to second)	Accession
					Group TC5[d]: t(14;16) or t(14;20) translocation ($n = 5$) V all others	7 miRNAs ↓[e]	
					MM with gain 1q ($n = 20$) V all others	4 miRNAs ↑[e]	
					MM with del 13q ($n = 22$) V all others	3 miRNAs ↑[e]	
					MM with del 17p ($n = 4$) V all others	5 miRNAs ↑[e]	
					Hyperdiploid ($n = 17$) V all others		
Wu et al. (2013)	Affymetrix GeneChip microRNA arrays v1.0	847 human miRNAs	Diagnosis	BMPC	Individual pairwise comparisons between:	Various	GSE41276[b]
					MM with del 4p16 ($n = 26$)		
					MM with dysregulation of MAF ($n = 7$)		
					MM with translocations involving 11q13 ($n = 26$)		
					Other MM ($n = 85$)		
Ronchetti et al. (2012)	Affymetrix GeneChip Human Gene 1.0 ST Array	215 snoRNAs and 17 scaRNAs	Various	BMPC	Disease [diagnosis MM ($n = 55$) and sPCL ($n = 8$)] V normal tonsil ($n = 4$)	General decrease	GSE39683[b]
					sPCL ($n = 8$) v MM at diagnosis ($n = 55$)	11 snoRNAs ↓	
					Hyperdiploid MM ($n = 48$) V nonhyperdiploid MM ($n = 81$)	11 snoRNAs ↑; 1 snoRNA ↓	

(Continued)

TABLE 4.1 (Continued)

Study	Experimental platform	Noncoding RNAs interrogated	Stage of disease	Sample type	Supervised comparisons/ analyses performed	Deregulated miRNAs (first group relative to second)	Accession
Huang et al. (2012)	Applied Biosystems TaqMan Low-Density Arrays	667 human miRNAs	Diagnosis	Plasma	MM (n = 12) V healthy control (n = 8)	6 miRNAs ↑	Contact authors
Kubiczkova et al. (2014)	Applied Biosystems TaqMan Low Density Arrays	667 human miRNAs	Various	Serum	MM (n = 4) V healthy control (n = 4)	7 miRNAs ↑; 7 miRNAs ↓	Contact authors
					MGUS (n = 5) V healthy control (n = 4)	1 miRNA ↑; 4 miRNAs ↓	
Wang et al. (2015)	Agilent Human miRNA Microarray	851 human miRNAs	Diagnosis	Extracellular supernatant of BM aspirates	MM (n = 20) V healthy control (n = 8)	42 miRNAs ↑; 69 miRNAs ↓	GSE49261[b]
Hao et al. (2016)	Exiqon MiRCURY LNA Array	1891 human miRNAs	Diagnosis	Serum	MM (n = 7) V healthy control (n = 5)	4 miRNAs ↑; 23 miRNAs ↓	Contact authors
Manier et al. (2017b)	Small RNA-seq	NA	Diagnosis	Exosomes	MM (n = 10) V healthy control (n = 5)	Various	GSE94564[b]
	Applied Biosystems custom TaqMan low-density array	22 human miRNAs				22 miRNAs ↓	

[a] *Available at Array Express: https://www.ebi.ac.uk/arrayexpress/.*
[b] *Available at Gene Expression Omnibus: https://www.ncbi.nlm.nih.gov/geo/.*
[c] *Deletion of Retinoblastoma gene (RB) considered a unique abnormality.*
[d] *TC (translocation/cyclin D expression) group according to Hideshima classification.*
[e] *Not within the involved chromosome arm.*

BM, bone marrow; BMPC, bone marrow plasma cells; MGUS, monoclonal gammopathy of undetermined significance; miRNA, microRNA; MM, multiple myeloma; NA, not applicable; scaRNA, small Cajal body-specific RNA; snoRNA, small nucleolar RNA; sPCL, secondary plasma cell leukemia.

TABLE 4.2 Datasets of long and circular noncoding RNA expression in multiple myeloma patients and cell lines.

Study	Experimental platform	Noncoding RNAs interrogated	Stage of disease	Sample type	Supervised comparisons/analyses performed	Deregulated ncRNAs (first group relative to second)	Accession
David et al. (2020)	Mining existing microarray dataset (Affymetrix Exon-1.0 ST Array)	9311 human lncRNAs	Diagnosis	BMPC	t(4;14)$^+$ MM ($n = 10$) V healthy control ($n = 6$)	32 lncRNAs ↑; 24 lncRNAs ↓	Mined from GSE39754[a]
					t(11;14)$^+$ MM ($n = 10$) V healthy control ($n = 6$)	28 lncRNAs ↑; 27 lncRNAs ↓	
					Other MM (t(4;14)$^-$ and t(11;14)$^-$) ($n = 10$) V healthy control ($n = 6$)	24 lncRNAs ↑; 23 lncRNAs ↓	
Ronchetti et al. (2016a)	Mining existing microarray dataset (Affymetrix Human Gene 1.0 ST Array)	1852 human lncRNAs	Various	BMPC	MGUS ($n = 20$) V healthy control ($n = 9$)	4 lncRNAs ↑; 57 lncRNAs ↓	Mined from GSE66293 and GSE47552[a]
					Smoldering MM ($n = 33$) V healthy control ($n = 9$)	42 lncRNAs ↑; 18 lncRNAs ↓	
					Symptomatic MM ($n = 170$) V healthy control ($n = 9$)	72 lncRNAs ↑; 9 lncRNAs ↓	
					Plasma cell leukemia ($n = 36$) V healthy control ($n = 9$)	54 lncRNAs ↑; 31 lncRNAs ↓	
					Progressive change during disease evolution	15 lncRNAs ↑; 6 lncRNAs ↓	
			Diagnosis	BMPC	Hyperdiploid ($n = 48$) V nonhyperdiploid patients ($n = 81$)	44 lncRNAs ↑; 3 lncRNAs ↓	
					t(11;14)$^+$ MM ($n = 34$) V all other MM ($n = 95$)	47 lncRNAs ↑; 22 lncRNAs ↓	
					t(4;14)$^+$ MM ($n = 19$) V all other MM ($n = 110$)	37 lncRNAs ↑; 7 lncRNAs ↓	
					MAF$^+$ MM ($n = 6$) V all other MM ($n = 123$)	5 lncRNAs ↑; 21 lncRNAs ↓	

(Continued)

TABLE 4.2 (Continued)

Study	Experimental platform	Noncoding RNAs interrogated	Stage of disease	Sample type	Supervised comparisons/ analyses performed	Deregulated ncRNAs (first group relative to second)	Accession
Ronchetti et al. (2018)	RNA-seq	9540 human lncRNAs	Diagnosis	BMPC	Hyperdiploid MM (n = 8) V other MM (n = 20)	34 lncRNAs ↑; 116 lncRNAs ↓	GSE109116[a]
					t(11;14)$^+$ MM (n = 8) V other MM (n = 22)	50 lncRNAs ↑; 68 lncRNAs ↓	
					t(4;14)$^+$ MM (n = 7) V other MM (n = 23)	62 lncRNAs ↑; 34 lncRNAs ↓	
					MAF$^+$ MM (n = 4) V other MM (n = 26)	16 lncRNAs ↑; 26 lncRNAs ↓	
					MM with gain 1q (n = 15) V other MM (n = 13)	8 lncRNAs ↑ (2 within 1q); 4 lncRNAs ↓	
					MM with del (13) (n = 18) V other MM (n = 12)	78 lncRNAs ↑; 31 lncRNAs ↓ (7 on chr 13)	
					MM with del (17) (n = 3) V other MM (n = 27)	2 lncRNAs ↓	
					MM with mutation in DIS3 (n = 6) V other MM (n = 17)	97 lncRNAs ↑	
					MM with MAPK mutations (n = 14) V other MM (n = 9)	6 lncRNAs ↑	
Samur et al. (2018)	RNA-seq	7277 human lincRNAs	Diagnosis	BMPC	MM (n = 308) V healthy control (n = 16)	474 lncRNAs ↑; 395 lncRNAs ↓	Contact authors
Dahl et al. (2018)	RNA-seq	CircRNAs	Cell line	NIH-929	NA (619 unique circRNAs detected)	NA	GSE108111[a]
Gao et al. (2019)	Arraystar Human circRNA Array V2	13617 circRNAs	Not specified	BMPC	MM (n = 3) V control (iron deficiency anemia) (n = 3)	131 circRNAs ↑; 16 circRNAs ↓	GSE133058[a]

[a] Available at Gene Expression Omnibus: https://www.ncbi.nlm.nih.gov/geo/.

BMPC, Bone marrow plasma cells; circRNA, circular RNA; lincRNA, long intergenic noncoding RNA; lncRNA, long noncoding RNA; MGUS, monoclonal gammopathy of undetermined significance; MM, multiple myeloma; NA, not applicable.

patient samples, allowing the prediction of miRNA/mRNA regulatory interactions (Corthals et al., 2011; Gutierrez et al., 2010; Lionetti et al., 2009; Zhou et al., 2010). More recently, secreted miRNAs have been profiled in either extracellular supernatant of bone marrow aspirates or circulating exosomal fractions and serum of MM patients and healthy individuals, with a particular interest in identifying easily accessible diagnostic or prognostic biomarkers (Table 4.1) (Hao et al., 2016; Huang, Yu, Li, Liu, & Zhong, 2012; Kubiczkova et al., 2014; Manier et al., 2017b; Wang et al., 2015). Other classes of small ncRNAs implicated in malignant disease include small nucleolar RNAs (snoRNAs) and small Cajal body-specific RNAs (scaRNAs), which play essential roles in processes of RNA maturation, including splicing and ribosomal ribonucleic acid (rRNA) processing (Cao et al., 2018). Microarray profiling of sno/scaRNA expression has been carried out in plasma cells from MM patients at diagnosis and following progression to secondary plasma cell leukemia (sPCL), compared to normal controls (Ronchetti et al., 2012). Expression of sno/scaRNAs was generally decreased in tumor plasma cells of patients at diagnosis and declined further upon progression to sPCL. This study also revealed changes in the expression of certain sno/scaRNA that were specific for different molecular disease subtypes (Ronchetti et al., 2012).

Taken altogether, while there are some clear consistent candidates that have been revealed in these datasets (discussed further in Section 4.4.1), there are also considerable differences between the sncRNAs identified in comparable studies (Chi et al., 2011; Pichiorri et al., 2008; Zhou et al., 2010). These differences are likely a reflection of the evolution of experimental platforms used over time, as well as differences in sample size and molecular breakdown, as well as the small number of normal samples included in some studies, making direct comparison of the findings problematic (Table 4.1). The application of small RNA-seq technology will undoubtedly extend the current state of knowledge beyond the known sncRNAs represented in existing microarrays and expression panels and will lead to the discovery of new molecules dysregulated in MM (Manier et al., 2017b).

4.3.2 Datasets profiling expression of long noncoding RNAs in multiple myeloma

In keeping with the emerging awareness of lncRNAs as an important class of regulatory molecules, interest in the contribution of lncRNAs to the pathophysiology of MM has increased dramatically over the last decade. There are several useful databases containing transcriptomic profiles of lncRNAs in large sets of MM patients, performed on either microarray or RNA-seq platforms (Table 4.2). While microarray platforms do not sample the long noncoding transcriptome completely, useful information on large sets of lncRNAs can still be mined from these microarray-based datasets and this approach has revealed distinct changes in lncRNA expression within specific molecular subtypes and progressive stages of the disease (David et al., 2020; Ronchetti et al., 2016a). Two studies have used RNA-seq to profile the expression of previously annotated lncRNA genes in MM patient cohorts. In 2018, Ronchetti et al. (2018) sequenced RNA from BMPC of 30 MM patients at diagnosis and analyzed the expression of 9540 lncRNAs across different molecular subtypes of the disease. Cross-comparison with the expression of 770 coding genes previously associated with

MM revealed 43 pairs of lncRNA /proximal MM-genes, the expression of which was positively correlated, suggesting potential *cis*-regulatory interactions important for the disease. In the same year, Samur et al. (2018) sequenced RNA from BMPC of 16 healthy donors and 308 newly diagnosed MM patients enrolled in the IFM-DFCI 2009 clinical trial (Attal et al., 2017) and analyzed the expression of 7277 long intergenic noncoding RNAs (lincRNA). This study, the most extensive to date, identified 869 lincRNAs differentially expressed in MM, with 474 increasing and 395 decreasing in tumor plasma cells relative to healthy controls. This study also integrated copy number data from Single Nucleotide Polymorphism arrays to identify changes in lincRNA expression that are driven by alterations of gene dosage.

CircRNAs, covalently closed RNA molecules formed by noncanonical back-splicing of linear pre-mRNA transcripts, are another important class of regulatory ncRNA that can modify expression of downstream genes by affecting transcription and splicing or by acting as sponges for miRNAs and other RNA-binding proteins (Holdt, Kohlmaier, & Teupser, 2018). CircRNAs have been implicated in various malignant diseases and their high stability and resistance to exonucleases in body fluids render them especially interesting as easily accessible biomarkers (Wang, Nazarali, & Ji, 2016). Despite this potential, at present circRNAs have not been extensively studied in large MM patient cohorts. One database contains 619 unique circRNAs that were detected in the MM cell line NIH-929 by RNA-seq (Dahl et al., 2018). Focused analysis of 52 cancer-associated circRNAs using Nanostring assays revealed that MM cell lines clustered separately from cell lines of other B-cell malignancies (Dahl et al., 2018). Another study used microarrays to profile expression of circRNAs in BMPC from three MM patients and three patients of iron deficiency anemia (used as control) (Gao et al., 2019). Of the approximately, 10000 circRNAs detected in these samples, 147 were differently expressed between the MM and control groups (Table 4.2).

4.4 Noncoding RNAs implicated in the etiology of multiple myeloma

Collectively, the above datasets and studies demonstrate widespread changes in the expression of ncRNAs in MM patients. As is also the case for coding genes (Mattioli et al., 2005; Zhan et al., 2006), unsupervised clustering based on the expression profiles of ncRNAs generally grouped MM patients according to the molecular disease subtype (David et al., 2020; Lionetti et al., 2009; Ronchetti et al., 2012, 2018), indicating that the underlying genetic alterations are a primary determinant of changes in the noncoding transcriptome. While the different comparisons, sample sizes, and experimental platforms used in these various studies make it difficult to draw direct comparison between them, a number of consistently altered ncRNAs have emerged and have been further validated and functionally tested. Here we highlight some of the best-characterized ncRNAs with a verified contribution to disease progression and/or outcome.

4.4.1 MicroRNAs implicated in the etiology of multiple myeloma

As described earlier, dynamic changes take place in the expression of several miRNAs during progression of MM. Alterations in the expression and activity of miRNAs in MM can be driven by response to upstream signaling pathways, changes in gene copy number, epigenetic deregulation of the miRNA-host gene, impairment of normal miRNA biosynthesis and processing, as well as changes in the levels of competing-endogenous RNAs or so-called miRNA sponges (Misiewicz-Krzeminska et al., 2019). Important candidate miRNAs that have emerged from the datasets described in Section 4.3.1 and have demonstrated functional and/or prognostic significance in preclinical studies are described next and summarized in Table 4.3. The reader is also referred to recent reviews specifically focused on miRNAs implicated in MM (Caracciolo et al., 2018; Chen, Yang, Liu, Zhang, & Xing, 2021).

4.4.1.1 *miR-21*

MiRNA-21, a well-established oncogenic miRNA in other cancers (Bautista-Sanchez et al., 2020), has been shown to be upregulated as of the premalignant MGUS stage of MM (Chi et al., 2011; Pichiorri et al., 2008). Inhibitors of miR-21 reduced proliferation of MM cells in vitro as well as in an in vivo xenograft model (Leone et al., 2013). Upregulation of miR-21, which is triggered when MM cells adhere to bone marrow stromal cells (BMSC), also increases resistance of the tumor to the commonly used drugs dexamethasone, bortezomib, or doxorubicin, suggesting an important role for this miRNA in growth and survival of MM cells within the bone marrow milieu (Wang et al., 2011). Mechanistically, it was shown that miR-21 inhibits expression of the tumor suppressor protein PTEN (phosphatase and tensin homolog), thereby upregulating prosurvival PI3K/AKT/mTOR signaling that is normally repressed by PTEN in cancer cells (Leone et al., 2013) (see also Section 4.5.3). MiRNA-21 is also implicated in the interleukin (IL)6/JAK/STAT3 signaling pathway in MM cells (see Section 4.5.1) and also acts in BMSC to influence bone homeostasis (see Section 4.6).

4.4.1.2 *miR-181a/b and the miR-106b~25 cluster*

Both the miR-181a/b family and the miR-106b~25 cluster (encoding mature miRNAs, miR-106b, -93, and -25) are upregulated in plasma cells in the early MGUS stage of the disease, as well as in symptomatic MM (Chi et al., 2011; Pichiorri et al., 2008; Roccaro et al., 2009; Zhou et al., 2010). For three of these miRNAs (miR-181b, miR-106b, and miR-25), their expression level was significantly associated with disease risk score (Zhou et al., 2010). MiRNA-181a was also detected at elevated levels in the plasma of MM patients, although plasma levels were not prognostically informative (Huang et al., 2012). Antagomirs directed against the *miR-106b~25* cluster suppressed MM cell viability and colony formation capacity in vitro (Gu et al., 2017a), while inhibition of the miR-181s significantly impaired in vivo tumor growth in an MM xenograft model (Pichiorri et al., 2008) confirming the oncogenic potential of these miRNAs in this disease. Mechanistically, it has been shown that miR-181s and members of the *miR-106b~25* cluster repress expression in MM cells of p300-CBP-associated factor (PCAF), an activator of p53 activity, thus implicating these miRNAs in suppression of p53 signaling in MM (Pichiorri et al., 2008) (see Section 4.5.2). MiR-106b~25 has also been shown to activate MAPK signaling in MM cells (see Section 4.5.3) (Gu et al., 2017a). In contrast to the above findings, miR-181a has been reported to exert a tumor-suppressive effect in MM cell lines in one study (Chen, Hu, Wang, Zhao, & Gu, 2018) (see Section 4.4.4).

TABLE 4.3 Major short noncoding RNAs (sncRNAs) implicated in multiple myeloma.

miRNA	Prognostic significance	Direction of change in disease	Downstream genes (direction of effect in MM)	Pathways affected (direction of effect in MM)	References	Effect of sncRNA on tumor cells and disease
miR-21	YES	↑ In MM tumor cells	↓ PTEN	↑ PI3K/AKT/mTOR signaling	Leone et al. (2013)	Oncogenic: promotes proliferation, drug resistance and bone destruction
			↓ PIAS3	↑ IL6/JAK/STAT signaling	Loffler et al. (2007), Xiong et al. (2012)	
		↑ In BMSC	↓ OPG ↑ RANKL	↑ Osteoclastogenic RANKL signaling (BMSC)	Pitari et al. (2015)	
miR-181a/b	YES	↑ In MM tumor cells[a]	↓ PCAF	↓ p53 signaling	Pichiorri et al. (2008)	Oncogenic: promotes tumorigenesis in vivo
			↓ HOXA11	ND	Shen et al. (2018)	
miR-106b~25 cluster	YES	↑ In MM tumor cells	↓ PCAF	↓ p53 signaling	Pichiorri et al. (2008)	oncogenic: increases viability and colony formation of MM cells
			↓ TP53	↓ p53 signaling	Kumar et al. (2011)	
			↓ CDKN1A/p21	↓ p53 signaling	Zhou et al. (2010)	
			?	↑ MAPK signaling	Gu et al. (2017a)	
miR-17~92 cluster	YES	↑ In MM tumor cells	↓ BCL2L11	↓ Apoptosis	Chen et al. (2011)	Oncogenic: reduces apoptosis, promotes proliferation and tumorigenesis
			↓ SOCS1	↑ IL6/JAK/STAT signaling	Pichiorri et al. (2008)	
			↓ CDKN1A/p21	↓ p53 signaling	Zhou et al. (2010)	
			↓ PTEN	↑ PI3K/AKT/mTOR signaling[b]	Morelli et al. (2018)	

(Continued)

TABLE 4.3 (Continued)

miRNA	Prognostic significance	Direction of change in disease	Downstream genes (direction of effect in MM)	Pathways affected (direction of effect in MM)	References	Effect of sncRNA on tumor cells and disease
miR-125a	ND	↑ In MM tumor cells	↓ TP53	↓ p53 signaling	Leotta et al. (2014)	Oncogenic: promotes proliferation, invasion, and survival
miR-125b	[c]	↓ In MM tumor cells[c]	↑ IRF4, BLIMP-1	↓ Apoptosis	Morelli et al. (2015)	Tumor-suppressive: promotes apoptosis and autophagy-mediated death
miR-15a/16 cluster	YES	↓ In MM tumor cells	↑ CCND1, CCND2, CDC25A	↑ Cell cycling	Roccaro et al. (2009)	Tumor-suppressive: blocks cell cycle and impedes proliferation; impedes angiogenesis in BMEC
			↑ AKT3	↑ PI3K/AKT/mTOR signaling		
			↑ TAB3	↑ NFkB signaling		
			↑ VEGF secretion	↑ Angiogenesis	Sun et al. (2013a)	
ACA11 snoRNA	[d]	↑ In t(4;14)+ MM tumor cells	↓ Other snoRNAs	↓ Oxidative stress	Chu et al. (2012)	Oncogenic: increased proliferation and drug resistance (but hypersensitivity to bortezomib); reduced oxidative stress
			↑ Pre-rRNA, ribosomes	↑ Protein synthesis	Oliveira et al. (2019)	

[a] MiR-181a reported to act as tumor-suppressor in one study (Chen et al., 2018).
[b] Presumptive but not directly demonstrated in MM.
[c] Upregulation of miR-125b in MM has also been reported (Chi et al., 2011; Lionetti et al., 2009; Jiang et al., 2018); high expression correlates with poor survival in (Jiang et al., 2018).
[d] Prognostic significance independent of t(4;14) translocation not known.

BMEC, bone marrow endothelial cells; BMSC, bone marrow stromal cells; miRNA, microRNA; MM, multiple myeloma; ND, not determined; PCAF, p300-CBP-associated factor; PTEN, phosphatase and tensin homolog; NFκB, nuclear factor-kappa B; VEGF, vascular endothelial growth factor

4.4.1.3 miR-17~92 cluster

MiRNA-17~92 is a polycistronic cluster located at human chromosome 13q31.3 that produces six mature miRNAs (miR-17, -18a, -19a, -19b-1, -20a, and -92a-1). Like its paralogous family cluster miR-106b~25 described earlier, miR-17~92 is a well-established oncomir, being implicated in various hematological and solid tumors (Mendell, 2008). Several studies have reported upregulation of the miR-17~92 cluster in tumor plasma cells of MM patients (Chi et al., 2011; Pichiorri et al., 2008; Zhou et al., 2010). Unlike other miRNAs discussed earlier, upregulation of miR-17~92 is not observed during the premalignant MGUS stage, suggesting that this cluster may play an important role during progression to symptomatic disease (Pichiorri et al., 2008). Elevated levels of members of this cluster in tumor plasma cells (Chen et al., 2011; Gao et al., 2012) and patient serum (Huang et al., 2012) are associated with shorter times of progression-free survival, suggesting that this cluster could be a useful noninvasive prognostic marker. In MM, as in other tumors, expression of miR-17~92 is induced by the oncoprotein MYC (Chen et al., 2011; Zhou et al., 2010). Since the miR-19 components of the cluster in turn downregulate the expression of the proapoptotic gene *BCL2L11* (Pichiorri et al., 2008), this cluster plays a central role in the tumor-promoting, antiapoptotic influence of MYC in MM cells (Chen et al., 2011). In addition to its involvement in the MYC pathway, miR-19a and -19b from this cluster also downregulate expression of SOCS1, an inhibitor of IL6 signaling, and so also contribute to the important growth-promoting IL6 signaling pathway in MM cells (Pichiorri et al., 2008) (see Section 4.5.1). MiRNA-17 also downregulates expression in MM cells of the cyclin-dependent kinase inhibitor CDKN1A/p21 that regulates cell-cycle entry into G_1 and is a highly significant indicator of OS in MM patients (Zhou et al., 2010) (see Section 4.5.2). Further, like miR-21 above, the miR-17~92 cluster also represses PTEN, the inhibitor of PI3K/AKT/mTOR signaling, and so may also contribute to this important pathway in MM (Morelli et al., 2018). Thus the miR-17~92 cluster is implicated in multiple signaling pathways controlling MM cell growth and survival.

4.4.1.4 miR-125 family

The miR-125 family, comprising miR-125a and -125b, has been implicated in several aspects of MM etiology. Upregulation of miR-125a has been observed in several molecular subgroups of MM (Gutierrez et al., 2010; Lionetti et al., 2013; Wu et al., 2013). Expression of this miRNA is also dramatically induced upon adhesion of MM cells to BMSCs in vitro (Leotta et al., 2014). Mechanistically, it appears that miR-125a contributes to tumor progression by dampening the p53 signaling pathway (Leotta et al., 2014) (see Section 4.5.2). Inhibition of miR-125a suppresses growth and migration and increases apoptosis of MM cells in a p53-dependent manner. Surprisingly, its homolog miR-125b has been shown in one study to exert a tumor suppressor activity in MM and is downregulated in MM cell lines and tumor cells from some patient subgroups (Morelli et al., 2015). Ectopic expression of miR-125b impaired growth and survival of MM cells and diminished tumor progression in an in vivo xenograft model. The tumor suppressor effect of miR-125b in MM was shown to be mediated by direct silencing of the lymphoid-specific transcription factor IRF4 (interferon regulatory factor 4) and BLIMP-1 (B-lymphocyte-induced maturation protein-1) (Morelli et al., 2015). It should be noted, however, that other studies have reported increased levels of miR-125b in MM patients and subsets (Chi et al., 2011; Jiang, Luan, Chang, & Chen, 2018; Lionetti et al., 2009), suggesting that miR-125b could also play a different tumor-promoting role in some patient subtypes. These findings are consistent with observations in other cancers, where miR-125b can exert distinct functional effects depending on the molecular context (Sun, Lin, & Chen 2013b).

4.4.1.5 miR-15a/16 cluster

The miR-15a/16 cluster at human chr13q14 is an important tumor suppressor, the reduced expression of which has been implicated in a variety of tumors (Liu, Xu, Ou, Liu, & Zhang, 2019). Reduced levels of miR-15a/16 are also observed in tumor plasma cells of symptomatic MM patients (Chi et al., 2011; Roccaro et al., 2009). Increasing miR-15a and miR-16−1 by transfecting MM cell lines with pre-miRNAs reduced proliferation and induced G_1 cell-cycle arrest. Overexpression of this cluster was associated with reduction of the cycle regulators cyclinD1, cyclinD2, and CDC25A together with a decrease of prosurvival PI3K/AKT/mTOR, MAP kinase, and NFκB signaling (Roccaro et al., 2009) (see Section 4.5.3). The 13q14 region containing the miR-15a/16 cluster is commonly deleted in MM patients, although it has been shown that downregulation of these miRNAs is not only determined by the chr13 deletion status (Corthals et al., 2010). In addition to the copy number of chr13q, the levels of these tumor suppressor miRNAs can also be influenced by interactions between MM tumor cells and stromal cells in the BM niche. Roccaro et al. (2013) demonstrated that exosomes produced from BM stromal cells are transferred to MM tumor plasma cells in vitro and in vivo and constitute an important means of signaling between the niche and the tumor. Importantly, while BMSC-derived exosomes from MM patients promoted growth of tumor cells, those from healthy individuals were inhibitory to growth. Among other molecules, BMSC-derived exosomes from healthy individuals contain higher levels of miR-15a and it was shown that this exosomally derived miRNA impairs growth of MM tumor cells (Roccaro et al., 2013). Further, BMSC of *miR-15a/16−1* knockout mice had no growth repressive activity, indicating that the absence of exosome-mediated transfer of this miRNA was permissive for tumor expansion (Roccaro et al., 2013).

4.4.1.6 ACA11

ACA11 (also called SCARNA22) is a 125 nt snoRNA implicated specifically in the etiology of t(4;14)[+] MM. ACA11 is encoded within intron 18−19 of the *MMSET* gene, which becomes upregulated in t(4;14)[+] MM due to its juxtaposition downstream of IgH enhancer regions at the translocation breakpoint (Chu et al., 2012). ACA11 is localized in the nucleolus, where it affects posttranscriptional processing of other snoRNAs, as well as rRNA levels, ribosome biogenesis and protein synthesis (Chu et al., 2012; Oliveira et al., 2019). As a result of these molecular effects, ACA11 increases proliferation of MM cells and reduces oxidative stress and sensitivity to cytotoxic chemotherapy but renders cells more sensitive to the PI bortezomib.

4.4.2 Long noncoding RNAs implicated in multiple myeloma

As described in Section 4.3.2 and Table 4.2, recent studies have documented deregulated expression of large numbers of lncRNAs in MM tumor samples. Although the functional significance of most of these changes remains to be investigated, several lncRNAs have now been shown to play important roles in disease progression and outcome. Many of these lncRNAs have also been implicated in other malignant diseases, although detailed studies suggest that their molecular mechanism(s) of action may often be disease and context dependent. Next, we describe the most important lncRNAs for MM that have emerged to date (summarized in Table 4.4). A more extensive list of lncRNAs implicated in MM is discussed in recent reviews (Butova, Vychytilova-Faltejskova, Souckova, Sevcikova, & Hajek, 2019; Cui, Song, & Fang, 2019).

4.4.2.1 *Metastasis-associated lung adenocarcinoma transcript 1*

The lncRNA MALAT1 (metastasis-associated lung adenocarcinoma transcript 1, also known as lnc-SCYL1−1) is an oncogenic lncRNA that has been implicated in a wide variety of solid and hematological malignancies (Goyal et al., 2021). In MM, increased levels of MALAT1 are detected in BMPC at all disease stages, and high expression of this lncRNA is a prognostic indicator of early progression and poor outcome (Cho et al., 2014; Handa et al., 2017; Ronchetti et al., 2016a). MALAT1 may also be implicated in extramedullary expansion of the tumor, with dramatically increased levels of MALAT1 detected in extramedullary plasma cells compared to intramedullary cells of the same patient (Handa et al., 2017). Knockdown of MALAT1 decreases MM cell proliferation and enhances apoptosis both in vitro and in vivo (Amodio et al., 2018b; Hu et al., 2018; Liu et al., 2017a). MALAT1 has been implicated in various cellular processes in MM cells. MALAT1 interacts with PARP1 and LIG3, components of the alternative nonhomologous end-joining complex and contributes to repair of DNA damage in MM cells (Hu et al., 2018). MALAT1 also regulates the proteasomal machinery in MM cells via an effect on expression of NRF1 and NRF2, transcriptional activators of proteasome subunit genes (Amodio et al., 2018b). MALAT1 has also been implicated in dampening the p53-mediated DNA damage response in MM (Ronchetti et al., 2016a) (see Section 4.5.2) and in inhibiting terminal differentiation of plasma cells by inducing the transcriptional repressor FOXP1 (Gu, Xiao, & Yang, 2017) (see Section 4.4.4). Finally, it has also been shown that this lncRNA interacts with and stabilizes the high mobility group box 1, leading to increased autophagy in MM cells and enhanced chemoresistance (Gao et al., 2017).

4.4.2.2 *Nuclear paraspeckle assembly transcript 1*

NEAT1 (nuclear paraspeckle assembly transcript 1) is a structural lncRNA that is essential for the formation of paraspeckles, ribonucleoprotein structures in the nucleus that regulate expression of certain genes by controlling nuclear retention of mRNA (Bond & Fox, 2009). NEAT1 was found to be upregulated in plasma cells of symptomatic MM patients (Taiana et al., 2019), although this finding has not been consistently observed (Sedlarikova et al., 2017). A follow-up study showed that knockdown of NEAT1 expression inhibited proliferation and triggered apoptosis of MM cells in vitro and in vivo (Taiana et al., 2020). Molecular analyses revealed that depletion of NEAT1 downregulated expression of several genes involved in homologous recombination and DNA repair, including *RAD51B* and *RAD51D*, and MM cells with depleted NEAT1 showed elevated signs of DNA damage. In MM cell lines, NEAT1 has also been implicated in activation of PI3K/AKT/mTOR signaling, an important pathway for MM cell growth and survival (Xu et al., 2018) (see

TABLE 4.4 Major long noncoding RNA (lncRNAs) implicated in multiple myeloma.

lncRNA	Prognostic significance	Direction of change in disease	Downstream genes/interacting partners (direction of effect in MM)	Pathways affected (direction of effect in MM)	References	Effect of lncRNA on tumor cells and disease
MALAT1	YES	↑ In MM tumor cells; ↑ extramedullary v intramedullary tumor cells	Interacts with PARP1 and LIG3	↑ A-NHEJ and DNA repair	Hu et al. (2018)	Oncogenic: promotes proliferation, tumorigenesis, autophagy/chemoresistance and extramedullary tumor expansion
			↓ KEAP1 then ↑ NRF1 and NRF2	↑ Proteasomal activity	Amodio et al. (2018b)	
			Stabilizes HMGB1	↑ Autophagy	Gao et al. (2017)	
			↓ miR-509 then ↑FOXP1	↓ Of plasma cell terminal differentiation[a]	Gu et al. (2017b)	
			?	↓ p53-mediated DNA damage response	Ronchetti et al. (2016a)	
NEAT1	YES	↑ In MM tumor cells	↑ RAD51B and RAD51D	↑ DNA repair	Taiana et al. (2020)	Oncogenic: promotes cell growth and drug resistance
			?	↑ PI3K/AKT/mTOR signaling	Xu et al. (2018)	
PVT1	YES	↑ In MM tumor cells	↑ c-MYC	↑ MYC-driven tumorigenesis[a]	Jin et al. (2019), Handa et al. (2020)	Oncogenic: not extensively characterized

(Continued)

TABLE 4.4 (Continued)

lncRNA	Prognostic significance	Direction of change in disease	Downstream genes/interacting partners (direction of effect in MM)	Pathways affected (direction of effect in MM)	References	Effect of lncRNA on tumor cells and disease
MEG3	YES	↓ In MM tumor cells	↑ miR-181 then ↓HOXA11	?	Shen et al. (2018)	Tumor-suppressive: represses proliferation and tumorigenesis; promotes osteogenic differentiation
		↓ In MSC of MM patients	↓ BMP4	↓ Osteogenic differentiation	Zhuang et al. (2015)	
CRNDE	YES	↑ In MM tumor cells	↑ IL6R	↑ IL6/JAK/STAT signaling	David et al. (2020)	Oncogenic: promotes proliferation, drug resistance, interaction with the niche and tumorigenesis
			↑ CDH2	↑ Adhesion		
			↓ miR-451	?	Meng et al. (2017)	

[a]*Presumptive but not directly demonstrated in MM.*
A-NHEJ, Alternative nonhomologous end joining; MM, multiple myeloma; MSC, mesenchymal stem cells.

Section 4.5.3). Knockdown of NEAT1 also sensitized cells to treatment by bortezomib, carfilzomib, and melphalan, standard agents used in the treatment of MM (Taiana et al., 2020). Thus NEAT1 contributes to MM cell growth and survival through effects on cell signaling and by promoting DNA repair and genome stability.

4.4.2.3 Plasmacytoma variant translocation 1

The gene encoding the lncRNA PVT1 (plasmacytoma variant translocation 1) is located 53 kb downstream of the MYC proto-oncogene at a region on human chr8q24 that is a fragile site frequently affected by genomic rearrangements in cancer (Jin et al., 2019). In other cancers, a complex regulatory cross talk has been observed between these two genes, both at the level of regulation of their expression and the stability of the c-Myc protein (Jin et al., 2019) and it has been shown that PVT1 is important for MYC-driven tumorigenesis in preclinical models (Tseng et al., 2014). In MM, genomic rearrangements affecting the *MYC* gene are detected in 36% of patients at diagnosis and frequently lead to upregulation of both *MYC* and the downstream *PVT1* gene (Mikulasova et al., 2020). At present, the contribution of PVT1 to the role of MYC during MM disease progression remains unclear and warrants further characterization (Handa et al., 2020; Mikulasova et al., 2020). Other rearrangements within the 8q24 region observed in MM occur within the PVT1 gene itself, forming chimaeras with the coding genes NBEA or WWOX (Nagoshi et al., 2012). These rearrangements retain only the first exon of PVT1 and express variant forms of NBEA or WWOX lacking their normal amino termini, suggesting that the PVT1 lncRNA does not contribute to their transforming potential. The pathological consequences of these PVT1 chimaeras for the disease are not clear (Nagoshi et al., 2012).

4.4.2.4 Maternally expressed gene 3

The imprinted lncRNA MEG3 (maternally expressed gene 3) is a well-established tumor-suppressor gene, the expression of which is decreased in a variety of malignant disorders (Ghafouri-Fard & Taheri, 2019). MEG3 influences cell growth and apoptosis via affects on a variety of downstream factors and pathways, including p53, MYC, retinoblastoma, and Transforming Growth Factor Beta (TGFβ). Expression of MEG3 is progressively decreased with disease evolution in MM patients, where lower levels of expression correlate with shorter progression-free and OS times (Shen et al., 2018). As in other cancers, decreased expression of MEG3 in MM is associated with the acquisition of DNA methylation at the MEG3 promoter on the normally active maternal allele (Benetatos et al., 2008; Yu et al., 2020). Overexpression of MEG3 inhibited proliferation of MM cells in vitro and reduced tumorigenesis in a xenograft model (Shen et al., 2018). In MM cells, MEG3 functions as a molecular sponge for the oncogenic miRNA miR-181a, the expression of which is upregulated in MM patients (Shen et al., 2018). Decreased activity of miR-181a leads in turn to upregulation of its target gene HOXA11, encoding a homeobox transcription factor with presumptive tumor suppressor activity. Interestingly, MEG3 is also decreased in bone marrow mesenchymal stem cells (MSC) of MM patients, where its downregulation contributes to impaired osteogenic potential (Zhuang et al., 2015) (see Section 4.6). Thus as well as tumor growth, the depletion of MEG3 may also contribute to the onset of osseous lesions that are a defining feature of this disease.

4.4.2.5 Colorectal neoplasia differentially expressed

Two studies have reported upregulation of the lncRNA CRNDE (Colorectal Neoplasia Differentially Expressed) in MM patient cohorts (David et al., 2020; Meng et al., 2017). CRNDE levels increase with disease progression and high expression of this lncRNA is an indicator of poor prognosis. CRNDE is a broadly expressed lncRNA gene localized on chr16q12, the oncogenic potential of which was first observed in colorectal cancer (Graham et al., 2011), but which has now been implicated in a range of solid and hematological malignancies (Lu et al., 2020). The CRNDE transcript is subject to extensive alternative splicing, with alternative spliceforms exhibiting distinct localizations within MM cells, suggesting that they may exert different molecular roles (David et al., 2020). Depleting CRNDE either by siRNA-mediated knockdown (Meng et al., 2017) or CRISPR-mediated deletion of the *CRNDE* gene (David et al., 2020) inhibited proliferation and increased apoptosis of MM cells. CRNDE deletion also impaired expansion of MM cells in bone marrow of transplanted mice and reduced the formation of extramedullary tumors (David et al., 2020). Molecular analyses indicated that CRNDE affects MM tumorigenesis by influencing the IL6 signaling pathway, as well as the interaction between MM cells and the bone marrow niche (see Sections 4.5.1 and 4.6).

4.4.3 Other ncRNAs implicated in multiple myeloma

Although less well studied, several circRNAs have also been explored for potential oncogenic or tumor suppressor activities in MM patients. Microarray-based profiling of circRNA expression led to the identification of hsa_circ_0007841 (Gao et al., 2019) and hsa_circRNA_101237 (Liu, Tang, Liu, & Wang, 2020), two circRNAs whose expression is increased in bone marrow cells of MM patients. Both of these circRNAs were markers of poor prognosis and hsa_circRNA_101237 was also higher in relapsed/refractory MM than in treatment-naive patients, suggesting a potential contribution to disease progression (Gao et al., 2019; Liu et al., 2020). Both of these circRNAs were implicated in resistance of primary tumors and MM cell lines to treatment with the PI bortezomib, although the mechanistic basis for this observation and indeed for the contribution of these circRNAs to the progression of MM has not yet been elucidated. Other circRNAs have been reported to exert antiproliferative and tumor suppressor effects in MM cells. Levels of hsa_circ_0000190 are decreased in bone marrow and peripheral blood of MM patients, with higher expression of this circRNA being associated with favorable prognosis among MM patients (Feng et al., 2019). Overexpression of hsa_circ_0000190 in two different MM cell lines led to a G_1 cell-cycle block and impaired proliferation in vitro as well as reduced tumorigenesis in murine xenografts. Another circRNA, hsa_circ_0069767, was investigated as a derivative of the *c-KIT* gene, the expression of which is a favorable prognostic marker in MM (Chen et al., 2020). Expression of hsa_circ_0069767 correlates with c-KIT in MM cells, and this circRNA is an independent indicator of longer progression-free and OS. Enforced expression of hsa_circ_0069767 increased apoptosis in an MM cell line and decreased migration and invasion in transwell assays, indicating that this circRNA exercises a tumor suppressor function independently of c-KIT (Chen et al., 2020).

4.4.4 Interactions between ncRNAs in multiple myeloma

As well as directly regulating the expression of downstream target genes, ncRNAs are themselves often involved in cross-regulatory interactive networks. The most common forms of these interactions are those that take place between miRNAs and competitive endogenous RNAs (ceRNAs), also called sponge RNAs, which are often lncRNAs, circRNAs, and transcribed pseudogenes, and which buffer the activity of miRNAs by competitively sequestering them away from their mRNA targets (Bak & Mikkelsen, 2014).

In MM, several antagonistic cross talk relationships between lncRNAs and miRNAs have been implicated in disease evolution and outcome, including the lncRNA CRNDE that can act as a sponge for miR-451 (Meng et al., 2017) and MALAT1 that can diminish the activity of miR-509 (Gu et al., 2017b). Dampening of miR-509 by MALAT1 upregulates expression of its target gene *FOXP1*, encoding a repressor of plasma cell terminal differentiation (van Keimpema et al., 2015), thereby adding to the mechanistic pathways by which MALAT1 is likely to influence disease progression in MM (Gu et al., 2017b) (see also Section 4.4.2.1). In another recent study, the lncRNA PCAT1 (prostate cancer-associated transcript 1) was shown to diminish the tumor suppressor miR-129, a miRNA that decreases in primary MM tumor cells and cell lines (Shen et al., 2020). The decrease of miR-129 and resulting activation of its downstream target gene *MAP3K7* promoted MM cell proliferation and survival via an effect on the NFκB pathway (see also Section 4.5.3). Another example of an important sponging interaction involves circRNA hsa_circ_0000190, the expression of which is decreased in MM and which inhibits MM proliferation and tumorigenicity (see Section 4.4.3) (Feng et al., 2019). This circRNA exerted its tumor-suppressor activity by sponging the miRNA miR-767 and thereby increasing expression of its downstream target gene MAPK4, which in turn blocked cell-cycle progression. The lncRNA CCAT1, which is upregulated in MM cell lines and primary tumor cells, and is associated with poor OS, can act as a ceRNA to repress the activity of the miRNA miR-181a (Chen et al., 2018). Knockdown of CCAT1 reduced the proliferation of MM cells in vitro and in vivo, and this effect was neutralized by concurrent inhibition of miR-181a, indicating that CCAT1 exerted its effects on MM cell growth by repression of this miRNA. However, this finding contrasts with a previous study that observed that miR-181a promotes MM tumorigenesis (Pichiorri et al., 2008), and the significance of this CCAT1/miR-181a interaction will require further exploration.

Expanding beyond the sponging interactions that have been identified using candidate approaches, Ronchetti, Manzoni, Todoerti, Neri, and Agnelli (2016) applied in silico target prediction algorithms to published lncRNA and miRNA expression datasets from 95 MM patients and 4 normal controls and identified 11 lncRNA/miRNA pairs predicted to be involved in ceRNA interactions. These interactions and their functional significance will require further investigation in MM cells.

4.5 Cell signaling pathways modulated by noncoding RNAs in multiple myeloma

Many of the ncRNAs discussed earlier impact the disease by affecting major signaling pathway(s) that control proliferation and survival of MM cells. Conversely, changes in ncRNA expression are frequently driven by upstream signaling pathways. Here we focus on the important interactions that have been observed in MM cells between various ncRNAs and the major signaling pathways controlling disease progression and outcome.

4.5.1 IL6/JAK/STAT signaling and noncoding RNAs in multiple myeloma

IL6 is a major promoter of MM cell growth and survival (Matthes, Manfroi, & Huard, 2016; Rosean et al., 2014). In addition to autocrine signaling, tumor plasma cells also receive IL6 from stromal cells in the surrounding bone marrow niche. Binding of IL6 to its receptor activates several downstream signaling pathways (Heinrich et al., 2003). In particular, activation of the JAK/STAT3 pathway triggers translocation of phosphorylated STAT3 into the nucleus where it upregulates genes critical for growth and survival of MM cells, including MCL1, BCL-xl, and MYC (Gupta et al., 2017; Puthier, Bataille, & Amiot, 1999) (Fig. 4.1). IL6/JAK/STAT3 signaling is repressed by proteins of the SOCS family that inhibit STAT3 phosphorylation (Kang, Tanaka, Narazaki, & Kishimoto, 2019), and epigenetic silencing of SOCS1 due to hypermethylation of its promoter has been observed in MM patients (Chim, Fung, Cheung, Liang, & Kwong, 2004; Galm, Yoshikawa, Esteller, Osieka, & Herman, 2003). Due to its critical role in MM disease progression and drug resistance, the IL6 signaling pathway has been the target of several therapeutic strategies (Matthes et al., 2016).

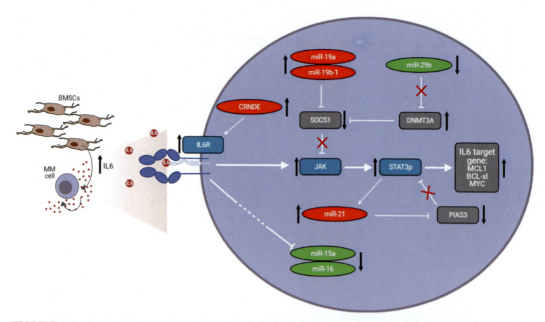

FIGURE 4.1 Interactions between ncRNAs and IL6/JAK/STAT signalling in multiple myeloma. Noncoding RNAs with tumor-promoting and tumor-suppressing activity in MM are indicated as red and green ovals respectively. Black arrows indicate the direction of change in expression in MM. Normal activating and inhibitory interactions between pathway components are shown in white, with red crosses indicating where these interactions are impaired in MM. Figure was created using bioRENDER (https://biorender.com/).

Noncoding RNAs have now been implicated at several points in the IL6 signaling pathway, both as downstream mediators and upstream regulators (Fig. 4.1). One of the important ncRNAs implicated in this pathway is the miRNA miR-21, which is upregulated in MM patients (see Section 4.4.1.1) and whose host gene is a direct transcriptional target of STAT3 in MM cell lines (Loffler et al., 2007). This study showed that the STAT3-mediated upregulation of miR-21 is an important component of the prosurvival influence of IL6 on MM cells. Interestingly, miR-21 directly silences PIAS3, encoding an inhibitor of STAT3, thus forming a positive feed-forward loop to enhance the effects of IL6/JAK/STAT signaling in MM cells (Xiong et al., 2012). Paracrine IL6 from BMSC has also been shown to drive silencing of the tumor suppressors miRNA-15a/-16 in MM cells, also contributing to the protective effects of IL6 (Hao et al., 2011). The mechanistic basis for repression of these miRNAs by IL6 has not yet been elucidated.

In addition to acting downstream of IL6 signals, ncRNAs can also act as upstream regulators of the IL6 signaling pathway. MiRNA miR-29b, the expression of which is reduced in MM patients and cell lines (Zhang et al., 2011), contributes to epigenetic regulation of the *SOCS1* gene encoding the repressor of JAK signaling (Amodio et al., 2013). Amodio et al. showed that mimics of miR-29b increased SOCS1 expression and inhibited STAT3 phosphorylation. These effects were due to silencing of *DNMT3A*, a miR-29b target gene encoding the DNA methyltransferase 3A, thereby inducing demethylation of the SOCS1 promoter (Amodio et al., 2013). In contrast to the tumor suppressor role of miR-29b in dampening IL6 signaling via derepression of SOCS1, the oncogenic miRNAs miR-19a and miR-19b-1, which are upregulated in MM as part of the miR-17~92 cluster (see Section 4.4.1.3), are thought to enhance IL6 signaling by silencing SOCS1 (Pichiorri et al., 2008). Thus the tumor-promoting and prosurvival effects of IL6 are likely to be mediated by multiple changes in miRNA expression (Fig. 4.1). A regulatory effect on the IL6 signaling pathway has also recently been demonstrated for the lncRNA CRNDE, which is upregulated in tumor plasma cells, and which promotes MM cell proliferation, survival, and tumorigenesis (David et al., 2020). Transcriptomic profiling revealed that CRNDE activates expression of the IL6 receptor and CRNDE-deleted MM cells displayed reduced phosphorylation of STAT3 and IL6-dependent growth (David et al., 2020). In addition, MM cells lacking CRNDE were also less able to induce IL6 secretion by BMSC in coculture assays, indicating that CRNDE influences IL6 signaling to tumor cells via both cell-intrinsic and cell-extrinsic mechanisms (Fig. 4.1).

4.5.2 Noncoding RNAs and the p53 pathway in multiple myeloma

Disruption or inactivation of the p53 pathway is one of the primary drivers of drug resistance in MM (Jovanovic et al., 2018). The pathway can be impaired by a variety of different mechanisms, including chromosomal deletion or mutations affecting the TP53 gene, epigenetic silencing of its promoter, and overexpression of the p53 inhibitor MDM2. Similar to the IL6 pathway, it is clear that ncRNAs are also implicated as both mediators and regulators of p53 activity, and that dysregulation of ncRNAs is another important mechanism for disruption of the p53 pathway in MM cells (Fig. 4.2).

Several miRNAs, in particular miR-34a, miRs-192, -194, and -215, are directly induced by the p53 pathway and contribute to its tumor-suppressing activity in MM cells (Di

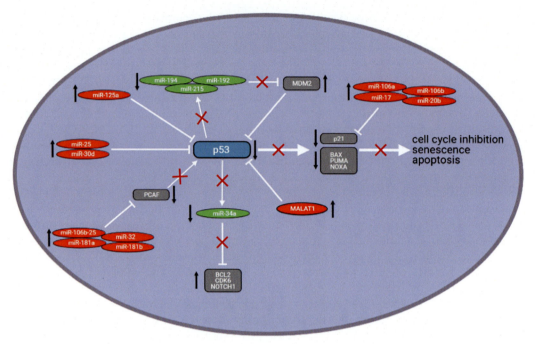

FIGURE 4.2 **Interactions between ncRNAs and p53 signalling in multiple myeloma.** Annotations are as described in Figure 4.1. Figure was created using bioRENDER (https://biorender.com/).

Martino et al., 2012; Pichiorri et al., 2010). Treatment with mimics of miR-34a reduced MM cell proliferation by silencing its downstream target genes *CDK6*, *BCL2*, and *NOTCH1*, even in a p53 mutant background (Di Martino et al., 2012). Pichiorri et al. showed that miRs-192, -194, and -215 are produced from two related clusters that are directly activated by p53 and also target and silence expression of the p53 inhibitor MDM2. These miRNAs therefore establish an important feed-forward regulatory loop that contributes to p53-mediated cell-cycle and growth arrest (Pichiorri et al., 2010). Interestingly, expression of all three of these miRNAs is reduced at the transition from MGUS to full MM, suggesting that decline of p53 signaling due to disruption of this feed-forward loop may be important for progression to symptomatic disease (Pichiorri et al., 2010).

Direct or indirect effects on p53 activity have also been demonstrated for several other miRNAs in MM cells. As discussed in Section 4.4.1.4, the miRNA miR-125a is upregulated in specific subtypes of MM, where it promotes cellular proliferation and survival (Leotta et al., 2014). This study demonstrated that miR-125a targets the 3′ untranslated region (UTR) of the p53 transcript, and that inhibition of miR-125a increased expression of p53 and its downstream target genes, inducing growth arrest and apoptosis in p53^{WT} MM cells. Two other miRNAs, miR-25 (from the miR-106b~25 cluster) and miR-30d, have also been shown to directly repress p53 expression in an MM cell line (Kumar et al., 2011). Expression of p53 is also silenced indirectly by several miRNAs (the miR-106b~25 cluster, miR-32, miR-181a, and -181b) that are upregulated in MM patients and that target PCAF, a positive regulator of p53 activity (Pichiorri et al., 2008). Inhibition of these miRNAs increased expression of PCAF and improved p53 responses in MM cell lines. MiRNAs also regulate expression of the *CDKN1A/p21* gene encoding the key cyclin-dependent kinase inhibitor responsible for cell-cycle arrest in response to p53 signaling as well as other cellular stresses. Zhou et al. (2010) identified four paralogous miRNAs expressed in MM cells, miR-106b (from the miR-106b~25 cluster), miR-17 (from the miR-17~92 cluster), as well as miR-106a and miR-20b that decrease p21 protein levels when overexpressed. All four of these miRNAs were associated with a high disease risk score.

Although less well studied, several lncRNAs have also been implicated in p53 responses in MM cells. Consistent with observations in other cancers, evidence suggests that the oncogenic lncRNA MALAT1 represses the p53 pathway in MM, since higher expression of MALAT1 in tumor plasma cells was associated with diminution of a p53-DNA damage response gene expression signature (Ronchetti et al., 2016a).

Conversely, the lncRNA MEG3, the expression of which is decreased in MM due to promoter hypermethylation (see Section 4.4.2.4), has been shown in other cancers to exert tumor-suppressor effects by activating p53 (Zhou et al., 2007), although this mechanism has not yet been directly demonstrated in MM.

4.5.3 The influence of noncoding RNAs on PI3K/AKT/mTOR, MAPK, and NFκB signaling in multiple myeloma

Noncoding RNAs are also implicated in several other major signaling pathways important in the pathogenesis of MM. Activation of the PI3K/AKT/mTOR signaling pathway is observed in a large proportion of MM patients, where it contributes to tumor growth, drug resistance, and disease relapse (Ramakrishnan & Kumar, 2018). Activation of this pathway can occur through numerous mechanisms, several of which involve the activity of short or long ncRNAs. The miRNA miR-21, which is expressed at high levels in MM cell lines and tumor plasma cells (see Section 4.4.1.1) represses expression of the tumor suppressor gene *PTEN*, an important inhibitor of the PI3K/AKT/mTOR signaling (Leone et al., 2013). Inhibition of miR-21 diminished tumorigenic growth in a xenograft model of MM, which was associated with increased levels of PTEN protein and reduced phosphorylation of AKT (Leone et al., 2013). PTEN is also a validated target of repression by the miR-17~92 cluster (Mogilyansky & Rigoutsos, 2013), which is also upregulated in MM, and inhibition of this cluster leads to activation of PTEN expression in MM cells (Morelli et al., 2018). As in other cancers (Guo et al., 2018), the protumorigenic effects of NEAT1 in MM are also associated with activation by this lncRNA of PI3K/AKT/mTOR signaling (Xu et al., 2018). Conversely, the tumor suppressor miRNAs miR-15a/16, which are decreased in MM plasma cells, can inhibit PI3K/AKT/mTOR signaling by repressing expression of AKT3 (Roccaro et al., 2009). The MAPK signaling pathway is affected by the miR-106b~25 cluster that is upregulated early in MM progression. Inhibition of this miRNA cluster impairs phosphorylation of p38-MAPK and reduces viability and colony-forming capacity of MM cells (Gu et al., 2017a). It should also be noted that since both PI3K/AKT/mTOR and MAPK pathways are downstream of IL6 activation, they could also be affected by the ncRNAs influencing the IL6 signaling pathway described earlier. Consistent with this, the reduced expression of the IL6 receptor observed in MM cells lacking the lncRNA CRNDE was associated with gene expression changes indicating downregulation of PI3K/AKT/mTOR and MAPK as well as JAK/STAT signaling (David et al., 2020). Finally, ncRNAs also impact NFκB signaling, another important pathway in the biology of MM (Dehghanifard et al., 2018). The tumor-suppressor miRNAs miR-15a/16 repress NFκB signaling in MM cells by silencing expression of the NFκB-activator, MAP3KIP3 (also called TAB3) (Roccaro et al., 2009). As described earlier (Section 4.4.4), it has also been reported that the lncRNA PCAT1 promotes MM cell proliferation and survival by activating the NFκB pathway, via an indirect effect on MAP3K7 (Shen et al., 2020).

4.6 Noncoding RNAs affecting interactions with the bone marrow niche

MM tumor plasma cells reside in the bone marrow within a highly heterogeneous niche that comprises both cellular components [including BMSC, osteoclasts, osteoblasts, bone marrow endothelial cells (BMEC), inflammatory, and other immune cells] and noncellular components (including the extracellular matrix, cytokines, chemokines, growth factors, and other signaling molecules). This microenvironment exerts a strong influence on progression of the disease and resistance to treatment (Hideshima, Mitsiades, Tonon, Richardson, & Anderson, 2007; Lomas, Tahri, & Ghobrial, 2020; Manier et al., 2012). Conversely, the tumor cells in turn remodel the surrounding microenvironment to establish a highly supportive neoplastic niche (Hideshima et al., 2007).

Several important ncRNAs have been shown to contribute to MM disease progression by affecting the niche and its interactions with tumor plasma cells. An important example is provided by the tumor suppressor miR-15a/16 cluster, the expression of which is reduced in tumor plasma cells. In addition to enhancing PI3K/AKT/mTOR and NFκB signaling as described in Section 4.5.3, reduction of miR-15a/16 also upregulates expression of *VEGF* (vascular endothelial growth factor), the transcript of which is a direct target of these miRNAs (Sun et al., 2013a). Elevated secretion of VEGF by MM plasma cells following inhibition of miR-15a/16 enhances the angiogenic potential of BMEC in cocultures in vitro and increases angiogenesis and tumor development in an in vivo model. In another example, it has been shown that deletion of the lncRNA CRNDE decreases the adhesive properties of MM cells and impairs their interaction with stromal cells in coculture experiments (David et al., 2020). Gene expression profiling revealed dysregulation of signatures related to cell–cell and cell–substrate adhesion in the absence of CRNDE. In particular, CRNDE activates expression of CDH2, encoding the cell adhesion molecule N-cadherin (David et al., 2020), which has previously been identified as a marker of poor prognosis in MM patients (Vandyke, Chow, Williams, To, & Zannettino, 2013). MM cells lacking CRNDE were less effective at inducing IL6 secretion from BMSCs than CRNDE^{WT} cells, suggesting that the influence of CRNDE on the adhesive properties of MM tumor cells could also feed-forward to further enhance IL6 signaling (David et al., 2020). Another important lncRNA within the bone marrow niche is the tumor-suppressing lncRNA MEG3, which has been shown to promote osteoblastic differentiation of MSC, by activating expression of BMP4 (bone morphogenetic protein 4) (Zhuang et al., 2015). In addition to being reduced in tumor plasma cells (see Section 4.4.2.4), expression of MEG3 is also decreased in MSC of MM patients, contributing to the impaired osteogenic differentiation that is an important clinical feature of this disease. Another ncRNA that regulates osteogenesis in MM patients is the miRNA miR-21. As described earlier, expression of miR-21 is increased in plasma cells of MM patients where it promotes growth and survival (Leone et al., 2013). Pitari et al. showed that expression of this miRNA is also strongly induced in BMSC when cocultured with MM cell lines (Pitari et al., 2015). In the stromal cells, miR-21 inhibits expression of the bone-protecting factor OPG and indirectly enhances the osteoclastogenic factor Receptor activator of nuclear factor kappa-B ligand (RANKL), with the resulting imbalance of these factors contributing to the characteristic bone destruction observed in MM.

In many other cases, interactions between ncRNAs and the niche are mediated via cell-extrinsic mechanisms, in which ncRNAs (particularly miRNAs) are expressed and released from one cell type to exert their activity in another (Raimondi et al., 2016). MiRNAs can be transmitted between cells either entrapped in microvesicles or other shedded bodies, or in a free form, complexed only with the AGO2 protein (Turchinovich, Samatov, Tonevitsky, & Burwinkel, 2013). One miRNA that is shed from MM cells, miR-135b, has been shown to promote angiogenesis in the hypoxic bone marrow environment by repressing expression in BMEC of FIH-1, an inhibitor of Hypoxia-inducible factor 1-alpha (HIF1α) (Umezu et al., 2014). Tumor-promoting miRNAs such as miR-181a/b and miR-21 may also be transferred to malignant plasma cells from cells of the surrounding niche, such as bone marrow adipocytes, via an exosome-mediated mechanism (Soley, Falank, & Reagan, 2017). Adding further complexity to the potential regulatory effects of extracellular miRNAs, studies in other cell types indicate that these molecules can exert noncanonical functions in the receiving cell, for example, by engaging Toll-like receptors to trigger proinflammatory signaling (Fabbri et al., 2012), although, at present, these noncanonical pathways have not been explored in MM.

4.7 Noncoding RNAs as diagnostic and prognostic biomarkers in multiple myeloma

Many ncRNAs stably persist in body fluids such as serum and urine, and there is considerable interest in the use of ncRNAs as easily accessible diagnostic biomarkers and prognostic predictors of disease evolution and outcome. Profiling of miRNA levels in serum samples from MM patients and healthy donors has identified a number of miRNAs which show high diagnostic power in receiver operating characteristic analyses (reviewed in Zhu et al., 2018). In particular, serum levels of miR-720 and -1308 (Jones et al., 2012), miR-29a (Sevcikova, Kubiczkova, Sedlarikova, Slaby, & Hajek, 2013) as well as a combination of miR-19a with miR-4254 (Hao et al., 2015) are able to discriminate MM patients from healthy donors with high sensitivity and specificity. The study by Hao et al. also revealed that the serum level of miR-135b was able to distinguish the presence or absence of bone disease in MM patients (Hao et al., 2015). Among lncRNAs, TUG1 (taurine-upregulated gene 1) (Yin, Shen, Cui, & Ju, 2019), PCAT1 (Shen et al., 2017), and H19 (Pan et al., 2018) are also be detected at elevated levels in the serum of MM patients and have been proposed as biomarkers that could improve MM diagnosis. It was found that the serum levels of TUG1 or PCAT1 provided higher diagnostic sensitivity and specificity than traditional biomarkers (serum β2 microglobulin and albumin) and significantly improved the diagnostic accuracy of these markers when used in combination (Shen et al., 2017; Yin et al., 2019). While these findings require validation in larger cohorts, these ncRNAs represent potentially useful auxiliary biomarkers to improve the accuracy of MM diagnosis.

Signatures based on the expression of either sncRNAs or lncRNAs have also been described that demonstrate significant power to predict disease prognosis and outcome. Based on miRNA expression profiles generated in BMPC of 153 MM patients at diagnosis, Wu et al. identified an "outcome classifier" based on the expression of only two miRNAs (miR-17 and miR-886), which was able to accurately predict disease risk in univariate analysis (Wu et al., 2013). Multivariate analysis revealed that this 2-miRNA-based classifier was independent of and more prognostically significant than other prognostic indicators based on gene expression profiling (Decaux et al., 2008; Dickens et al., 2010), International Staging Scheme (ISS) classification or cytogenetics (Wu et al., 2013). Several studies have also reported prognostically predictive changes in miRNAs that can be detected outside tumor plasma cells, either in extracellular supernatants from bone marrow aspirates or in circulating exosomal isolates or whole serum (Hao et al., 2016; Huang et al., 2012; Kubiczkova et al., 2014; Manier et al., 2017b; Wang et al., 2015). Although the prognostic signatures identified by these studies were largely distinct, some consistent changes observed include decreased circulating levels of miR-106b, miR-20a, miR-16, let-7b, and let-7e (Kubiczkova et al., 2014; Manier et al., 2017b; Wang et al., 2015). The prognostic utility of these biomarkers will need to be verified in larger patient cohorts.

The expression of lncRNAs in BMPC has also been investigated as potential biomarkers of disease risk and prognosis. Some prognostically informative lncRNAs are discussed in Section 4.4.2 and include MALAT1 (Cho et al., 2014), CRNDE (David et al., 2020; Meng et al., 2017), and NEAT1 (Yu, Peng, Chen, Han, & Luo, 2020), all associated with poor prognosis, and MEG3 (Shen et al., 2018), for which high expression in MM tumor cells is a good prognostic indicator. Four other studies have extracted multicomponent lncRNA signatures from transcriptomic datasets that are reported to accurately and sensitively distinguish poor and good prognosis outcomes among MM patients (Hu, Huang, Zhang, & Shen, 2017; Zhong, Liu, Li, Liao, & Li, 2020; Zhou et al., 2015) (Table 4.5). By mining an existing microarray transcriptomic dataset, Hu et al. extracted a signature of 176 lncRNAs, the expression levels of which could accurately distinguish between favorable and poor outcome in an independent patient set (Hu et al., 2017). Taking a similar approach, the studies of Zhou et al. (2015) and Zhong et al. (2020) identified more refined signatures comprising 4 or 7 lncRNAs, respectively, which could accurately predict survival in MM patients. The 7 lncRNA signature predicted disease risk with higher accuracy than two previously published risk models based on the expression of coding genes (Zhong et al., 2020). More recently, an RNA-seq study focusing on lincRNA expression identified a signature comprising 14 lincRNAs, the expression levels of which distinguished patients with poor versus favorable prognosis with high accuracy (Samur et al., 2018). This signature was able to distinguish high- and low-risk subgroups even within risk groups previously defined based on other prognostic indicators, such as ISS status, high-risk cytogenetic status, and minimum residual disease status. As above, it was found that the lincRNA signature added additional prognostic power when used in combination with a signature (EMC92) (Kuiper et al., 2012) based on expression of coding genes (Samur et al., 2018).

TABLE 4.5 Prognostic long noncoding RNA (lncRNA) signatures in multiple myeloma.

Study	Experimental platform	Ensembl gene ID	Gene symbol	Genomic coordinates (hg19)	Correlation with disease risk
Zhou et al. (2015)	Affymetrix HG-U133 Plus 2.0 array	ENSG00000237481	RP4–803 J11.2/AL117350	chr1:229455150–229458834	Correlate
		ENSG00000230424	RP1–43E13.2/EMC1-AS1	chr1:19536995–19567198	Correlate
		ENSG00000259976	RP11–553 L6.5/AC093010.2	chr3:114033348–114048561	Anticorrelate
		ENSG00000233070	ZFY-AS1	chrY:2834885–2870667	Anticorrelate
Hu et al. (2017)[a]	Affymetrix HG-U133 Plus 2.0 array	ENSG00000235169	SMIM1[b]	chr 1: 3,689,352 3,692,546	Correlate
		ENSG00000286656	AC008875.2	chr5:42,985,503–42,993,435	Correlate
		ENSG00000220785	MTMR9L	chr1:32,697,259–32,707,282	Correlate
		ENSG00000247271	AC069360.2/BED5-AS1	chr11:10,879,806–10,900,823	Correlate
		ENSG00000275198	AL512791.2	chr14:90,849,868–90,854,251	Correlate
		ENSG00000223975	AP001048.1	chr21:44,885,189–44,887,178	Correlate
		ENSG00000230006	ANKRD36BP2	chr2:89,065,324–89,106,126	Correlate
		ENSG00000132832	AL139352.1	chr20:43,285,092–43,324,737	Correlate
		ENSG00000176728	TTTY14	chrY:21,034,387–21,239,302	Correlate
		ENSG00000177989	ODF3B[b]	chr22:50,968,838–50,970,543	Correlate
		ENSG00000235590	GNAS-AS1	chr20:57,393,974–57,425,958	Anticorrelate
		ENSG00000240204	SMKR1[b]	chr7:129,142,320–129,152,759	Anticorrelate
		ENSG00000244879	GABPB1-AS1	chr15:50,655,998–50,660,476	Anticorrelate
		ENSG00000185477	GPRIN3[b]	chr4:90,166,086–90,172,345	Anticorrelate

(Continued)

TABLE 4.5 (Continued)

| Study | Experimental platform | lncRNAs included in the signature ||| Correlation with disease risk |
		Ensembl gene ID	Gene symbol	Genomic coordinates (hg19)	
		ENSG00000228649	SNHG26	chr7:22,897,143–22,898,161	Anticorrelate
		ENSG00000265148	BZRAP1-AS1/TSPOAP1-AS1	chr17:56,402,811–56,431,077	Anticorrelate
		ENSG00000238083	LRRC37A2[b]	chr17:44,620,700–44,622,797	Anticorrelate
		ENSG00000143401	ANP32E[b]	chr1:150,190,861–150,192,882	Anticorrelate
		ENSG00000244879	GABPB1-AS1	chr15:50,647,664–50,650,501	Anticorrelate
		ENSG00000261716	RP11–196G18.22	chr1:149,816,066–149,820,591	Anticorrelate
Zhong et al. (2020)	Affymetrix HG-U133 Plus 2.0 array	ENSG00000260213	AC092718.2	chr16:81,050,397–81,058,149	Correlate
		ENSG00000277332	AC108002.2	chr8:143,719,949–143,722,027	Correlate
		ENSG00000285407	AL033530.1	chr1:69,144,884–69,409,074	Correlate
		ENSG00000269621	AL589765.7	chr1:151,728,017–151,732,387	Correlate
		ENSG00000248874	C5orf17	chr5:23,951,457–24,178,374	Correlate
		ENSG00000265148	BZRAP1-AS1 / TSPOAP1-AS1	chr17:56402811–56430647	Correlate
		ENSG00000229719	MIR194–2HG	chr11:64,655,934–64,660,921	Anticorrelate
Samur et al. (2018)	RNA-seq	ENSG00000255284	AP006621.5/LOC171391	chr11:777578–784297	Correlate
		ENSG00000213468	FIRRE	chrX:130822807–130964632	Correlate
		ENSG00000267321	RP11–1094M14.11/SNHG30	chr17:33895128–33901862	Correlate
		ENSG00000231551	RP11–495P10.1	chr1:147887350–147901125	Correlate

(Continued)

TABLE 4.5 (Continued)

Study	Experimental platform	lncRNAs included in the signature			
		Ensembl gene ID	Gene symbol	Genomic coordinates (hg19)	Correlation with disease risk
		ENSG00000225975	LINC01534	chr19:37176341–37178351	Correlate
		ENSG00000203356	LINC01562	chr1:51660767–51700768	Correlate
		ENSG00000249550	LINC01234	chr12:114182382–114211488	Correlate
		ENSG00000178977	LINC00324	chr17:8,123,948–8,127,361	Anticorrelate
		ENSG00000259673	IQCH-AS1	chr15:67695957–67814182	Anticorrelate
		ENSG00000270641	TSIX	chrX:73012040–73049066	Anticorrelate
		ENSG00000274561	RP11–147L13.12/ AC005332.3	chr17:66127603–66128048	Anticorrelate
		ENSG00000242258	LINC00996	chr7:150130742–150145228	Anticorrelate
		ENSG00000278175	GLIDR	chr9:41,948,514–41,955,177	Anticorrelate
		ENSG00000260923	AC137934.1/LINC02193	chr16:90239625–90289081	Anticorrelate

[a]Showing the top 20 most significant (based on Z score) of a total of 176 lncRNAs.
[b]Indicates likely coding gene (based on coordinates published in the study).

Surprisingly, there is very little overlap between the individual elements of the predictive lncRNA signatures identified in these four studies (Zhong et al., 2020) (Table 4.5). These variable findings may reflect differences in the patient cohorts examined and experimental platforms used, as well as different statistical analyses (Zhong et al., 2020). Further, at present, the biological function and mechanism(s) of action for most of the lncRNAs identified in these studies remain to be determined. It will therefore be important to validate these prognostic signatures in larger and independent patient cohorts and to carry out thorough functional characterization of ncRNAs validated as prognostically informative. Moreover, these existing signatures are based only on progression-free or OS and do not consider different treatment regimes. Malek, Kim, and Driscoll (2016) specifically explored changes in the expression of ncRNAs in an MM cell line upon acquisition of resistance to bortezomib, carfilzomib, or ixazomib, three PIs used in the treatment of MM. This study identified changes in both miRNAs and lncRNAs that were consistently altered in PI-resistant cells, many of which showed similar changes of expression in primary MM tumor samples. Identifying ncRNAs implicated in the response and outcome following specific treatments will improve the use of these molecules as predictive biomarkers and could eventually lead to more personalized therapeutic regimes.

4.8 Therapeutic potential of noncoding RNAs in multiple myeloma patients

Beyond their utility as diagnostic and prognostic biomarkers, ncRNAs also present new possibilities for the development of novel therapeutic molecules to be used in combination with established regimes for the treatment of MM. Although, at the time of writing, ncRNA-based therapies for MM have not yet been developed as far as clinical trials, there are many encouraging findings from preclinical studies, including some in animal models. Among other notable findings, Morelli et al. describe a Locked nucleic acid (LNA)-gapmeR reagent (MIR17PTi) that inhibits the six mature miRNAs of the miR-17~92 cluster implicated in several signaling pathways important for disease progression (Morelli et al., 2018) (see Sections 4.4.1.3 and 4.5). MIR17PTi displayed favorable pharmacokinetic and safety profiles in nonhuman primates and strongly inhibited the growth of human MM cells in four clinically relevant mouse models. Another promising therapeutic target is miR-34a, which is downstream of p53 activation and represses the growth of MM tumor cells (Di Martino et al., 2012) (see Fig. 4.2). Effective delivery of miR-34a mimics to MM cells has been achieved using both lipid vesicle- (Di Martino et al., 2014) and nanoparticle-based approaches (Cosco et al., 2015), impairing tumor growth and improving survival in mouse xenograft models of human MM. Similarly, intratumor or systemic delivery of mimics of the tumor suppressor miRNA miR-125b exerted a significant antitumor effect and improved survival of Nonobese Diabetic / Severe Combined Immunodeficiency (NOD/SCID) mice engrafted with human MM cells (Morelli et al., 2015). These effects were due to silencing by miR-125b of the lymphocyte-specific transcription factor IRF4, which forms part of a critical autoregulatory loop with the MYC proto-oncogene in MM cells (Shaffer et al., 2008). The lncRNA MALAT1, which has been implicated in a variety of malignant diseases including MM (see Section 4.4.2.1), has been the subject of extensive efforts to develop therapeutic reagents (Amodio et al., 2018a). Targeting MALAT1 using an LNA-gapmeR antisense oligonucleotide impaired growth of MM tumor cells in a xenograft model (Amodio et al., 2018b). Several nonnucleic acid inhibitors of MALAT1 are also under development, in particular targeting a conserved triple helix structure at the 3′ end of the MALAT1 transcript that is critical for its stability (Brown et al., 2014; Wilusz et al., 2012). Small molecules targeting this triple helix structure have been identified in library screens (Abulwerdi et al., 2019; Donlic, Morgan, Xu, Liu, Roble, Jr., Hargrove, 2018), at least one of which was shown to reduce MALAT1 and impair expression of its downstream target genes in cellular models (Abulwerdi et al., 2019). These molecules have not yet been tested in MM cells. Opportunities for therapeutic intervention on lncRNAs also lie in the identification or design of small molecules that inhibit their interactions with the partner proteins through which they exert their biological activity (Pedram Fatemi et al., 2015). Advances in this direction will require thorough molecular characterization of the lncRNAs implicated functionally in the etiology of MM and identification of their critical partner proteins.

It is clear that therapeutic targeting of ncRNAs can improve the efficacy of conventional agents when used in combinations. MIR17PTi, the LNA inhibitor of the miR-17~92 cluster discussed earlier, which was effective when used as a single agent, exhibited further synergistic effects when used in combination with the conventional agents, dexamethasone and melphalan (Morelli et al., 2018). Systemic delivery of antagonists of miR-221/222, a family of miRNAs implicated in the acquisition of drug resistance phenotypes in MM cells, improved sensitivity of MM tumor growth to dexamethasone (Zhao et al., 2015) and melphalan (Gulla et al., 2016) in mouse xenotransplants. In another example of therapeutic synergy, it has been shown that ectopic reexpression of miRNAs miR-*192*, *-194* and *-215* that target MDM2, the repressor of p53 signaling (see Fig. 4.2), increased the sensitivity of MM tumor xenotransplants to MI-219, an inhibitor of MDM2, due to synergistic activation of the p53 pathway (Pichiorri et al., 2010). In in vitro assays, augmentation of the tumor suppressor miR-29b has been shown to increase sensitivity of MM cell lines and primary tumor cells to the PI bortezomib (Jagannathan et al., 2015). Conversely, inhibition of the oncogenic miRNA miR-21 increased the sensitivity of MM cell lines to bortezomib as well as the chemotherapeutic agent doxorubicin, and the glucocorticoid dexamethasone (Wang et al., 2011). Similarly, inhibition of the lncRNA NEAT1 has been shown to increase the sensitivity of MM cell lines to bortezomib, carfilzomib, melphalan, and dexamethasone (Che, Ye, Wang, Ma, & Wang, 2021; Taiana et al., 2020; Wu & Wang, 2018). Many of these synergistic interactions between modifiers of ncRNAs and conventional therapies will need to be validated in preclinical animal models before possible clinical trials. However, taken together, these findings suggest that in the future, novel reagents targeting ncRNAs will form an important complement for existing therapies.

4.9 Summary and conclusion

As in other malignant diseases, recent years have observed a rapid advance in our understanding of the contribution of ncRNAs to the onset and progression of MM and suggest that targeting these molecules could lead to improved treatment of this currently incurable disease. NcRNAs are likely to form an important component of the personalization of MM treatment, in which complete molecular profiling will facilitate the selection of specific reagents, including those targeting appropriate ncRNAs, to complement existing immunomodulatory drugs, PIs, and other emerging immunotherapies. Challenges that will need to be addressed to fulfill the potential of ncRNA-based therapies in MM include improving the pharmacokinetics and stability of ncRNA-based reagents in the face of innate immune responses and cellular nuclease, as well as optimizing delivery of ncRNA-based reagents to MM tumor in the bone marrow niche, while limiting on-target but off-tumor effects in healthy tissues (Arun, Diermeier, & Spector, 2018).

Despite much recent progress, at present few of the ncRNAs the expression of which is perturbed in MM have been characterized in detail at the molecular and functional level. The field will advance rapidly by moving beyond analyses of single candidates and toward more holistic approaches to identify functionally relevant ncRNAs, including appropriate CRISPR-based strategies to carry out high-throughput functional screens of ncRNAs (Aquino-Jarquin, 2017; Liu et al., 2017b) both in vitro and in clinically relevant mouse models. The wide-scale identification of functionally important ncRNAs, together with detailed characterization of their molecular mechanisms of action, their interacting partner proteins, the downstream signaling pathways, and upstream genetic, epigenetic, or signaling events driving their dysregulated expression should provide a wealth of potential new targets for therapeutic intervention in the future.

Acknowledgments

Research in our laboratory on ncRNAs has been supported by the Association pour la Recherche sur le Cancer (Fondation ARC, France), the Fondation Française pour la Recherche contre le Myélome et les Gammapathies (FFRMG, France), the Cancéropôle (Ile-de-France), and the Institut National du Cancer (INCa), France. SZ was supported by the Initiative d'Excellence Program (IDEX, Université de Paris). We thank Jean-Christophe Bories for helpful comments on the manuscript.

References

Amodio, N., Bellizzi, D., Leotta, M., Raimondi, L., Biamonte, L., D'Aquila, P., ... Tassone, P. (2013). miR-29b induces SOCS-1 expression by promoter demethylation and negatively regulates migration of multiple myeloma and endothelial cells. *Cell Cycle, 12*, 3650–3662. Available from https://doi.org/10.4161/cc.26585.

Amodio, N., et al. (2018a). MALAT1: A druggable long non-coding RNA for targeted anti-cancer approaches. *Journal of Hematology & Oncology, 11*, 63. Available from https://doi.org/10.1186/s13045-018-0606-4.

Amodio, N., et al. (2018b). Drugging the lncRNA MALAT1 via LNA gapmeR ASO inhibits gene expression of proteasome subunits and triggers anti-multiple myeloma activity. *Leukemia, 32*, 1948–1957. Available from https://doi.org/10.1038/s41375-018-0067-3.

Abulwerdi, F. A., Xu, W., Ageeli, A. A., Yonkunas, M. J., Arun, G., Nam, H., ... Le Grice, S. F. J. (2019). Selective small-molecule targeting of a triple helix encoded by the long noncoding RNA, MALAT1. *ACS Chemical Biology, 14*, 223–235. Available from https://doi.org/10.1021/acschembio.8b00807.

Aquino-Jarquin, G. (2017). Emerging role of CRISPR/Cas9 technology for microRNAs editing in cancer research. *Cancer Research, 77*, 6812–6817. Available from https://doi.org/10.1158/0008-5472.CAN-17-2142.

Arun, G., Diermeier, S. D., & Spector, D. L. (2018). Therapeutic targeting of long non-coding RNAs in cancer. *Trends in Molecular Medicine, 24*, 257–277. Available from https://doi.org/10.1016/j.molmed.2018.01.001.

Attal, M., et al. (2017). Lenalidomide, bortezomib, and dexamethasone with transplantation for myeloma. *New England Journal of Medicine, 376*, 1311–1320. Available from https://doi.org/10.1056/NEJMoa1611750.

Bak, R. O., & Mikkelsen, J. G. (2014). miRNA sponges: Soaking up miRNAs for regulation of gene expression. *Wiley Interdisciplinary Reviews: RNA, 5*, 317–333. Available from https://doi.org/10.1002/wrna.1213.

References

Bautista-Sanchez, D., et al. (2020). The promising role of miR-21 as a cancer biomarker and its importance in RNA-based therapeutics. *Molecular Therapy – Nucleic Acids, 20*, 409–420. Available from https://doi.org/10.1016/j.omtn.2020.03.003.

Benetatos, L., et al. (2008). Promoter hypermethylation of the MEG3 (DLK1/MEG3) imprinted gene in multiple myeloma. *Clinical Lymphoma, Myeloma & Leukemia, 8*, 171–175. Available from https://doi.org/10.3816/CLM.2008.n.021.

Bolli, N., et al. (2014). Heterogeneity of genomic evolution and mutational profiles in multiple myeloma. *Nature Communications, 5*, 2997. Available from https://doi.org/10.1038/ncomms3997.

Bond, C. S., & Fox, A. H. (2009). Paraspeckles: Nuclear bodies built on long noncoding RNA. *Journal of Cell Biology, 186*, 637–644. Available from https://doi.org/10.1083/jcb.200906113.

Boyle, E. M., et al. (2021). The molecular make up of smoldering myeloma highlights the evolutionary pathways leading to multiple myeloma. *Nature Communications, 12*, 293. Available from https://doi.org/10.1038/s41467-020-20524-2.

Brown, J. A., et al. (2014). Structural insights into the stabilization of MALAT1 noncoding RNA by a bipartite triple helix. *Nature Structural & Molecular Biology, 21*, 633–640. Available from https://doi.org/10.1038/nsmb.2844.

Butova, R., Vychytilova-Faltejskova, P., Souckova, A., Sevcikova, S., & Hajek, R. (2019). Long non-coding RNAs in multiple myeloma. *Noncoding RNA, 5*. Available from https://doi.org/10.3390/ncrna5010013.

Cao, T., et al. (2018). Biology and clinical relevance of noncoding sno/scaRNAs. *Trends in Cardiovascular Medicine, 28*, 81–90. Available from https://doi.org/10.1016/j.tcm.2017.08.002.

Caprio, C., Sacco, A., Giustini, V., & Roccaro, A. M. (2020). Epigenetic aberrations in multiple myeloma. *Cancers (Basel), 12*. Available from https://doi.org/10.3390/cancers12102996.

Caracciolo, D., et al. (2018). The potential role of miRNAs in multiple myeloma therapy. *Expert Review of Hematology, 11*, 793–803. Available from https://doi.org/10.1080/17474086.2018.1517041.

Che, F., Chen, J., Wan, C., & Huang, X. (2020). MicroRNA-27 inhibits autophagy and promotes proliferation of multiple myeloma cells by targeting the NEDD4/Notch1 axis. *Frontiers in Oncology, 10*, 571914. Available from https://doi.org/10.3389/fonc.2020.571914.

Che, F., Ye, X., Wang, Y., Ma, S., & Wang, X. (2021). Lnc NEAT1/miR-29b-3p/Sp1 form a positive feedback loop and modulate bortezomib resistance in human multiple myeloma cells. *European Journal of Pharmacology, 891*, 173752. Available from https://doi.org/10.1016/j.ejphar.2020.173752.

Chen, D., Yang, X., Liu, M., Zhang, Z., & Xing, E. (2021). Roles of miRNA dysregulation in the pathogenesis of multiple myeloma. *Cancer Gene Therapy*. Available from https://doi.org/10.1038/s41417-020-00291-4.

Chen, F., et al. (2020). Effect of the up-regulation of circular RNA Hsa_circ_0069767 derived from C-KIT on the biological behavior of multiple myeloma cells. *Cancer Management and Research, 12*, 11321–11331. Available from https://doi.org/10.2147/CMAR.S259393.

Chen, L., et al. (2011). miR-17–92 cluster microRNAs confers tumorigenicity in multiple myeloma. *Cancer Letters, 309*, 62–70. Available from https://doi.org/10.1016/j.canlet.2011.05.017.

Chen, L., Hu, N., Wang, C., Zhao, H., & Gu, Y. (2018). Long non-coding RNA CCAT1 promotes multiple myeloma progression by acting as a molecular sponge of miR-181a-5p to modulate HOXA1 expression. *Cell Cycle, 17*, 319–329. Available from https://doi.org/10.1080/15384101.2017.1407893.

Chi, J., et al. (2011). MicroRNA expression in multiple myeloma is associated with genetic subtype, isotype and survival. *Biology Direct, 6*, 23. Available from https://doi.org/10.1186/1745-6150-6-23.

Chilosi, M., et al. (1999). CD138/syndecan-1: A useful immunohistochemical marker of normal and neoplastic plasma cells on routine trephine bone marrow biopsies. *Modern Pathology, 12*, 1101–1106.

Chim, C. S., Fung, T. K., Cheung, W. C., Liang, R., & Kwong, Y. L. (2004). SOCS1 and SHP1 hypermethylation in multiple myeloma: Implications for epigenetic activation of the Jak/STAT pathway. *Blood, 103*, 4630–4635. Available from https://doi.org/10.1182/blood-2003-06-2007.

Cho, S. F., et al. (2014). MALAT1 long non-coding RNA is overexpressed in multiple myeloma and may serve as a marker to predict disease progression. *BMC Cancer, 14*, 809. Available from https://doi.org/10.1186/1471-2407-14-809.

Chretien, M. L., et al. (2015). Understanding the role of hyperdiploidy in myeloma prognosis: Which trisomies really matter? *Blood, 126*, 2713–2719. Available from https://doi.org/10.1182/blood-2015-06-650242.

Chu, L., et al. (2012). Multiple myeloma-associated chromosomal translocation activates orphan snoRNA ACA11 to suppress oxidative stress. *The Journal of Clinical Investigation, 122*, 2793–2806. Available from https://doi.org/10.1172/JCI63051.

Corthals, S. L., et al. (2011). MicroRNA signatures characterize multiple myeloma patients. *Leukemia, 25*, 1784–1789. Available from https://doi.org/10.1038/leu.2011.147.

Corthals, S. L., et al. (2010). Micro-RNA-15a and micro-RNA-16 expression and chromosome 13 deletions in multiple myeloma. *Leukemia Research, 34*, 677–681. Available from https://doi.org/10.1016/j.leukres.2009.10.026.

Cosco, D., et al. (2015). Delivery of miR-34a by chitosan/PLGA nanoplexes for the anticancer treatment of multiple myeloma. *Scientific Reports, 5*, 17579. Available from https://doi.org/10.1038/srep17579.

Cui, Y. S., Song, Y. P., & Fang, B. J. (2019). The role of long non-coding RNAs in multiple myeloma. *European Journal of Haematology, 103*, 3–9. Available from https://doi.org/10.1111/ejh.13237.

Dahl, M., et al. (2018). Enzyme-free digital counting of endogenous circular RNA molecules in B-cell malignancies. *Laboratory Investigation, 98*, 1657–1669. Available from https://doi.org/10.1038/s41374-018-0108-6.

David, A., et al. (2020). The long non-coding RNA CRNDE regulates growth of multiple myeloma cells via an effect on IL6 signalling. *Leukemia*. Available from https://doi.org/10.1038/s41375-020-01034-y.

Decaux, O., et al. (2008). Prediction of survival in multiple myeloma based on gene expression profiles reveals cell cycle and chromosomal instability signatures in high-risk patients and hyperdiploid signatures in low-risk patients: A study of the Intergroupe Francophone du Myelome. *Journal of Clinical Oncology: Official Journal of the American Society of Clinical Oncology, 26*, 4798–4805. Available from https://doi.org/10.1200/JCO.2007.13.8545.

Dehghanifard, A., et al. (2018). Various signaling pathways in multiple myeloma cells and effects of treatment on these pathways. *Clinical Lymphoma, Myeloma & Leukemia, 18*, 311–320. Available from https://doi.org/10.1016/j.clml.2018.03.007.

Di Martino, M. T., et al. (2014). In vivo activity of miR-34a mimics delivered by stable nucleic acid lipid particles (SNALPs) against multiple myeloma. *PLoS One, 9*, e90005. Available from https://doi.org/10.1371/journal.pone.0090005.

Di Martino, M. T., et al. (2012). Synthetic miR-34a mimics as a novel therapeutic agent for multiple myeloma: In vitro and in vivo evidence. *Clinical Cancer Research, 18*, 6260–6270. Available from https://doi.org/10.1158/1078-0432.CCR-12-1708.

Dickens, N. J., et al. (2010). Homozygous deletion mapping in myeloma samples identifies genes and an expression signature relevant to pathogenesis and outcome. *Clinical Cancer Research, 16*, 1856–1864. Available from https://doi.org/10.1158/1078-0432.CCR-09-2831.

Donlic, A., Morgan, B. S., Xu, J. L., Liu, A., Roble, C. Jr., & Hargrove, A. E. (2018). Discovery of small molecule ligands for MALAT1 by tuning an RNA-binding scaffold. *Angewandte Chemie International Edition, 57*, 13242–13247. Available from https://doi.org/10.1002/anie.201808823.

Fabbri, M., et al. (2012). MicroRNAs bind to toll-like receptors to induce prometastatic inflammatory response. *Proceedings of the National Academy of Sciences of the United States of America, 109*, E2110–E2116. Available from https://doi.org/10.1073/pnas.1209414109.

Feng, Y., et al. (2019). CircRNA circ_0000190 inhibits the progression of multiple myeloma through modulating miR-767-5p/MAPK4 pathway. *Journal of Experimental & Clinical Cancer Research, 38*, 54. Available from https://doi.org/10.1186/s13046-019-1071-9.

Galm, O., Yoshikawa, H., Esteller, M., Osieka, R., & Herman, J. G. (2003). SOCS-1, a negative regulator of cytokine signaling, is frequently silenced by methylation in multiple myeloma. *Blood, 101*, 2784–2788. Available from https://doi.org/10.1182/blood-2002-06-1735.

Gao, D., Lv, A. E., Li, H. P., Han, D. H., & Zhang, Y. P. (2017). LncRNA MALAT-1 elevates HMGB1 to promote autophagy resulting in inhibition of tumor cell apoptosis in multiple myeloma. *Journal of Cellular Biochemistry, 118*, 3341–3348. Available from https://doi.org/10.1002/jcb.25987.

Gao, M., et al. (2019). hsa_circ_0007841: A novel potential biomarker and drug resistance for multiple myeloma. *Frontiers in Oncology, 9*, 1261. Available from https://doi.org/10.3389/fonc.2019.01261.

Gao, X., et al. (2012). MiR-15a, miR-16-1 and miR-17–92 cluster expression are linked to poor prognosis in multiple myeloma. *Leukemia Research, 36*, 1505–1509. Available from https://doi.org/10.1016/j.leukres.2012.08.021.

Ghafouri-Fard, S., & Taheri, M. (2019). Maternally expressed gene 3 (MEG3): A tumor suppressor long non coding RNA. *Biomedicine & Pharmacotherapy, 118*, 109129. Available from https://doi.org/10.1016/j.biopha.2019.109129.

References

Gowda, P. S., et al. (2018). Runx2 suppression by miR-342 and miR-363 inhibits multiple myeloma progression. *Molecular Cancer Research, 16,* 1138−1148. Available from https://doi.org/10.1158/1541-7786.MCR-17-0606.

Goyal, B., et al. (2021). Diagnostic, prognostic, and therapeutic significance of long non-coding RNA MALAT1 in cancer. *Biochimica et Biophysica Acta — Reviews on Cancer, 1875,* 188502. Available from https://doi.org/10.1016/j.bbcan.2021.188502.

Graham, L. D., et al. (2011). Colorectal neoplasia differentially expressed (CRNDE), a novel gene with elevated expression in colorectal adenomas and adenocarcinomas. *Genes Cancer, 2,* 829−840. Available from https://doi.org/10.1177/1947601911431081.

Gu, C., et al. (2017a). Integrative analysis of signaling pathways and diseases associated with the miR-106b/25 cluster and their function study in berberine-induced multiple myeloma cells. *Functional & Integrative Genomics, 17,* 253−262. Available from https://doi.org/10.1007/s10142-016-0519-7.

Gu, Y., Xiao, X., & Yang, S. (2017b). LncRNA MALAT1 acts as an oncogene in multiple myeloma through sponging miR-509-5p to modulate FOXP1 expression. *Oncotarget, 8,* 101984−101993. Available from https://doi.org/10.18632/oncotarget.21957.

Gulla, A., et al. (2016). A 13 mer LNA-i-miR-221 inhibitor restores drug sensitivity in melphalan-refractory multiple myeloma cells. *Clinical Cancer Research, 22,* 1222−1233. Available from https://doi.org/10.1158/1078-0432.CCR-15-0489.

Guo, H. M., et al. (2018). LncRNA NEAT1 regulates cervical carcinoma proliferation and invasion by targeting AKT/PI3K. *European Review for Medical and Pharmacological Sciences, 22,* 4090−4097. Available from https://doi.org/10.26355/eurrev_201807_15400.

Gupta, V. A., et al. (2017). Bone marrow microenvironment-derived signals induce Mcl-1 dependence in multiple myeloma. *Blood, 129,* 1969−1979. Available from https://doi.org/10.1182/blood-2016-10-745059.

Gutierrez, N. C., et al. (2010). Deregulation of microRNA expression in the different genetic subtypes of multiple myeloma and correlation with gene expression profiling. *Leukemia, 24,* 629−637. Available from https://doi.org/10.1038/leu.2009.274.

Haematological Malignancy Research Network (2020). <https://www.hmrn.org/statistics/incidence>.

Handa, H., et al. (2017). Long non-coding RNA MALAT1 is an inducible stress response gene associated with extramedullary spread and poor prognosis of multiple myeloma. *British Journal of Haematology, 179,* 449−460. Available from https://doi.org/10.1111/bjh.14882.

Handa, H., et al. (2020). Long noncoding RNA PVT1 is regulated by bromodomain protein BRD4 in multiple myeloma and is associated with disease progression. *International Journal of Molecular Sciences, 21.* Available from https://doi.org/10.3390/ijms21197121.

Hao, M., et al. (2015). Low serum miR-19a expression as a novel poor prognostic indicator in multiple myeloma. *International Journal of Cancer, 136,* 1835−1844. Available from https://doi.org/10.1002/ijc.29199.

Hao, M., et al. (2016). Serum high expression of miR-214 and miR-135b as novel predictor for myeloma bone disease development and prognosis. *Oncotarget, 7,* 19589−19600. Available from https://doi.org/10.18632/oncotarget.7319.

Hao, M., et al. (2011). Suppressing miRNA-15a/-16 expression by interleukin-6 enhances drug-resistance in myeloma cells. *Journal of Hematology & Oncology, 4,* 37. Available from https://doi.org/10.1186/1756-8722-4-37.

Heinrich, P. C., et al. (2003). Principles of interleukin (IL)-6-type cytokine signalling and its regulation. *Biochemical Journal, 374,* 1−20. Available from https://doi.org/10.1042/BJ20030407.

Hideshima, T., Mitsiades, C., Tonon, G., Richardson, P. G., & Anderson, K. C. (2007). Understanding multiple myeloma pathogenesis in the bone marrow to identify new therapeutic targets. *Nature Reviews Cancer, 7,* 585−598. Available from https://doi.org/10.1038/nrc2189.

Holdt, L. M., Kohlmaier, A., & Teupser, D. (2018). Molecular roles and function of circular RNAs in eukaryotic cells. *Cellular and Molecular Life Sciences: CMLS, 75,* 1071−1098. Available from https://doi.org/10.1007/s00018-017-2688-5.

Hu, A. X., Huang, Z. Y., Zhang, L., & Shen, J. (2017). Potential prognostic long non-coding RNA identification and their validation in predicting survival of patients with multiple myeloma. *Tumor Biology, 39.* Available from https://doi.org/10.1177/1010428317694563, 1010428317694563.

Hu, Y., et al. (2018). Targeting the MALAT1/PARP1/LIG3 complex induces DNA damage and apoptosis in multiple myeloma. *Leukemia, 32,* 2250−2262. Available from https://doi.org/10.1038/s41375-018-0104-2.

Huang, J. J., Yu, J., Li, J. Y., Liu, Y. T., & Zhong, R. Q. (2012). Circulating microRNA expression is associated with genetic subtype and survival of multiple myeloma. *Medical Oncology, 29,* 2402–2408. Available from https://doi.org/10.1007/s12032-012-0210-3.

Isin, M., et al. (2014). Investigation of circulating lncRNAs in B-cell neoplasms. *Clinica Chimica Acta, 431,* 255–259. Available from https://doi.org/10.1016/j.cca.2014.02.010.

Jagannathan, S., et al. (2015). MiR-29b replacement inhibits proteasomes and disrupts aggresome + autophagosome formation to enhance the antimyeloma benefit of bortezomib. *Leukemia, 29,* 727–738. Available from https://doi.org/10.1038/leu.2014.279.

Jiang, Y., Luan, Y., Chang, H., & Chen, G. (2018). The diagnostic and prognostic value of plasma microRNA-125b-5p in patients with multiple myeloma. *Oncology Letters, 16,* 4001–4007. Available from https://doi.org/10.3892/ol.2018.9128.

Jin, K., et al. (2019). Long non-coding RNA PVT1 interacts with MYC and its downstream molecules to synergistically promote tumorigenesis. *Cellular and Molecular Life Sciences: CMLS, 76,* 4275–4289. Available from https://doi.org/10.1007/s00018-019-03222-1.

Jones, C. I., et al. (2012). Identification of circulating microRNAs as diagnostic biomarkers for use in multiple myeloma. *British Journal of Cancer, 107,* 1987–1996. Available from https://doi.org/10.1038/bjc.2012.525.

Jovanovic, K. K., et al. (2018). Deregulation and targeting of TP53 pathway in multiple myeloma. *Frontiers in Oncology, 8,* 665. Available from https://doi.org/10.3389/fonc.2018.00665.

Kang, S., Tanaka, T., Narazaki, M., & Kishimoto, T. (2019). Targeting interleukin-6 signaling in clinic. *Immunity, 50,* 1007–1023. Available from https://doi.org/10.1016/j.immuni.2019.03.026.

Kubiczkova, L., et al. (2014). Circulating serum microRNAs as novel diagnostic and prognostic biomarkers for multiple myeloma and monoclonal gammopathy of undetermined significance. *Haematologica, 99,* 511–518. Available from https://doi.org/10.3324/haematol.2013.093500.

Kuiper, R., et al. (2012). A gene expression signature for high-risk multiple myeloma. *Leukemia, 26,* 2406–2413. Available from https://doi.org/10.1038/leu.2012.127.

Kumar, M., et al. (2011). Negative regulation of the tumor suppressor p53 gene by microRNAs. *Oncogene, 30,* 843–853. Available from https://doi.org/10.1038/onc.2010.457.

Kumar, S. K., et al. (2017). Multiple myeloma. *Nature Reviews Disease Primers, 3,* 17046. Available from https://doi.org/10.1038/nrdp.2017.46.

Landgren, O., & Iskander, K. (2017). Modern multiple myeloma therapy: Deep, sustained treatment response and good clinical outcomes. *Journal of Internal Medicine, 281,* 365–382. Available from https://doi.org/10.1111/joim.12590.

Lennox, K. A., & Behlke, M. A. (2016). Cellular localization of long non-coding RNAs affects silencing by RNAi more than by antisense oligonucleotides. *Nucleic Acids Research, 44,* 863–877. Available from https://doi.org/10.1093/nar/gkv1206.

Leone, E., et al. (2013). Targeting miR-21 inhibits in vitro and in vivo multiple myeloma cell growth. *Clinical Cancer Research, 19,* 2096–2106. Available from https://doi.org/10.1158/1078-0432.CCR-12-3325.

Leotta, M., et al. (2014). A p53-dependent tumor suppressor network is induced by selective miR-125a-5p inhibition in multiple myeloma cells. *Journal of Cellular Physiology, 229,* 2106–2116. Available from https://doi.org/10.1002/jcp.24669.

Lionetti, M., et al. (2013). Biological and clinical relevance of miRNA expression signatures in primary plasma cell leukemia. *Clinical Cancer Research, 19,* 3130–3142. Available from https://doi.org/10.1158/1078-0432.CCR-12-2043.

Lionetti, M., et al. (2009). Identification of microRNA expression patterns and definition of a microRNA/mRNA regulatory network in distinct molecular groups of multiple myeloma. *Blood, 114,* e20–e26. Available from https://doi.org/10.1182/blood-2009-08-237495.

Liu, H., et al. (2017a). Down-regulation of long non-coding RNA MALAT1 by RNA interference inhibits proliferation and induces apoptosis in multiple myeloma. *Clinical and Experimental Pharmacology & Physiology, 44,* 1032–1041. Available from https://doi.org/10.1111/1440-1681.12804.

Liu, S. J., et al. (2017b). CRISPRi-based genome-scale identification of functional long noncoding RNA loci in human cells. *Science, 355.* Available from https://doi.org/10.1126/science.aah7111.

Liu, T., Xu, Z., Ou, D., Liu, J., & Zhang, J. (2019). The miR-15a/16 gene cluster in human cancer: A systematic review. *Journal of Cellular Physiology, 234,* 5496–5506. Available from https://doi.org/10.1002/jcp.27342.

Liu, X., Tang, H., Liu, J., & Wang, X. (2020). hsa_circRNA_101237: A novel diagnostic and prognostic biomarker and potential therapeutic target for multiple myeloma. *Cancer Management and Research, 12,* 2109–2118. Available from https://doi.org/10.2147/CMAR.S241089.

Loffler, D., et al. (2007). Interleukin-6 dependent survival of multiple myeloma cells involves the Stat3-mediated induction of microRNA-21 through a highly conserved enhancer. *Blood, 110,* 1330–1333. Available from https://doi.org/10.1182/blood-2007-03-081133.

Lomas, O. C., Tahri, S., & Ghobrial, I. M. (2020). The microenvironment in myeloma. *Current Opinion in Oncology, 32,* 170–175. Available from https://doi.org/10.1097/CCO.0000000000000615.

Lu, Y., et al. (2020). CRNDE: An oncogenic long non-coding RNA in cancers. *Cancer Cell International, 20,* 162. Available from https://doi.org/10.1186/s12935-020-01246-3.

Malek, E., Kim, B. G., & Driscoll, J. J. (2016). Identification of long non-coding RNAs deregulated in multiple myeloma cells resistant to proteasome inhibitors. *Genes (Basel), 7.* Available from https://doi.org/10.3390/genes7100084.

Manier, S., Salem, K. Z., Park, J., Landau, D. A., Getz, G., & Ghobrial, I. M. (2017a). Genomic complexity of multiple myeloma and its clinical implications. *Nature Reviews Clinical Oncology, 14,* 100–113. Available from https://doi.org/10.1038/nrclinonc.2016.122.

Manier, S., Liu, C. J., Avet-Loiseau, H., Park, J., Shi, J., Campigotto, F., . . . Ghobrial, I. M. (2017b). Prognostic role of circulating exosomal miRNAs in multiple myeloma. *Blood, 129,* 2429–2436. Available from https://doi.org/10.1182/blood-2016-09-742296.

Manier, S., Sacco, A., Leleu, X., Ghobrial, I. M., & Roccaro, A. M. (2012). Bone marrow microenvironment in multiple myeloma progression. *Journal of Biomedicine and Biotechnology, 2012,* 157496. Available from https://doi.org/10.1155/2012/157496.

Matthes, T., Manfroi, B., & Huard, B. (2016). Revisiting IL-6 antagonism in multiple myeloma. *Critical Reviews in Oncology/Hematology, 105,* 1–4. Available from https://doi.org/10.1016/j.critrevonc.2016.07.006.

Mattioli, M., et al. (2005). Gene expression profiling of plasma cell dyscrasias reveals molecular patterns associated with distinct IGH translocations in multiple myeloma. *Oncogene, 24,* 2461–2473. Available from https://doi.org/10.1038/sj.onc.1208447.

Mendell, J. T. (2008). miRiad roles for the miR-17–92 cluster in development and disease. *Cell, 133,* 217–222. Available from https://doi.org/10.1016/j.cell.2008.04.001.

Meng, Y. B., et al. (2017). Long noncoding RNA CRNDE promotes multiple myeloma cell growth by suppressing miR-451. *Oncology Research, 25,* 1207–1214. Available from https://doi.org/10.3727/096504017X14886679715637.

Mikulasova, A., et al. (2020). Microhomology-mediated end joining drives complex rearrangements and overexpression of MYC and PVT1 in multiple myeloma. *Haematologica, 105,* 1055–1066. Available from https://doi.org/10.3324/haematol.2019.217927.

Misiewicz-Krzeminska, I., et al. (2019). Factors regulating microRNA expression and function in multiple myeloma. *Noncoding RNA, 5.* Available from https://doi.org/10.3390/ncrna5010009.

Mogilyansky, E., & Rigoutsos, I. (2013). The miR-17/92 cluster: A comprehensive update on its genomics, genetics, functions and increasingly important and numerous roles in health and disease. *Cell Death & Differentiation, 20,* 1603–1614. Available from https://doi.org/10.1038/cdd.2013.125.

Morelli, E., et al. (2015). Selective targeting of IRF4 by synthetic microRNA-125b-5p mimics induces anti-multiple myeloma activity in vitro and in vivo. *Leukemia, 29,* 2173–2183. Available from https://doi.org/10.1038/leu.2015.124.

Morelli, E., et al. (2018). Therapeutic vulnerability of multiple myeloma to MIR17PTi, a first-in-class inhibitor of pri-miR-17–92. *Blood, 132,* 1050–1063. Available from https://doi.org/10.1182/blood-2018-03-836601.

Nagoshi, H., et al. (2012). Frequent PVT1 rearrangement and novel chimeric genes PVT1-NBEA and PVT1-WWOX occur in multiple myeloma with 8q24 abnormality. *Cancer Research, 72,* 4954–4962. Available from https://doi.org/10.1158/0008-5472.CAN-12-0213.

National Cancer Institute USA. (2020). *SEER cancer stat facts: Myeloma,* <https://seer.cancer.gov/statfacts/html/mulmy.html>.

Oliveira, V., et al. (2019). The snoRNA target of t(4;14) in multiple myeloma regulates ribosome biogenesis. *FASEB BioAdvances, 1,* 404–414. Available from https://doi.org/10.1096/fba.2018-00075.

Pan, Y., et al. (2018). Serum level of long noncoding RNA H19 as a diagnostic biomarker of multiple myeloma. *Clinica Chimica Acta, 480,* 199–205. Available from https://doi.org/10.1016/j.cca.2018.02.019.

Pedram Fatemi, R., et al. (2015). Screening for small-molecule modulators of long noncoding RNA-protein interactions using alphascreen. *Journal of Biomolecular Screening*, 20, 1132−1141. Available from https://doi.org/10.1177/1087057115594187.

Pichiorri, F., et al. (2010). Downregulation of p53-inducible microRNAs 192, 194, and 215 impairs the p53/MDM2 autoregulatory loop in multiple myeloma development. *Cancer Cell*, 18, 367−381. Available from https://doi.org/10.1016/j.ccr.2010.09.005.

Pichiorri, F., et al. (2008). MicroRNAs regulate critical genes associated with multiple myeloma pathogenesis. *Proceedings of the National Academy of Sciences of the United States of America*, 105, 12885−12890. Available from https://doi.org/10.1073/pnas.0806202105.

Pitari, M. R., et al. (2015). Inhibition of miR-21 restores RANKL/OPG ratio in multiple myeloma-derived bone marrow stromal cells and impairs the resorbing activity of mature osteoclasts. *Oncotarget*, 6, 27343−27358. Available from https://doi.org/10.18632/oncotarget.4398.

Puthier, D., Bataille, R., & Amiot, M. (1999). IL-6 up-regulates mcl-1 in human myeloma cells through JAK / STAT rather than ras/MAP kinase pathway. *European Journal of Immunology*, 29, 3945−3950, doi:10.1002/(SICI)1521-4141(199912)29:12 < 3945::AID-IMMU3945 > 3.0.CO;2-O.

Raimondi, L., et al. (2016). MicroRNAs: Novel crossroads between myeloma cells and the bone marrow microenvironment. *BioMed Research International*, 2016, 6504593. Available from https://doi.org/10.1155/2016/6504593.

Rajkumar, S. V., et al. (2013). Impact of primary molecular cytogenetic abnormalities and risk of progression in smoldering multiple myeloma. *Leukemia*, 27, 1738−1744. Available from https://doi.org/10.1038/leu.2013.86.

Rajkumar, S. V., et al. (2014). International Myeloma Working Group updated criteria for the diagnosis of multiple myeloma. *Lancet Oncology*, 15, e538−e548. Available from https://doi.org/10.1016/S1470-2045(14)70442-5.

Ramakrishnan, V., & Kumar, S. (2018). PI3K/AKT/mTOR pathway in multiple myeloma: From basic biology to clinical promise. *Leukemia & Lymphoma*, 59, 2524−2534. Available from https://doi.org/10.1080/10428194.2017.1421760.

Ravi, P., et al. (2018). Defining cure in multiple myeloma: A comparative study of outcomes of young individuals with myeloma and curable hematologic malignancies. *Blood Cancer Journal*, 8, 26. Available from https://doi.org/10.1038/s41408-018-0065-8.

Roccaro, A. M., et al. (2013). BM mesenchymal stromal cell-derived exosomes facilitate multiple myeloma progression. *The Journal of Clinical Investigation*, 123, 1542−1555. Available from https://doi.org/10.1172/JCI66517.

Roccaro, A. M., et al. (2009). MicroRNAs 15a and 16 regulate tumor proliferation in multiple myeloma. *Blood*, 113, 6669−6680. Available from https://doi.org/10.1182/blood-2009-01-198408.

Ronchetti, D., et al. (2018). A compendium of long non-coding RNAs transcriptional fingerprint in multiple myeloma. *Scientific Reports*, 8, 6557. Available from https://doi.org/10.1038/s41598-018-24701-8.

Ronchetti, D., et al. (2016a). Distinct lncRNA transcriptional fingerprints characterize progressive stages of multiple myeloma. *Oncotarget*, 7, 14814−14830. Available from https://doi.org/10.18632/oncotarget.7442.

Ronchetti, D., et al. (2012). The expression pattern of small nucleolar and small Cajal body-specific RNAs characterizes distinct molecular subtypes of multiple myeloma. *Blood Cancer Journal*, 2, e96. Available from https://doi.org/10.1038/bcj.2012.41.

Ronchetti, D., Manzoni, M., Todoerti, K., Neri, A., & Agnelli, L. (2016). In silico characterization of miRNA and long non-coding RNA interplay in multiple myeloma. *Genes (Basel)*, 7. Available from https://doi.org/10.3390/genes7120107.

Rosean, T. R., et al. (2014). Preclinical validation of interleukin 6 as a therapeutic target in multiple myeloma. *Immunologic Research*, 59, 188−202. Available from https://doi.org/10.1007/s12026-014-8528-x.

Samur, M. K., et al. (2018). Long intergenic non-coding RNAs have an independent impact on survival in multiple myeloma. *Leukemia*, 32, 2626−2635. Available from https://doi.org/10.1038/s41375-018-0116-y.

Sant, M., et al. (2010). Incidence of hematologic malignancies in Europe by morphologic subtype: Results of the HAEMACARE project. *Blood*, 116, 3724−3734. Available from https://doi.org/10.1182/blood-2010-05-282632.

Sedlarikova, L., et al. (2018). Circulating exosomal long noncoding RNA PRINS-First findings in monoclonal gammopathies. *Hematology & Oncology*, 36, 786−791. Available from https://doi.org/10.1002/hon.2554.

Sedlarikova, L., et al. (2017). Deregulated expression of long non-coding RNA UCA1 in multiple myeloma. *European Journal of Haematology*, 99, 223−233. Available from https://doi.org/10.1111/ejh.12908.

Sevcikova, S., Kubiczkova, L., Sedlarikova, L., Slaby, O., & Hajek, R. (2013). Serum miR-29a as a marker of multiple myeloma. *Leukemia & Lymphoma, 54*, 189−191. Available from https://doi.org/10.3109/10428194.2012.704030.

Shaffer, A. L., et al. (2008). IRF4 addiction in multiple myeloma. *Nature, 454*, 226−231. Available from https://doi.org/10.1038/nature07064.

Shen, X., et al. (2018). Long non-coding RNA MEG3 functions as a competing endogenous RNA to regulate HOXA11 expression by sponging miR-181a in multiple myeloma. *Cellular Physiology and Biochemistry: International Journal of Experimental Cellular Physiology, Biochemistry, and Pharmacology, 49*, 87−100. Available from https://doi.org/10.1159/000492846.

Shen, X., et al. (2020). PCAT-1 promotes cell growth by sponging miR-129 via MAP3K7/NF-kappaB pathway in multiple myeloma. *Journal of Cellular and Molecular Medicine, 24*, 3492−3503. Available from https://doi.org/10.1111/jcmm.15035.

Shen, X., et al. (2017). Upregulated lncRNA-PCAT1 is closely related to clinical diagnosis of multiple myeloma as a predictive biomarker in serum. *Cancer Biomarkers, 18*, 257−263. Available from https://doi.org/10.3233/CBM-160158.

Soley, L., Falank, C., & Reagan, M. R. (2017). MicroRNA transfer between bone marrow adipose and multiple myeloma cells. *Current Osteoporosis Reports, 15*, 162−170. Available from https://doi.org/10.1007/s11914-017-0360-5.

Sun, C. Y., She, X. M., Qin, Y., Chu, Z. B., Chen, L., Ai, L. S., ... Hu, Y. (2013a). miR-15a and miR-16 affect the angiogenesis of multiple myeloma by targeting VEGF. *Carcinogenesis, 34*, 426−435. Available from https://doi.org/10.1093/carcin/bgs333.

Sun, Y. M., Lin, K. Y., & Chen, Y. Q. (2013b). Diverse functions of miR-125 family in different cell contexts. *Journal of Hematology & Oncology, 6*, 6. Available from https://doi.org/10.1186/1756-8722-6-6.

Taiana, E., et al. (2019). Long non-coding RNA NEAT1 shows high expression unrelated to molecular features and clinical outcome in multiple myeloma. *Haematologica, 104*, e72−e76. Available from https://doi.org/10.3324/haematol.2018.201301.

Taiana, E., et al. (2020). Long non-coding RNA NEAT1 targeting impairs the DNA repair machinery and triggers anti-tumor activity in multiple myeloma. *Leukemia, 34*, 234−244. Available from https://doi.org/10.1038/s41375-019-0542-5.

Tseng, Y. Y., et al. (2014). PVT1 dependence in cancer with MYC copy-number increase. *Nature, 512*, 82−86. Available from https://doi.org/10.1038/nature13311.

Turchinovich, A., Samatov, T. R., Tonevitsky, A. G., & Burwinkel, B. (2013). Circulating miRNAs: Cell-cell communication function? *Frontiers in Genetics, 4*, 119. Available from https://doi.org/10.3389/fgene.2013.00119.

Turesson, I., et al. (2018). Rapidly changing myeloma epidemiology in the general population: Increased incidence, older patients, and longer survival. *European Journal of Haematology, 101*. Available from https://doi.org/10.1111/ejh.13083.

Umezu, T., et al. (2014). Exosomal miR-135b shed from hypoxic multiple myeloma cells enhances angiogenesis by targeting factor-inhibiting HIF-1. *Blood, 124*, 3748−3757. Available from https://doi.org/10.1182/blood-2014-05-576116.

van Keimpema, M., et al. (2015). The forkhead transcription factor FOXP1 represses human plasma cell differentiation. *Blood, 126*, 2098−2109. Available from https://doi.org/10.1182/blood-2015-02-626176.

van Nieuwenhuijzen, N., Spaan, I., Raymakers, R., & Peperzak, V. (2018). From MGUS to multiple myeloma, a paradigm for clonal evolution of premalignant cells. *Cancer Research, 78*, 2449−2456. Available from https://doi.org/10.1158/0008-5472.CAN-17-3115.

Vandyke, K., Chow, A. W., Williams, S. A., To, L. B., & Zannettino, A. C. (2013). Circulating N-cadherin levels are a negative prognostic indicator in patients with multiple myeloma. *British Journal of Haematology, 161*, 499−507. Available from https://doi.org/10.1111/bjh.12280.

Walker, B. A., et al. (2015). Mutational spectrum, copy number changes, and outcome: Results of a sequencing study of patients with newly diagnosed myeloma. *Journal of Clinical Oncology: Official Journal of the American Society of Clinical Oncology, 33*, 3911−3920. Available from https://doi.org/10.1200/JCO.2014.59.1503.

Wang, F., Nazarali, A. J., & Ji, S. (2016). Circular RNAs as potential biomarkers for cancer diagnosis and therapy. *American Journal of Cancer Research, 6*, 1167−1176.

Wang, W., et al. (2015). Aberrant levels of miRNAs in bone marrow microenvironment and peripheral blood of myeloma patients and disease progression. *Journal of Molecular Diagnostics, 17*, 669−678. Available from https://doi.org/10.1016/j.jmoldx.2015.06.006.

Wang, X., et al. (2011). Myeloma cell adhesion to bone marrow stromal cells confers drug resistance by microRNA-21 up-regulation. *Leukemia & Lymphoma, 52*, 1991−1998. Available from https://doi.org/10.3109/10428194.2011.591004.

Wilusz, J. E., et al. (2012). A triple helix stabilizes the 3′ ends of long noncoding RNAs that lack poly(A) tails. *Genes & Development, 26*, 2392−2407. Available from https://doi.org/10.1101/gad.204438.112.

Wu, P., et al. (2013). Improved risk stratification in myeloma using a microRNA-based classifier. *British Journal of Haematology, 162*, 348−359. Available from https://doi.org/10.1111/bjh.12394.

Wu, Y., & Wang, H. (2018). LncRNA NEAT1 promotes dexamethasone resistance in multiple myeloma by targeting miR-193a/MCL1 pathway. *Journal of Biochemical and Molecular Toxicology, 32*. Available from https://doi.org/10.1002/jbt.22008.

Xiong, Q., et al. (2012). Identification of novel miR-21 target proteins in multiple myeloma cells by quantitative proteomics. *Journal of Proteome Research, 11*, 2078−2090. Available from https://doi.org/10.1021/pr201079y.

Xu, H., Li, J., & Zhou, Z. G. (2018). NEAT1 promotes cell proliferation in multiple myeloma by activating PI3K/AKT pathway. *European Review for Medical and Pharmacological Sciences, 22*, 6403−6411. Available from https://doi.org/10.26355/eurrev_201810_16053.

Yadav, P., Cook, M., & Cockwell, P. (2016). Current trends of renal impairment in multiple myeloma. *Kidney Diseases (Basel), 1*, 241−257. Available from https://doi.org/10.1159/000442511.

Yang, Y., Li, Y., Gu, H., Dong, M., & Cai, Z. (2020). Emerging agents and regimens for multiple myeloma. *Journal of Hematology & Oncology, 13*, 150. Available from https://doi.org/10.1186/s13045-020-00980-5.

Yin, Q., Shen, X., Cui, X., & Ju, S. (2019). Elevated serum lncRNA TUG1 levels are a potential diagnostic biomarker of multiple myeloma. *Experimental Hematology, 79*(47−55), e42. Available from https://doi.org/10.1016/j.exphem.2019.10.002.

Yu, H., Peng, S., Chen, X., Han, S., & Luo, J. (2020). Long non-coding RNA NEAT1 serves as a novel biomarker for treatment response and survival profiles via microRNA-125a in multiple myeloma. *Journal of Clinical Laboratory Analysis, 34*, e23399. Available from https://doi.org/10.1002/jcla.23399.

Yu, W., et al. (2020). Promoter hypermethylation influences the suppressive role of long non-coding RNA MEG3 in the development of multiple myeloma. *Experimental and Therapeutic Medicine, 20*, 637−645. Available from https://doi.org/10.3892/etm.2020.8723.

Zhan, F., et al. (2006). The molecular classification of multiple myeloma. *Blood, 108*, 2020−2028. Available from https://doi.org/10.1182/blood-2005-11-013458.

Zhang, Y. K., et al. (2011). Overexpression of microRNA-29b induces apoptosis of multiple myeloma cells through down regulating Mcl-1. *Biochemical and Biophysical Research Communications, 414*, 233−239. Available from https://doi.org/10.1016/j.bbrc.2011.09.063.

Zhao, J. J., et al. (2015). Targeting the miR-221−222/PUMA/BAK/BAX pathway abrogates dexamethasone resistance in multiple myeloma. *Cancer Research, 75*, 4384−4397. Available from https://doi.org/10.1158/0008-5472.CAN-15-0457.

Zhong, Y., Liu, Z., Li, D., Liao, Q., & Li, J. (2020). Identification and validation of a potential prognostic 7-lncRNA signature for predicting survival in patients with multiple myeloma. *BioMed Research International, 2020*, 3813546. Available from https://doi.org/10.1155/2020/3813546.

Zhou, M., et al. (2015). Identification and validation of potential prognostic lncRNA biomarkers for predicting survival in patients with multiple myeloma. *Journal of Experimental & Clinical Cancer Research, 34*, 102. Available from https://doi.org/10.1186/s13046-015-0219-5.

Zhou, Y., et al. (2007). Activation of p53 by MEG3 non-coding RNA. *The Journal of Biological Chemistry, 282*, 24731−24742. Available from https://doi.org/10.1074/jbc.M702029200.

Zhou, Y., et al. (2010). High-risk myeloma is associated with global elevation of miRNAs and overexpression of EIF2C2/AGO2. *Proceedings of the National Academy of Sciences of the United States of America, 107*, 7904−7909. Available from https://doi.org/10.1073/pnas.0908441107.

Zhu, B., et al. (2018). The potential function of microRNAs as biomarkers and therapeutic targets in multiple myeloma. *Oncology Letters, 15*, 6094−6106. Available from https://doi.org/10.3892/ol.2018.8157.

Zhuang, W., et al. (2015). Upregulation of lncRNA MEG3 promotes osteogenic differentiation of mesenchymal stem cells from multiple myeloma patients by targeting BMP4 transcription. . *Stem Cells, 33*, 1985−1997. Available from https://doi.org/10.1002/stem.1989.

CHAPTER 5

Clinical applications of noncoding RNAs in lung cancer patients

Santosh Kumar[1], Naveen Kumar Vishvakarma[2] and Ajay Kumar[3]

[1]Department of Life Science, National Institute of Technology, Rourkela, India [2]Department of Biotechnology, Guru Ghasidas Vishwavidyalaya, Bilaspur, India [3]Department of Zoology, Institute of Science, Banaras Hindu University, Varanasi, India

Abbreviations

ANRIL	antisense noncoding RNA in the INK4 locus
BALF	bronchoalveolar lavage fluid
BANCR	BRAF-activated noncoding RNA
CCAT2	colon cancer–associated transcript 2
CLL	chronic lymphocytic leukemia
DLL3	delta-like protein 3
EBUS-TBNA	endobronchial untrasonography-guided transbronchial needle aspiration
EGFR	epidermal growth factor receptor
EMT	epithelial–mesenchymal transition
FAM83A-AS1	family with sequence similarity 83 member A antisense RNA 1
GAS5	growth arrest–specific transcript 5
HOTAIR	HOX transcript antisense intergenic RNA
let-7	lethal-7
lncRNA	long noncoding RNA
LUAD	lung adenocarcinoma
MALAT1	metastasis-associated lung adenocarcinoma transcript 1
MEG3	maternal-expressed gene 3
miRNA	micro-RNA
NAT	natural antisense transcript
ncRNA	noncoding RNA
NSCLC	non–small cell lung carcinoma
OMCD	OncomiR Cancer Database
PEAK1	pseudopodium-enriched atypical kinase 1
piRNA	PIWI-interacting RNA

PRNCR1	prostate cancer–associated ncRNA 1
PVT1	plasmacytoma variant translocation 1
rtqPCR	real-time quantitative polymerase chain reaction
SCC	squamous-cell carcinoma
SCLC	small-cell lung carcinoma
shRNA	small hairpin RNA
snoRNA	small nucleolar RNA
SP1	specificity protein 1
SPRY4-IT1	SPRY4 intronic transcript 1
tsRNA	tRNA-derived small RNA
TTNA	transthoracic needle aspiration
T-UCR	transcribed ultraconserved region

5.1 Introduction

Lung cancer is the foremost cause of cancer-related deaths worldwide and accounts for approximately 20% of all cancer deaths (Bray et al., 2018). Based on histopathological features, lung cancer is mainly classified into two categories: non–small cell lung carcinoma (NSCLC) and small-cell lung carcinoma (SCLC). NSCLC comprises 85% of total lung cancer cases, whereas the rest of the 15% cases belong to the SCLC. Despite recent advances in the understanding of lung cancer–associated mechanistic pathways and advancement in therapeutic approaches, the 5-year survival rate for lung cancer remains less than 15% (Tao et al., 2016). The poor 5-year survival rate is largely due to late diagnosis, chemo-, and radioresistance which strongly emphasizes identifying the novel early diagnostic and prognostic markers along with the factors responsible for chemo- and radioresistance. Remarkably, several recent evidences strongly suggest the diagnostic, prognostic, and therapeutic potential of noncoding RNAs (ncRNAs) for lung cancer.

The surge in high-throughput RNA-sequencing technique advancement has led to the discovery of a completely unexplored field of gene regulation mediated by ncRNAs. These are the RNA transcripts that lack functional protein-coding capability. Only a meager 1.9% of total genome is transcribed to produce messenger RNA (mRNA), rest of the genome gets transcribed into ncRNAs with relatively low expression (Palazzo and Koonin, 2020). Several different classes of ncRNA are known to regulate molecular processes either at transcription, translation, or functional protein levels, affecting a diverse range of cellular and pathological processes. Moreover, dysregulated expression of ncRNA is a predominant characteristic of many malignancies, cardiovascular, and brain-related disorders (Braicu et al., 2019; Grillone et al., 2020; Kumar and Suryawanshi, 2019; Mishra and Kumar, 2021) with major implication in diagnosis and therapy (Redis, Berindan-Neagoe, Pop, & Calin, 2012). Different classes of ncRNA deregulated in cancer include micro-RNA (miRNA), long noncoding RNA (lncRNA), circular RNA (circRNA), PIWI-interacting RNA (piRNA), transcribed ultraconserved region (T-UCR), tRNA-derived small RNA (tsRNA), small nucleolar RNA (snoRNA), and natural antisense transcripts (NATs) (Braicu et al., 2019). The clinical significance of these noncoding transcripts is emphasized by their potential as biomarkers for diagnosis, and therapeutic usage. In the present chapter, we have discussed about the role of dysregulated levels of ncRNAs, namely, lncRNAs (ncRNAs more than 200 nucleotides), and miRNAs (ncRNAs composed of 19–25 nucleotides) in the development and progression of lung cancer along with their clinical applications.

5.2 Experimental methods and tools for analyzing ncRNAs in lung cancer patients

The experimental methods used for the analysis of ncRNA in cancer can be broadly divided into two approaches, high-throughput approach giving holistic view, and candidate gene approach for a more focused view. The high-throughput approaches utilize microarray and RNA-sequencing techniques to understand the role of ncRNA in lung cancer patients. The comparative analysis of transcript expression levels in lung cancer patients versus normal individuals can give us insight into the ncRNAs involved in the pathogenesis of lung cancer. Similarly, high-throughput sequencing data are also useful for the analysis of dysregulated expression of ncRNAs, including miRNAs, lncRNAs, and circRNAs, to ascertain their role in the pathogenesis of lung cancer. Furthermore, comparison of mRNA expression data with ncRNA expression could be useful for the predictive information and classification of cancer types.

Real-time quantitative polymerase chain reaction (RTqPCR) is a useful experimental method for the analysis of differential expression of ncRNAs in cancer patients and healthy individuals. Moreover, RTqPCR can also be utilized for the quantification of ncRNAs under different physiological conditions or drug treatments. The RTqPCR panels for miRNA (Zaporozhchenko et al., 2018), lncRNA (Acha-Sagredo et al., 2020), and circRNA (Liu et al., 2019) are used for the profiling of these ncRNA expression levels in lung cancer patients. In addition, Northern blotting assays and in situ hybridization assays are also useful for detection and quantification of ncRNAs. Several small hairpin RNA (shRNA) libraries, targeting specific or a set of lncRNAs, are used to study the downstream mechanistic details of the lncRNAs by which they affect phenotypic behavior of tumor cells. Many ncRNAs also interact with a plethora of RNA-binding proteins (RBPs) to perform their functions. The direct binders of the ncRNA can be identified using pull-down assays followed by mass spectrometric analysis (Sun et al., 2016). Another method to analyze the proteins to lncRNA is RNA antisense purification approach (McHugh et al., 2015). In this method, lncRNA is cross-linked using ultraviolet (UV) light to the directly bound protein in cells followed by mass spectrometry. This method can be performed using stable isotope labeling by/with amino acids in cell culture medium in combination with quantitative proteomics to improve the accuracy and sensitivity of the analysis of lncRNA-bound proteins (Cheung and Juan, 2017).

5.3 Datasets and informatics for analyzing ncRNAs in lung cancer patients

Recent advances in the field of high-throughput next-generation sequencing have led to the development of several datasets and bioinformatics tools for the analysis of ncRNA in lung cancer patients. Several genomic, RNA-sequencing, and expression microarray datasets have been generated to study the regulation of gene expression and cellular pathways mediated in cancer patients (Weinstein et al., 2013). Recently, a high-resolution genomics dataset for NSCLC subtypes, lung adenocarcinoma (LUAD), and squamous-cell carcinoma (SCC) has been profiled. This dataset contains lung cancer–related genes/loci, copy-number alterations, regional amplifications, and deletions, along with gene-expression profiling data for human lung cancer (Tonon et al., 2005). Similarly, genome-wide methylation dataset for

LUAD cells provides the information about the methylation status of the tumor tissue. Thus insight into the epigenetic landscape of LUAD cell can help us to understand its role in remodeling various signaling pathways (Bjaanæs et al., 2016). In recent times the analysis of transcriptomic data has completely transformed our understanding of the interplay between the coding as well as noncoding transcripts to modulate the cellular physiology under pathological conditions. Numerous datasets for the expression analysis of coding and ncRNAs are available on gene-expression omnibus (https://www.ncbi.nlm.nih.gov/geo/), and The Cancer Genome Atlas (TCGA, https://portal.gdc.cancer.gov). These datasets include the RNA-sequencing, single-cell RNA-sequencing, and gene-expression microarray datasets (Jin, Guan, Zhang, & Wang, 2020; Seo et al., 2012; LAzzawi et al., 2017, Lim, Tan, Lim, & Lim, 2018, Kim et al., 2020, Smolander, Stupnikov, Glazko, Dehmer, & Emmert-Streib, 2019).

Moreover, numerous databases compile a comprehensive set of ncRNAs, including miRNAs, lncRNAs, and circRNAs. These databases can be utilized to understand the sequence, structure, and functional relationship of these ncRNAs in lung cancer along with other tumors and diseases. OncomiR Cancer Database (OMCD) (http://www.oncomir.umn.edu/omcd/) contains miRNA sequencing data derived from different types of cancer patients, including LUAD, and SCC (OMCD). Recently, updated Lnc2cancer3.0 database is a good resource of RNA-sequencing and single-cell RNA-sequencing data for the expression analysis of lncRNAs and circRNAs across different human cancer types. This database includes experimentally supported data for understanding the complex interplay of cancer-associated lncRNAs, miRNAs, circRNAs, transcription factors, and enhancers and their regulatory mechanisms (Gao et al., 2021). Another database for analyzing the lncRNA targets in human cancers is LncTarD (http://biocc.hrbmu.edu.cn/LncTarD/ or http://bio-bigdata.hrbmu.edu.cn/LncTarD/). It contains the manually curated experimental data for the lncRNA—target regulation, lncRNA-influenced biological functions, lncRNA-mediated regulatory mechanisms, and lncRNA—target regulations responsible for drug resistance or sensitivity. Notably, this database contains maximum entries (320 entries) for the lncRNA—target regulations in lung cancer (Zhao et al., 2020).

Furthermore, many ncRNAs also perform their functions through interaction with transacting RBPs. Many databases used for the RNA—protein interaction analysis used experimentally validated data (Lewis et al., 2010; Stark et al., 2006) and, thus, can only be used for limited purpose specially they are not useful for the analysis of recently discovered classes of ncRNA, including lncRNA and circRNA. Therefore machine learning algorithm-based prediction using software like catRAPID (Bellucci, Agostini, Masin, & Tartaglia, 2011) and RPISeq (Muppirala, Honavar, & Dobbs, 2011) is more successful in predicting ncRNA interactions with RBPs.

5.4 Expression of ncRNAs in lung cancer patients

The expression of several ncRNAs is found extensively dysregulated in a variety of cancers, leading to altered cellular physiology and signaling pathways (Grillone et al., 2020). The altered expression of ncRNAs (Fig. 5.1) also contributes an essential role in the pathogenesis of lung cancer (Braicu et al., 2019). Here, we will discuss about the expression patterns of important classes of ncRNAs in lung cancer.

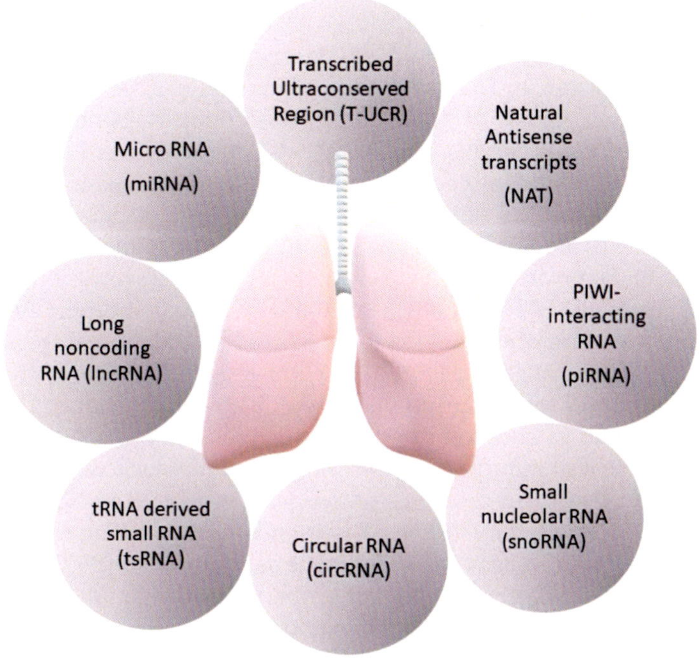

FIGURE 5.1 Different types of ncRNA expressing in lung cancer.

5.4.1 Micro-RNA

A very large fraction (about 65% of all lung cancer—related studies) of literature available for lung neoplasms are linked with the investigation of miRNAs implicated in lung cancer. They are involved in the regulation of each of the hallmarks of cancer. The expression analysis using miRNA microarray revealed a large set of miRNAs with dysregulated expression patterns. Among these miRNAs an increased expression of particularly hsa-mir-155 and lower expression of hsa-let-7a-2 were found correlated to poor survival of lung cancer patients (Yanaihara et al., 2006). One of the important tumor suppressor miRNAs, miR-218-5p, is downregulated in NSCLC cells. It binds to the 3'-untranslated region (UTR) of epidermal growth factor receptor (EGFR) and downregulates its expression leading to the suppression of cell proliferation and migration in lung cancer cells (Zhu et al., 2016). Two miRNAs, miR-15a and miR-16, are downregulated in SCC and LUAD tissue samples. They inversely regulate the expression of cyclin-D1 thereby controlling the cell cycle progression and tumorigenesis (Bandi et al., 2009). One of the important tumor suppressors and cell-autonomous suppressors of epithelial—mesenchymal transition (EMT) and metastasis, miRNA-200, is downregulated in lung cancer (Chen et al., 2014). Several other profiling experiments with miRNA expression analysis in blood plasma (Zaporozhchenko et al., 2018) and meta-analysis of lung cancer datasets (Vosa et al., 2013, Wang, Chen, & Wu, 2017) have revealed the expression patterns of miRNAs with high prognostic and diagnostic value in lung cancer patients (Wu, Tsai, Lien, Kuo, & Hung, 2019).

5.4.2 Long noncoding RNA

Increasing number of evidences are supporting lncRNA as emerging global regulators of the biological processes, including several cancer hallmarks. One of the best studied lncRNA, metastasis-associated lung adenocarcinoma transcript 1 (MALAT1), was originally identified in the NSCLC with upregulated expression and strong association with metastasis in NSCLC patients (Ji et al., 2003). MALAT1 directly binds to the C-terminal domain of the transcription factor specificity protein 1 (SP1) and increases its stability in LUAD cells. This stabilized SP1 leads to the overexpression of target genes along with a transcription of MALAT1 in a positive feedback loop manner (Li et al., 2018). Hypoxia in NSCLC cells induces the overexpression of lncRNA HOX transcript antisense intergenic RNA (HOTAIR) resulting in increased cancer cell proliferation, migration, and invasion. This upregulation is mediated by the binding of hypoxia-inducible factor-1α to the promoter region of HOTAIR (Zhou et al., 2015). One of the most studied and deregulated transcription factors in tumors, c-Myc, is known to induce the expression of lncRNA H19 in lung cancer in addition to several other cancer types (Barsyte-Lovejoy et al., 2006). Additionally, c-Myc also induces the expression of another lncRNA called antisense ncRNA in the INK4 locus (ANRIL) and promotes tumorigenic properties in NSCLC cells (Lu et al., 2016). Similarly, transcription factor Yin Yang-1 is overexpressed in lung cancer tissues and induces the expression of lncRNA plasmacytoma variant translocation 1 (PVT1) by directly binding to the promoter region and activating its transcription (Huang et al., 2017). The downregulation of tumor suppressor lncRNA SPRY4 intronic transcript 1 (SPRY4-IT1) is mediated by the polycomb group protein enhancer of zeste homolog 2 in NSCLC cells. Repression of lncRNA SPRY4-IT1 affects the EMT leading to enhanced cell proliferation and metastasis and reduced survival of NSCLC patients (Sun et al., 2014a).

5.4.3 tRNA-derived small RNA

Conventionally, transfer RNAs are supposed to be only involved in the protein translation process by transferring the activated amino acids to the ribosome. However, recent discoveries of tRNA-derived ncRNAs in cancer (Huang et al., 2018) have revealed their regulatory capacity too. Notably, tsRNA-3676 and tsRNA-4521 were identified in chronic lymphocytic leukemia (CLL) and lung cancer samples with downregulated expression and mutations (Pekarsky et al., 2016). Another study found that in addition to tsRNA-3676 and tsRNA-4521 (also known as ts-101 and ts-53, respectively), two more tsRNAs, ts-46, and ts-47, are downregulated in CLL and lung cancer (Balatti et al., 2017).

5.4.4 Circular RNA

circRNA perform their regulatory function predominantly through miRNA sponging activity. Many studies of circRNA function focus on the circRNA—miRNA—mRNA ternary complex formation. The expression of circRNA 100146 is upregulated in NSCLC cells and directly binds with miR-361-3p and miR-615-5p along with splicing factor SF3B3. The sponging of these miRNAs and splicing factors affects the expression of several downstream targets involved in cancer cell proliferation, invasion, and apoptosis (Chen et al., 2019). Similarly, circRNA PVT1 sponges miR-497 and promotes cellular processes involved in cancer progression in NSCLC (Qin et al., 2019) Another circRNA, hsa_circ_0013958, is found to be upregulated in LUAD patients. It acts as a sponge for miRNA-134, leading to the upregulation of cyclin-D1, and promotes cancer progression in LUAD (Zhu et al., 2017). The overexpression of circ_0000735 in NSCLC cells and tissue enhances the cancer-related cellular activities by sponging miR-1179 and miR-1182 (Li, Jiang, Liu, Lv, & Guan, 2019).

5.4.5 Small nucleolar RNA

Another class of ncRNA with an established role in lung tumorigenesis is snoRNA. The expression profiling study of lung tumor—initiating cells revealed that there are 22 snoRNAs showing altered expression patterns. Among these 22 snoRNAs, particularly two snoRNAs (snoRA3 and snoRA42) are associated with increased mortality of NSCLC patients (Mannoor, Shen, Liao, Liu, & Jiang, 2014). Additionally, the expression of SNORA42 is also upregulated in NSCLC tissues. The overexpression of SNORA42 is caused by the genomic amplification, leading to an increase in cell growth and colony formation in NSCLC and bronchial epithelial cell lines (Mei et al., 2012). One more snoRNA, overexpressed in NSCLC, is SNORD78. The oncogenic activity of SNORD78 is associated with EMT leading to increased cellular invasion in NSCLC cells (Zheng et al., 2015).

5.4.6 PIWI-interacting RNA

In general, piRNAs are thought to be expressed in germline cells with roles in the suppression of transposon activity and gene expression during the developmental processes. However, many piRNAs are emerging as important regulators of cellular processes during cancer conditions. The expression of 555 piRNA was measured using RNA-sequencing experiments with lung bronchial epithelial and NSCLC cell lines. Among these 555 piRNAs, the one with the most downregulated expression is piRNA-like-163 in NSCLC cells (Mei et al., 2015). Moreover, the microarray expression analysis has revealed that piRNA-34871 and piRNA-52200 are upregulated and piRNA-35127 and piRNA-46545 are downregulated in NSCLC cells (Reeves, Firek, Jliedi, & Amaar, 2017). Similarly, piRNA-651 is also overexpressed in NSCLC along with other cancer cells (Zhang et al., 2018). These piRNAs are involved in the regulation of the tumorigenic behavior of cancer cells.

5.4.7 Natural antisense transcripts

Pervasive transcription of the human genome produces a large number of NATs that regulate the expression of various genes in cells. The strand-specific paired-end RNA-sequencing experiments have revealed many important NATs with functional roles in cancer. One such NAT identified in lung cancer patient samples is NKX2–1-AS1 which regulates the expression of oncogene NKX2 and regulates cell proliferation (Balbin et al., 2015). The expression of another lncRNA NAT, family with sequence similarity 83 member A antisense RNA 1 (FAM83A-AS1), is also upregulated in NSCLC cells. FAM83A-AS1 overexpression leads to the upregulation of FAM83A expression and thereby promotes cellular proliferation, and invasion (Shi, Jiao, Yu, & Wang, 2019).

5.4.8 Transcribed ultraconserved region

These regions encode for the lncRNAs with regulatory roles in human cancers. One such T-UCR with prooncogenic potential, ultraconserved element 338 (uc.338), is highly expressed in lung cancer patients. It regulates the migration and invasion properties of lung cancer cells (Gao, Gao, Li, Zhang, & Gao, 2016). Another T-UCR, uc.339, which acts as a decoy for the miR-339-3p, -663b-3p, and -95-5p, is overexpressed in NSCLC samples. Thus uc.339 affects the expression of miRNA target and regulates the tumorigenesis in lung cancer (Vannini et al., 2017). On the contrary, the expression of uc.454 in lung cancer cells is downregulated with a potential tumor-suppressive role in lung cancer (Zhou et al., 2018).

5.5 Sample types used for analyzing ncRNAs

The differential expression of ncRNAs among malignant and nonmalignant lung tissues/cell lines makes them a credible biomarker for the detection of lung neoplasm. Fortunately, investigations focused on pathological consequences of ncRNAs are available in decent numbers and they collectively suggest their mechanistic involvement in the manifestation of diverse forms of lung diseases, including nonmalignant as well as of malignant nature. Appropriateness of ncRNAs has also been braced by various investigations. A large effort in the characterization of diverse forms of ncRNAs in varied malignant conditions of lungs leads to the identification of associated ncRNA signatures.

Apart from its molecular expression profile, suitability for discrimination of disease conditions, measurability with samples obtained with marginal invasiveness, distress, and risk of injury to patients are aided advantages. Samples acquired through noninvasive or minimally invasive interventions include biofluids. Amusingly, few forms of ncRNAs (majorly miRNAs) have been detected in all types of body fluids, including blood and urine. It is interesting to note that few of body fluids have inimical circumstances that favor the degradation of RNAs. Moreover, biofluids derived from respiratory tracts such as sputum also have the presence of ncRNAs which can be exploited for diagnosis. A relatively more invasive fluid sampling of bronchoalveolar lavages has been shown to provide better specificity and sensitivity.

5.5.1 Blood

Peripheral blood is among the most frequent samples to be analyzed for disease diagnosis purposes. Detection and profiling of ncRNAs in the circulation for discrimination between healthy and neoplastic conditions are also no exception. Moreover, profiling of ncRNAs in blood has the power to categorize nonmalignant conditions such as pulmonary fibrosis, pulmonary tuberculosis, chronic obstructive pulmonary disease (COPD), and cystic fibrosis, where the clinical exhibitions and radiological features mostly overlap. Accurate categorization of disease conditions is prime-most for rightful and effective therapeutic management. Using more than one ncRNAs preferably from different classes/origins can improve the accuracy of diagnosis. Among the ncRNAs, miRNA hold the prevalent acceptability for the diagnosis of different lung tumor stages (Le, Romano,

Nana-Sinkam, & Acunzo, 2021). A study on profiling of cell-free circulating miRNAs from blood plasma has identified 179 miRNAs for the detection of lung cancer (Zaporozhchenko et al., 2018). Intriguingly, the levels of miR-15/16, miR-29, miR-34, miR-200 family, let-7, miR-21, miR-155, and miR-17−92 cluster were found significantly altered in lung cancers. Increased or decreased expression level in biofluids correlates to the either potential of miRNA, that is, tumor suppressor or oncogenic (Le et al., 2021). Apart from miRNA, other small ncRNAs also hold great promise in diagnosis of lung malignancies. Gu et al. (2020) demonstrated the potential of circulating small ncRNA other than miRNA, which includes tsRNA, rRNA (rsRNA), and YRNA (ysRNA), collectively termed TRY-RNA. Moreover, blood-derived TRY-RNA leaves only miRNA profiling behind the classification of normal, nonmalignant, and malignant lung conditions (Gu et al., 2020). It is also suggested that combining TRY-RNA with circulating miRNA profiling provides additional authority in the discrimination of pathological conditions. Several lines of investigations collectively suggest the potential of small ncRNA for lung cancer diagnosis. Moreover, blood samples also serve as a source of lncRNA biomarkers in the diagnosis of respiratory organ malignancies (Ginn, Shi, Montagna, & Garofalo, 2020). High specificity of MALAT1 in blood samples was reported for the diagnosis of NSCLC (Ginn et al., 2020; Weber et al., 2013). However, low sensitivity of blood MALAT1 restricts its alone use for diagnostic detection of lung cancers. Nevertheless, a meta-analysis demonstrates that despite shortcomings as diagnostic marker, blood MALAT1 expression in NSCLC alone can serve as a very efficient prognostic factor (Ginn et al., 2020; Li et al., 2016). MALAT1 has also been found overexpressed in blood as well as other samples derived from patients/animal models of malignancies other than lung cancer, including colon, prostate, and breast (Sun and Ma, 2019). Serum also serves as a source for the detection of HIF1A-AS1 and XIST as lung cancer biomarkers (Tantai, Hu, Yang, & Geng, 2015); and combined detection of both (HIF1A-AS1 and XIST) has added accuracy in the diagnosis of lung cancer as compared to their individual detection (Ginn et al., 2020; Tantai et al., 2015). Lin et al. found reliable measurement of 7 lncRNAs in plasma in a cohort study with lung cancer patients for certifying the diagnostic potential. Furthermore, they detected augmented levels of SNHG1 and RMRP lncRNAs in plasma of patients suffering from lung malignancies as compared to healthy individuals and suggested a high specificity ($>84\%$) and sensitivity ($>87\%$) when used in combination (Lin, Leng, Zhan, & Jiang, 2018). An investigation using the murine model of lung carcinogenesis confirms the comparative potential of blood samples for the detection of NR_026689 lncRNA (Wu, Li et al., 2016; Wu, Lin et al., 2016). Long intragenic nonprotein-coding RNA p53-induced transcript is another lncRNA of which low serum level correlates to low tissue expression in NSCLC and associated with metastasis (Zhang, Gong, Li, & Tang, 2021; Zhang, Liu, Zhang, & Zhang, 2021). These investigations collectively suggest that blood samples hold sturdy contest with tissue samples for the detection of both small and long noncoding RNAs for lung cancer diagnosis.

Although circulating ncRNAs show stability in blood samples, various investigations suggest for quick and fast processing of sample without any delay (Wu, Li et al., 2016; Wu, Lin et al., 2016) for high sensitivity. Investigation on stability of ncRNAs in blood samples based on their collection, processing, and storage (Glinge et al., 2017) indicates differential susceptibility. Immediate processing of blood samples for ncRNA favors the optimal use in diagnostic purposes as this prevents the changes over time. Low-temperature storage with minimal disturbances was found less deteriorative for ncRNAs (Felekkis and Papaneophytou, 2020, Glinge et al., 2017). The type of anticoagulant for the collection of blood samples also affects the processing outcome of the samples. Heparin can interfere with the activity of reverse transcriptase and citrate can hinder the PCR outcomes (Felekkis and Papaneophytou, 2020). Therefore, ethylenediaminetetraacetic acid (EDTA) is considered as the most suitable anticoagulant for blood sample collection for ncRNA. Moreover, heparinase treatment before processing has been shown to prevent the hindrance with reverse transcriptase activity by heparin.

Storage of whole blood is not preferred for ncRNA detection as it causes qualitative and quantitative changes. Interestingly, ncRNAs are more stable in plasma as compared to serum during low-temperature storage and freeze–thaw cycles (Farina et al., 2014; Felekkis and Papaneophytou, 2020; Glinge et al., 2017). Plasma sample storage at ultralow temperature and freeze–thaw cycles demonstrates the differential susceptibility among a panel of eight circulating miRNAs (Matias-Garcia et al., 2020). However, ultralow temperature for storage of plasma samples maintains the stability of almost all of the miRNAs analyzed for decades (Matias-Garcia et al., 2020).

5.5.2 Bronchoalveolar lavage fluid

In lung cancer, bronchoalveolar lavage fluid (BALF) has the secretory components from neoplastic cells and hence serves as a prospective sample for the detection of biomarkers. Few of the very earlier studies indicate the presence of coding RNA entities specific to the tumor in BALF (Engel et al., 2004). The presence of ncRNAs in BALF has also been confirmed with their diagnostic implication in a variety of lung disorders (Li, Yin, Fan, Zhang, & Yang, 2019). MiRNA profile in BALF differs from that of blood samples in patients of respiratory pathology (Molina-Pinelo et al., 2012) that suggests the necessity of careful corroboration of ncRNA sourced to BALF for clinical implications. Rodríguez et al. demonstrated that BALF has less amount and diversity of miRNA as compared to plasma miRNAs. Augmented miRNA levels in tumor plasma miRNA pool more intensely correlate to occurrence of malignancies. However, BALF of tumor patients was found devoid of many miRNAs that were present only in the plasma pool of nonmalignant patients (Rodríguez et al., 2014). The differential profile of BALF miRNAs (miR-15b, miR-192, and miR-221) is reported in inflammatory lug disorder sarcoidosis (Kiszałkiewicz et al., 2016). Sarcoidosis refers to the formation of granulomata of inflammatory cells and correlates to increased risk of lung malignancy (El Jammal, Pavic, Gerfaud-Valentin, Jamilloux, & Sève, 2020). Using a panel of miRNAs, namely, miR-21, miR-143, miR-155, miR-210, and miR-372, in BALF and sputum, a high level of diagnostic sensitivity (>85%) and specificity (100%) was achieved by Kim et al. (2015). Without BALF miRNA profile a cluster analysis of only sputum has a relatively low sensitivity and specificity of ~68% and 90%, respectively. Schmidt et al. also reported differential expression of eight miRNAs in BALF of lung cancer patients when compared to normal individuals. Among these eight, five upregulated miRNAs (U6 snRNA, hsa-miR 1285, 1303, 29a-5p, 650) were suggested as promising biomarkers for lung cancer (Schmidt, Rehbein, & Fleischhacker, 2016). Among these miRNAs, hsa-miR 1285 potentiates lung cancer progression via enhancing proliferation and metastasis through the downregulation of CDH1 and Smad4 (Zhou, Zhang, Zheng, Zhao, & Han, 2017). This indicates that BALF may have the functional ncRNA biomarkers that have better diagnostic precision along with potential for therapeutic targeting.

BALF miRNA profile modulates the events favorable to carcinogenesis. With a murine experimental model, cigarette smoke exposure modulated the miRNA in BALF (Izzotti et al., 2018). These altered profiles of miRNA in BALF were detected before the occurrence of any detectable pathological changes in histology. This suggests that along with miRNA profiles of other body fluids, BALF also holds promise for the early diagnosis of lung cancer. The levels of miR-126 and Let-7a are reported higher in exosomes derived from BALF of LUAD patients as compared to control subjects in a pilot study (Kim et al., 2018). BALF exosomal miRNA levels correlate to miRNA levels in tumor tissue samples. Another pilot study confirms the diagnostic potential of combined mRNA levels of fragile histidine triad, and retinoic acid receptor-β and their regulatory miRNAs (miRNA-34a, miRNA-141, miRNA-143, and miRNA-217) profile in BALF for SCC (Dutkowska, Antczak, Domańska-Senderowska, & Brzeziańska-Lasota, 2019). Thus available evidence suggests that BALF ncRNAs might also help in the early detection of lung cancer.

5.5.3 Sputum

Sputum is one of the samples which can be collected without much discomfort to the patient. As per its origin, it has molecular as well as cytological features that can be exploited for the diagnosis of respiratory tract pathologies, including the diagnosis of lung cancer (Gyoba, Shan, Roa, & Bédard, 2016; Liao et al., 2014). In a review, Gyoba et al. (2016) reported that sputum miRNA analysis has a sensitivity and specificity of more than 61% and 80%, respectively. Apart from miRNAs in sputum, other ncRNAs have also shown their presence with diagnostic potential for lung cancer. The presence of snoRNA along with miRNA was confirmed (Su, Fang, & Jiang, 2016; Su, Guarnera, Fang, & Jiang, 2016). Combined sputum snoRNA along with miRNA presented high sensitivity and specificity (~89%), suggesting their implication in early detection of lung cancer. Sputum ncRNA analyses not only provide the early detection of lung malignancies but also assist in the identification of therapeutic targets/interventions. One such study demonstrated that the methylated form of miR-196b upregulates in sputum samples during events of lung carcinogenesis (Tellez et al., 2016). Through sputum miRNA analysis, this study demonstrated the epigenetic correlations of ncRNAs in lung pathology. Moreover, integration of miRNA detection in sputum and methylated DNA biomarkers (*RASSF1A*, *PRDM14*, and *3OST2*) significantly improved the sensitivity and specificity of NSCLC diagnosis (Su, Fang et al., 2016; Su, Guarnera et al., 2016). Similarly, combination of different classes of ncRNAs can be optimized for a better and accurate diagnosis approach. Kim et al. (2015) indicated that sputum miRNA profile alone has the diagnostic potential for NSCLC but combining with plasma miRNA profile significantly improves the degree of sensitivity and specificity. A meta-analysis to assess the diagnostic potential of miRNA biomarker detection in sputum for NSCLC included 19 investigations (Zhang, Wang, & Zhang, 2019), which demonstrates the consistency of sputum miRNA detection for diagnosis with pooled sensitivity and specificity of 75% and 88%, respectively. This meta-analysis also confirmed that analysis of multiple miRNA biomarkers has high sensitivity (84%) as compared to single miRNA. A clinical trial for clustering and profiling of sputum miRNAs for the detection of NSCLC confirmed that a panel of five miRNAs (miR-21, miR-143, miR-155, miR-210, and miR-372) has the potential of prospectively diagnose with high sensitivity (83.3%) and specificity (100%) (Roa et al., 2012). Collectively, it can be inferred that sputum miRNAs hold high potential in the diagnosis of lung neoplastic disorder with the added advantage of the ease of sampling. Moreover, sputum ncRNA profile correlates to other biomarkers, and combining them further improves the accuracy of diagnosis.

5.5.4 Tumor biopsies

Tissue samples of suspicious malignancies will have the actual molecular alteration existent in the neoplasm. As it is established that ncRNAs have the mechanistic role as protooncogene and tumor suppressor, the alterations in ncRNA is reflected in tumor tissues. A large number of ncRNAs in tumor tissues have been identified as biomarkers for lung cancer (Ginn et al., 2020; Le et al., 2021; Li et al., 2016; Sun and Ma, 2019). Different methods of tissue sampling from the suspicious area can be implemented as the necessity of downstream processing, degree of clinical manifestations, etc. Therefore despite being an invasive technique, tissue samples provide a better diagnosis with an improved degree of accuracy. The methods of lung tissue sampling include needle biopsy, transbronchial biopsy, thoracoscopic biopsy, and open biopsy.

Needle biopsy from the lung is done under local anesthesia while putting a needle through the chest. Placing the needle in the right place is insured by guidance through computational tomography or fluoroscopy. Alternatively, such a type of biopsy method is also called percutaneous biopsy or transthoracic needle aspiration (TTNA). Fassina, Cappellesso, and Fassan (2011) reported that miRNAs of hsa-let family were upregulated in adenocarcinoma, while has-miR-205 was found upregulated in SCC in TTNA samples of NSCLC. This indicates the potential of ncRNAs as biomarkers as well as markers for the classification of diseases.

For the transbronchial biopsy, endobronchial ultrasonography-guided transbronchial needle aspiration (EBUS-TBNA) is the most commonly used technique. Microarray-based expression analysis of RNAs in 24 EBUS-TBNA samples of lung cancer reported the feasibility of whole transcript and miRNA analysis for the identification of disease and hierarchical clustering (Nakajima et al., 2012). Another study by Petriella et al. (2013) also indicates the potential of fine needle aspirate samples' miRNA analysis in discrimination of disease and healthy patients compared to that of open biopsy. Needle biopsy's miRNA analysis was found to have a specificity of more than 95% (Nakajima et al., 2012). Analysis of miRNA panel in EBUS-TBNA samples also suggests that miR-200c can be exploited as a biomarker for enhanced nodal staging in lung cancer (Inage et al., 2018).

Thoracoscopic biopsies are done under general anesthesia using an endoscope through the middle of the ribs to take tissue samples. This method is relatively more invasive than the previously discussed TTNA and TBNA. However, therapeutic intervention in the form of removal of tissue/nodule can be achieved by this sampling method. With miRNA analysis in thoracoscopic samples, miR-486-5p downregulation is found, which is correlated to stages of lung cancer (Mohamed, Mohamed, El-Kaream, Badawi, & Darwish, 2018). Hence, mir-486-5p can serve as a biomarker for local and systemic spread of lung cancer and can provide additional confidence when used along with other more established techniques for staging.

Open biopsies provide the possibility of large tissue samples which can be processed to meet the heterogeneity in tumor tissues (core and peripheral tissues). This biopsy method assists the preclinical research for the identification of ncRNA profiles in lung cancer staging and their mechanistic involvement.

5.6 Cell signaling pathways modulated by ncRNAs in lung cancer patients

Gene-expression patterns, activation of enzymes/pathways, and phenotypic consequences in cells are regulated by various signaling pathways. A fault in mediators and effectors of cell signaling leads to various pathological conditions. In malignancies, changes at the genetic level are mostly associated with mediators of cell signaling, including the receptors, intracellular signal mediators, and transcription factors. ncRNAs also contribute to the regulation of gene expression (transcription/translation) and/or availability of functional proteins (Lin et al., 2018). Most of the small ncRNAs are coded from chromosomal regions having the highest degree of instability which makes them susceptible to a probable collateral change during chromosomal alterations. This leads to a change in the copy number of ncRNAs (Lin et al., 2018; Ramón y Cajal, Segura, & Hümmer, 2019), and their interactivity to RNA and other biomolecules (Ramón y Cajal et al., 2019). The most studied mechanism is transcriptional regulation by ncRNAs. Sequence-specific interaction of miRNA with mRNA halts the transcription of mRNA leading to the prevention of protein formation. Additionally, lncRNAs have the ability to interact with miRNA and restrict their translation inhibitory effects. It will be interesting to note that lncRNAs have the tissue specificity for their expression. Nevertheless, they also have specificity for their expression at different stages of development. This indicates that lncRNAs regulate the availability of functional proteins, by regulating their inhibitor miRNA. LncRNAs also have the ability to directly interact with mediators of cell signaling (such as receptors and signal mediators) (Lin et al., 2018). Moreover, in extracellular spaces, miRNAs are detected with significant level of stability. This suggests the regulation of cellular phenotype by neighboring or distant cells through the production and secretion of miRNA (Ramón y Cajal et al., 2019).

LncRNAs have the ability to modulate the cell growth regulatory pathways (Lin et al., 2018). The modulation of AKT/ERK pathways by lncRNA SOX2-OT leads to proliferation and poor prognosis in lung malignancies (Herrera-Solorio et al., 2021). MALAT1 is one of few first ncRNAs, overexpression of which favors lung cancer progression via inducing invasion through modulated ERK/MAPK pathway (Liu et al., 2019). Myc signaling was also demonstrated to favor oncogenic events driven by MALAT1 involving the regulation of miR-204 (Ye, Dong, Hou, Zhang, & Shen, 2021). Modulation of IGF2BP2-mediated signaling has been observed in response to MALAT1-mediated progression (Ye et al., 2021). Amendments in cell signaling by MALAT1 are also evident in malignancies of origins other than lung (Ramón y Cajal et al., 2019; Lin et al., 2018). STAT-3 also serves as a frequent target of ncRNAs in the oncogenesis and progression of cancer. LncRNA LEISA augments STAT-3's ability to interact with the promoter site in LUAD and favors the progression (Wu et al., 2021). In NSCLC, Yu et al. (2021) reported that SNORA47 favors the pathogenesis of lung malignancy through modulated PI3K/Akt signaling. Knockdown of SNORA47 significantly inhibits proliferation, and invasion and migration in NSCLC. Activation of Wnt/β-catenin signaling by long intergenic ncRNA LINC01006 augments proliferation, migratory potential, and mesenchymal transition in LUAD cells (Zhang, Gong et al., 2021; Zhang, Liu et al., 2021). LINC01006 targets miR-129−2-3p miRNA for activating the Wnt/β-catenin signaling (Zhang, Gong et al., 2021; Zhang, Liu et al., 2021). Wnt/β-catenin signaling is linked with augmented proliferation and metastatic behavior in malignancies of the lung and other origins (Koni, Pinnarò, & Brizzi, 2020). miR-29b-3p regulates the expression of MMP9 and TLR4 in lung cancer cells by directly interacting with HMGB1 (Zhong et al., 2021). Using shRNA-mediated silencing, lncRNA H19 was found as hierarchical regulator of miR-29b-3p. Aberrant expression of MMP9 and TLR4 in silenced cells indicates that H19 inhibits miR-29b-3p and prevents to suppress the expression of MMP9 and metastasis (Zhong et al., 2021). Tumor suppressor function of lncRNA Fer-1-like protein 4 drives through the activation of p53 pathway of apoptosis along with the inhibition of AKT phosphorylation in NSCLC (Ouyang et al., 2021).

Among the ncRNAs, miRNAs directly influence the translational availability of proteins. Various investigations have demonstrated that these small ncRNAs also influence the cellular signaling governing either oncogenic or tumor suppressor consequences (Ramón y Cajal et al., 2019; Ahn and Ko, 2020). Protumor miRNAs can induce the cell proliferative signaling, angiogenic regulators, EMT driving cascades on their own or can enhance their consequences (Ramón y Cajal et al., 2019). MiRNAs that have the tumor suppressor function either inhibit the translation of protooncogene or factor promoting their transcription (Ramón y Cajal et al., 2019; Ahn and Ko, 2020; Du and Pertsemlidis, 2012). Wnt signaling cascade regulates the fate of lung carcinogenesis and ncRNAs affect this axis in either way (Ahn and Ko, 2020; Chen, Cai, Gu, Yang, & Fan, 2021; Xie, Xue, Guo, & Yang, 2021). Inhibition of Wnt signaling by miRNA-520a culminated in the suppression of onset and progression of NSCLC (Xie et al., 2021). In vitro and in silico analysis demonstrated that miRNA-520a targets ribonucleotide reductase subunit 2 mRNA to inactivate Wnt/β-catenin signaling causing inhibited tumor growth and metastasis in an animal model (Xie et al., 2021). In contrast, another miRNA miR-665 modulates Wnt/β-catenin axis to facilitate the proliferation of lung cancer cells (Chen et al., 2021). The miR-665 inhibits apoptosis induction through the downregulation of caspases-3 signaling. Wnt/β-catenin was stimulated by miR-665 by targeting Tripartite motif 8 mRNA (Chen et al., 2021). Another miRNA, miR-22-3p, is reportedly downregulated in various cancer types by suppressing the cell proliferation and colony-forming ability of lung cancer cells. miR-22-3p inhibits the expression of STAT-3 and MET (MNNG HOS transforming gene) proteins in various cell lines of lung cancer origin (Chen et al., 2021). NcRNAs have their intervening regulatory network. Even the processing control of miRNA governs downstream signaling cascade outcome. Processing of oncogenic miRNA by DHX9 and NPM1 stabilization by lncRNA KIMAT1 involves KRAS and myc signaling in lung cancer (Shi et al., 2021). Augmented myc level also downregulates the expression of tumor suppressor miRNA, further strengthening the KRAS-mediated lung cell oncogenic transformation (Shi et al., 2021). The regulatory event among ncRNAs can be seen with miR-363-3p, a tumor-suppressing miRNAs, which has the suppressing effect on metastasis through the inhibition of pseudopodium-enriched atypical kinase 1 (PEAK1) (Geng, Li, Li, Wu, & Chen, 2021). RNA pull-down assay verifies its inhibition by lncRNA NORAD (noncoding RNA activated by DNA damage) leading to upregulation of PEAK1, which consequently favors metastasis (Geng et al., 2021). The miR-363-3p also has an inhibitory effect on ERK signaling in NSCLC cells (Geng et al., 2021). EMT has been linked with aggressive behavior of cancer cells, including resistance to therapy. The miR-183 augments EMT in LUAD H1299 cells through modulated expression of zinc finger E-box-binding homeobox 1 along with resistance to radiotherapy (Huang et al., 2021). Genetic transformation causing over expression of miR-183 favors the malignant cell proliferation while inhibition using shRNA endorse radiosensitivity in lung neoplastic cells (Huang et al., 2021). MiRNAs also show chemo-sensitizing ability. mir-200c, an miRNA from the extracellular vesicle, augments NSCLC cell sensitivity toward gefitinib via inhibiting downstream signaling of EGFR (Lin et al., 2021). Inactivation of EGFR signaling by miR-200c also prevents EMT in NSCLC (Lin et al., 2021). miR-518d-5p modulates chemotherapy response in SCLC through regulating Delta-like protein 3 (DLL3) and RBP lin-28 homolog B (LIN28B) (Huang, Cao, Sha, Zhu, & Han, 2019). The augmented expression of DLL3 correlates to severity in lung malignancies and serves as a biomarker for poor prognosis in SCLC patients (Huang et al., 2019). Another miRNA, miR-483-3p, is found suppressed in EGFR-mutant NSCLC cells and has shown counteracting ability against chemoresistance (Yue et al., 2018). The overexpression of miR-483-3p favors the sensitivity of lung cancer cells toward tyrosine kinase inhibitor gefitinib. Targeting β3 integrin by miR-483-3p suppresses the ERK signaling and prevents invasion and metastasis of NSCLC cells (Yue et al., 2018). Epigenetic alteration in the miR-483-3p promoter region (hypermethylation) prevents the expression of this tumor-suppressing miRNA in lung cancer cells (Yue et al., 2018).

The altered levels of ncRNAs modulate various cell signaling pathways in lung cancer through preventing the expression of signal mediators by miRNAs and through the regulation of miRNAs by lncRNAs. In lung malignancies, ncRNAs affect various signaling pathways through modulating expression/activity of various signaling molecules, including EGFR, ERK, PI3K, AKT, PEAK1, SNAIL, and DLL3. These ncRNA-driven alterations in cell signaling events modulate the phenotypic behavior of lung cancer cells. EMT and chemoresistance in lung neoplasm can be regulated by ncRNAs in either direction favoring aggression or repression. Careful exploration of ncRNA-mediated amendment in cell signaling in oncogenic transformation or maintaining the normal cell phenotype can provide various therapeutic targets.

5.7 NcRNAs as predictive markers for lung cancer patients

NcRNAs can affect the maintenance of cancer hallmarks and aggressive behavior of transformed cells. Different ncRNAs belong either to protumor (protooncogene) or tumor suppressor categories. Moreover, they also play an instrumental role before the onset of oncogenesis transformation governing the susceptibility for malignant transformation (Chen, Zheng, Zhuo, & Wang, 2017; Chen, Zhao et al., 2017; Chen, Min et al., 2017; Srivastava and Srivastava, 2012). Various ncRNAs have been linked with the susceptibility of individuals for malignancies originated in the lung. The expression of miRNAs, their single nucleotide polymorphism (SNP) variants, or epigenetically altered forms show either increased or decreased susceptibility for lung cancer. One of the first pieces of evidence of its kind by Tian et al. reported that rs11614913 SNP-associated functional variant of miR-196 (miR-196a2) correlates significantly to a higher risk of lung cancer in the Chinese population. SNP variant miR-196a2 was found to increase the risk of lung cancer by 25% in a case–control study on more than 1000 lung cancer patients as well as cancer-free individuals (Tian et al., 2009). A meta-analysis of various miRNA polymorphisms also suggested mir-196a2 rs11614913 as a risk factor for overall cancer risk in the Asian population especially lung and colorectal cancer (Srivastava and Srivastava, 2012). However, an SNP of miR-135a/b (rs2240688A > C) decreases the risk of lung cancer in Chinese individuals (Cheng et al., 2013). rs2240688A > C SNP of miR-135a/b augments its binding ability and prevents the translation of mRNA of CD133. Moreover, the presence of rs2240688A > C predicts a good prognosis in lung cancer patients as observed in the analyses of different case–control studies involving more than 4500 individuals (Cheng et al., 2013). Genetic variants of miRNA also affect the risk based on other factors. A polymorphism CG or GG at rs2910164 (in miR-146a coding region) has a decreased risk of lung cancer as compared to CC polymorphism (Jeon et al., 2014). While comparing in a stratified statistical analysis with gender, and age, CG/GG variants do not have a significant difference with CC variant in risk level. However, the risk for lung cancer is higher with CC variants in never smokers, while no significant difference exists among smokers (Jeon et al., 2014). This indicates that predictive values of ncRNA in the incidence of malignancies must be considered with careful inspection of other factors like exposure to carcinogens. Specific polymorphism variants of miR-146a rs2910164, miR-196a2 rs11614913, miR-608 rs4919510, miR-27a rs895819, and miR-423 rs6505162 and cooking oil exposure were analyzed in Chinese females for incidence of lung cancer by Yin et al. (2015). Although no biological interaction between exposure to cooking oil fume and polymorphism variants was detected in multivariate analysis, females having specific risk genotype and cooking oil fume exposure have a significantly high risk of lung cancer development (Yin et al., 2015). Differential expression of genetic variants of miRNA in LUAD tumors of TCGA is observed (Xie et al., 2016). The quantitative trait locus analysis reveals that specific variants of regulatory regions of miR-200b/200a/429 cluster and miR-30a are linked with risk for NSCLC (Xie et al., 2016).

The predictive value of miRNA varies for different cancer types, populations, and exposure levels of carcinogenic factors. Polymorphism of miR-432 can modify the risk level for cancer. A stratified analysis of cancer type with rs6505162 polymorphism variant of miR-432 indicates a decreased risk of lung cancer along with gastrointestinal cancer and colorectal cancer (Moazeni-Roodi, Ghavami, & Hashemi, 2019). However, the variant is not associated with risk modification for gastric, esophageal, or breast cancer. As compared to the non-Asian population, the Asian population having the miR-432 variant hence shows relatively less susceptibility toward the onset of cancer development (Moazeni-Roodi et al., 2019). Predictive estimates of miRNAs in cancer are also affected by variations in sequences of their target gene/mRNA affecting their binding. An SNP (rs9224) in the untranslated region of FAM13A affects the binding of specific miRNA miRNA-22-5p and consequently affects the susceptibility for oncogenic transformation (Yu et al., 2019). Moreover, it also modifies the prognosis of patients suffering from squamous carcinoma of the lung (Yu et al., 2019). Using more than 11,000 subjects of different case–control studies, a meta-analysis indicates that miR-4293 rs12220909 variant heterozygous polymorphism could be a shielding factor especially against lung cancer (Ji, An, & Fang, 2020). Using KEGG pathway analysis, the protective role of the miR-4293 rs12220909 variant was traced down to the regulation of transcription factor and RNA polymerase binding with DNA (Ji et al., 2020). Epigenetic alterations in miRNA are also modifying their predictive values. The methylated form of miR-9 family members predicts better overall survival among NSCLC patients (Muraoka et al., 2012). Methylation causes the expressional silencing of miRNA which indicates that miR-9 family members have a role in oncogenic transformation and metastatic events (Muraoka et al., 2012).

Similar to miRNA, expression/polymorphism of lncRNA also has predictive implication for lung cancer. Expression of MALAT1 is associated with lung cancer as a diagnostic as well as a predictive biomarker. MALAT1 overexpression in lung cancer patients predicts a significantly high degree of metastasis both to nearby or distant tissues (Zhu, Liu, Shenglin, & Zhang, 2015). MALAT1 expression level also correlates to poor overall survival among patients affected by lung cancer (Zhu et al., 2015). However, up recent Chen et al. (2021) demonstrated a significantly diminished risk of lung cancer with rs619586 A/G polymorphisms of MALAT1. A meta-analysis of lncRNA polymorphism and risk of cancer indicate that specific variants of MALAT1, ANRIL, HOXA distal transcript antisense RNA, and prostate cancer-associated ncRNA 1 (PRNCR1) correlate to the risk of cancers of various origins, including lung malignancies (Huang, Zhang, & Shao, 2018). PRNCR1 polymorphism variants also correlate to incidences of NSCLC in a meta-analysis of case—control study with Chinese individuals (Li et al., 2021). Among the Chinese population, polymorphism in another lncRNA prostate cancer-associated transcript 1 correlates to the risk of lung cancer independent of environmental factors (Bi et al., 2019). A hospital-based case—control study reveals that two polymorphisms of maternal-expressed gene 3 (MEG3) determine the susceptibility for lung cancer (Yang et al., 2018). Stratified analysis suggests no significant interaction between polymorphism and smoking in lung cancer susceptibility. Various polymorphism variations affect the secondary structure and consequently modulate their role either in the prevention or promotion of malignant transformation of cells. In silico analysis indicates that risk SNP rs114020893 polymorphism in the lncRNA NEXN-AS1 drives the changes in the overall structure of this lncRNA (Yuan et al., 2016). This polymorphism was found significantly associated with lung cancer risk in a meta-analysis of genome-wide association studies, including more than 250,000 subjects (Yuan et al., 2016).

Taken together, miRNA and lncRNA expression levels and their polymorphic variants correlate to risk level for lung carcinogenesis. Epigenetic modifications in the genes of ncRNA or target gene promoter regions also affect their predictive values. Few of these predictive markers of ncRNA also predict the aggressive phenotypes of transformed cells, including the degree of metastasis, and therapy resistance. Predictive values of these ncRNAs are affected by the presence/exposure to carcinogenic factors like smoking, oil fume exposure, race, age, and gender. Therefore it can be suggested that a careful evaluation of ncRNAs as risk predictive factors and their use along with other molecular markers will provide a better strategy for efficient prediction of individuals with a high risk of lung cancer. This will pave the path for designing the preventive strategies or avoiding the risk factors exposure on individuals having a specific profile of predictive ncRNA markers for lung cancer.

5.8 Diagnostic potential of ncRNAs in lung cancer patients

Accumulating evidence strongly suggests the implication of dysregulated levels of ncRNAs in the development and progression of lung cancer. The aberrant expression/level of ncRNAs, namely, miRNAs and lncRNAs, has been reported in the alteration of various crucial cellular processes, including proliferation, apoptosis, angiogenesis, invasion, and metastasis (Enfield, Pikor, Martinez, & Lam, 2012; Ghafouri-Fard, Shoorei, Branicki, & Taheri, 2020). In recent few years, several ncRNAs have been shown a great diagnostic potential for the diagnosis of different stages of lung cancers.

5.8.1 LncRNAs as diagnostic markers

Various lncRNAs are identified as crucial regulatory molecules for the initiation and progression of lung cancer due to their relative stability in plasma and hence have a great potential to be used as a suitable diagnostic marker for lung cancer. More so, various serum exosome-associated lncRNAs are also reported as potential clinical biomarkers for staging lung cancer.

Studies have reported an elevated level of HOTAIR in patients with lung cancer. The plasma level of HOTAIR has also been used for the diagnosis and monitoring of lung cancer. Likewise, the increased level of H19 is also observed in the lung cancer tissues and is negatively associated with lung cancer patient survival. A high level of ANRIL is detected in the serum of lung cancer patients, which supports its diagnostic value for the early detection of lung cancer. MALAT1 (also known as nuclear-enriched transcript 2), a highly conserved lncRNA, is found highly overexpressed in a wide variety of cancers, including lung cancer during the early metastasizing stage (Ji et al., 2003). The role of MALAT1 has been reported in the progression of lung cancer via inducing migration and invasion possibly via mediating abnormal alternative splicing and promoting altered expression of few oncogenic transcription factors such as B-MYB (Schmidt et al., 2011; Tripathi et al., 2013). Further, a study has shown a significantly elevated level of MALAT1 in the blood of lung cancer patients (Weber et al., 2013). Together, the abovementioned reports strongly suggest that MALAT1 can be a potential diagnostic biomarker for the detection of lung cancer. However, some of the drawbacks such as unsatisfactory sensitivity prevent MALAT1 from being considered a single diagnostic marker but can be used as a complementary marker. A study by Zhang et al. (2017) has suggested the utilization of serum exosomal MALAT1 as a noninvasive diagnostic marker. In this study, they noticed a significantly elevated level of MALAT1 in lung cancer patients, which was associated with tumor stage and lymphatic metastasis. Colon cancer–associated transcript 2 (CCAT2) has also been identified as a diagnostic marker due to its significantly higher level of expression in lung cancer (Qiu et al., 2014). Further, Qiu et al. have shown the crucial role of CCAT2 in the proliferation and invasion of lung cancer cells. They have also suggested that the detection of CCAT2 in combination with serum carcinoembryonic antigen could be a better choice for the prediction of lymph node metastasis in lung cancer patients. Liang et al. assessed the level of lncRNA-growth arrest–specific transcript 5 (GAS5) in the blood samples of lung cancer patients along with healthy controls and observed significantly decreased GAS5 levels in lung cancer patients. Further, they suggested that GAS5 may be a promising biomarker for the diagnosis of lung cancer. The detection of two or more lncRNAs is considered a better method for the diagnosis of lung cancer. Hu et al. (2016) have suggested to diagnose lung cancer by detecting the plasma levels of SPRY4-IT1, ANRIL, and nuclear paraspeckle assembly transcript (NEAT). In addition, the detection of plasma levels of XIST and HIF1A-AS1 in combination is identified as a much better choice for the diagnosis of lung cancer than XIST or HIF1A-AS1 alone (Tantai et al., 2015).

5.8.2 miRNAs as diagnostic markers

The altered levels of miRNAs are found at different stages of lung cancer in the tissues as well as in the blood of cancer patients with respect to their normal counterparts. Therefore a growing number of evidence suggests the great diagnostic potential of miRNAs for the detection of lung tumor stages.

To identify the diagnostic significance of miRNAs for lung cancer, Fan et al. analyzed the expression profile of 12 miRNAs in the serum of lung cancer patients and healthy controls. Remarkably, they found significantly lower expression of five miRNAs, miR-16-5p, miR-17b-5p, miR-19-3p, miR-20a-5p, and miR-92-3p, whereas the high expression level of miR-15b-5p (Fan et al., 2016). However, out of these only three miRNAs (miR-15b-5p, miR-16-5p, and miR-20a-5p) were identified as promising biomarkers for the diagnosis of lung cancer. A study conducted by Chen et al. (2021) identified 10 miRNAs (miR20a, miR24, miR25, miR-145, miR152, miR199a-5p, miR-221, miR222, miR-223, and miR320), expression levels of which were found drastically altered in the serum samples of lung cancer patients than controls. Further, the expression pattern of these miRNAs showed a correlation with the tumor stage of the cancer patients. An investigation has shown the diagnostic potential of five plasma miRNAs, miR-20a, miR-145, miR-21, miR-223, and miR-221, for the screening of early-stage lung cancer (Geng et al., 2014). Various tumor-derived exosomal miRNAs are also identified as promising biomarkers for the diagnosis of early-stage lung cancer (Jin et al., 2017).

5.9 Prognostic potential of ncRNAs in lung cancer patients

Convincing evidence indicates a direct correlation between the expression levels of various ncRNAs and patients' prognosis. Studies suggest that depending on the regulatory actions of ncRNAs, their up- or downregulated expressions are associated with the poor prognosis of lung cancer patients.

5.9.1 LncRNAs as prognostic biomarkers

The level of PVT is found upregulated in the tissues of lung cancer and considered an independent prognostic factor for lung cancer. Further, an upregulated expression of ANRIL is detected in the tissues and cell clines of lung cancer, which is correlated to the poor prognosis of lung cancer patients. MALAT1 is also being considered a prognostic marker for lung cancer as its expression is directly correlated to the poor prognosis in lung cancer patients (Schmidt et al., 2011). A recent study by Zhou et al. (2019) has shown the lower expression of LOC285194 in the lung tumor tissues compared to adjacent tumor-free tissues. Moreover, the findings of this study suggested that the tumor-suppressive nature of LOC285194 depends on its ability to target p53. Further, this investigation demonstrated that the lower expression of LOC285194 is associated with the poor survival of lung cancer patients and suggested LOC285194 as a novel prognostic biomarker in patients with lung cancer. In another recent study, the expression level of lncRNA SLC16A1-AS1 is found significantly lower in the tumor tissues of lung cancer than corresponding nontumor tissues (Liu et al., 2020). The expression analysis also showed downregulated expression of lncRNA SLC16A1-AS1 in various cell lines of lung cancer. Interestingly, the overexpression of lncRNA SLC16A1-AS1 is reported in the inhibition of survival and proliferation of lung cancer cells via preventing the phosphorylation of RAS/RAF/MEK/ERK pathway. Further, the results of this study suggested that higher expression of lncRNA SLC16A1-AS1 is associated with better overall survival rate and disease-free survival of lung cancer patients and, hence, revealed the prognostic value of lncRNA SLC16A1-AS1 in lung cancer. Shi et al. (2015) have reported the downregulated expression of lncRNA GAS5 in lung cancer tissues compared to corresponding nontumor tissues and have shown a correlation with tumor size and TNM (tumor—node—metastasis) stage. A study conducted by Sun et al. (2014b) has shown decreased expression of BRAF-activated noncoding RNA (BANCR) in lung cancer tissues and cell lines and established an association between the downregulated expression of BANCR with tumor size, TNM stage, and decreased survival of cancer patients. The experimental findings of this study demonstrated that BANCR may be a potential prognostic biomarker of lung cancer.

5.9.2 miRNAs as prognostic biomarkers

To date, several miRNAs are identified as prognostic biomarkers for lung cancer. But in this section, we have discussed some of the well-established prognostic biomarkers of lung cancer. The expression of miR21 has been found highly elevated in a variety of cancers, including lung cancer (Zheng et al., 2018). The upregulated expression of miR21 is correlated to the overall poor survival of lung cancer patients (Markou et al., 2008). Further, convincing reports suggest that miR21 may be a potential prognostic biomarker for lung cancer (Saito et al., 2011; Zheng et al., 2018). miR-210, a hypoxia regulatory miRNA, has been found significantly increased in the serum as well as tissue samples of lung cancer patients compared to healthy control subjects (Eilertsen et al., 2014; Li et al., 2013). Further, miR-210 level has shown an association with the clinical stages and, hence, is predicted as one of the promising prognostic biomarkers for lung cancer. Chen, Min et al. (2017) have shown a strong correlation between the downregulated expression of miR148a with higher tumor grade and risk of death in lung cancer patients. Further, the low level of miR148b has also shown a direct association with high tumor grade, lymph node metastasis, and poor patient survival (Ge et al., 2015). Therefore downregulated expression of miR148a and miR148b has prognostic significance for lung cancer.

5.10 Therapeutic potential of ncRNAs in lung cancer patients

The recent understanding of ncRNAs indicates that ncRNAs could be potential cancer therapeutic targets due to their ability to regulate the expression of a vast number of coding and noncoding genes, which directly or indirectly facilitate the development and progression of cancer by altering various crucial cellular processes, including survival, proliferation, apoptosis, angiogenesis, invasion and metastasis, and chemo- and radioresistance.

5.10.1 Therapeutic potential of lncRNAs

The role of aberrant expression of lncRNAs has now been well established in the development and progression of various cancers, including lung cancer owing to their crucial involvement in the regulation of various cancer-promoting activities such as evasion of apoptosis, cell proliferation, invasion, and metastasis (Tao et al., 2016). Reports suggest that lncRNAs exhibit oncogenic as well as tumor-suppressive properties (Khandelwal, Bacolla, Vasquez, & Jain, 2015).

The elevated level of MALAT1 is widely reported in lung tumorigenesis via facilitating proliferation, invasion, and metastasis (Schmidt et al., 2011; Weber et al., 2013). Further, various in vitro and in vivo studies have demonstrated the metastatic inhibitory ability of MALAT1 in lung cancer (Eißmann et al., 2012; Gutschner et al., 2013) and suggested that MALAT1 could be a potential therapeutic target for the antimetastasis therapy. Further, the upregulated expression of HOTAIR is also reported in proliferation, migration, invasion, and metastasis in lung cancer (Nakagawa et al., 2013; Ono et al., 2014; Zhao, An, Liang, & Xie, 2014). In vitro and in vivo studies have shown the inhibition of migration, invasion, and metastasis by the silencing of HOTAIR (Ono et al., 2014; Zhao et al., 2014). These shreds of evidence suggest that HOTAIR exhibits a promising therapeutic potential for lung cancer treatment. Available reports have shown the downregulated expression of MEG3 in lung cancer tissues compared to corresponding nontumor tissues (Lu et al., 2013). In addition, the suppressed expression of MEG3 has shown an association with pathological stages, and shorter overall survival of lung cancer patients (Lu et al., 2013). Hence, MEG3 could be a promising therapeutic candidate for the treatment of lung cancer. Further, experimental data indicate an important role of the downregulated expression of SPRY4-IT1, BANCR, and taurine-upregulated gene 1 lncRNAs in the lung tumorigenesis by supporting cancer-inducing cellular processes such as cell proliferation, migration, and invasion (Sun et al., 2014a,b; Zhang et al., 2014). Moreover, their downregulated expression is correlated to poor overall survival, and hence they are considered useful prognostic biomarkers and therapeutic targets of lung cancer.

5.10.2 Therapeutic potential of miRNAs

Convincing reports indicate the tumor-suppressive and oncogenic activity of various miRNAs and their involvement in the development and progression of lung cancer. Lethal-7 (let-7) is reported as a tumor suppressor miRNA in lung cancer due to the association of its suppressed expression with poor prognosis (Landi et al., 2010; Xia, Zhu, Zhou, & Chen,

2014; Yanaihara et al., 2006). Let-7 has been identified as a negative regulator of crucial cell cycle oncogenes such as KRAS, MYC, and HMGA2 (Johnson et al., 2005; Lee and Dutta, 2007; Sampson et al., 2007). Further, the overexpression of let-7 is reported in the inhibition of growth of lung cancer cells by targeting cell cycle regulatory molecules such as CDC25A, CDK6, and cyclin D2 (Johnson et al., 2007). In vivo experiments also suggest tumor inhibitory ability of let-7 against lung cancer (Esquela-Kerscher et al., 2008). Hence, let-7 could be a better therapeutic option for the treatment of lung cancer. The expression of members of the miR-34 family, particularly miR-34a and miR34c, is found downregulated in lung cancer and has shown a correlation with the progression of lung cancer (Garofalo et al., 2013; Zhao et al., 2017). The miR-34 replacement has been observed in the inhibition of tumorigenicity. So, miR-34 could be a promising therapeutic agent for lung cancer treatment. The downregulated expression of miR-145 and miR-200c has also been reported in the progression of lung cancer (Ling et al., 2015; Liu et al., 2017, 2018), which suggests that their ectopic expression might be an effective therapeutic strategy for the retardation of lung cancer progression (Liu et al., 2017; Sadeghiyeh et al., 2019). The oncogenic property of miR21 is reported in various cancers, including lung cancer (Volinia et al., 2006). Available reports suggest the critical role of elevated levels of miR21 in lung tumorigenesis. Further, in vitro and in vivo experiments have shown the inhibition of lung cancer growth by the downregulation of miR21 and, thus, suggest the therapeutic potential of miR21 against lung cancer (Hatley et al., 2010; Markou, Zavridou, & Lianidou, 2016; Zhang et al., 2010).

As discussed earlier, ncRNAs (lncRNAs or miRNAs) are emerged as novel therapeutic targets for designing effective therapeutic approaches against lung cancer (Table 5.1).

5.11 Potential of ncRNAs in predicting chemoresistance and radioresistance in lung cancer patients

5.11.1 NcRNAs and their role in chemoresistance

Surgery is the best choice for the treatment of lung cancer if it is detected at an early stage but unfortunately, in a majority of cases, lung cancer is usually detected at advanced stages. Hence, chemotherapy plays an essentially important role in the treatment and life quality of lung cancer patients. But unfortunately, chemoresistance has emerged as a major hindrance to the treatment of cancer by chemotherapeutic strategies. Further, recent findings suggest the crucial involvement of ncRNAs in the development of chemoresistance in lung cancer as altered expression of various ncRNAs has been reported in lung cancer.

5.11.1.1 Role of lncRNAs in chemoresistance

HOTAIR, an lncRNA, is reported to be involved in the chemoresistance to cisplatin in lung cancer. A study has shown a significantly elevated expression of HOTAIR in cisplatin-resistant A549/DDP cells compared to parental ones (Liu et al., 2013). Further, this investigation suggested that HOTAIR mediates cisplatin resistance possibly through downregulating the expression of p21. Interestingly, restoration of cisplatin sensitivity was observed in A549/DDP cells by the knockdown of HOTAIR. The role of MALAT1 lncRNA has also been seen in the development of chemoresistance in lung cancer. A study by

TABLE 5.1 Therapeutic potential of noncoding RNAs (ncRNAs) for lung cancer.

NcRNAs (LncRNAs/miRNAs)	Expression level	Affected cellular processes	References
LncRNAs			
MALAT1	Upregulated	Increased proliferation, and invasion and metastasis	Weber et al. (2013), Schmidt et al. (2011)
HOTAIR	Upregulated	Increased proliferation, migration, invasion, and metastasis	Nakagawa et al. (2013), Zhao et al. (2014), Ono et al. (2014)
MEG3	Downregulated	Increased cell proliferation and inhibited apoptosis	Lu et al. (2013)
SPRY4-IT1	Downregulated	Increased cell proliferation and metastasis	Sun et al. (2014a)
BANCR	Downregulated	Increased proliferation, migration, invasion, and metastasis	Sun et al. (2014b)
TUG1	Downregulated	Increased proliferation	Zhang et al. (2014)
miRNAs			
Let-7 family	Downregulated	Increased cell proliferation	Johnson et al. (2007)
miR-34 family	Downregulated	Inhibited apoptosis and increased migration and invasion	Garofalo et al. (2013)
miR-145	Downregulated	Increased survival, migration, invasion, and metastasis	Ling et al. (2015), Liu et al. (2018)
miR-200c	Downregulated	Increased migration, invasion	Liu et al. (2017)
miR21	Upregulated	Increased proliferation, migration, invasion and inhibited apoptosis	Zhang et al. (2010), Markou et al. (2016), Hatley et al. (2010)

BANCR, BRAF-activated noncoding RNA; *HOTAIR*, HOX transcript antisense intergenic RNA; *Let-7*, lethal-7; *LncRNAs*, ncRNAs more than 200 nucleotides; *MALAT1*, metastasis-associated lung adenocarcinoma transcript 1; *MEG3*, maternal-expressed gene 3; *miRNAs*, microRNAs; *SPRY4-IT1*, SPRY4 intronic transcript 1; *TUG1*, taurine-upregulated gene 1.

Chen, Zhao et al. (2017) has shown a significant role of MALAT1 in the cisplatin resistance in lung cancer. They observed increased sensitivity in MALAT1 knockdown A549/DDP cells for cisplatin. The findings of their study showed a crucial role of the MALAT1-miR-101-SOX9 feedback loop in the chemoresistance of lung cancer to cisplatin and suggested MALAT1 as a potential target for the chemosensitization of lung cancer. A report by Fang, Chen, Yuan, Liu, and Jiang (2018) has also indicated a pivotal role of MALAT1 in the chemoresistance of lung cancer to cisplatin. This study explored that MALAT1 mediates cisplatin resistance in lung cancer via upregulating MDR1 and MRP1 through the activation of STAT-3. Further, the higher expression of MALAT1 is also reported in increasing the resistance to anti-PD1 immunotherapy against lung cancer (Wei, Wang, Huang, Zhao, & Zhao, 2019).

Lung cancer also exhibits chemoresistance to paclitaxel, another promising chemotherapeutic drug against lung cancer. Numerous studies have shown the role of lncRNAs in the promotion of drug resistance to paclitaxel in lung cancer. The upregulated expression of lncRNA ANRIL is reported in paclitaxel-resistant lung cancer cells (A549/PTX cells) by apoptosis inhibition through up- and downregulating the expression of Bcl2 and cleaved PARP, respectively (Xu et al., 2017). The expression of lncRNA KCNQ1OT1 is also found elevated in A549/PTX cells and its knockdown is reported to enhance the chemosensitivity of A549/PTX cells to paclitaxel possibly through suppressing MDR1 expression (Ren, Xu, Huang, Zhao, & Shi, 2017). LncRNAs CCAT1 and ROR mediate chemoresistance in lung cancer to paclitaxel via modulating vital cellular processes such as apoptosis and cell proliferation (Chen, Xu et al., 2016; Chen, Zhang et al., 2016; Pan et al., 2017).

5.11.1.2 *miRNAs in chemoresistance*

A large body of evidence indicates an association between the aberrant expression of miRNAs and chemoresistance in various cancers, including lung cancer. Further, reports suggest that miRNAs not only affect the chemotherapeutic response but their levels are also modulated by the chemotherapeutic drugs. Several studies show the changes in the expression profiling of miRNAs in response to chemotherapeutic drugs, which clearly suggest the implication of miRNAs in the development of drug resistance.

A study has shown the crucial involvement of miR-125b in cisplatin resistance in lung cancer. In this study, Cui et al. (2013) examined the serum levels of four different miRNAs, miR-125b, miR-10b, miR-34a, and miR-155, and observed dramatically increased expression of miR-125b in lung cancer patients not responding to cisplatin. Further, they observed a correlation between the miR-125b level and poor patient survival and, thus, suggested miR-125b as a potential therapeutic target to reduce cisplatin resistance in lung cancer. A report by Gao et al. (2012) has shown the predictive biomarker potential of miR-21 for platinum-based chemotherapy response as they found a significantly high level of miR-21 in the tissues as well as plasma of chemotherapy-resistant patients. In an investigation, Rui, Bing, Hai-Zhu, Wei, and Long-Bang (2010) examined the expression profile of six miRNAs (miR-192, 200b, 194, 424, 98, and 212) in docetaxel-resistant lung cancer cells (SPC-A1/docetaxel cells) and showed an association between the decreased levels of miR-200b, 194, and 212, and increased levels of miR-192, 424, and 98 with docetaxel resistance in lung cancer. Further, a study has shown a correlation between the increased level of miR-135a and paclitaxel resistance in lung cancer (Holleman et al., 2011). In this study, resensitization was noticed in paclitaxel-resistant lung cancer cells by the blockage of miR-135a.

Taken together, abovementioned reports suggest that ncRNAs might be among crucial players in predicting the chemotherapeutic response of chemotherapeutic drugs against lung cancer.

5.11.2 NcRNAs and their role in radioresistance

Radiotherapy is considered among one of the most effective therapeutic strategies for various cancers, including lung cancer due to the advantage of the localized application. In radiotherapy, ionizing radiation is being used to induce cell death in cancer cells.

Ionizing radiation triggers oxidative stress and induces cell death in cancer cells by causing damages to the plasma membrane, subcellular organelles, and DNA. Like other cancers, radiotherapy is among the best treatment of choice for lung cancer but resistance to ionizing radiation is identified as a major reason for the radiotherapy failure, which contributes a significant role in the poor prognosis in lung cancer patients (Gomez-Casal et al., 2015). Recent accumulating evidence suggests the involvement of various factors, including ncRNAs, in the development of radioresistance.

5.11.2.1 LncRNAs in radioresistance

To date, only a few studies have reported the role of lncRNAs in the development of radioresistance in lung cancer. Accumulating experimental data suggests the association of altered expression of HOTAIR, PVT, and BANCR in inducing resistance to ionizing radiation in lung cancer. The elevated level of lncRNA HOTAIR induces radioresistance in lung cancer by activating the Wnt signaling pathway (Chen, Chen, Wang, & Shen, 2015). A study by Wu, Li, Zhang, and Hu (2017) has indicated the role of PVT1 inhibition in the radiosensitization of lung cancer by sponging miR195. Besides, the upregulated expression of BANCR has shown a crucial role in the radiotherapy of lung cancer (Chen, Shen, Zheng, Wang, & Mao, 2015).

5.11.2.2 miRNAs in radioresistance

Besides chemoresistance, miRNAs are also implicated in radioresistance in lung cancer via altering various radiotherapy-mediated mechanisms. In an investigation, Wang et al. (2011) detected differential expression of 12 miRNAs in radiosensitive and radioresistant lung cancer patients. This study showed upregulated expression of five miRNAs (miR-126, let-7a, miR495, miR451, and miR128b), whereas downregulated expression of seven miRNAs (miR130a, miR106b, miR19b, miR-22, miR-15b, miR17-5p, and miR21) in radiosensitive lung cancer patients compared to radioresistant malignancies. In another study, Chen, Xu et al. (2016), reported 14 miRNAs (miR153-3p, miR1-3p, miR613, miR-372-3p, miR302e, miR495-3p, miR206, miR520a-3p, miR328-3p, miR520b, miR1297, miR520d-3p, miR193a-3p, and miR520e) as radioresistant genes, while five (let7c-5p, miR98-5p, miR203a-3p, miR137, and miR34c-5p) as radiosensitive genes by examining expression profiles of miRNAs in radioresistant and radiosensitive lung cancer cells. Further, they demonstrated significantly elevated expression of four miRNAs (miR98-5p, miR302e, miR495-3p, and miR613) in responding group (complete response + partial response) than nonresponding group (stable disease + progressive disease). The role of the upregulated expression of miR21, miR-210, and miR1323 has also been reported in the development of radioresistant in lung cancer (Grosso et al., 2013; Jiang et al., 2016; Li et al., 2015).

5.12 Summary and conclusion

NcRNAs, namely, lncRNAs and miRNAs, play an indispensible role in the lung carcinogenesis through altering various crucial cellular and pathological processes. Further, various ncRNAs are identified as diagnostic and prognostic markers for lung cancer. More importantly, a growing number of evidence suggests that ncRNAs could be promising therapeutic targets to weaken chemo- and radioresistance of lung cancer. Taken together, in-depth understanding about the role of aberrant expression of ncRNAs in the pathogenesis of lung cancer might play a significant role in developing novel diagnostic and prognostic tools for lung cancer along with designing highly effective therapeutic approaches with least or no side effects.

Acknowledgments

SK acknowledges the funding support from SERB (EEQ/2018/000022), New Delhi, India. NKV acknowledges the DBT-BUILDER scheme (grant no. BT/PR7020/INF/22/172/2012) and UGC-Special Assistance Program (UGC-SAP) at the Department of Biotechnology, Guru Ghasidas Vishwavidyalaya. AK acknowledges the University Grants Commission and SERB, New Delhi, India, for providing financial support in the form of UGC-Start-Up Research [F. No. 30–370/2017 (BSR)] and Early Career Research Award (ECR/2016/001117), respectively. Financial support of Institute of Eminence (IoE) (6031) from Banaras Hindu University, India is highly acknowledged. AK also acknowledges UGC-CAS and DST-FIST program to the Department of Zoology, Banaras Hindu University, India.

References

Acha-Sagredo, A., Uko, B., Pantazi, P., Bediaga, N. G., Moschandrea, C., Rainbow, L., ... Liloglou, T. (2020). Long non-coding RNA dysregulation is a frequent event in non-small cell lung carcinoma pathogenesis. *British Journal of Cancer*, 122(7), 1050–1058.

Ahn, Y. H., & Ko, Y. H. (2020). Diagnostic and therapeutic implications of microRNAs in non-small cell lung cancer. *International Journal of Molecular Sciences*, 21(22), 8782.

Azzawi, H., Hou, J., Alanni, R., Xiang, Y., Abdu-Aljabar, R., & Azzawi, A. (2017). Multiclass lung cancer diagnosis by gene expression programming and microarray datasets. In: *International conference on advanced data mining and applications* (pp. 541–553). Cham: Springer.

Balatti, V., Nigita, G., Veneziano, D., Drusco, A., Stein, G. S., Messier, T. L., ... Croce, C. M. (2017). tsRNA signatures in cancer. *Proceedings of the National Academy of Sciences*, 114(30), 8071–8076.

Balbin, O. A., Malik, R., Dhanasekaran, S. M., Prensner, J. R., Cao, X., Wu, Y. M., ... Chinnaiyan, A. M. (2015). The landscape of antisense gene expression in human cancers. *Genome Research*, 25(7), 1068–1079.

Bandi, N., Zbinden, S., Gugger, M., Arnold, M., Kocher, V., Hasan, L., ... Vassella, E. (2009). miR-15a and miR-16 are implicated in cell cycle regulation in a Rb-dependent manner and are frequently deleted or downregulated in non–small cell lung cancer. *Cancer Research*, 69(13), 5553–5559.

Barsyte-Lovejoy, D., Lau, S. K., Boutros, P. C., Khosravi, F., Jurisica, I., Andrulis, I. L., ... Penn, L. Z. (2006). The c-Myc oncogene directly induces the H19 noncoding RNA by allele-specific binding to potentiate tumorigenesis. *Cancer Research*, 66(10), 5330–5337.

Bellucci, M., Agostini, F., Masin, M., & Tartaglia, G. G. (2011). Predicting protein associations with long noncoding RNAs. *Nature Methods*, 8(6), 444–445.

Bi, Y., Cui, Z., Li, H., Lv, X., Li, J., Yang, Z., ... Yin, Z. (2019). Polymorphisms in long noncoding RNA-prostate cancer-associated transcript 1 are associated with lung cancer susceptibility in a northeastern Chinese population. *DNA and Cell Biology*, 38(11), 1357–1365.

Bjaanæs, M. M., Fleischer, T., Halvorsen, A. R., Daunay, A., Busato, F., Solberg, S., ... Helland, Å. (2016). Genome-wide DNA methylation analyses in lung adenocarcinomas: Association with EGFR, KRAS and TP53 mutation status, gene expression and prognosis. *Molecular Oncology*, 10(2), 330–343.

Braicu, C., Zimta, A. A., Harangus, A., Iurca, I., Irimie, A., Coza, O., ... Berindan-Neagoe, I. (2019). The function of non-coding RNAs in lung cancer tumorigenesis. *Cancers*, 11(5), 605.

Bray, F., Ferlay, J., Soerjomataram, I., Siegel, R. L., Torre, L. A., ... Jemal, A. (2018). Global cancer statistics 2018: GLOBOCAN estimates of incidence and mortality worldwide for 36 cancers in 185 countries. *CA: A Cancer Journal for Clinicians*, 68(6), 394–424.

Chen, J. X., Chen, M., Wang, S. Y., & Shen, Z. P. (2015). Up-regulation of BRAF activated non-coding RNA is associated with radiation therapy for lung cancer. *Biomedicine & Pharmacotherapy, 71*, 79–83.

Chen, J., Shen, Z., Zheng, Y., Wang, S., & Mao, W. (2015). Radiotherapy induced Lewis lung cancer cell apoptosis via inactivating β-catenin mediated by upregulated HOTAIR. *International Journal of Clinical and Experimental Pathology, 8*(7), 7878.

Chen, J., Zhang, K., Song, H., Wang, R., Chu, X., & Chen, L. (2016). Long noncoding RNA CCAT1 acts as an oncogene and promotes chemoresistance in docetaxel-resistant lung adenocarcinoma cells. *Oncotarget, 7*(38), 62474.

Chen, L., Gibbons, D. L., Goswami, S., Cortez, M. A., Ahn, Y. H., Byers, L. A., ... Qin, F. X. F. (2014). Metastasis is regulated via microRNA-200/ZEB1 axis control of tumour cell PD-L1 expression and intratumoral immunosuppression. *Nature Communications, 5*(1), 1–12.

Chen, L., Nan, A., Zhang, N., Jia, Y., Li, X., Ling, Y., ... Jiang, Y. (2019). Circular RNA 100146 functions as an oncogene through direct binding to miR-361-3p and miR-615-5p in non-small cell lung cancer. *Molecular Cancer, 18*(1), 1–8.

Chen, M., Cai, D., Gu, H., Yang, J., & Fan, L. (2021). MALAT1 rs619586 A/G polymorphisms are associated with decreased risk of lung cancer. *Medicine, 100*(12), e23716.

Chen, R., Zheng, Y., Zhuo, L., & Wang, S. (2017). The association between miR-423 rs6505162 polymorphism and cancer susceptibility: A systematic review and meta-analysis. *Oncotarget, 8*(25), 40204.

Chen, T. J., Zheng, Q., Gao, F., Yang, T., Ren, H., Li, Y., ... Chen, M. W. (2021). MicroRNA-665 facilitates cell proliferation and represses apoptosis through modulating Wnt5a/β-Catenin and Caspase-3 signaling pathways by targeting TRIM8 in LUSC. *Cancer Cell International, 21*(1), 1–15.

Chen, W., Zhao, W., Zhang, L., Wang, L., Wang, J., Wan, Z., ... Yu, L. (2017). MALAT1-miR-101-SOX9 feedback loop modulates the chemo-resistance of lung cancer cell to DDP via Wnt signaling pathway. *Oncotarget, 8*(55), 94317.

Chen, X., Xu, Y., Liao, X., Liao, R., Zhang, L., Niu, K., ... Sun, J. (2016). Plasma miRNAs in predicting radiosensitivity in non-small cell lung cancer. *Tumor Biology, 37*(9), 11927–11936.

Chen, Y., Min, L., Ren, C., Xu, X., Yang, J., Sun, X., ... Zhang, X. (2017). miRNA-148a serves as a prognostic factor and suppresses migration and invasion through Wnt1 in non-small cell lung cancer. *PLoS One, 12*(2), e0171751.

Cheng, M., Yang, L., Yang, R., Yang, X., Deng, J., Yu, B., ... Lu, J. (2013). A microRNA-135a/b binding polymorphism in CD133 confers decreased risk and favorable prognosis of lung cancer in Chinese by reducing CD133 expression. *Carcinogenesis, 34*(10), 2292–2299.

Cheung, C. H. Y., & Juan, H. F. (2017). Quantitative proteomics in lung cancer. *Journal of Biomedical Science, 24*(1), 1–11.

Cui, E. H., Li, H. J., Hua, F., Wang, B., Mao, W., Feng, X. R., ... Wang, X. (2013). Serum microRNA 125b as a diagnostic or prognostic biomarker for advanced NSCLC patients receiving cisplatin-based chemotherapy. *Acta Pharmacologica Sinica, 34*(2), 309–313.

Du, L., & Pertsemlidis, A. (2012). microRNA regulation of cell viability and drug sensitivity in lung cancer. *Expert Opinion on Biological Therapy, 12*(9), 1221–1239.

Dutkowska, A., Antczak, A., Domańska-Senderowska, D., & Brzeziańska-Lasota, E. (2019). Expression of selected miRNA, RARβ and FHIT genes in BALf of squamous cell lung cancer (squamous-cell carcinoma, SCC) patients: A pilot study. *Molecular Biology Reports, 46*(6), 6593–6597.

Eilertsen, M., Andersen, S., Al-Saad, S., Richardsen, E., Stenvold, H., Hald, S. M., ... Bremnes, R. M. (2014). Positive prognostic impact of miR-210 in non-small cell lung cancer. *Lung Cancer, 83*(2), 272–278.

Eißmann, M., Gutschner, T., Hämmerle, M., Günther, S., Caudron-Herger, M., Groß, M., ... Diederichs, S. (2012). Loss of the abundant nuclear non-coding RNA MALAT1 is compatible with life and development. *RNA Biology, 9*(8), 1076–1087.

El Jammal, T., Pavic, M., Gerfaud-Valentin, M., Jamilloux, Y., ... Sève, P. (2020). Sarcoidosis and cancer: A complex relationship. *Frontiers in Medicine, 7*.

Enfield, K. S., Pikor, L. A., Martinez, V. D., & Lam, W. L. (2012). Mechanistic roles of noncoding RNAs in lung cancer biology and their clinical implications. *Genetics Research International, 2012*.

Engel, E., Schmidt, B., Carstensen, T., Weickmann, S., Jandrig, B., Witt, C., ... Fleischhacker, M. (2004). Detection of tumor-specific mRNA in cell-free bronchial lavage supernatant in patients with lung cancer. *Annals of the New York Academy of Sciences, 1022*(1), 140–146.

Esquela-Kerscher, A., Trang, P., Wiggins, J. F., Patrawala, L., Cheng, A., Ford, L., ... Slack, F. J. (2008). The let-7 microRNA reduces tumor growth in mouse models of lung cancer. *Cell Cycle, 7*(6), 759–764.

Fan, L., Qi, H., Teng, J., Su, B., Chen, H., Wang, C., ... Xia, Q. (2016). Identification of serum miRNAs by nano-quantum dots microarray as diagnostic biomarkers for early detection of non-small cell lung cancer. *Tumor Biology, 37*(6), 7777–7784.

Fang, Z., Chen, W., Yuan, Z., Liu, X., & Jiang, H. (2018). LncRNA-MALAT1 contributes to the cisplatin-resistance of lung cancer by upregulating MRP1 and MDR1 via STAT3 activation. *Biomedicine & Pharmacotherapy, 101*, 536–542.

Farina, N. H., Wood, M. E., Perrapato, S. D., Francklyn, C. S., Stein, G. S., Stein, J. L., ... Lian, J. B. (2014). Standardizing analysis of circulating microRNA: Clinical and biological relevance. *Journal of Cellular Biochemistry, 115*(5), 805–811.

Fassina, A., Cappellesso, R., & Fassan, M. (2011). Classification of non-small cell lung carcinoma in transthoracic needle specimens using microRNA expression profiling. *Chest, 140*(5), 1305–1311.

Felekkis, K., & Papaneophytou, C. (2020). Challenges in using circulating micro-RNAs as biomarkers for cardiovascular diseases. *International Journal of Molecular Sciences, 21*(2), 561.

Gao, W., Lu, X., Liu, L., Xu, J., Feng, D., & Shu, Y. (2012). MiRNA-21: A biomarker predictive for platinum-based adjuvant chemotherapy response in patients with non-small cell lung cancer. *Cancer Biology & Therapy, 13*(5), 330–340.

Gao, X., Gao, X., Li, C., Zhang, Y., & Gao, L. (2016). Knockdown of long noncoding RNA uc. 338 by siRNA inhibits cellular migration and invasion in human lung cancer cells. *Oncology Research, 24*(5), 337.

Gao, Y., Shang, S., Guo, S., Li, X., Zhou, H., Liu, H., ... Zhang, Y. (2021). Lnc2Cancer 3.0: An updated resource for experimentally supported lncRNA/circRNA cancer associations and web tools based on RNA-seq and scRNA-seq data. *Nucleic Acids Research, 49*(D1), D1251–D1258.

Garofalo, M., Jeon, Y. J., Nuovo, G. J., Middleton, J., Secchiero, P., Joshi, P., ... Croce, C. M. (2013). MiR-34a/c-dependent PDGFR-α/β downregulation inhibits tumorigenesis and enhances TRAIL-induced apoptosis in lung cancer. *PLoS One, 8*(6), e67581.

Ge, H., Li, B., Hu, W. X., Li, R. J., Jin, H., Gao, M. M., ... Ding, C. M. (2015). MicroRNA-148b is down-regulated in non-small cell lung cancer and associated with poor survival. *International Journal of Clinical and Experimental Pathology, 8*(1), 800.

Geng, Q., Fan, T., Zhang, B., Wang, W., Xu, Y., & Hu, H. (2014). Five microRNAs in plasma as novel biomarkers for screening of early-stage non-small cell lung cancer. *Respiratory Research, 15*(1), 1–9.

Geng, Q., Li, Z., Li, X., Wu, Y., & Chen, N. (2021). LncRNA NORAD, sponging miR-363-3p, promotes invasion and EMT by upregulating PEAK1 and activating the ERK signaling pathway in NSCLC cells. *Journal of Bioenergetics and Biomembranes, 53*(3), 321–332.

Ghafouri-Fard, S., Shoorei, H., Branicki, W., & Taheri, M. (2020). Non-coding RNA profile in lung cancer. *Experimental and Molecular Pathology, 114*, 104411.

Ginn, L., Shi, L., Montagna, M. L., & Garofalo, M. (2020). LncRNAs in non-small-cell lung cancer. *Non-coding RNA, 6*(3), 25.

Glinge, C., Clauss, S., Boddum, K., Jabbari, R., Jabbari, J., Risgaard, B., ... Tfelt-Hansen, J. (2017). Stability of circulating blood-based microRNAs—pre-analytic methodological considerations. *PLoS One, 12*(2), e0167969.

Gomez-Casal, R., Epperly, M. W., Wang, H., Proia, D. A., Greenberger, J. S., ... Levina, V. (2015). Radioresistant human lung adenocarcinoma cells that survived multiple fractions of ionizing radiation are sensitive to HSP90 inhibition. *Oncotarget, 6*(42), 44306.

Grillone, K., Riillo, C., Scionti, F., Rocca, R., Tradigo, G., Guzzi, P. H., ... Tassone, P. (2020). Non-coding RNAs in cancer: Platforms and strategies for investigating the genomic "dark matter.". *Journal of Experimental & Clinical Cancer Research, 39*(1), 1–19.

Grosso, S., Doyen, J., Parks, S. K., Bertero, T., Paye, A., Cardinaud, B., ... Mari, B. (2013). MiR-210 promotes a hypoxic phenotype and increases radioresistance in human lung cancer cell lines. *Cell Death & Disease, 4*(3), e544.

Gu, W., Shi, J., Liu, H., Zhang, X., Zhou, J. J., Li, M., ... Zhou, T. (2020). Peripheral blood non-canonical small non-coding RNAs as novel biomarkers in lung cancer. *Molecular Cancer, 19*(1), 1–6.

Gutschner, T., Hämmerle, M., Eißmann, M., Hsu, J., Kim, Y., Hung, G., ... Diederichs, S. (2013). The noncoding RNA MALAT1 is a critical regulator of the metastasis phenotype of lung cancer cells. *Cancer Research, 73*(3), 1180–1189.

Gyoba, J., Shan, S., Roa, W., & Bédard, E. L. (2016). Diagnosing lung cancers through examination of micro-RNA biomarkers in blood, plasma, serum and sputum: A review and summary of current literature. *International Journal of Molecular Sciences, 17*(4), 494.

References

Hatley, M. E., Patrick, D. M., Garcia, M. R., Richardson, J. A., Bassel-Duby, R., Van Rooij, E., ... Olson, E. N. (2010). Modulation of K-Ras-dependent lung tumorigenesis by MicroRNA-21. *Cancer Cell, 18*(3), 282–293.

Herrera-Solorio, A. M., Peralta-Arrieta, I., Armas López, L., Hernández-Cigala, N., Mendoza Milla, C., Ortiz Quintero, B., ... Ávila-Moreno, F. (2021). LncRNA SOX2-OT regulates AKT/ERK and SOX2/GLI-1 expression, hinders therapy, and worsens clinical prognosis in malignant lung diseases. *Molecular Oncology, 15*(4), 1110–1129.

Holleman, A., Chung, I., Olsen, R. R., Kwak, B., Mizokami, A., Saijo, N., ... Zetter, B. R. (2011). miR-135a contributes to paclitaxel resistance in tumor cells both in vitro and in vivo. *Oncogene, 30*(43), 4386–4398.

Hu, X., Bao, J., Wang, Z., Zhang, Z., Gu, P., Tao, F., ... Jiang, W. (2016). The plasma lncRNA acting as fingerprint in non-small-cell lung cancer. *Tumor Biology, 37*(3), 3497–3504.

Huang, J., Cao, D., Sha, J., Zhu, X., & Han, S. (2019). DLL3 is regulated by LIN28B and miR-518d-5p and regulates cell proliferation, migration and chemotherapy response in advanced small cell lung cancer. *Biochemical and Biophysical Research Communications, 514*(3), 853–860.

Huang, S. Q., Sun, B., Xiong, Z. P., Shu, Y., Zhou, H. H., Zhang, W., ... Li, Q. (2018). The dysregulation of tRNAs and tRNA derivatives in cancer. *Journal of Experimental & Clinical Cancer Research, 37*(1), 1–11.

Huang, T., Wang, G., Yang, L., Peng, B., Wen, Y., Ding, G., ... Wang, Z. (2017). Transcription factor YY1 modulates lung cancer progression by activating lncRNA-PVT1. *DNA and Cell Biology, 36*(11), 947–958.

Huang, X., Zhang, W., & Shao, Z. (2018). Association between long non-coding RNA polymorphisms and cancer risk: A meta-analysis. *Bioscience Reports, 38*(4).

Huang, Y., Zhang, M., Li, Y., Luo, J., Wang, Y., Geng, W., ... Bai, Y. (2021). miR-183 promotes radioresistance of lung adenocarcinoma H1299 cells via epithelial-mesenchymal transition. *Brazilian Journal of Medical and Biological Research, 54*.

Inage, T., Nakajima, T., Itoga, S., Ishige, T., Fujiwara, T., Sakairi, Y., ... Yoshino, I. (2018). Molecular nodal staging using miRNA expression in lung cancer patients by endobronchial ultrasound-guided transbronchial needle aspiration. *Respiration, 96*(3), 267–274.

Izzotti, A., Longobardi, M., La Maestra, S., Micale, R. T., Pulliero, A., Camoirano, A., ... De Flora, S. (2018). Release of MicroRNAs into body fluids from ten organs of mice exposed to cigarette smoke. *Theranostics, 8*(8), 2147.

Jeon, H. S., Lee, Y. H., Lee, S. Y., Jang, J. A., Choi, Y. Y., Yoo, S. S., ... Park, J. Y. (2014). A common polymorphism in pre-microRNA-146a is associated with lung cancer risk in a Korean population. *Gene, 534*(1), 66–71.

Ji, D., An, M., & Fang, Q. (2020). Whether miR-4293 rs12220909 variant affects cancer susceptibility: Evidence from 11255 subjects. *Artificial Cells, Nanomedicine, and Biotechnology, 48*(1), 933–938.

Ji, P., Diederichs, S., Wang, W., Böing, S., Metzger, R., Schneider, P. M., ... Müller-Tidow, C. (2003). MALAT-1, a novel noncoding RNA, and thymosin beta4 predict metastasis and survival in early-stage non-small cell lung cancer. *Oncogene, 22*(39), 8031–8041.

Jiang, S., Wang, R., Yan, H., Jin, L., Dou, X., & Chen, D. (2016). MicroRNA-21 modulates radiation resistance through upregulation of hypoxia-inducible factor-1α-promoted glycolysis in non-small cell lung cancer cells. *Molecular Medicine Reports, 13*(5), 4101–4107.

Jin, X., Chen, Y., Chen, H., Fei, S., Chen, D., Cai, X., ... Xie, C. (2017). Evaluation of tumor-derived exosomal miRNA as potential diagnostic biomarkers for early-stage non–small cell lung cancer using next-generation sequencing. *Clinical Cancer Research, 23*(17), 5311–5319.

Jin, X., Guan, Y., Zhang, Z., & Wang, H. (2020). Microarray data analysis on gene and miRNA expression to identify biomarkers in non-small cell lung cancer. *BMC Cancer, 20*, 1–10.

Johnson, C. D., Esquela-Kerscher, A., Stefani, G., Byrom, M., Kelnar, K., Ovcharenko, D., ... Slack, F. J. (2007). The let-7 microRNA represses cell proliferation pathways in human cells. *Cancer Research, 67*(16), 7713–7722.

Johnson, S. M., Grosshans, H., Shingara, J., Byrom, M., Jarvis, R., Cheng, A., ... Slack, F. J. (2005). RAS is regulated by the let-7 microRNA family. *Cell, 120*(5), 635–647.

Khandelwal, A., Bacolla, A., Vasquez, K. M., & Jain, A. (2015). Long non-coding RNA: A new paradigm for lung cancer. *Molecular Carcinogenesis, 54*(11), 1235–1251.

Kim, J. E., Eom, J. S., Kim, W. Y., Jo, E. J., Mok, J., Lee, K., ... Kim, M. H. (2018). Diagnostic value of microRNAs derived from exosomes in bronchoalveolar lavage fluid of early-stage lung adenocarcinoma: A pilot study. *Thoracic Cancer, 9*(8), 911–915.

Kim, J. O., Gazala, S., Razzak, R., Guo, L., Ghosh, S., Roa, W. H., ... Bedard, E. L. (2015). Non-small cell lung cancer detection using microRNA expression profiling of bronchoalveolar lavage fluid and sputum. *Anticancer Research, 35*(4), 1873–1880.

Kim, N., Kim, H. K., Lee, K., Hong, Y., Cho, J. H., Choi, J. W., ... Lee, H. O. (2020). Single-cell RNA sequencing demonstrates the molecular and cellular reprogramming of metastatic lung adenocarcinoma. *Nature Communications, 11*(1), 1−15.

Kiszałkiewicz, J., Piotrowski, W. J., Pastuszak-Lewandoska, D., Górski, P., Antczak, A., Górski, W., ... Brzeziańska-Lasota, E. (2016). Altered miRNA expression in pulmonary sarcoidosis. *BMC Medical Genetics, 17*(1), 1−12.

Koni, M., Pinnarò, V., & Brizzi, M. F. (2020). The Wnt signalling pathway: A tailored target in cancer. *International Journal of Molecular Sciences, 21*(20), 7697.

Kumar, S., & Suryawanshi, H. (2019). Role of microRNAs in cardiovascular diseases and their therapeutic implications. *AGO-driven non-coding RNAs* (pp. 233−259). Academic Press.

Landi, M. T., Zhao, Y., Rotunno, M., Koshiol, J., Liu, H., Bergen, A. W., ... Wang, E. (2010). MicroRNA expression differentiates histology and predicts survival of lung cancer. *Clinical Cancer Research, 16*(2), 430−441.

Le, P., Romano, G., Nana-Sinkam, P., & Acunzo, M. (2021). Non-coding RNAs in cancer diagnosis and therapy: Focus on lung cancer. *Cancers, 13*(6), 1372.

Lee, Y. S., & Dutta, A. (2007). The tumor suppressor microRNA let-7 represses the HMGA2 oncogene. *Genes & Development, 21*(9), 1025−1030.

Lewis, B. A., Walia, R. R., Terribilini, M., Ferguson, J., Zheng, C., Honavar, V., ... Dobbs, D. (2010). PRIDB: A protein−RNA interface database. *Nucleic Acids Research, 39*(suppl_1), D277−D282.

Li, N., Cui, Z., Gao, M., Li, S., Song, M., Wang, Y., ... Yin, Z. (2021). Genetic polymorphisms of PRNCR1 and lung cancer risk in Chinese northeast population: A case−control study and meta-analysis. *DNA and Cell Biology, 40*(1), 132−144.

Li, S., Ma, F., Jiang, K., Shan, H., Shi, M., ... Chen, B. (2018). Long non-coding RNA metastasis-associated lung adenocarcinoma transcript 1 promotes lung adenocarcinoma by directly interacting with specificity protein 1. *Cancer Science, 109*(5), 1346−1356.

Li, W., Jiang, W., Liu, T., Lv, J., & Guan, J. (2019). Enhanced expression of circ_0000735 forecasts clinical severity in NSCLC and promotes cell progression via sponging miR-1179 and miR-1182. *Biochemical and Biophysical Research Communications, 510*(3), 467−471.

Li, Y., Han, W., Ni, T. T., Lu, L., Huang, M., Zhang, Y., ... Li, H. (2015). Knockdown of microRNA-1323 restores sensitivity to radiation by suppression of PRKDC activity in radiation-resistant lung cancer cells. *Oncology Reports, 33*(6), 2821−2828.

Li, Y., Yang, Z., Wan, X., Zhou, J., Zhang, Y., Ma, H., ... Bai, Y. (2016). Clinical prognostic value of metastasis-associated lung adenocarcinoma transcript 1 in various human cancers: An updated meta-analysis. *The International Journal of Biological Markers, 31*(2), 173−182.

Li, Y., Yin, Z., Fan, J., Zhang, S., & Yang, W. (2019). The roles of exosomal miRNAs and lncRNAs in lung diseases. *Signal Transduction and Targeted Therapy, 4*(1), 1−12.

Li, Z. H., Zhang, H., Yang, Z. G., Wen, G. Q., Cui, Y. B., ... Shao, G. G. (2013). Prognostic significance of serum microRNA-210 levels in nonsmall-cell lung cancer. *Journal of International Medical Research, 41*(5), 1437−1444.

Liao, Q. B., Guo, J. Q., Zheng, X. Y., Zhou, Z. F., Li, H., Lai, X. Y., ... Ye, J. F. (2014). Test performance of sputum microRNAs for lung cancer: A meta-analysis. *Genetic Testing and Molecular Biomarkers, 18*(8), 562−567.

Lim, S. B., Tan, S. J., Lim, W. T., & Lim, C. T. (2018). A merged lung cancer transcriptome dataset for clinical predictive modeling. *Scientific Data, 5*(1), 1−8.

Lin, C. C., Wu, C. Y., Tseng, J. T., Hung, C. H., Wu, S. Y., Huang, Y. T., ... Su, W. C. (2021). Extracellular vesicle miR-200c enhances gefitinib sensitivity in heterogeneous EGFR-mutant NSCLC. *Biomedicines, 9*(3), 243.

Lin, Y., Leng, Q., Zhan, M., & Jiang, F. (2018). A plasma long noncoding RNA signature for early detection of lung cancer. *Translational Oncology, 11*(5), 1225−1231.

Ling, D. J., Chen, Z. S., Zhang, Y. D., Liao, Q. D., Feng, J. X., Zhang, X. Y., ... Shi, T. S. (2015). MicroRNA-145 inhibits lung cancer cell metastasis. *Molecular Medicine Reports, 11*(4), 3108−3114.

Liu, C., Li, H., Jia, J., Ruan, X., Liu, Y., & Zhang, X. (2019). High metastasis-associated lung adenocarcinoma transcript 1 (MALAT1) expression promotes proliferation, migration, and invasion of non-small cell lung cancer via ERK/mitogen-activated protein kinase (MAPK) signaling pathway. *Medical Science Monitor: International Medical Journal of Experimental and Clinical Research, 25*, 5143.

Liu, H. Y., Lu, S. R., Guo, Z. H., Zhang, Z. S., Ye, X., Du, Q., ... Liu, J. L. (2020). lncRNA SLC16A1-AS1 as a novel prognostic biomarker in non-small cell lung cancer. *Journal of Investigative Medicine, 68*(1), 52−59.

Liu, K., Chen, H., You, Q., Ye, Q., Wang, F., Wang, S., ... Gu, M. (2018). miR-145 inhibits human non-small-cell lung cancer growth by dual-targeting RIOK2 and NOB1. *International Journal of Oncology, 53*(1), 257–265.

Liu, P. L., Liu, W. L., Chang, J. M., Chen, Y. H., Liu, Y. P., Kuo, H. F., ... Chong, I. W. (2017). MicroRNA-200c inhibits epithelial-mesenchymal transition, invasion, and migration of lung cancer by targeting HMGB1. *PLoS One, 12*(7), e0180844.

Liu, X. X., Yang, Y. E., Liu, X., Zhang, M. Y., Li, R., Yin, Y. H., ... Qu, Y. Q. (2019). A two-circular RNA signature as a noninvasive diagnostic biomarker for lung adenocarcinoma. *Journal of Translational Medicine, 17*(1), 1–13.

Liu, Z., Sun, M., Lu, K., Liu, J., Zhang, M., Wu, W., ... Wang, R. (2013). The long noncoding RNA HOTAIR contributes to cisplatin resistance of human lung adenocarcinoma cells via downregualtion of p21 WAF1/CIP1 expression. *PLoS One, 8*(10), e77293.

Lu, K. H., Li, W., Liu, X. H., Sun, M., Zhang, M. L., Wu, W. Q., ... Hou, Y. Y. (2013). Long non-coding RNA MEG3 inhibits NSCLC cells proliferation and induces apoptosis by affecting p53 expression. *BMC Cancer, 13*(1), 1–11.

Lu, Y., Zhou, X., Xu, L., Rong, C., Shen, C., ... Bian, W. (2016). Long noncoding RNA ANRIL could be transactivated by c-Myc and promote tumor progression of non-small-cell lung cancer. *OncoTargets and Therapy, 9*, 3077.

Mannoor, K., Shen, J., Liao, J., Liu, Z., & Jiang, F. (2014). Small nucleolar RNA signatures of lung tumor-initiating cells. *Molecular Cancer, 13*(1), 1–12.

Markou, A., Tsaroucha, E. G., Kaklamanis, L., Fotinou, M., Georgoulias, V., ... Lianidou, E. S. (2008). Prognostic value of mature microRNA-21 and microRNA-205 overexpression in non–small cell lung cancer by quantitative real-time RT-PCR. *Clinical Chemistry, 54*(10), 1696–1704.

Markou, A., Zavridou, M., & Lianidou, E. S. (2016). miRNA-21 as a novel therapeutic target in lung cancer. *Lung Cancer: Targets and Therapy, 7*, 19.

Matias-Garcia, P. R., Wilson, R., Mussack, V., Reischl, E., Waldenberger, M., Gieger, C., ... Kuehn-Steven, A. (2020). Impact of long-term storage and freeze-thawing on eight circulating microRNAs in plasma samples. *PLoS One, 15*(1), e0227648.

McHugh, C. A., Chen, C. K., Chow, A., Surka, C. F., Tran, C., McDonel, P., ... Guttman, M. (2015). The Xist lncRNA interacts directly with SHARP to silence transcription through HDAC3. *Nature, 521*(7551), 232–236.

Mei, Y. P., Liao, J. P., Shen, J., Yu, L., Liu, B. L., Liu, L., ... Jiang, F. (2012). Small nucleolar RNA 42 acts as an oncogene in lung tumorigenesis. *Oncogene, 31*(22), 2794–2804.

Mei, Y., Wang, Y., Kumari, P., Shetty, A. C., Clark, D., Gable, T., ... Mao, L. (2015). A piRNA-like small RNA interacts with and modulates p-ERM proteins in human somatic cells. *Nature Communications, 6*(1), 1–12.

Mishra, P., & Kumar, S. (2021). Association of lncRNA with regulatory molecular factors in brain and their role in the pathophysiology of schizophrenia. *Metabolic Brain Disease, 36*, 1–10.

Moazeni-Roodi, A., Ghavami, S., & Hashemi, M. (2019). Association between miR-423 rs6505162 polymorphism and susceptibility to cancer. *Archives of Medical Research, 50*(1), 21–30.

Mohamed, M. A., Mohamed, E. I., El-Kaream, S. A. A., Badawi, M. I., ... Darwish, S. H. (2018). Underexpression of miR-486-5p but not overexpression of miR-155 is associated with lung cancer stages. *Microrna, 7*(2), 120–127.

Molina-Pinelo, S., Suárez, R., Pastor, M. D., Nogal, A., Márquez-Martín, E., Martín-Juan, J., ... Paz-Ares, L. (2012). Association between the miRNA signatures in plasma and bronchoalveolar fluid in respiratory pathologies. *Disease Markers, 32*(4), 221–230.

Muppirala, U. K., Honavar, V. G., & Dobbs, D. (2011). Predicting RNA-protein interactions using only sequence information. *BMC Bioinformatics, 12*(1), 1–11.

Muraoka, T., Soh, J., Toyooka, S., Maki, Y., Shien, K., Furukawa, M., ... Miyoshi, S. (2012). Impact of aberrant methylation of microRNA-9 family members on non-small cell lung cancers. *Molecular and Clinical Oncology, 1*(1), 185–189.

Nakagawa, T., Endo, H., Yokoyama, M., Abe, J., Tamai, K., Tanaka, N., ... Satoh, K. (2013). Large noncoding RNA HOTAIR enhances aggressive biological behavior and is associated with short disease-free survival in human non-small cell lung cancer. *Biochemical and Biophysical Research Communications, 436*(2), 319–324.

Nakajima, T., Zamel, R., Anayama, T., Kimura, H., Yoshino, I., Keshavjee, S., ... Yasufuku, K. (2012). Ribonucleic acid microarray analysis from lymph node samples obtained by endobronchial ultrasonography-guided transbronchial needle aspiration. *The Annals of Thoracic Surgery, 94*(6), 2097–2101.

Ono, H., Motoi, N., Nagano, H., Miyauchi, E., Ushijima, M., Matsuura, M., ... Ishikawa, Y. (2014). Long noncoding RNA HOTAIR is relevant to cellular proliferation, invasiveness, and clinical relapse in small-cell lung cancer. *Cancer Medicine, 3*(3), 632–642.

Ouyang, L., Yang, M., Wang, X., Fan, J., Liu, X., Zhang, Y., ... Shu, Y. (2021). Long non-coding RNA FER1L4 inhibits cell proliferation and promotes cell apoptosis via the PTEN/AKT/p53 signaling pathway in lung cancer. *Oncology Reports, 45*(1), 359−367.

Palazzo, A. F., & Koonin, E. V. (2020). Functional long non-coding RNAs evolve from junk transcripts. *Cell, 183* (5), 1151−1161.

Pan, Y., Chen, J., Tao, L., Zhang, K., Wang, R., Chu, X., ... Chen, L. (2017). Long noncoding RNA ROR regulates chemoresistance in docetaxel-resistant lung adenocarcinoma cells via epithelial mesenchymal transition pathway. *Oncotarget, 8*(20), 33144−33158.

Pekarsky, Y., Balatti, V., Palamarchuk, A., Rizzotto, L., Veneziano, D., Nigita, G., ... Croce, C. M. (2016). Dysregulation of a family of short noncoding RNAs, tsRNAs, in human cancer. *Proceedings of the National Academy of Sciences, 113*(18), 5071−5076.

Petriella, D., Galetta, D., Rubini, V., Savino, E., Paradiso, A., Simone, G., ... Tommasi, S. (2013). Molecular profiling of thin-prep FNA samples in assisting clinical management of non-small-cell lung cancer. *Molecular Biotechnology, 54*(3), 913−919.

Qin, S., Zhao, Y., Lim, G., Lin, H., Zhang, X., ... Zhang, X. (2019). Circular RNA PVT1 acts as a competing endogenous RNA for miR-497 in promoting non-small cell lung cancer progression. *Biomedicine & Pharmacotherapy, 111*, 244−250.

Qiu, M., Xu, Y., Yang, X., Wang, J., Hu, J., Xu, L., ... Yin, R. (2014). CCAT2 is a lung adenocarcinoma-specific long non-coding RNA and promotes invasion of non-small cell lung cancer. *Tumor Biology, 35*(6), 5375−5380.

Ramón y Cajal, S., Segura, M. F., & Hümmer, S. (2019). Interplay between ncRNAs and cellular communication: A proposal for understanding cell-specific signaling pathways. *Frontiers in Genetics, 10*, 281.

Redis, R. S., Berindan-Neagoe, I., Pop, V. I., & Calin, G. A. (2012). Non-coding RNAs as theranostics in human cancers. *Journal of Cellular Biochemistry, 113*(5), 1451−1459.

Reeves, M. E., Firek, M., Jliedi, A., & Amaar, Y. G. (2017). Identification and characterization of RASSF1C piRNA target genes in lung cancer cells. *Oncotarget, 8*(21), 34268.

Ren, K., Xu, R., Huang, J., Zhao, J., & Shi, W. (2017). Knockdown of long non-coding RNA KCNQ1OT1 depressed chemoresistance to paclitaxel in lung adenocarcinoma. *Cancer Chemotherapy and Pharmacology, 80*(2), 243−250.

Roa, W. H., Kim, J. O., Razzak, R., Du, H., Guo, L., Singh, R., ... Bedard, E. L. (2012). Sputum microRNA profiling: A novel approach for the early detection of non-small cell lung cancer. *Clinical and Investigative Medicine, 35*, E271−E281.

Rodríguez, M., Silva, J., López-Alfonso, A., López-Muñiz, M. B., Peña, C., Domínguez, G., ... Bonilla, F. (2014). Different exosome cargo from plasma/bronchoalveolar lavage in non-small-cell lung cancer. *Genes, Chromosomes and Cancer, 53*(9), 713−724.

Rui, W., Bing, F., Hai-Zhu, S., Wei, D., & Long-Bang, C. (2010). Identification of microRNA profiles in docetaxel-resistant human non-small cell lung carcinoma cells (SPC-A1). *Journal of Cellular and Molecular Medicine, 14*(1-2), 206−214.

Sadeghiyeh, N., Sehati, N., Mansoori, B., Mohammadi, A., Shanehbandi, D., Khaze, V., ... Baradaran, B. (2019). MicroRNA-145 replacement effect on growth and migration inhibition in lung cancer cell line. *Biomedicine & Pharmacotherapy, 111*, 460−467.

Saito, M., Schetter, A. J., Mollerup, S., Kohno, T., Skaug, V., Bowman, E. D., ... Harris, C. C. (2011). The association of microRNA expression with prognosis and progression in early-stage, non−small cell lung adenocarcinoma: A retrospective analysis of three cohorts. *Clinical Cancer Research, 17*(7), 1875−1882.

Sampson, V. B., Rong, N. H., Han, J., Yang, Q., Aris, V., Soteropoulos, P., ... Krueger, L. J. (2007). MicroRNA let-7a down-regulates MYC and reverts MYC-induced growth in Burkitt lymphoma cells. *Cancer Research, 67*(20), 9762−9770.

Schmidt, B., Rehbein, G., & Fleischhacker, M. (2016). Liquid profiling in lung cancer−quantification of extracellular miRNAs in bronchial lavage. *Circulating nucleic acids in serum and plasma−CNAPS IX* (pp. 33−37). Cham: Springer.

Schmidt, L. H., Spieker, T., Koschmieder, S., Humberg, J., Jungen, D., Bulk, E., ... Muller-Tidow, C. (2011). The long noncoding MALAT-1 RNA indicates a poor prognosis in non-small cell lung cancer and induces migration and tumor growth. *Journal of Thoracic Oncology, 6*(12), 1984−1992.

Seo, J. S., Ju, Y. S., Lee, W. C., Shin, J. Y., Lee, J. K., Bleazard, T., ... Kim, Y. T. (2012). The transcriptional landscape and mutational profile of lung adenocarcinoma. *Genome Research, 22*(11), 2109−2119.

Shi, L., Magee, P., Fassan, M., Sahoo, S., Leong, H. S., Lee, D., ... Garofalo, M. (2021). A KRAS-responsive long non-coding RNA controls microRNA processing. *Nature Communications, 12*(1), 1−19.

Shi, R., Jiao, Z., Yu, A., & Wang, T. (2019). Long noncoding antisense RNA FAM83A-AS1 promotes lung cancer cell progression by increasing FAM83A. *Journal of Cellular Biochemistry, 120*(6), 10505−10512.

Shi, X., Sun, M., Liu, H., Yao, Y., Kong, R., Chen, F., & Song, Y. (2015). A critical role for the long non-coding RNA GAS5 in proliferation and apoptosis in non-small-cell lung cancer. *Molecular Carcinogenesis, 54*(S1), E1−E12.

Smolander, J., Stupnikov, A., Glazko, G., Dehmer, M., & Emmert-Streib, F. (2019). Comparing biological information contained in mRNA and non-coding RNAs for classification of lung cancer patients. *BMC Cancer, 19*(1), 1−15.

Srivastava, K., & Srivastava, A. (2012). Comprehensive review of genetic association studies and *meta*-analyses on miRNA polymorphisms and cancer risk. *PLoS One, 7*(11), e50966.

Stark, C., Breitkreutz, B. J., Reguly, T., Boucher, L., Breitkreutz, A., & Tyers, M. (2006). BioGRID: A general repository for interaction datasets. *Nucleic Acids Research, 34*(suppl_1), D535−D539.

Su, Y., Fang, H., & Jiang, F. (2016). Integrating DNA methylation and microRNA biomarkers in sputum for lung cancer detection. *Clinical Epigenetics, 8*(1), 1−9.

Su, Y., Guarnera, M. A., Fang, H., & Jiang, F. (2016). Small non-coding RNA biomarkers in sputum for lung cancer diagnosis. *Molecular Cancer, 15*(1), 1−4.

Sun, C., Li, S., Zhang, F., Xi, Y., Wang, L., Bi, Y., ... Li, D. (2016). Long non-coding RNA NEAT1 promotes non-small cell lung cancer progression through regulation of miR-377-3p-E2F3 pathway. *Oncotarget, 7*(32), 51784.

Sun, M., Liu, X. H., Lu, K. H., Nie, F. Q., Xia, R., Kong, R., ... Wang, Z. X. (2014a). EZH2-mediated epigenetic suppression of long noncoding RNA SPRY4-IT1 promote s NSCLC cell proliferation and metastasis by affecting the epithelial−mesenchymal transition. *Cell Death & Disease, 5*(6), e1298.

Sun, M., Liu, X. H., Wang, K. M., Nie, F. Q., Kong, R., Yang, J. S., ... Wang, Z. X. (2014b). Downregulation of BRAF activated non-coding RNA is associated with poor prognosis for non-small cell lung cancer and promotes metastasis by affecting epithelial-mesenchymal transition. *Molecular Cancer, 13*(1), 1−12.

Sun, Y., & Ma, L. (2019). New insights into long non-coding RNA MALAT1 in cancer and metastasis. *Cancers, 11*(2), 216.

Tantai, J., Hu, D., Yang, Y., & Geng, J. (2015). Combined identification of long non-coding RNA XIST and HIF1A-AS1 in serum as an effective screening for non-small cell lung cancer. *International Journal of Clinical and Experimental Pathology, 8*(7), 7887−7895.

Tao, H., Yang, J. J., Zhou, X., Deng, Z. Y., Shi, K. H., & Li, J. (2016). Emerging role of long noncoding RNAs in lung cancer: Current status and future prospects. *Respiratory Medicine, 110*, 12−19.

Tellez, C. S., Juri, D. E., Do, K., Picchi, M. A., Wang, T., Liu, G., ... Belinsky, S. A. (2016). miR-196b is epigenetically silenced during the premalignant stage of lung carcinogenesis. *Cancer Research, 76*(16), 4741−4751.

Tian, T., Shu, Y., Chen, J., Hu, Z., Xu, L., Jin, G., ... Shen, H. (2009). A functional genetic variant in microRNA-196a2 is associated with increased susceptibility of lung cancer in Chinese. *Cancer Epidemiology and Prevention Biomarkers, 18*(4), 1183−1187.

Tonon, G., Wong, K. K., Maulik, G., Brennan, C., Feng, B., Zhang, Y., ... DePinho, R. A. (2005). High-resolution genomic profiles of human lung cancer. *Proceedings of the National Academy of Sciences, 102*(27), 9625−9630.

Tripathi, V., Shen, Z., Chakraborty, A., Giri, S., Freier, S. M., Wu, X., ... Prasanth, K. V. (2013). Long noncoding RNA MALAT1 controls cell cycle progression by regulating the expression of oncogenic transcription factor B-MYB. *PLoS Genet, 9*(3), e1003368.

Vannini, I., Wise, P. M., Challagundla, K. B., Plousiou, M., Raffini, M., Bandini, E., ... Fabbri, M. (2017). Transcribed ultraconserved region 339 promotes carcinogenesis by modulating tumor suppressor microRNAs. *Nature Communications, 8*(1), 1−19.

Volinia, S., Calin, G. A., Liu, C. G., Ambs, S., Cimmino, A., Petrocca, F., ... Croce, C. M. (2006). A microRNA expression signature of human solid tumors defines cancer gene targets. *Proceedings of the National Academy of Sciences, 103*(7), 2257−2261.

Vosa, U., Vooder, T., Kolde, R., Vilo, J., Metspalu, A., & Annilo, T. (2013). Meta-analysis of microRNA expression in lung cancer. *International Journal of Cancer, 132*(12), 2884−2893.

Wang, K., Chen, M., & Wu, W. (2017). Analysis of microRNA (miRNA) expression profiles reveals 11 key biomarkers associated with non-small cell lung cancer. *World Journal of Surgical Oncology, 15*(1), 1−10.

Wang, X. C., Du, L. Q., Tian, L. L., Wu, H. L., Jiang, X. Y., Zhang, H., ... Meng, A. M. (2011). Expression and function of miRNA in postoperative radiotherapy sensitive and resistant patients of non-small cell lung cancer. *Lung Cancer, 72*(1), 92–99.

Weber, D. G., Johnen, G., Casjens, S., Bryk, O., Pesch, B., Jöckel, K. H., ... Brüning, T. (2013). Evaluation of long noncoding RNA MALAT1 as a candidate blood-based biomarker for the diagnosis of non-small cell lung cancer. *BMC Research Notes, 6*(1), 1–9.

Wei, S., Wang, K., Huang, X., Zhao, Z., & Zhao, Z. (2019). LncRNA MALAT1 contributes to non-small cell lung cancer progression via modulating miR-200a-3p/programmed death-ligand 1 axis. *International Journal of Immunopathology and Pharmacology, 33*, 2058738419859699.

Weinstein, J. N., Collisson, E. A., Mills, G. B., Shaw, K. R. M., Ozenberger, B. A., Ellrott, K., ... Stuart, J. M. (2013). The cancer genome atlas pan-cancer analysis project. *Nature Genetics, 45*(10), 1113–1120.

Wu, C. S., Lin, F. C., Chen, S. J., Chen, Y. L., Chung, W. J., & Cheng, C. I. (2016). Optimized collection protocol for plasma microRNA measurement in patients with cardiovascular disease. *BioMed Research International, 2016*.

Wu, D., Li, Y., Zhang, H., & Hu, X. (2017). Knockdown of Lncrna PVT1 enhances radiosensitivity in non-small cell lung cancer by sponging Mir-195. *Cellular Physiology and Biochemistry, 42*(6), 2453–2466.

Wu, J., Li, X., Xu, Y., Yang, T., Yang, Q., Yang, C., ... Jiang, Y. (2016). Identification of a long non-coding RNA NR_026689 associated with lung carcinogenesis induced by NNK. *Oncotarget, 7*(12), 14486–14498.

Wu, K. L., Tsai, Y. M., Lien, C. T., Kuo, P. L., & Hung, J. Y. (2019). The roles of MicroRNA in lung cancer. *International Journal of Molecular Sciences, 20*(7), 1611.

Wu, S., Liu, B., Zhang, Y., Hong, R., Liu, S., Xiang, T., ... Guan, H. (2021). Long non-coding RNA LEISA promotes progression of lung adenocarcinoma via enhancing interaction between STAT3 and IL-6 promoter. *Oncogene*, 1–11.

Xia, Y., Zhu, Y., Zhou, X., & Chen, Y. (2014). Low expression of let-7 predicts poor prognosis in patients with multiple cancers: A meta-analysis. *Tumor Biology, 35*(6), 5143–5148.

Xie, K., Wang, C., Qin, N., Yang, J., Zhu, M., Dai, J., ... Hu, Z. (2016). Genetic variants in regulatory regions of microRNAs are associated with lung cancer risk. *Oncotarget, 7*(30), 47966–47974.

Xie, Y., Xue, C., Guo, S., & Yang, L. (2021). MicroRNA-520a suppresses pathogenesis and progression of non-small-cell lung cancer through targeting the RRM2/Wnt axis. *Analytical Cellular Pathology, 2021*.

Xu, R., Mao, Y., Chen, K., He, W., Shi, W., & Han, Y. (2017). The long noncoding RNA ANRIL acts as an oncogene and contributes to paclitaxel resistance of lung adenocarcinoma A549 cells. *Oncotarget, 8*(24), 39177–39184.

Yanaihara, N., Caplen, N., Bowman, E., Seike, M., Kumamoto, K., Yi, M., ... Harris, C. C. (2006). Unique microRNA molecular profiles in lung cancer diagnosis and prognosis. *Cancer Cell, 9*(3), 189–198.

Yang, Z., Li, H., Li, J., Lv, X., Gao, M., Bi, Y., ... Yin, Z. (2018). Association between long noncoding RNA MEG3 polymorphisms and lung cancer susceptibility in Chinese northeast population. *DNA and Cell Biology, 37*(10), 812–820.

Ye, M., Dong, S., Hou, H., Zhang, T., & Shen, M. (2021). Oncogenic role of long noncoding RNAMALAT1 in thyroid cancer progression through regulation of the miR-204/IGF2BP2/m6A-MYC signaling. *Molecular Therapy-Nucleic Acids, 23*, 1–12.

Yin, Z., Cui, Z., Guan, P., Li, X., Wu, W., Ren, Y., ... Zhou, B. (2015). Interaction between polymorphisms in pre-MiRNA genes and cooking oil fume exposure on the risk of lung cancer in Chinese non-smoking female population. *PLoS One, 10*(6), e0128572.

Yu, H., Tian, L., Yang, L., Liu, S., Wang, S., & Gong, J. (2021). Knockdown of SNORA47 Inhibits the tumorigenesis of NSCLC via Mediation of PI3K/Akt Signaling Pathway. *Frontiers in Oncology, 11*, 181.

Yu, Y., Mao, L., Lu, X., Yuan, W., Chen, Y., Jiang, L., ... Chu, M. (2019). Functional variant in 3′ UTR of FAM13A is potentially associated with susceptibility and survival of lung squamous carcinoma. *DNA and Cell Biology, 38*(11), 1269–1277.

Yuan, H., Liu, H., Liu, Z., Owzar, K., Han, Y., Su, L., ... Wei, Q. (2016). A novel genetic variant in long non-coding RNA gene NEXN-AS1 is associated with risk of lung cancer. *Scientific Reports, 6*(1), 1–8.

Yue, J., Lv, D., Wang, C., Li, L., Zhao, Q., Chen, H., ... Xu, L. (2018). Epigenetic silencing of miR-483-3p promotes acquired gefitinib resistance and EMT in EGFR-mutant NSCLC by targeting integrin β3. *Oncogene, 37*(31), 4300–4312.

Zaporozhchenko, I. A., Morozkin, E. S., Ponomaryova, A. A., Rykova, E. Y., Cherdyntseva, N. V., Zheravin, A. A., ... Laktionov, P. P. (2018). Profiling of 179 miRNA expression in blood plasma of lung cancer patients and cancer-free individuals. *Scientific Reports, 8*(1), 1–13.

Zhang, C., Gong, C., Li, J., & Tang, J. (2021). Downregulation of long non-coding RNA LINC-PINT serves as a diagnostic and prognostic biomarker in patients with non-small cell lung cancer. *Oncology Letters, 21*(3), 1.

Zhang, E. B., Yin, D. D., Sun, M., Kong, R., Liu, X. H., You, L. H., ... Wang, Z. X. (2014). P53-regulated long noncoding RNA TUG1 affects cell proliferation in human non-small cell lung cancer, partly through epigenetically regulating HOXB7 expression. *Cell Death & Disease, 5*(5), e1243.

Zhang, J. G., Wang, J. J., Zhao, F., Liu, Q., Jiang, K., & Yang, G. H. (2010). MicroRNA-21 (miR-21) represses tumor suppressor PTEN and promotes growth and invasion in non-small cell lung cancer (NSCLC). *Clinica Chimica Acta, 411*(11−12), 846−852.

Zhang, R., Xia, Y., Wang, Z., Zheng, J., Chen, Y., Li, X., ... Ming, H. (2017). Serum long non coding RNA MALAT-1 protected by exosomes is up-regulated and promotes cell proliferation and migration in non-small cell lung cancer. *Biochemical and Biophysical Research Communications, 490*(2), 406−414.

Zhang, S. J., Yao, J., Shen, B. Z., Li, G. B., Kong, S. S., Bi, D. D., ... Cheng, B. L. (2018). Role of piwi-interacting RNA-651 in the carcinogenesis of non-small cell lung cancer. *Oncology Letters, 15*(1), 940−946.

Zhang, X., Wang, Q., & Zhang, S. (2019). MicroRNAs in sputum specimen as noninvasive biomarkers for the diagnosis of nonsmall cell lung cancer: An updated *meta*-analysis. *Medicine, 98*(6).

Zhang, Y., Liu, H., Zhang, Q., & Zhang, Z. (2021). LncRNA LINC01006 facilitates cell proliferation, migration and EMT in lung adenocarcinoma via targeting miR-129-2-3p/CTNNB1 axis and activating Wnt/β-catenin signaling pathway. *Molecular and Cellular Biology*.

Zhao, H., Shi, J., Zhang, Y., Xie, A., Yu, L., Zhang, C., ... Li, X. (2020). LncTarD: A manually-curated database of experimentally-supported functional lncRNA−target regulations in human diseases. *Nucleic Acids Research, 48*(D1), D118−D126.

Zhao, K., Cheng, J., Chen, B., Liu, Q., Xu, D., & Zhang, Y. (2017). Circulating microRNA-34 family low expression correlates with poor prognosis in patients with non-small cell lung cancer. *Journal of Thoracic Disease, 9*(10), 3735−3746.

Zhao, W., An, Y., Liang, Y., & Xie, X. W. (2014). Role of HOTAIR long noncoding RNA in *meta*static progression of lung cancer. *European Review for Medical and Pharmacological Sciences, 18*(13), 1930−1936.

Zheng, D., Zhang, J., Ni, J., Luo, J., Wang, J., Tang, L., ... Chen, G. (2015). Small nucleolar RNA 78 promotes the tumorigenesis in non-small cell lung cancer. *Journal of Experimental & Clinical Cancer Research, 34*(1), 1−15.

Zheng, W., Zhao, J., Tao, Y., Guo, M., Ya, Z., Chen, C., ... Xu, L. (2018). MicroRNA-21: A promising biomarker for the prognosis and diagnosis of non-small cell lung cancer. *Oncology Letters, 16*(3), 2777−2782.

Zhong, X., Zhang, C., Diao, Y., Liao, S., Ling, Q., Zhang, Z., ... Long, P. (2021). The LncRNA H19/microRNA-29b-3p/HMGB1 signaling axis contributes to the regulation of lung cancer cell growth. *FEBS Open Bio*.

Zhou, C., Ye, L., Jiang, C., Bai, J., Chi, Y., & Zhang, H. (2015). Long noncoding RNA HOTAIR, a hypoxia-inducible factor-1α activated driver of malignancy, enhances hypoxic cancer cell proliferation, migration, and invasion in non-small cell lung cancer. *Tumor Biology, 36*(12), 9179−9188.

Zhou, H., Chen, A., Shen, J., Zhang, X., Hou, M., Li, J., ... He, J. (2019). Long non-coding RNA LOC285194 functions as a tumor suppressor by targeting p53 in non-small cell lung cancer. *Oncology Reports, 41*(1), 15−26.

Zhou, J., Wang, C., Gong, W., Wu, Y., Xue, H., Jiang, Z., ... Shi, M. (2018). Uc. 454 inhibited growth by targeting heat shock protein family a member 12B in non-small-cell lung cancer. *Molecular Therapy-Nucleic Acids, 12*, 174−183.

Zhou, S., Zhang, Z., Zheng, P., Zhao, W., & Han, N. (2017). MicroRNA-1285-5p influences the proliferation and metastasis of non-small-cell lung carcinoma cells via downregulating CDH1 and Smad4. *Tumor Biology, 39*(6), 1010428317705513.

Zhu, K., Ding, H., Wang, W., Liao, Z., Fu, Z., Hong, Y., ... Chen, X. (2016). Tumor-suppressive miR-218-5p inhibits cancer cell proliferation and migration via EGFR in non-small cell lung cancer. *Oncotarget, 7*(19), 28075−28085.

Zhu, L., Liu, J., Shenglin, M. A., & Zhang, S. (2015). Long noncoding RNA MALAT-1 can predict metastasis and a poor prognosis: A meta-analysis. *Pathology & Oncology Research, 21*(4), 1259−1264.

Zhu, X., Wang, X., Wei, S., Chen, Y., Chen, Y., Fan, X., ... Wu, G. (2017). hsa_circ_0013958: A circular RNA and potential novel biomarker for lung adenocarcinoma. *The FEBS Journal, 284*(14), 2170−2182.

CHAPTER 6

Noncoding RNAs in intraocular tumor patients

Daniel Fernandez-Diaz[1,2], Beatriz Fernandez-Marta[1], Nerea Lago-Baameiro[3], Paula Silva-Rodríguez[2,4], Laura Paniagua[5], María José Blanco-Teijeiro[1,2], María Pardo[2,3], Antonio Piñeiro[1,2] and Manuel F. Bande[1,2]

[1]Department of Ophthalmology, University Hospital of Santiago de Compostela, Santiago de Compostela, Spain [2]Intraocular Tumors in Adults, Health Research Institute of Santiago de Compostela (IDIS), Santiago de Compostela, Spain [3]Obesidomics Group, Health Research Institute of Santiago de Compostela (IDIS), Santiago de Compostela, Spain [4]Galician Public Foundation of Genomic Medicine, University Hospital of Santiago de Compostela, Santiago de Compostela, Spain [5]Department of Ophthalmology, University Hospital of Coruña, La Coruña, Spain

6.1 Introduction

It is well known that epigenetics play a fundamental role in ocular pathologies, encompassing, among many other mechanisms, regulation through noncoding RNAs (ncRNAs) (Lee, 2012; Schmitz, Grote, & Herrmann, 2016). These molecules constitute the vast majority of the human genome and, although they lack the capacity for translation into proteins, their influence on numerous biological functions is greater than that of coding RNAs (Guzel et al., 2020).

Different classes of ncRNAs, including long noncoding RNA (lncRNA), microRNA (miRNA), or circular RNA (circRNA), contribute to the appearance, pathogenesis, and evolution of various eye diseases, such as cataracts, proliferative vitreoretinopathy and diabetic retinopathy, corneal neovascularization, premature retinopathy, age-related macular degeneration, high myopia, primary open-angle glaucoma, pterygium, uveitis, strabismus, ophthalmological disorders induced by hyperhomocysteinemia, and of course eye tumors (Guo, Liu, et al., 2019; Li, Wen, Zhang, & Fan, 2016; Wawrzyniak, Zarębska, Rolle, & Gotz-Więckowska, 2018; Yan, Yao,

Tao, & Jiang, 2014; Zhang et al., 2019). Retinoblastoma (RB) and uveal melanoma (UM) are the most common primary intraocular malignancies in childhood and adulthood, respectively, and are not only a major cause of visual function loss but also a significant cause of mortality (Global Retinoblastoma Study Group et al., 2020; Kaliki & Shields, 2017). Given that the influence of ncRNAs on the carcinogenic process of these tumors has been widely demonstrated, these epigenetic regulators are considered relevant and innovative biomarkers for both diagnosis and prognosis, and their use as therapeutic targets could become a promising new tool for clinical treatment.

6.2 Retinoblastoma

6.2.1 Introduction

RB is a neoplasm of retinal origin that constitutes the most frequent primary intraocular malignancy in childhood (Dimaras et al., 2012). Its incidence is one case per 15,000−20,000 live births, resulting in approximately 9000 new cases per year (Kivelä, 2009) without significant differences according to sex or race (Rao & Honavar, 2017). Although there are several therapeutic modalities that allow local control of the tumor, the development of metastatic disease remains the most daunting adverse effect (AlAli, Kletke, Gallie, & Lam, 2018). Both the conservation of the eyeball and the survival of the patient will depend largely on the tumor stage at the time of diagnosis. In developing regions, detection is carried out at a later stage, resulting in a much lower survival rate and a mortality rate of 50%−70%.

RB is a genetic disease in which the loss of the tumor suppressor gene *RB1* plays a major role (Knudson, 1971). There is a nonhereditary form with monocular involvement and a hereditary form in which the involvement may be single or bilateral. At present, more precise genomic studies are broadening the field of knowledge about this neoplasm, so it is increasingly clear that the presence of other genetic and epigenetic events is necessary for the tumorigenesis process (Thériault, Dimaras, Gallie, & Corson, 2014).

For these reasons, there is a growing interest in ncRNA molecules because it is vitally important to have markers that not only allow early diagnosis of the disease but also provide a therapeutic and prognostic approach to improve the quality of life and survival of patients with RB.

6.2.2 Long noncoding RNAs and circular RNAs in retinoblastoma

lncRNAs are ncRNAs with a length of more than 200 nucleotides and are crucial in neoplastic processes (Chu, Qu, Zhong, Artandi, & Chang, 2011; Gibb, Brown, & Lam, 2011; Prensner & Chinnaiyan, 2011) including RB (Li et al., 2016; Yang & Wei, 2019; Zhang et al., 2019).

To date, a few tumor suppressor lncRNAs have been described. Shang, Yang, Zhang, and Wu (2018) demonstrated that BDNF (Brain Derived Neurotrophic Factor) antisense RNA was downregulated in RB samples and correlated to a more advanced tumor stage and lower overall survival (OS). Similarly, lncRNA maternally expressed gene 3 showed decreased expression levels in RB tissues and cell lines, which was negatively associated with metastatic disease development, the International Intraocular Retinoblastoma Classification (IIRC) stage, and patient survival, owing to regulation of the *Wnt/β-catenin* and *p53* pathways (Gao & Lu, 2016, 2017). Similarly, it was described that the metallothionein 1J pseudogene was underexpressed and negatively modulated the activity of the *Wnt/β-catenin* axis, correlating to the invasion of the optic nerve, the IIRC stage, and the appearance of metastasis in cases of RB (Bi, Han, Zhang, & Li, 2018).

In the field of lncRNAs with oncogenic activity, an enormous variety of molecules has been discovered. Hao, Mou, Zhang, Wang, and Yang (2018) and Su et al. (2015) observed that actin filament–associated protein 1-antisense RNA 1 and BRAF(B-Raf proto-oncogene)-activated ncRNA (BANCR) were overexpressed and were independent unfavorable prognostic factors associated with tumor size, choroidal infiltration, and optic nerve invasion. The antisense ncRNA in the INK4 locus (ANRIL), which was upregulated in the RB samples, favored proliferation, migration, and cell invasion and reduced apoptosis by, among other effects, inhibiting the *ATM-E2F1* pathway (Yang & Peng, 2018). The lncRNA CCAT1 induced similar actions at the cellular level, in this case through negative modulation of miR-218-5p (Zhang et al., 2017). Wang, Yang, and Li (2018) demonstrated that differentiation-antagonizing nonprotein-coding RNA was overexpressed in RB cell lines and tissues and worsens disease-free survival (DFS) and OS by stimulating proliferation, migration, invasion, and epithelial–mesenchymal transition (EMT) by acting as a competing endogenous RNA (ceRNA) to miR-34c and miR-613. In line with this, tumoral elevation of HOX antisense intergenic RNA favors tumorigenesis and reduces OS through the *miR-613/c-Met* axis and the *Notch1* pathway (Dong et al., 2016; Yang et al., 2018). Sheng, Wu, Gong, Dong, and Sun (2018) indicated that increased levels of the promoter of CDKN1A antisense DNA damage-activated RNA, a novel lncRNA, were associated with unfavorable clinicopathological characteristics, such as optic nerve invasion and advanced IIRC stage, in part by inhibiting cell apoptosis when interacting with the *Bcl-2/caspase-3* pathway. Similarly, there is evidence of the clinical role of testis-associated highly conserved oncogenic lncRNA, the overexpression of which potentiates the malignant phenotype of RB in relation to the *c-Myc* oncogene and *insulin-like growth factor 2 mRNA-binding protein 1* (Shang, 2018). Finally, two other lncRNAs that promote tumor aggressiveness and lower OS are the small nucleolar RNA host gene 14 and plasmacytoma variant translocation 1 (PVT1), which are sponging miRNAs of miR-124 and miR-488-3p, respectively (Sun, Shen, Liu, Gao, & Zhang, 2020; Wu, Cui, Lv, & Feng, 2019).

Regarding lncRNA H19, Zhang, Shang, Nie, Li, and Li (2018) found that it showed decreased expression in RB and exerted its tumor-suppressive action at the level of proliferation, cell cycle, and apoptosis through interaction with the miR-17–92 cluster. However, in other publications, it was demonstrated that this lncRNA was overexpressed in tumor samples, that is, it was a carcinogenic molecule favoring proliferation, migration, invasion, and reduced cell apoptosis, which ultimately translated into a larger tumor size, invasion of the optic nerve, choroidal infiltration, and shorter survival (Li, Chen, Wang, Tang, & Han, 2018; Qi, Wang, & Yu, 2019).

Table 6.1 describes in detail all main deregulated lncRNAs, including their chromosomal location, molecular targets, signaling pathways in which they intervene, the type of study sample, and the technical validation methods used for their characterization.

Fig. 6.1 is a schematic summarizing the expression levels of the different lncRNAs, as well as the cellular processes in which they participate.

TABLE 6.1 Long noncoding RNAs (lncRNAs) involved in retinoblastoma (RB) pathogenesis.

lncRNA	Location[a]	Expression	Target mechanism	Sample type	Validation method	References
AFAP1-AS1	4p16.1	Upregulated	—	RB cell lines (Weri-Rb1, Y79) and tumor tissues	RT-PCR, PA, MA, IA	Hao et al. (2018)
ANRIL	9p21.3	Upregulated	ATM-E2F1, miR-99a/c-Myc, miR-99a/JAK/STAT, and miR-99a/PI3K/AKT pathways	RB cell lines (HXO-RB44, Y79) and tumor tissues	RT-PCR, PA, MA, IA, AA, WB, LUCA, IF	Wang, Zhang, et al. (2019), Yang and Peng (2018)
BANCR	9q21.12	Upregulated	MAPK and NF-kB pathways	RB cell lines (Weri-Rb1, Y79) and tumor tissues	RT-PCR, PA, MA, IA	Su et al. (2015)
BDNF-AS	11p14.1	Downregulated	CDC42, cyclin E and BDNF	RB cell lines (Weri-Rb1, Y79, SO-RB50, HXO-RB44, Rb116, Rb143) and tumor tissues	RT-PCR, PA, MA, WB	Shang et al. (2018)
CCAT1	8q24.21	Upregulated	miR-218-5p	RB cell lines (Weri-Rb1, Y79, SO-RB50) and tumor tissues	RT-PCR, PA, MA, IA, AA, WB	Zhang et al. (2017)
CYTOR (LINC00152)	2p11.2	Upregulated	Caspase-3, caspase-8, ki-67, Bcl-2, and MMP-9	RB cell lines (SO-RB50, Y79) and tumor tissue	RT-PCR, PA, MA, IA, AA, WB, ATM	Li, Wen, et al. (2018)
DANCR	4q12	Upregulated	Notch pathway, miR-34c, and miR-613/MMP-9 axis	RB cell lines (Weri-Rb1, Y79, SO-RB50, HXO-RB44) and tumor tissue	RT-PCR, PA, MA, IA, WB, LUCA, ATM	Wang, Yang, et al. (2018)
FAM238B (LINC00202)	10p12.1	Upregulated	miR-3619-5p/RIN1 axis	RB cell lines (Weri-Rb1, Y79, SO-RB50, HXO-RB44) and tumor tissue	RT-PCR, PA, MA, IA, WB, LUCA	Yan, Su, Ma, Yu, and Chen (2019)
FEZF1-AS1	7q31.32	Upregulated	miR-1236-3p	RB cell lines (Weri-Rb1, Y79, SO-RB50, RBL-13)	RT-PCR, PA, MA, IA, WB, LUCA	Zhang et al. (2020)

(Continued)

TABLE 6.1 (Continued)

lncRNA	Location[a]	Expression	Target mechanism	Sample type	Validation method	References
H19	11p15.5	Upregulated Downregulated	miR-143/RUNX2 axis and PI3K/AKT/mTOR pathways miR-17–92 cluster, p21, and STAT3 pathways	RB cell lines (Weri-Rb1, Y79) and tumor tissues RB cell lines (Weri-Rb1, Y79, SO-RB50) and tumor tissues	RT-PCR, PA, MA, IA, AA, WB, LUCA, ATM RT-PCR, PA, AA, WB, IF	Li, Chen, et al. (2018), Qi et al. (2019), Zhang, Shang, et al. (2018)
HOTAIR	12q13.13	Upregulated	Notch1 pathway and miR-613/c-Met axis	RB cell lines (HXO-RB44, Y79, SO-RB50, Weri-Rb1) and tumor tissue	RT-PCR, PA, IA, AA, WB, LUCA, ATM	Dong et al. (2016), Yang et al. (2018)
HOXA11-AS	7p15.2	Upregulated	miR-506-3p/NEK3 axis	RB cell lines (HXO-RB44, Y79, SO-RB50, Weri-Rb1) and tumor tissue	RT-PCR, PA, AA, LUCA	Han et al. (2019)
MALAT1	11q13.1	Upregulated	miR-124/STX17 and miR-124/Slug axis	RB cell lines (HXO-RB44, Y79, SO-RB50, Weri-Rb1)	RT-PCR, PA, MA, IA, AA, WB, LUCA	Huang, Yang, Fang, and Liu (2018), Liu, Yan, Zhang, and Yu (2018)
MEG3	14q32.2	Downregulated	Wnt/β-catenin and p53 pathways	RB cell lines (HXO-RB44, Y79, SO-RB50, Weri-Rb1) and tumor tissue	RT-PCR, PA, AA, WB, LUCA	Gao and Lu (2016, 2017)
MIR7-3HG	19p13.3	Upregulated	miR-27a-3p/PEG10 network	RB cell lines (Weri-Rb1, Y79) and tumor tissue	RT-PCR, PA, AA, WB, LUCA	Ding, Jiang, Sheng, Li, and Zhu (2020)
MT1JP	16q13	Downregulated	Wnt/β-catenin pathway, cyclin D1 and c-Myc	RB cell lines (Weri-Rb1, Y79) and tumor tissue	RT-PCR, PA, MA, IA, AA, WB, LUCA	Bi et al. (2018)
NEAT1	11q13.1	Upregulated	miR-204/CXCR4 axis and miR-124	RB cell lines (Weri-Rb1, Y79, SO-RB50) and tumor tissue	RT-PCR, PA, MA, AA, WB, LUCA, ATM	Wang, Yang, Tian, and Zhang (2019), Zhong, Yang, Li, Li, and Li (2019)

(Continued)

TABLE 6.1 (Continued)

lncRNA	Location[a]	Expression	Target mechanism	Sample type	Validation method	References
PANDAR	6p21.2	Upregulated	Bcl-2/caspase-3 pathway	RB cell lines (HXO-RB44, Y79, SO-RB50, Weri-Rb1) and tumor tissue	RT-PCR, PA, AA, WB, LUCA, IHC, ATM	Sheng et al. (2018)
PlncRNA-1	21q22.12	Upregulated	CBR3	RB cell lines (Weri-Rb1, Y79) and tumor tissue	RT-PCR, PA, MA, IA, WB	Wang, Liu, Yang, Hao, and Zhang (2018)
PVT1	8q24.21	Upregulated	miR-488-3p/Notch2 pathway	RB cell lines (HXO-RB44, Y79, SO-RB50, Weri-Rb1) and tumor tissue	RT-PCR, PA, MA, IA, AA, WB, LUCA, ATM	Wu, Cui, et al. (2019)
SNHG14	15q11.2	Upregulated	miR-124/STAT3 pathway	RB cell lines (Weri-Rb1, Y79, SO-RB50) and tumor tissue	RT-PCR, PA, MA, IA, AA, WB, LUCA, ATM	Sun et al. (2020)
SNHG16	17q25.1	Upregulated	miR-140-5p	RB cell lines (Weri-Rb1, Y79, SO-RB50) and tumor tissue	RT-PCR, PA, AA, LUCA, ATM	Xu, Hu, Wang, and Liu (2019)
TCL6	14q32.13	Downregulated	miR-21/PTEN/PI3K/AKT pathway	RB cell lines (Weri-Rb1, Y79) and tumor tissue	RT-PCR, PA, AA, WB, LUCA	Tao, Wang, Liu, Wang, & Chen, 2019
THOR	2q14.2	Upregulated	c-Myc/IGF2BP1 interaction	RB cell lines (Weri-Rb1, Y79) and tumor tissue	RT-PCR, PA, MA, AA, WB, ATM	Shang (2018)
TP73-AS1	1p36.32	Upregulated	miR-139-3p and miR-874-3p/TFAP2B/Wnt/β-catenin pathway	RB cell lines (HXO-RB44, Y79, SO-RB50, Weri-Rb1) and tumor tissue	RT-PCR, PA, MA, IA, AA, WB, LUCA	Wang, Wang, Wu, and Sun (2020), Xia, Yang, Wu, et al. (2019)
UCA1	19p13.12	Upregulated	miR-513a-5p/STMN1 axis	RB cell lines (SO-RB50) and tumor tissue	RT-PCR, PA, WB, LUCA	Yang et al. (2020b)
XIST	Xq13.2	Upregulated	miR-101/ZEB1-ZEB2, miR-124/STAT3, and miR-140-5p/SOX4 axis	RB cell lines (HXO-RB44, Y79, SO-RB50, Weri-Rb1) and tumor tissue	RT-PCR, PA, MA, IA, AA, WB, LUCA	Cheng, Chang, Zheng, et al. (2019), Hu, Liu, Han, Wang, and Xu (2018), Wang, Sun, et al. (2020)

(Continued)

TABLE 6.1 (Continued)

lncRNA	Location[a]	Expression	Target mechanism	Sample type	Validation method	References
ZFPM2-AS1	8q23.1	Upregulated	miR-515/HOXA1/Wnt/β-catenin axis	RB cell lines (Weri-Rb1, Y79, SO-RB50) and tumor tissue	RT-PCR, PA, MA, IA, AA, WB, LUCA, IF, IHC, ATM	Lyv et al. (2020)

[a]*Information obtained from GeneCards database (http://www.genecards.org).*
AA, Apoptosis in vitro assay; *ATM*, animal tumor models; *IA*, invasion in vitro assay; *IF*, immunofluorescence; *IHC*, immunohistochemistry; *LUCA*, luciferase expression assays; *MA*, migration in vitro assay; *PA*, proliferation in vitro assay; *RT-PCR*, real-time polymerase chain reaction; *WB*, Western blot immunodetection.

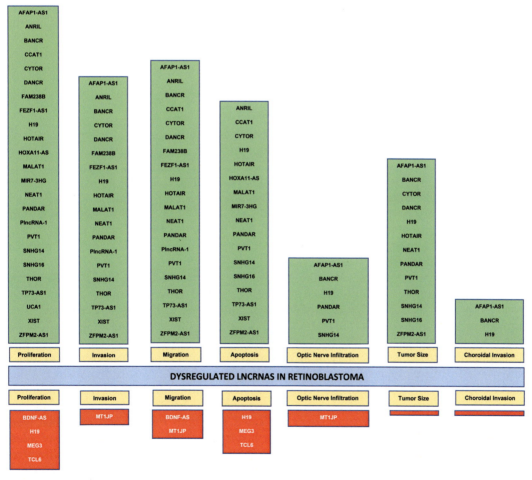

FIGURE 6.1 Events associated with dysregulated lncRNAs in RB. Red and green boxes indicate, respectively, underexpressed and overexpressed lncRNAs in RB. *lncRNAs*, Long noncoding RNAs; *RB*, retinoblastoma.

circRNAs are a novel molecular group of noncoding endogenous RNAs with a loop structure and lower susceptibility to degradation (Jeck & Sharpless, 2014; Vicens & Westhof, 2014). Their functions include acting as miRNA sponges, removing the initiation codons of mature mRNAs to reduce protein translation, regulating alternative splicing, and binding to *RNA binding proteins* (Salzman, 2016; Salzman, Gawad, Wang, Lacayo, & Brown, 2012; Zhang et al., 2013). Current evidence has shown that they play key roles in the cellular processes of many ocular diseases, including RB (Guo, Liu, et al., 2019; Wawrzyniak et al., 2018). Thus an overall decrease in circRNA expression levels has been demonstrated in this type of neoplasm (Lyu, Wang, Zheng, et al., 2019). Moreover, Lyu et al. found that TET1-hsa_circ_0093996 was underregulated in both primary and recurrent RB samples, leading to an increase in miR-183 and ultimately a reduction in the tumor suppressor *programmed cell death 4*. Xing, Zhang, Feng, Cui, and Ding (2018) demonstrated that lower levels of hsa_circ_0001649 was also expressed in RB and favored tumorigenesis by regulating cell proliferation and apoptosis through the *AKT/mTOR* signaling pathway, and that it was correlated to larger tumor size, more advanced clinical stage, and shorter survival.

6.2.3 MicroRNAs in retinoblastoma

6.2.3.1 Dysregulated expression of microRNAs

The aberrant expression of miRNAs, a type of small ncRNA with a length of 17–22 nucleotides, plays a crucial role in the pathophysiology and progression of RB (Delsin, Salomao, Pezuk, & Brassesco, 2019; Singh, Malik, Goswami, Shukla, & Kaur, 2016). The study of these molecules represents the most extensive field of knowledge among ncRNAs, with new molecules constantly and rapidly emerging. In Tables 6.2 and 6.3 are shown a wide variety of miRNAs discovered to date, stratified into two groups (tumor suppressor and oncogenic action), as well as their molecular targets, axes of action, and functions at the cellular level, in addition to the type of study sample and the technical validation methods used.

Among tumor suppressors, Guo, Bai, Ji, and Ma (2019) and Lei et al. (2014) described that the downregulation of miR-98 and miR-101 in the RB samples implied a worse prognosis as it was associated with unfavorable clinical–histopathological characteristics (tumor size, choroidal/orbital infiltration, and optic nerve invasion). Although both molecules had similar repercussions, they act through different pathways: miR-98 influences the *insulin-like growth factor-1 receptor*, while miR-101 inhibits the expression of the oncogenic *enhancer of zeste homolog 2 (EZH2)* protein. Zhao and Cui (2019) demonstrated that miR-361-3p is underregulated in RB cells, tissue, and serum samples, and that its action in suppressing tumorigenesis and progression is carried out by acting on the *sonic hedgehog* signaling pathway. An interesting study indicated that both miR-31 and miR-200a were underexpressed and that the elevation of their levels failed to limit the expansion of a less aggressive RB cell line (Weri-Rb1), but it did restrict the growth of the highly proliferative Y79 cell line (Montoya et al., 2015). Many other miRNAs have also shown low levels of expression. For example, Martin et al. (2013) confirmed the deregulation of miR-22, miR-129-3p, miR-129-5p, miR-382, and miR-504, among others.

TABLE 6.2 MicroRNAs (miRNAs) with retinoblastoma (RB) tumor suppressor action.

miRNA	Location[a]	Expression	Target mechanism	Suppressive events	Sample type	Validation method	Reference
miR-23a	19p13.12	Downregulated	Zeb1	Cell migration	RB cell lines (Weri-Rb1, Y79)	RT-PCR, MA, IF, WB	Wang, Luo, et al. (2018)
mir-29a	7q32.3	Downregulated	STAT3	Cell proliferation, migration, and invasion Induces apoptosis	RB cell lines (SO-RB50, Y79) and tumor tissue	RT-PCR, PA, MA, IA, AA, WB, LUCA, ATM	Liu, Zhang, Hu, Wang, and Xu (2018)
miR-31	9p21.3	Downregulated	STK40, PPP6C and DLL3	Cell proliferation (Y79) Induces apoptosis (Y79)	RB cell lines (Weri-Rb1, Y79) and tumor tissue	RT-PCR, PA, AA, WB, LUCA, IF, IHC	Montoya et al. (2015)
miR-34a	1p36.22	Downregulated	HMGB1	Cell autophagy Induces apoptosis	RB cell lines (Weri-Rb1, Y79)	RT-PCR, AA, WB, LUCA	Liu, Huang, et al. (2014)
miR-98	Xp11.22	Downregulated	IGF1R/k-Ras/Raf/MEK/ERK pathway	Cell proliferation, migration, and invasion Induces apoptosis	RB cell lines (Weri-Rb1, Y79, SO-RB50) and tumor tissue	RT-PCR, PA, MA, IA, AA, WB, LUCA, IHC	Guo, Bai, et al. (2019)
miR-101	1p31.3	Downregulated	EZH2	Cell proliferation Induces apoptosis	RB cell lines (Weri-Rb1, Y79) and tumor tissues	RT-PCR, PA, AA, LUCA	Lei et al. (2014)
miR-125a-5p	19q13.41	Downregulated	TAZ-EGFR pathway	Cell proliferation	RB cell lines (Weri-Rb1, Y79) and tumor tissues	RT-PCR, PA, WB, LUCA, IHC, ATM	Zhang et al. (2016)
miR-138-5p	3p21.32	Downregulated	PDK1	Cell proliferation, migration, and invasion Induces apoptosis	RB cell lines (Y79)	RT-PCR, PA, MA, IA, AA, WB, LUCA	Wang et al. (2017)
miR-183	7q32.2	Downregulated	LRP6	Cell proliferation, migration, and invasion Induces apoptosis	RB cell lines (Weri-Rb1, Y79, SO-RB50) and tumor tissues	RT-PCR, PA, MA, IA, AA, WB, LUCA, IF, ATM	Wang, Wang, Li, Liu, and Teng (2014)

(Continued)

TABLE 6.2 (Continued)

miRNA	Location[a]	Expression	Target mechanism	Suppressive events	Sample type	Validation method	Reference
miR-184	15q25.1	Downregulated	SLC7A5/ ATR/ATM pathway	Cell proliferation, migration, and invasion Induces apoptosis	RB cell lines (Weri-Rb1, Y79) and tumor tissue	RT-PCR, PA, MA, IA, AA, WB, LUCA, IF, IHC	He et al. (2019)
miR-186	1p31.1	Downregulated	DIXDC1	Cell proliferation and invasion	RB cell lines (Weri-Rb1, Y79, SO-RB50, HXO-RB44) and tumor tissue	RT-PCR, PA, IA, WB, LUCA	Che, Qian, and Li (2018)
miR-200a	1p36.33	Downregulated	ACOT7 and DLL3	Cell proliferation (Y79) Induces apoptosis (Y79)	RB cell lines (Weri-Rb1, Y79) and tumor tissue	RT-PCR, PA, AA, WB, LUCA, IF, IHC	Montoya et al. (2015)
miR-204	9q21.12	Downregulated	Cyclin D2 and MMP-9	Cell proliferation, migration, and invasion	RB cell lines (Weri-Rb1, Y79, SO-RB50) and tumor tissue	RT-PCR, PA, MA, IA, WB, LUCA, ATM	Wu et al. (2015)
miR-214-3p	1q24.3	Downregulated	ABCB1 and XIAP	Cell proliferation Induces apoptosis	RB cell lines (Weri-Rb1, Y79, SO-RB50) and tumor tissues	RT-PCR, PA, AA, WB, LUCA, ATM	Yang et al. (2020a)
miR-218-5p	4p15.31	Downregulated	NACC1/ AKT/mTOR axis	Cell proliferation Induces apoptosis	RB cell lines (Weri-Rb1)	RT-PCR, PA, AA, WB	Li, Yu, and Ren (2020)
miR-330	19q13.32	Downregulated	ROCK1	Cell proliferation and invasion	RB cell lines (Weri-Rb1, Y79, SO-RB50) and tumor tissue	RT-PCR, PA, IA, WB, LUCA	Wang, Wang, Li, Zhang, and Lyu (2019)
miR-361-3p	Xq21.2	Downregulated	GLI1, GLI3 and SHH pathway	Cell proliferation	RB cell lines (Weri-Rb1, Y79), tumor tissues and serum samples	RT-PCR, PA, WB, LUCA	Zhao and Cui (2019)
miR-365b-3p	17q11.2	Downregulated	PAX6	Cell proliferation Induces apoptosis	RB cell lines (Weri-Rb1, Y79, SO-RB50) and tumor tissues	RT-PCR, PA, AA, WB, IF, ATM	Wang, Wang, Wu, Hou, and Hu (2013)

(Continued)

TABLE 6.2 (Continued)

miRNA	Location[a]	Expression	Target mechanism	Suppressive events	Sample type	Validation method	Reference
miR-485	14q32.31	Downregulated	Wnt3a and Wnt/β-catenin pathway	Cell proliferation, migration, and invasion Induces apoptosis	RB cell lines (Weri-Rb1, Y79, SO-RB50) and tumor tissue	RT-PCR, PA, MA, IA, AA, WB, LUCA, ATM	Lyu, Wang, Lu, Zhang, and Wang (2019)
miR-504	Xq26.3	Downregulated	AEG-1	Cell proliferation and invasion	RB cell lines (Weri-Rb1, Y79, SO-RB50) and tumor tissues	RT-PCR, PA, IA, WB, LUCA	Wang, Lyu, Ma, Wu, and Wang (2019)

[a]Information obtained from GeneCards database (http://www.genecards.org).
AA, Apoptosis in vitro assay; ATM, animal tumor models.; IA, invasion in vitro assay; IF, immunofluorescence; IHC, immunohistochemistry; LUCA, luciferase expression assays; MA, migration in vitro assay; PA, proliferation in vitro assay; RT-PCR, real-time polymerase chain reaction; WB, Western blot immunodetection.

TABLE 6.3 MicroRNAs (miRNAs) with retinoblastoma (RB) carcinogenic action.

miRNA	Location[a]	Expression	Target mechanism	Oncogenic events	Sample type	Validation method	References
miR-17–92 cluster	13q31.3	Upregulated	EpCAM and STAT3	Cell proliferation and invasion Represses apoptosis	RB cell lines (Y79) and tumor tissue	RT-PCR, PA, IA, AA, WB, IHC, ATM	Jo et al. (2014), Kandalam et al. (2012)
miR-21	17q23.1	Upregulated	PDCD4	Cell proliferation and invasion Represses apoptosis	RB cell lines (Weri-Rb1, Y79, Rb355)	RT-PCR, WB	Shen et al. (2014)
miR-25-3p	7q22.1	Upregulated	PTEN/AKT pathway	Cell proliferation, migration, and invasion Represses apoptosis	RB cell lines (Weri-Rb1, Y79, SO-RB50) and tumor tissue	RT-PCR, PA, MA, IA, AA, WB, LUCA, IF, IHC, ATM	Wan et al. (2019)
miR-106b	7q22.1	Upregulated	RUNX3	Cell proliferation and migration Represses apoptosis	RB cell lines (Y79)	RT-PCR, PA, MA, AA, WB	Yang, Fu, Zhang, Lu, and Li (2017)
miR-125b	11q24.1	Upregulated	DRAM2	Cell proliferation, migration, and invasion Represses apoptosis	RB cell lines (HXO-RB44, Y79, SO-RB50) and tumor tissue	RT-PCR, PA, MA, IA, AA, WB, LUCA	Bai et al. (2016)

(Continued)

TABLE 6.3 (Continued)

miRNA	Location[a]	Expression	Target mechanism	Oncogenic events	Sample type	Validation method	References
miR-130b	22q11.21	Upregulated	EpCAM	Cell proliferation and invasionRepresses apoptosis	RB cell lines (Weri-Rb1, Y79) and tumor tissues	RT-PCR, PA, IA, AA, WB	Beta et al. (2014)
miR-181c	19p13.12	Upregulated	EpCAM	Cell proliferation and invasion Represses apoptosis	RB cell lines (Weri-Rb1, Y79) and tumor tissues	RT-PCR, PA, IA, AA, WB	Beta et al. (2014)
miR-224-3p	Xq28	Upregulated	LATS2/Hippo-YAP axis	Cell proliferation and angiogenesis Represses apoptosis	RB cell lines (Y79) and tumor tissues	RT-PCR, PA, AA, WB, LUCA, IHC, ATM	Song et al. (2020)
miR-492	12q22	Upregulated	LATS2	Cell proliferation and invasion	RB cell lines (Weri-Rb1, Y79, SO-RB50) and tumor tissues	RT-PCR, PA, IA, WB, LUCA	Sun, Zhang, and Zhang (2019)
miR-494	14q32.31	Upregulated	PTEN/PI3K/AKT pathway	Cell proliferation, migration, and invasion	RB cell lines (SO-RB50, Y79) and tumor tissues	RT-PCR, PA, MA, IA, WB, LUCA	Xu et al. (2020)

[a]Information obtained from GeneCards database (http://www.genecards.org).
AA, Apoptosis in vitro assay; ATM, animal tumor models; IA, invasion in vitro assay; IF, immunofluorescence; IHC, immunohistochemistry; LUCA, luciferase expression assays; MA, migration in vitro assay; PA, proliferation in vitro assay; RT-PCR, real-time polymerase chain reaction; WB, Western blot immunodetection.

As for miRNAs with tumorigenic action, in 2009, Zhao et al. (2009) identified a range of miRNAs with elevated expression levels in RB, including miR-129−1, miR-129−2, miR-198, miR-373, miR-492, miR-494, miR-498, miR-503, miR-513−1, miR-513−2, and miR-518c. One of the best known (Jo et al., 2014; Kandalam, Beta, Maheswari, Swaminathan, & Krishnakumar, 2012) is the miR-17−92 cluster that promotes cell proliferation and invasion while suppressing apoptosis; its action is modulated by interactions with *epithelial cell adhesion molecule (EpCAM)* and *signal transducer and activator of transcription 3*. The study by Beta et al. (2014) similarly focused on the influence of *EpCAM* on miRNAs, concluding that it is able to interact with several oncogenic miRNAs, such as miR-130b and miR-181c. Wan et al. (2019) described that miR-25-3p promoted tumor malignancy by promoting proliferation, migration, EMT, in vitro cell invasion, and in vivo tumor formation, all of which increased *Akt* phosphorylation by suppressing *phosphatase and tensin homolog (PTEN)*.

6.2.3.2 miRNAs as potential biomarkers for retinoblastoma

As previously mentioned, a delay in diagnosis will negatively influence the ocular and vital prognosis of patients with RB. For this reason the early detection of biomarkers is considered a crucial milestone, and it is in this area where the study of miRNAs in body fluids, such as plasma or serum, is of great interest. These circulating miRNAs serve as mediators of intercellular communication and are transported by ribonucleoproteins or wrapped in exosomes for protection against degradation (Bail et al., 2010; Hunter et al., 2008). Therefore liquid biopsy has become an accessible, noninvasive tool that provides considerable information on diagnosis, therapeutic response, and tumor prognosis (Delsin et al., 2019; Heitzer, Ulz, & Geigl, 2015; Lande et al., 2020; Mitchell et al., 2008).

Beta et al. (2013) compared 14 serum samples from healthy subjects with a similar number of serum samples from patients with RB, noting that 24 miRNAs were underexpressed and 21 were overexpressed in the latter group. They also compared the serum miRNA profiles with those of published tumor samples, describing 33 coinciding miRNAs (8 downregulated and 25 upregulated), of which only the upregulation of miR-17, miR-18a, and miR-20a could be validated. Liu, Wang, Sun et al. (2014) studied 65 plasma samples from patients with RB and healthy individuals, confirming the downregulation of miR-21, miR-320, and let-7e and their association with tumor progression. In a more recent study (Castro-Magdonel et al., 2020), the circulating miRNome was analyzed in 12 children with RB and 12 healthy controls of the same age, and a plasma signature of 19 miRNAs was detected (miR-378h, miR-455-3p, miR-1281, miR-3201, miR-3613-3p, miR-3921, miR-4507, miR-4508, miR-4529-3p, miR-4668-5p, miR-4707-5p, miR-4750-3p, miR-4763-3p, miR-6069, miR-6085, miR-6511b-5p, miR-6777-5p, miR-6794-5p, and miR-8084) in all RB patients with discriminatory capacity against controls; of these, 14 were also detected in the corresponding primary tumors. Similarly, nine miRNAs were present in all samples (particularly miR-638, miR-5787, and miR-6732-5p), which can serve as a control plasma reference. Moreover, in the same study, four miRNAs (miR-378h, miR-4706, miR-4763-3p, and miR-6511b-5p) associated with male sex and five miRNAs (miR-1469, miR-3620-5p, miR-6088, miR-6089, and miR-6794-5p) associated with female sex were revealed, revealing a specific hormonal functional axis. In this line, Zheng et al. (2020) demonstrated reduced expression of miR-144 in both tissue and serum samples of patients with RB, which were positively correlated to one another and associated with tumor size, metastasis development, and lower OS and DFS.

Finally, it should also be mentioned that research is emerging to assess the presence of nucleic acids and other tumor markers at the aqueous humor level (Berry et al., 2017; Shehata, Abou Ghalia, Elsayed, Ahmed Said, & Mahmoud, 2016).

6.2.4 Therapeutic potential of noncoding RNAs in retinoblastoma

The manipulation of ncRNAs and their regulators is considered a novel approach in the treatment of RB (Golabchi et al., 2018; Mirakholi, Mahmoudi, & Heidari, 2013). Examples of this include the use of agents that promote the expression of ncRNAs with tumor-suppressing action or molecules that silence the expression of those with oncogenic functions, such as antagonists against miR-130b and miR-181c (Beta et al., 2014). Regulation of the expression of ncRNAs involved in the physiology of chemoresistance would allow a more personalized tumor treatment, thus avoiding the use of ineffective drugs from the outset (Plousiou & Vannini, 2019). In this line, it was found that the underexpression of miR-15a, miR-16, miR-34a, miR-3163, and the let-7 family and overregulation of miR-18a, miR-19b, miR-106a, and miR-198 correlated to resistance to chemotherapy (Jia, Wei, Liu, &

Zhao, 2016; Liu, Huang, Xie, et al., 2014; Mitra et al., 2012). In addition, He et al. (2019) and Yang, Zhang, Lu, and Wang (2020a) described that the overexpression of miR-184 and miR-214-3p enhanced tumor chemosensitivity by inducing apoptosis. It was also observed that silencing the onco-lncRNA urothelial cancer associated 1 sensitized cells with multiple chemotherapies (Yang, Zhang, Lu, & Wang, 2020b).

For their part, Xu et al. (2011) showed that hypoxia is a key factor in the tumorigenic microenvironment and is associated with chemoresistance. They also observed that hypoxia-inducible factor (HIF) was capable of modulating a series of miRNAs known as hypoxia-regulated miRNAs, including miR-30c-2, miR-125a-3p, miR-181b, miR-491-3p, and miR-497. Similarly, miR-320 was found to regulate the autophagy of RB cells through *HIF-1α* under hypoxic conditions (Liang, Chen, & Liang, 2017). Considering its relevance, hypoxia has been proposed as a therapeutic target in advanced cases of RB (Boutrid et al., 2008). Currently, another field of action is the study of targeted therapies against miRNAs in cancer stem cells, a tumor cell subgroup with a higher risk of resistance to cytotoxins and, therefore, of tumor recurrence (Mirakholi et al., 2013).

The use of ncRNAs as therapeutic tools is mainly limited by inefficient systemic administration; therefore new molecular encapsulation mechanisms (Wang et al., 2015) and the use of alternative administration routes are being investigated. Thus in the same way that studies are examining the injection of chemotherapeutic agents into the vitreous cavity (Ghassemi, Shields, Ghadimi, Khodabandeh, & Roohipoor, 2014), research is also being carried out on the direct intravitreal injection of ncRNA (Bai et al., 2011; Chung, Gillies, Yam, Wang, & Shen, 2016). This therapeutic modality has two advantages: the direct action on tumor cells and the low endonuclease activity present in the vitreous humor (de Fougerolles, Vornlocher, Maraganore, & Lieberman, 2007).

In short, although the results are very promising, clinical applications are a great medical challenge due to the complexity of the mechanisms surrounding ncRNAs, such as the absence of target specificity, in addition to other aspects, such as the fact that the drug in question selectively reaches the axis or route of action of ncRNA at the level of the tumor tissue (Adams, Parsons, Walker, Zhang, & Slack, 2017).

6.2.5 Conclusion

Based on this analysis, it is clear that the different classes of ncRNAs are a fundamental part of the oncogenic process of RB. Although a wide variety of these molecules have been studied, the function and expression of many ncRNAs remain unknown. The research and identification of the different types of ncRNAs, as well as their underlying mechanisms of action, will allow a better understanding of the physiopathological processes involved as well as significant advances from the diagnostic, therapeutic, and prognostic point of view.

6.3 Uveal melanoma

6.3.1 Introduction

UM is the most frequent primary intraocular tumor in adulthood. Although current treatment options have improved the local control of the disease, metastasis occurs in more than 50% of patients within 15 years of diagnosis, resulting in a very poor prognosis (Rodriguez-Vidal et al., 2020).

Some prognostic markers related to the development of the metastatic process have been identified, including tumor height, presence of monosomy 3 (M3), gain of chromosome 8 (Damato et al., 2007), or tumor-specific mutations in the genes *GNAQ*, *GNA11*, *EIF1AX*, *SF3B1*, and *BAP1* (Rodrigues & Stern, 2015; Shain et al., 2019). Of these, cytogenetic factors are the most relevant prognostic markers. In this context, many studies have investigated the role of ncRNAs in the progression of UM (Aughton, Kalirai, & Coupland, 2020; Bande et al., 2020).

6.3.2 Long noncoding RNAs in uveal melanoma

As in RB, lncRNAs also play crucial roles in the development and progression of UM. Among lncRNAs with tumor suppressor action in UM, CASC15-new-transcript 1 is downregulated and inhibits cell migration and tumorigenesis by modulating the *JPX/FTX-XIST* noncoding cascade. Therefore inhibition of *JPX/FTX* transmits the signal to *XIST*, decreasing its expression levels (Xing et al., 2017). Wu, Yuan, Ma, Xu, and Zhang (2020) described that the small nucleolar RNA host gene 7 is downregulated in UM, inversely associated with *EZH2*, and correlates to a higher proportion of epithelioid cells, poor clinical staging, low tumor-free survival, and a high mortality rate. For its part, Pax6 upstream antisense RNA acts to silence *HES1*, a mediator of the *Notch* signaling pathway, resulting in increased migration and cell colony formation (Ding et al., 2016). Another lncRNA with a tumor suppressor role is lncRNA-numb (Cheng et al., 2016).

However, several lncRNAs with carcinogenic action have been described in the literature, such as rhophilin RHO Gtpase—binding protein 1-antisense 1 (Lu, Yu, Zhang, et al., 2017), purinergic receptor P2X7 variant3 (Pan et al., 2016), and ferritin heavy-chain 1 pseudogene 3 (Zheng et al., 2017). The following lncRNAs are among the most studied and share aberrant expression with RB, since both types of neoplasms are overregulated. Homeobox A11 antisense promotes tumorigenesis by interacting with *EZH2* to silence the transcription of *p21*, inhibiting miR-124 expression, and acting as a ceRNA of miR-124 (Lu, Zhao, Zha, et al., 2017). Metastasis-associated lung adenocarcinoma transcript 1 (MALAT1) acts in an oncogenic manner through two axes; Sun, Sun, Zhou, Gao, and Han (2016) demonstrated that it inhibits miR-140, thus favoring the expression of *Slug* and *ADAM10* and producing an increase in proliferation, migration, colony formation, and invasion of UM cell lines. Wu, Chen, Zuo, Jiang, and Yan (2020) described that MALAT1 inhibits miR-608 to enhance *HOXC4* expression, leading to UM cell proliferation, invasion, and migration. Lastly, PVT1 acts through inhibiting miR-17-3p, thereby promoting the expression of *MDM2* and inactivation of *p53*. There is also a correlation between high expression of PVT1 and a number of parameters of poor prognosis in UM, such as advanced age, histological epithelioid type, presence of extrascleral extension, and metastasis (Wu, Chen, Han, Zhang, & Yan, 2019; Xu, Gong, & Liu, 2017).

Table 6.4 describes in detail all main deregulated lncRNAs, including their chromosomal location, molecular targets, signaling pathways, cellular events in which they intervene, the type of study sample, and the technical validation methods used for their characterization.

TABLE 6.4 Long noncoding RNAs (lncRNAs) involved in uveal melanoma (UM) pathogenesis.

lncRNA	Location[a]	Expression	Target mechanism	Cellular events	Sample type	Validation method	References
CANT1 (CASC15-NT1)	17q25.3	Downregulated	JPX/FTX-XIST lncing cascade	Represses cell proliferation and migration	UM cell lines (MUM-2B, OCM-1, OCM-1A, OM431, SP6.5, 92.1) and tumor tissue	RT-PCR, PA, MA, IF, ATM	Xing et al. (2017)
FTH1P3	2p23.3	Upregulated	miR-224-5p/Rac1/Fizzled5 axis	Induces cell proliferation and migration	UM cell lines (C918, MUM-2B, MUM-2C, OCM-1A) and tumor tissue	RT-PCR, PA, MA, WB, LUCA	Zheng et al. (2017)
HOXA11-AS	7p15.2	Upregulated	EZH2 and miR-124	Induces cell proliferation and invasion Represses apoptosis	UM cell lines (C918, MUM-2B, MUM-2C, OCM-1A) and tumor tissue	RT-PCR, PA, IA, AA, WB, LUCA	Lu, Zhao, et al. (2017)
lncRNA-numb	14q24.3	Downregulated	—	Represses cell proliferation and invasion	UM cell lines (MUM-2B, OCM-1a, SP6.5)	RT-PCR, PA, IA, WB, IHC	Cheng et al. (2016)
MALAT1	11q13.1	Upregulated	miR-140-Slug/ADAM10 and miR-608/HOXC4 axis	Induces cell proliferation, migration, and invasion	UM cell lines (C918, MUM-2B, MUM-2C, M619, OCM-1A) and tumor tissue	RT-PCR, PA, MA, IA, WB, LUCA, IHC, ATM	Sun et al. (2016), Wu, Chen, et al. (2020)
PAUPAR	11p13	Downregulated	HES1	Represses cell proliferation and migration	UM cell lines (MUM-2B, OCM-1, OM431) and tumor tissue	RT-PCR, PA, MA, WB, ATM	Ding et al. (2016)
PVT1	8q24.21	Upregulated	EZH2 and miR-17-3p/MDM2 pathway	Induces cell proliferation, migration, and invasion Represses apoptosis	UM cell lines (C918, MUM-2B, MUM-2C, OCM-1, OCM-1A) and tumor tissue	RT-PCR, PA, MA, IA, AA, WB, LUCA, IF, IHC, ATM	Huang et al. (2019), Wu, Chen, et al. (2019), Xu et al. (2017)
P2RX7-V3	12q24.31	Upregulated	PI3K/AKT pathway	Induces cell proliferation and migration	UM cell lines (MUM-2B, OM431)	RT-PCR, PA, MA, IHC, ATM	Pan et al. (2016)

RHPN1-AS1	8q24.3	Upregulated	Nicotinate–nicotinamide metabolism and TGF-β	Induces cell proliferation, migration, and invasion	UM cell lines (OCM-1, OM431)	RT-PCR, PA, MA, IA, IF, ATM	Lu, Yu, et al. (2017)
SNHG7	9q34.3	Downregulated	EZH2	Represses cell proliferation Induces apoptosis	UM cell lines (MEL202, MEL270, MEL290, OMM2.3, OMM2.5, 92.1) and tumor tissue	RT-PCR, PA, AA, WB	Wu, Yuan, et al. (2020)
SNHG15	7p13	Upregulated	-	Poor prognosis clinicopathological characteristic	TCGA UM database	GSEA	Wu, Li, et al. (2020)

[a]Information obtained from GeneCards database (http://www.genecards.org).
AA, Apoptosis in vitro assay; ATM, animal tumor models; GSEA, gene set enrichment analysis; IA, invasion in vitro assay; IF, immunofluorescence; IHC, immunohistochemistry; LUCA, luciferase expression assays; MA, migration in vitro assay; PA, proliferation in vitro assay; RT-PCR, real-time polymerase chain reaction; WB, Western blot immunodetection.

6.3.3 MicroRNAs in uveal melanoma

6.3.3.1 Dysregulated expression of microRNAs

miRNAs are a type of ncRNAs that have been demonstrated to be involved in the progression of UM, being able to act as oncogenic factors or tumor suppressors. Oncogenic miRNAs downregulate the transcription of tumor suppressor genes, whereas tumor-suppressing miRNAs inhibit the transcription of oncogenic genes (Liz & Esteller, 2016).

In Tables 6.5 and 6.6 are shown a wide variety of miRNAs stratified into two groups (tumor suppressor and oncogenic action), as well as their molecular targets, axes of action, and functions at the cellular level, in addition to the type of study sample and the technical validation methods used.

TABLE 6.5 MicroRNAs (miRNAs) with uveal melanoma (UM) tumor suppressor action.

miRNA	Location[a]	Expression	Target mechanism	Suppressive events	Sample type	Validation method	References
miR-9	1q22	Downregulated	NF-Kb1 pathway	Cell proliferation, migration, and invasion	UM cell lines (C918, MUM-2B, MUM-2C, OCM-1A)	RT-PCR, PA, MA, IA, FRA, WB	Liu et al. (2012)
miR-15a	13q14.2	Downregulated	IL-10Rα	Cell proliferation	UM cell lines (OCM-1) and tumor tissues	RT-PCR, PA, WB, LUCA	Venza et al. (2015)
miR-23a	19p13.12	Downregulated	Zeb1	Cell migration	UM cell lines (OCM-1)	RT-PCR, MA, IF, WB	Wang, Luo, et al. (2018)
miR-34a	1p36.22	Downregulated	c-Met and LGR4	Cell proliferation, migration, and invasion	UM cell lines (M17, M21, M23, SP6.5) and tumor tissues	PA, MA, IA, WB, LUCA, IF	Hou et al. (2019), Yan et al. (2009)
miR-34b/c	11q23.1	Downregulated	c-Met and p-AKT	Cell proliferation and migration Induces apoptosis	UM cell lines (SP6.5) and tumor tissues	RT-PCR, PA, MA, AA, WB, LUCA	Dong and Lou (2012)
miR-122	18q21.31	Downregulated	c-Met and ADAM10	Cell proliferation and migration	UM cell lines (MEL270, OMM2.5, UPMM2, UPMM3, 92.1), tumor tissues and TCGA UM database	RT-PCR, PA, MA, AA, WB, LUCA	Amaro et al. (2020)
miR-124a	3 loci: 8p23.1, 8q12.3, 20q13.33	Downregulated	CDK4, CDK6, EZH2, and cyclin D2	Cell proliferation, migration, and invasion	UM cell lines (M17, M21, M23, SP6.5) and tumor tissues	RT-PCR, PA, MA, IA, WB, LUCA, ATM	Chen et al. (2013)

(Continued)

TABLE 6.5 (Continued)

miRNA	Location[a]	Expression	Target mechanism	Suppressive events	Sample type	Validation method	References
miR-137	1p21.3	Downregulated	c-Met, CDK6 and MITF	Cell proliferation	UM cell lines (M17, M23, SP6.5)	RT-PCR, PA, WB, LUCA	Chen et al. (2011)
miR-142-3p	17q22	Downregulated	CDC25C, GNAQ, RAC1, TGFβR1, and WASL	Cell proliferation, migration, and invasion	UM cell lines (M17, SP6.5) and tumor tissues	RT-PCR, PA, MA, IA, AA, WB, LUCA, ATM	Peng et al. (2019)
miR-144	17q11.2	Downregulated	c-Met and ADAM10	Cell proliferation, migration, and invasion Induces apoptosis	UM cell lines (C918, MEL270, MUM-2B, MUM-2C, OCM-1A, OMM2.5, UPMM2, UPMM3, 92.1), tumor tissues and TCGA UM database	RT-PCR, PA, MA, IA, AA, WB, LUCA	Amaro et al. (2020), Sun, Bian, et al. (2015)
miR-145	5q32	Downregulated	NRP1/CDC42	Cell proliferation, migration, and invasion	UM cell lines (C918, OCM-1A), tumor tissues, and TCGA UM database	RT-PCR, PA, MA, IA, WB, LUCA	Li, Luo, et al. (2020)
miR-182	7q32.2	Downregulated	Bcl-2, c-Met, cyclin D2, and MITF	Cell proliferation, migration, and invasion Induces apoptosis	UM cell lines (M23, SP6.5) and tumor tissues	RT-PCR, PA, MA, IA, AA, WB, LUCA, ATM	Yan et al. (2012)
miR-185	22q11.21	Downregulated	IL-10Rα	Cell proliferation	UM cell lines (OCM-1) and tumor tissues	RT-PCR, PA, WB, LUCA	Venza et al. (2015)
miR-205	1q32.2	Downregulated	NRP1/CDC42	Cell proliferation, migration, and invasion	UM cell lines (C918, OCM-1A), tumor tissues, and TCGA UM database	RT-PCR, PA, MA, IA, WB, LUCA	Li, Luo, et al. (2020)
miR-211	15q13.3	Downregulated	IL-10Rα	Cell proliferation	UM cell lines (OCM-1) and tumor tissues	RT-PCR, PA, WB, LUCA	Venza et al. (2015)
miR-216a-5p	2p16.1	Downregulated	HK2	Cell proliferation	UM cell lines (MUM-2B) and tumor tissues	RT-PCR, PA, WB, LUCA, GA, IHC, ATM	Liu, Huo, et al. (2018)

(*Continued*)

TABLE 6.5 (Continued)

miRNA	Location[a]	Expression	Target mechanism	Suppressive events	Sample type	Validation method	References
miR-224-5p	Xq28	Downregulated	PIK3R3/PI3K/AKT3 axis	Cell proliferation, migration, and invasion	UM cell lines (OCM-1A) and tumor tissues	RT-PCR, PA, MA, IA, WB, LUCA	Li et al. (2019)

[a]*Information obtained from GeneCards database (http://www.genecards.org).*
AA, Apoptosis in vitro assay; *ATM*, animal tumor models; *FRA*, fluorescence reporter assay; *GA*, glycolysis assay; *IA*, invasion in vitro assay; *IF*, immunofluorescence; *IHC*, immunohistochemistry; *LUCA*, luciferase expression assays; *MA*, migration in vitro assay; *PA*, proliferation in vitro assay; *RT-PCR*, real-time polymerase chain reaction; *WB*, Western blot immunodetection.

TABLE 6.6 MicroRNAs (miRNAs) with uveal melanoma (UM) carcinogenic action.

miRNA	Location[a]	Expression	Target mechanism	Oncogenic events	Sample type	Validation method	Reference
miR-20a	13q31.3	Upregulated	–	Cell proliferation, migration, and invasion	UM cell lines (MUM-2B, MUM-2C) and tumor tissues	RT-PCR, PA, MA, IA	Zhou et al. (2016)
miR-21	17q23.1	Upregulated	p53/LASP1 axis	Cell proliferation, migration, and invasion	UM cell lines (MUM-2B, M619, OCM-1)	RT-PCR, PA, MA, IA, AA, WB, LUCA, ATM	Wang, Yang, Wei, and Xu (2018)
miR-92a-3p	2 loci: 13q31.3, Xq26.2	Upregulated	MYCBP2	Represses apoptosis	UM cell lines (OCM-1, OCM-3, 92.1) and tumor tissues	RT-PCR, AA, WB, LUCA	Venza et al. (2016)
miR-155	21q21.3	Upregulated	NDFIP1	Cell proliferation and invasion	UM cell lines (C918, MUM-2B, MUM-2C, OCM-1A) and tumor tissues	RT-PCR, PA, IA, WB, LUCA	Peng et al. (2017)
miR-181 family	Chr 1, 9, 19	Upregulated	CTDSPL/pRB/E2F1 pathway	Cell proliferation	UM cell lines (MUM-2B, OCM-1, OCM-1A SP6.5, VUP, 92.1) and tumor tissues	RT-PCR, WB, LUCA	Zhang, He, et al. (2018)
miR-222	Xp11.3	Upregulated	PI3K/AKT/MMP-9 pathway	Cell proliferation and migration Represses apoptosis	UM cell lines (C918, MUM-2B)	RT-PCR, PA, MA, AA, WB, IHC, ATM	Cheng, Cheng, et al. (2019)
miR-367	4q25	Upregulated	PTEN	Cell proliferation and migration	UM cell lines (C918, MUM-2B, M17, M23) and tumor tissues	RT-PCR, PA, MA, WB, LUCA	Ling, Lu, et al. (2017)

(Continued)

TABLE 6.6 (Continued)

miRNA	Location[a]	Expression	Target mechanism	Oncogenic events	Sample type	Validation method	Reference
miR-454	17q22	Upregulated	*PTEN, AKT* and *mTOR*	Cell proliferation and invasion	UM cell lines (C918, MUM-2B, MUM-2C, OCM-1A) and tumor tissues	RT-PCR, PA, IA, WB, LUCA	Sun, Wang, Gao, et al. (2015)
miR-652	Xq23	Upregulated	*HOXA9/HIF-1* pathway	Cell proliferation and migration	UM cell lines (MEL270, MUM-2B) and tumor tissues	RT-PCR, PA, MA, WB, LUCA	Xia, Yang, Yang, et al. (2019)

[a]*Information obtained from GeneCards database (http://www.genecards.org).*
AA, Apoptosis in vitro assay; *ATM*, animal tumor models; *IA*, invasion in vitro assay; *IF*, immunofluorescence; *IHC*, immunohistochemistry; *LUCA*, luciferase expression assays; *MA*, migration in vitro assay; *PA*, proliferation in vitro assay; *RT-PCR*, real-time polymerase chain reaction; *WB*, Western blot immunodetection.

Within the tumor suppressor group, miR-34a was identified as a transcriptional target of *p53* and inhibited the expression of *c-Met, Rb, cdc2*, and *E2F3*; additionally, *leucine-rich repeat-containing G-protein coupled receptor 4*, which is implicated in embryological development, cell mobility, and metastasis, was identified as another of its targets (Hou et al., 2019; Yan et al., 2009). Similarly, Dong and Lou (2012) described that miR-34b and miR-34c expression reduced cell growth and migration in UM cell lines, and the *c-Met* signaling pathway (*c-Met, p-Akt, CDK4*, and *CDK6*) was identified as an miR-34b/c target. Moreover, miR-122 and miR-144 decreased the expression of *c-Met* and *ADAM10* proteins, thereby reducing proliferation, migration, and cell cycle progression in UM (Amaro et al., 2020).

Chen et al. (2011) revealed lower expression of miR-137 in UM cell lines compared to that in normal melanocytes, which was associated decreases in *MITF, c-Met, CDK2*, and *CDK6*. In addition, Eedunuri et al. (2015) reported the downregulation of *steroid receptor coactivators* induced by miR-137. Similarly, miR-182 was found to suppress the expression of *MITF* and cell cycle—related genes (Yan et al., 2012).

Transfection of miR-15a, miR-185, and miR-211 decreased cell proliferation in UM cell lines by targeting *IL-10Rα*, thus favoring the inflammatory response and immune reactions against this tumor (Venza et al., 2015). As in RB, EMT is responsible for facilitating the spread of tumor cells through the action of matrix metallopeptidases (MMPs). Furthermore, miR-9 suppresses cell migration and invasion by targeting *NF-kb1*, resulting in the downregulation of proteins related to this pathway, including *VEGFA, MMP-2*, and *MMP-9* (Liu et al., 2012). Likewise, miR-296-3p inhibits the expression of *MMP-2/MMP-9* (Wang, Hu, Cui, Zhou, & Chen, 2018), and miR-23a targets *Zeb1* (Wang, Luo, Guan, & Zhao, 2018).

Chen et al. (2013) discussed the relationship between miR-124a and its targets *CDK4, CDK6, cyclin D2*, and *EZH2* and stated that the restoration of miR-124a was possible through treatment with hypomethylating factors. In addition, miR-216a-5p targets *hexokinase 2*, an aerobic glycolysis enzyme overexpressed in UM (Liu, Huo, et al., 2018). Furthermore, miRNA-145/miRNA-205 inhibits the proliferation and invasion of UM cells by targeting *NRP1/CDC42* (Li, Luo, Liu, & Wei, 2020). Finally, miR-142-3p (Peng et al., 2019) and miR-224-5p (Li, Liu, Li, & Wang, 2019) are involved in tumor suppression and inhibition of cell growth in UM.

Among miRNAs with oncogenic action, miR-222 was identified by Cheng, Cheng, Zhao, and Qu (2019) as a promoter of the *high-mobility group A1* signaling pathway, and its introduction to cell lines increased the expression of *PI3K*, *p-Akt*, and *MMP-9*, leading to EMT. Zhang, He, Li et al. (2018) described high levels of miR-181 family members (miR-181a, miR-181b, and miR-181c) in UM cell lines; in particular, miR-181b was found to inhibit the expression of the tumor-suppressing phosphatase enzyme *CTDSPL*, which dephosphorylates the tumor suppressor *Rb*.

miR-367 and miR-454 target *PTEN*, a tumor suppressor involved in the *PI3K/Akt* pathway, and introduction of these miRNAs into cultures resulted in an increase in cell proliferation, progression, and migration (Ling, Lu, Zhang, Jiang, & Zhang, 2017; Sun et al., 2015). Furthermore, miR-155 exerts its oncogenic action by inhibiting *protein I related to the Nedd4 family*, which is involved in the ubiquitination and nuclear translocation of *PTEN* (Peng, Liu, & Liu, 2017).

In addition, miR-652 inhibits the expression of *HOXA9*, leading to the *HIF-1* signaling pathway and an increase in cell migration and proliferation, thus favoring the tumorigenic process (Xia, Yang, Yang, et al., 2019). Venza et al. (2016) reported that miR-92a-3p has also been linked to an increase in cell proliferation by targeting *MYC-binding protein 2*. Similarly, the introduction of miR-20a led to an increase in cell invasion and migration in UM, but its cellular target has not yet been determined (Zhou, Jiang, Wang, & Xia, 2016).

6.3.3.2 miRNAs as potential biomarkers for uveal melanoma

As mentioned, the presence of miRNAs in multiple body fluids, their high degree of stability, and their long half-lives (up to 24 h) make them excellent potential biomarkers that could be combined with possible protein biomarkers in serum (Bande et al., 2015; Reiniger et al., 2005).

Numerous studies have been conducted on samples of serum, plasma, or blood, some of which are described here. Stark et al. (2019) studied 17 miRNAs present in blood to differentiate UM from benign melanocytic lesions such as nevi. Using six of these (miR-16, miR-145, miR-146a, miR-204, miR-211, and miR-363-3p), UM was identified with 93% sensitivity and 100% specificity. Russo et al. (2016) found eight miRNAs with differential expression in the serum of UM patients and healthy controls, but only miR-146a showed significant differences after validation. Similarly, Achberger et al. (2014) identified increased plasma levels of miR-20a, miR-125b, miR-146a, miR-155, miR-181a, and miR-223 in UM patients compared to those in controls; of these the levels of miR-20a, miR-125b, miR-146a, miR-155, and miR-223 were found to be increased in metastatic stages, while miR-181a was decreased. Ragusa et al. (2015) described miRNA expression in the vitreous humor, vitreous humor exosomes, and serum of six patients with UM compared to healthy controls. They found that 90% of miRNA expression was shared between the vitreous humor and vitreous exosomes. miR-146a was upregulated in UM patients' serum and serum exosomes. Levels of miR-21, miR-34a, and miR-146a were increased in the vitreous humor, vitreous exosomes, and histological samples. Moreover, Triozzi et al. (2016) proposed a relationship between miRNA plasma levels and treatment with interferon-α-2b, so measuring blood levels of specific miRNAs implicated in angiogenesis, including miR-16, miR-106a, miR-126, and miR-199a, may have clinical utility in monitoring antiangiogenic therapy. They also found overexpression of miR-92b, miR-199-5p, and miR-223 in plasma samples of UM patients with M3.

Among the mentioned miRNAs, miR-146a levels have been shown to be increased in several studies. This fact suggests that miR-146a is a promising serum biomarker of UM (Bande et al., 2020).

From the point of view of the cell lines and tissue samples, Worley, Long, Onken, and Harbour (2008) determined the expression of miRNAs through microarray analysis in 24 primary frozen UM samples and related the results to the risk of metastasis based on chromosome 3 abnormalities. In this manner the combination of six miRNAs (miR-143, miR-193b, miR-199a, miR-199*, miR-652, and let-7b) was established to distinguish between class 1 and class 2 UM. miR-199a and let-7b showed the greatest discriminatory potential. Venkatesan et al. (2016) identified five miRNAs (miR-134, miR-143, miR-146b, miR-199a, and miR-214) differentially expressed in M3 and disomy 3 UM tumors. They also found a strong correlation between liver metastasis and the expression of miR-134 and miR-149. Robertson et al. (2017) conducted an extensive study analyzing 80 primary UM samples and identified four clusters of miRNAs related to chromosome M3, metastasis, and DNA methylation profiles.

Smit et al. (2019) identified 13 miRNAs that were differentially expressed depending on the level of genetic mutations linked to oncological progression in UM (*GNAQ*, *GNA11*, *EIF1AX*, *SF3B1*, *BAP1*, and M3). Five were overexpressed in the high-risk group: miR-16-5p, miR-17-5p, miR-21-5p, miR-132-5p, and miR-151a-3p, while eight were underexpressed: miR-99a-3p, miR-99a-5p, miR-101-3p, miR-181a-2-3p, miR-181b-5p, miR-378, miR-1537-3p, and let-7c-5p. Wróblewska et al. (2020) established six miRNAs to identify primary and metastatic UM; this second group found overexpression of miR-346, miR-592, and miR-1247 and underexpression of miR-506 and miR-513c.

Through the analysis of miRNA expression data in the TCGA (The Cancer Genome Atlas) UM database, Falzone et al. (2019) determined seven miRNAs associated with tumor stage, vital status, and survival. Six miRNAs were downregulated: miR-211-5p, miR-508-3p, miR-509–3-5p, miR-513a-5p, miR-513c-5p, and miR-514a-3p, while six were upregulated: miR-199a-5p, miR-224, miR-452, miR-592, let-7b-3p, and let-7b-5p. Similarly, Xin et al. (2019) inferred nine miRNAs with differential expression between the high- and low-risk UM groups: miR-195, miR-224, miR-365a, miR-365b, miR-452, miR-513c, miR-873, miR-4709, and miR-7702. However, in spite of the abovementioned results, Larsen et al. (2014) did not find associations between miRNA expression and metastasis and survival in UM.

6.3.4 Conclusion

ncRNAs have become very important in UM research. In recent years the number of studies related to ncRNAs has multiplied exponentially. Moreover, ncRNAs regulate multiple biological processes in tumors, and it seems clear that they will play an important role in the diagnosis, prognosis, and treatment of UM in the coming decades.

ncRNAs could serve as excellent prognostic markers, with the advantage that they can be obtained noninvasively. However, multiple studies have produced very few ncRNAs that appear to be common biomarkers. The reasons for this could be the use of small samples and heterogeneity in the inclusion processes or quantification methods (Bande Rodríguez et al., 2020).

Future studies will permit a better understanding of ncRNAs in UM. New studies should include a greater number of samples and apply uniform techniques to clarify and validate the obtained information.

6.4 Conclusion

In this review, we aimed to provide a comprehensive clinical overview of the current knowledge about ncRNAs in relation to the two most common primary intraocular malignancies, UM and RB (Fig. 6.2). Thanks to the development of bioinformatics and genomic technologies, new perspectives and research are emerging that highlight the importance of the various groups of ncRNAs as innovative tools in the management of these ophthalmic neoplasms. Despite the fact that these molecules are already considered potential diagnostic biomarkers, therapeutic targets, and prognostic factors, further studies are needed to thoroughly investigate the underlying mechanisms and to precisely define the role they can play in routine clinical practice with the ultimate goal of controlling the disease in early stages and improving patients' OS.

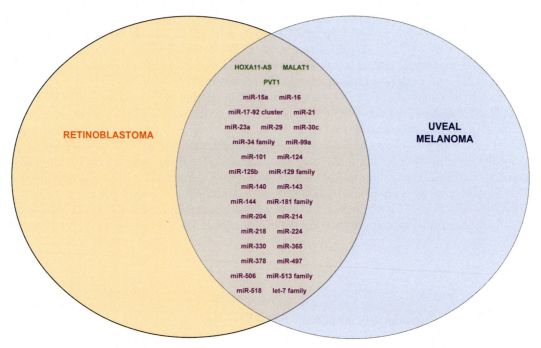

FIGURE 6.2 Venn diagram representing the lncRNAs and miRNAs that share deregulation in RB and UM. *lncRNAs*, Long noncoding RNAs; *miRNAs*, microRNAs; *RB*, retinoblastoma; *UM*, uveal melanoma.

References

Achberger, S., Aldrich, W., Tubbs, R., Crabb, J. W., Singh, A. D., & Triozzi, P. L. (2014). Circulating immune cell and microRNA in patients with uveal melanoma developing metastatic disease. *Molecular Immunology*, 58(2), 182−186. Available from https://doi.org/10.1016/j.molimm.2013.11.018.

Adams, B. D., Parsons, C., Walker, L., Zhang, W. C., & Slack, F. J. (2017). Targeting noncoding RNAs in disease. *The Journal of Clinical Investigation*, 127(3), 761−771. Available from https://doi.org/10.1172/JCI84424.

AlAli, A., Kletke, S., Gallie, B., & Lam, W.-C. (2018). Retinoblastoma for pediatric ophthalmologists. *Asia-Pacific Journal of Ophthalmology (Philadelphia, PA)*, 7(3), 160−168. Available from https://doi.org/10.22608/APO.201870.

Amaro, A., Croce, M., Ferrini, S., Barisione, G., Gualco, M., Perri, P., ... Gangemi, R. (2020). Potential oncosuppressive role of miR122 and miR144 in uveal melanoma through ADAM10 and C-Met inhibition. *Cancers*, 12(6). Available from https://doi.org/10.3390/cancers12061468.

Aughton, K., Kalirai, H., & Coupland, S. E. (2020). MicroRNAs and uveal melanoma: Understanding the diverse role of these small molecular regulators. *International Journal of Molecular Sciences*, 21(16). Available from https://doi.org/10.3390/ijms21165648.

Bai, S., Tian, B., Li, A., Yao, Q., Zhang, G., & Li, F. (2016). MicroRNA-125b promotes tumor growth and suppresses apoptosis by targeting DRAM2 in retinoblastoma. *Eye (London, England)*, 30(12), 1630−1638.

Bai, Y., Bai, X., Wang, Z., Zhang, X., Ruan, C., & Miao, J. (2011). MicroRNA-126 inhibits ischemia-induced retinal neovascularization via regulating angiogenic growth factors. *Experimental and Molecular Pathology*, 91(1), 471−477. Available from https://doi.org/10.1016/j.yexmp.2011.04.016.

Bail, S., Swerdel, M., Liu, H., Jiao, X., Goff, L. A., Hart, R. P., & Kiledjian, M. (2010). Differential regulation of microRNA stability. *RNA (New York, N.Y.)*, 16(5), 1032−1039. Available from https://doi.org/10.1261/rna.1851510.

Bande Rodríguez, M. F., Fernandez Marta, B., Lago Baameiro, N., Santiago-Varela, M., Silva-Rodríguez, P., Blanco-Teijeiro, M. J., ... Piñeiro Ces, A. (2020). Blood Biomarkers of uveal melanoma: Current perspectives. *Clinical Ophthalmology (Auckland, N.Z.)*, 14, 157−169. Available from https://doi.org/10.2147/OPTH.S199064.

Bande, M., Fernandez-Diaz, D., Fernandez-Marta, B., Rodriguez-Vidal, C., Lago-Baameiro, N., Silva-Rodríguez, P., ... Piñeiro, A. (2020). The role of non-coding RNAs in uveal melanoma. *Cancers*, 12(10). Available from https://doi.org/10.3390/cancers12102944.

Bande, M. F., Santiago, M., Mera, P., Piulats, J. M., Blanco, M. J., Rodríguez-Álvarez, M. X., ... Pardo, M. (2015). ME20-S as a potential biomarker for the evaluation of uveal melanoma. *Investigative Ophthalmology & Visual Science*, 56(12), 7007−7011. Available from https://doi.org/10.1167/iovs.15-17183.

Berry, J. L., Xu, L., Murphree, A. L., Krishnan, S., Stachelek, K., Zolfaghari, E., ... Hicks, J. (2017). Potential of aqueous humor as a surrogate tumor biopsy for retinoblastoma. *JAMA Ophthalmology*, 135(11), 1221−1230. Available from https://doi.org/10.1001/jamaophthalmol.2017.4097.

Beta, M., Khetan, V., Chatterjee, N., Suganeswari, G., Rishi, P., Biswas, J., & Krishnakumar, S. (2014). EpCAM knockdown alters microRNA expression in retinoblastoma—Functional implication of EpCAM regulated miRNA in tumor progression. *PLoS One*, 9(12), e114800. Available from https://doi.org/10.1371/journal.pone.0114800.

Beta, M., Venkatesan, N., Vasudevan, M., Vetrivel, U., Khetan, V., & Krishnakumar, S. (2013). Identification and in silico analysis of retinoblastoma serum microRNA profile and gene targets towards prediction of novel serum biomarkers. *Bioinformatics and Biology Insights*, 7, 21−34. Available from https://doi.org/10.4137/BBI.S10501.

Bi, L.-L., Han, F., Zhang, X.-M., & Li, Y.-Y. (2018). LncRNA MT1JP acts as a tumor inhibitor via reciprocally regulating Wnt/β-Catenin pathway in retinoblastoma. *European Review for Medical and Pharmacological Sciences*, 22(13), 4204−4214. Available from https://doi.org/10.26355/eurrev_201807_15414.

Boutrid, H., Jockovich, M.-E., Murray, T. G., Piña, Y., Feuer, W. J., Lampidis, T. J., & Cebulla, C. M. (2008). Targeting hypoxia, a novel treatment for advanced retinoblastoma. *Investigative Ophthalmology & Visual Science*, 49(7), 2799−2805. Available from https://doi.org/10.1167/iovs.08-1751.

Castro-Magdonel, B. E., Orjuela, M., Alvarez-Suarez, D. E., Camacho, J., Cabrera-Muñoz, L., Sadowinski-Pine, S., ... Ponce-Castañeda, M. V. (2020). Circulating miRNome detection analysis reveals 537 miRNAS in plasma, 625 in extracellular vesicles and a discriminant plasma signature of 19 miRNAs in children with retinoblastoma from which 14 are also detected in corresponding primary tumors. *PLoS One, 15*(4), e0231394. Available from https://doi.org/10.1371/journal.pone.0231394.

Che, X., Qian, Y., & Li, D. (2018). Suppression of disheveled-axin domain containing 1 (DIXDC1) by microRNA-186 inhibits the proliferation and invasion of retinoblastoma cells. *Journal of Molecular Neuroscience, 64*(2), 252–261.

Chen, X., He, D., Dong, X. D., Dong, F., Wang, J., Wang, L., . . . Tu, L. (2013). MicroRNA-124a is epigenetically regulated and acts as a tumor suppressor by controlling multiple targets in uveal melanoma. *Investigative Ophthalmology & Visual Science, 54*(3), 2248–2256. Available from https://doi.org/10.1167/iovs.12-10977.

Chen, X., Wang, J., Shen, H., Lu, J., Li, C., Hu, D.-N., . . . Tu, L. (2011). Epigenetics, microRNAs, and carcinogenesis: Functional role of microRNA-137 in uveal melanoma. *Investigative Ophthalmology & Visual Science, 52*(3), 1193–1199. Available from https://doi.org/10.1167/iovs.10-5272.

Cheng, Y., Chang, Q., Zheng, B., Xu, J., Li, H., & Wang, R. (2019). LncRNA XIST promotes the epithelial to mesenchymal transition of retinoblastoma via sponging miR-101. *European Journal of Pharmacology, 843*, 210–216.

Cheng, G., He, J., Zhang, L., Ge, S., Zhang, H., & Fan, X. (2016). HIC1 modulates uveal melanoma progression by activating lncRNA-numb. *Tumour Biology: The Journal of the International Society for Oncodevelopmental Biology and Medicine, 37*(9), 12779–12789. Available from https://doi.org/10.1007/s13277-016-5243-3.

Cheng, Y., Cheng, T., Zhao, Y., & Qu, Y. (2019). HMGA1 exacerbates tumor progression by activating miR-222 through PI3K/Akt/MMP-9 signaling pathway in uveal melanoma. *Cellular Signalling, 63*, 109386. Available from https://doi.org/10.1016/j.cellsig.2019.109386.

Chu, C., Qu, K., Zhong, F. L., Artandi, S. E., & Chang, H. Y. (2011). Genomic maps of long noncoding RNA occupancy reveal principles of RNA-chromatin interactions. *Molecular Cell, 44*(4), 667–678. Available from https://doi.org/10.1016/j.molcel.2011.08.027.

Chung, S. H., Gillies, M., Yam, M., Wang, Y., & Shen, W. (2016). Differential expression of microRNAs in retinal vasculopathy caused by selective Müller cell disruption. *Scientific Reports, 6*, 28993. Available from https://doi.org/10.1038/srep28993.

Damato, B., Duke, C., Coupland, S. E., Hiscott, P., Smith, P. A., Campbell, I., . . . Howard, P. (2007). Cytogenetics of uveal melanoma: A 7-year clinical experience. *Ophthalmology, 114*(10), 1925–1931. Available from https://doi.org/10.1016/j.ophtha.2007.06.012.

de Fougerolles, A., Vornlocher, H.-P., Maraganore, J., & Lieberman, J. (2007). Interfering with disease: A progress report on siRNA-based therapeutics. *Nature Reviews Drug Discovery, 6*(6), 443–453. Available from https://doi.org/10.1038/nrd2310.

Delsin, L. E. A., Salomao, K. B., Pezuk, J. A., & Brassesco, M. S. (2019). Expression profiles and prognostic value of miRNAs in retinoblastoma. *Journal of Cancer Research and Clinical Oncology, 145*(1), 1–10. Available from https://doi.org/10.1007/s00432-018-2773-7.

Dimaras, H., Kimani, K., Dimba, E. A. O., Gronsdahl, P., White, A., Chan, H. S. L., & Gallie, B. L. (2012). Retinoblastoma. *Lancet (London, England), 379*(9824), 1436–1446. Available from https://doi.org/10.1016/S0140-6736(11)61137-9.

Ding, F., Jiang, K., Sheng, Y., Li, C., & Zhu, H. (2020). LncRNA MIR7-3HG executes a positive role in retinoblastoma progression via modulating miR-27a-3p/PEG10 axis. *Experimental Eye Research, 193*, 107960.

Ding, X., Wang, X., Lin, M., Xing, Y., Ge, S., Jia, R., . . . Li, J. (2016). PAUPAR lncRNA suppresses tumourigenesis by H3K4 demethylation in uveal melanoma. *FEBS Letters, 590*(12), 1729–1738. Available from https://doi.org/10.1002/1873-3468.12220.

Dong, C., Liu, S., Lv, Y., Zhang, C., Gao, H., Tan, L., & Wang, H. (2016). Long non-coding RNA HOTAIR regulates proliferation and invasion via activating Notch signalling pathway in retinoblastoma. *Journal of Biosciences, 41*(4), 677–687. Available from https://doi.org/10.1007/s12038-016-9636-7.

Dong, F., & Lou, D. (2012). MicroRNA-34b/c suppresses uveal melanoma cell proliferation and migration through multiple targets. *Molecular Vision, 18*, 537–546.

Eedunuri, V. K., Rajapakshe, K., Fiskus, W., Geng, C., Chew, S. A., Foley, C., . . . Mitsiades, N. (2015). MiR-137 targets p160 steroid receptor coactivators SRC1, SRC2, and SRC3 and inhibits cell proliferation. *Molecular Endocrinology (Baltimore, MD), 29*(8), 1170–1183. Available from https://doi.org/10.1210/me.2015-1080.

Falzone, L., Romano, G. L., Salemi, R., Bucolo, C., Tomasello, B., Lupo, G., . . . Candido, S. (2019). Prognostic significance of deregulated microRNAs in uveal melanomas. *Molecular Medicine Reports, 19*(4), 2599–2610. Available from https://doi.org/10.3892/mmr.2019.9949.

Gao, Y., & Lu, X. (2016). Decreased expression of MEG3 contributes to retinoblastoma progression and affects retinoblastoma cell growth by regulating the activity of Wnt/β-catenin pathway. *Tumour Biology: The Journal of the International Society for Oncodevelopmental Biology and Medicine, 37*(2), 1461−1469. Available from https://doi.org/10.1007/s13277-015-4564-y.

Gao, Y., & Lu, X. (2017). LncRNA-MEG3 mediated apoptosis of retinoblastoma by regulating P53 pathway. *Recent Advances in Ophthalmology, 37*(4), 301−304.

Ghassemi, F., Shields, C. L., Ghadimi, H., Khodabandeh, A., & Roohipoor, R. (2014). Combined intravitreal melphalan and topotecan for refractory or recurrent vitreous seeding from retinoblastoma. *JAMA Ophthalmology, 132*(8), 936−941. Available from https://doi.org/10.1001/jamaophthalmol.2014.414.

Gibb, E. A., Brown, C. J., & Lam, W. L. (2011). The functional role of long non-coding RNA in human carcinomas. *Molecular Cancer, 10*, 38. Available from https://doi.org/10.1186/1476-4598-10-38.

Global Retinoblastoma Study GroupFabian, I. D., Abdallah, E., Abdullahi, S. U., Abdulqader, R. A., Adamou Boubacar, S., ... Bowman, R. (2020). Global Retinoblastoma presentation and analysis by national income level. *JAMA Oncology, 6*(5), 685−695. Available from https://doi.org/10.1001/jamaoncol.2019.6716.

Golabchi, K., Soleimani-Jelodar, R., Aghadoost, N., Momeni, F., Moridikia, A., Nahand, J. S., ... Mirzaei, H. (2018). MicroRNAs in retinoblastoma: Potential diagnostic and therapeutic biomarkers. *Journal of Cellular Physiology, 233*(4), 3016−3023. Available from https://doi.org/10.1002/jcp.26070.

Guo, L., Bai, Y., Ji, S., & Ma, H. (2019). MicroRNA-98 suppresses cell growth and invasion of retinoblastoma via targeting the IGF1R/k-Ras/Raf/MEK/ERK signaling pathway. *International Journal of Oncology, 54*(3), 807−820. Available from https://doi.org/10.3892/ijo.2019.4689.

Guo, N., Liu, X.-F., Pant, O. P., Zhou, D.-D., Hao, J.-L., & Lu, C.-W. (2019). Circular RNAs: Novel promising biomarkers in ocular diseases. *International Journal of Medical Sciences, 16*(4), 513−518. Available from https://doi.org/10.7150/ijms.29750.

Guzel, E., Okyay, T. M., Yalcinkaya, B., Karacaoglu, S., Gocmen, M., & Akcakuyu, M. H. (2020). Tumor suppressor and oncogenic role of long non-coding RNAs in cancer. *Northern Clinics of Istanbul, 7*(1), 81−86. Available from https://doi.org/10.14744/nci.2019.46873.

Han, N., Zuo, L., Chen, H., Zhang, C., He, P., & Yan, H. (2019). Long non-coding RNA homeobox A11 antisense RNA (HOXA11-AS) promotes retinoblastoma progression via sponging miR-506-3p. *OncoTargets and Therapy, 12*, 3509−3517.

Hao, F., Mou, Y., Zhang, L., Wang, S., & Yang, Y. (2018). LncRNA AFAP1-AS1 is a prognostic biomarker and serves as oncogenic role in retinoblastoma. *Bioscience Reports, 38*(3). Available from https://doi.org/10.1042/BSR20180384.

He, T.-G., Xiao, Z.-Y., Xing, Y.-Q., Yang, H.-J., Qiu, H., & Chen, J.-B. (2019). Tumor suppressor miR-184 enhances chemosensitivity by directly inhibiting SLC7A5 in retinoblastoma. *Frontiers in Oncology, 9*, 1163. Available from https://doi.org/10.3389/fonc.2019.01163.

Heitzer, E., Ulz, P., & Geigl, J. B. (2015). Circulating tumor DNA as a liquid biopsy for cancer. *Clinical Chemistry, 61*(1), 112−123. Available from https://doi.org/10.1373/clinchem.2014.222679.

Hou, Q., Han, S., Yang, L., Chen, S., Chen, J., Ma, N., ... Tu, L. (2019). The interplay of microRNA-34a, LGR4, EMT-associated factors, and MMP2 in regulating uveal melanoma cells. *Investigative Ophthalmology & Visual Science, 60*(13), 4503−4510. Available from https://doi.org/10.1167/iovs.18-26477.

Hu, C., Liu, S., Han, M., Wang, Y., & Xu, C. (2018). Knockdown of lncRNA XIST inhibits retinoblastoma progression by modulating the miR-124/STAT3 axis. *Biomedicine & Pharmacotherapy, 107*, 547−554.

Huang, J., Yang, Y., Fang, F., & Liu, K. (2018). MALAT1 modulates the autophagy of retinoblastoma cell through miR-124-mediated stx17 regulation. *Journal of Cellular Biochemistry, 119*(5), 3853−3863.

Huang, X.-M., Shi, S.-S., Jian, T.-M., Tang, D.-R., Wu, T., & Sun, F.-Y. (2019). LncRNA PVT1 knockdown affects proliferation and apoptosis of uveal melanoma cells by inhibiting EZH2. *European Review for Medical and Pharmacological Sciences, 23*(7), 2880−2887.

Hunter, M. P., Ismail, N., Zhang, X., Aguda, B. D., Lee, E. J., Yu, L., ... Marsh, C. B. (2008). Detection of microRNA expression in human peripheral blood microvesicles. *PLoS One, 3*(11), e3694. Available from https://doi.org/10.1371/journal.pone.0003694.

Jeck, W. R., & Sharpless, N. E. (2014). Detecting and characterizing circular RNAs. *Nature Biotechnology, 32*(5), 453−461. Available from https://doi.org/10.1038/nbt.2890.

Jia, M., Wei, Z., Liu, P., & Zhao, X. (2016). Silencing of ABCG2 by microRNA-3163 inhibits multidrug resistance in retinoblastoma cancer stem cells. *Journal of Korean Medical Science, 31*(6), 836−842. Available from https://doi.org/10.3346/jkms.2016.31.6.836.

Jo, D. H., Kim, J. H., Cho, C. S., Cho, Y.-L., Jun, H. O., Yu, Y. S., ... Kim, J. H. (2014). STAT3 inhibition suppresses proliferation of retinoblastoma through down-regulation of positive feedback loop of STAT3/miR-17−92 clusters. *Oncotarget, 5*(22), 11513−11525. Available from https://doi.org/10.18632/oncotarget.2546.

Kaliki, S., & Shields, C. L. (2017). Uveal melanoma: Relatively rare but deadly cancer. *Eye (London, England), 31*(2), 241−257. Available from https://doi.org/10.1038/eye.2016.275.

Kandalam, M. M., Beta, M., Maheswari, U. K., Swaminathan, S., & Krishnakumar, S. (2012). Oncogenic microRNA 17−92 cluster is regulated by epithelial cell adhesion molecule and could be a potential therapeutic target in retinoblastoma. *Molecular Vision, 18*, 2279−2287.

Kivelä, T. (2009). The epidemiological challenge of the most frequent eye cancer: Retinoblastoma, an issue of birth and death. *The British Journal of Ophthalmology, 93*(9), 1129−1131. Available from https://doi.org/10.1136/bjo.2008.150292.

Knudson, A. G. (1971). Mutation and cancer: Statistical study of retinoblastoma. *Proceedings of the National Academy of Sciences of the United States of America, 68*(4), 820−823. Available from https://doi.org/10.1073/pnas.68.4.820.

Lande, K., Gupta, J., Ranjan, R., Kiran, M., Torres Solis, L. F., Solís Herrera, A., ... Karnati, R. (2020). Exosomes: Insights from retinoblastoma and other eye cancers. *International Journal of Molecular Sciences, 21*(19). Available from https://doi.org/10.3390/ijms21197055.

Larsen, A.-C., Holst, L., Kaczkowski, B., Andersen, M. T., Manfé, V., Siersma, V. D., ... Heegaard, S. (2014). MicroRNA expression analysis and multiplex ligation-dependent probe amplification in metastatic and non-metastatic uveal melanoma. *Acta Ophthalmologica, 92*(6), 541−549. Available from https://doi.org/10.1111/aos.12322.

Lee, J. T. (2012). Epigenetic regulation by long noncoding RNAs. *Science (New York, N.Y.), 338*(6113), 1435−1439. Available from https://doi.org/10.1126/science.1231776.

Lei, Q., Shen, F., Wu, J., Zhang, W., Wang, J., & Zhang, L. (2014). MiR-101, downregulated in retinoblastoma, functions as a tumor suppressor in human retinoblastoma cells by targeting EZH2. *Oncology Reports, 32*(1), 261−269. Available from https://doi.org/10.3892/or.2014.3167.

Li, L., Yu, H., & Ren, Q. (2020). MiR-218-5p suppresses the progression of retinoblastoma through targeting NACC1 and inhibiting the AKT/mTOR signaling pathway. *Cancer Management and Research, 12*, 6959−6967.

Li, S., Wen, D., Che, S., Cui, Z., Sun, Y., Ren, H., et al. (2018). Knockdown of long noncoding RNA 00152 (LINC00152) inhibits human retinoblastoma progression. *OncoTargets and Therapy, 11*, 3215−3223.

Li, F., Wen, X., Zhang, H., & Fan, X. (2016). Novel insights into the role of long noncoding RNA in ocular diseases. *International Journal of Molecular Sciences, 17*(4), 478. Available from https://doi.org/10.3390/ijms17040478.

Li, J., Liu, X., Li, C., & Wang, W. (2019). MiR-224-5p inhibits proliferation, migration, and invasion by targeting PIK3R3/AKT3 in uveal melanoma. *Journal of Cellular Biochemistry, 120*(8), 12412−12421. Available from https://doi.org/10.1002/jcb.28507.

Li, L., Chen, W., Wang, Y., Tang, L., & Han, M. (2018). Long non-coding RNA H19 regulates viability and metastasis, and is upregulated in retinoblastoma. *Oncology Letters, 15*(6), 8424−8432. Available from https://doi.org/10.3892/ol.2018.8385.

Li, Y., Luo, J.-T., Liu, Y.-M., & Wei, W.-B. (2020). MiRNA-145/miRNA-205 inhibits proliferation and invasion of uveal melanoma cells by targeting NPR1/CDC42. *International Journal of Ophthalmology, 13*(5), 718−724. Available from https://doi.org/10.18240/ijo.2020.05.04.

Liang, Y., Chen, X., & Liang, Z. (2017). MicroRNA-320 regulates autophagy in retinoblastoma by targeting hypoxia inducible factor-1α. *Experimental and Therapeutic Medicine, 14*(3), 2367−2372. Available from https://doi.org/10.3892/etm.2017.4779.

Ling, J. W., Lu, P. R., Zhang, Y. B., Jiang, S., & Zhang, Z. C. (2017). MiR-367 promotes uveal melanoma cell proliferation and migration by regulating PTEN. *Genetics and Molecular Research: GMR, 16*(3). Available from https://doi.org/10.4238/gmr16039067.

Liu, S., Yan, G., Zhang, J., & Yu, L. (2018). Knockdown of long noncoding RNA (lncRNA) metastasis-associated lung adenocarcinoma transcript 1 (MALAT1) inhibits proliferation, migration, and invasion and promotes apoptosis by targeting miR-124 in retinoblastoma. *Oncology Research, 26*(4), 581−591.

References

Liu, S., Zhang, X., Hu, C., Wang, Y., & Xu, C. (2018). miR-29a inhibits human retinoblastoma progression by targeting STAT3. *Oncology Reports*, *39*(2), 739−746.

Liu, K., Huang, J., Xie, M., Yu, Y., Zhu, S., Kang, R., ... Duan, X. (2014). MIR34A regulates autophagy and apoptosis by targeting HMGB1 in the retinoblastoma cell. *Autophagy*, *10*(3), 442−452. Available from https://doi.org/10.4161/auto.27418.

Liu, N., Sun, Q., Chen, J., Li, J., Zeng, Y., Zhai, S., ... Wang, X. (2012). MicroRNA-9 suppresses uveal melanoma cell migration and invasion through the NF-κB1 pathway. *Oncology Reports*, *28*(3), 961−968. Available from https://doi.org/10.3892/or.2012.1905.

Liu, S.-S., Wang, Y.-S., Sun, Y.-F., Miao, L.-X., Wang, J., Li, Y.-S., ... Liu, Q.-L. (2014). Plasma microRNA-320, microRNA-let-7e and microRNA-21 as novel potential biomarkers for the detection of retinoblastoma. *Biomedical Reports*, *2*(3), 424−428. Available from https://doi.org/10.3892/br.2014.246.

Liu, Y., Huo, Y., Wang, D., Tai, Y., Li, J., Pang, D., ... Huang, Y. (2018). MiR-216a-5p/Hexokinase 2 axis regulates uveal melanoma growth through modulation of Warburg effect. *Biochemical and Biophysical Research Communications*, *501*(4), 885−892. Available from https://doi.org/10.1016/j.bbrc.2018.05.069.

Liz, J., & Esteller, M. (2016). LncRNAs and microRNAs with a role in cancer development. *Biochimica et Biophysica Acta*, *1859*(1), 169−176. Available from https://doi.org/10.1016/j.bbagrm.2015.06.015.

Lu, L., Yu, X., Zhang, L., Ding, X., Pan, H., Wen, X., ... Fan, X. (2017). The long non-coding RNA RHPN1-AS1 promotes uveal melanoma progression. *International Journal of Molecular Sciences*, *18*(1). Available from https://doi.org/10.3390/ijms18010226.

Lu, Q., Zhao, N., Zha, G., Wang, H., Tong, Q., & Xin, S. (2017). LncRNA HOXA11-AS exerts oncogenic functions by repressing p21 and miR-124 in uveal melanoma. *DNA and Cell Biology*, *36*(10), 837−844. Available from https://doi.org/10.1089/dna.2017.3808.

Lyu, X., Wang, L., Lu, J., Zhang, H., & Wang, L. (2019). microRNA-485 inhibits the malignant behaviors of retinoblastoma by directly targeting Wnt3a. *Oncology Reports*, *41*(5), 3137−3147.

Lyu, J., Wang, Y., Zheng, Q., Hua, P., Zhu, X., Li, J., ... Zhao, P. (2019). Reduction of circular RNA expression associated with human retinoblastoma. *Experimental Eye Research*, *184*, 278−285. Available from https://doi.org/10.1016/j.exer.2019.03.017.

Lyv, X., Wu, F., Zhang, H., Lu, J., Wang, L., & Ma, Y. (2020). Long noncoding RNA ZFPM2-AS1 knockdown restrains the development of retinoblastoma by modulating the microRNA-515/HOXA1/Wnt/β-Catenin Axis. *Investigative Ophthalmology & Visual Science*, *61*(6), 41.

Martin, J., Bryar, P., Mets, M., Weinstein, J., Jones, A., Martin, A., ... Laurie, N. A. (2013). Differentially expressed miRNAs in retinoblastoma. *Gene*, *512*(2), 294−299. Available from https://doi.org/10.1016/j.gene.2012.09.129.

Mirakholi, M., Mahmoudi, T., & Heidari, M. (2013). MicroRNAs horizon in retinoblastoma. *Acta Medica Iranica*, *51*(12), 823−829.

Mitchell, P. S., Parkin, R. K., Kroh, E. M., Fritz, B. R., Wyman, S. K., Pogosova-Agadjanyan, E. L., ... Tewari, M. (2008). Circulating microRNAs as stable blood-based markers for cancer detection. *Proceedings of the National Academy of Sciences of the United States of America*, *105*(30), 10513−10518. Available from https://doi.org/10.1073/pnas.0804549105.

Mitra, M., Mohanty, C., Harilal, A., Maheswari, U. K., Sahoo, S. K., & Krishnakumar, S. (2012). A novel in vitro three-dimensional retinoblastoma model for evaluating chemotherapeutic drugs. *Molecular Vision*, *18*, 1361−1378.

Montoya, V., Fan, H., Bryar, P. J., Weinstein, J. L., Mets, M. B., Feng, G., ... Laurie, N. A. (2015). Novel miRNA-31 and miRNA-200a-mediated regulation of retinoblastoma proliferation. *PLoS One*, *10*(9), e0138366. Available from https://doi.org/10.1371/journal.pone.0138366.

Pan, H., Ni, H., Zhang, L., Xing, Y., Fan, J., Li, P., ... Fan, X. (2016). P2RX7-V3 is a novel oncogene that promotes tumorigenesis in uveal melanoma. *Tumour Biology: The Journal of the International Society for Oncodevelopmental Biology and Medicine*, *37*(10), 13533−13543. Available from https://doi.org/10.1007/s13277-016-5141-8.

Peng, D., Dong, J., Zhao, Y., Peng, X., Tang, J., Chen, X., ... Yan, D. (2019). MiR-142-3p suppresses uveal melanoma by targeting CDC25C, TGFβR1, GNAQ, WASL, and RAC1. *Cancer Management and Research*, *11*, 4729−4742. Available from https://doi.org/10.2147/CMAR.S206461.

Peng, J., Liu, H., & Liu, C. (2017). MiR-155 promotes uveal melanoma cell proliferation and invasion by regulating NDFIP1 expression. *Technology in Cancer Research & Treatment*, *16*(6), 1160−1167. Available from https://doi.org/10.1177/1533034617737923.

Plousiou, M., & Vannini, I. (2019). Non-coding RNAs in retinoblastoma. *Frontiers in Genetics, 10,* 1155. Available from https://doi.org/10.3389/fgene.2019.01155.

Prensner, J. R., & Chinnaiyan, A. M. (2011). The emergence of lncRNAs in cancer biology. *Cancer Discovery, 1*(5), 391−407. Available from https://doi.org/10.1158/2159-8290.CD-11-0209.

Qi, D., Wang, M., & Yu, F. (2019). Knockdown of lncRNA-H19 inhibits cell viability, migration and invasion while promotes apoptosis via microRNA-143/RUNX2 axis in retinoblastoma. *Biomedicine & Pharmacotherapy = Biomedecine & Pharmacotherapie, 109,* 798−805. Available from https://doi.org/10.1016/j.biopha.2018.10.096.

Ragusa, M., Barbagallo, C., Statello, L., Caltabiano, R., Russo, A., Puzzo, L., ... Reibaldi, M. (2015). miRNA profiling in vitreous humor, vitreal exosomes and serum from uveal melanoma patients: Pathological and diagnostic implications. *Cancer Biology & Therapy, 16*(9), 1387−1396. Available from https://doi.org/10.1080/15384047.2015.1046021.

Rao, R., & Honavar, S. G. (2017). Retinoblastoma. *Indian Journal of Pediatrics, 84*(12), 937−944. Available from https://doi.org/10.1007/s12098-017-2395-0.

Reiniger, I. W., Schaller, U. C., Haritoglou, C., Hein, R., Bosserhoff, A. K., Kampik, A., & Mueller, A. J. (2005). "Melanoma inhibitory activity" (MIA): A promising serological tumour marker in metastatic uveal melanoma. *Graefe's Archive for Clinical and Experimental Ophthalmology = Albrecht Von Graefes Archiv Fur Klinische Und Experimentelle Ophthalmologie, 243*(11), 1161−1166. Available from https://doi.org/10.1007/s00417-005-1171-4.

Robertson, A. G., Shih, J., Yau, C., Gibb, E. A., Oba, J., Mungall, K. L., ... Woodman, S. E. (2017). Integrative analysis identifies four molecular and clinical subsets in uveal melanoma. *Cancer Cell, 32*(2), 204−220. Available from https://doi.org/10.1016/j.ccell.2017.07.003, e15.

Rodrigues, M. J., & Stern, M.-H. (2015). Genetic landscape of uveal melanoma. *Journal Francais D'ophtalmologie, 38* (6), 522−525. Available from https://doi.org/10.1016/j.jfo.2015.04.004.

Rodriguez-Vidal, C., Fernandez-Diaz, D., Fernandez-Marta, B., Lago-Baameiro, N., Pardo, M., Silva, P., ... Bande, M. (2020). Treatment of metastatic uveal melanoma: systematic review. *Cancers, 12*(9), 2557. Available from https://doi.org/10.3390/cancers12092557.

Russo, A., Caltabiano, R., Longo, A., Avitabile, T., Franco, L. M., Bonfiglio, V., ... Reibaldi, M. (2016). Increased levels of miRNA-146a in serum and histologic samples of patients with uveal melanoma. *Frontiers in Pharmacology, 7,* 424. Available from https://doi.org/10.3389/fphar.2016.00424.

Salzman, J. (2016). Circular RNA expression: its potential regulation and function. *Trends in Genetics: TIG, 32*(5), 309−316. Available from https://doi.org/10.1016/j.tig.2016.03.002.

Salzman, J., Gawad, C., Wang, P. L., Lacayo, N., & Brown, P. O. (2012). Circular RNAs are the predominant transcript isoform from hundreds of human genes in diverse cell types. *PLoS One, 7*(2), e30733. Available from https://doi.org/10.1371/journal.pone.0030733.

Schmitz, S. U., Grote, P., & Herrmann, B. G. (2016). Mechanisms of long noncoding RNA function in development and disease. *Cellular and Molecular Life Sciences: CMLS, 73*(13), 2491−2509. Available from https://doi.org/10.1007/s00018-016-2174-5.

Shain, A. H., Bagger, M. M., Yu, R., Chang, D., Liu, S., Vemula, S., ... Kiilgaard, J. F. (2019). The genetic evolution of metastatic uveal melanoma. *Nature Genetics, 51*(7), 1123−1130. Available from https://doi.org/10.1038/s41588-019-0440-9.

Shang, W., Yang, Y., Zhang, J., & Wu, Q. (2018). Long noncoding RNA BDNF-AS is a potential biomarker and regulates cancer development in human retinoblastoma. *Biochemical and Biophysical Research Communications, 497*(4), 1142−1148. Available from https://doi.org/10.1016/j.bbrc.2017.01.134.

Shang, Y. (2018). LncRNA THOR acts as a retinoblastoma promoter through enhancing the combination of c-myc mRNA and IGF2BP1 protein. *Biomedicine & Pharmacotherapy, 106,* 1243−1249. Available from https://doi.org/10.1016/j.biopha.2018.07.052.

Shehata, H. H., Abou Ghalia, A. H., Elsayed, E. K., Ahmed Said, A. M., & Mahmoud, S. S. (2016). Clinical significance of high levels of survivin and transforming growth factor beta-1 proteins in aqueous humor and serum of retinoblastoma patients. *Journal of AAPOS: The Official Publication of the American Association for Pediatric Ophthalmology and Strabismus, 20*(5), 444.e1−444.e9. Available from https://doi.org/10.1016/j.jaapos.2016.07.223.

Shen, F., Mo, M.-H., Chen, L., An, S., Tan, X., Fu, Y., et al. (2014). MicroRNA-21 Down-regulates Rb1 Expression by Targeting PDCD4 in Retinoblastoma. *Journal of Cancer, 5*(9), 804−812.

Sheng, L., Wu, J., Gong, X., Dong, D., & Sun, X. (2018). SP1-induced upregulation of lncRNA PANDAR predicts adverse phenotypes in retinoblastoma and regulates cell growth and apoptosis in vitro and in vivo. *Gene, 668*, 140−145. Available from https://doi.org/10.1016/j.gene.2018.05.065.

Singh, U., Malik, M. A., Goswami, S., Shukla, S., & Kaur, J. (2016). Epigenetic regulation of human retinoblastoma. *Tumour Biology: The Journal of the International Society for Oncodevelopmental Biology and Medicine, 37*(11), 14427−14441. Available from https://doi.org/10.1007/s13277-016-5308-3.

Smit, K. N., Chang, J., Derks, K., Vaarwater, J., Brands, T., Verdijk, R. M., ... Kilic, E. (2019). Aberrant microRNA expression and its implications for uveal melanoma metastasis. *Cancers, 11*(6). Available from https://doi.org/10.3390/cancers11060815.

Song, L., Huang, Y., Zhang, X., Han, S., Hou, M., & Li, H. (2020). Downregulation of microRNA-224-3p hampers retinoblastoma progression via activation of the hippo-YAP signaling pathway by increasing LATS2. *Investigative Ophthalmology & Visual Science, 61*(3), 32.

Stark, M. S., Gray, E. S., Isaacs, T., Chen, F. K., Millward, M., McEvoy, A., ... Hayward, N. K. (2019). A panel of circulating microRNAs detects uveal melanoma with high precision. *Translational Vision Science & Technology, 8*(6), 12. Available from https://doi.org/10.1167/tvst.8.6.12.

Su, S., Gao, J., Wang, T., Wang, J., Li, H., & Wang, Z. (2015). Long non-coding RNA BANCR regulates growth and metastasis and is associated with poor prognosis in retinoblastoma. *Tumour Biology: The Journal of the International Society for Oncodevelopmental Biology and Medicine, 36*(9), 7205−7211. Available from https://doi.org/10.1007/s13277-015-3413-3.

Sun, L., Bian, G., Meng, Z., Dang, G., Shi, D., & Mi, S. (2015). MiR-144 inhibits uveal melanoma cell proliferation and invasion by regulating c-Met expression. *PLoS One, 10*(5), e0124428.

Sun, Z., Zhang, A., & Zhang, L. (2019). Inhibition of microRNA-492 attenuates cell proliferation and invasion in retinoblastoma via directly targeting LATS2. *Molecular Medicine Reports, 19*(3), 1965−1971.

Sun, L., Sun, P., Zhou, Q.-Y., Gao, X., & Han, Q. (2016). Long noncoding RNA MALAT1 promotes uveal melanoma cell growth and invasion by silencing of miR-140. *American Journal of Translational Research, 8*(9), 3939−3946.

Sun, L., Wang, Q., Gao, X., Shi, D., Mi, S., & Han, Q. (2015). MicroRNA-454 functions as an oncogene by regulating PTEN in uveal melanoma. *FEBS Letters, 589*(19 Pt B), 2791−2796. Available from https://doi.org/10.1016/j.febslet.2015.08.007.

Sun, X., Shen, H., Liu, S., Gao, J., & Zhang, S. (2020). Long noncoding RNA SNHG14 promotes the aggressiveness of retinoblastoma by sponging microRNA-124 and thereby upregulating STAT3. *International Journal of Molecular Medicine, 45*(6), 1685−1696. Available from https://doi.org/10.3892/ijmm.2020.4547.

Tao, S., Wang, W., Liu, P., Wang, H., & Chen, W. (2019). Long non-coding RNA T-cell leukemia/lymphoma 6 serves as a sponge for miR-21 modulating the cell proliferation of retinoblastoma through PTEN. *The Korean Journal of Physiology & Pharmacology: Official Journal of the Korean Physiological Society and the Korean Society of Pharmacology, 23*(6), 449−458. Available from https://doi.org/10.4196/kjpp.2019.23.6.449.

Thériault, B. L., Dimaras, H., Gallie, B. L., & Corson, T. W. (2014). The genomic landscape of retinoblastoma: A review. *Clinical & Experimental Ophthalmology, 42*(1), 33−52. Available from https://doi.org/10.1111/ceo.12132.

Triozzi, P. L., Achberger, S., Aldrich, W., Crabb, J. W., Saunthararajah, Y., & Singh, A. D. (2016). Association of tumor and plasma microRNA expression with tumor monosomy-3 in patients with uveal melanoma. *Clinical Epigenetics, 8*, 80. Available from https://doi.org/10.1186/s13148-016-0243-0.

Venkatesan, N., Kanwar, J., Deepa, P. R., Khetan, V., Crowley, T. M., Raguraman, R., ... Krishnakumar, S. (2016). Clinico-pathological association of delineated miRNAs in uveal melanoma with monosomy 3/disomy 3 chromosomal aberrations. *PLoS One, 11*(1), e0146128. Available from https://doi.org/10.1371/journal.pone.0146128.

Venza, I., Visalli, M., Beninati, C., Benfatto, S., Teti, D., & Venza, M. (2015). IL-10Rα expression is post-transcriptionally regulated by miR-15a, miR-185, and miR-211 in melanoma. *BMC Medical Genomics, 8*, 81. Available from https://doi.org/10.1186/s12920-015-0156-3.

Venza, M., Visalli, M., Beninati, C., Benfatto, S., Teti, D., & Venza, I. (2016). MiR-92a-3p and MYCBP2 are involved in MS-275-induced and c-myc-mediated TRAIL-sensitivity in melanoma cells. *International Immunopharmacology, 40*, 235−243. Available from https://doi.org/10.1016/j.intimp.2016.09.004.

Vicens, Q., & Westhof, E. (2014). Biogenesis of circular RNAs. *Cell, 159*(1), 13−14. Available from https://doi.org/10.1016/j.cell.2014.09.005.

Wan, W., Wan, W., Long, Y., Li, Q., Jin, X., Wan, G., ... Zhu, Y. (2019). MiR-25-3p promotes malignant phenotypes of retinoblastoma by regulating PTEN/Akt pathway. *Biomedicine & Pharmacotherapy = Biomedecine & Pharmacotherapie, 118*, 109111. Available from https://doi.org/10.1016/j.biopha.2019.109111.

Wang, J., Wang, X., Li, Z., Liu, H., & Teng, Y. (2014). MicroRNA-183 suppresses retinoblastoma cell growth, invasion and migration by targeting LRP6. *The FEBS Journal, 281*(5), 1355−1365.

Wang, J., Wang, X., Wu, G., Hou, D., & Hu, Q. (2013). MiR-365b-3p, down-regulated in retinoblastoma, regulates cell cycle progression and apoptosis of human retinoblastoma cells by targeting PAX6. *FEBS Letters, 587*(12), 1779−1786.

Wang, L., Lyu, X., Ma, Y., Wu, F., & Wang, L. (2019). MicroRNA-504 targets AEG-1 and inhibits cell proliferation and invasion in retinoblastoma. *Molecular Medicine Reports, 19*(4), 2935−2942.

Wang, L., Wang, C., Wu, T., & Sun, F. (2020). Long non-coding RNA TP73-AS1 promotes TFAP2B-mediated proliferation, metastasis and invasion in retinoblastoma via decoying of miRNA-874-3p. *Journal of Cell Communication and Signaling, 14*(2), 193−205.

Wang, L., Wang, L., Li, L., Zhang, H., & Lyu, X. (2019). MicroRNA-330 is downregulated in retinoblastoma and suppresses cell viability and invasion by directly targeting ROCK1. *Molecular Medicine Reports, 20*(4), 3440−3447.

Wang, L., Yang, D., Tian, R., & Zhang, H. (2019). NEAT1 promotes retinoblastoma progression via modulating miR-124. *Journal of Cellular Biochemistry, 120*(9), 15585−15593.

Wang, S., Liu, J., Yang, Y., Hao, F., & Zhang, L. (2018). PlncRNA-1 is overexpressed in retinoblastoma and regulates retinoblastoma cell proliferation and motility through modulating CBR3. *IUBMB Life, 70*(10), 969−975.

Wang, X., Zhang, X., Han, Y., Wang, Q., Ren, Y., Wang, B., et al. (2019). Silence of lncRNA ANRIL represses cell growth and promotes apoptosis in retinoblastoma cells through regulating miR-99a and c-Myc. *Artificial Cells, Nanomedicine, and Biotechnology, 47*(1), 2265−2273.

Wang, Y., Sun, D., Sheng, Y., Guo, H., Meng, F., & Song, T. (2020). XIST promotes cell proliferation and invasion by regulating miR-140-5p and SOX4 in retinoblastoma. *World Journal of Surgical Oncology, 18*(1), 49.

Wang, Y.-C., Yang, X., Wei, W.-B., & Xu, X.-L. (2018). Role of microRNA-21 in uveal melanoma cell invasion and metastasis by regulating p53 and its downstream protein. *International Journal of Ophthalmology, 11*(8), 1258−1268.

Wang, Z., Yao, Y.-J., Zheng, F., Guan, Z., Zhang, L., Dong, N., et al. (2017). Mir-138-5p acts as a tumor suppressor by targeting pyruvate dehydrogenase kinase 1 in human retinoblastoma. *European Review for Medical and Pharmacological Sciences, 21*(24), 5624−5629.

Wang, J.-X., Yang, Y., & Li, K. (2018). Long noncoding RNA DANCR aggravates retinoblastoma through miR-34c and miR-613 by targeting MMP-9. *Journal of Cellular Physiology, 233*(10), 6986−6995. Available from https://doi.org/10.1002/jcp.26621.

Wang, S., Cao, M., Deng, X., Xiao, X., Yin, Z., Hu, Q., ... Zeng, Y. (2015). Degradable hyaluronic acid/protamine sulfate interpolyelectrolyte complexes as miRNA-delivery nanocapsules for triple-negative breast cancer therapy. *Advanced Healthcare Materials, 4*(2), 281−290. Available from https://doi.org/10.1002/adhm.201400222.

Wang, X., Hu, Y., Cui, J., Zhou, Y., & Chen, L. (2018). Coordinated targeting of MMP-2/MMP-9 by miR-296-3p/FOXCUT exerts tumor-suppressing effects in choroidal malignant melanoma. *Molecular and Cellular Biochemistry, 445*(1−2), 25−33. Available from https://doi.org/10.1007/s11010-017-3248-x.

Wang, Y., Luo, Y., Guan, W., & Zhao, H. (2018). Role of miR-23a/Zeb1 negative feedback loop in regulating epithelial-mesenchymal transition and tumorigenicity of intraocular tumors. *Oncology Letters, 16*(2), 2462−2470. Available from https://doi.org/10.3892/ol.2018.8940.

Wawrzyniak, O., Zarębska, Ż., Rolle, K., & Gotz-Więckowska, A. (2018). Circular and long non-coding RNAs and their role in ophthalmologic diseases. *Acta Biochimica Polonica, 65*(4), 497−508. Available from https://doi.org/10.18388/abp.2018_2639.

Worley, L. A., Long, M. D., Onken, M. D., & Harbour, J. W. (2008). Micro-RNAs associated with metastasis in uveal melanoma identified by multiplexed microarray profiling. *Melanoma Research, 18*(3), 184−190. Available from https://doi.org/10.1097/CMR.0b013e3282feeac6.

Wróblewska, J. P., Lach, M. S., Ustaszewski, A., Kulcenty, K., Ibbs, M., Jagiełło, I., ... Marszałek, A. (2020). The potential role of selected miRNA in uveal melanoma primary tumors as early biomarkers of disease progression. *Genes, 11*(3). Available from https://doi.org/10.3390/genes11030271.

Wu, X., Li, X.-F., Wu, Q., Ma, R.-Q., Qian, J., & Zhang, R. (2020). LncRNA SNHG15 predicts poor prognosis in uveal melanoma and its potential pathways. *Journal of Ophthalmology, 13*(8), 1195—1201.

Wu, X., Zeng, Y., Wu, S., Zhong, J., Wang, Y., & Xu, J. (2015). MiR-204, down-regulated in retinoblastoma, regulates proliferation and invasion of human retinoblastoma cells by targeting CyclinD2 and MMP-9. *FEBS Letters, 589*(5), 645—650.

Wu, S., Chen, H., Han, N., Zhang, C., & Yan, H. (2019). Long noncoding RNA PVT1 silencing prevents the development of uveal melanoma by impairing microRNA-17-3p-dependent MDM2 upregulation. *Investigative Ophthalmology & Visual Science, 60*(14), 4904—4914. Available from https://doi.org/10.1167/iovs.19-27704.

Wu, S., Chen, H., Zuo, L., Jiang, H., & Yan, H. (2020). Suppression of long noncoding RNA MALAT1 inhibits the development of uveal melanoma via microRNA-608-mediated inhibition of HOXC4. *American Journal of Physiology. Cell Physiology, 318*(5), C903—C912. Available from https://doi.org/10.1152/ajpcell.00262.2019.

Wu, X., Yuan, Y., Ma, R., Xu, B., & Zhang, R. (2020). LncRNA SNHG7 affects malignant tumor behaviors through downregulation of EZH2 in uveal melanoma cell lines. *Oncology Letters, 19*(2), 1505—1515. Available from https://doi.org/10.3892/ol.2019.11240.

Wu, X.-Z., Cui, H.-P., Lv, H.-J., & Feng, L. (2019). Knockdown of lncRNA PVT1 inhibits retinoblastoma progression by sponging miR-488-3p. *Biomedicine & Pharmacotherapy = Biomedecine & Pharmacotherapie, 112*, 108627. Available from https://doi.org/10.1016/j.biopha.2019.108627.

Xia, Z., Yang, X., Wu, S., Feng, Z., Qu, L., Chen, X., et al. (2019). LncRNA TP73-AS1 down-regulates miR-139-3p to promote retinoblastoma cell proliferation. *Bioscience Reports, 39*(5).

Xia, Z., Yang, C., Yang, X., Wu, S., Feng, Z., Qu, L., … Ma, Y. (2019). MiR-652 promotes proliferation and migration of uveal melanoma cells by targeting HOXA9. *Medical Science Monitor: International Medical Journal of Experimental and Clinical Research, 25*, 8722—8732. Available from https://doi.org/10.12659/MSM.917099.

Xin, X., Zhang, Y., Ling, F., Wang, L., Sheng, X., Qin, L., & Zhao, X. (2019). Identification of a nine-miRNA signature for the prognosis of uveal melanoma. *Experimental Eye Research, 180*, 242—249. Available from https://doi.org/10.1016/j.exer.2019.01.004.

Xing, L., Zhang, L., Feng, Y., Cui, Z., & Ding, L. (2018). Downregulation of circular RNA hsa_circ_0001649 indicates poor prognosis for retinoblastoma and regulates cell proliferation and apoptosis via AKT/mTOR signaling pathway. *Biomedicine & Pharmacotherapy = Biomedecine & Pharmacotherapie, 105*, 326—333. Available from https://doi.org/10.1016/j.biopha.2018.05.141.

Xing, Y., Wen, X., Ding, X., Fan, J., Chai, P., Jia, R., … Fan, X. (2017). CANT1 lncRNA triggers efficient therapeutic efficacy by correcting aberrant lncing cascade in malignant uveal melanoma. *Molecular Therapy: The Journal of the American Society of Gene Therapy, 25*(5), 1209—1221. Available from https://doi.org/10.1016/j.ymthe.2017.02.016.

Xu, C., Hu, C., Wang, Y., & Liu, S. (2019). Long noncoding RNA SNHG16 promotes human retinoblastoma progression via sponging miR-140-5p. *Biomed Pharmacother Biomedecine Pharmacother, 117*, 109153.

Xu, F., Liu, G., Wang, L., Wang, X., Jin, X., & Bo, W. (2020). miR-494 promotes progression of retinoblastoma via PTEN through PI3K/AKT signaling pathway. *Oncology Letters, 20*(2), 1952—1960.

Xu, H., Gong, J., & Liu, H. (2017). High expression of lncRNA PVT1 independently predicts poor overall survival in patients with primary uveal melanoma. *PLoS One, 12*(12), e0189675. Available from https://doi.org/10.1371/journal.pone.0189675.

Xu, X., Jia, R., Zhou, Y., Song, X., Wang, J., Qian, G., … Fan, X. (2011). Microarray-based analysis: Identification of hypoxia-regulated microRNAs in retinoblastoma cells. *International Journal of Oncology, 38*(5), 1385—1393. Available from https://doi.org/10.3892/ijo.2011.961.

Yan, G., Su, Y., Ma, Z., Yu, L., & Chen, N. (2019). Long noncoding RNA LINC00202 promotes tumor progression by sponging miR-3619-5p in retinoblastoma. *Cell Structure and Function, 44*(1), 51—60.

Yan, B., Yao, J., Tao, Z.-F., & Jiang, Q. (2014). Epigenetics and ocular diseases: From basic biology to clinical study. *Journal of Cellular Physiology, 229*(7), 825—833. Available from https://doi.org/10.1002/jcp.24522.

Yan, D., Dong, X. D., Chen, X., Yao, S., Wang, L., Wang, J., … Tu, L. (2012). Role of microRNA-182 in posterior uveal melanoma: Regulation of tumor development through MITF, BCL2 and cyclin D2. *PLoS One, 7*(7), e40967. Available from https://doi.org/10.1371/journal.pone.0040967.

Yan, D., Zhou, X., Chen, X., Hu, D.-N., Dong, X. D., Wang, J., … Qu, J. (2009). MicroRNA-34a inhibits uveal melanoma cell proliferation and migration through downregulation of c-Met. *Investigative Ophthalmology & Visual Science, 50*(4), 1559—1565. Available from https://doi.org/10.1167/iovs.08-2681.

Yang, G., Fu, Y., Zhang, L., Lu, X., & Li, Q. (2017). miR106b regulates retinoblastoma Y79 cells through Runx3. *Oncology Reports, 38*(5), 3039–3043.

Yang, G., Fu, Y., Lu, X., Wang, M., Dong, H., & Li, Q. (2018). LncRNA HOTAIR/miR-613/c-met axis modulated epithelial-mesenchymal transition of retinoblastoma cells. *Journal of Cellular and Molecular Medicine, 22*(10), 5083–5096. Available from https://doi.org/10.1111/jcmm.13796.

Yang, L., Zhang, L., Lu, L., & Wang, Y. (2020a). MiR-214-3p regulates multi-drug resistance and apoptosis in retinoblastoma cells by targeting ABCB1 and XIAP. *OncoTargets and Therapy, 13*, 803–811. Available from https://doi.org/10.2147/OTT.S235862.

Yang, L., Zhang, L., Lu, L., & Wang, Y. (2020b). LncRNA UCA1 increases proliferation and multidrug resistance of retinoblastoma cells through downregulating miR-513a-5p. *DNA and Cell Biology, 39*(1), 69–77. Available from https://doi.org/10.1089/dna.2019.5063.

Yang, M., & Wei, W. (2019). Long non-coding RNAs in retinoblastoma. *Pathology, Research and Practice, 215*(8), 152435. Available from https://doi.org/10.1016/j.prp.2019.152435.

Yang, Y., & Peng, X.-W. (2018). The silencing of long non-coding RNA ANRIL suppresses invasion, and promotes apoptosis of retinoblastoma cells through the ATM-E2F1 signaling pathway. *Bioscience Reports, 38*(6). Available from https://doi.org/10.1042/BSR20180558.

Zhang, G., Yang, W., Li, D., Li, X., Huang, J., Huang, R., et al. (2020). lncRNA FEZF1-AS1 promotes migration, invasion and epithelial-mesenchymal transition of retinoblastoma cells by targeting miR-1236-3p. *Molecular Medicine Reports, 22*(5), 3635–3644.

Zhang, Y., Xue, C., Zhu, X., Zhu, X., Xian, H., & Huang, Z. (2016). Suppression of microRNA-125a-5p upregulates the TAZ-EGFR signaling pathway and promotes retinoblastoma proliferation. *Cellular Signalling, 28*(8), 850–860.

Zhang, A., Shang, W., Nie, Q., Li, T., & Li, S. (2018). Long non-coding RNA H19 suppresses retinoblastoma progression via counteracting miR-17–92 cluster. *Journal of Cellular Biochemistry, 119*(4), 3497–3509. Available from https://doi.org/10.1002/jcb.26521.

Zhang, H., Zhong, J., Bian, Z., Fang, X., Peng, Y., & Hu, Y. (2017). Long non-coding RNA CCAT1 promotes human retinoblastoma SO-RB50 and Y79 cells through negative regulation of miR-218-5p. *Biomedicine & Pharmacotherapy = Biomedecine & Pharmacotherapie, 87*, 683–691. Available from https://doi.org/10.1016/j.biopha.2017.01.004.

Zhang, L., Dong, Y., Wang, Y., Gao, J., Lv, J., Sun, J., . . . Xu, W. (2019). Long non-coding RNAs in ocular diseases: New and potential therapeutic targets. *The FEBS Journal, 286*(12), 2261–2272. Available from https://doi.org/10.1111/febs.14827.

Zhang, L., He, X., Li, F., Pan, H., Huang, X., Wen, X., . . . Fan, X. (2018). The miR-181 family promotes cell cycle by targeting CTDSPL, a phosphatase-like tumor suppressor in uveal melanoma. *Journal of Experimental & Clinical Cancer Research: CR, 37*(1), 15. Available from https://doi.org/10.1186/s13046-018-0679-5.

Zhang, Y., Zhang, X.-O., Chen, T., Xiang, J.-F., Yin, Q.-F., Xing, Y.-H., . . . Chen, L.-L. (2013). Circular intronic long noncoding RNAs. *Molecular Cell, 51*(6), 792–806. Available from https://doi.org/10.1016/j.molcel.2013.08.017.

Zhao, D., & Cui, Z. (2019). MicroRNA-361-3p regulates retinoblastoma cell proliferation and stemness by targeting hedgehog signaling. *Experimental and Therapeutic Medicine, 17*(2), 1154–1162. Available from https://doi.org/10.3892/etm.2018.7062.

Zhao, J.-J., Yang, J., Lin, J., Yao, N., Zhu, Y., Zheng, J., . . . Ma, X. (2009). Identification of miRNAs associated with tumorigenesis of retinoblastoma by miRNA microarray analysis. *Child's Nervous System: ChNS: Official Journal of the International Society for Pediatric Neurosurgery, 25*(1), 13–20. Available from https://doi.org/10.1007/s00381-008-0701-x.

Zheng, Q., Zhu, Q., Li, C., Hao, S., Li, J., Yu, X., . . . Pan, Y. (2020). MicroRNA-144 functions as a diagnostic and prognostic marker for retinoblastoma. *Clinics (Sao Paulo, Brazil), 75*, e1804. Available from https://doi.org/10.6061/clinics/2020/e1804.

Zheng, X., Tang, H., Zhao, X., Sun, Y., Jiang, Y., & Liu, Y. (2017). Long non-coding RNA FTH1P3 facilitates uveal melanoma cell growth and invasion through miR-224-5p. *PLoS One, 12*(11), e0184746. Available from https://doi.org/10.1371/journal.pone.0184746.

Zhong, W., Yang, J., Li, M., Li, L., & Li, A. (2019). Long noncoding RNA NEAT1 promotes the growth of human retinoblastoma cells via regulation of miR-204/CXCR4 axis. *Journal of Cellular Physiology, 234*(7), 11567–11576.

Zhou, J., Jiang, J., Wang, S., & Xia, X. (2016). Oncogenic role of microRNA-20a in human uveal melanoma. *Molecular Medicine Reports, 14*(2), 1560–1566. Available from https://doi.org/10.3892/mmr.2016.5433.

Applications of noncoding RNAs in renal cancer patients

Eman A. Toraih[1,2], Jessica A. Sedhom[1], Muhib Haidari[1] and Manal S. Fawzy[3,4]

[1]Department of Surgery, Tulane University School of Medicine, New Orleans, LA, United States [2]Genetics Unit, Department of Histology and Cell Biology, Faculty of Medicine, Suez Canal University, Ismailia, Egypt [3]Department of Medical Biochemistry and Molecular Biology, Faculty of Medicine, Suez Canal University, Ismailia, Egypt [4]Department of Biochemistry, Faculty of Medicine, Northern Border University, Arar, Saudi Arabia

7.1 Introduction

Renal cell carcinoma (RCC), which originates in renal tubular epithelial cells, is one of the 10 most common types of cancer worldwide (Siegel, Miller, Fuchs, & Jemal, 2021). It can be classified into clear cell RCC (ccRCC), which accounts for 70%–80% of renal cancer cases, papillary RCC (pRCC), and chromophobe RCC subtypes, among others (Prasad et al., 2006). Each subtype displays a distinct pathological picture, unique genetic/epigenetic signatures, and variable disease outcomes (Bhan, Soleimani, & Mandal, 2017). The present noninvasive biomarkers have been reported to lack predictive values for early-stage RCC detection (Cimadamore et al., 2020). Also, although surgical resection is the preferred treatment modality for localized RCC, postoperative local recurrence or distant metastasis still occurs in some patients after several years (Athar & Gentile, 2008; Graham et al., 2019; Lieder, Guenzel, Lebentrau, Schneider, & Franzen, 2017; Pal & Agarwal, 2016). Furthermore, the 5-year disease-specific survival for ccRCC and pRCC is 10.5% and 10.3%, respectively, once metastasis occurs (Patard et al., 2005). Also, metastatic RCC has been found to resist chemotherapy and radiotherapy treatment. Therefore the development of novel diagnostic and/or prognostic biomarkers for early detection of recurrence and cancer metastasis, and identification of new therapeutic targets to overcome drug resistance are important issues to improve RCC patient's prognosis.

Protein-related biomarkers and therapeutics have clear roles in this area; however, some showed inadequate sensitivity and/or predictive values in RCC early detection, cancer prognosis, patient's survival, or response to treatment (Wang, Zheng, Chen, Shi, & Chen, 2020). Therefore searching for other biomolecules implicated in oncogenic processes and/or enhancing response to therapy for successful RCC management is necessary.

Accumulating evidence shows most of the cancer-related genomic deregulations, including mutations or copy number variations located in noncoding DNA elements (Guttman & Rinn, 2012). These elements have been identified to cover almost 95% of human genome DNA sequences and transcribed into many functional noncoding RNAs (ncRNAs) implicated in several biological processes, including tumor development, progression, and metastasis (Goodall & Wickramasinghe, 2021; Guttman & Rinn, 2012; Matsui & Corey, 2017). Hence, ncRNAs may be attractive diagnostic/prognostic biomarkers and therapeutic molecular targets (Wang et al., 2019).

The most extensively studied microRNAs (miRNAs) family is a type of small ncRNAs (<200 nucleotides) involved in many aspects of renal development, physiological functions, homeostasis, and abnormal growth (Bhatt, Mi, & Dong, 2011; Chandrasekaran et al., 2012; Grange, Brossa, & Bussolati, 2019; Hayes, Peruzzi, & Lawler, 2014; Ma & Qu, 2013; Wei, Mi, & Dong, 2013). The canonical pathway of their biosynthesis includes several "miRNA machinery genes" (Fawzy, Abu AlSel, & Toraih, 2020) and occurs in sequential steps (Fig. 7.1).

Moreover, another class of regulating ncRNAs is the long noncoding RNAs (lncRNAs), a highly heterogenous family of RNAs, which can be classified into various subtypes, including the sense/antisense transcripts intergenic lncRNAs, and intronic RNAs (Boon, Jaé, Holdt, & Dimmeler, 2016) (Fig. 7.2). Consistent with their heterogeneity, lncRNAs have diverse regulatory roles and functions in gene expression as signaling, decoys, guiding, scaffolding, and RNA enhancers, among other molecular functions (Fang & Fullwood, 2016) (Fig. 7.3). LncRNAs are demonstrated to regulate carcinogenesis and metastasis of RCC and could serve as biological markers and therapeutic targets (Li et al., 2017; Liu et al., 2018).

Compared to the previously mentioned classes of linear RNAs, the circular RNAs (circRNAs), which have no 3′ or 5′ terminals, have also drawn considerable research interest. They have been identified to resist degradation by exonucleases that support their emerging roles as potential biomarkers due to their exceptional stability (Patop, Wüst, & Kadener, 2019) (Fig. 7.4). This class of ncRNAs acts through several pleiotropic mechanisms in tumorigenesis, including cellular proliferation, migration, invasion, and apoptosis (Dong et al., 2017; Matsui & Corey, 2017), and may have the advantage to act as molecular therapeutic targets (Li, Yang, & Chen, 2018).

Besides miRNAs, lncRNAs, and circular intronic RNAs (ciRNAs), the authors note that there exist other families of well-known housekeeping ncRNAs and newly emerged regulatory ones, such as small nuclear RNA, small nucleolar RNA, small interfering RNA (siRNA), Piwi-interacting RNA, enhancer RNA, and YRNA, among others, [reviewed in Zhang, Wu, Chen, and Chen (2019)]. Due to space limitations, this chapter will focus on and summarize the putative roles of miRNAs, lncRNAs, and ciRNAs in RCC as diagnostic/prognostic biomarkers and therapeutic targets for their potential clinical translation and align with the rapid nucleic acid therapeutics evolution.

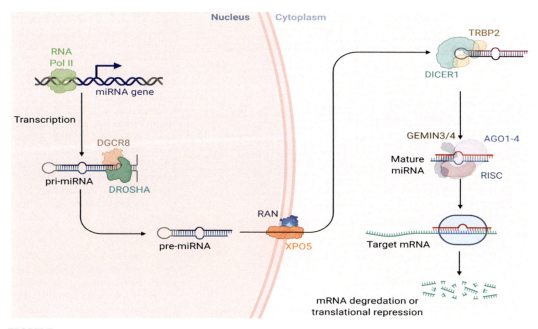

FIGURE 7.1 miRNA biogenesis and mechanism of action. Canonical miRNA biogenesis begins with the generation of the pri-miRNA transcript. The microprocessor complex, which comprises DGCR8, cleaves the pri-miRNA to produce the pre-miRNA. The pre-miRNA is exported to the cytoplasm in an Exportin5-/RanGTP-dependent manner and processed to produce the mature miRNA duplex. Finally, either the 5p or 3p strands of the mature miRNA duplex are loaded into the AGO family of proteins to form an miRISC. miRISC binds to target mRNAs to induce translational inhibition. Decapped mRNA may then undergo 5′ − 3′ degradation. *AGO*, Argonaute; *DGCR8*, Drosha and DiGeorge syndrome critical region 8; *miRISC*, miRNA-induced silencing complex; *miRNA*, microRNA; *mRNA*, messenger RNA; *pre-miRNA*, precursor-miRNA. *Source: Created by Biorender.com.*

7.2 Datasets and informatics for analyzing noncoding RNAs in renal cancer patients

Given the diverse functions and the large data volume of ncRNAs, their analysis has been verified to be challenging (He, Liu, & Sun, 2018). In this section the commonly used datasets and computational tools for identifying and analyzing ncRNAs in renal cancer patients will be summarized. This can help investigators quickly find the appropriate tools for their genomic data analyses and identify the gap areas that will require further analysis. It is worth noting that due to the rapidly expanding field of ncRNAs, this summary is by no means to cover all the datasets and informatics related to ncRNAs but rather to demonstrate the common and user-friendly datasets for analyzing ncRNAs.

7.2.1 Datasets and informatics for analyzing microRNAs in renal cancers

Table 7.1 lists several databases for miRNA, including miRNAs-related information, gene targets prediction, gene interactions, enrichment analysis, miRNA/targets gene variation assignments, and miRNA−disease interactions. According to prediction methods, publicly available human miRNA−target prediction resources are summarized in Fig. 7.5.

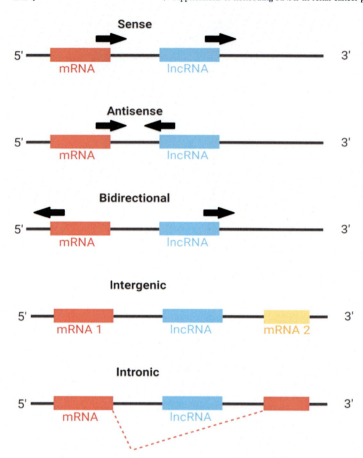

FIGURE 7.2 LncRNAs classification. LncRNAs can be classified into (1) sense or (2) antisense categories when overlapping one or more exons of another transcript on the same strand or opposite strand, respectively; they may also be (3) bidirectional, where a neighboring coding transcript on the opposite strand is initiated in close genomic proximity to its expression; (4) intergenic, derived within the genomic interval between two genes; or (5) intronic, that is, derived completely from an intron of a second transcript. *LncRNAs*, Long noncoding RNAs. *Source: Created by Biorender.com; Derived from Boon, R. A., Jaé, N., Holdt, L., & Dimmeler, S. (2016). Long noncoding RNAs: From clinical genetics to therapeutic targets? Journal of the American College of Cardiology, 67(10), 1214–1226. https://doi.org/10.1016/j.jacc.2015.12.051.*

7.2.2 Datasets and informatics for analyzing long noncoding RNAs in renal cancers

As depicted in Table 7.2, numerous lncRNA-predictive tools and databases are available for experimental scientists. These relevant databases compile and integrate different types of lncRNA-related information, for example, expression, functionality, subcellular localization, transcriptional regulation, and interactions.

7.2.3 Datasets and informatics for analyzing circular RNAs in renal cancers

Several circRNA-related databases have been established to store circRNA-related data and regulatory networks. CircAtlas v2.0, CIRCpedia v2, and TCSD all report circRNA expression levels across various human tissues and cell lines. In addition, CircRiC focuses on circRNAs in cancer cell lines and MiOncoCirc v2.0 on circRNAs in clinical human cancer samples. CircAtlas v2.0, circBase, and circRNADisease also provide circRNA–miRNA interactions. CircInteractome, CircRiC, CSCD, TSCD, CircFunBase, and Circ2Disease

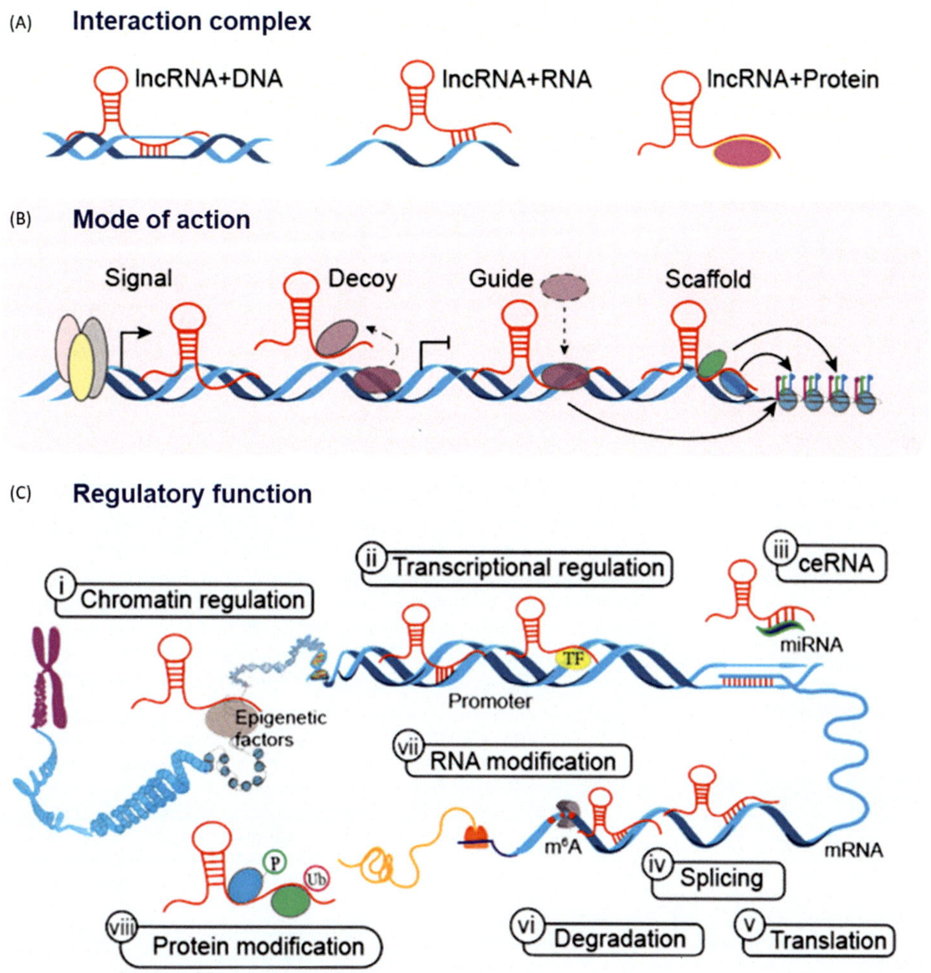

FIGURE 7.3 Mode of action and functions of lncRNAs in cancer. (A) lncRNAs can bind DNA, RNA, and protein molecules to regulate gene expression at multiple levels via base pairing or secondary structure formation. (B) LncRNAs have four primary roles as signals, decoys, guides, and scaffolds. Signal: lncRNAs may serve as molecular signaling mediators on signaling pathways to modulate a certain set of gene expressions. Decoy: lncRNAs may function as molecular decoys that take proteins or RNAs away from specific locations. Guide: lncRNAs can play a role as molecular guides, locating certain ribonucleoprotein complexes to a specific target site on the chromatin. Scaffold: lncRNAs can support the assembly of protein complexes that link factors together in the proper complex formation to generate brand new functions. (C) Mechanism of action for lncRNAs in the nucleus (i and ii) and cytoplasm (iii–viii). (i) LncRNAs can recruit epigenetic factors to change patterns of chromatin organization, (ii) activate or repress the transcription of certain genes by interacting with DNA sequences or TFs, (iii) act as ceRNAs by base pairing with miRNA and diminishing its inhibitory effects and manipulate mRNA function by base pairing to (iv) regulate alternative splicing (e.g., MALAT 1), (v) affect mRNA translation (e.g., TTN-AS1 and AC132217.4), and (vi) mRNA degradation (e.g., CASC11). (vii–viii) lncRNAs can modify mRNA and proteins, playing regulatory roles in methylation, phosphorylation, and ubiquitination (Cheng, Wang, Wang, et al., 2019). *ceRNAs*, Competing endogenous RNA; *lncRNAs*, long noncoding RNAs; *miRNA*, microRNA; *mRNA*, messenger RNA. *Source: Created by Biorender.com.*

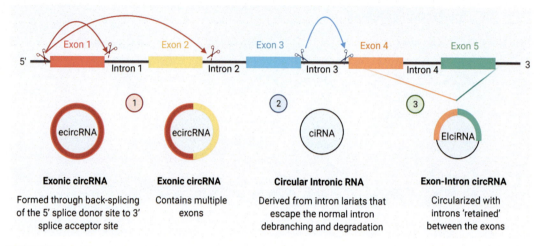

FIGURE 7.4 Formation of three types of circRNAs. (1) ecircRNA is formed through backsplicing of the 5'splice site (splice donor site) to a 3'splice site (splice acceptor site). The Intron 1 is removed and brings the 5' splice site of Exon 2 close to the 3' splice site of Exon 1 to form an ecircRNA that contains multiple exons. Exons can also skip splicing; Exon 1 can also link with Exon 3; (2) ciRNAs are derived from intron lariats (Intron 3) that escape the normal intron debranching and degradation. Reverse complementary sequences of lariat intron excised from pre-mRNA can pair to produce a close loop structure termed ciRNA; (3) EIciRNAs are circularized with introns "retained" between the exons. Intron 4 retains with Exons 4 and 5 to form an EIciRNA. *circRNAs*, Circular RNAs; *ciRNAs*, circular intronic RNAs; *ecircRNA*, Exonic circular RNA; EIciRNAs, exon–intron circRNAs. *Source: Created by Biorender.com.*

TABLE 7.1 MicroRNAs (miRNAs) and miRNAs gene targets databases.

Name	Description	Address
miRBase	Provide the users the information of target information, primary evidence, and known miRNA sequences for recorded miRNA	http://www.mirbase.org/
miRGator	Compile miRNA–mRNA–target interactions and inverse correlation between gene expression and the miRNA–mRNA relationships	http://mirgator.kobic.re.kr/
miRGen	Study the associations between miRNA function and miRNA genomic organization, and offering useful tools for studies in miRNA genomic organization, cotranscription, and targeting	http://www.microrna.gr/mirgen/
IntmiR	Focus on intronic miRNAs of human and mouse genome and provide the users with comprehensive miRNA-associated information, including target genes, pathways, and diseases	http://rgcb.res.in/intmir/
mirDIP	miRNA data integration portal. Find miRNAs that target a gene, or genes targeted by a miRNA, in *Homo sapiens*	http://ophid.utoronto.ca/mirDIP/

(Continued)

TABLE 7.1 (Continued)

Name	Description	Address
StarBase	starBase is designed for decoding miRNA−lncRNA, miRNA−mRNA, miRNA−circRNA, miRNA−pseudogene, miRNA−sncRNA, protein−lncRNA, protein−sncRNA, protein−mRNA, and protein−pseudogene interactions and ceRNA networks from 108 CLIP-Seq (HITS-CLIP, PAR-CLIP, iCLIP, CLASH) datasets	http:/starbase.sysu.edu.cn/
StarScan	StarScan is developed for scanning small RNA (miRNA, piRNA, and siRNA)-mediated RNA cleavage events in lncRNA, circRNA, mRNA, and pseudogenes from degradome sequencing data	http://mirlab.sysu.edu.cn/starscan/
Cupid	Cupid is a method for simultaneous prediction of miRNA−target interactions and their mediated ceRNA interactions	cupidtool.sourceforge.net
TargetScan	Predicts biological targets of miRNAs by searching for the presence of sites that match the seed region of each miRNA	http://www.targetscan.org/vert_72/
TarBase	Indexed more than half a million experimentally validated miRNA−gene interactions covering 356 distinct cell types from 24 species	http://www.microrna.gr/tarbase
Diana-microT	DIANA-microT 3.0 is an algorithm based on several parameters calculated individually for each miRNA, and it combines conserved and nonconserved miRNA recognition elements into a final prediction score	DIANA LAB—DNA Intelligent Analysis—microT v3.0 Web Server (archive.org)
PicTar	PicTar is combinatorial miRNA−target predictions	PicTar (archive.org)
PITA	PITA incorporates the role of target-site accessibility, as determined by base−pairing interactions within the mRNA, in miRNA−target recognition	https://genie.weizmann.ac.il/pubs/mir07/mir07_data.html
RepTar	A database of inverse miRNA−target predictions based on the RepTar algorithm that is independent of evolutionary conservation considerations and is not limited to seed pairing sites	http://reptar.ekmd.huji.ac.il/
RNA22	Visualize the predictions within a cDNA map and also find transcripts where multiple miR's of interest target. The second website link (custom) first finds putative miRNA-binding sites in the sequence of interest then identifies the targeted miRNA	http://cm.jefferson.edu/rna22/Interactive/
MBSTAR	Multiple instance approach for finding out true or functional miRNA-binding sites	https://www.isical.ac.in/~bioinfo_miu/MBStar30.htm
miRTarBase	The experimentally validated miRNA−target interaction database. As a database, miRTarBase has accumulated more than 360,000 MTIs, which are collected by manually surveying pertinent literature after NLP of the text systematically to filter research articles related to functional studies of miRNAs	http://miRTarBase.mbc.nctu.edu.tw/

(Continued)

TABLE 7.1 (Continued)

Name	Description	Address
miReg	Curate the regulatory relationships among miRNA-associated elements, including validated upstream regulators, downstream targets, associated biological process, experimental condition, and disease state	http://iioab-mireg.webs.com/
miRecords	Provide the users with predicted miRNA targets produced by 11 previously proposed miRNA–target prediction algorithms	http://miRecords.umn.edu/miRecordshttp://c1.accurascience.com/
miRwalk	Aggregates and compares results from other miRNA-to-mRNA databases. Stores the largest amount of predicted and experimentally verified miRNA–target interactions	zmf.umm.uni-heidelberg.de/apps/zmf/mirwalk2/
miRNAMap	Collect experimentally verified interactions between miRNAs and miRNA target and provide three established computational tools for identifying miRNA targets in 30-UTR of genes	http://mirnamap.mbc.nctu.edu.tw/
microRNA.org	Provide miRNA expression profiles derived from tissues and cell lines	http://www.microrna.org
miRPathDB	Provide researchers easy access to the information about target pathways regulated by miRNAs	https://mpd.bioinf.uni-sb.de/
PEMDAM	Construct three bipartite association networks: EF–miRNA network, EF–disease network, and miRNA–disease network	http://lmmd.ecust.edu.cn/database/pemdam/
SomamiR 2.0	A better platform for functional analysis of somatic mutations altering miRNA–ceRNA interactions	http://compbio.uthsc.edu/SomamiR
MiREnvironment	Users can also perform bioinformatics analysis to predict cancer treatment as well as EF–disease associations through this web resource	http://210.73.221.6/miren
MiRCancer	Provide 878 miRNA–cancer associations, which includes 236 miRNAs and 79 human cancers	http://mircancer.ecu.edu/
MiR2Disease	Provide a data resource of miRNA deregulation in various human diseases	http://www.mir2disease.org/
HMDD	Detailed and comprehensive annotations for miRNA–disease associations are presented in each entity	http://www.cuilab.cn/hmdd
MiREC	Focus on the miRNA–disease associations that are specific for endometrial cancer	http://www.mirecdb.org
DbDEMC	Record the miRNA expression information in 14 kinds of cancers	http://www.picb.ac.cn/dbDEMC/
OncomiRDB	Store information on experimentally validated oncogenic and tumor-suppressive miRNAs	http://bioinfo.au.tsinghua.edu.cn/oncomirdb/

ceRNA, Competing endogenous RNA; *circRNA*, circular RNA; *lncRNA*, long noncoding RNA; *mRNA*, messenger RNA; *MTIs*, miRNA–target interactions; *piRNA*, Piwi-interacting RNA; *siRNA*, small interfering RNA; *EF*, environmental factor; *NLP*, natural language processing; *sncRNA*, small noncoding RNA; *UTR*, untranslated region

7.2 Datasets and informatics for analyzing noncoding RNAs in renal cancer patients

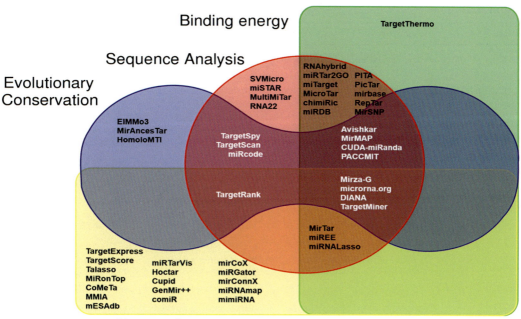

FIGURE 7.5 Publicly available human miRNA–target prediction resources according to methods of prediction. These included assessment of evolutionary conservation of the putative binding region; target sequence analysis—referring to methods considering explicit properties of the putative binding region, such as miRNA sequence complementarity, G–C content, and accessibility to RISC complex; calculation of binding energy between miRNA and its putative target sequence; and those which use of miRNA/mRNA expression profiles. *RISC*, RNA-induced silencing complex; *miRNA*, microRNA; *mRNA*, messenger RNA

TABLE 7.2 Web-based databases for long noncoding RNAs.

Name	Description	Address	Refs.
General lncRNA databases			
LNCipedia	General lncRNA database, expertly curated	https://lncipedia.org/	Zhang, Yao, et al. (2020)
LNCBook	General lncRNA database, some community curation	http://bigd.big.ac.cn/lncbook/index	Derrien et al. (2012)
LncRNAWiki	A wiki-based, publicly editable, and open-content platform for community curation of human lncRNAs	http://lncrna.big.ac.cn/index.php/Main_Page	Ma et al. (2019)
Expression databases			
GTEx	A comprehensive public resource to study tissue-specific gene expression and regulation	https://gtexportal.org/home/	The GTEx Consortium Atlas of Genetic Regulatory Effects Across Human Tissues (2020)

(Continued)

TABLE 7.2 (Continued)

Name	Description	Address	Refs.
NRED	A database of long noncoding RNA expression	http://jsm-research.imb.uq.edu.au/NRED	Dinger et al. (2009)
LncExpDB	An expression database of human long ncRNAs	https://bigd.big.ac.cn/lncexpdb	Li et al. (2021)
MiTranscriptome	A catalog of human long poly-adenylated RNA transcripts derived from computational analysis of high-throughput RNA-Seq data from over 6500 samples, spanning diverse cancer and tissue types	http://mitranscriptome.org/	Iyer et al. (2015)
TANRIC	Database of noncoding RNAs in cancer	http://www.tanric.org	Li et al. (2015)
CANTATAdb	Database of plant lncRNAs	http://cantata.amu.edu.pl/	Szcześniak, Bryzghalov, Ciomborowska-Basheer, and Makałowska (2019)
NRED	Gene expression information on lncRNAs	http://nred.matticklab.com/cgi-bin/ncrnadb.pl	Dinger et al. (2009)
lncRNAtor	lncRNA coding potential and phylogenetic conservation	http://lncrnator.ewha.ac.kr/index.htm	Park, Yu, Choi, Kim, and Lee (2014)
FARNA	Knowledgebase of inferred functions of noncoding RNA transcripts	http://cbrc.kaust.edu.sa/farna	Alam et al. (2017)
deepBase	Identification, expression, evolution, and function of lncRNAs, small RNAs, and circular RNAs from deep-sequencing data	http://deepbase.sysu.edu.cn/	Yang, Li, Jiang, Zhou, and Qu (2013)
Protein-coding potential			
CPC	Protein-coding potential prediction	http://cpc.gao-lab.org/	Anderson et al. (2015)
CPPred	Protein-coding potential prediction	http://www.rnabinding.com/CPPred/	Kang et al. (2017)
cncRNAdb	Manually curated resource of experimentally supported RNAs with both protein-coding and noncoding function	http://www.rna-society.org/cncrnadb/	Huang et al. (2021)
TransCirc	Protein-coding potential of circRNAs	https://www.biosino.org/transcirc	Huang et al. (2021)
ncEP	Manually curated database for experimentally validated ncRNA-encoded proteins or peptides	http://www.jianglab.cn/ncEP/	Liu et al. (2020)
CNIT	Protein-coding potential prediction	http://cnit.noncode.org/CNIT/	Sun et al. (2013)

(Continued)

TABLE 7.2 (Continued)

Name	Description	Address	Refs.
Subcellular localization			
LncSLdb	Database of lncRNA subcellular localization	http://bioinformatics.xidian.edu.cn/lncSLdb/	Cesana et al. (2011)
LncATLAS	Database of lncRNA subcellular localization	https://lncatlas.crg.eu/	Mas-Ponte et al. (2017)
LncLocator	LncRNA subcellular localization prediction	http://www.csbio.sjtu.edu.cn/bioinf/lncLocator/	Cao, Pan, Yang, Huang, and Shen (2018)
Structural conformation and sequence			
RMDB	Database of RNA structures	https://rmdb.stanford.edu	Yesselman et al. (2018)
RNAfold	RNA structure prediction	http://rna.tbi.univie.ac.at/cgi-bin/RNAWebSuite/RNAfold.cgi	Gruber, Lorenz, Bernhart, Neuböck, and Hofacker (2008)
DMfold	RNA structure prediction with pseudoknots	https://github.com/linyuwangPHD/RNA-Secondary-Structure-Database	Wang et al. (2019)
lncEvo	Automated identification and conservation study of lncRNAs	https://gitlab.com/spirit678/lncrna_conservation_nf	Bryzghalov, Makałowska, and Szcześniak (2021)
LNCipedia 3.0	Structure, protein-coding potential, miR-binding sites	http://www.lncipedia.org	Volders et al. (2013)
slncky Evolution Browser	This site contains alignments and evolutionary metrics of conserved lncRNAs	https://slncky.github.io/	Chen et al. (2016)
Interactions			
RNAInter	Database of RNA interactions	http://www.rna-society.org/rnainter/	Yi et al. (2017)
DIANA-LncBase	miR targets of lncRNAs	http://diana.imis.athena-innovation.gr/DianaTools/index.php?r=lncBase/index	Paraskevopoulou et al. (2013)
DES-ncRNA	Knowledgebase for exploring information about human micro- and lncRNAs based on literature mining	http://www.cbrc.kaust.edu.sa/des_ncrna	Salhi et al. (2017)
lncRNAMap	Regulatory function exploration of lncRNAs via siRNAs and miRNA	http://lncrnamap.mbc.nctu.edu.tw/php/	Chan, Huang, and Chang (2014)

(Continued)

TABLE 7.2 (Continued)

Name	Description	Address	Refs.
LncTarD	A manually curated database of experimentally supported functional lncRNA–target regulations in human diseases	http://bio-bigdata.hrbmu.edu.cn/LncTarD	Zhao et al. (2020)
LncRNA2Target v2.0	A comprehensive database for target genes of lncRNAs in human and mouse	http://123.59.132.21/lncrna2target	Cheng, Wang, Tian, et al. (2019)
RISE	Database of RNA interactions	http://rise.life.tsinghua.edu.cn	Gong, Shao, et al. (2018)
IntaRNA 2.0	RNA–RNA interaction prediction	http://rna.informatik.uni-freiburg.de/IntaRNA/Input.jsp	Mann, Wright, and Backofen (2017)
LncRRISearch	LncRNA–RNA interaction prediction	http://rtools.cbrc.jp/LncRRIsearch/	Fukunaga and Hamada (2017)
LnChrom	Database of lncRNA–chromatin interactions	http://biocc.hrbmu.edu.cn/LnChrom/	Yu et al. (2018)
lncRNABase (starBase v2.0)	lncRNA functional genomic annotations and their coordinated regulatory networks	http://starbase.sysu.edu.cn/mirLncRNA.php	
lncPro	lncRNA–protein association predictions	http://www.bioinfo.org/NPInter/lncPro.htm	Lu et al. (2013)
Triplexator	RNA–DNA interaction prediction	http://bioinformatics.org.au/tools/triplexator/	Papatheodorou et al. (2020)
SFPEL-LPI	LncRNA–protein interaction prediction	http://bioinfotech.cn/SFPEL-LPI/	Zhang et al. (2018)
Function prediction and enrichment analysis			
SEEKR	K-mer similarity predictor	http://seekr.org/home	Kirk et al. (2018)
NONCODE 6.0	Detailed annotation and potential function	http://www.noncode.org	Zhao et al. (2021)
LncRBase	Genomic and molecular features of lncRNAs for the analysis of functional and behavioral complexities of lncRNAs	http://bicresources.jcbose.ac.in/zhumur/lncrbase/	Chakraborty, Deb, Maji, Saha, and Ghosh (2014)
LncSEA	Platform for lncRNA-related sets and enrichment analysis	http://bio.liclab.net/LncSEA/index.php	Chen et al. (2021)
Co-LncRNA	GO annotations and KEGG pathways of lncRNAs	http://www.bio-bigdata.com/Co-LncRNA/	Zhao et al. (2015)
lncRNA2function	RNA-seq-based functional annotation of lncRNAs	http://mlg.hit.edu.cn/lncrna2function//	Jiang et al. (2015)

(Continued)

TABLE 7.2 (Continued)

Name	Description	Address	Refs.
lncRNAdb 2.0	Detailed annotation and information	http://www.lncrnadb.org	Quek et al. (2015)
fRNAdb	Cellular function, miR precursor, repeats	http://valadkhanlab.org/database	Kin et al. (2007)
Linc2Go	Functional annotations for human lincRNA	http://www.bioinfo.tsinghua.edu.cn/~liuke/Linc2GO/	Liu, Yan, Li, and Sun (2013)
LnCompare	Gene set feature analysis for human lncRNAs	http://www.rnanut.net/lncompare/	Carlevaro-Fita et al. (2019)
ncFANs v2	Detailed functional annotation based on a coding–noncoding gene coexpression network	http://www.bioinfo.org/ncfans/	Liao et al. (2011)
Transcription regulation			
lncRNome	Transcriptional regulation, structure, disease, and genomic variations	http://genome.igib.res.in/lncRNome/	Bhartiya et al. (2013)
ChIPBase	Transcriptional regulation map of lncRNAs	http://deepbase.sysu.edu.cn/chipbase/	Yang et al. (2013)
TF2lncRNA	ChIP-seq data-based lncRNA-TF associations	http://mlg.hit.edu.cn/tf2lncrna/index.jsp	Jiang et al. (2014)
LongTarget	lncRNA DNA-binding motifs and binding sites	http://lncrna.smu.edu.cn	He, Zhang, Liu, and Zhu (2015)
BIGTranscriptome	High confidence of coding and noncoding transcriptomes assembled with hundreds of pseudo-stranded and stranded RNA-seq datasets		You, Yoon, and Nam (2017)
Tissue specificity			
C-It-Loci	Positional/chronological locations in tissue-specific lncRNA transcripts. A tool to explore and to compare the expression profiles of conserved loci among various tissues in three organisms	http://c-it-loci.uni-frankfurt.de	Weirick, John, Dimmeler, and Uchida (2015)
lncRNAKB	Exploring lncRNA biology in the context of tissue specificity and disease association	http://www.lncrnakb.org/	Seifuddin et al. (2020)
Genetic variants			
ncRNAVar	Manually curated database for identification of ncRNA variants associated with human diseases	http://www.liwzlab.cn/ncrnavar/	Zhang, Zhen, et al. (2020)

(Continued)

TABLE 7.2 (Continued)

Name	Description	Address	Refs.
LnCeVar	A comprehensive database of genomic variations that disturb ceRNA network regulation	http://www.bio-bigdata.net/LnCeVar/	Wang et al. (2020)
lncRNASNP v2	Analysis of SNPs in lncRNAs with their potential impacts on structure and function	http://bioinfo.life.hust.edu.cn/lncRNASNP/	Gong, Liu, Zhang, Miao, and Guo (2015)
lincSNP v3.0	Database of linking disease−associated SNPs to human large intergenic noncoding RNAs	http://bioinfo.hrbmu.edu.cn/LincSNP	Ning et al. (2014)
AnnoLncr webtool	One-stop portal for systematically annotating novel human lncRNAs	http://annolnc1.gao-lab.org/ http://annolnc.cbi.pku.edu.cn	Yang, Ke, Ding, and Gao (2021)
PancancerQTL	GWAS + eQTL	http://bioinfo.life.hust.edu.cn/PancanQTL/	Gong, Mei, et al. (2018)
Lnc2Catlas	GWAS	https://lnc2catlas.bioinfotech.org/	Ren et al. (2018)
RMVar	An updated database of functional variants involved in RNA modifications	http://rmvar.renlab.org	Luo et al. (2021)

Disease association

Name	Description	Address	Refs.
LncRNADisease	lncRNA associations from 166 diseases	http://www.cuilab.cn/lncrnadisease	Chen, Wang, et al. (2013)
Lnc2Cancer 3.0	lncRNA expression associations of from 86 cancers	http://www.bio-bigdata.net/lnc2cancer	Ning et al. (2016)
TANRIC	Functional and clinical relevance of lncRNAs in cancer	http://ibl.mdanderson.org/tanric/_design/basic/index.html	Li et al. (2015)
CLC	Database of lncRNAs causally implicated in cancer through in vivo, in vitro, and other evidence	https://www.gold-lab.org/clc	Carlevaro-Fita et al. (2020)
CNCDatabase	Cancer drivers at noncoding regions	https://cncdatabase.med.cornell.edu/	Liu, Martinez-Fundichely, Bollapragada, Spiewack, and Khurana (2021)
MONOCLdb	The MOuse NOnCode Lung database provides the annotations and expression profiles of mouse lncRNAs (lncRNAs) involved in influenza and SARS-CoV infections	https://www.monocldb.org	Josset et al. (2014)
MNDR v2.0	Comprehensive annotations for lncRNA−disease associations	http://www.rna-society.org/mndr/	Cui et al. (2018)

(Continued)

TABLE 7.2 (Continued)

Name	Description	Address	Refs.
EVLncRNAs	lncRNA–disease associations that are validated by low-throughput experiments	http://biophy.dzu.edu.cn/EVLncRNAs2	Zhou et al. (2021), Zhou et al. (2018)
NSDNA	ncRNA–NSD associations	http://www.bio-bigdata.net/nsdna/	Wang et al. (2017)
Nc2Eye	First high-quality manually curated ncRNAomics knowledge base associated with eye disease	http://nc2eye.bio-data.cn/	Wang et al. (2017)
HDncRNA	A comprehensive database of noncoding RNAs associated with heart diseases	http://hdncrna.cardiacdev.com	Wang et al. (2018)
ncRI	Manually curated database for experimentally validated ncRNAs in inflammation	http://www.jianglab.cn/ncRI/	Wang, Zhou, et al. (2020)
ncRPheno	A comprehensive database platform for identification and validation of disease-related ncRNAs	http://lilab2.sysu.edu.cn/ncrpheno	Zhang, Yao, et al. (2020)
Drug targets			
noncoRNA	Database of experimentally supported ncRNAs and drug targets in cancer	http://www.ncdtcdb.cn:8080/NoncoRNA/	Li et al. (2020)

ceRNAs, Competing endogenous RNA; *circRNAs*, circular RNAs; *CLC*, cancer LncRNA census; *lncRNA*, long noncoding RNA; *miRNA*, microRNA; *NSD*, nervous system disease; *siRNAs*, small interfering RNAs; *eQTL*, expression quantitative trait locus; *GWAS*, genome-wide association studies; *SNP*, single-nucleotide polymorphism.

predict circRNA–RBP interactions based on cross-linking immunoprecipitation sequencing (CLIP-seq) data from starBase v2.0. For projects involving a cross-species comparison, CIRCpedia, circFunbase, Circad, NPInter (v4), and CircAtlas (v2) databases contain circRNA information on several different species (Table 7.3).

In 2011 the first study reported that the concept of "competing endogenous RNAs (ceRNAs)," also refer to as "miRNA sponges," was hypothesized (Salmena, Poliseno, Tay, Kats, & Pandolfi, 2011). According to this hypothesis, coding/ncRNAs s compete for miRNAs as they share one or more "miRNA response elements," acting in this way as "ceRNAs" (Zhao et al., 2020). Several studies have underscored the significant roles that ceRNAs play in tumorigenesis and cancer progression, including RCC (Jiang & Ye, 2019; Wang et al., 2016; Zhao et al., 2020). For instance, one study has identified the lncRNA "HOX transcript antisense RNA; HOTAIR" acting as miR-217-ceRNA with target gene "hypoxia-inducible factor 1 subunit alpha; HIF-1α" upregulation which promotes renal carcinogenesis, progression, and epithelial–mesenchymal transition (EMT) (Hong et al., 2017). Also, it has been found that the "HOXA transcript at the distal tip (HOTTIP)/miR-615/insulin-like growth factor-2 (IGF-2)" network promotes RCC progression (Wang et al., 2018). Recently, Zhao et al. (2020) constructed and analyzed a metastatic RCC-related ceRNA network that included 11 lncRNAs, 2 miRNAs, and 20 mRNAs, from which 7/11 and 3/20 were identified to be associated with the overall survival (OS) of cancer patients.

TABLE 7.3 Web-based resources and databases for circRNAs.

Name	Description	Address	Refs.
Circ2Traits	CircRNAs potentially associated with disease and traits and SNP	http://gyanxet-beta.com/circdb	Ghosal, Das, Sen, Basak, and Chakrabarti (2013)
CircInteractome	Interacting miRNAs and RNA-binding proteins of circRNAs	https://circinteractome.nia.nih.gov	Dudekula et al. (2016)
CircNet	CircRNA–miRNA–mRNA interaction networks	http://syslab5.nchu.edu.tw/CircNet	Liu et al. (2016)
circRNADb	CircRNAs with protein-coding potential	http://reprod.njmu.edu.cn/circrnadb	Chen et al. (2016)
CSCD	Cancer-specific circRNAs	http://gb.whu.edu.cn/CSCD	Xia et al. (2018)
TSCD	Tissue-specific circRNAs	http://gb.whu.edu.cn/TSCD	Xia et al. (2017)
circBase	A unified circRNA dataset from previous publications	http://circbase.org	Memczak et al. (2013)
CIRCpedia v2	CircRNA annotations from over 180 RNA-seq datasets across six different species	http://www.picb.ac.cn/rnomics/circpedia	Dong, Ma, Li, and Yang (2018)
CircR2Disease	Experimentally supported associations between circRNAs and diseases	http://bioinfo.snnu.edu.cn/CircR2Disease	Yao et al. (2018)
Circ2Disease	Manually curated database of experimentally validated circRNAs in human disease	http://bioinformatics.zju.edu.cn/Circ2Disease/index.html	Yao et al. (2018)
circRNADisease	A curated database of experimentally supported circRNA–disease associations	http://cgga.org.cn:9091/circRNADisease/	Zhao et al. (2018)
CircFunBase	Functional circRNAs	http://bis.zju.edu.cn/CircFunBase	Meng, Hu, Zhang, Chen, and Chen (2019)
circRNABase	circRNABase is designed for decoding miRNA–circRNA interaction networks from thousands of circRNAs and 108 CLIP-Seq (HITS-CLIP, PAR-CLIP, iCLIP, CLASH)	circRNABase	Li, Liu, Zhou, Qu, and Yang (2014)
Gokool, Anwar, and Voineagu (2020)	The Landscape of Circular RNA Expression in the Human Brain	http://www.voineagulab.unsw.edu.au/circ_rna	Gokool et al. (2020)
circBank	Comprehensive database for circRNA with standard nomenclature	http://www.circbank.cn	Liu, Wang, Shen, Yang, and Ding (2019)
CircPro	An integrated tool for the identification of circRNAs with protein-coding potential	http://bis.zju.edu.cn/CircPro	Meng, Chen, Zhang, and Chen (2017)
Cerina	Systematic circRNA functional annotation based on integrative analysis of ceRNA interactions	https://www.bswhealth.med/research/Pages/biostat-software.aspx	Cardenas, Balaji, and Gu (2020)

(Continued)

TABLE 7.3 (Continued)

Name	Description	Address	Refs.
RefCirc	A reference database for circRNAs and annotated disease-associated circRNAs from independent research groups, validated by experiments	http://www.ncvar.org/RefCirc/index.php	
CircAtlas	Coexpression, and regulatory networks for circRNA annotation	http://circatlas.biols.ac.cn/	Wu, Ji, and Zhao (2020)
deepBase	Identification, expression, evolution, and function of small RNAs, LncRNAs, and circRNAs from deep-sequencing data	http://biocenter.sysu.edu.cn/deepBase/	Zheng et al. (2016)
starBase	Decoding miRNA–ceRNA, miRNA–ncRNA, and protein–RNA interaction networks from large-scale CLIP-Seq data	http://starbase.sysu.edu.cn/	Li et al. (2014)
CircR2Cancer	Manually curated database of associations between circRNAs and cancers	http://www.biobdlab.cn:8000	Lan et al. (2020)
Circad	A comprehensive manually curated resource of circRNA associated with diseases	http://clingen.igib.res.in/circad/	Rophina, Sharma, Poojary, and Scaria (2020)
exorbase	Database of circRNA, lncRNA, and mRNA in human blood exosomes	http://www.exorbase.org	Li et al. (2018)
MiOncoCirc	A comprehensive catalog of cancer-based circRNA species	https://mioncocirc.github.io/	Vo, Cieslik, Zhang, Shukla, Xiao, Zhang, et al. (2019)
isoCirc	Catalogs full-length circRNA isoforms in human transcriptomes	https://github.com/Xinglab/isoCirc	Xin et al. (2021)
CircAST	Full-length assembly and quantification of alternatively spliced isoforms in circRNAs	https://github.com/xiaofengsong/CircAST	Wu et al. (2019)
CircRiC	Detection of lineage-specific circRNAs in 935 cancer cell lines; reports drug response, biogenesis, interactions between circRNAs and mRNA (including miRNAs), proteins, or mutations	https://hanlab.uth.edu/cRic	Ruan et al. (2019)
TRCirc	A database for providing information on transcriptional regulation of circRNAs	http://www.licpathway.net/TRCirc/view/index	Tang et al. (2019)
LncACTdb 2.0	A database of endogenous RNAs, including circRNAs	http://www.bio-bigdata.net/LncACTdb/	Wang et al. (2019)
ncrpheno	A database that integrates and annotates ncRNA–disease association data	http://lilab2.sysu.edu.cn/ncrpheno	Zhang, Yao, et al. (2020)
NPInter v4	An integrated database of ncRNA interaction, including circRNA interaction	http://bigdata.ibp.ac.cn/npinter4	Teng et al. (2020)

ceRNAs, Competing endogenous RNA; *circRNAs*, circular RNAs; *lncRNAs*, long noncoding RNAs; *miRNAs*, microRNAs; *mRNA*, messenger RNA; *ncRNA*, noncoding RNAs; *SNP*, single-nucleotide polymorphism

The important databases that are related to ceRNAs network construction, functional and interaction prediction, annotation, and disease associations, among other functions, are summarized in Table 7.4.

7.3 Expression of noncoding RNAs in renal cancer patients

7.3.1 Expression of microRNAs in renal cancer patients

As ncRNAs have been identified to participate in multiple cellular processes, including proliferation, differentiation, survival and apoptosis, migration, invasion, and EMT, dysregulation of ncRNAs expression can be implicated in RCC pathogenesis and progression (Barth et al., 2020). The deregulated expression of miRNAs and lncRNAs in RCC are summarized in Tables 7.5—7.8.

7.3.2 Expression of long noncoding RNAs in renal cancer patients

Aberrant expression of lncRNAs has been implicated in RCC initiation, progression, prognosis, metastasis, and recurrence (Liu et al., 2018; Martens-Uzunova et al., 2014; Shi, Sun, Liu, Yao, & Song, 2013; Zhou, Wang, & Zhang, 2014). For instance, the oncogenic lncRNA HOTAIR promotes RCC tumorigenesis through AXL signaling by acting as a ceRNA sequestrating the tumor suppressor miR-217, with subsequent HIF-1 overexpression and AXL upregulation (Hong et al., 2017). The lncRNA-MRCCAT1 (metastatic RCC-associated transcript 1) upregulation promotes ccRCC metastasis via NPR3 inhibition and p38-MAPK signaling activation (Li et al., 2017). Further, overexpression of the lncRNA-UCA1 (urothelial carcinoma associated 1) (Li, Wang, Chen, et al., 2016), lncRNA-ATB (Xiong, Liu, Jiang, Zeng, & Tang, 2016), lncRNA-H19 (Wang et al., 2015), and lncRNA-FTX (He et al., 2017) are also involved in RCC tumorigenesis and proposed to be important biomarkers for RCC.

On the other hand, lncRNAs can also act as tumor suppressor roles. For example, lncRNA-SARCC (suppressing androgen receptor in RCC) is differentially modulated in a VHL (von Hippel—Lindau)-dependent way under hypoxia, suppressing the proliferation of VHL-mutant RCC cell yet promoting the growth of VHL-normal RCC cell (Zhai et al., 2016). Decreases of lncRNAs such as neuroblastoma-associated transcript-1 (Xue et al., 2015) and cancer susceptibility candidate 2 (Cao et al., 2016) are associated with poor prognosis in patients with RCC. To date, novel lncRNAs continue to be identified, such as lnc-ACACA-1, lnc-FOXG1-2, lnc-CPN2-1, lnc-BMP2-2, and lnc-TTC34-3, which were predicted by computational analyses to participate in RNA—protein interaction networks, including spliceosome and other complexes in RCC (Blondeau et al., 2015). Furthermore, the evolutionarily conserved intronic antisense lncRNAs are commonly expressed in RCC and possibly modulated by epigenetic modifiers (Fachel et al., 2013). Besides, a recent study shows that lncRNA can also be packaged into exosomes and function critically to promote RCC progress. Le et al. identified lncARSR (lncRNA activated in RCC with sunitinib resistance), which acts as ceRNA for miR-34 and miR-449 to promote c-MET and AXL expression. In addition, lncARSR can be packaged into exosomes and transmitted to sensitive

TABLE 7.4 Competing endogenous RNA databases.

Name	Description	Type	Address	Refs.
ceRNABase	ceRNABase is designed for decoding pan-cancer ceRNA networks involving lncRNAs and mRNAs by analyzing 5599 tumor and normal samples and 108 CLIP-Seq (HITS-CLIP, PAR-CLIP, iCLIP, CLASH) datasets	Database and server	http://starbase.sysu.edu.cn/	Li et al. (2014)
cefinder	Competing endogenous RNA database: predicted ceRNA candidates from the genome	Database	http://www.oncomir.umn.edu/cefinder/	Sarver and Subramanian (2012)
ceRNAFunction	ceRNAFunction is a web server to predict lncRNA and protein functions from pan-cancer ceRNA networks using 13 functional terms (including GO, KEGG, and BIOCARTA)	Webserver		Yang et al. (2011)
Cupid	Prediction of miRNA–target interactions and their mediated ceRNA interactions	Software (MATLAB)	http://cupidtool.sourceforge.net/	Chiu et al. (2015)
lnCeDB	Database of human long ncRNA acting as ceRNA		http://gyanxet-beta.com/lncedb	Das, Ghosal, Sen, and Chakrabarti (2014)
LnCeCell	Provide functional ceRNA network in a single cell and subcellular locations		http://www.bio-bigdata.net/LnCeCell/	Wang et al. (2021)
Hermes	Hermes predicts ceRNA interactions from expression profiles of candidate RNAs and their common miRNA regulators using conditional mutual information	Software (MATLAB)		Sumazin et al. (2011)
Linc2GO	A human LincRNA function annotation resource based on ceRNA webserver	Database	http://www.bioinfo.tsinghua.edu.cn/~liuke/Linc2GO/index.html	Liu et al. (2013)
starBase	Decoding miRNA–ceRNA, miRNA–ncRNA, and protein–RNA interaction networks from large-scale CLIP-Seq data	Database	http://starbase.sysu.edu.cn/	Li et al. (2014)
miRTissue ce	Extending miRTissue web service with the analysis of ceRNA–ceRNA interactions	Database	http://tblab.pa.icar.cnr.it/mirtissue.html	Fiannaca, Paglia, Rosa, Rizzo, and Urso (2020)
LncACTdb 2.0	A database of endogenous RNAs, including circRNAs	Database	http://www.bio-bigdata.net/LncACTdb/	Wang et al. (2019)
ncrpheno	A database that integrates and annotates ncRNA–disease association data	Database	http://lilab2.sysu.edu.cn/ncrpheno	Zhang, Yao, et al. (2020)
NPInter v4	An integrated database of ncRNA interaction, including circRNA interaction	Database	http://bigdata.ibp.ac.cn/npinter4	Teng et al. (2020)

ceRNAs, Competing endogenous RNA; *circRNAs*, circular RNAs; *lncRNAs*, long noncoding RNAs; *miRNAs*, microRNAs; *mRNAs*, messenger RNAs; *ncRNA*, noncoding RNA.

TABLE 7.5 Upregulated microRNAs in renal cell carcinoma.

MicroRNAs	Specimen	Function	Target genes	Pathways/mechanisms involved	Refs.
miR-7	RCC tissues	Oncomir		Migration, proliferation, and apoptosis	Yu et al. (2013)
miR-21	RCC tissues, metastatic RCC, cell lines	Tumor marker	PDCD4, TPM1, PTEN, FASL, TIMP3	Growth, apoptosis, cell cycle, invasion, and migration	Dey et al. (2012), Li et al. (2014), Lv et al. (2013)
miR-23b-3p	RCC tissues, cell lines		PTEN	Cell cycle, apoptosis, invasion, and migration	Zaman, Thamminana, et al. (2012)
miR-23b	Cell lines, RCC tissues	Oncomir	POX	HIF and apoptosis	Liu et al. (2010b)
miR-155	RCC tissues, cell lines	Oncomir	BACH1	Proliferation, migration, and apoptosis	Li et al. (2012)
miR-210	Cell line		E2F3	Cell cycle, migration, and invasion	Nakada et al. (2011)
miR-224/383	ccRCC tissues		DIO1	Tissue hypothyroidism	Boguslawska, Wojcicka, Piekielko-Witkowska, Master, and Nauman (2011)
miR-590-5p	ACHN, 786-O cells	Oncomir	PBRM1	Proliferation, invasion, and cell cycle	Xiao, Tang, Xiao, Fu, and Yu (2013)
miR-301a	RCC tissues		PTEN	Cell cycle G1/S transition	Li et al. (2020)
miR-429	RCC tissues		CRKL	TGF-β'SOS1/MEK/ERK/MMP2/MMP9 pathway, role in migration and invasion	Wang et al. (2020)
miR-92a-3p	RCC tissues		FBXW7	Silencing suppressed cell proliferation and reduced colony number	Zeng, Huang, Sun, and Luo (2020)
miR-1293	RCC tissues		HAO2	EMT, cell viability, invasion, and migration	Liu et al. (2020)
miR-210-3p	RCC tissues, urine		VEGFR	VHL/hypoxia	Petrozza et al. (2020)
miR-671-5p	RCC tissues		APC	Has a role in invasion and migration, Wnt signaling	Chi et al. (2020)
miR-592	RCC tissues, urine		SPRY2	Has a role in proliferation, migration, and invasion	Lv et al. (2019)
miR-22	RCC tissues, urine		PTEN	Has a role in the invasion	Gong, Zhao, Saar, Peehl, and Brooks (2019)
miR-720	RCC tissues		E-cadherin and E-catenin	Has a role in EMT and metastasis	Bhat et al. (2017a)
miR-210, miR-218, and miR-1233	RCC tissues, plasma			Upregulation had a higher risk of specific death by RCC	Dias et al. (2017)

(Continued)

TABLE 7.5 (Continued)

MicroRNAs	Specimen	Function	Target genes	Pathways/mechanisms involved	Refs.
miR-122	RCC tissues		miR-122/Dicer pathway	Induces EMT, migration, and invasion in RCC	Fan et al. (2018a)
miR-125b	RCC tissues			Forecasts recurrence and outcome of ccRCC after surgical resection	Fu et al. (2014)
miR-378 and miR-210	Serum			Serve as powerful noninvasive detection in RCC	Fedorko et al. (2015)
miR-224	RCC tissues, serum			Increased cell viability and invasion ability reduced apoptosis	Fujii et al. (2017)
miR-7	RCC tissues		MEG3, RASL11B	Induces progression of ccRCC	He et al. (2018)
miR-203a	RCC tissues		GSK-3β	Wnt/β-catenin pathway induces cell proliferation, migration, cell cycle and suppresses apoptosis of RCC cells	Hu et al. (2014)
miR-155	RCC tissues		FOXO3a	miR-155 increased the proliferation and inhibited apoptosis and cell cycle arrest	Ji et al. (2017)
miR-125b	RCC tissues			miR-125b induced cell mobility and inhibited apoptosis	Jin et al. (2017)
miR-122	RCC tissues		Occludin	MAPK pathway enhanced cell proliferation, migration, and invasion	Jingushi et al. (2017)
miR-221/222	RCC tissues		KDR	Angiogenesis pathways enhance tumor cell proliferation	Khella et al. (2015)
miR-223-3p	RCC tissues			Levels of miR-223-3p may be a biomarker for ccRCC, and it was correlated to cancer-specific survival	Kowalik et al. (2017)
miR-21, miR-155, and/or miR-142-5p	RCC tissues	—	—	Three-miRNA combination is a potential predictor of renal cancer in patients	Lokeshwar et al. (2018)
miR-193a-3p	RCC tissues		PTEN	PI3K signaling pathway induces cell proliferation, cell migration, and the cell cycle	Liu et al. (2017)
miR-193a-3p and miR-224	RCC tissues		ST3GalIV	PI3K/Akt pathway enhanced RCC cell proliferation and migration by directly suppressing ST3GalIV	Pan et al. (2018)
miR-99, miR–miR-200b, miR-106a, miR-106b	RCC tissues		mTOR, VHL	These miRNAs increased the aggressiveness of RCC	Oliveira et al. (2017)

(Continued)

TABLE 7.5 (Continued)

MicroRNAs	Specimen	Function	Target genes	Pathways/mechanisms involved	Refs.
miR-106b-5p	RCC tissues		β-catenin, LZTFL1, SFRP1, and DKK2	Induces tumor growth and metastasis through the induction of Wnt/β-catenin signaling	Lu et al. (2017)
hsa-miR-27b, hsa-miR-23b and hsa-miR-628-5p	RCC tissues		c-Met and Notch1	These miRNAs may be biomarkers of sunitinib response	Puente et al. (2017)
miR-9-1	RCC tissues			Induces ccRCC progression	Pronina et al. (2017)
miR-193a-3p, miR-362, and miR-572	Serum			Diagnostic biomarker for RCC	Wang et al. (2015)
miR-34a	RCC tissues		MET, E2F3, TP53INP2, and SOX2	Promotes RCC tumorigenesis and progression	Toraih et al. (2017)
miR-210	Metastatic RCC tissues			Induces aggressive behavior in ccRCC	Samaan et al. (2015)
miR-122	RCC tissues		Sprouty2	Induces cell proliferation by targeting Sprouty2	Wang et al. (2017)
miR-146a-5p, miR-128a-3p, and miR-17-5p	RCC tissues		CXCL8/IL8, UHRF1, MCM10, and CDKN3	Induce the evolution from primary RCC without metastases into metastatic form	Wotschofsky et al. (2016)
miR-106b-5p	RCC tissues		SETD2	The P53 pathway induces cell proliferation and inhibits apoptosis by reducing SETD2 expression	Xiang et al. (2015)
miR-144-3p	RCC tissues		ARID1A	mir-144-3p induces cell proliferation and metastasis in ccRCC by reducing ARID1A expression	Xiao et al. (2017)
miR-1233	RCC tissues			Diagnostic biomarkers	Yadav et al. (2017)
miR-29b	RCC tissues		KIF1B	Increases cell proliferation, and invasion suppresses apoptosis	Xu et al. (2015)
miR-210-3p	RCC tissues		TWIST1	Promotes cell proliferation, EMT, and tumorigenesis	Yoshino et al. (2017)
miR-210 and miR-1233	Serum			It might be useful as liquid biopsies for diagnosing RCC patients	Xu, Deng, and Zhang (2018)
miR-18a-5p	RCC tissues			Enhances cell proliferation and cell mobility reduces cell apoptosis	Zhou et al. (2018)
miR-489-3p and miR-630	RCC tissues		OCT2/c-Myc	Induced chemoresistance to oxaliplatin	Chen et al. (2019)

(Continued)

TABLE 7.5 (Continued)

MicroRNAs	Specimen	Function	Target genes	Pathways/mechanisms involved	Refs.
miR-21	RCC tissues		Cell cycle	Has a role in migration, invasion, proliferation, and resistance to apoptosis	Gaudelot et al. (2017)
miR-21	RCC tissues		TIMP3	Decreased miR-21 expression, decreased cell invasion and migration, and inhibited cell apoptosis	Chen, Gu, and Shen (2017)
miR-210			ISCU1/2	VHL/HIF-1α/centrosome amplification/migratory and invasive potential of ACHN metastatic RCC cells	Petrozza et al. (2020)
miR-217				HIF-1α/AXL HOTAIR/proliferation, migration, EMT process, and apoptosis	Hong et al. (2017)
miR-224			VHL, SMAD4, SMAD5, DIO1	VHL/HIF-1α/tissue	Fujii et al. (2017)

ccRCC, Clear cell RCC; *EMT*, epithelial–mesenchymal transition; *HIF*, hypoxia-inducible factor; *HOTAIR*, HOX transcript antisense RNA; *miRNA*, microRNA; *RCC*, renal cell carcinoma; *VHL*, von Hippel–Lindau; *ACHL*, renal adenocarcinoma cell line; *AXL*, cell surface receptor tyrosine kinase; *MAPK*, mitogen-activated protein kinase.

cells to disseminate sunitinib resistance (Qu et al., 2016). Although studies are accumulating over the past decade, our knowledge of the underlying mechanisms of lncRNAs' role in renal malignancies is still preliminary, which needs further research to be explored.

7.3.3 Expression of circular RNAs in renal cancer patients

The role of circRNAs remains unclear; however, initial efforts in understanding the genome-wide transcriptional profiles of circRNAs in RCC are well underway. By using microarray technology, Ma et al. (2020) found that 542 circRNAs were aberrantly expressed in ccRCC, wherein 324 circRNAs were significantly downregulated and 218 were upregulated in ccRCC tumors. Furthermore, using the Arraystar microarray to profile 7 matched ccRCC samples, Franz et al. detected a total of 13,261 circRNAs, of which 78 circRNAs were upregulated and 91 were downregulated more than twofold (Franz et al., 2019). Past the identification process, the next step is beginning to understand precisely how circRNAs act on a molecular level.

Several studies indicate that circRNAs normally function as tumor suppressors and are aberrantly promoting tumorigenesis in RCC. One such circRNA of this class is circ-ZNF609, which was highly expressed in multiple RCC cell lines compared with the normal renal epithelial cell line KiMA (Xiong et al., 2019). Mechanistically, circ-ZNF609 acts as an miR-138-5p molecular sponge, subsequently increasing levels of FOXP4, the known target of miR-138-5p. This study revealed a critical role of the circ-ZNF609/miR-138-5p/FOXP4 axis in RCC in vitro; however, its function in RCC progression in vivo awaits

TABLE 7.6 Downregulated microRNAs in renal cell carcinoma.

MicroRNAs	Specimen	Function	Target genes	Pathways/mechanisms involved	Refs.
miR-1/133a	RCC tissues, cell lines	TS	TAGLN2	Proliferation, invasion, apoptosis, and cell cycle	Kawakami et al. (2012)
miR-30c	RCC tissues, cell lines		Slug	Hypoxia, EMT	Huang et al. (2013)
miR-30d	RCC tissues	TS	Cyclin E2	Proliferation, colony formation, and cell cycle	Yu et al. (2014)
miR-34a	RCC tissues	TS	c-Met, c-MYC, and Notch1	Cell growth, cell cycle	Yamamura et al. (2012), Zhang et al. (2014)
miR-99a	RCC tissues		mTOR	Cells growth, clonality, migration, invasion, and cell cycle	Cui et al. (2012)
miR-133b	Cell lines	TS	MMP-9	Proliferation, migration, and invasion	Wu, Pan, et al. (2014)
miR-135a	RCC tissues	TS	c-MYC	Cell proliferation and cell cycle	Yamada et al. (2013)
miR-138	Cell lines	TS	VIM, HIF-1α, EZH2	Migration, invasion, and senescence	Liang et al. (2013), Song et al. (2011), Yamasaki et al. (2012)
miR-141	RCC tissues		CDC25B	Cell growth, metastasis, EphA2/p-FAK/p-AKT/MMPs signaling	Chen, Wang, et al. (2014), Yu, Zhang, Liu, Zhan, and Kong (2013)
miR-143/145 cluster	RCC tissues		HK2	Cell proliferation and invasion	Yoshino, Enokida, Itesako, Kojima, et al. (2013)
miR-145	RCC tissues	TS	ADAM17, ANGPT2, NEDD9	Proliferation, migration	Doberstein et al. (2013), Lu et al. (2014)
miR-182-5p	RCC tissues	TS	FLOT1	AKT/FOXO3a signaling	Xu et al. (2014)
miR-187	Tumor tissue, plasma	TS	B7-H3	Proliferation, tumor growth, and motility	Zhao et al. (2013)
miR-192/194/215	Metastatic tumors		ZEB2, MDM2, TYMS	Migration, invasion, and proliferation	Khella et al. (2013)
miR-199a-3p	RCC tissues, cell lines		c-Met	HGF/c-Met signaling	Huang et al. (2014)
miR-200c	RCC tissues		ZEB1	EMT, p-Akt, and Akt	Wang et al. (2013)
miR-205	RCC tissue, cell line		SFKs	Ras/Raf/ERK1/2 signaling, cell cycle, apoptosis, proliferation, colony formation, migration, and invasion	Majid et al. (2011)

(Continued)

TABLE 7.6 (Continued)

MicroRNAs	Specimen	Function	Target genes	Pathways/mechanisms involved	Refs.
miR-217	ccRCC tissues	TS		Proliferation, motility	Li et al. (2013)
miR-218	RCC tissues	TS	CAV2	Migration, invasion, and focal adhesion	Yamasaki, Seki, Yoshino, Itesako, Hidaka, et al. (2013)
miR-508-3p/509-3p	RCC tissues, plasma	TS		Proliferation, apoptosis, and migration	Zhai et al. (2012)
miR-509-5p	RCC tissues, plasma			Proliferation, migration, and apoptosis	Zhang, Pan, Yang, and Zheng (2013)
miR-584	Cell lines	TS	ROCK-1	Cell motility	Ueno et al. (2011)
miR-708	RCC tissues	TS	Survivin, ZEB2, and BMI1	Cell growth, clonality, invasion, migration, and apoptosis	Saini et al. (2011)
miR-1285	RCC tissues	TS	TGM2	Proliferation, invasion, and migration	Hidaka et al. (2012)
miR-1291	RCC tissues	TS	SLC2A1/GLUT1	Cell proliferation, migration, and invasion	Yamasaki, Seki, Yoshino, Itesako, Yamada, et al. (2013)
miR-1826	RCC tissues	TS	β-catenin, MEK1	Proliferation, invasion, migration, apoptosis, and cell cycle	Hirata et al. (2012)
hsa-miR-30c-5p	RCC tissues	–	–	Inhibits proliferation and tumor formation	Onyshchenko et al. (2020)
hsa-miR-138-1	RCC tissues	–	–	Associated with an unfavorable course of the disease	Onyshchenko et al. (2020)
miR-363	RCC tissues		S1PR1, ERK, including PDGF-A, PDGF-B, EMT	Inhibited the proliferation, migration, and invasive capacity of ccRCC cells	Xie et al. (2020)
miR-362-3p	RCC tissues, cell lines		SP1/AKT/FOXO3	Inhibited the proliferation of RCC cells	Zhu, Wang, Shen, Zheng, and Xu (2020)
miR-214	–		LIVIN	Reduces cell proliferation and tumorigenesis	Xu et al. (2020)
miR-133b	RCC tissues	–	ERK	Suppresses cell proliferation, migration, and invasion while inducing apoptosis	Xu, Ma, Liu, and Gao (2020)
miR-206	RCC tissues		CDK6	Effectively caused apoptosis and cell cycle arrest at G0/G1 phase	Guo, Jia, and Ge (2020)
miR-143	RCC tissues		ABL2	Decreases cell adhesion, migration, and EMT	Xu et al. (2020)
miR-124 and miR-203	RCC tissues		ZEB2	Inhibit cell proliferation and migration	Chen, Zhong, and Li (2020)

(Continued)

TABLE 7.6 (Continued)

MicroRNAs	Specimen	Function	Target genes	Pathways/mechanisms involved	Refs.
miR-101-5p and miR-101-3p	RCC tissues		DONSON	G2/M checkpoint, EMT, induced cell cycle arrest, and apoptosis	Yamada et al. (2020)
miR-765	Plasma		PLP2	Inhibited cell proliferation and metastasis	Xiao et al. (2020)
miR-212-5p	RCC tissues		TBX15	Acted as a tumor suppressor gene in ccRCC	Deng, Zheng, Li, and Ji (2019)
miR-200 family	RCC tissues, urine			Affects the carcinogenic potential of malignant cells	Gilyazova et al. (2020)
miR-135a-5p	RCC tissues	—	—	Identify renal carcinogenesis and metachronous metastasis in ccRCCs	Shiomi et al. (2019)
miR-141	RCC tissues		ZEB2	Decreased cell proliferation	Berkers et al. (2013)
miR-124-3p, -30a-5p and -200c-3p	RCC tissues		CAV1 and FLOT1	Decreased migration and invasion in ccRCC cell lines	Butz et al. (2015)
miR-148a	RCC tissues		AKT2	Cell proliferation, colony formation, migration, and invasion	Cao et al. (2017)
miR-766-3p	RCC tissues		SF2	Suppresses cell-cycle progression. SF2/P-AKT/P-ERK signaling pathway	Chen et al. (2017)
miR-30a-5p	RCC tissues		ZEB2	Inhibits cell growth, migration, and invasion	Chen et al. (2017)
miR-129-3p	RCC tissues		SOX4 and MMP-2/9	Inhibits migration and invasion in RCC	Chen, Ruan, et al. (2014)
miR-99a	RCC tissues		mTOR	Inhibits tumorigenicity, and tumor growth promotes G1-phase cell cycle arrest	Cui et al. (2012)
miR-203	RCC tissues		HOTAIR	Upregulation reduces cell proliferation, migration, and invasion and induces apoptosis and cell-cycle arrest. PTEN pathway	Dasgupta et al. (2018)
miR-145	RCC tissues		ADAM17	Suppresses proliferation and promotes cell apoptosis in RCC	Doberstein et al. (2013)
miR-22	RCC tissues		PTEN	Inhibits cell proliferation, migration, and invasion. Ras/mitogen-activated protein kinase pathway	Fan, Huang, Xiao, and Liang (2016)
miR-217	RCC tissues		HIF-1α/AXL signaling, HOTAIR	Reduces proliferation, migration, and EMT and increases apoptosis	Hong et al. (2017)
miR-122-5p and miR-206	Serum		—	Biomarkers for patients with ccRCC	Heinemann et al. (2018)
miR-199a-5p	RCC tissues		TGFBR1 and JunB	Reduce the invasion of ccRCC cells	He et al. (2015)

(Continued)

TABLE 7.6 (Continued)

MicroRNAs	Specimen	Function	Target genes	Pathways/mechanisms involved	Refs.
miR-10b	RCC tissues		–	Inhibits cell proliferation, invasive ability, and migration and induces cell cycle arrest	He, Zhao, Jiang, Zhong, and Xu (2015)
miR-30c	RCC tissues		Slug	Suppresses EMT	Huang et al. (2013)
miR-372	RCC tissues	TS	IGF 2BP 1	Inhibits tumor progression, cell proliferation, and cell invasion	Huang, Huang, Kong, and Li (2015)
miR-186	RCC tissues		SENP1	NF-κB signaling pathway, cell proliferation and invasion, and induces apoptosis	Jiao, Wu, Ji, Liu, and Liu (2018)
miR-126	RCC tissues		EGFL7, PIK3CD, VEGFA, and PIK3R2	Reduced cell proliferation and migration in RCC cells. HIF-1, VEGF, mTOR, and PI3K–Akt signaling pathways	Khella et al. (2015)
miR-10b	RCC tissues		PDGFB, ETS1, GRB2, PIK3CA, PIK3R3, CRK, BCL2 and MDM2	Prognostic significance in ccRCC and its overexpression is associated with PDF and OS through MAPK, Wnt, and p53 signaling pathways	Kulkarni et al. (2018)
miR-10a-5p, -miR-10b-5p	RCC tissues		–	Biomarkers for ccRCC and were correlated to cancer-specific survival	Kowalik et al. (2017)
miR-182-5p	RCC tissues		MALAT-1	Inhibits tumorigenicity and enhances apoptosis	Kulkarni et al. (2018)
miR-144-3p	RCC tissues		MAP3K8	MAP3K8 pathway suppresses EMT, viability, and metastasis	Liu, Chen, Xiao, Wang, and Pan (2016)
miR-138	RCC tissues		SOX4	MiR-138 inhibits EMT, tumor growth, cell proliferation, migration, and invasion	Liu et al. (2017)
miR-192 and miR-194	RCC tissues		–	Two-miRNA combination is a potential predictor of renal cancer in patients	Lokeshwar et al. (2018)
miR-124	RCC tissues		HOTAIR	miR-124 inhibits RCC cell proliferation and metastasis	Zhou et al. (2018)
miR-149-5p	RCC tissues		FOXM1	miR-149-5p suppresses cell migration and invasion through targeting FOXM1	Okato et al. (2017)
miR-194	RCC tissues		HIF1A, MDM2, PIK3R2, MAPK1, IGF1R, BCL2, ITGB1, and CRK	A biomarker for prognosis in ccRCC. HIF-hypoxia pathway, VEGF, mTOR, Wnt, TGF-beta, and MAPK signaling pathways	Nofech-Mozes et al. (2016)
miR-429	RCC tissues		E-cadherin	Inhibits cellular migration and cell motility	Machackova et al. (2016)
miR-199a	RCC tissues		ROCK1	MiR-199a inhibits cell proliferation, migration, and invasion	Qin et al. (2018)
miR-106a-5p	RCC tissues		PAK5	Inhibits RCC progression and metastasis via PAK5	Pan et al. (2017)
miR-129-2	RCC tissues		NKIRAS1, RARB, CHL1, and RHOA	Suppresses ccRCC progression	Pronina et al. (2017)

(Continued)

TABLE 7.6 (Continued)

MicroRNAs	Specimen	Function	Target genes	Pathways/mechanisms involved	Refs.
miR-28-5p and miR-378	Serum		–	Diagnostic biomarker for RCC	Wang et al. (2015)
miR-30a-5p	RCC tissues		GRP78 signaling pathway	Suppresses cell growth and induces apoptosis in RCC	Wang et al. (2017)
miR-28-5p	RCC tissues		RAP1B, p38, and Erk1/2 pathways	Suppress tumorigenesis, cell proliferation, cell migration, and invasion	Wang, Wu, et al. (2016)
miR-30e-3p	RCC tissues		Snail1	Reduces cell invasion and migration	Wang et al. (2017)
miR-492	RCC tissues		–	Induces apoptosis and suppresses cell proliferation and invasion	Wu et al. (2015)
miR-137	RCC tissues		RLIP76	Inhibits cell growth and metastasis, induces apoptosis	Wang et al. (2018)
miR-144	RCC tissues		MTOR/PI3K/AKT signaling pathway	Inhibits cell proliferation and cell viability and promotes cell cycle arrest	Xiang, Cui, and Ke (2016)
miR-34a, miR-200c and miR-141	Serum		–	Diagnostic biomarkers	Long, Du, Liu, Li, and He (2015)
miR-203	RCC tissues		FGF2	Inhibits cell proliferation, migration, and invasion of RCC via the inhibiting of FGF2	Xu et al. (2015)
miR-101	RCC tissues		TIGAR	Induces glycolysis and cell proliferation	Xu, Liu, and Bao (2017)
miR-137	RCC tissues		PI3K, p-AKT	Decreases cell proliferation, migration, and invasion and induces cell apoptosis. PI3 K/AKT signaling pathway	Zhang and Li (2016)
miR-451	RCC tissues		PSMB8	Promotes cell apoptosis and suppresses cell proliferation and growth of RCC. Inflammation pathway	Zhu, Huang, and Su (2016)
miR-497	RCC tissues	–	–	Reduces cell proliferation, migration, and invasion of RCC	Zhao, Zhao, Xu, Hou, and Du (2015)
miR-375	RCC tissues		YWHAZ	Inhibits cell proliferation, migration, and invasion	Zhang et al. (2018)
miR-451	–		ATF-2	Enhanced drug resistance and cell apoptosis and reduced cell viability	Sun, Lou, Zhong, and Wan (2017)
miR-381	RCC tissues		–	Enhances cell apoptosis and inhibits cell proliferation and chemoresistance	Chan et al. (2019)
miR-124	–		FZD5, P-gp	Promotes cell apoptosis and inhibits chemoresistance. Wnt signaling pathway	Long et al. (2015)

B7-H3, B7 homolog 3; *ccRCC*, clear cell RCC; *EMT*, epithelial–mesenchymal transition; *HIF*, hypoxia-inducible factor; *HK2*, Hexokinase-2; *HOTAIR*, HOX transcript antisense RNA; *OS*, overall survival; *RCC*, renal cell carcinoma; *TS*, tumor suppressor.

TABLE 7.7 Upregulated long noncoding RNAs (lncRNAs) in renal cell carcinoma.

LncRNAs	Specimen	Function	Target genes	Pathways/mechanisms involved	Refs.
SPRY4-IT1	RCC tissues, cell lines	Oncogene		Proliferation, migration, and invasion	Zhang, Yang, Yan, Che, and Zheng (2014)
HOTAIR	Cell lines	Oncogene	H3K27me, EZH2, miR-141, and Ago2	Proliferation, invasion, and cell cycle	Chiyomaru et al. (2014)
UCA1	RCC tissues, cell lines	Oncogene		Proliferation, migration, and apoptosis	Li, Wang, Li, et al. (2016)
lncRNA-ATB	RCC tissues, cell lines	Oncogene	EMT	Proliferation, apoptosis, migration, and invasion	Xiong et al. (2016)
RCCRT1	RCC tissues	Biomarker		Migration and invasion	Song et al. (2014)
MALAT1	RCC tissues, cell lines	Oncogene	Ezh2, miR-205	Proliferation, migration, and invasion	Hirata et al. (2015), Zhang, Yang, Chen, Che, and Zheng (2015)
Linc00152	RCC tissues, cell lines	Oncogene, biomarker		Proliferation, invasion, apoptosis, and cell cycle	Wu et al. (2016)
TUG1	RCC tissues, cell lines	Oncogene		Migration, invasion, proliferation, and apoptosis	Zhang et al. (2016)
H19	RCC tissues, cell lines	Biomarker		Proliferation, invasion, and migration	Wang et al. (2015)
HEIRCC	RCC tissues, cell lines	Oncogene	EMT	Proliferation, apoptosis, migration, and invasion	Xiong et al. (2017)
CRNDE	RCC tissues, cell lines	Oncogene	Wnt/β-catenin signaling	Proliferation, growth, and cell cycle	Shao et al. (2016)
uc009yby.1	RCC tissues	Oncogene		Proliferation	Ren et al. (2016)
FTX	RCC tissues, cell lines	Oncogene		Proliferation, cell cycle, migration, and invasion	He et al. (2017)
PVT1	RCC tissues	Oncogene	MYC	Promoter hypomethylation	Posa, Carvalho, Tavares, and Grosso (2016)

EMT, Epithelial−mesenchymal transition; *HOTAIR*, HOX transcript antisense RNA; *RCC*, renal cell carcinoma.

TABLE 7.8 Downregulated long noncoding RNAs (lncRNAs) in renal cell carcinoma.

LncRNAs	Specimen	Function	Target genes	Pathways/mechanisms involved	Refs.
CADM1-AS1	RCC tissues	TS	CADM1	Cell proliferation, apoptosis, and migration	Yao et al. (2014)
TRIM52-AS1	RCC tissues	TS		Proliferation, cell migration, and apoptosis	Liu, Yan, Xia, Zhang, and Xiu (2016)
GAS-5	RCC tissues	TS		Proliferation, apoptosis, cell cycle, migration, and invasion	Qiao, Gao, Huo, and Yang (2013)
MEG3	RCC tissues, cell lines	TS	Bcl-2, rocaspase-9, leaved caspase-9, cytochrome c	Apoptosis, mitochondrial pathway	Wang et al. (2015)
BX357664	RCC tissues, cell lines	TS	EMT, MMP2, MMP9, TGF-β1/p38/HSP27	Proliferation, migration, invasion, and cell cycle	Liu et al. (2016)
TCL6	RCC tissues	TS		Proliferation and apoptosis	Su et al. (2017)
CASC2	RCC tissues, cell lines	TS	miR-21	Proliferation and migration	Cao et al. (2016)
NBAT-1	RCC tissues, cell lines	Prognostic biomarker		Proliferation, migration, and invasion	Xue et al. (2015)

CASC2, Cancer susceptibility candidate 2; *EMT*, epithelial–mesenchymal transition; *RCC*, renal cell carcinoma; *TS*, tumor suppressor.

further investigation. Although no in vivo studies have been conducted yet, the in vitro role of the circ-ZNF609/miR-138-5p/FOXP4 axis appears critical in RCC. Further, circ-ZNF609 can itself be translated into a functional protein in a process that may be regulated by m6A modification (Di Timoteo et al., 2020). The versatility of this circRNA family may also be contributing to RCC (Fig. 7.6) and altogether merits further investigation.

7.4 Cell signaling pathways modulated by noncoding RNAs in renal cancer patients

RCC tumorigenesis is a complex process that involves multiple genetic/epigenetic mutations and molecular pathways dysregulation (Di Cristofano et al., 2007; Ricketts et al., 2018; Shang, Liu, Ito, Kamoto, & Ogawa, 2007; Xu, Krause, Samoylenko, & Vainio, 2016). Next are several cell signaling pathways and molecules that have been identified to be modulated by ncRNAs in RCC patients.

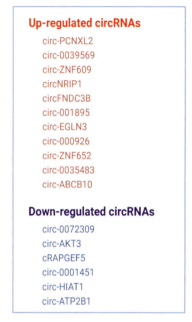
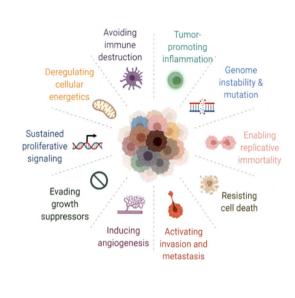

FIGURE 7.6 Deregulated circRNAs family is involved in hallmarks of cancer in RCC. *RCC*, Renal cell carcinoma *Source: Created by Biorender.com; Chen et al. (2020); Chen, Shao, Li, Liu, and Cao (2019); Chen, Yu, Shao, and Guo (2020); Dong et al. (2020); Han et al. (2018); Huang, Zhang, Jia, Liu, and Xu (2019); Jin et al. (2019); Lin and Cai (2020); Wang et al. (2018); Wang, Sun, Tao, Fei, and Chang (2017); Xiong, Zhang, and Song (2019); Xue et al. (2019); Yan, Liu, Cao, Zhang, and Shao (2019); Zhang and Guo (2020); Zhang et al. (2019); Zhou et al. (2018).*

7.4.1 Cell signaling pathways modulated by microRNAs in renal cancer patients

The tumor suppressor "von Hippel−Lindau (VHL)" gene is one of the most frequently mutated genes in RCC and plays a central role in the VHL/HIFs pathway (Zhai et al., 2016). It has been found that the main cause of most cases of familial ccRCC and 67% of sporadic ccRCC remains VHL gene inactivation (Banumathy & Cairns, 2010). This inactivation leads to higher tumorigenic target protein levels such as VEGF, PDGF, TGFα, and Glut 1b by upregulating HIF-1α and HIF-2α (Gläsker, Vergauwen, Koch, Kutikov, & Vortmeyer, 2020; Kim, Kim, Kim, & Zang, 2017). MiR-210 and many other miRNAs have shown this relationship by increasing the expression of HIF-1α and HIF-2α, directly by inactivating the VHL gene or indirectly (McCormick et al., 2013; Neal, Michael, Rawlings, Van der Hoek, & Gleadle, 2010). MiR-210 allows for anaerobic respiration of tumor cells by targeting the iron−sulfur cluster protein (ISCUI 1,2) and increases migration and invasion of tumor cells (Redova et al., 2013). Other miRNAs also have a role in this pathway. Like miR-210, miR-92a, and miR-23b expression are also positively correlated to migration and invasion capacity of tumor cells (Liu et al., 2010a; Valera, Walter, Linehan, & Merino, 2011). On the contrary, miR-138 and miR-200 expression may induce apoptosis and reduce the migratory properties of tumor cells (Liu et al., 2010; Song et al., 2011).

Several miRNAs play a role in the EMT. Tumor cells with EMT usually have a higher potential for invasion and metastasis. EMT is driven by the loss of adhesion mainly acquired by decreasing E-cadherin or increasing vimentin expression. MiR-200 gene family may have a regulatory function by inhibiting the ZEB1, which suppresses E-cadherin (Wang et al., 2013; Yoshino, Enokida, Itesako, Tatarano, et al., 2013). MiR-141 is also a tumor suppressor gene, and it functions by regulating the EphA2/p-FAK/p-Akt/MMP2/9 signaling cascade (Chen et al., 2014). Honokiol, *Magnolia* spp., bark may exhibit its antimetastatic activity by increasing the miR-141 expression (Li et al., 2014). Other miRNAs such as MiR-215, MiR-708, and miR-318 all have antitumor activities as well, and lower expressions of them lead to high migration, invasion, and vimentin levels (Huang et al., 2013; Liang et al., 2013; Saini et al., 2011; White et al., 2011; Yamasaki et al., 2012).

Many miRNAs have oncogenic roles in cell proliferation, invasion, apoptosis, and angiogenesis in RCC. MiR-21 is upregulated in RCC and induces invasion and antiapoptotic activity (Liu et al., 2010). In a study a knockout of this gene resulted in higher apoptosis by activating the caspase pathway. This gene also may regulate Akt kinase/TORC1, PTEN, KISS1, and cyclin D1 (Bera et al., 2013; Dey et al., 2012; Zhang, Guo, Shang, Song, & Wu, 2012). MiR-155 also inhibits apoptosis while increasing cell invasion and migration through SOCC and BACHI 1 (Li et al., 2012). Another miRNA, miR-23b-3p, achieves its oncogenic qualities by suppressing PTEN and upregulating PI3-kinase, total Akt, and IL-32 (Zaman, Thamminana, et al., 2012). MiR-17-92 cluster act on MYC oncogene and PTEN to induce cell proliferation (33) (Chow et al., 2010). Other miRNAs that have prooncogenic properties in RCC are miR-122 and miR-183 (Lian, Wang, Wang, Zhang, & Li, 2013; Qiu et al., 2014).

Moreover, numerous miRNAs play antioncogenic roles. MiR-205 is downregulated in RCC, and overexpression of this gene leads to cell apoptosis and decreased cell invasion and migration (Majid et al., 2011). This gene may also have a role in EMT by inhibiting the ZEB2 genes (Chen et al., 2014; Majid et al., 2011). MiR-1 and miR-133a may decrease the expression of transgelin-2 and induce cell death (Kawakami et al., 2012). Additionally, miR-146 blocks the expression of ADAM17, ANGPT2, and NEDD9 to lower cell proliferation (Doberstein et al., 2013; Lu et al., 2014; Wu, Li, et al., 2014). MiR-99a downregulates cell proliferation by regulating the mTOR pathway (Cui et al., 2012). MiR-30a regulates the DLL4 endothelial ligand, and lack of it results in abundant tumor angiogenesis (Huang et al., 2013). MiR-281 decreases cell mitosis and makes cells more susceptible to 5-FU chemotherapy (Chen, Duan, Yin, Tan, & Jiang, 2013).

7.4.2 Cell signaling pathways modulated by long noncoding RNAs in renal cancer patients

lncRNAs have been implicated in dysregulated intracellular signaling networks in RCC to maintain cell proliferation, enhance viability, impair differentiation, and promote motility (Liu et al., 2018). For example, the lncRNA HOTAIR knockdown significantly affects the G0/G1 cell cycle phase and decrease cell proliferation of RCC (Wu et al., 2014). The lncRNA-SARCC can interact with and decrease the stability of the androgen receptor (AR), altering the HIF-2a/c-myc signals under hypoxia (Zhai et al., 2016). LncRNA CRNDE upregulation promotes RCC cell proliferation by activating "Wnt/β-catenin" signaling (Shao et al., 2016). Also, BX357664 regulates RCC cell proliferation and the EMT via "TGF-β1/p38/HSP27" signaling inhibition (Liu et al., 2016). The lncRNAs TRIM52-AS1, GAS5, and DRAIC/PCAT29 have been found to play important roles in modulating growth-arrest and tumor-suppressor signaling pathways (Liu et al., 2016; Qiao et al., 2013; Sakurai, Reon, Anaya, & Dutta, 2015).

7.4.3 Cell signaling pathways modulated by circular RNAs in renal cancer patients

In addition to miRNA regulation, recent evidence has demonstrated that some circRNAs act through cancer-associated signaling pathways to promote RCC progression. For example, circNRIP1 plays an inhibitory role in RCC by regulating the "adenosine monophosphate-activated protein kinase" and "PI3K/AKT/mTOR pathways" (Dong et al., 2020) and "signaling pathways" are further involved in RCC development. The circZNF652 increased proliferation and EMT of RCC cells by regulating the "Ras/Raf/mitogen-activated kinase/extracellular signal-regulated kinase" and "Janus kinase/signal transducer and activator of transcription 3" signaling pathways (Zhang & Guo, 2020). Although few circRNAs have been reported in the signaling pathway of RCC thus far, further identification is anticipated as more light is shed on the pathogenesis of RCC. The urgency is to understand the rapidly evolving role another circRNAs play in RCC continues today (Chen et al., 2019, 2020; Han et al., 2018; Wang et al., 2017, 2018; Xue et al., 2019).

7.5 Diagnostic potential of noncoding RNAs in renal cancer patients

The screening and early diagnosis of RCC are essential for improved treatment efficacy and improving patient outcomes (Porta et al., 2016). Not only do RCC patients remain undetected for an extended period, but only a small proportion present with classical symptoms (i.e., hematuria, flank pain, a palpable abdominal mass), adding to the challenge of RCC diagnosis (Rossi, Klatte, Usher-Smith, & Stewart, 2018). Owing to the widespread use of computed tomography (CT) and magnetic resonance imaging (MRI), incidental detection of renal masses has increased; nonetheless, specific biomarkers are necessary, particularly in the setting of postoperative RCC recurrence and metastasis. The identification of the diagnostic utility of ncRNAs for RCC detection in the early stages of cancer has been reported in several studies and these may serve as potential biomarkers for the diagnosis of RCC as will be detailed in the following sections.

7.5.1 Diagnostic potential of microRNAs in renal cancer patients

There have been studies that have performed experiments to see if there are any miRNAs that could be used as a diagnostic marker. Unfortunately, there has not been much success in using miRNAs as a primary diagnostic marker due to their various limitations. However, it could be used in patients for follow-up. Even with the limitation, some miRNAs are associated with RCC (Table 7.9). One study concluded that

TABLE 7.9 Diagnostic role of microRNAs in renal cell carcinoma.

Type of samples	Comparison samples	Sample size (cancer/normal)	MicroRNA	AUC	Sensitivity (%)	Specificity (%)	Refs.
Tissue	Paired	96/96	miR-135a-5p	0.675	45.5	81.1	Shiomi et al. (2019)
Tissue	Paired	30/30	miR-720	0.905	80.0	100.0	Bhat et al. (2017b)
Tissue	Paired	53/53	miR-182-5p	0.954	90.0	97.0	Kulkarni et al. (2018)
Tissue	Paired	69/69	miR-129-3p	0.735	75.9	62.1	Chen, Ruan, et al. (2014)
Tissue	Cancer and biopsy tissues	156/46	Combined miR-10a-5p, miR-10b-5p, miR-223-3p	0.895	86.7	75.0	Kowalik et al. (2017)
Tissue	Paired	38/38	Combined miR-21 and miR-194	—	80.0	97.5	Lokeshwar et al. (2018)
Plasma	RCC and healthy	54/50	miR-210	0.7	60.9	73.1	Dias et al. (2017)
			miR-221	0.62	71.4	65.0	
			miR-1233	0.61	39.1	92.6	
Serum	RCC and healthy	195/100	Combined miR-378 and miR-210	0.85	80.0	78.0	Fedorko et al. (2015)
Serum	RCC and healthy	68/28	Combined miR-122-5p and miR-206	0.733	83.8	57.1	Heinemann et al. (2018)
Serum	RCC and healthy	82/80	miR-210	0.69	70.0	62.2	Xu et al. (2018)
			miR-1233	0.82	81.0	76.0	

AUC, Area under the curve, Paired: cancer and adjacent tissues; RCC, renal cell carcinoma.

miR-28, miR-185, miR-27, and let-7f-2 might be used in RCC patients as they may regulate PTEN expression (Gottardo et al., 2007). Moreover, miR-15a may be used to differentiate RCC from oncocytoma (von Brandenstein et al., 2012; Youssef et al., 2011). MiR-301a and miR-1293 expression has been positively correlated to the cancer stage (Li et al., 2020; Liu et al., 2020).

MiR-141 and miR-200c have been found to act as tumor suppressor genes. They were severely underexpressed in RCC and may function as E-cadherin regulators (Nakada et al., 2008). MiR-30c-5p and miR-138-1 also function as tumor suppressors as their expression was downregulated in RCC (Onyshchenko et al., 2020). Overexpression of MiR-200b resulted in lower migration and invasion (Li et al., 2019). Thus it may be used as a diagnostic marker. Lastly, in a study, miR-35 and miR-37 were shown to be useful diagnostic markers as well (Nakada et al., 2008).

7.5.2 Diagnostic potential of long noncoding RNAs in renal cancer patients

Many studies of determining aberrant lncRNAs expression for diagnostic purposes and identifying novel deregulated lncRNAs as biomarkers for RCC have been carried out. A recent study found that the expression level of CYP4A22-2/3 can discriminate ccRCCs from normal kidney tissues (Ellinger et al., 2015). Also, LncRNA LINC00887 has been evaluated as a potential diagnostic biomarker for RCC in a study (Xie et al., 2020). The expression of this lncRNA has been higher in RCC than in normal tissues. The study further analyzed this by performing an receiver operating characteristic (ROC) curve. The result of the area was 0.8001 with a 71.05% sensitivity and 89.87% specificity. They also showed that this lncRNA enhances cell proliferation in vivo.

7.5.3 Diagnostic potential of circular RNAs in renal cancer patients

The screening and early diagnosis of RCC are essential for improved treatment efficacy and reducing the mortality of patients, further prioritizing RCC research (Porta et al., 2016). Not only do RCC patients remain undetected for an extended period, but only a small proportion present with classical symptoms (i.e., hematuria, flank pain, a palpable abdominal mass), adding to the challenge of RCC diagnosis (Rossi et al., 2018). Owing to the now-widespread use of CT and MRI, incidental detection of renal masses has increased; nonetheless, specific biomarkers are necessary, particularly in the setting of postoperative RCC recurrence and metastasis. Serum and urine analyses serve as promising diagnostic tools in large part owing to their noninvasive and cost-effective nature.

circRNAs have great potential as a novel and attractive class of ncRNA biomarkers in liquid biopsy. Due to their unique structure, they are resistant to RNase R digestion and remain stable in the circulation. circRNAs are abundant in body fluids, such as saliva, blood, and urine (Fanale, Taverna, Russo, & Bazan, 2018; Jafari Ghods, 2018; Kölling et al., 2019; Vea, Llorente-Cortes, & de Gonzalo-Calvo, 2018). Additionally, they can be detected using relatively inexpensive quantitative reverse transcriptase–polymerase chain reaction (qRT-PCR) assays (Vo, Cieslik, Zhang, Shukla, Xiao, Wu, et al., 2019). Studies using clinical RCC samples have demonstrated the abnormal expression, higher disease specificity, and clinical relevance of specific circRNAs, further promoting them as ideal biomarker candidates. One such circRNA is hsa_circ_0001451, which is downregulated in ccRCC. Its diminished presence in RCC was linked to clinicopathological features and OS, indicating its potential as a diagnostic marker for ccRCC. hsa_circ_0001451 had an area under the receiver operating characteristic curve (AUC-ROC) of 0.704 for discriminating ccRCC from normal controls, with a sensitivity and specificity of 0.755, and 0.608, respectively (Wang et al., 2018). Franz et al. (2019) identified dysregulated circRNAs in ccRCC tissues by using a whole-genome microarray. Three circRNAs (circEGLN3, circNOX4, and circRHOBTB3) were identified and clinically validated by qRT-PCR. The AUC-ROC of circNOX4 and circRHOBTB3 in RCC tissues were 0.81 and 0.82, suggesting that both could be used as potential diagnostic biomarkers. Most importantly, circEGLN3 had AUC-ROC of 0.98, a remarkably reliable diagnostic value. Further, the combined detection of circEGLN3 and linEGLN3 increased the AUC-ROC to 0.99, with 95% sensitivity and 99% specificity. Collectively, the data indicate that the diagnostic value of a combination of circRNAs is higher than any single individual circRNA.

7.6 Prognostic potential of noncoding RNAs in renal cancer patients

Despite recent advances in understanding RCC, patient prognosis remains poor. Increasing studies showed that aberrant ncRNAs expression is associated with 5-year survival, OS, disease grade and stage, recurrence, and metastasis. In this sense, ncRNAs show promise as prognostic molecular biomarkers.

7.6.1 Prognostic potential of miRNAs in RCC

Several miRNAs' signatures have been identified for the potential prognostic utility in RCC, including association with favorable clinicopathological features. For example, elevated miR-210 expression is associated with lower tumor stage and grade (McCormick et al., 2013). The latter researchers found that in ccRCC tissue from three patient groups, miR-210 overexpression was associated with better OS. In accordance with this, Samaan et al. (2015) reported similar preliminary findings regarding miR-210 overexpression, although statistical significance was lost after adjusting for tumor characteristics. Inversely, another study found that increased miR-210 had a direct correlation with poor patient outcomes (Dang & Myers, 2015). Altogether, the role of miR-210 in RCC remains unclear with studies reporting contradictory findings.

Another miRNA, miR-21, has a more clearly outlined relationship with RCC. In a study of miR-21 expression and OS, only 50% of patients with high miR-21 expression survived at 5 years, while all patients with low miR-21 survived. Furthermore, high miR-21 expression is associated with advanced RCC stage (Zaman, Thamminana, et al., 2012). In another study involving 121 patients, the highest levels of miR-21 were in ccRCC and pRCC. In this investigation, high miR-21 expression was associated with shorter disease-free survival (DFS) and OS (Faragalla et al., 2012). Using bioinformatic approach, Vergho et al. (2014) concluded a significant correlation: miR-21 upregulation and miR-126 downregulation resulted in metastasis and cancer-specific survival (CSS). The ratio of miR-21 to 10b has further been shown to be an independent prognostic factor for ccRCC that has not metastasized to distant sites. This combined with the poor prognosis of patients with high miR21:10b levels means such patients require stricter surveillance. Another independent prognostic factor is miR-100, high expression of which in RCC tissues is associated with poor prognosis. Another miRNA, low expression of miR-155, is associated with poor prognosis in stage III and IV ccRCC (Shinmei et al., 2013). Yet another miRNA, miR-106b, is low in patients who develop metastasis. As such, the current literature suggests that miR-115 may be a predictive marker for early metastasis after nephrectomy (Slaby et al., 2010a). Ultimately, the established association between low expression of several miRNAs and poor prognosis testifies to the role miRNAs play as tumor suppressors (Li et al., 2013; Zhao et al., 2013, 2015).

Among 33 miRNAs analyzed in metastasis versus nonmetastatic ccRCC, the following miRNAs showed significant downregulation: miR-451, miR-221, miR-30a, miR-10b, and miR-29a (Fedorko et al., 2016). Further, a group of four miRNAs, which included let-7a, let-7c, miR-26a, miR-30c, were associated with either primary metastatic, late metastatic, or nonmetastatic disease processes (Fedorko et al., 2016). Additionally, low expression of several miRNAs (miR-26a, miR-10a, miR-143, miR-19b, and let-7c) showed a significant association with poor progression-free survival (PFS) and CSS (Heinzelmann et al., 2011). PFS and CSS were associated with miR-30c downregulation and miR-126 and miR-451 upregulation (Heinzelmann et al., 2014). In comparing metastatic and nonrecurrent patients, Osanto et al. (2012) identified 12 miRNAs with differential expression. In a further investigation of metastatic versus localized tumor, two miRNAs showed upregulation (miR-130b and miR-199b-5p) and two showed downregulation (miR-10b and miR-139-5p). Combined, these four miRNAs were used to build a metastatic tumor signature in RCC (Wu et al., 2012). The validation test showed that the signature could identify concurrent metastases (81%) more than subsequent/future metastasis of the primary tumors (69%) with a 76% sensitivity and 100% specificity. In another study comparing metastatic ccRCC and matched primary tumors from the same patients, the following miRNAs were differentially expressed: miR-10b, miR-126, miR-196a, miR-204, miR-215, miR-192, and miR-194 (Khella et al., 2012). Further investigation revealed that miR-215, miR-192, and miR-215 had both a convergent "the same molecule can be targeted by several miRNAs" and divergent "the same miRNA can control multiple targets" properties (Khella et al., 2013). This point in particular testifies to the complex function of miRNAs.

For the detection of early relapse after nephrectomy, miR-143, miR-26a, miR-145, miR-10b, miR-195, and miR-126 were confirmed as being downregulated in RCC patients who developed tumor relapse. In nonmetastatic patients, miR-127-3p, miR-145, and miR-126 significantly correlated to relapse-free survival (Slaby et al., 2012). In another study by Khella et al. (2015), higher miR-126 levels were associated with DFS. For patients with ccRCC, high miR-125b expression levels indicated poor survival and early recurrence after nephrectomy particularly in the setting of advanced-stage disease (Fu et al., 2014).

7.6.1.1 Association with renal cancer recurrence

Several miRNAs have been implicated in the recurrence of RCC. One study found higher levels of methylated miRNA-9-1 and miRNA-9-3 from the DNA of primary RCC recurrent patients when compared to those whose cancer did not reoccur (P-values = .012 and .009 for miR-9-1 and miR-9-3, respectively). Furthermore, increased levels of methylation of miRNA-9-1 and miRNA-9-3 were associated with decreased recurrence-free survival (RFS) time for about 30 months (P-values .034 for miR-9-1 and .007 for miR-9-3) (Hildebrandt et al., 2010). miR-9-3 methylation alone was associated with a higher risk of recurrence [hazard ratio (HR) 5.85, 95% confidence interval 1.30–26.35]. In a separate study from Nakata et al. (2015), miR-27a-3p levels and miR-193a-3p levels showed significant association with cancer progression. Lastly, in a study of in ccRCCs tissues, higher relative methylation of miRNA-124-3s was associated with worse RFS (HR = 9.37, P = .0005) (Gebauer et al., 2013).

7.6.1.2 Association with renal cancer metastasis

Several studies have identified the role miRNAs play in metastasis of RCC (Bhat et al., 2017a; Fan et al., 2018b; Ghafouri-Fard, Shirvani-Farsani, Branicki, & Taheri, 2020; Lu et al., 2017; Mytsyk et al., 2018; Samaan et al., 2015; Shiomi et al., 2019; Slaby et al., 2010a; Wotschofsky et al., 2016; Xu et al., 2015; Yang, Zhang, Chen, Yan, & Zheng, 2015). In one such study, miRNA-106b expression levels were lower in patients with metastatic RCC when compared to nonmetastatic patients. In postnephretomy renal cancer patients, miR-106b expression level was predictive for early metastasis (long-rank $P = .032$) (Slaby et al., 2010a). Investigation of another miRNA, miR-506, revealed its down-regulation in ccRCC. It is believed that miR-506 exerts an antioncogenic affect by acting upon *FLOT1*, a nearby target gene (Yang et al., 2015). Downregulation of miRNA combinations "miR-21 + 194 miR-21 + 142-5p + 194" was significantly associated with higher risk for metastasis (Lokeshwar et al., 2018).

7.6.1.3 Association with renal cancer patient's survival

Currently, the results of several studies suggest the potential of miRNAs as RCC survival predictors (Fedorko et al., 2016; Ghafouri-Fard et al., 2020; Mytsyk et al., 2018). For example, one study found that RCCs with higher stage and grade were associated with significantly higher miRNA-21 levels in tissue samples (Faragalla et al., 2012). miRNA-21-positive patients had a reliably shorter disease-free survival (Redova et al., 2012). In 46 cases, stages III and IV of RCCs, miR-486 overexpression was associated with TNM staging and poor cancer-specific mortality independent of other covariates (Goto et al., 2013). Low miR-129-3p level has been associated with short DFS and OS (Chen et al., 2014). Levels of miR-10a-5p, -10b-5p, and miR-223-3p were correlated to RCC-specific survival (Kowalik et al., 2017). Upregulation of miR-301a has been associated with advanced stage and poor survival of RCC patients (Li et al., 2020). Also, miR-21 and miR-210 upregulations, while miR-141, miR-200c, and miR-429 downregulations, have been related to poor CSS after tumor excision as evidenced by previous metaanalysis (Tang & Xu, 2015). The available data on miRNAs' impact on RCC oncological manifestations, which may play an important role in predicting disease outcome and can be directly associated with RCC prognosis, are summarized in Table 7.10.

7.6.2 Prognostic potential of long noncoding RNAs in RCC

Several studies have demonstrated that dysregulated lncRNas signatures are associated with various prognostic indices, including disease grade/stage, recurrence, cancer metastasis, disease OS, the 5-year survival, and/or the DFS (Table 7.11). For instance, high levels of LINC00887 in sera are associated with poor survival of RCC patients, and it also promotes RCC cell proliferation (Xie et al., 2020). HOTAIR upregulation showed a positive correlation with RCC cancer migration and metastasis (Pei et al., 2014). Tumor cells with nearby lymph node spread were found to have higher expression of HOTAIR than localized RCC. HOTAIR upregulates insulin growth factor-binding protein 2 and downregulates miR-217. Both in vitro and in vivo experiments have proved that the inhibition of HOTAIR expression leads to higher G0/G1 cell cycle arrest and lower tumor cell invasion and proliferation (Hong et al., 2017; Katayama et al., 2017; Wu et al., 2014; Xia et al., 2017). Overexpression of GIHCG in RCC tissues is positively correlated to advanced TNM stages, Fuhrman grades, and poor prognosis (He, Qin, Zhang, Yi, & Han, 2018). Serum GIHCG level is also significantly upregulated in RCC patients and correlated to advanced TNM stages (He et al., 2018). The lncRNA-ATB was found to be elevated in metastatic RCC patients, and higher expression of lncRNA-ATB correlated to

TABLE 7.10 Prognostic role of microRNAs in renal cancer patients.

MicroRNA	Expression	M	Stage	Grade	CSS	OS	Rec	Prog	Refs.
miR-100	H	(+)	(+)		(+)	(+)			Wang et al. (2013)
miR-101-5p/3p	L		(+)			(+)			Nakamura et al. (2020)
miR-106b	L	(+)							Slaby et al. (2010b)
miR-106b-5p	H					(+)			Puente et al. (2017)
miR-10a-5p/miR-10b-5p	H					(+)			Kowalik et al. (2017)
miR-10b	H					(+)			Khella et al. (2017)
miR-122	H					(+)			Fan et al. (2018a)
miR-122-5p/miR-206	H					(+)			Heinemann et al. (2018)
miR-1236	H					(−)			Wang, Tang, et al. (2016)
miR-124-3p	H		(+)			(−)	(+)		Butz et al. (2015)
miR-125b	H				(+)	(+)	(+)		Fu et al. (2014)
miR-126	H				(+)	(−)			Samaan et al. (2015)
miR-1293	H					(+)			Liu et al. (2020)
miR-129-3p	H					(+)			Chen, Ruan, et al. (2014)
miR-135a-5p	L		(+)						Shiomi et al. (2019)
miR-138	L					(+)			Liu et al. (2017)
miR-143	L					(+)			Xu et al. (2020)
miR-144-3	L					(+)			Liu et al. (2016)
miR-155	L						(+)	(+)	Shinmei et al. (2013)
miR-187	L		(+)	(+)		(+)			Zhao et al. (2013)
miR-18a-5p	H					(+)			Zhou et al. (2018)
miR-194	H					(−)			Nofech-Mozes et al. (2016)
miR-19a	H							(+)	Ma et al. (2016)
miR-200c-3p	H					(+)			Butz et al. (2015)
miR-203a	L					(+)	(+)	(+)	Hu et al. (2014)
miR-206	L					(+)			Guo et al. (2020)
miR-21	H		(+)	(+)		(+)		(+)	Zaman et al. (2012)
miR-21/10b ratio	H					(+)		(+)	Fritz, Lindgren, Ljungberg, Axelson, and Dahlbäck (2014)
miR-21/126 ratio	H				(+)				Vergho et al. (2014)

(Continued)

TABLE 7.10 (Continued)

MicroRNA	Expression	M	Stage	Grade	CSS	OS	Rec	Prog	Refs.
miR-21/miR-194	L	(+)							Lokeshwar et al. (2018)
miR-210	H		(−)	(−)		(−)			Vergho et al. (2014)
miR-210	H					(+)			Samaan et al. (2015)
miR-210/miR-1233	H				(+)				Dias et al. (2017)
miR-210-3p	L					(+)			Yoshino et al. (2017)
miR-217	L		(+)	(+)		(+)			Li et al. (2013)
miR-22	H					(+)			Gong et al. (2019)
miR-221	H					(+)			Khella et al. (2015)
miR-223-3p	L					(+)			Kowalik et al. (2017)
miR-29b	H		(+)			(+)			Xu et al. (2015)
miR-30a-5p/30e-3p	L					(+)			Chen et al. (2017), Wang et al. (2017)
miR-310a	H					(+)			Li et al. (2020)
miR-378	H		(+)			(+)			Fedorko et al. (2015)
miR-429	H	(+)				(−)		(+)	Machackova et al. (2016)
miR-451	H	(+)							Heinzelmann et al. (2011)
miR-451	L					(+)			Zhu et al. (2016)
miR-497	L					(+)		(+)	Zhao, Zhao et al. (2015)
miR-514	L	(+)					(+)		Wocschofsky (2013)
miR-592	H					(+)			Lv et al. (2019)
miR-628	H					(−)			Puente et al. (2017)
miR-671-5p	H					(+)			Chi et al. (2020)
miR-720	L					(−)			Bhat et al. (2017b)
miR-766-3p	H					(−)			Khella et al. (2017)
miR-99a	L					(+)			Cui et al. (2012)

(+): worse outcome, (−): good outcome. CSS, Cancer-specific survival; M, metastasis; OS, overall survival; Prog, prognosis; Rec, recurrence.

disease progression and a more invasive feature, as well as metastasis (Xiong et al., 2016). In addition, RCCRT1 has a significant relationship with clinicopathologic features of ccRCC patients, including the size of the tumor, the pathological staging, and tumor grade (Song et al., 2014). It is also associated with metastasis of lymph nodes and distal

TABLE 7.11 Prognostic role of long noncoding RNAs in renal cancer patients.

lncRNAs	Expression	Gene targets	Cellular functions	Prognostic information	Refs.
CADM1-AS1	Low	CADM1	Proliferation	Poor prognosis (OS)	Yao et al. (2014)
TCL6	Low			Poor prognosis (OS)	Su et al. (2017)
BX357664		TGF-beta, p38	Proliferation		Liu et al. (2016)
NONHSAT123350	Low			Poor prognosis (DFS, OS)	Liu et al. (2016)
lnc-ZNF180-2	Low			Poor prognosis (PFS, CSS, OS)	Ellinger et al. (2015)
CRNDE		Wnt/beta-catenin	Apoptosis		Shao et al. (2016)
SPRY4-IT1	High			Poor prognosis (OS)	Zhang et al. (2014)
RCCRT1	High			Poor prognosis	Song et al. (2014)
MALAT1	High	miR-200, HIF2A	Proliferation, migration	Poor prognosis (OS)	Zhang et al. (2015)
Linc00152	High			Poor prognosis (OS)	Wu et al. (2016)
lncARSR		miR-34, miR-449, YAP	Stem cell–like, chemoresistance		Qu et al. (2016)
DRAIC	High			Good prognosis	Sakurai et al. (2015)
PVT1	High	MYC	Proliferation	Poor prognosis (OS)	Posa et al. (2016)
MEG3		BCL2, procaspase 9	Apoptosis		Wang et al. (2015)
HOTAIR		miR-141, miR-217, JMJD3	Proliferation		Wu et al. (2014)
lncRNA-SARCC		AR, HIF2A, miR-143-3p, MYC	Proliferation, migration		Zhai et al. (2017)

CSS, Cancer-specific survival; *DFS*, disease-free survival; *lncRNAs*, long noncoding RNAs; *OS*, overall survival; *PFS*, progression-free survival; *SARCC*, suppressing androgen receptor in RCC.

metastasis in ccRCC (Song et al., 2014). Moreover, TCL6 expression was an independent predictor of ccRCC aggressiveness and was negatively correlated to the TNM stage (Su et al., 2017).

A multitude of studies has shown that aberrant lncRNA expressions are associated with the disease OS, the 5-year survival, the DFS, the disease grade and stage, recurrence, metastasis, etc. As shown in Table 7.11, decreased expressions of lncRNA

NONHSAT123350, CADM1-AS1, TCL6, and lnc-ZNF180-2 are correlated to a poorer prognosis of RCC patients (Ellinger et al., 2015; Liu et al., 2016; Su et al., 2017; Yao et al., 2014). On the other hand, overexpression of SPRY4-IT1, RCCRT1, MALAT1, Linc00152, and PVT1 shows a poor prognosis (Hirata et al., 2015; Posa et al., 2016; Song et al., 2014; Wu et al., 2016; Zhang et al., 2014, 2015).

7.6.3 Prognostic potential of circular RNAs in RCC

Accumulating evidence suggests that circRNA may be contributing to RCC prognosis via the dysregulation of several related cellular and molecular mechanisms (Table 7.12). Cadherin 2 (CDH2, also known as N-cadherin) is a marker of EMT and a major contributor to RCC aggressiveness that can be directly targeted by miR-411 (Alimperti &

TABLE 7.12 Prognostic role of some circular RNAs in renal cancer patients.

CircRNA	Apop	Prolif.	Inv.	Mig.	Auto	EMT	Target miRNA	miRNA—target genes/protein	Refs.
Upregulated circRNAs									
circ-PCNXL2		(+)	(+)				miR-153	ZEB2	Zhou et al. (2018)
circ-0039569		(+)	(+)	(+)			miR-34a-5p	CCL22	Jin et al. (2019)
circ-ZNF609		(+)	(+)				miR-138-5p	FOXP4	Xiong et al. (2019)
circNRIP1		(+)		(+)			miR-505	AMPK and PI3K/AKT/mTOR	Dong et al. (2020)
circ-001895	(−)	(+)	(+)	(+)			miR-296-5p	SOX12	Chen et al. (2020)
circ-EGLN3	(−)	(+)	(+)	(+)			miR-1299	IRF7	Lin and Cai (2020)
circ-000926		(+)	(+)	(+)		(+)	miR-411	CDH2	Zhang et al. (2019)
circ-ZNF652	(−)	(+)				(+)	miR-205	Ras/Raf/MEK/ERK and JAK1/STAT3	Zhang and Guo (2020)
circ-0035483		(+)			(+)		miR-335	CCNB1	Yan et al. (2019)
circ-ABCB10	(−)	(+)		(+)					Huang et al. (2019)
Downregulated circRNAs									
circ-0072309	(+)	(−)	(−)	(−)			miR-100	PI3K/AKT/mTOR	Chen et al. (2019)
circ-AKT3			(−)	(−)		(−)	miR-296-3p	E-cadherin	Xue et al. (2019)
cRAPGEF5		(−)	(−)	(−)			miR-27a-3p	TXNIP	Chen et al. (2020)
circ-0001451	(+)	(−)					—	—	Wang et al. (2018)
circ-HIAT1			(−)	(−)			miR-195-5p	CDC42	Wang et al. (2017)
circ-ATP2B1			(−)	(−)			miR-204-3p	FN1	Han et al. (2018)

Apop, Apoptosis; *Auto*, autophagy; *CDH2*, cadherin 2; *EMT*, epithelial—mesenchymal transition; *Inv.*, invasion; *Mig.*, migration; *miRNA*, microRNA; *Prolif*, proliferation.

Andreadis, 2015; Chen et al., 2019; Han et al., 2019). Another study reported that a novel circRNA (circ_000926) increased CDH2 expression by inhibiting miR-411 in RCC cells (Zhang et al., 2019). Moreover, silencing circ_000926-reduced cell proliferation, migration, and invasion in vitro and markedly suppressed the growth and metastasis of RCC tumors in vivo. These results reveal a novel circ_000926/miR-411/CDH2 regulatory axis in RCC growth and metastasis.

The circ-ABCB10 could be a prognostic marker as it has been shown to promote RCC cell growth and suppress apoptosis in vitro (Huang et al., 2019). Moreover, circ-ABCB10 was markedly increased in RCC tissues, and it was associated with pathologic grade and TNM stage. In the study of another potential prognostic marker, cRAPGEF5, 245 RCC cases were included (Chen et al., 2020). Among these cases, cRAPGEF5 downregulation was significantly associated with large tumor size, advanced TNM stage, and distant metastasis. Moreover, the level of cRAPGEF5 was correlated to aggressive tumor characteristics, poor OS, and RFS in patients, further implicating it as a prognostic biomarker for RCC. Further RNA immunoprecipitation and in vitro biochemical assays showed that cRAPGEF5 functioned as an oncogenic miR-27a-3p sponge further targeting the tumor suppressor gene TXNIP (Chen et al., 2020).

In a study of circPRRC2A, expression was increased in RCC tissues. Similar to the abovementioned circRNAs, circPRRC2A was positively correlated to advanced TNM stage and lymph node metastasis in RCC patients. Further multivariate analyses of circPRRC2A expression level identified it as an independent risk factor for OS alongside RCC tumor size, T stage, and Fuhrman grade (Li et al., 2020).

Another circRNA of importance, the AKT3-derived circAKT3, is significantly reduced in RCC tumors and cell lines (Xue et al., 2019). Overexpression of circAKT3 significantly inhibited ccRCC cell migration and invasion by sponging miR-296-3p, which induced E-cadherin expression. Several other regulatory cascades comprising circRNA, miRNA, and mRNA have been reported in RCC including "circPCNXL2/miR-153/ZEB2" (Zhou et al., 2018), "circ_0039569/miR-34a-5p/CCL22" (Jin et al., 2019), "hsa_circ_001895/miR-296-5p/SOX12" (Chen et al., 2020), and "circEGLN/miR-1299/IRF7" (Lin & Cai, 2020).

Some circRNAs play an integral role as miRNA sponges or reservoirs, eliciting a regulatory effect on RCC progression. This regulation can be accomplished by modulating steroid receptors such as AR or estrogen receptor, both of which play a critical role in the progression and metastasis of hormone-regulated cancers such as RCC. Wang et al. (2017) investigated the potential effects of functional interactions between AR and circRNAs in ccRCC. In this study, circHIAT1 increased "miR-195-5p/29a-3p/29c-3p" activity by acting as an miRNA sponge, inhibiting AR-dependent tumorigenesis of ccRCC cells. In another study the novel circATP2B1 was also shown to regulate RCC progression by acting as an miRNA sponge (Han et al., 2018). Data indicate that the Erβ can promote ccRCC invasion by regulating the "circATP2B1/miR-204-3p/FN1" signaling pathway. Collectively, a handful of studies testify to the complex role circRNAs play in steroid receptor modulation.

7.7 Therapeutic potential of noncoding RNAs in renal cancer patients

In this section, we discuss recent advances in the therapeutic potential of ncRNAs in RCC treatment. ncRNAs are considered newly emerged molecular targets for RCC treatment, and several therapeutic strategies, such as gene silencing, immunotherapy, and small molecule inhibition, have been designed and tested in different preclinical/clinical studies. In the future the application of ncRNAs-targeted agents alone or in combination with conventional therapies may open a new era for novel therapeutic strategies against RCC.

7.7.1 Therapeutic potential of microRNAs in renal cancer patients

Another possible application of miRNAs in the RCC patients rises from their importance in determining patients' response to chemotherapy. Therefore prior identification of miRNA profiles in the biopsy samples might facilitate selecting the most appropriate therapeutic regimen in a personalized manner. Moreover, targeted suppression of certain miRNAs is a possible modality to enhance patients' response to chemotherapy. miR-21 represents a promising candidate in this regard since it is overexpressed in RCC samples in independent studies, and its silencing has enhanced response to multiple anticancer drugs such as paclitaxel 5-fluorouracil, oxaliplatin, and dovitinib. However, miRNA-based therapies face several challenges, such as designing specific formulations to avoid off-target effects and the low efficacy of delivery methods (Baumann & Winkler, 2014).

7.7.2 Therapeutic potential of long noncoding RNAs in renal cancer patients

The current loss/gain of function work, genetic loss/overexpression of function models, or RNA interfering technology are good strategies to explore the therapeutic potential of lncRNAs in cancer (Gutschner & Diederichs, 2012). LncRNAs might be a promising therapeutic option to attenuate the growth/metastatic potential of RCC and/or improve the sensitivity to anticancer treatment via therapeutic silencing/overexpression of this family of ncRNAs (Li et al., 2017). For instance, the lncRNA HOTAIR knockdown can decrease RCC proliferation, invasion and suppress the tumorigenesis in vivo in the xenograft experiments (Wu et al., 2014). Suppression of MALAT1 not only inhibited RCC proliferation and promoted apoptosis but also inhibited migration and invasion of cancer cells (Hirata et al., 2015; Zhang et al., 2015). Indeed, H19 inhibition can suppress RCC proliferation, invasion, and migration (Wang et al., 2015). Using RNAi to knock down the lncRNA RCCRT1 suppressed RCC migration and invasion in the cell lines (Song et al., 2014). On the other hand, overexpression of the LncRNA CADM1-AS1 can decrease RCC growth and migration and increase apoptosis (Yao et al., 2014). The lncRNA-sorafenib resistance-associated lncRNA in RCC (SRLR) was found upregulated in "intrinsically sorafenib resistant RCCs," and its knockdown sensitized nonresponsive RCC cells to sorafenib treatment (Xu et al., 2017). Also, the activated in RCC with sunitinib resistance (ARSR) lncRNA promotes the sunitinib resistance of RCC and its targeting can restore the sensitivity of this type of RCC to sunitinib (Qu et al., 2016). Indeed, lncARSR knockdown attenuated the self-renewal, tumorigenicity, and metastasis of renal cancer stem cells and was associated with a poor prognosis of ccRCC (Qu et al., 2016).

7.7.3 Therapeutic potential of circular RNAs in renal cancer patients

In terms of RCC therapeutic targets, circRNAs continue to show promise (Jiang & Ye, 2019). In one study, circHIAT1 and circATP2B1 acted as metastatic inhibitors, preventing AR- or ERβ-induced ccRCC cell migration and invasion. Targeting these newly identified AR-circHIAT1-miR-195-5p/29a-3p/29c-3p/CDC42 and ERβ-circATP2B1-mediated miR-204-3p/FN1 signaling cascades offers a novel approach toward the treatment of metastatic ccRCC (Han et al., 2018; Wang et al., 2017). In addition to this, circRNAs may serve a role in resensitizing cancer cells that have developed resistance to current therapies. Gemcitabine, a tyrosine kinase inhibitor used to treat RCC, develops resistance in many patients over time (Yang et al., 2018). Recently, Yan et al. (2019) found that hsa-circ_0035483 was aberrantly expressed in RCC and enhanced gemcitabine (GEM) resistance via the hsa-miR-335/CCNB1 signaling pathway. The study results suggest that hsa_circ_0035483 may be a promising target for preventing GEM resistance in RCC therapy, although the role circRNAs play in RCC drug resistance remains in its nascent stage.

With an emphasis on the role of circRNAs as miRNA sponges, several RCC strategies have been proposed. Either inhibiting or restoring circRNA function to regulate miRNAs and, in turn, their downstream targets. Thus far, several ways of circRNAs function inhibition have emerged, such as the antisense oligonucleotide, siRNA, CRISPR/Cas9-mediated knockout, and circRNAs association interruption with miRNAs or RNA-binding proteins (RBPs) by saturating the conserved binding sites on circRNAs (Zhang & Xin, 2018). In contrast, when circRNAs function as sponges for RBPs or miRNAs, restoring circRNA expression may induce apoptosis or inhibit abnormal proliferation (Wang, Zhang, Wang, Fu, & Lin, 2020). Altogether, artificial circRNA sponges may be an effective strategy to restore circRNAs functions, providing a new approach for drug design based on miRNAs target therapy (Holdt, Kohlmaier, & Teupser, 2018; Kristensen, Hansen, Venø, & Kjems, 2018; Rossbach, 2019).

Given the cell/tissue specificity and stability of endogenous circRNAs, targeting this class of ncRNAs can be a promising strategy for gene-based therapy with potentially lower toxicity than synthetic molecules (e.g., synthetic drugs and siRNAs) (Wang et al., 2020). "Engineered circRNAs" could be an effective and promising technology for delivering therapeutic agents. For instance, high-quality protein translation was achieved using an artificial circRNA with a threefold extension of the protein half-life (Wesselhoeft, Kowalski, & Anderson, 2018).

7.8 Potential of noncoding RNAs in predicting chemoresistance and radioresistance in renal cancer patients

ncRNAs have been studied extensively for their role in RCC chemoresistance or chemosensitivity (Corrà, Agnoletto, Minotti, Baldassari, & Volinia, 2018; Zhang et al., 2020), and some ncRNAs play double-faced roles in enhancing the cancer sensitivity to various therapeutic modalities (May, Bylicky, Chopra, Coleman, & Aryankalayil, 2021; Zhang et al., 2020). However, it should be noted that the clinical correlation of ncRNAs with chemosensitivity does not necessarily guarantee the functional relevance of ncRNAs in drug resistance. Functional experiments assessing the effects of ncRNA modulation are essential (Wang et al., 2019).

7.8.1 Potential of microRNAs in predicting chemoresistance/radioresistance in renal cancer patients

miRNAs may play a vital role in sunitinib treatment in RCC. One study showed that underexpression of miR-141 led to poor sunitinib response in RCC. Moreover, overexpression of this gene overcame the therapy resistance (Berkers et al., 2013). Another very predictor of sunitinib efficacy in metastatic RCC is miR-942 (Prior et al., 2014). The latter investigators found that miR-942, among other upregulated miRNAs (i.e., miR-133a, miR-628-5p, and miR-48) in tumors from sunitinib-resistant patients, could accurately classify extreme phenotypes of marked sensitivity versus resistance to sunitinib. This finding was replicated in sunitinib resistant cell line "Caki-2" compared to the sensitive cell line (Prior et al., 2014). Exogenously expressing miR-30a in RCC "786-0 or A489 cells" inhibited *Beclin-1* expression and enhanced sorafenib-induced cytotoxicity. On the contrary, introducing antagomiR-30a inhibited "sorafenib-induced cytotoxicity" against RCC cells (Zheng et al., 2015).

Other miRNAs such as miR-192, miR-193a-3p, and miR-501-3p may play a role in sunitinib treatment (Gámez-Pozo et al., 2012). MiR-183 increases the sensitivity of tumor cells to the toxic effect of natural killer cells (Zhang et al., 2015). When it comes to 5-FU treatment, miR-381 targets *WWE1* and improves the response to the drug (Chen, Duan, et al., 2013). MiRp-451 may also affect multidrug resistance cell lines as it suppresses *ATF-2* (Sun et al., 2017). The downregulated miR-99a-3p in RCC, in addition, is implicated in sunitinib-resistant RCC (Osako et al., 2019). Similarly, the downregulated miR-101 in sunitinib-treated RCC tissues has been associated with *UHRF1* overexpression, contributing to drug resistance (Goto et al., 2016). Indeed, seven miRNAs are differentially expressed in sunitinib-resistant cells compared to control ones. Among these miRNAs, miR-18a has been correlated to HIF1A, miR-29b-1-5p with the PI3K/Akt and HIF1 pathways, miR-4430 with the PTEN and mTOR pathways; and miR-4521 with ZEB2 and the mTOR pathways, which could be potential mechanisms for drug resistance of RCC (Katayama et al., 2017). Taken collectively, the previous examples of deregulated miRNAs associated with sunitinib resistance could help in the development of therapeutic strategies to reduce such resistance and/or improve the prognosis of patients with RCC (Yamaguchi et al., 2017).

7.8.2 Potential of long noncoding RNAs in predicting chemoresistance/radioresistance in renal cancer patients

lncRNAs have been implicated in different mechanisms of chemoresistance, such as altering drug efflux, triggering apoptosis, interfering DNA damage repair, or inducing mutations of drug targets, among others (Chen, Wei, Wang, & Sun, 2017; Zhao et al., 2019). These biomolecules may improve the management of RCC patients receiving targeted therapy by acting as predictors and/or potential candidates for target therapy resistance (Li et al., 2017). The general mechanisms that underlie the role of lncRNAs in cancer therapy resistance have been summarized in recent reviews (Zhang et al., 2020; Zhao et al., 2019). The earliest identified lncRNA "ARSR" is implicated in the development of resistance to sunitinib in RCC tissues/cell lines, and targeting it could restore sunitinib sensitivity in vivo (Qu et al., 2016). Also, the lncRNA-SRLR elicits intrinsic sorafenib resistance via promoting the "IL-6/STAT3" axis in RCC (Xu et al., 2017). MALAT1 expression is dramatically upregulated in sunitinib resistance RCC tissues and cell lines, and its knockdown induced RCC apoptosis and inhibited proliferation, invasion, and the chemoresistance through modulating "RasGAP SH3-domain-Binding Protein 1" via sponging miR-362-3p in RCC cells (Wang, Chang, Zhu, Gao, & Chang, 2020). Also, the lncRNA SNHG12 promotes RCC growth, migration, invasion, and sunitinib resistance via SNHG12/SP1/CDCA3 axis, and knocking down of this lncRNA could reverse RCC sunitinib resistance (Liu et al., 2020). Meanwhile, Jiang et al. find that the lncRNA LINC01094 expression is increased with the radiation exposure time to reach eight times the initial level. Knocking down of LINC01094 results in enhanced radiosensitivity of ccRCC cells through targeting the miR-577/CHEK2/FOXM1 axis (Jiang et al., 2020). In contrast, Liu et al. (2019) find that GAS5 is downregulated in sorafenib nonresponsive RCCs, and GAS5 overexpression can promote responsiveness to sorafenib in RCC cells via the "miR-21/SOX5 pathway". Lastly, Shi et al. (2020) find that targeting the nuclear receptor (TR4) can modulate the "TR4-lncTASR/AXL" signaling and increase the sunitinib sensitivity to suppress the RCC progression.

7.8.3 Potential of ciRNAs in predicting chemoresistance/radioresistance in renal cancer patients

CircRNAs could play both oncogenic and antitumor functions when it comes to treatment. Has-circ-0000338 promotes apoptosis, while exosomal has-circ-0000338 also has a carcinogenic property (Ma, Kong, Wang, & Ju, 2020). In RCC, has-circ-0035483 has a vital role in GEM treatment. Upregulated circ-0035483 can facilitate GEM-induced autophagy and enhance the resistance through regulating hsa-miR-335/cyclin B1 (a checkpoint molecule involved in GEM resistance). In this sense, silencing this ciRNA can enhance GEM sensitivity in vivo (Yan et al., 2019) (Fig. 7.7). The field of exploring the roles of ciRNAs in RCC chemoresistance/radioresistance prediction is still in the infancy stage, and more research is required to give new insights into this area.

It is worth noting that future clinical trials are warranted to confirm the abovementioned ncRNAs-mediated mechanisms explored by in vivo/in vitro experiments to help in the development of novel therapeutic strategies to sensitize RCC to chemotherapy/radiotherapy treatment and suppress the tumor progress.

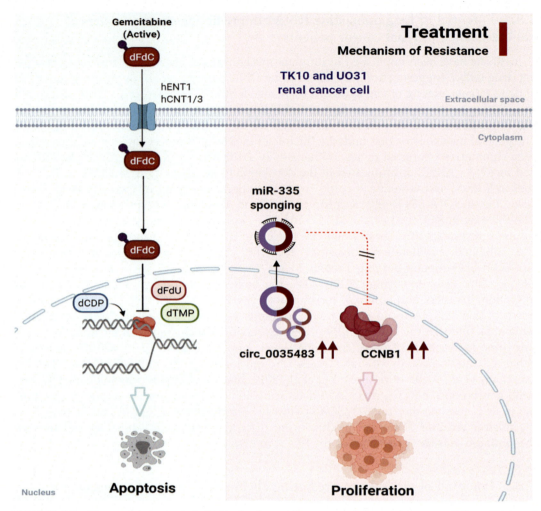

FIGURE 7.7 The mechanisms of circRNAs in drug-resistant tumors. In renal cancer cell lines, TX10 and UO31, circ_0035483 is significantly increased after treatment with gemcitabine, an S-phase antimetabolite drug. This could enhance cell viability to induce gemcitabine resistance. The results of RNA pull-down experiments revealed that circ_0035483 could bind with miR-335 as a sponge. Promotion of circ_0035483 expression increased the expression of CCNB1 and inhibited cell apoptosis. CCNB1 is an important checkpoint molecule in the cell cycle and is deeply involved in cell cycle–specific gemcitabine resistance. Thus circ_0035483 promotes gemcitabine resistance in human renal cancer cells by sponging miR-335 to upregulate CCNB1 (Yan et al., 2019). Source: Created by Biorender.com.

7.9 Summary and conclusion

ncRNAs are essential epigenetic regulators of gene expression, affecting all stages of renal cancer development, prognosis, and therapeutic response. The discovery of ncRNAs adds a new layer of complexity to the molecular landscape of renal cancer.

Several ncRNAs have been identified and implicated in the renal cancer hallmarks. This chapter discussed and summarized the research progress and recent discoveries on RCC-related ncRNAs and further emphasized their potential clinical applications, which provides a new direction for developing RCC theranostics. However, several aspects related to the contribution of the newly emerged ncRNAs to RCC initiation, progression, and response to therapeutic interventions are still not fully understood. In this sense, further studies are required to explore new ncRNAs related to RCC that are not defined yet and confirm its diagnostic, prognostic, and/or clinical therapeutic utility and validate the currently available ones to pave the way for early diagnosis and open a new era of individualized therapy.

References

Alam, T., Uludag, M., Essack, M., Salhi, A., Ashoor, H., Hanks, J. B., ... Bajic, V. B. (2017). FARNA: Knowledgebase of inferred functions of non-coding RNA transcripts. *Nucleic Acids Research*, 45(5), 2838–2848. Available from https://doi.org/10.1093/nar/gkw973.

Alimperti, S., & Andreadis, S. T. (2015). CDH2 and CDH11 act as regulators of stem cell fate decisions. *Stem Cell Research*, 14(3), 270–282.

Anderson, D. M., Anderson, K. M., Chang, C. L., Makarewich, C. A., Nelson, B. R., McAnally, J. R., ... Olson, E. N. (2015). A micropeptide encoded by a putative long noncoding RNA regulates muscle performance. *Cell*, 160(4), 595–606. Available from https://doi.org/10.1016/j.cell.2015.01.009.

Athar, U., & Gentile, T. C. (2008). Treatment options for metastatic renal cell carcinoma: A review. *The Canadian Journal of Urology*, 15(2), 3954–3966. Available from https://www.ncbi.nlm.nih.gov/pubmed/18405442.

Banumathy, G., & Cairns, P. (2010). Signaling pathways in renal cell carcinoma. *Cancer Biology & Therapy*, 10(7), 658–664. Available from https://doi.org/10.4161/cbt.10.7.13247.

Barth, D. A., Drula, R., Ott, L., Fabris, L., Slaby, O., Calin, G. A., & Pichler, M. (2020). Circulating non-coding RNAs in renal cell carcinoma—pathogenesis and potential implications as clinical biomarkers. *Frontiers in Cell and Developmental Biology*, 8, 828. Available from https://www.frontiersin.org/article/10.3389/fcell.2020.00828.

Baumann, V., & Winkler, J. (2014). miRNA-based therapies: Strategies and delivery platforms for oligonucleotide and non-oligonucleotide agents. *Future Medicinal Chemistry*, 6(17), 1967–1984. Available from https://doi.org/10.4155/fmc.14.116.

Bera, A., Ghosh-Choudhury, N., Dey, N., Das, F., Kasinath, B. S., Abboud, H. E., & Choudhury, G. G. (2013). NFκB-mediated cyclin D1 expression by microRNA-21 influences renal cancer cell proliferation. *Cellular Signalling*, 25(12), 2575–2586. Available from https://doi.org/10.1016/j.cellsig.2013.08.005.

Berkers, J., Govaere, O., Wolter, P., Beuselinck, B., Schöffski, P., van Kempen, L. C., ... Lerut, E. (2013). A possible role for microRNA-141 down-regulation in sunitinib resistant metastatic clear cell renal cell carcinoma through induction of epithelial-to-mesenchymal transition and hypoxia resistance. *The Journal of Urology*, 189(5), 1930–1938. Available from https://doi.org/10.1016/j.juro.2012.11.133.

Bhan, A., Soleimani, M., & Mandal, S. S. (2017). Long noncoding RNA and cancer: A new paradigm. *Cancer Research*, 77(15), 3965–3981. Available from https://doi.org/10.1158/0008-5472.CAN-16-2634.

Bhartiya, D., Pal, K., Ghosh, S., Kapoor, S., Jalali, S., Panwar, B., ... Scaria, V. (2013). lncRNome: A comprehensive knowledgebase of human long noncoding RNAs. *Database*, 2013. Available from https://doi.org/10.1093/database/bat034.

Bhat, N. S., Colden, M., Dar, A. A., Saini, S., Arora, P., Shahryari, V., ... Majid, S. (2017a). MicroRNA-720 regulates E-cadherin–αE-catenin complex and promotes renal cell carcinoma. *Molecular Cancer Therapeutics*, 16(12), 2840–2848.

Bhat, N. S., Colden, M., Dar, A. A., Saini, S., Arora, P., Shahryari, V., ... Dahiya, R. (2017b). MicroRNA-720 regulates E-cadherin-αE-catenin complex and promotes renal cell carcinoma. *Molecular Cancer Therapeutics*, 16(12), 2840–2848. Available from https://doi.org/10.1158/1535-7163.MCT-17-0400.

Bhatt, K., Mi, Q.-S., & Dong, Z. (2011). microRNAs in kidneys: Biogenesis, regulation, and pathophysiological roles. *American Journal of Physiology. Renal Physiology*, 300(3), F602−F610. Available from https://doi.org/10.1152/ajprenal.00727.2010.

Blondeau, J. J., Deng, M., Syring, I., Schrödter, S., Schmidt, D., Perner, S., ... Ellinger, J. (2015). Identification of novel long non-coding RNAs in clear cell renal cell carcinoma. *Clinical Epigenetics*, 7, 10. Available from https://doi.org/10.1186/s13148-015-0047-7.

Boguslawska, J., Wojcicka, A., Piekielko-Witkowska, A., Master, A., & Nauman, A. (2011). MiR-224 targets the 3′UTR of type 1 5′-iodothyronine deiodinase possibly contributing to tissue hypothyroidism in renal cancer. *PLoS One*, 6(9), e24541. Available from https://doi.org/10.1371/journal.pone.0024541.

Boon, R. A., Jaé, N., Holdt, L., & Dimmeler, S. (2016). Long noncoding RNAs: From clinical genetics to therapeutic targets? *Journal of the American College of Cardiology*, 67(10), 1214−1226. Available from https://doi.org/10.1016/j.jacc.2015.12.051.

Bryzghalov, O., Makałowska, I., & Szcześniak, M. W. (2021). lncEvo: Automated identification and conservation study of long noncoding RNAs. *BMC Bioinformatics*, 22(1), 59. Available from https://doi.org/10.1186/s12859-021-03991-2.

Butz, H., Szabó, P. M., Khella, H. W., Nofech-Mozes, R., Patocs, A., & Yousef, G. M. (2015). miRNA-target network reveals miR-124as a key miRNA contributing to clear cell renal cell carcinoma aggressive behaviour by targeting CAV1 and FLOT1. *Oncotarget*, 6(14), 12543−12557. Available from https://doi.org/10.18632/oncotarget.3815.

Cao, H., Liu, Z., Wang, R., Zhang, X., Yi, W., Nie, G., ... Zhu, M. (2017). miR-148a suppresses human renal cell carcinoma malignancy by targeting AKT2. *Oncology Reports*, 37(1), 147−154. Available from https://doi.org/10.3892/or.2016.5257.

Cao, Y., Xu, R., Xu, X., Zhou, Y., Cui, L., & He, X. (2016). Downregulation of lncRNA CASC2 by microRNA-21 increases the proliferation and migration of renal cell carcinoma cells. *Molecular Medicine Reports*, 14(1), 1019−1025. Available from https://doi.org/10.3892/mmr.2016.5337.

Cao, Z., Pan, X., Yang, Y., Huang, Y., & Shen, H. B. (2018). The lncLocator: A subcellular localization predictor for long non-coding RNAs based on a stacked ensemble classifier. *Bioinformatics (Oxford, England)*, 34(13), 2185−2194. Available from https://doi.org/10.1093/bioinformatics/bty085.

Cardenas, J., Balaji, U., & Gu, J. (2020). Cerina: Systematic circRNA functional annotation based on integrative analysis of ceRNA interactions. *Scientific Reports*, 10(1), 22165. Available from https://doi.org/10.1038/s41598-020-78469-x.

Carlevaro-Fita, J., Lanzós, A., Feuerbach, L., Hong, C., Mas-Ponte, D., Pedersen, J. S., ... Consortium, P. (2020). Cancer LncRNA census reveals evidence for deep functional conservation of long noncoding RNAs in tumorigenesis. *Communications Biology*, 3(1), 56. Available from https://doi.org/10.1038/s42003-019-0741-7.

Carlevaro-Fita, J., Liu, L., Zhou, Y., Zhang, S., Chouvardas, P., Johnson, R., & Li, J. (2019). LnCompare: Gene set feature analysis for human long non-coding RNAs. *Nucleic Acids Research*, 47(W1), W523−w529. Available from https://doi.org/10.1093/nar/gkz410.

Cesana, M., Cacchiarelli, D., Legnini, I., Santini, T., Sthandier, O., Chinappi, M., ... Bozzoni, I. (2011). A long noncoding RNA controls muscle differentiation by functioning as a competing endogenous RNA. *Cell*, 147(2), 358−369. Available from https://doi.org/10.1016/j.cell.2011.09.028.

Chakraborty, S., Deb, A., Maji, R. K., Saha, S., & Ghosh, Z. (2014). LncRBase: An enriched resource for lncRNA information. *PLoS One*, 9(9), e108010. Available from https://doi.org/10.1371/journal.pone.0108010.

Chan, W. L., Huang, H. D., & Chang, J. G. (2014). lncRNAMap: A map of putative regulatory functions in the long non-coding transcriptome. *Computational Biology and Chemistry*, 50, 41−49. Available from https://doi.org/10.1016/j.compbiolchem.2014.01.003.

Chan, Y., Yu, Y., Wang, G., Wang, C., Zhang, D., Wang, X., ... Zhang, C. (2019). Inhibition of microRNA-381 promotes tumor cell growth and chemoresistance in clear-cell renal cell carcinoma. *Medical Science Monitor: International Medical Journal of Experimental and Clinical Research*, 25, 5181−5190. Available from https://doi.org/10.12659/MSM.915524.

Chandrasekaran, K., Karolina, D. S., Sepramaniam, S., Armugam, A., Wintour, E. M., Bertram, J. F., & Jeyaseelan, K. (2012). Role of microRNAs in kidney homeostasis and disease. *Kidney International*, 81, 617−627, United States.

Chen, B., Duan, L., Yin, G., Tan, J., & Jiang, X. (2013). miR-381, a novel intrinsic WEE1 inhibitor, sensitizes renal cancer cells to 5-FU by up-regulation of Cdc2 activities in 786-O. *Journal of Chemotherapy (Florence, Italy)*, 25(4), 229–238. Available from https://doi.org/10.1179/1973947813Y.0000000092.

Chen, C., Xue, S., Zhang, J., Chen, W., Gong, D., Zheng, J., ... Zhai, W. (2017). DNA-methylation-mediated repression of miR-766-3p promotes cell proliferation via targeting SF2 expression in renal cell carcinoma. *International Journal of Cancer*, 141(9), 1867–1878. Available from https://doi.org/10.1002/ijc.30853.

Chen, D., Zhu, M., Su, H., Chen, J., Xu, X., & Cao, C. (2019). LINC00961 restrains cancer progression via modulating epithelial-mesenchymal transition in renal cell carcinoma. *Journal of Cellular Physiology*, 234(5), 7257–7265. Available from https://doi.org/10.1002/jcp.27483.

Chen, G., Wang, Z., Wang, D., Qiu, C., Liu, M., Chen, X., ... Cui, Q. (2013). LncRNADisease: A database for long-non-coding RNA-associated diseases. *Nucleic Acids Research*, 41(Database issue), D983–D986. Available from https://doi.org/10.1093/nar/gks1099.

Chen, J., Gu, Y., & Shen, W. (2017). MicroRNA-21 functions as an oncogene and promotes cell proliferation and invasion via TIMP3 in renal cancer. *European Review for Medical and Pharmacological Sciences*, 21(20), 4566–4576.

Chen, J., Shishkin, A. A., Zhu, X., Kadri, S., Maza, I., Guttman, M., ... Garber, M. (2016). Evolutionary analysis across mammals reveals distinct classes of long non-coding RNAs. *Genome Biology*, 17, 19. Available from https://doi.org/10.1186/s13059-016-0880-9.

Chen, J., Zhang, J., Gao, Y., Li, Y., Feng, C., Song, C., ... Li, C. (2021). LncSEA: A platform for long non-coding RNA related sets and enrichment analysis. *Nucleic Acids Research*, 49(D1), D969–d980. Available from https://doi.org/10.1093/nar/gkaa806.

Chen, J., Zhong, Y., & Li, L. (2020). miR-124 and miR-203 synergistically inactivate EMT pathway via coregulation of ZEB2 in clear cell renal cell carcinoma (ccRCC). *Journal of Translational Medicine*, 18(1), 69. Available from https://doi.org/10.1186/s12967-020-02242-x.

Chen, L., Qin, Z., Lei, J., Ye, S., Zeng, K., Wang, H., ... Yu, L. (2019). Upregulation of miR-489-3p and miR-630 inhibits oxaliplatin uptake in renal cell carcinoma by targeting OCT2. *Acta Pharmaceutica Sinica B*, 9(5), 1008–1020. Available from https://doi.org/10.1016/j.apsb.2019.01.002.

Chen, Q., Liu, T., Bao, Y., Zhao, T., Wang, J., Wang, H., ... Wang, L. (2020). CircRNA cRAPGEF5 inhibits the growth and metastasis of renal cell carcinoma via the miR-27a-3p/TXNIP pathway. *Cancer Letters*, 469, 68–77. Available from https://doi.org/10.1016/j.canlet.2019.10.017.

Chen, Q. N., Wei, C. C., Wang, Z. X., & Sun, M. (2017). Long non-coding RNAs in anti-cancer drug resistance. *Oncotarget*, 8(1), 1925–1936. Available from https://doi.org/10.18632/oncotarget.12461.

Chen, T., Shao, S., Li, W., Liu, Y., & Cao, Y. (2019). The circular RNA hsa-circ-0072309 plays anti-tumour roles by sponging miR-100 through the deactivation of PI3K/AKT and mTOR pathways in the renal carcinoma cell lines. *Artificial Cells, Nanomedicine, and Biotechnology*, 47(1), 3638–3648. Available from https://doi.org/10.1080/21691401.2019.1657873.

Chen, T., Yu, Q., Shao, S., & Guo, L. (2020). Circular RNA circFNDC3B protects renal carcinoma by miR-99a downregulation. *Journal of Cellular Physiology*, 235(5), 4399–4406. Available from https://doi.org/10.1002/jcp.29316.

Chen, X., Han, P., Zhou, T., Guo, X., Song, X., & Li, Y. (2016). circRNADb: A comprehensive database for human circular RNAs with protein-coding annotations. *Scientific Reports*, 6, 34985. Available from https://doi.org/10.1038/srep34985.

Chen, X., Ruan, A., Wang, X., Han, W., Wang, R., Lou, N., ... Zhang, X. (2014). miR-129-3p, as a diagnostic and prognostic biomarker for renal cell carcinoma, attenuates cell migration and invasion via downregulating multiple metastasis-related genes. *Journal of Cancer Research and Clinical Oncology*, 140(8), 1295–1304. Available from https://doi.org/10.1007/s00432-014-1690-7.

Chen, X., Wang, X., Ruan, A., Han, W., Zhao, Y., Lu, X., ... Zhang, X. (2014). miR-141 is a key regulator of renal cell carcinoma proliferation and metastasis by controlling EphA2 expression. *Clinical Cancer Research: An Official Journal of the American Association for Cancer Research*, 20(10), 2617–2630. Available from https://doi.org/10.1158/1078-0432.Ccr-13-3224.

Chen, Z., Tang, Z. Y., He, Y., Liu, L. F., Li, D. J., & Chen, X. (2014). miRNA-205 is a candidate tumor suppressor that targets ZEB2 in renal cell carcinoma. *Oncology Research and Treatment*, 37(11), 658–664. Available from https://doi.org/10.1159/000368792.

Chen, Z., Xiao, K., Chen, S., Huang, Z., Ye, Y., & Chen, T. (2020). Circular RNA hsa_circ_001895 serves as a sponge of microRNA-296-5p to promote clear cell renal cell carcinoma progression by regulating SOX12. *Cancer Science, 111*(2), 713−726. Available from https://doi.org/10.1111/cas.14261.

Chen, Z., Zhang, J., Zhang, Z., Feng, Z., Wei, J., Lu, J., ... Luo, J. (2017). The putative tumor suppressor microRNA-30a-5p modulates clear cell renal cell carcinoma aggressiveness through repression of ZEB2. *Cell Death and Disease, 8*(6), e2859. Available from https://doi.org/10.1038/cddis.2017.252.

Cheng, J. T., Wang, L., Wang, H., Tang, F. R., Cai, W. Q., Sethi, G., ... Ma, Z. (2019). Insights into biological role of LncRNAs in epithelial-mesenchymal transition. *Cells, 8*(10). Available from https://doi.org/10.3390/cells8101178.

Cheng, L., Wang, P., Tian, R., Wang, S., Guo, Q., Luo, M., ... Jiang, Q. (2019). LncRNA2Target v2.0: A comprehensive database for target genes of lncRNAs in human and mouse. *Nucleic Acids Research, 47*(D1), D140−d144. Available from https://doi.org/10.1093/nar/gky1051.

Chi, X. G., Meng, X. X., Ding, D. L., Xuan, X. H., Chen, Y. Z., Cai, Q., & Wang, A. (2020). HMGA1-mediated miR-671-5p targets APC to promote metastasis of clear cell renal cell carcinoma through Wnt signaling. *Neoplasma, 67*(1), 46−53. Available from https://doi.org/10.4149/neo_2019_190217N135.

Chiu, H. S., Llobet-Navas, D., Yang, X., Chung, W. J., Ambesi-Impiombato, A., Iyer, A., ... Sumazin, P. (2015). Cupid: Simultaneous reconstruction of microRNA-target and ceRNA networks. *Genome Res, 25*(2), 257−267. Available from https://doi.org/10.1101/gr.178194.114.

Chiyomaru, T., Fukuhara, S., Saini, S., Majid, S., Deng, G., Shahryari, V., ... Yamamura, S. (2014). Long noncoding RNA HOTAIR is targeted and regulated by miR-141 in human cancer cells. *Journal of Biological Chemistry, 289*(18), 12550−12565. Available from https://doi.org/10.1074/jbc.M113.488593.

Chow, T. F., Mankaruos, M., Scorilas, A., Youssef, Y., Girgis, A., Mossad, S., ... Yousef, G. M. (2010). The miR-17−92 cluster is over expressed in and has an oncogenic effect on renal cell carcinoma. *Journal of Urology, 183*(2), 743−751. Available from https://doi.org/10.1016/j.juro.2009.09.086.

Cimadamore, A., Massari, F., Santoni, M., Mollica, V., Di Nunno, V., Cheng, L., ... Moch, H. (2020). Molecular characterization and diagnostic criteria of renal cell carcinoma with emphasis on liquid biopsies. *Expert Review of Molecular Diagnostics, 20*(2), 141−150. Available from https://doi.org/10.1080/14737159.2019.1665510.

Corrà, F., Agnoletto, C., Minotti, L., Baldassari, F., & Volinia, S. (2018). The network of non-coding RNAs in cancer drug resistance. *Frontiers in Oncology, 8*, 327. Available from https://doi.org/10.3389/fonc.2018.00327.

Cui, L., Zhou, H., Zhao, H., Zhou, Y., Xu, R., Xu, X., ... He, X. (2012). MicroRNA-99a induces G1-phase cell cycle arrest and suppresses tumorigenicity in renal cell carcinoma. *BMC Cancer, 12*, 546. Available from https://doi.org/10.1186/1471-2407-12-546.

Cui, T., Zhang, L., Huang, Y., Yi, Y., Tan, P., Zhao, Y., ... Wang, D. (2018). MNDR v2.0: An updated resource of ncRNA-disease associations in mammals. *Nucleic Acids Research, 46*(D1), D371−d374. Available from https://doi.org/10.1093/nar/gkx1025.

Dang, Kyvan, & Myers, Kenneth A (2015). The role of hypoxia-induced miR-210 in cancer progression. *Int J Mol Sci, 16*(3), 6353−6372. Available from https://doi.org/10.3390/ijms16036353.

Das, S., Ghosal, S., Sen, R., & Chakrabarti, J. (2014). lnCeDB: Database of human long noncoding RNA acting as competing endogenous RNA. *PLoS One, 9*(6), e98965. Available from https://doi.org/10.1371/journal.pone.0098965.

Dasgupta, P., Kulkarni, P., Majid, S., Shahryari, V., Hashimoto, Y., Bhat, N. S., ... Dahiya, R. (2018). MicroRNA-203 inhibits long noncoding RNA HOTAIR and regulates tumorigenesis through epithelial-to-mesenchymal transition pathway in renal cell carcinoma. *Molecular Cancer Therapeutics, 17*(5), 1061−1069. Available from https://doi.org/10.1158/1535-7163.Mct-17-0925.

Deng, J. H., Zheng, G. Y., Li, H. Z., & Ji, Z. G. (2019). MiR-212-5p inhibits the malignant behavior of clear cell renal cell carcinoma cells by targeting TBX15. *European Review for Medical and Pharmacological Sciences, 23*(24), 10699−10707. Available from https://doi.org/10.26355/eurrev_201912_19770.

Derrien, T., Johnson, R., Bussotti, G., Tanzer, A., Djebali, S., Tilgner, H., ... Guigó, R. (2012). The GENCODE v7 catalog of human long noncoding RNAs: Analysis of their gene structure, evolution, and expression. *Genome Research, 22*(9), 1775−1789. Available from https://doi.org/10.1101/gr.132159.111.

Dey, N., Das, F., Ghosh-Choudhury, N., Mandal, C. C., Parekh, D. J., Block, K., ... Choudhury, G. G. (2012). microRNA-21 governs TORC1 activation in renal cancer cell proliferation and invasion. *PLoS One, 7*(6), e37366. Available from https://doi.org/10.1371/journal.pone.0037366.

Di Cristofano, C., Minervini, A., Menicagli, M., Salinitri, G., Bertacca, G., Pefanis, G., ... Cavazzana, A. (2007). Nuclear expression of hypoxia-inducible factor-1alpha in clear cell renal cell carcinoma is involved in tumor progression. *The American Journal of Surgical Pathology, 31*, 1875−1881, United States.

Di Timoteo, G., Dattilo, D., Centrón-Broco, A., Colantoni, A., Guarnacci, M., Rossi, F., ... Bozzoni, I. (2020). Modulation of circRNA metabolism by m6A modification. *Cell Reports, 31*(6), 107641. Available from https://doi.org/10.1016/j.celrep.2020.107641.

Dias, F., Teixeira, A. L., Ferreira, M., Adem, B., Bastos, N., Vieira, J., ... Medeiros, R. (2017). Plasmatic miR-210, miR-221 and miR-1233 profile: Potential liquid biopsies candidates for renal cell carcinoma. *Oncotarget, 8*(61), 103315−103326. Available from https://doi.org/10.18632/oncotarget.21733.

Dinger, M. E., Pang, K. C., Mercer, T. R., Crowe, M. L., Grimmond, S. M., & Mattick, J. S. (2009). NRED: A database of long noncoding RNA expression. *Nucleic Acids Research, 37*(Database issue), D122−D126. Available from https://doi.org/10.1093/nar/gkn617.

Doberstein, K., Steinmeyer, N., Hartmetz, A. K., Eberhardt, W., Mittelbronn, M., Harter, P. N., ... Gutwein, P. (2013). MicroRNA-145 targets the metalloprotease ADAM17 and is suppressed in renal cell carcinoma patients. *Neoplasia, 15*(2), 218−230. Available from https://doi.org/10.1593/neo.121222.

Dong, R., Ma, X. K., Li, G. W., & Yang, L. (2018). CIRCpedia v2: an updated database for comprehensive circular RNA annotation and expression comparison. *Genomics, Proteomics & Bioinformatics/Beijing Genomics Institute, 16*(4), 226−233. Available from https://doi.org/10.1016/j.gpb.2018.08.001.

Dong, Y., He, D., Peng, Z., Peng, W., Shi, W., Wang, J., ... Duan, C. (2017). Circular RNAs in cancer: An emerging key player. *Journal of hematology & oncology, 10*(1), 2. Available from https://doi.org/10.1186/s13045-016-0370-2.

Dong, Z., Liu, Y., Wang, Q., Wang, H., Ji, J., Huang, T., ... Cao, Y. (2020). The circular RNA-NRIP1 plays oncogenic roles by targeting microRNA-505 in the renal carcinoma cell lines. *Journal of Cellular Biochemistry, 121*(3), 2236−2246. Available from https://doi.org/10.1002/jcb.29446.

Dudekula, D. B., Panda, A. C., Grammatikakis, I., De, S., Abdelmohsen, K., & Gorospe, M. (2016). CircInteractome: A web tool for exploring circular RNAs and their interacting proteins and microRNAs. *RNA Biology, 13*(1), 34−42. Available from https://doi.org/10.1080/15476286.2015.1128065.

Ellinger, J., Alam, J., Rothenburg, J., Deng, M., Schmidt, D., Syring, I., ... Müller, S. C. (2015). The long noncoding RNA lnc-ZNF180-2 is a prognostic biomarker in patients with clear cell renal cell carcinoma. *American Journal of Cancer Research, 5*(9), 2799−2807.

Fachel, A. A., Tahira, A. C., Vilella-Arias, S. A., Maracaja-Coutinho, V., Gimba, E. R., Vignal, G. M., ... Verjovski-Almeida, S. (2013). Expression analysis and in silico characterization of intronic long noncoding RNAs in renal cell carcinoma: Emerging functional associations. *Molecular Cancer, 12*(1), 140. Available from https://doi.org/10.1186/1476-4598-12-140.

Fan, W., Huang, J., Xiao, H., & Liang, Z. (2016). MicroRNA-22 is downregulated in clear cell renal cell carcinoma, and inhibits cell growth, migration and invasion by targeting PTEN. *Molecular Medicine Reports, 13*(6), 4800−4806. Available from https://doi.org/10.3892/mmr.2016.5101.

Fan, Y., Ma, X., Li, H., Gao, Y., Huang, Q., Zhang, Y., ... Liu, K. (2018a). miR-122 promotes metastasis of clear-cell renal cell carcinoma by downregulating Dicer. *International Journal of Cancer, 142*(3), 547−560.

Fan, Y., Ma, X., Li, H., Gao, Y., Huang, Q., Zhang, Y., ... Zhang, X. (2018b). miR-122 promotes metastasis of clear-cell renal cell carcinoma by downregulating Dicer. *International Journal of Cancer. Journal International du Cancer, 142*(3), 547−560. Available from https://doi.org/10.1002/ijc.31050.

Fanale, D., Taverna, S., Russo, A., & Bazan, V. (2018). Circular RNA in exosomes. *Advances in Experimental Medicine and Biology, 1087*, 109−117. Available from https://doi.org/10.1007/978-981-13-1426-1_9.

Fang, Y., & Fullwood, M. J. (2016). Roles, functions, and mechanisms of long non-coding RNAs in cancer. *Genomics, Proteomics & Bioinformatics/Beijing Genomics Institute, 14*(1), 42−54. Available from https://doi.org/10.1016/j.gpb.2015.09.006.

Faragalla, H., Youssef, Y. M., Scorilas, A., Khalil, B., White, N. M., Mejia-Guerrero, S., ... Yousef, G. M. (2012). The clinical utility of miR-21 as a diagnostic and prognostic marker for renal cell carcinoma. *The Journal of Molecular Diagnostics: JMD, 14*(4), 385−392. Available from https://doi.org/10.1016/j.jmoldx.2012.02.003.

Fawzy, M. S., Abu AlSel, B. T., & Toraih, E. A. (2020). Analysis of microRNA processing machinery gene (DROSHA, DICER1, RAN, and XPO5) variants association with end-stage renal disease. *Journal of Clinical Laboratory Analysis, 34*(12), e23520. Available from https://doi.org/10.1002/jcla.23520.

Fedorko, M., Pacik, D., Wasserbauer, R., Juracek, J., Varga, G., Ghazal, M., & Nussir, M. I. (2016). MicroRNAs in the pathogenesis of renal cell carcinoma and their diagnostic and prognostic utility as cancer biomarkers. *The International Journal of Biological Markers, 31*(1), e26−e37. Available from https://doi.org/10.5301/jbm.5000174.

Fedorko, M., Stanik, M., Iliev, R., Redova-Lojova, M., Machackova, T., Svoboda, M., ... Slaby, O. (2015). Combination of MiR-378 and MiR-210 serum levels enables sensitive detection of renal cell carcinoma. *International Journal of Molecular Sciences*, *16*(10), 23382−23389. Available from https://doi.org/10.3390/ijms161023382.

Fiannaca, A., Paglia, L., Rosa, M., Rizzo, R., & Urso, A. (2020). miRTissue (ce): Extending miRTissue web service with the analysis of ceRNA-ceRNA interactions. *BMC Bioinformatics*, *21*(Suppl 8), 199. Available from https://doi.org/10.1186/s12859-020-3520-z.

Franz, A., Ralla, B., Weickmann, S., Jung, M., Rochow, H., Stephan, C., ... Jung, K. (2019). Circular RNAs in clear cell renal cell carcinoma: Their microarray-based identification, analytical validation, and potential use in a clinico-genomic model to improve prognostic accuracy. *Cancers*, *11*(10), 1473. Available from https://doi.org/10.3390/cancers11101473.

Fritz, H. K. M., Lindgren, D., Ljungberg, B., Axelson, H., & Dahlbäck, B. (2014). The miR(21/10b) ratio as a prognostic marker in clear cell renal cell carcinoma. *European Journal of Cancer*, *50*(10), 1758−1765. Available from https://doi.org/10.1016/j.ejca.2014.03.281.

Fu, Q., Liu, Z., Pan, D., Zhang, W., Xu, L., Zhu, Y., ... Xu, J. (2014). Tumor miR-125b predicts recurrence and survival of patients with clear-cell renal cell carcinoma after surgical resection. *Cancer Science*, *105*(11), 1427−1434. Available from https://doi.org/10.1111/cas.12507.

Fujii, N., Hirata, H., Ueno, K., Mori, J., Oka, S., Shimizu, K., ... Matsuyama, H. (2017). Extracellular miR-224 as a prognostic marker for clear cell renal cell carcinoma. *Oncotarget*, *8*(66), 109877−109888. Available from https://doi.org/10.18632/oncotarget.22436.

Fukunaga, T., & Hamada, M. (2017). RIblast: An ultrafast RNA-RNA interaction prediction system based on a seed-and-extension approach. *Bioinformatics (Oxford, England)*, *33*(17), 2666−2674. Available from https://doi.org/10.1093/bioinformatics/btx287.

Gámez-Pozo, A., Antón-Aparicio, L. M., Bayona, C., Borrega, P., Gallegos Sancho, M. I., García-Domínguez, R., ... Espinosa Arranz, E. (2012). MicroRNA expression profiling of peripheral blood samples predicts resistance to first-line sunitinib in advanced renal cell carcinoma patients. *Neoplasia*, *14*(12), 1144−1152. Available from https://doi.org/10.1593/neo.12734.

Gaudelot, K., Gibier, J. B., Pottier, N., Hémon, B., Van Seuningen, I., Glowacki, F., ... Perrais, M. (2017). Targeting miR-21 decreases expression of multi-drug resistant genes and promotes chemosensitivity of renal carcinoma. *Tumour Biology*, *39*(7), 1010428317707372. Available from https://doi.org/10.1177/1010428317707372.

Gebauer, K., Peters, I., Dubrowinskaja, N., Hennenlotter, J., Abbas, M., Scherer, R., ... Serth, J. (2013). Hsa-mir-124-3 CpG island methylation is associated with advanced tumours and disease recurrence of patients with clear cell renal cell carcinoma. *British Journal of Cancer*, *108*(1), 131−138. Available from https://doi.org/10.1038/bjc.2012.537.

Ghafouri-Fard, S., Shirvani-Farsani, Z., Branicki, W., & Taheri, M. (2020). MicroRNA signature in renal cell carcinoma. *Frontiers in Oncology*, *10*, 596359. Available from https://doi.org/10.3389/fonc.2020.596359.

Ghosal, S., Das, S., Sen, R., Basak, P., & Chakrabarti, J. (2013). Circ2Traits: A comprehensive database for circular RNA potentially associated with disease and traits. *Frontiers in Genetics*, *4*(283). Available from https://doi.org/10.3389/fgene.2013.00283.

Gilyazova, I. R., Klimentova, E. A., Bulygin, K. V., Izmailov, A. A., Bermisheva, M. A., Galimova, E. F., ... Khusnutdinova, E. K. (2020). MicroRNA-200 family expression analysis in metastatic clear cell renal cell carcinoma patients. *Cancer Gene Therapy*, *27*(10−11), 768−772. Available from https://doi.org/10.1038/s41417-019-0149-z.

Gläsker, S., Vergauwen, E., Koch, C. A., Kutikov, A., & Vortmeyer, A. O. (2020). Von Hippel−Lindau disease: Current challenges and future prospects. *OncoTargets and Therapy*, *13*, 5669−5690. Available from https://doi.org/10.2147/OTT.S190753.

Gokool, A., Anwar, F., & Voineagu, I. (2020). The landscape of circular RNA expression in the human brain. *Biological Psychiatry*, *87*(3), 294−304. Available from https://doi.org/10.1016/j.biopsych.2019.07.029.

Gong, J., Liu, W., Zhang, J., Miao, X., & Guo, A. Y. (2015). lncRNASNP: A database of SNPs in lncRNAs and their potential functions in human and mouse. *Nucleic Acids Research*, *43*(Database issue), D181−D186. Available from https://doi.org/10.1093/nar/gku1000.

Gong, J., Mei, S., Liu, C., Xiang, Y., Ye, Y., Zhang, Z., ... Han, L. (2018). PancanQTL: Systematic identification of cis-eQTLs and trans-eQTLs in 33 cancer types. *Nucleic Acids Research*, *46*(D1), D971−d976. Available from https://doi.org/10.1093/nar/gkx861.

Gong, J., Shao, D., Xu, K., Lu, Z., Lu, Z. J., Yang, Y. T., & Zhang, Q. C. (2018). RISE: A database of RNA interactome from sequencing experiments. *Nucleic Acids Research, 46*(D1), D194−d201. Available from https://doi.org/10.1093/nar/gkx864.

Gong, X., Zhao, H., Saar, M., Peehl, D. M., & Brooks, J. D. (2019). miR-22 Regulates invasion, gene expression and predicts overall survival in patients with clear cell renal cell carcinoma. *Kidney Cancer, 3*(2), 119−132. Available from https://doi.org/10.3233/kca-190051.

Goodall, G. J., & Wickramasinghe, V. O. (2021). RNA in cancer. *Nature Reviews Cancer, 21*, 22−36, England.

Goto, K., Oue, N., Shinmei, S., Sentani, K., Sakamoto, N., Naito, Y., ... Yasui, W. (2013). Expression of. *Molecular and Clinical Oncology, 1*(2), 235−240. Available from https://doi.org/10.3892/mco.2012.46.

Goto, Y., Kurozumi, A., Nohata, N., Kojima, S., Matsushita, R., Yoshino, H., ... Seki, N. (2016). The microRNA signature of patients with sunitinib failure: Regulation of UHRF1 pathways by microRNA-101 in renal cell carcinoma. *Oncotarget, 7*(37), 59070−59086. Available from https://doi.org/10.18632/oncotarget.10887.

Gottardo, F., Liu, C. G., Ferracin, M., Calin, G. A., Fassan, M., Bassi, P., ... Baffa, R. (2007). Micro-RNA profiling in kidney and bladder cancers. *Urologic Oncology, 25*(5), 387−392. Available from https://doi.org/10.1016/j.urolonc.2007.01.019.

Graham, J., Wells, J. C., Donskov, F., Lee, J. L., Fraccon, A., Pasini, F., ... Heng, D. Y. C. (2019). Cytoreductive nephrectomy in metastatic papillary renal cell carcinoma: Results from the international metastatic renal cell carcinoma database consortium. *European Urology Oncology, 2*(6), 643−648.

Grange, C., Brossa, A., & Bussolati, B. (2019). Extracellular vesicles and carried miRNAs in the progression of renal cell carcinoma. *International Journal of Molecular Sciences, 20*(8).

Gruber, A. R., Lorenz, R., Bernhart, S. H., Neuböck, R., & Hofacker, I. L. (2008). The Vienna RNA websuite. *Nucleic Acids Research, 36*(Web Server issue), W70−W74. Available from https://doi.org/10.1093/nar/gkn188.

Guo, Z., Jia, H., & Ge, J. (2020). MiR-206 suppresses proliferation and epithelial-mesenchymal transition of renal cell carcinoma by inhibiting CDK6 expression. *Human Cell: Official Journal of Human Cell Research Society, 33*(3), 750−758. Available from https://doi.org/10.1007/s13577-020-00355-5.

Gutschner, T., & Diederichs, S. (2012). The hallmarks of cancer: A long non-coding RNA point of view. *RNA Biology, 9*(6), 703−719. Available from https://doi.org/10.4161/rna.20481.

Guttman, M., & Rinn, J. L. (2012). Modular regulatory principles of large non-coding RNAs. *Nature, 482*(7385), 339−346.

Han, X., Piao, L., Yuan, X., Wang, L., Liu, Z., & He, X. (2019). Knockdown of NSD2 suppresses renal cell carcinoma metastasis by inhibiting epithelial−mesenchymal transition. *International Journal of Medical Sciences, 16*(10), 1404−1411. Available from https://doi.org/10.7150/ijms.36128.

Han, Z., Zhang, Y., Sun, Y., Chen, J., Chang, C., Wang, X., & Yeh, S. (2018). ERβ-mediated alteration of circATP2B1 and miR-204-3p signaling promotes invasion of clear cell renal cell carcinoma. *Cancer Research, 78*(10), 2550−2563. Available from https://doi.org/10.1158/0008-5472.Can-17-1575.

Hayes, J., Peruzzi, P. P., & Lawler, S. (2014). MicroRNAs in cancer: Biomarkers, functions and therapy. *Trends in Molecular Medicine, 20*, 460−469.

He, C., Zhao, X., Jiang, H., Zhong, Z., & Xu, R. (2015). Demethylation of miR-10b plays a suppressive role in ccRCC cells. *International Journal of Clinical and Experimental Pathology, 8*(9), 10595−10604.

He, H., Dai, J., Zhuo, R., Zhao, J., Wang, H., Sun, F., ... Xu, D. (2018). Study on the mechanism behind lncRNA MEG3 affecting clear cell renal cell carcinoma by regulating miR-7/RASL11B signaling. *Journal of Cellular Physiology, 233*(12), 9503−9515. Available from https://doi.org/10.1002/jcp.26849.

He, H., Wang, L., Zhou, W., Zhang, Z., Xu, S., Wang, D., ... Ge, J. (2015). MicroRNA expression profiling in clear cell renal cell carcinoma: identification and functional validation of key miRNAs. *PLoS One, 10*(5), e0125672. Available from https://doi.org/10.1371/journal.pone.0125672.

He, Q., Liu, Y., & Sun, W. (2018). Statistical analysis of non-coding RNA data. *Cancer Letters, 417*, 161−167. Available from https://doi.org/10.1016/j.canlet.2017.12.029.

He, S., Zhang, H., Liu, H., & Zhu, H. (2015). LongTarget: A tool to predict lncRNA DNA-binding motifs and binding sites via Hoogsteen base-pairing analysis. *Bioinformatics (Oxford, England), 31*(2), 178−186. Available from https://doi.org/10.1093/bioinformatics/btu643.

He, X., Sun, F., Guo, F., Wang, K., Gao, Y., Feng, Y., ... Li, Y. (2017). Knockdown of long noncoding RNA FTX inhibits proliferation, migration, and invasion in renal cell carcinoma cells. *Oncology Research, 25*(2), 157−166. Available from https://doi.org/10.3727/096504016x14719078133203.

He, Z. H., Qin, X. H., Zhang, X. L., Yi, J. W., & Han, J. Y. (2018). Long noncoding RNA GIHCG is a potential diagnostic and prognostic biomarker and therapeutic target for renal cell carcinoma. *European Review for Medical and Pharmacological Sciences*, 22(1), 46−54. Available from https://doi.org/10.26355/eurrev_201801_14099.

Heinemann, F. G., Tolkach, Y., Deng, M., Schmidt, D., Perner, S., Kristiansen, G., ... Ellinger, J. (2018). Serum miR-122-5p and miR-206 expression: Non-invasive prognostic biomarkers for renal cell carcinoma. *Clinical Epigenetics*, 10, 11. Available from https://doi.org/10.1186/s13148-018-0444-9.

Heinzelmann, J., Henning, B., Sanjmyatav, J., Posorski, N., Steiner, T., Wunderlich, H., ... Junker, K. (2011). Specific miRNA signatures are associated with metastasis and poor prognosis in clear cell renal cell carcinoma. *World Journal of Urology*, 29(3), 367−373. Available from https://doi.org/10.1007/s00345-010-0633-4.

Heinzelmann, J., Unrein, A., Wickmann, U., Baumgart, S., Stapf, M., Szendroi, A., ... Junker, K. (2014). MicroRNAs with prognostic potential for metastasis in clear cell renal cell carcinoma: A comparison of primary tumors and distant metastases. *Annals of Surgical Oncology*, 21(3), 1046−1054. Available from https://doi.org/10.1245/s10434-013-3361-3.

Hidaka, H., Seki, N., Yoshino, H., Yamasaki, T., Yamada, Y., Nohata, N., ... Enokida, H. (2012). Tumor suppressive microRNA-1285 regulates novel molecular targets: Aberrant expression and functional significance in renal cell carcinoma. *Oncotarget*, 3(1), 44−57. Available from https://doi.org/10.18632/oncotarget.417.

Hildebrandt, M. A., Gu, J., Lin, J., Ye, Y., Tan, W., Tamboli, P., ... Wu, X. (2010). Hsa-miR-9 methylation status is associated with cancer development and metastatic recurrence in patients with clear cell renal cell carcinoma. *Oncogene*, 29(42), 5724−5728. Available from https://doi.org/10.1038/onc.2010.305.

Hirata, H., Hinoda, Y., Shahryari, V., Deng, G., Nakajima, K., Tabatabai, Z. L., ... Dahiya, R. (2015). Long noncoding RNA MALAT1 promotes aggressive renal cell carcinoma through Ezh2 and interacts with miR-205. *Cancer Research*, 75(7), 1322−1331. Available from https://doi.org/10.1158/0008-5472.Can-14-2931.

Hirata, H., Hinoda, Y., Ueno, K., Nakajima, K., Ishii, N., & Dahiya, R. (2012). MicroRNA-1826 directly targets beta-catenin (CTNNB1) and MEK1 (MAP2K1) in VHL-inactivated renal cancer. *Carcinogenesis*, 33(3), 501−508. Available from https://doi.org/10.1093/carcin/bgr302.

Holdt, L. M., Kohlmaier, A., & Teupser, D. (2018). Circular RNAs as therapeutic agents and targets. *Frontiers in Physiology*, 9, 1262. Available from https://doi.org/10.3389/fphys.2018.01262.

Hong, Q., Li, O., Zheng, W., Xiao, W.-Z., Zhang, L., Wu, D., ... Chen, X.-M. (2017). LncRNA HOTAIR regulates HIF-1α/AXL signaling through inhibition of miR-217 in renal cell carcinoma. *Cell Death & Disease*, 8(5), e2772. Available from https://doi.org/10.1038/cddis.2017.181. (Accession No. 28492542).

Hu, G., Lai, P., Liu, M., Xu, L., Guo, Z., Liu, H., ... Xu, Y. (2014). miR-203a regulates proliferation, migration, and apoptosis by targeting glycogen synthase kinase-3β in human renal cell carcinoma. *Tumour Biology: The Journal of the International Society for Oncodevelopmental Biology and Medicine*, 35(11), 11443−11453. Available from https://doi.org/10.1007/s13277-014-2476-x.

Huang, J., Dong, B., Zhang, J., Kong, W., Chen, Y., Xue, W., ... Huang, Y. (2014). miR-199a-3p inhibits hepatocyte growth factor/c-Met signaling in renal cancer carcinoma. *Tumour Biology: The Journal of the International Society for Oncodevelopmental Biology and Medicine*, 35(6), 5833−5843. Available from https://doi.org/10.1007/s13277-014-1774-7.

Huang, J., Yao, X., Zhang, J., Dong, B., Chen, Q., Xue, W., ... Huang, Y. (2013). Hypoxia-induced downregulation of miR-30c promotes epithelial-mesenchymal transition in human renal cell carcinoma. *Cancer Science*, 104(12), 1609−1617. Available from https://doi.org/10.1111/cas.12291.

Huang, Q. B., Ma, X., Zhang, X., Liu, S. W., Ai, Q., Shi, T. P., ... Zheng, T. (2013). Down-regulated miR-30a in clear cell renal cell carcinoma correlated with tumor hematogenous metastasis by targeting angiogenesis-specific DLL4. *PLoS One*, 8(6), e67294. Available from https://doi.org/10.1371/journal.pone.0067294.

Huang, W., Ling, Y., Zhang, S., Xia, Q., Cao, R., Fan, X., ... Zhang, G. (2021). TransCirc: An interactive database for translatable circular RNAs based on multi-omics evidence. *Nucleic Acids Research*, 49(D1), D236−D242. Available from https://doi.org/10.1093/nar/gkaa823.

Huang, X., Huang, M., Kong, L., & Li, Y. (2015). miR-372 suppresses tumour proliferation and invasion by targeting IGF2BP1 in renal cell carcinoma. *Cell Proliferation*, 48(5), 593−599. Available from https://doi.org/10.1111/cpr.12207.

Huang, Y., Wang, J., Zhao, Y., Wang, H., Liu, T., Li, Y., ... Zhang, Y. (2021). cncRNAdb: A manually curated resource of experimentally supported RNAs with both protein-coding and noncoding function. *Nucleic Acids Research*, 49(D1), D65−D70. Available from https://doi.org/10.1093/nar/gkaa791.

Huang, Y., Zhang, Y., Jia, L., Liu, C., & Xu, F. (2019). Circular RNA ABCB10 promotes tumor progression and correlates with pejorative prognosis in clear cell renal cell carcinoma. *The International Journal of Biological Markers, 34*(2), 176−183. Available from https://doi.org/10.1177/1724600819842279.

Iyer, M. K., Niknafs, Y. S., Malik, R., Singhal, U., Sahu, A., Hosono, Y., ... Chinnaiyan, A. M. (2015). The landscape of long noncoding RNAs in the human transcriptome. *Nature Genetics, 47*(3), 199−208. Available from https://doi.org/10.1038/ng.3192.

Jafari Ghods, F. (2018). Circular RNA in saliva. *Advances in Experimental Medicine and Biology, 1087*, 131−139. Available from https://doi.org/10.1007/978-981-13-1426-1_11.

Ji, H., Tian, D., Zhang, B., Zhang, Y., Yan, D., & Wu, S. (2017). Overexpression of miR-155 in clear-cell renal cell carcinoma and its oncogenic effect through targeting FOXO3a. *Experimental and Therapeutic Medicine, 13*(5), 2286−2292. Available from https://doi.org/10.3892/etm.2017.4263.

Jiang, Q., Ma, R., Wang, J., Wu, X., Jin, S., Peng, J., ... Wang, Y. (2015). LncRNA2Function: A comprehensive resource for functional investigation of human lncRNAs based on RNA-seq data. *BMC Genomics, 16*(Suppl 3), S2. Available from https://doi.org/10.1186/1471-2164-16-s3-s2.

Jiang, Q., Wang, J., Wang, Y., Ma, R., Wu, X., & Li, Y. (2014). TF2LncRNA: Identifying common transcription factors for a list of lncRNA genes from ChIP-Seq data. *BioMed Research International, 2014*, 317642. Available from https://doi.org/10.1155/2014/317642.

Jiang, W.-D., & Ye, Z.-H. (2019). Integrated analysis of a competing endogenous RNA network in renal cell carcinoma using bioinformatics tools. *Bioscience Reports, 39*(7). Available from https://doi.org/10.1042/bsr20190996.

Jiang, Y., Li, W., Yan, Y., Yao, X., Gu, W., & Zhang, H. (2020). LINC01094 triggers radio-resistance in clear cell renal cell carcinoma via miR-577/CHEK2/FOXM1 axis. *Cancer Cell International, 20*, 274. Available from https://doi.org/10.1186/s12935-020-01306-8.

Jiao, D., Wu, M., Ji, L., Liu, F., & Liu, Y. (2018). MicroRNA-186 suppresses cell proliferation and metastasis through targeting sentrin-specific protease 1 in renal cell carcinoma. *Oncology Research, 26*(2), 249−259. Available from https://doi.org/10.3727/096504017x14953948675430.

Jin, C., Shi, L., Li, Z., Liu, W., Zhao, B., Qiu, Y., ... Zhu, Q. (2019). Circ_0039569 promotes renal cell carcinoma growth and metastasis by regulating miR-34a-5p/CCL22. *American Journal of Translational Research, 11*(8), 4935−4945.

Jin, L., Zhang, Z., Li, Y., He, T., Hu, J., Liu, J., ... Lai, Y. (2017). miR-125b is associated with renal cell carcinoma cell migration, invasion and apoptosis. *Oncology Letters, 13*(6), 4512−4520. Available from https://doi.org/10.3892/ol.2017.5985.

Jingushi, K., Kashiwagi, Y., Ueda, Y., Kitae, K., Hase, H., Nakata, W., ... Tsujikawa, K. (2017). High miR-122 expression promotes malignant phenotypes in ccRCC by targeting occludin. *International Journal of Oncology, 51*(1), 289−297. Available from https://doi.org/10.3892/ijo.2017.4016.

Josset, L., Tchitchek, N., Gralinski, L. E., Ferris, M. T., Eisfeld, A. J., Green, R. R., ... Katze, M. G. (2014). Annotation of long non-coding RNAs expressed in collaborative cross founder mice in response to respiratory virus infection reveals a new class of interferon-stimulated transcripts. *RNA Biology, 11*(7), 875−890. Available from https://doi.org/10.4161/rna.29442.

Kang, Y. J., Yang, D. C., Kong, L., Hou, M., Meng, Y. Q., Wei, L., & Gao, G. (2017). CPC2: A fast and accurate coding potential calculator based on sequence intrinsic features. *Nucleic Acids Research, 45*(W1), W12−W16. Available from https://doi.org/10.1093/nar/gkx428.

Katayama, H., Tamai, K., Shibuya, R., Nakamura, M., Mochizuki, M., Yamaguchi, K., ... Satoh, K. (2017). Long non-coding RNA HOTAIR promotes cell migration by upregulating insulin growth factor-binding protein 2 in renal cell carcinoma. *Scientific Reports, 7*(1), 12016. Available from https://doi.org/10.1038/s41598-017-12191-z.

Kawakami, K., Enokida, H., Chiyomaru, T., Tatarano, S., Yoshino, H., Kagara, I., ... Nakagawa, M. (2012). The functional significance of miR-1 and miR-133a in renal cell carcinoma. *European Journal of Cancer, 48*(6), 827−836. Available from https://doi.org/10.1016/j.ejca.2011.06.030.

Khella, H. W., Bakhet, M., Allo, G., Jewett, M. A., Girgis, A. H., Latif, A., ... Yousef, G. M. (2013). miR-192, miR-194 and miR-215: A convergent microRNA network suppressing tumor progression in renal cell carcinoma. *Carcinogenesis, 34*(10), 2231−2239. Available from https://doi.org/10.1093/carcin/bgt184.

Khella, H. W., Scorilas, A., Mozes, R., Mirham, L., Lianidou, E., Krylov, S. N., ... Yousef, G. M. (2015). Low expression of miR-126 is a prognostic marker for metastatic clear cell renal cell carcinoma. *The American Journal of Pathology, 185*(3), 693−703. Available from https://doi.org/10.1016/j.ajpath.2014.11.017.

Khella, H. W., White, N. M., Faragalla, H., Gabril, M., Boazak, M., Dorian, D., ... Yousef, G. M. (2012). Exploring the role of miRNAs in renal cell carcinoma progression and metastasis through bioinformatic and experimental analyses. *Tumour Biology: The Journal of the International Society for Oncodevelopmental Biology and Medicine*, *33*(1), 131–140. Available from https://doi.org/10.1007/s13277-011-0255-5.

Khella, H. W. Z., Butz, H., Ding, Q., Rotondo, F., Evans, K. R., Kupchak, P., ... Yousef, G. M. (2015). miR-221/222 are involved in response to sunitinib treatment in metastatic renal cell carcinoma. *Molecular Therapy: The Journal of the American Society of Gene Therapy*, *23*(11), 1748–1758. Available from https://doi.org/10.1038/mt.2015.129.

Khella, H. W. Z., Daniel, N., Youssef, L., Scorilas, A., Nofech-Mozes, R., Mirham, L., ... Yousef, G. M. (2017). miR-10b is a prognostic marker in clear cell renal cell carcinoma. *Journal of Clinical Pathology*, *70*(10), 854–859. Available from https://doi.org/10.1136/jclinpath-2017-204341.

Kim, B. J., Kim, J. H., Kim, H. S., & Zang, D. Y. (2017). Prognostic and predictive value of VHL gene alteration in renal cell carcinoma: A *meta*-analysis and review. *Oncotarget*, *8*(8), 13979–13985. Available from https://doi.org/10.18632/oncotarget.14704.

Kin, T., Yamada, K., Terai, G., Okida, H., Yoshinari, Y., Ono, Y., ... Asai, K. (2007). fRNAdb: A platform for mining/annotating functional RNA candidates from non-coding RNA sequences. *Nucleic Acids Research*, *35* (Database issue), D145–D148. Available from https://doi.org/10.1093/nar/gkl837.

Kirk, J. M., Kim, S. O., Inoue, K., Smola, M. J., Lee, D. M., Schertzer, M. D., ... Calabrese, J. M. (2018). Functional classification of long non-coding RNAs by k-mer content. *Nature Genetics*, *50*(10), 1474–1482. Available from https://doi.org/10.1038/s41588-018-0207-8.

Kölling, M., Haddad, G., Wegmann, U., Kistler, A., Bosakova, A., Seeger, H., ... Lorenzen, J. M. (2019). Circular RNAs in urine of kidney transplant patients with acute T cell-mediated allograft rejection. *Clinical Chemistry*, *65*(10), 1287–1294. Available from https://doi.org/10.1373/clinchem.2019.305854.

Kowalik, C. G., Palmer, D. A., Sullivan, T. B., Teebagy, P. A., Dugan, J. M., Libertino, J. A., ... Rieger-Christ, K. M. (2017). Profiling microRNA from nephrectomy and biopsy specimens: Predictors of progression and survival in clear cell renal cell carcinoma. *BJU International*, *120*(3), 428–440. Available from https://doi.org/10.1111/bju.13886.

Kristensen, L. S., Hansen, T. B., Venø, M. T., & Kjems, J. (2018). Circular RNAs in cancer: Opportunities and challenges in the field. *Oncogene*, *37*(5), 555–565. Available from https://doi.org/10.1038/onc.2017.361.

Kulkarni, P., Dasgupta, P., Bhat, N. S., Shahryari, V., Shiina, M., Hashimoto, Y., ... Dahiya, R. (2018). Elevated miR-182-5p associates with renal cancer cell mitotic arrest through diminished. *Molecular Cancer Research: MCR*, *16*(11), 1750–1760. Available from https://doi.org/10.1158/1541-7786.MCR-17-0762.

Lan, W., Zhu, M., Chen, Q., Chen, B., Liu, J., Li, M., & Chen, Y.-P. P. (2020). CircR2Cancer: A manually curated database of associations between circRNAs and cancers. *Database: The Journal of Biological Databases and Curation*, *2020*, baaa085. Available from https://doi.org/10.1093/database/baaa085.

Li, H., Zhao, J., Zhang, J. W., Huang, Q. Y., Huang, J. Z., Chi, L. S., ... Ma, W. M. (2013). MicroRNA-217, downregulated in clear cell renal cell carcinoma and associated with lower survival, suppresses cell proliferation and migration. *Neoplasma*, *60*(5), 511–515. Available from https://doi.org/10.4149/neo_2013_066.

Li, J., Han, L., Roebuck, P., Diao, L., Liu, L., Yuan, Y., ... Liang, H. (2015). TANRIC: An interactive open platform to explore the function of lncRNAs in cancer. *Cancer Research*, *75*(18), 3728–3737. Available from https://doi.org/10.1158/0008-5472.Can-15-0273.

Li, J., Jiang, D., Zhang, Q., Peng, S., Liao, G., Yang, X., ... Pang, J. (2020). MiR-301a promotes cell proliferation by repressing PTEN in renal cell carcinoma. *Cancer Management and Research*, *12*, 4309–4320. Available from https://doi.org/10.2147/cmar.S253533.

Li, J.-H., Liu, S., Zhou, H., Qu, L.-H., & Yang, J.-H. (2014). starBase v2.0: Decoding miRNA-ceRNA, miRNA-ncRNA and protein-RNA interaction networks from large-scale CLIP-Seq data. *Nucleic Acids Research*, *42* (Database issue), D92–D97. Available from https://doi.org/10.1093/nar/gkt1248.

Li, J. K., Chen, C., Liu, J. Y., Shi, J. Z., Liu, S. P., Liu, B., ... Wang, L. H. (2017). Long noncoding RNA MRCCAT1 promotes metastasis of clear cell renal cell carcinoma via inhibiting NPR3 and activating p38-MAPK signaling. *Molecular Cancer*, *16*(1), 111. Available from https://doi.org/10.1186/s12943-017-0681-0.

Li, L., Wu, P., Wang, Z., Meng, X., Zha, C., Li, Z., ... Cai, J. (2020). NoncoRNA: A database of experimentally supported non-coding RNAs and drug targets in cancer. *J Hematol Oncol*, *13*(1), 15. Available from https://doi.org/10.1186/s13045-020-00849-7.

Li, M., Wang, Y., Cheng, L., Niu, W., Zhao, G., Raju, J. K., ... Bu, R. (2017). Long non-coding RNAs in renal cell carcinoma: A systematic review and clinical implications. *Oncotarget*, 8(29), 48424–48435. Available from https://doi.org/10.18632/oncotarget.17053.

Li, S., Chen, T., Zhong, Z., Wang, Y., Li, Y., & Zhao, X. (2012). microRNA-155 silencing inhibits proliferation and migration and induces apoptosis by upregulating BACH1 in renal cancer cells. *Molecular Medicine Reports*, 5(4), 949–954. Available from https://doi.org/10.3892/mmr.2012.779.

Li, S., Li, Y., Chen, B., Zhao, J., Yu, S., Tang, Y., ... Huang, S. (2018). exoRBase: A database of circRNA, lncRNA and mRNA in human blood exosomes. *Nucleic Acids Research*, 46(D1), D106–D112. Available from https://doi.org/10.1093/nar/gkx891.

Li, W., Wang, Q., Su, Q., Ma, D., An, C., Ma, L., & Liang, H. (2014). Honokiol suppresses renal cancer cells' metastasis via dual-blocking epithelial-mesenchymal transition and cancer stem cell properties through modulating miR-141/ZEB2 signaling. *Molecules and Cells*, 37(5), 383–388. Available from https://doi.org/10.14348/molcells.2014.0009.

Li, W., Yang, F. Q., Sun, C. M., Huang, J. H., Zhang, H. M., Li, X., ... Liu, M. (2020). circPRRC2A promotes angiogenesis and metastasis through epithelial-mesenchymal transition and upregulates TRPM3 in renal cell carcinoma. *Theranostics*, 10(10), 4395–4409. Available from https://doi.org/10.7150/thno.43239.

Li, X., Xin, S., He, Z., Che, X., Wang, J., Xiao, X., ... Song, X. (2014). MicroRNA-21 (miR-21) post-transcriptionally downregulates tumor suppressor PDCD4 and promotes cell transformation, proliferation, and metastasis in renal cell carcinoma. *Cellular Physiology and Biochemistry: International Journal of Experimental Cellular Physiology, Biochemistry, and Pharmacology*, 33(6), 1631–1642. Available from https://doi.org/10.1159/000362946.

Li, X., Yang, L., & Chen, L. L. (2018). The biogenesis, functions, and challenges of circular RNAs. *Molecular Cell*, 71, 428–442.

Li, Y., Guan, B., Liu, J., Zhang, Z., He, S., Zhan, Y., ... Zhao, W. (2019). MicroRNA-200b is downregulated and suppresses metastasis by targeting LAMA4 in renal cell carcinoma. *EBioMedicine*, 44, 439–451. Available from https://doi.org/10.1016/j.ebiom.2019.05.041.

Li, Y., Wang, T., Chen, D., Yu, Z., Jin, L., Ni, L., ... Lai, Y. (2016). Identification of long-non coding RNA UCA1 as an oncogene in renal cell carcinoma. *Molecular Medicine Reports*, 13(4), 3326–3334. Available from https://doi.org/10.3892/mmr.2016.4894.

Li, Y., Wang, T., Li, Y., Chen, D., Yu, Z., Jin, L., ... Lai, Y. (2016). Identification of long-non coding RNA UCA1 as an oncogene in renal cell carcinoma. *Molecular Medicine Reports*, 13(4), 3326–3334. Available from https://doi.org/10.3892/mmr.2016.4894.

Li, Z., Liu, L., Jiang, S., Li, Q., Feng, C., Du, Q., ... Ma, L. (2021). LncExpDB: An expression database of human long non-coding RNAs. *Nucleic Acids Research*, 49(D1), D962–d968. Available from https://doi.org/10.1093/nar/gkaa850.

Lian, J. H., Wang, W. H., Wang, J. Q., Zhang, Y. H., & Li, Y. (2013). MicroRNA-122 promotes proliferation, invasion and migration of renal cell carcinoma cells through the PI3K/Akt signaling pathway. *Asian Pacific Journal of Cancer Prevention: APJCP*, 14(9), 5017–5021. Available from https://doi.org/10.7314/apjcp.2013.14.9.5017.

Liang, J., Zhang, Y., Jiang, G., Liu, Z., Xiang, W., Chen, X., ... Zhao, J. (2013). MiR-138 induces renal carcinoma cell senescence by targeting EZH2 and is downregulated in human clear cell renal cell carcinoma. *Oncology Research*, 21(2), 83–91. Available from https://doi.org/10.3727/096504013x13775486749218.

Liao, Q., Xiao, H., Bu, D., Xie, C., Miao, R., Luo, H., ... Zhao, Y. (2011). ncFANs: A web server for functional annotation of long non-coding RNAs. *Nucleic Acids Research*, 39(Web Server issue), W118–W124. Available from https://doi.org/10.1093/nar/gkr432.

Lieder, A., Guenzel, T., Lebentrau, S., Schneider, C., & Franzen, A. (2017). Diagnostic relevance of metastatic renal cell carcinoma in the head and neck: An evaluation of 22 cases in 671 patients. *International Brazilian Journal of Urology: Official journal of the Brazilian Society of Urology*, 43(2), 202–208. Available from https://doi.org/10.1590/S1677-5538.IBJU.2015.0665.

Lin, L., & Cai, J. (2020). Circular RNA circ-EGLN3 promotes renal cell carcinoma proliferation and aggressiveness via miR-1299-mediated IRF7 activation. *Journal of Cellular Biochemistry*, 121(11), 4377–4385. Available from https://doi.org/10.1002/jcb.29620.

Liu, E. M., Martinez-Fundichely, A., Bollapragada, R., Spiewack, M., & Khurana, E. (2021). CNCDatabase: A database of non-coding cancer drivers. *Nucleic Acids Research*, 49(D1), D1094–D1101. Available from https://doi.org/10.1093/nar/gkaa915.

Liu, F., Chen, N., Xiao, R., Wang, W., & Pan, Z. (2016). miR-144-3p serves as a tumor suppressor for renal cell carcinoma and inhibits its invasion and metastasis by targeting MAP3K8. *Biochemical and Biophysical Research Communications*, 480(1), 87−93. Available from https://doi.org/10.1016/j.bbrc.2016.10.004.

Liu, F., Wu, L., Wang, A., Xu, Y., Luo, X., Liu, X., ... Chen, S. (2017). MicroRNA-138 attenuates epithelial-to-mesenchymal transition by targeting SOX4 in clear cell renal cell carcinoma. *American Journal of Translational Research*, 9(8), 3611−3622.

Liu, H., Brannon, A. R., Reddy, A. R., Alexe, G., Seiler, M. W., Arreola, A., ... Bhanot, G. V. (2010). Identifying mRNA targets of microRNA dysregulated in cancer: With application to clear cell renal cell carcinoma. *BMC Systems Biology*, 4, 51. Available from https://doi.org/10.1186/1752-0509-4-51.

Liu, H., Chen, P., Jiang, C., Han, J., Zhao, B., Ma, Y., & Mardan, M. (2016). Screening for the key lncRNA targets associated with metastasis of renal clear cell carcinoma. *Medicine(Baltimore)*, 95(2), e2507. Available from https://doi.org/10.1097/md.0000000000002507.

Liu, H., Zhou, X., Yuan, M., Zhou, S., Huang, Y. E., Hou, F., ... Jiang, W. (2020). ncEP: A manually curated database for experimentally validated ncRNA-encoded proteins or peptides. *Journal of Molecular Biology*, 432(11), 3364−3368. Available from https://doi.org/10.1016/j.jmb.2020.02.022.

Liu, K., Yan, Z., Li, Y., & Sun, Z. (2013). Linc2GO: A human LincRNA function annotation resource based on ceRNA hypothesis. *Bioinformatics (Oxford, England)*, 29(17), 2221−2222. Available from https://doi.org/10.1093/bioinformatics/btt361.

Liu, L., Li, Y., Liu, S., Duan, Q., Chen, L., Wu, T., ... Xin, D. (2017). Downregulation of miR-193a-3p inhibits cell growth and migration in renal cell carcinoma by targeting PTEN. *Tumour Biology: The Journal of the International Society for Oncodevelopmental Biology and Medicine*, 39(6). Available from https://doi.org/10.1177/1010428317711951, 1010428317711951.

Liu, L., Pang, X., Shang, W., Xie, H., Feng, Y., & Feng, G. (2019). Long non-coding RNA GAS5 sensitizes renal cell carcinoma to sorafenib via miR-21/SOX5 pathway. *Cell Cycle (Georgetown, Tex.)*, 18(3), 257−263. Available from https://doi.org/10.1080/15384101.2018.1475826.

Liu, M., Wang, Q., Shen, J., Yang, B. B., & Ding, X. (2019). Circbank: A comprehensive database for circRNA with standard nomenclature. *RNA Biology*, 16(7), 899−905. Available from https://doi.org/10.1080/15476286.2019.1600395.

Liu, W., Zabirnyk, O., Wang, H., Shiao, Y. H., Nickerson, M. L., Khalil, S., ... Phang, J. M. (2010a). miR-23b targets proline oxidase, a novel tumor suppressor protein in renal cancer. *Oncogene*, 29(35), 4914−4924. Available from https://doi.org/10.1038/onc.2010.237.

Liu, W., Zabirnyk, O., Wang, H., Shiao, Y. H., Nickerson, M. L., Khalil, S., ... Phang, J. M. (2010b). miR-23b* targets proline oxidase, a novel tumor suppressor protein in renal cancer. *Oncogene*, 29(35), 4914−4924.

Liu, X., Hao, Y., Yu, W., Yang, X., Luo, X., Zhao, J., ... Li, L. (2018). Long non-coding RNA emergence during renal cell carcinoma tumorigenesis. *Cellular Physiology and Biochemistry: International Journal of Experimental Cellular Physiology, Biochemistry, and Pharmacology*, 47(2), 735−746. Available from https://doi.org/10.1159/000490026.

Liu, X. L., Pan, W. G., Li, K. L., Mao, Y. J., Liu, S. D., & Zhang, R. M. (2020). miR-1293 suppresses tumor malignancy by targeting hydrocyanic oxidase 2: Therapeutic potential of a miR-1293/hydrocyanic oxidase 2 axis in renal cell carcinoma. *Cancer Biotherapy & Radiopharmaceuticals*, 35(5), 377−386. Available from https://doi.org/10.1089/cbr.2019.2957.

Liu, Y., Cheng, G., Huang, Z., Bao, L., Liu, J., Wang, C., ... Zhang, X. (2020). Long noncoding RNA SNHG12 promotes tumour progression and sunitinib resistance by upregulating CDCA3 in renal cell carcinoma. *Cell Death and Disease*, 11(7), 515. Available from https://doi.org/10.1038/s41419-020-2713-8.

Liu, Y., Qian, J., Li, X., Chen, W., Xu, A., Zhao, K., ... Wang, Z. (2016). Long noncoding RNA BX357664 regulates cell proliferation and epithelial-to-mesenchymal transition via inhibition of TGF-β1/p38/HSP27 signaling in renal cell carcinoma. *Oncotarget*, 7(49), 81410−81422. Available from https://doi.org/10.18632/oncotarget.12937.

Liu, Y.-C., Li, J.-R., Sun, C.-H., Andrews, E., Chao, R.-F., Lin, F.-M., ... Huang, H.-D. (2016). CircNet: A database of circular RNAs derived from transcriptome sequencing data. *Nucleic Acids Research*, 44(D1), D209−D215. Available from https://doi.org/10.1093/nar/gkv940.

Liu, Z., Yan, H. Y., Xia, S. Y., Zhang, C., & Xiu, Y. C. (2016). Downregulation of long non-coding RNA TRIM52-AS1 functions as a tumor suppressor in renal cell carcinoma. *Molecular Medicine Reports*, 13(4), 3206−3212. Available from https://doi.org/10.3892/mmr.2016.4908.

Lokeshwar, S. D., Talukder, A., Yates, T. J., Hennig, M. J. P., Garcia-Roig, M., Lahorewala, S. S., ... Lokeshwar, V. B. (2018). Molecular characterization of renal cell carcinoma: A potential three-microRNA prognostic signature. *Cancer Epidemiology, Biomarkers & Prevention: A Publication of the American Association for Cancer Research, Cosponsored by the American Society of Preventive Oncology, 27*(4), 464−472. Available from https://doi.org/10.1158/1055-9965.EPI-17-0700.

Long, Q. Z., Du, Y. F., Liu, X. G., Li, X., & He, D. L. (2015). miR-124 represses FZD5 to attenuate P-glycoprotein-mediated chemo-resistance in renal cell carcinoma. *Tumour Biology: The Journal of the International Society for Oncodevelopmental Biology and Medicine, 36*(9), 7017−7026. Available from https://doi.org/10.1007/s13277-015-3369-3.

Lu, J., Wei, J. H., Feng, Z. H., Chen, Z. H., Wang, Y. Q., Huang, Y., ... Luo, J. H. (2017). miR-106b-5p promotes renal cell carcinoma aggressiveness and stem-cell-like phenotype by activating Wnt/β-catenin signalling. *Oncotarget, 8*(13), 21461−21471. Available from https://doi.org/10.18632/oncotarget.15591.

Lu, Q., Ren, S., Lu, M., Zhang, Y., Zhu, D., Zhang, X., & Li, T. (2013). Computational prediction of associations between long non-coding RNAs and proteins. *BMC Genomics, 14*(1), 651. Available from https://doi.org/10.1186/1471-2164-14-651.

Lu, R., Ji, Z., Li, X., Zhai, Q., Zhao, C., Jiang, Z., ... Yu, Z. (2014). miR-145 functions as tumor suppressor and targets two oncogenes, ANGPT2 and NEDD9, in renal cell carcinoma. *Journal of Cancer Research and Clinical Oncology, 140*(3), 387−397. Available from https://doi.org/10.1007/s00432-013-1577-z.

Luo, X., Li, H., Liang, J., Zhao, Q., Xie, Y., Ren, J., & Zuo, Z. (2021). RMVar: An updated database of functional variants involved in RNA modifications. *Nucleic Acids Research, 49*(D1), D1405−d1412. Available from https://doi.org/10.1093/nar/gkaa811.

Lv, L., Huang, F., Mao, H., Li, M., Li, X., Yang, M., & Yu, X. (2013). MicroRNA-21 is overexpressed in renal cell carcinoma. *The International Journal of Biological Markers, 28*(2), 201−207. Available from https://doi.org/10.5301/jbm.2013.10831.

Lv, X., Shen, J., Guo, Z., Kong, L., Zhou, G., & Ning, H. (2019). Aberrant expression of miR-592 is associated with prognosis and progression of renal cell carcinoma. *OncoTargets and Therapy, 12*, 11231−11239. Available from https://doi.org/10.2147/ott.S227834.

Ma, C., Qin, J., Zhang, J., Wang, X., Wu, D., & Li, X. (2020). Construction and analysis of circular RNA molecular regulatory networks in clear cell renal cell carcinoma. *Molecular Medicine Reports, 21*(1), 141−150. Available from https://doi.org/10.3892/mmr.2019.10811.

Ma, L., & Qu, L. (2013). The function of microRNAs in renal development and pathophysiology. *Journal of Genetics and Genomics, 40*, 143−152.

Ma, L., Cao, J., Liu, L., Li, Z., Shireen, H., Pervaiz, N., ... Zhang, Z. (2019). Community curation and expert curation of human long noncoding RNAs with LncRNAWiki and LncBook. *Current Protocols in Bioinformatics, 67*(1), e82. Available from https://doi.org/10.1002/cpbi.82.

Ma, Q., Peng, Z., Wang, L., Li, Y., Wang, K., Zheng, J., ... Liu, T. (2016). miR-19a correlates with poor prognosis of clear cell renal cell carcinoma patients via promoting cell proliferation and suppressing PTEN/SMAD4 expression. *International Journal of Oncology, 49*(6), 2589−2599. Available from https://doi.org/10.3892/ijo.2016.3746.

Ma, S., Kong, S., Wang, F., & Ju, S. (2020). CircRNAs: Biogenesis, functions, and role in drug-resistant tumours. *Molecular Cancer, 19*(1), 119. Available from https://doi.org/10.1186/s12943-020-01231-4.

Machackova, T., Mlcochova, H., Stanik, M., Dolezel, J., Fedorko, M., Pacik, D., ... Slaby, O. (2016). MiR-429 is linked to metastasis and poor prognosis in renal cell carcinoma by affecting epithelial-mesenchymal transition. *Tumour Biology: The Journal of the International Society for Oncodevelopmental Biology and Medicine, 37*(11), 14653−14658. Available from https://doi.org/10.1007/s13277-016-5310-9.

Majid, S., Saini, S., Dar, A. A., Hirata, H., Shahryari, V., Tanaka, Y., ... Dahiya, R. (2011). MicroRNA-205 inhibits Src-mediated oncogenic pathways in renal cancer. *Cancer Research, 71*(7), 2611−2621. Available from https://doi.org/10.1158/0008-5472.CAN-10-3666.

Mann, M., Wright, P. R., & Backofen, R. (2017). IntaRNA 2.0: Enhanced and customizable prediction of RNA−RNA interactions. *Nucleic Acids Research, 45*(W1), W435−w439. Available from https://doi.org/10.1093/nar/gkx279.

Martens-Uzunova, E. S., Böttcher, R., Croce, C. M., Jenster, G., Visakorpi, T., & Calin, G. A. (2014). Long noncoding RNA in prostate, bladder, and kidney cancer. *European Urology, 65*(6), 1140−1151. Available from https://doi.org/10.1016/j.eururo.2013.12.003.

Mas-Ponte, D., Carlevaro-Fita, J., Palumbo, E., Hermoso Pulido, T., Guigo, R., & Johnson, R. (2017). LncATLAS database for subcellular localization of long noncoding RNAs. *RNA (New York, N.Y.), 23*(7), 1080–1087. Available from https://doi.org/10.1261/rna.060814.117.

Matsui, M., & Corey, D. R. (2017). Non-coding RNAs as drug targets. *Nature Reviews Drug Discovery, 16*(3), 167–179.

May, J. M., Bylicky, M., Chopra, S., Coleman, C. N., & Aryankalayil, M. J. (2021). Long and short non-coding RNA and radiation response: A review. *Translational Research: The Journal of Laboratory and Clinical Medicine, 233*, 162–179. Available from https://doi.org/10.1016/j.trsl.2021.02.005.

McCormick, R. I., Blick, C., Ragoussis, J., Schoedel, J., Mole, D. R., Young, A. C., ... Harris, A. L. (2013). miR-210 is a target of hypoxia-inducible factors 1 and 2 in renal cancer, regulates ISCU and correlates with good prognosis. *British Journal of Cancer, 108*(5), 1133–1142. Available from https://doi.org/10.1038/bjc.2013.56.

Memczak, S., Jens, M., Elefsinioti, A., Torti, F., Krueger, J., Rybak, A., ... Rajewsky, N. (2013). Circular RNAs are a large class of animal RNAs with regulatory potency. *Nature, 495*(7441), 333–338. Available from https://doi.org/10.1038/nature11928.

Meng, X., Chen, Q., Zhang, P., & Chen, M. (2017). CircPro: An integrated tool for the identification of circRNAs with protein-coding potential. *Bioinformatics (Oxford, England), 33*(20), 3314–3316. Available from https://doi.org/10.1093/bioinformatics/btx446.

Meng, X., Hu, D., Zhang, P., Chen, Q., & Chen, M. (2019). CircFunBase: A database for functional circular RNAs. *Database (Oxford), 2019*. Available from https://doi.org/10.1093/database/baz003.

Mytsyk, Y., Dosenko, V., Skrzypczyk, M. A., Borys, Y., Diychuk, Y., Kucher, A., ... Manyuk, L. (2018). Potential clinical applications of microRNAs as biomarkers for renal cell carcinoma. *Central European Journal of Urology, 71*(3), 295–303. Available from https://doi.org/10.5173/ceju.2018.1618.

Nakada, C., Matsuura, K., Tsukamoto, Y., Tanigawa, M., Yoshimoto, T., Narimatsu, T., ... Moriyama, M. (2008). Genome-wide microRNA expression profiling in renal cell carcinoma: Significant down-regulation of miR-141 and miR-200c. *The Journal of Pathology, 216*(4), 418–427. Available from https://doi.org/10.1002/path.2437.

Nakada, C., Tsukamoto, Y., Matsuura, K., Nguyen, T. L., Hijiya, N., Uchida, T., ... Moriyama, M. (2011). Overexpression of miR-210, a downstream target of HIF1α, causes centrosome amplification in renal carcinoma cells. *The Journal of Pathology, 224*(2), 280–288. Available from https://doi.org/10.1002/path.2860.

Nakamura, T., Iwamoto, T., Nakamura, H. M., Shindo, Y., Saito, K., Yamada, A., ... Fukumoto, S. (2020). Regulation of miR-1-mediated connexin 43 expression and cell proliferation in dental epithelial cells. *Frontiers in Cell and Developmental Biology, 8*, 156. Available from https://doi.org/10.3389/fcell.2020.00156.

Nakata, W., Uemura, M., Sato, M., Fujita, K., Jingushi, K., Ueda, Y., ... Nonomura, N. (2015). Expression of miR-27a-3p is an independent predictive factor for recurrence in clear cell renal cell carcinoma. *Oncotarget, 6*(25), 21645–21654. Available from https://doi.org/10.18632/oncotarget.4064.

Neal, C. S., Michael, M. Z., Rawlings, L. H., Van der Hoek, M. B., & Gleadle, J. M. (2010). The VHL-dependent regulation of microRNAs in renal cancer. *BMC Medicine, 8*, 64. Available from https://doi.org/10.1186/1741-7015-8-64.

Ning, S., Zhang, J., Wang, P., Zhi, H., Wang, J., Liu, Y., ... Li, X. (2016). Lnc2Cancer: A manually curated database of experimentally supported lncRNAs associated with various human cancers. *Nucleic Acids Research, 44*(D1), D980–D985. Available from https://doi.org/10.1093/nar/gkv1094.

Ning, S., Zhao, Z., Ye, J., Wang, P., Zhi, H., Li, R., ... Li, X. (2014). LincSNP: A database of linking disease-associated SNPs to human large intergenic non-coding RNAs. *BMC Bioinformatics, 15*, 152. Available from https://doi.org/10.1186/1471-2105-15-152.

Nofech-Mozes, R., Khella, H. W., Scorilas, A., Youssef, L., Krylov, S. N., Lianidou, E., ... Yousef, G. M. (2016). MicroRNA-194 is a marker for good prognosis in clear cell renal cell carcinoma. *Cancer Med, 5*(4), 656–664. Available from https://doi.org/10.1002/cam4.631.

Okato, A., Arai, T., Yamada, Y., Sugawara, S., Koshizuka, K., Fujimura, L., ... Seki, N. (2017). Dual strands of pre-miR-149 inhibit cancer cell migration and invasion through targeting FOXM1 in renal cell carcinoma. *International Journal of Molecular Sciences, 18*(9). Available from https://doi.org/10.3390/ijms18091969.

Oliveira, R. C., Ivanovic, R. F., Leite, K. R. M., Viana, N. I., Pimenta, R. C. A., Junior, J. P., ... Reis, S. T. (2017). Expression of micro-RNAs and genes related to angiogenesis in ccRCC and associations with tumor characteristics. *BMC Urology, 17*(1), 113. Available from https://doi.org/10.1186/s12894-017-0306-3.

Onyshchenko, K. V., Voitsitskyi, T. V., Grygorenko, V. M., Saidakova, N. O., Pereta, L. V., Onyschuk, A. P., & Skrypkina, I. Y. (2020). Expression of micro-RNA hsa-miR-30c-5p and hsa-miR-138-1 in renal cell carcinoma.

Experimental Oncology, 42(2), 115−119. Available from https://doi.org/10.32471/exp-oncology.2312-8852.vol-42-no-2.14632.

Osako, Y., Yoshino, H., Sakaguchi, T., Sugita, S., Yonemori, M., Nakagawa, M., & Enokida, H. (2019). Potential tumor-suppressive role of microRNA-99a-3p in sunitinib-resistant renal cell carcinoma cells through the regulation of RRM2. *International Journal of Oncology, 54*(5), 1759−1770. Available from https://doi.org/10.3892/ijo.2019.4736.

Osanto, S., Qin, Y., Buermans, H. P., Berkers, J., Lerut, E., Goeman, J. J., & van Poppel, H. (2012). Genome-wide microRNA expression analysis of clear cell renal cell carcinoma by next generation deep sequencing. *PLoS One, 7*(6), e38298. Available from https://doi.org/10.1371/journal.pone.0038298.

Pal, S. K., & Agarwal, N. (2016). Kidney cancer: Finding a niche for girentuximab in metastatic renal cell carcinoma. *Nature Reviews Urology, 13*, 442−443.

Pan, Y., Hu, J., Ma, J., Qi, X., Zhou, H., Miao, X., ... Jia, L. (2018). MiR-193a-3p and miR-224 mediate renal cell carcinoma progression by targeting alpha-2,3-sialyltransferase IV and the phosphatidylinositol 3 kinase/Akt pathway. *Molecular Carcinogenesis, 57*(8), 1067−1077. Available from https://doi.org/10.1002/mc.22826.

Pan, Y. J., Wei, L. L., Wu, X. J., Huo, F. C., Mou, J., & Pei, D. S. (2017). MiR-106a-5p inhibits the cell migration and invasion of renal cell carcinoma through targeting PAK5. *Cell Death and Disease, 8*(10), e3155. Available from https://doi.org/10.1038/cddis.2017.561.

Papatheodorou, I., Moreno, P., Manning, J., Fuentes, A. M., George, N., Fexova, S., ... Brazma, A. (2020). Expression Atlas update: From tissues to single cells. *Nucleic Acids Research, 48*(D1), D77−d83. Available from https://doi.org/10.1093/nar/gkz947.

Paraskevopoulou, M. D., Georgakilas, G., Kostoulas, N., Reczko, M., Maragkakis, M., Dalamagas, T. M., & Hatzigeorgiou, A. G. (2013). DIANA-LncBase: Experimentally verified and computationally predicted microRNA targets on long non-coding RNAs. *Nucleic Acids Research, 41*(Database issue), D239−D245. Available from https://doi.org/10.1093/nar/gks1246.

Park, C., Yu, N., Choi, I., Kim, W., & Lee, S. (2014). lncRNAtor: A comprehensive resource for functional investigation of long non-coding RNAs. *Bioinformatics (Oxford, England), 30*(17), 2480−2485. Available from https://doi.org/10.1093/bioinformatics/btu325.

Patard, J. J., Leray, E., Rioux-Leclercq, N., Cindolo, L., Ficarra, V., Zisman, A., ... Pantuck, A. J. (2005). Prognostic value of histologic subtypes in renal cell carcinoma: A multicenter experience. *Journal of Clinical Oncology: Official Journal of the American Society of Clinical Oncology, 23*, 2763−2771.

Patop, I. L., Wüst, S., & Kadener, S. (2019). Past, present, and future of circRNAs. *The EMBO Journal, 38*(16), e100836.

Pei, C. S., Wu, H. Y., Fan, F. T., Wu, Y., Shen, C. S., & Pan, L. Q. (2014). Influence of curcumin on HOTAIR-mediated migration of human renal cell carcinoma cells. *Asian Pacific Journal of Cancer Prevention: APJCP, 15*(10), 4239−4243. Available from https://doi.org/10.7314/apjcp.2014.15.10.4239.

Petrozza, V., Costantini, M., Tito, C., Giammusso, L. M., Sorrentino, V., Cacciotti, J., ... Fazi, F. (2020). Emerging role of secreted miR-210-3p as potential biomarker for clear cell renal cell carcinoma metastasis. *Cancer Biomarkers: Section A of Disease Markers, 27*(2), 181−188. Available from https://doi.org/10.3233/cbm-190242.

Porta, C., Gore, M. E., Rini, B. I., Escudier, B., Hariharan, S., Charles, L. P., ... Motzer, R. J. (2016). Long-term safety of sunitinib in metastatic renal cell carcinoma. *European Urology, 69*(2), 345−351. Available from https://doi.org/10.1016/j.eururo.2015.07.006.

Posa, I., Carvalho, S., Tavares, J., & Grosso, A. R. (2016). A pan-cancer analysis of MYC-PVT1 reveals CNV-unmediated deregulation and poor prognosis in renal carcinoma. *Oncotarget, 7*(30), 47033−47041. Available from https://doi.org/10.18632/oncotarget.9487.

Prasad, S. R., Humphrey, P. A., Catena, J. R., Narra, V. R., Srigley, J. R., Cortez, A. D., ... Chintapalli, K. N. (2006). Common and uncommon histologic subtypes of renal cell carcinoma: Imaging spectrum with pathologic correlation. *Radiographics: A Review Publication of the Radiological Society of North America, Inc, 26*(6), 1795−1806, discussion 1806-1710. Available from https://doi.org/10.1148/rg.266065010.

Prior, C., Perez-Gracia, J. L., Garcia-Donas, J., Rodriguez-Antona, C., Gurucega, E., Esteban, E., ... Calvo, A. (2014). Identification of tissue microRNAs predictive of sunitinib activity in patients with metastatic renal cell carcinoma. *PLoS One, 9*(1), e86263. Available from https://doi.org/10.1371/journal.pone.0086263.

Pronina, I. V., Klimov, E. A., Burdennyy, A. M., Beresneva, E. V., Fridman, M. V., Ermilova, V. D., ... Loginov, V. I. (2017). Methylation of the genes for the microRNAs miR-129-2 and miR-9-1, changes in their expression, and activation of their potential target genes in clear cell renal cell carcinoma. *Molecular Biology (Mosk)*, *51*(1), 73−84. Available from https://doi.org/10.7868/s0026898416060161.

Puente, J., Laínez, N., Dueñas, M., Méndez-Vidal, M. J., Esteban, E., Castellano, D., ... Group, S. S. O. G. (2017). Novel potential predictive markers of sunitinib outcomes in long-term responders vs primary refractory patients with metastatic clear-cell renal cell carcinoma. *Oncotarget*, *8*(18), 30410−30421. Available from https://doi.org/10.18632/oncotarget.16494.

Qiao, H. P., Gao, W. S., Huo, J. X., & Yang, Z. S. (2013). Long non-coding RNA GAS5 functions as a tumor suppressor in renal cell carcinoma. *Asian Pacific Journal of Cancer Prevention: APJCP*, *14*(2), 1077−1082. Available from https://doi.org/10.7314/apjcp.2013.14.2.1077.

Qin, Z., Wei, X., Jin, N., Wang, Y., Zhao, R., Hu, Y., ... Zhou, Q. (2018). MiR-199a targeting ROCK1 to affect kidney cell proliferation, invasion and apoptosis. *Artificial Cells, Nanomedicine, and Biotechnology*, *46*(8), 1920−1925. Available from https://doi.org/10.1080/21691401.2017.1396224.

Qiu, M., Liu, L., Chen, L., Tan, G., Liang, Z., Wang, K., ... Chen, H. (2014). microRNA-183 plays as oncogenes by increasing cell proliferation, migration and invasion via targeting protein phosphatase 2A in renal cancer cells. *Biochemical and Biophysical Research Communications*, *452*(1), 163−169. Available from https://doi.org/10.1016/j.bbrc.2014.08.067.

Qu, L., Ding, J., Chen, C., Wu, Z. J., Liu, B., Gao, Y., ... Wang, L. H. (2016). Exosome-transmitted lncARSR promotes sunitinib resistance in renal cancer by acting as a competing endogenous RNA. *Cancer Cell*, *29*(5), 653−668. Available from https://doi.org/10.1016/j.ccell.2016.03.004.

Quek, X. C., Thomson, D. W., Maag., Jesper, L. V., Bartonicek, N., Signal, B., Clark, M. B., ... Dinger, M. E. (2015). lncRNAdb v2.0: Expanding the reference database for functional long noncoding RNAs. *Nucleic Acids Research*, *43*(D1), D168−D173. Available from https://doi.org/10.1093/nar/gku988.

Redova, M., Poprach, A., Besse, A., Iliev, R., Nekvindova, J., Lakomy, R., ... Slaby, O. (2013). MiR-210 expression in tumor tissue and in vitro effects of its silencing in renal cell carcinoma. *Tumour Biology: The Journal of the International Society for Oncodevelopmental Biology and Medicine*, *34*(1), 481−491. Available from https://doi.org/10.1007/s13277-012-0573-2.

Redova, M., Poprach, A., Nekvindova, J., Iliev, R., Radova, L., Lakomy, R., ... Slaby, O. (2012). Circulating miR-378 and miR-451 in serum are potential biomarkers for renal cell carcinoma. *Journal of Translational Medicine*, *10*, 55. Available from https://doi.org/10.1186/1479-5876-10-55.

Ren, C., An, G., Zhao, C., Ouyang, Z., Bo, X., & Shu, W. (2018). Lnc2Catlas: An atlas of long noncoding RNAs associated with risk of cancers. *Scientific Reports*, *8*(1), 1909. Available from https://doi.org/10.1038/s41598-018-20232-4.

Ren, X., Lan, T., Chen, Y., Shao, Z., Yang, C., & Peng, J. (2016). lncRNA uc009yby.1 promotes renal cell proliferation and is associated with poor survival in patients with clear cell renal cell carcinomas. *Oncology Letters*, *12*(3), 1929−1934. Available from https://doi.org/10.3892/ol.2016.4856.

Ricketts, C. J., De Cubas, A. A., Fan, H., Smith, C. C., Lang, M., Reznik, E., ... Network, C. G. A. R. (2018). The cancer genome atlas comprehensive molecular characterization of renal cell carcinoma. *Cell Rep*, *23*(1), 313−326, e315. Available from https://doi.org/10.1016/j.celrep.2018.03.075.

Rophina, M., Sharma, D., Poojary, M., & Scaria, V. (2020). Circad: A comprehensive manually curated resource of circular RNA associated with diseases. *Database (Oxford)*, *2020*. Available from https://doi.org/10.1093/database/baaa019.

Rossbach, O. (2019). Artificial circular RNA sponges targeting microRNAs as a novel tool in molecular biology. *Molecular Therapy − Nucleic Acids*, *17*, 452−454. Available from https://doi.org/10.1016/j.omtn.2019.06.021.

Rossi, S. H., Klatte, T., Usher-Smith, J., & Stewart, G. D. (2018). Epidemiology and screening for renal cancer. *World Journal of Urology*, *36*(9), 1341−1353. Available from https://doi.org/10.1007/s00345-018-2286-7.

Ruan, H., Xiang, Y., Ko, J., Li, S., Jing, Y., Zhu, X., ... Han, L. (2019). Comprehensive characterization of circular RNAs in ~ 1000 human cancer cell lines. *Genome Medicine*, *11*(1), 55. Available from https://doi.org/10.1186/s13073-019-0663-5.

Saini, S., Yamamura, S., Majid, S., Shahryari, V., Hirata, H., Tanaka, Y., & Dahiya, R. (2011). MicroRNA-708 induces apoptosis and suppresses tumorigenicity in renal cancer cells. *Cancer Research*, *71*(19), 6208−6219. Available from https://doi.org/10.1158/0008-5472.Can-11-0073.

Sakurai, K., Reon, B. J., Anaya, J., & Dutta, A. (2015). The lncRNA DRAIC/PCAT29 locus constitutes a tumor-suppressive nexus. *Molecular Cancer Research: MCR*, *13*(5), 828−838. Available from https://doi.org/10.1158/1541-7786.Mcr-15-0016-t.

Salhi, A., Essack, M., Alam, T., Bajic, V. P., Ma, L., Radovanovic, A., ... Bajic, V. B. (2017). DES-ncRNA: A knowledgebase for exploring information about human micro and long noncoding RNAs based on literature-mining. *RNA Biology, 14*(7), 963–971. Available from https://doi.org/10.1080/15476286.2017.1312243.

Salmena, L., Poliseno, L., Tay, Y., Kats, L., & Pandolfi, P. P. (2011). A ceRNA hypothesis: The rosetta stone of a hidden RNA language? *Cell, 146*(3), 353–358.

Samaan, S., Khella, H. W., Girgis, A., Scorilas, A., Lianidou, E., Gabril, M., ... Yousef, G. M. (2015). miR-210 is a prognostic marker in clear cell renal cell carcinoma. *The Journal of Molecular Diagnostics: JMD, 17*(2), 136–144. Available from https://doi.org/10.1016/j.jmoldx.2014.10.005.

Sarver, A. L., & Subramanian, S. (2012). Competing endogenous RNA database. *Bioinformation, 8*(15), 731–733. Available from https://doi.org/10.6026/97320630008731.

Seifuddin, F., Singh, K., Suresh, A., Judy, J. T., Chen, Y. C., Chaitankar, V., ... Pirooznia, M. (2020). lncRNAKB, a knowledgebase of tissue-specific functional annotation and trait association of long noncoding RNA. *Scientific Data, 7*(1), 326. Available from https://doi.org/10.1038/s41597-020-00659-z.

Shang, D., Liu, Y., Ito, N., Kamoto, T., & Ogawa, O. (2007). Defective Jak-Stat activation in renal cell carcinoma is associated with interferon-alpha resistance. *Cancer Science, 98*, 1259–1264, England.

Shao, K., Shi, T., Yang, Y., Wang, X., Xu, D., & Zhou, P. (2016). Highly expressed lncRNA CRNDE promotes cell proliferation through Wnt/β-catenin signaling in renal cell carcinoma. *Tumour Biology: The Journal of the International Society for Oncodevelopmental Biology and Medicine*. Available from https://doi.org/10.1007/s13277-016-5440-0.

Shi, H., Sun, Y., He, M., Yang, X., Hamada, M., Fukunaga, T., ... Chang, C. (2020). Targeting the TR4 nuclear receptor-mediated lncTASR/AXL signaling with tretinoin increases the sunitinib sensitivity to better suppress the RCC progression. *Oncogene, 39*(3), 530–545. Available from https://doi.org/10.1038/s41388-019-0962-8.

Shi, X., Sun, M., Liu, H., Yao, Y., & Song, Y. (2013). Long non-coding RNAs: A new frontier in the study of human diseases. *Cancer Letters, 339*(2), 159–166. Available from https://doi.org/10.1016/j.canlet.2013.06.013.

Shinmei, S., Sakamoto, N., Goto, K., Sentani, K., Anami, K., Hayashi, T., ... Yasui, W. (2013). MicroRNA-155 is a predictive marker for survival in patients with clear cell renal cell carcinoma. *International Journal of Urology, 20*(5), 468–477. Available from https://doi.org/10.1111/j.1442-2042.2012.03182.x.

Shiomi, E., Sugai, T., Ishida, K., Osakabe, M., Tsuyukubo, T., Kato, Y., ... Obara, W. (2019). Analysis of expression patterns of microRNAs that are closely associated with renal carcinogenesis. *Frontiers in Oncology, 9*, 431. Available from https://doi.org/10.3389/fonc.2019.00431.

Siegel, R. L., Miller, K. D., Fuchs, H. E., & Jemal, A. (2021). Cancer statistics, 2021. *CA: A Cancer Journal for Clinicians, 71*(1), 7–33. Available from https://doi.org/10.3322/caac.21654.

Slaby, O., Jancovicova, J., Lakomy, R., Svoboda, M., Poprach, A., Fabian, P., ... Vyzula, R. (2010a). Expression of miRNA-106b in conventional renal cell carcinoma is a potential marker for prediction of early metastasis after nephrectomy. *Journal of Experimental & Clinical Cancer Research: CR, 29*, 90. Available from https://doi.org/10.1186/1756-9966-29-90.

Slaby, O., Jancovicova, J., Lakomy, R., Svoboda, M., Poprach, A., Fabian, P., ... Vyzula, R. (2010b).). Expression of miRNA-106b in conventional renal cell carcinoma is a potential marker for prediction of early metastasis after nephrectomy. *Journal of Experimental & Clinical Cancer Research: CR, 29*(1), 90. Available from https://doi.org/10.1186/1756-9966-29-90.

Slaby, O., Redova, M., Poprach, A., Nekvindova, J., Iliev, R., Radova, L., ... Vyzula, R. (2012). Identification of MicroRNAs associated with early relapse after nephrectomy in renal cell carcinoma patients. *Genes, Chromosomes & Cancer, 51*(7), 707–716. Available from https://doi.org/10.1002/gcc.21957.

Song, S., Wu, Z., Wang, C., Liu, B., Ye, X., Chen, J., ... Wang, L. (2014). RCCRT1 is correlated with prognosis and promotes cell migration and invasion in renal cell carcinoma. *Urology, 84*(3), 730, e731-737. Available from https://doi.org/10.1016/j.urology.2014.05.033.

Song, T., Zhang, X., Wang, C., Wu, Y., Cai, W., Gao, J., & Hong, B. (2011). MiR-138 suppresses expression of hypoxia-inducible factor 1α (HIF-1α) in clear cell renal cell carcinoma 786-O cells. *Asian Pacific Journal of Cancer Prevention: APJCP, 12*(5), 1307–1311. Retrieved from https://www.ncbi.nlm.nih.gov/pubmed/21875287.

Su, H., Sun, T., Wang, H., Shi, G., Zhang, H., Sun, F., & Ye, D. (2017). Decreased TCL6 expression is associated with poor prognosis in patients with clear cell renal cell carcinoma. *Oncotarget, 8*(4), 5789–5799. Available from https://doi.org/10.18632/oncotarget.11011.

Sumazin, P., Yang, X., Chiu, H.-S., Chung, W.-J., Iyer, A., Llobet-Navas, D., ... Califano, A. (2011). An extensive microRNA-mediated network of RNA–RNA interactions regulates established oncogenic pathways in glioblastoma. *Cell*, *147*(2), 370–381. Available from https://doi.org/10.1016/j.cell.2011.09.041.

Sun, L., Luo, H., Bu, D., Zhao, G., Yu, K., Zhang, C., ... Zhao, Y. (2013). Utilizing sequence intrinsic composition to classify protein-coding and long non-coding transcripts. *Nucleic Acids Research*, *41*(17), e166. Available from https://doi.org/10.1093/nar/gkt646.

Sun, X., Lou, L., Zhong, K., & Wan, L. (2017). MicroRNA-451 regulates chemoresistance in renal cell carcinoma by targeting ATF-2 gene. *Experimental Biology and Medicine (Maywood)*, *242*(12), 1299–1305. Available from https://doi.org/10.1177/1535370217701625.

Szcześniak, M. W., Bryzghalov, O., Ciomborowska-Basheer, J., & Makałowska, I. (2019). CANTATAdb 2.0: Expanding the collection of plant long noncoding RNAs. *Methods in Molecular Biology*, *1933*, 415–429. Available from https://doi.org/10.1007/978-1-4939-9045-0_26.

Tang, K., & Xu, H. (2015). Prognostic value of *meta*-signature miRNAs in renal cell carcinoma: An integrated miRNA expression profiling analysis. *Scientific Reports*, *5*, 10272. Available from https://doi.org/10.1038/srep10272.

Tang, Z., Li, X., Zhao, J., Qian, F., Feng, C., Li, Y., ... Li, C. (2019). TRCirc: A resource for transcriptional regulation information of circRNAs. *Briefings in Bioinformatics*, *20*(6), 2327–2333. Available from https://doi.org/10.1093/bib/bby083.

Teng, X., Chen, X., Xue, H., Tang, Y., Zhang, P., Kang, Q., ... He, S. (2020). NPInter v4.0: An integrated database of ncRNA interactions. *Nucleic Acids Research*, *48*(D1), D160–d165. Available from https://doi.org/10.1093/nar/gkz969.

The GTEx Consortium Atlas of Genetic Regulatory Effects Across Human Tissues. (2020). *Science (New York, N.Y.)*, *369*(6509), 1318–1330. Available from https://doi.org/10.1126/science.aaz1776.

Toraih, E. A., Ibrahiem, A. T., Fawzy, M. S., Hussein, M. H., Al-Qahtani, S. A. M., & Shaalan, A. A. M. (2017). MicroRNA-34a: A key regulator in the hallmarks of renal cell carcinoma. *Oxidative Medicine and Cellular Longevity*, *2017*, 3269379. Available from https://doi.org/10.1155/2017/3269379.

Ueno, K., Hirata, H., Shahryari, V., Chen, Y., Zaman, M. S., Singh, K., ... Dahiya, R. (2011). Tumour suppressor microRNA-584 directly targets oncogene Rock-1 and decreases invasion ability in human clear cell renal cell carcinoma. *British Journal of Cancer*, *104*(2), 308–315. Available from https://doi.org/10.1038/sj.bjc.6606028.

Valera, V. A., Walter, B. A., Linehan, W. M., & Merino, M. J. (2011). Regulatory effects of microRNA-92 (miR-92) on VHL gene expression and the hypoxic activation of miR-210 in clear cell renal cell carcinoma. *Journal of Cancer*, *2*, 515–526. Available from https://doi.org/10.7150/jca.2.515.

Vea, A., Llorente-Cortes, V., & de Gonzalo-Calvo, D. (2018). Circular RNAs in blood. *Advances in Experimental Medicine and Biology*, *1087*, 119–130. Available from https://doi.org/10.1007/978-981-13-1426-1_10.

Vergho, D., Kneitz, S., Rosenwald, A., Scherer, C., Spahn, M., Burger, M., ... Kneitz, B. (2014). Combination of expression levels of miR-21 and miR-126 is associated with cancer-specific survival in clear-cell renal cell carcinoma. *BMC Cancer*, *14*, 25. Available from https://doi.org/10.1186/1471-2407-14-25.

Vo, J. N., Cieslik, M., Zhang, Y., Shukla, S., Xiao, L., Wu, Y. M., ... Chinnaiyan, A. M. (2019). The landscape of circular RNA in cancer. *Cell*, *176*(4), 869–881, e813. Available from https://doi.org/10.1016/j.cell.2018.12.021.

Vo, J. N., Cieslik, M., Zhang, Y., Shukla, S., Xiao, L., Zhang, Y., ... Chinnaiyan, A. M. (2019). The landscape of circular RNA in cancer. *Cell*, *176*(4), 869–881, e813. Available from https://doi.org/10.1016/j.cell.2018.12.021.

Volders, P. J., Helsens, K., Wang, X., Menten, B., Martens, L., Gevaert, K., ... Mestdagh, P. (2013). LNCipedia: A database for annotated human lncRNA transcript sequences and structures. *Nucleic Acids Research*, *41*(Database issue), D246–D251. Available from https://doi.org/10.1093/nar/gks915.

von Brandenstein, M., Pandarakalam, J. J., Kroon, L., Loeser, H., Herden, J., Braun, G., ... Fries, J. W. (2012). MicroRNA 15a, inversely correlated to PKCα, is a potential marker to differentiate between benign and malignant renal tumors in biopsy and urine samples. *The American Journal of Pathology*, *180*(5), 1787–1797. Available from https://doi.org/10.1016/j.ajpath.2012.01.014.

Wang, C., Hu, J., Lu, M., Gu, H., Zhou, X., Chen, X., ... Zhang, C. (2015). A panel of five serum miRNAs as a potential diagnostic tool for early-stage renal cell carcinoma. *Scientific Reports*, *5*, 7610. Available from https://doi.org/10.1038/srep07610.

Wang, C., Tang, K., Li, Z., Chen, Z., Xu, H., & Ye, Z. (2016). Targeted p21(WAF1/CIP1) activation by miR-1236 inhibits cell proliferation and correlates with favorable survival in renal cell carcinoma. *Urologic Oncology*, *34*(2), 59, e23-34. Available from https://doi.org/10.1016/j.urolonc.2015.08.014.

Wang, C., Wu, C., Yang, Q., Ding, M., Zhong, J., Zhang, C. Y., ... Zhang, C. (2016). miR-28-5p acts as a tumor suppressor in renal cell carcinoma for multiple antitumor effects by targeting RAP1B. *Oncotarget, 7*(45), 73888−73902. Available from https://doi.org/10.18632/oncotarget.12516.

Wang, D., Zhu, C., Zhang, Y., Zheng, Y., Ma, F., Su, L., & Shao, G. (2017). MicroRNA-30e-3p inhibits cell invasion and migration in clear cell renal cell carcinoma by targeting Snail1. *Oncology Letters, 13*(4), 2053−2058. Available from https://doi.org/10.3892/ol.2017.5690.

Wang, G., Chen, L., Meng, J., Chen, M., Zhuang, L., & Zhang, L. (2013). Overexpression of microRNA-100 predicts an unfavorable prognosis in renal cell carcinoma. *International Urology and Nephrology, 45*(2), 373−379. Available from https://doi.org/10.1007/s11255-012-0374-y.

Wang, G., Xue, W., Jian, W., Liu, P., Wang, Z., Wang, C., ... Zhang, C. (2018). The effect of Hsa_circ_0001451 in clear cell renal cell carcinoma cells and its relationship with clinicopathological features. *Journal of Cancer, 9*(18), 3269−3277. Available from https://doi.org/10.7150/jca.25902.

Wang, J., Cao, Y., Zhang, H., Wang, T., Tian, Q., Lu, X., ... Wang, L. (2017). NSDNA: A manually curated database of experimentally supported ncRNAs associated with nervous system diseases. *Nucleic Acids Research, 45*(D1), D902−d907. Available from https://doi.org/10.1093/nar/gkw1038.

Wang, J., Wang, C., Li, Q., Guo, C., Sun, W., Zhao, D., ... Sun, M. Z. (2020). miR-429-CRKL axis regulates clear cell renal cell carcinoma malignant progression through SOS1/MEK/ERK/MMP2/MMP9 pathway. *Biomedicine & Pharmacotherapy = Biomedecine & Pharmacotherapie, 127*, 110215. Available from https://doi.org/10.1016/j.biopha.2020.110215.

Wang, J., Zhu, S., Meng, N., He, Y., Lu, R., & Yan, G. R. (2019). ncRNA-Encoded peptides or proteins and cancer. *Molecular Therapy: The Journal of the American Society of Gene Therapy, 27*(10), 1718−1725.

Wang, K., Sun, Y., Tao, W., Fei, X., & Chang, C. (2017). Androgen receptor (AR) promotes clear cell renal cell carcinoma (ccRCC) migration and invasion via altering the circHIAT1/miR-195-5p/29a-3p/29c-3p/CDC42 signals. *Cancer Letters, 394*, 1−12. Available from https://doi.org/10.1016/j.canlet.2016.12.036.

Wang, L., Cai, Y., Zhao, X., Jia, X., Zhang, J., Liu, J., ... Wang, J. (2015). Down-regulated long non-coding RNA H19 inhibits carcinogenesis of renal cell carcinoma. *Neoplasma, 62*(3), 412−418. Available from https://doi.org/10.4149/neo_2015_049.

Wang, L., Liu, Y., Zhong, X., Liu, H., Lu, C., Li, C., & Zhang, H. (2019). DMfold: A novel method to predict RNA secondary structure with pseudoknots based on deep learning and improved base pair maximization principle. *Frontiers in Genetics, 10*, 143. Available from https://doi.org/10.3389/fgene.2019.00143.

Wang, M., Gao, H., Qu, H., Li, J., Liu, K., & Han, Z. (2018). MiR-137 suppresses tumor growth and metastasis in clear cell renal cell carcinoma. *Pharmacological Reports: PR, 70*(5), 963−971. Available from https://doi.org/10.1016/j.pharep.2018.04.006.

Wang, M., Huang, T., Luo, G., Huang, C., Xiao, X. Y., Wang, L., ... Zeng, F. Q. (2015). Long non-coding RNA MEG3 induces renal cell carcinoma cells apoptosis by activating the mitochondrial pathway. *Journal of Huazhong University of Science and Technology. Medical Sciences = Hua Zhong ke ji da xue xue bao. Yi xue Ying De wen ban = Huazhong Keji Daxue Xuebao. Yixue Yingdewen ban, 35*(4), 541−545. Available from https://doi.org/10.1007/s11596-015-1467-5.

Wang, P., Guo, Q., Hao, Y., Liu, Q., Gao, Y., Zhi, H., ... Li, X. (2021). LnCeCell: A comprehensive database of predicted lncRNA-associated ceRNA networks at single-cell resolution. *Nucleic Acids Research, 49*(D1), D125−d133. Available from https://doi.org/10.1093/nar/gkaa1017.

Wang, P., Li, X., Gao, Y., Guo, Q., Ning, S., Zhang, Y., ... Li, X. (2020). LnCeVar: A comprehensive database of genomic variations that disturb ceRNA network regulation. *Nucleic Acids Research, 48*(D1), D111−d117. Available from https://doi.org/10.1093/nar/gkz887.

Wang, P., Li, X., Gao, Y., Guo, Q., Wang, Y., Fang, Y., ... Li, X. (2019). LncACTdb 2.0: An updated database of experimentally supported ceRNA interactions curated from low- and high-throughput experiments. *Nucleic Acids Research, 47*(D1), D121−d127. Available from https://doi.org/10.1093/nar/gky1144.

Wang, Q., Wu, G., Zhang, Z., Tang, Q., Zheng, W., Chen, X., ... Che, X. (2018). Long non-coding RNA HOTTIP promotes renal cell carcinoma progression through the regulation of the miR-615/IGF-2 pathway. *International Journal of Oncology, 53*(5), 2278−2288. Available from https://doi.org/10.3892/ijo.2018.4539.

Wang, S., Zheng, X., Chen, X., Shi, X., & Chen, S. (2020). Prognostic and predictive value of immune/stromal-related gene biomarkers in renal cell carcinoma. *Oncology Letters, 20*(1), 308−316.

Wang, S., Zhou, S., Liu, H., Meng, Q., Ma, X., Liu, H., ... Jiang, W. (2020). ncRI: A manually curated database for experimentally validated non-coding RNAs in inflammation. *BMC Genomics*, *21*(1), 380. Available from https://doi.org/10.1186/s12864-020-06794-6.

Wang, W. J., Wang, Y. M., Hu, Y., Lin, Q., Chen, R., Liu, H., ... Peng, L. Y. (2018). HDncRNA: A comprehensive database of non-coding RNAs associated with heart diseases. *Database (Oxford)*, *2018*. Available from https://doi.org/10.1093/database/bay067.

Wang, X., Chen, X., Wang, R., Xiao, P., Xu, Z., Chen, L., ... Zhang, X. (2013). microRNA-200c modulates the epithelial-to-mesenchymal transition in human renal cell carcinoma metastasis. *Oncology Reports*, *30*(2), 643−650. Available from https://doi.org/10.3892/or.2013.2530.

Wang, Y., Hou, J., He, D., Sun, M., Zhang, P., Yu, Y., & Chen, Y. (2016). The emerging function and mechanism of ceRNAs in cancer. *Trends in Genetics: TIG*, *32*(4), 211−224.

Wang, Y., Zhang, Y., Wang, P., Fu, X., & Lin, W. (2020). Circular RNAs in renal cell carcinoma: Implications for tumorigenesis, diagnosis, and therapy. *Molecular Cancer*, *19*(1), 149. Available from https://doi.org/10.1186/s12943-020-01266-7.

Wang, Z., Chang, X., Zhu, G., Gao, X., & Chang, L. (2020). Depletion of lncRNA MALAT1 inhibited sunitinib resistance through regulating miR-362-3p-mediated G3BP1 in renal cell carcinoma. *Cell Cycle (Georgetown, Tex.)*, *19*(16), 2054−2062. Available from https://doi.org/10.1080/15384101.2020.1792667.

Wang, Z., Qin, C., Zhang, J., Han, Z., Tao, J., Cao, Q., ... Gu, M. (2017). MiR-122 promotes renal cancer cell proliferation by targeting Sprouty2. *Tumour Biology: The Journal of the International Society for Oncodevelopmental Biology and Medicine*, *39*(2), 1010428317691184. Available from https://doi.org/10.1177/1010428317691184.

Wei, Q., Mi, Q.-S., & Dong, Z. (2013). The regulation and function of microRNAs in kidney diseases. *IUBMB Life*, *65*(7), 602−614. Available from https://doi.org/10.1002/iub.1174.

Weirick, T., John, D., Dimmeler, S., & Uchida, S. (2015). C-It-Loci: A knowledge database for tissue-enriched loci. *Bioinformatics (Oxford, England)*, *31*(21), 3537−3543. Available from https://doi.org/10.1093/bioinformatics/btv410.

Wesselhoeft, R. A., Kowalski, P. S., & Anderson, D. G. (2018). Engineering circular RNA for potent and stable translation in eukaryotic cells. *Nature Communications*, *9*(1), 2629. Available from https://doi.org/10.1038/s41467-018-05096-6.

White, N. M., Khella, H. W., Grigull, J., Adzovic, S., Youssef, Y. M., Honey, R. J., ... Yousef, G. M. (2011). miRNA profiling in metastatic renal cell carcinoma reveals a tumour-suppressor effect for miR-215. *British Journal of Cancer*, *105*(11), 1741−1749. Available from https://doi.org/10.1038/bjc.2011.401.

Wotschofsky, Z., Gummlich, L., Liep, J., Stephan, C., Kilic, E., Jung, K., ... Meyer, H. A. (2016). Integrated microRNA and mRNA signature associated with the transition from the locally confined to the metastasized clear cell renal cell carcinoma exemplified by miR-146-5p. *PLoS One*, *11*(2), e0148746. Available from https://doi.org/10.1371/journal.pone.0148746.

Wu, A., Wu, K., Li, M., Bao, L., Shen, X., Li, S., ... Yang, Z. (2015). Upregulation of microRNA-492 induced by epigenetic drug treatment inhibits the malignant phenotype of clear cell renal cell carcinoma in vitro. *Molecular Medicine Reports*, *12*(1), 1413−1420. Available from https://doi.org/10.3892/mmr.2015.3550.

Wu, D., Li, M., Wang, L., Zhou, Y., Zhou, J., Pan, H., & Qu, P. (2014). microRNA-145 inhibits cell proliferation, migration and invasion by targeting matrix metallopeptidase-11 in renal cell carcinoma. *Molecular Medicine Reports*, *10*(1), 393−398. Available from https://doi.org/10.3892/mmr.2014.2149.

Wu, D., Pan, H., Zhou, Y., Zhou, J., Fan, Y., & Qu, P. (2014). microRNA-133b downregulation and inhibition of cell proliferation, migration and invasion by targeting matrix metallopeptidase-9 in renal cell carcinoma. *Molecular Medicine Reports*, *9*(6), 2491−2498. Available from https://doi.org/10.3892/mmr.2014.2116.

Wu, J., Li, Y., Wang, C., Cui, Y., Xu, T., Wang, C., ... Song, X. (2019). CircAST: Full-length assembly and quantification of alternatively spliced isoforms in circular RNAs. *Genomics, Proteomics & Bioinformatics/Beijing Genomics Institute*, *17*(5), 522−534. Available from https://doi.org/10.1016/j.gpb.2019.03.004.

Wu, W., Ji, P., & Zhao, F. (2020). CircAtlas: An integrated resource of one million highly accurate circular RNAs from 1070 vertebrate transcriptomes. *Genome Biology*, *21*(1), 101. Available from https://doi.org/10.1186/s13059-020-02018-y.

Wu, X., Weng, L., Li, X., Guo, C., Pal, S. K., Jin, J. M., ... Wu, H. (2012). Identification of a 4-microRNA signature for clear cell renal cell carcinoma metastasis and prognosis. *PLoS One*, *7*(5), e35661. Available from https://doi.org/10.1371/journal.pone.0035661.

Wu, Y., Liu, J., Zheng, Y., You, L., Kuang, D., & Liu, T. (2014). Suppressed expression of long non-coding RNA HOTAIR inhibits proliferation and tumourigenicity of renal carcinoma cells. *Tumour Biology: The Journal of the International Society for Oncodevelopmental Biology and Medicine, 35*(12), 11887−11894. Available from https://doi.org/10.1007/s13277-014-2453-4.

Wu, Y., Tan, C., Weng, W. W., Deng, Y., Zhang, Q. Y., Yang, X. Q., . . . Wang, C. F. (2016). Long non-coding RNA Linc00152 is a positive prognostic factor for and demonstrates malignant biological behavior in clear cell renal cell carcinoma. *American Journal of Cancer Research, 6*(2), 285−299.

Xia, M., Yao, L., Zhang, Q., Wang, F., Mei, H., Guo, X., & Huang, W. (2017). Long noncoding RNA HOTAIR promotes metastasis of renal cell carcinoma by up-regulating histone H3K27 demethylase JMJD3. *Oncotarget, 8*(12), 19795−19802. Available from https://doi.org/10.18632/oncotarget.15047.

Xia, S., Feng, J., Chen, K., Ma, Y., Gong, J., Cai, F., . . . He, C. (2018). CSCD: A database for cancer-specific circular RNAs. *Nucleic Acids Research, 46*(D1), D925−d929. Available from https://doi.org/10.1093/nar/gkx863.

Xia, S., Feng, J., Lei, L., Hu, J., Xia, L., Wang, J., . . . He, C. (2017). Comprehensive characterization of tissue-specific circular RNAs in the human and mouse genomes. *Briefings in Bioinformatics, 18*(6), 984−992. Available from https://doi.org/10.1093/bib/bbw081.

Xiang, C., Cui, S. P., & Ke, Y. (2016). MiR-144 inhibits cell proliferation of renal cell carcinoma by targeting MTOR. *Journal of Huazhong University of Science and Technology. Medical Sciences = Hua Zhong ke ji da xue xue bao. Yi xue Ying De wen ban = Huazhong Keji Daxue Xuebao. Yixue Yingdewen ban, 36*(2), 186−192. Available from https://doi.org/10.1007/s11596-016-1564-0.

Xiang, W., He, J., Huang, C., Chen, L., Tao, D., Wu, X., . . . Jiang, G. (2015). miR-106b-5p targets tumor suppressor gene SETD2 to inactive its function in clear cell renal cell carcinoma. *Oncotarget, 6*(6), 4066−4079. Available from https://doi.org/10.18632/oncotarget.2926.

Xiao, W., Lou, N., Ruan, H., Bao, L., Xiong, Z., Yuan, C., . . . Zhang, X. (2017). Mir-144-3p promotes cell proliferation, metastasis, sunitinib resistance in clear cell renal cell carcinoma by downregulating ARID1A. *Cellular Physiology and Biochemistry: International Journal of Experimental Cellular Physiology, Biochemistry, and Pharmacology, 43*(6), 2420−2433. Available from https://doi.org/10.1159/000484395.

Xiao, W., Wang, C., Chen, K., Wang, T., Xing, J., Zhang, X., & Wang, X. (2020). MiR-765 functions as a tumour suppressor and eliminates lipids in clear cell renal cell carcinoma by downregulating PLP2. *EBioMedicine, 51*, 102622. Available from https://doi.org/10.1016/j.ebiom.2019.102622.

Xiao, X., Tang, C., Xiao, S., Fu, C., & Yu, P. (2013). Enhancement of proliferation and invasion by MicroRNA-590-5p via targeting PBRM1 in clear cell renal cell carcinoma cells. *Oncology Research, 20*(11), 537−544. Available from https://doi.org/10.3727/096504013x13775486749335.

Xie, J., Zhong, Y., Chen, R., Li, G., Luo, Y., Yang, J., . . . Song, Y. (2020). Serum long non-coding RNA LINC00887 as a potential biomarker for diagnosis of renal cell carcinoma. *FEBS Open Bio, 10*(9), 1802−1809. Available from https://doi.org/10.1002/2211-5463.12930.

Xie, Y., Chen, L., Gao, Y., Ma, X., He, W., Zhang, Y., . . . Gou, X. (2020). miR-363 suppresses the proliferation, migration and invasion of clear cell renal cell carcinoma by downregulating S1PR1. *Cancer Cell International, 20*, 227. Available from https://doi.org/10.1186/s12935-020-01313-9.

Xin, R., Gao, Y., Gao, Y., Wang, R., Kadash-Edmondson, K. E., Liu, B., . . . Xing, Y. (2021). isoCirc catalogs full-length circular RNA isoforms in human transcriptomes. *Nature Communications, 12*(1), 266. Available from https://doi.org/10.1038/s41467-020-20459-8.

Xiong, J., Liu, Y., Jiang, L., Zeng, Y., & Tang, W. (2016). High expression of long non-coding RNA lncRNA-ATB is correlated with metastases and promotes cell migration and invasion in renal cell carcinoma. *Japanese Journal of Clinical Oncology, 46*(4), 378−384. Available from https://doi.org/10.1093/jjco/hyv214.

Xiong, J., Liu, Y., Luo, S., Jiang, L., Zeng, Y., Chen, Z., . . . Tang, W. (2017). High expression of the long non-coding RNA HEIRCC promotes renal cell carcinoma metastasis by inducing epithelial-mesenchymal transition. *Oncotarget, 8*(4), 6555−6563. Available from https://doi.org/10.18632/oncotarget.14149.

Xiong, Y., Zhang, J., & Song, C. (2019). CircRNA ZNF609 functions as a competitive endogenous RNA to regulate FOXP4 expression by sponging miR-138-5p in renal carcinoma. *Journal of Cellular Physiology, 234*(7), 10646−10654. Available from https://doi.org/10.1002/jcp.27744.

Xu, B., Wang, C., Wang, Y. L., Chen, S. Q., Wu, J. P., Zhu, W. D., . . . Chen, M. (2020). miR-143 inhibits renal cell carcinoma cells metastatic potential by suppressing ABL2. *The Kaohsiung Journal of Medical Sciences, 36*(8), 592−598. Available from https://doi.org/10.1002/kjm2.12207.

Xu, H., Wu, S., Shen, X., Shi, Z., Wu, D., Yuan, Y., ... Chen, X. (2020). Methylation-mediated miR-214 regulates proliferation and drug sensitivity of renal cell carcinoma cells through targeting LIVIN. *Journal of Cellular and Molecular Medicine, 24*(11), 6410−6425. Available from https://doi.org/10.1111/jcmm.15287.

Xu, M., Gu, M., Zhang, K., Zhou, J., Wang, Z., & Da, J. (2015). miR-203 inhibition of renal cancer cell proliferation, migration and invasion by targeting of FGF2. *Diagnostic Pathology, 10*, 24. Available from https://doi.org/10.1186/s13000-015-0255-7.

Xu, Q., Krause, M., Samoylenko, A., & Vainio, S. (2016). Wnt signaling in renal cell carcinoma. *Cancers, 8*(6), 57. Available from https://doi.org/10.3390/cancers8060057.

Xu, X., Liu, C., & Bao, J. (2017). Hypoxia-induced hsa-miR-101 promotes glycolysis by targeting TIGAR mRNA in clear cell renal cell carcinoma. *Molecular Medicine Reports, 15*(3), 1373−1378. Available from https://doi.org/10.3892/mmr.2017.6139.

Xu, X., Wu, J., Li, S., Hu, Z., Zhu, Y., Liang, Z., ... Xie, L. (2014). Downregulation of microRNA-182-5p contributes to renal cell carcinoma proliferation via activating the AKT/FOXO3a signaling pathway. *Molecular Cancer, 13*, 109. Available from https://doi.org/10.1186/1476-4598-13-109.

Xu, Y., Deng, W., & Zhang, W. (2018). Long non-coding RNA TUG1 protects renal tubular epithelial cells against injury induced by lipopolysaccharide via regulating microRNA-223. *Biomedicine & Pharmacotherapy = Biomedecine & Pharmacotherapie, 104*, 509−519. Available from https://doi.org/10.1016/j.biopha.2018.05.069.

Xu, Y., Ma, Y., Liu, X. L., & Gao, S. L. (2020). miR-133b affects cell proliferation, invasion and chemosensitivity in renal cell carcinoma by inhibiting the ERK signaling pathway. *Molecular Medicine Reports, 22*(1), 67−76. Available from https://doi.org/10.3892/mmr.2020.11125.

Xu, Y., Zhu, J., Lei, Z., Wan, L., Zhu, X., Ye, F., & Tong, Y. (2015). Expression and functional role of miR-29b in renal cell carcinoma. *International Journal of Clinical and Experimental Pathology, 8*(11), 14161−14170.

Xu, Z., Yang, F., Wei, D., Liu, B., Chen, C., Bao, Y., ... Wang, L. (2017). Long noncoding RNA-SRLR elicits intrinsic sorafenib resistance via evoking IL-6/STAT3 axis in renal cell carcinoma. *Oncogene, 36*(14), 1965−1977. Available from https://doi.org/10.1038/onc.2016.356.

Xue, D., Wang, H., Chen, Y., Shen, D., Lu, J., Wang, M., ... Xia, L. (2019). Circ-AKT3 inhibits clear cell renal cell carcinoma metastasis via altering miR-296-3p/E-cadherin signals. *Molecular Cancer, 18*(1), 151. Available from https://doi.org/10.1186/s12943-019-1072-5.

Xue, S., Li, Q. W., Che, J. P., Guo, Y., Yang, F. Q., & Zheng, J. H. (2015). Decreased expression of long non-coding RNA NBAT-1 is associated with poor prognosis in patients with clear cell renal cell carcinoma. *International Journal of Clinical and Experimental Pathology, 8*(4), 3765−3774.

Yadav, S., Khandelwal, M., Seth, A., Saini, A. K., Dogra, P. N., & Sharma, A. (2017). Serum microRNA expression profiling: Potential diagnostic implications of a panel of serum microRNAs for clear cell renal cell cancer. *Urology, 104*, 64−69. Available from https://doi.org/10.1016/j.urology.2017.03.013.

Yamada, Y., Hidaka, H., Seki, N., Yoshino, H., Yamasaki, T., Itesako, T., ... Enokida, H. (2013). Tumor-suppressive microRNA-135a inhibits cancer cell proliferation by targeting the c-MYC oncogene in renal cell carcinoma. *Cancer Science, 104*(3), 304−312. Available from https://doi.org/10.1111/cas.12072.

Yamada, Y., Nohata, N., Uchida, A., Kato, M., Arai, T., Moriya, S., ... Seki, N. (2020). Replisome genes regulation by antitumor miR-101-5p in clear cell renal cell carcinoma. *Cancer Science, 111*(4), 1392−1406. Available from https://doi.org/10.1111/cas.14327.

Yamaguchi, N., Osaki, M., Onuma, K., Yumioka, T., Iwamoto, H., Sejima, T., ... Okada, F. (2017). Identification of MicroRNAs involved in resistance to sunitinib in renal cell carcinoma cells. *Anticancer Research, 37*(6), 2985−2992. Available from https://doi.org/10.21873/anticanres.11652.

Yamamura, S., Saini, S., Majid, S., Hirata, H., Ueno, K., Chang, I., ... Dahiya, R. (2012). MicroRNA-34a suppresses malignant transformation by targeting c-Myc transcriptional complexes in human renal cell carcinoma. *Carcinogenesis, 33*(2), 294−300. Available from https://doi.org/10.1093/carcin/bgr286.

Yamasaki, T., Seki, N., Yamada, Y., Yoshino, H., Hidaka, H., Chiyomaru, T., ... Enokida, H. (2012). Tumor suppressive microRNA-138 contributes to cell migration and invasion through its targeting of vimentin in renal cell carcinoma. *International Journal of Oncology, 41*(3), 805−817. Available from https://doi.org/10.3892/ijo.2012.1543.

Yamasaki, T., Seki, N., Yoshino, H., Itesako, T., Hidaka, H., Yamada, Y., ... Enokida, H. (2013). MicroRNA-218 inhibits cell migration and invasion in renal cell carcinoma through targeting caveolin-2 involved in focal adhesion pathway. *The Journal of Urology, 190*(3), 1059−1068. Available from https://doi.org/10.1016/j.juro.2013.02.089.

Yamasaki, T., Seki, N., Yoshino, H., Itesako, T., Yamada, Y., Tatarano, S., ... Enokida, H. (2013). Tumor-suppressive microRNA-1291 directly regulates glucose transporter 1 in renal cell carcinoma. *Cancer Science*, *104*(11), 1411–1419. Available from https://doi.org/10.1111/cas.12240.

Yan, L., Liu, G., Cao, H., Zhang, H., & Shao, F. (2019). Hsa_circ_0035483 sponges hsa-miR-335 to promote the gemcitabine-resistance of human renal cancer cells by autophagy regulation. *Biochemical and Biophysical Research Communications*, *519*(1), 172–178. Available from https://doi.org/10.1016/j.bbrc.2019.08.093.

Yang, D., Tang, Y., Fu, H., Xu, J., Hu, Z., Zhang, Y., & Cai, Q. (2018). Integrin β1 promotes gemcitabine resistance in pancreatic cancer through Cdc42 activation of PI3K p110β signaling. *Biochemical and Biophysical Research Communications*, *505*(1), 215–221. Available from https://doi.org/10.1016/j.bbrc.2018.09.061.

Yang, D. C., Ke, L., Ding, Y., & Gao, G. (2021). AnnoLnc: A one-stop portal to systematically annotate novel human long noncoding RNAs. *Methods in Molecular Biology*, *2254*, 111–131. Available from https://doi.org/10.1007/978-1-0716-1158-6_8.

Yang, F. Q., Zhang, H. M., Chen, S. J., Yan, Y., & Zheng, J. H. (2015). MiR-506 is down-regulated in clear cell renal cell carcinoma and inhibits cell growth and metastasis via targeting FLOT1. *PLoS One*, *10*(3), e0120258. Available from https://doi.org/10.1371/journal.pone.0120258.

Yang, J. H., Li, J. H., Jiang, S., Zhou, H., & Qu, L. H. (2013). ChIPBase: A database for decoding the transcriptional regulation of long non-coding RNA and microRNA genes from ChIP-Seq data. *Nucleic Acids Research*, *41*(Database issue), D177–D187. Available from https://doi.org/10.1093/nar/gks1060.

Yang, J. H., Li, J. H., Shao, P., Zhou, H., Chen, Y. Q., & Qu, L. H. (2011). starBase: A database for exploring microRNA-mRNA interaction maps from Argonaute CLIP-Seq and Degradome-Seq data. *Nucleic Acids Research*, *39*(Database issue), D202–D209. Available from https://doi.org/10.1093/nar/gkq1056.

Yao, D., Zhang, L., Zheng, M., Sun, X., Lu, Y., & Liu, P. (2018). Circ2Disease: A manually curated database of experimentally validated circRNAs in human disease. *Scientific Reports*, *8*(1), 11018. Available from https://doi.org/10.1038/s41598-018-29360-3.

Yao, J., Chen, Y., Wang, Y., Liu, S., Yuan, X., Pan, F., & Geng, P. (2014). Decreased expression of a novel lncRNA CADM1-AS1 is associated with poor prognosis in patients with clear cell renal cell carcinomas. *International Journal of Clinical and Experimental Pathology*, *7*(6), 2758–2767. Available from https://www.ncbi.nlm.nih.gov/pubmed/25031695.

Yesselman, J. D., Tian, S., Liu, X., Shi, L., Li, J. B., & Das, R. (2018). Updates to the RNA mapping database (RMDB), version 2. *Nucleic Acids Research*, *46*(D1), D375–d379. Available from https://doi.org/10.1093/nar/gkx873.

Yi, Y., Zhao, Y., Li, C., Zhang, L., Huang, H., Li, Y., ... Wang, D. (2017). RAID v2.0: An updated resource of RNA-associated interactions across organisms. *Nucleic Acids Research*, *45*(D1), D115–D118. Available from https://doi.org/10.1093/nar/gkw1052.

Yoshino, H., Enokida, H., Itesako, T., Kojima, S., Kinoshita, T., Tatarano, S., ... Seki, N. (2013). Tumor-suppressive microRNA-143/145 cluster targets hexokinase-2 in renal cell carcinoma. *Cancer Science*, *104*(12), 1567–1574. Available from https://doi.org/10.1111/cas.12280.

Yoshino, H., Enokida, H., Itesako, T., Tatarano, S., Kinoshita, T., Fuse, M., ... Seki, N. (2013). Epithelial-mesenchymal transition-related microRNA-200s regulate molecular targets and pathways in renal cell carcinoma. *Journal of Human Genetics*, *58*(8), 508–516. Available from https://doi.org/10.1038/jhg.2013.31.

Yoshino, H., Yonemori, M., Miyamoto, K., Tatarano, S., Kofuji, S., Nohata, N., ... Enokida, H. (2017). microRNA-210-3p depletion by CRISPR/Cas9 promoted tumorigenesis through revival of TWIST1 in renal cell carcinoma. *Oncotarget*, *8*(13), 20881–20894. Available from https://doi.org/10.18632/oncotarget.14930.

You, B. H., Yoon, S. H., & Nam, J. W. (2017). High-confidence coding and noncoding transcriptome maps. *Genome Research*, *27*(6), 1050–1062. Available from https://doi.org/10.1101/gr.214288.116.

Youssef, Y. M., White, N. M., Grigull, J., Krizova, A., Samy, C., Mejia-Guerrero, S., ... Yousef, G. M. (2011). Accurate molecular classification of kidney cancer subtypes using microRNA signature. *European Urology*, *59*(5), 721–730. Available from https://doi.org/10.1016/j.eururo.2011.01.004.

Yu, F., Zhang, G., Shi, A., Hu, J., Li, F., Zhang, X., ... Cheng, S. (2018). LnChrom: A resource of experimentally validated lncRNA-chromatin interactions in human and mouse. *Database (Oxford)*, *2018*. Available from https://doi.org/10.1093/database/bay039.

Yu, H., Lin, X., Wang, F., Zhang, B., Wang, W., Shi, H., ... Zhao, J. (2014). Proliferation inhibition and the underlying molecular mechanisms of microRNA-30d in renal carcinoma cells. *Oncology Letters*, *7*(3), 799–804. Available from https://doi.org/10.3892/ol.2013.1754.

Yu, X. Y., Zhang, Z., Liu, J., Zhan, B., & Kong, C. Z. (2013). MicroRNA-141 is downregulated in human renal cell carcinoma and regulates cell survival by targeting CDC25B. *OncoTargets and Therapy, 6*, 349−354. Available from https://doi.org/10.2147/ott.S41343.

Yu, Z., Ni, L., Chen, D., Zhang, Q., Su, Z., Wang, Y., . . . Li, X. (2013). Identification of miR-7 as an oncogene in renal cell carcinoma. *Journal of Molecular Histology, 44*(6), 669−677. Available from https://doi.org/10.1007/s10735-013-9516-5.

Zaman, M. S., Shahryari, V., Deng, G., Thamminana, S., Saini, S., Majid, S., . . . Dahiya, R. (2012). Up-regulation of microRNA-21 correlates with lower kidney cancer survival. *PLoS One, 7*(2), e31060. Available from https://doi.org/10.1371/journal.pone.0031060.

Zaman, M. S., Thamminana, S., Shahryari, V., Chiyomaru, T., Deng, G., Saini, S., . . . Dahiya, R. (2012). Inhibition of PTEN gene expression by oncogenic miR-23b-3p in renal cancer. *PLoS One, 7*(11), e50203. Available from https://doi.org/10.1371/journal.pone.0050203.

Zeng, R., Huang, J., Sun, Y., & Luo, J. (2020). Cell proliferation is induced in renal cell carcinoma through miR-92a-3p upregulation by targeting FBXW7. *Oncology Letters, 19*(4), 3258−3268. Available from https://doi.org/10.3892/ol.2020.11443.

Zhai, Q., Zhou, L., Zhao, C., Wan, J., Yu, Z., Guo, X., . . . Lu, R. (2012). Identification of miR-508-3p and miR-509-3p that are associated with cell invasion and migration and involved in the apoptosis of renal cell carcinoma. *Biochemical and Biophysical Research Communications, 419*(4), 621−626. Available from https://doi.org/10.1016/j.bbrc.2012.02.060.

Zhai, W., Sun, Y., Guo, C., Hu, G., Wang, M., Zheng, J., . . . Chang, C. (2017). LncRNA-SARCC suppresses renal cell carcinoma (RCC) progression via altering the androgen receptor(AR)/miRNA-143-3p signals. *Cell Death and Differentiation, 24*(9), 1502−1517. Available from https://doi.org/10.1038/cdd.2017.74.

Zhai, W., Sun, Y., Jiang, M., Wang, M., Gasiewicz, T. A., Zheng, J., & Chang, C. (2016). Differential regulation of LncRNA-SARCC suppresses VHL-mutant RCC cell proliferation yet promotes VHL-normal RCC cell proliferation via modulating androgen receptor/HIF-2α/C-MYC axis under hypoxia. *Oncogene, 35*, 4866−4880.

Zhang, C., Mo, R., Yin, B., Zhou, L., Liu, Y., & Fan, J. (2014). Tumor suppressor microRNA-34a inhibits cell proliferation by targeting Notch1 in renal cell carcinoma. *Oncology Letters, 7*(5), 1689−1694. Available from https://doi.org/10.3892/ol.2014.1931.

Zhang, D., Yang, X. J., Luo, Q. D., Fu, D. L., Li, Z. L., Zhang, P., & Chong, T. (2019). Down-regulation of circular RNA_000926 attenuates renal cell carcinoma progression through miRNA-411-dependent CDH2 inhibition. *The American Journal of Pathology, 189*(12), 2469−2486. Available from https://doi.org/10.1016/j.ajpath.2019.06.016.

Zhang, H., & Li, H. (2016). miR-137 inhibits renal cell carcinoma growth. *Oncology Letters, 12*(1), 715−720. Available from https://doi.org/10.3892/ol.2016.4616.

Zhang, H., Guo, Y., Shang, C., Song, Y., & Wu, B. (2012). miR-21 downregulated TCF21 to inhibit KISS1 in renal cancer. *Urology, 80*(6), 1298−1302, e1291. Available from https://doi.org/10.1016/j.urology.2012.08.013.

Zhang, H. M., Yang, F. Q., Chen, S. J., Che, J., & Zheng, J. H. (2015). Upregulation of long non-coding RNA MALAT1 correlates with tumor progression and poor prognosis in clear cell renal cell carcinoma. *Tumour Biology: The Journal of the International Society for Oncodevelopmental Biology and Medicine, 36*(4), 2947−2955. Available from https://doi.org/10.1007/s13277-014-2925-6.

Zhang, H. M., Yang, F. Q., Yan, Y., Che, J. P., & Zheng, J. H. (2014). High expression of long non-coding RNA SPRY4-IT1 predicts poor prognosis of clear cell renal cell carcinoma. *International Journal of Clinical and Experimental Pathology, 7*(9), 5801−5809.

Zhang, L., & Guo, Y. (2020). Silencing circular RNA-ZNF652 represses proliferation and EMT process of renal carcinoma cells via raising miR-205. *Artificial Cells, Nanomedicine, and Biotechnology, 48*(1), 648−655. Available from https://doi.org/10.1080/21691401.2020.1725532.

Zhang, M., & Xin, Y. (2018). Circular RNAs: A new frontier for cancer diagnosis and therapy. *Journal of Hematology & Oncology, 11*(1), 21. Available from https://doi.org/10.1186/s13045-018-0569-5.

Zhang, M., Lu, W., Huang, Y., Shi, J., Wu, X., Zhang, X., . . . Wu, S. (2016). Downregulation of the long noncoding RNA TUG1 inhibits the proliferation, migration, invasion and promotes apoptosis of renal cell carcinoma. *Journal of Molecular Histology, 47*(4), 421−428. Available from https://doi.org/10.1007/s10735-016-9683-2.

Zhang, P., Wu, W., Chen, Q., & Chen, M. (2019). Non-coding RNAs and their integrated networks. *Journal of Integrative Bioinformatics, 16*(3), 20190027. Available from https://doi.org/10.1515/jib-2019-0027.

Zhang, Q., Di, W., Dong, Y., Lu, G., Yu, J., Li, J., & Li, P. (2015). High serum miR-183 level is associated with poor responsiveness of renal cancer to natural killer cells. *Tumour Biology: The Journal of the International Society for Oncodevelopmental Biology and Medicine*, 36(12), 9245−9249. Available from https://doi.org/10.1007/s13277-015-3604-y.

Zhang, W., Yao, G., Wang, J., Yang, M., Wang, J., Zhang, H., & Li, W. (2020). ncRPheno: A comprehensive database platform for identification and validation of disease related noncoding RNAs. *RNA Biology*, 17(7), 943−955. Available from https://doi.org/10.1080/15476286.2020.1737441.

Zhang, W., Yue, X., Tang, G., Wu, W., Huang, F., & Zhang, X. (2018). SFPEL-LPI: Sequence-based feature projection ensemble learning for predicting LncRNA-protein interactions. *PLoS Computational Biology*, 14(12), e1006616. Available from https://doi.org/10.1371/journal.pcbi.1006616.

Zhang, W., Zeng, B., Yang, M., Yang, H., Wang, J., Deng, Y., ... Li, W. (2020). ncRNAVar: A manually curated database for identification of noncoding RNA variants associated with human diseases. *Journal of Molecular Biology*, 433, 166727. Available from https://doi.org/10.1016/j.jmb.2020.166727.

Zhang, W. B., Pan, Z. Q., Yang, Q. S., & Zheng, X. M. (2013). Tumor suppressive miR-509-5p contributes to cell migration, proliferation and antiapoptosis in renal cell carcinoma. *Irish Journal of Medical Science*, 182(4), 621−627. Available from https://doi.org/10.1007/s11845-013-0941-y.

Zhang, X., Xie, K., Zhou, H., Wu, Y., Li, C., Liu, Y., ... Tao, Y. (2020). Role of non-coding RNAs and RNA modifiers in cancer therapy resistance. *Molecular Cancer*, 19(1), 47. Available from https://doi.org/10.1186/s12943-020-01171-z.

Zhang, X., Xing, N. D., Lai, C. J., Liu, R., Jiao, W., Wang, J., ... Xu, Z. H. (2018). MicroRNA-375 suppresses the tumor aggressive phenotypes of clear cell renal cell carcinomas through regulating YWHAZ. *Chinese Medical Journal (Engl)*, 131(16), 1944−1950. Available from https://doi.org/10.4103/0366-6999.238153.

Zhao, H., Shi, J., Zhang, Y., Xie, A., Yu, L., Zhang, C., ... Li, X. (2020). LncTarD: A manually-curated database of experimentally-supported functional lncRNA-target regulations in human diseases. *Nucleic Acids Research*, 48(D1), D118−d126. Available from https://doi.org/10.1093/nar/gkz985.

Zhao, J., Lei, T., Xu, C., Li, H., Ma, W., Yang, Y., ... Liu, Y. (2013). MicroRNA-187, down-regulated in clear cell renal cell carcinoma and associated with lower survival, inhibits cell growth and migration though targeting B7-H3. *Biochemical and Biophysical Research Communications*, 438(2), 439−444. Available from https://doi.org/10.1016/j.bbrc.2013.07.095.

Zhao, K., Zhang, Q., Wang, Y., Zhang, J., Cong, R., Song, N., & Wang, Z. (2020). The construction and analysis of competitive endogenous RNA (ceRNA) networks in metastatic renal cell carcinoma: A study based on The Cancer Genome Atlas. *Translational Andrology and Urology*, 9(2), 303−311. Available from https://doi.org/10.21037/tau.2020.02.17.

Zhao, L., Wang, J., Li, Y., Song, T., Wu, Y., Fang, S., ... He, S. (2021). NONCODEV6: An updated database dedicated to long non-coding RNA annotation in both animals and plants. *Nucleic Acids Research*, 49(D1), D165−d171. Available from https://doi.org/10.1093/nar/gkaa1046.

Zhao, W., Shan, B., He, D., Cheng, Y., Li, B., Zhang, C., & Duan, C. (2019). Recent progress in characterizing long noncoding RNAs in cancer drug resistance. *Journal of Cancer*, 10(26), 6693−6702. Available from https://doi.org/10.7150/jca.30877.

Zhao, X., Zhao, Z., Xu, W., Hou, J., & Du, X. (2015). Down-regulation of miR-497 is associated with poor prognosis in renal cancer. *International journal of clinical and experimental pathology*, 8(1), 758−764. Retrieved from https://pubmed.ncbi.nlm.nih.gov/25755771. https://www.ncbi.nlm.nih.gov/pmc/articles/PMC4348935/.

Zhao, Z., Bai, J., Wu, A., Wang, Y., Zhang, J., Wang, Z., ... Li, X. (2015). Co-LncRNA: Investigating the lncRNA combinatorial effects in GO annotations and KEGG pathways based on human RNA-Seq data. *Database (Oxford)*, 2015. Available from https://doi.org/10.1093/database/bav082.

Zhao, Z., Wang, K., Wu, F., Wang, W., Zhang, K., Hu, H., ... Jiang, T. (2018). circRNA disease: A manually curated database of experimentally supported circRNA-disease associations. *Cell Death and Disease*, 9(5), 475. Available from https://doi.org/10.1038/s41419-018-0503-3.

Zheng, B., Zhu, H., Gu, D., Pan, X., Qian, L., Xue, B., ... Shan, Y. (2015). MiRNA-30a-mediated autophagy inhibition sensitizes renal cell carcinoma cells to sorafenib. *Biochemical and Biophysical Research Communications*, 459(2), 234−239. Available from https://doi.org/10.1016/j.bbrc.2015.02.084.

Zheng, L. L., Li, J. H., Wu, J., Sun, W. J., Liu, S., Wang, Z. L., ... Qu, L. H. (2016). DeepBase v2.0: Identification, expression, evolution and function of small RNAs, LncRNAs and circular RNAs from deep-sequencing data. *Nucleic Acids Research*, 44(D1), D196−D202. Available from https://doi.org/10.1093/nar/gkv1273.

Zhou, B., Ji, B., Liu, K., Hu, G., Wang, F., Chen, Q., ... Wang, J. (2021). EVLncRNAs 2.0: An updated database of manually curated functional long non-coding RNAs validated by low-throughput experiments. *Nucleic Acids Research*, 49(D1), D86–d91. Available from https://doi.org/10.1093/nar/gkaa1076.

Zhou, B., Zhao, H., Yu, J., Guo, C., Dou, X., Song, F., ... Wang, J. (2018). EVLncRNAs: A manually curated database for long non-coding RNAs validated by low-throughput experiments. *Nucleic Acids Research*, 46(D1), D100–d105. Available from https://doi.org/10.1093/nar/gkx677.

Zhou, B., Zheng, P., Li, Z., Li, H., Wang, X., Shi, Z., & Han, Q. (2018). CircPCNXL2 sponges miR-153 to promote the proliferation and invasion of renal cancer cells through upregulating ZEB2. *Cell Cycle (Georgetown, Tex.)*, 17(23), 2644–2654. Available from https://doi.org/10.1080/15384101.2018.1553354.

Zhou, L., Li, Z., Pan, X., Lai, Y., Quan, J., Zhao, L., ... Gui, Y. (2018). Identification of miR-18a-5p as an oncogene and prognostic biomarker in RCC. *American Journal of Translational Research*, 10(6), 1874–1886. Available from https://www.ncbi.nlm.nih.gov/pubmed/30018727.

Zhou, S., Wang, J., & Zhang, Z. (2014). An emerging understanding of long noncoding RNAs in kidney cancer. *Journal of Cancer Research and Clinical Oncology*, 140(12), 1989–1995. Available from https://doi.org/10.1007/s00432-014-1699-y.

Zhu, H., Wang, S., Shen, H., Zheng, X., & Xu, X. (2020). SP1/AKT/FOXO3 signaling is involved in miR-362-3p-mediated inhibition of cell-cycle pathway and EMT progression in renal cell carcinoma. *Frontiers in Cell and Developmental Biology*, 8, 297. Available from https://doi.org/10.3389/fcell.2020.00297.

Zhu, S., Huang, Y., & Su, X. (2016). Mir-451 correlates with prognosis of renal cell carcinoma patients and inhibits cellular proliferation of renal cell carcinoma. *Medical Science Monitor: International Medical Journal of Experimental and Clinical Research*, 22, 183–190. Available from https://doi.org/10.12659/msm.896792.

Clinical significance of long noncoding RNAs in breast cancer patients

Nikee Awasthee[1], Anusmita Shekher[1], Vipin Rai[1], Pranjal K. Baruah[2], Anurag Sharma[3], Kishore B. Challagundla[4,5] and Subash C. Gupta[1,6]

[1]Department of Biochemistry, Institute of Science, Banaras Hindu University, Varanasi, India [2]Department of Applied Sciences, GUIST, Gauhati University, Guwahati, India [3]Division of Environmental Health and Toxicology, Nitte (Deemed to Be University), Nitte University Centre for Science Education and Research (NUCSER), Mangalore, India [4]Department of Biochemistry and Molecular Biology, The Fred and Pamela Buffett Cancer Center; University of Nebraska Medical Center, Omaha, NE, United States [5]The Children's Health Research Institute, University of Nebraska Medical Center, Omaha, NE, United States [6]Department of Biochemistry, All India Institute of Medical Sciences, Guwahati, India

Abbreviations

AFAP1-AS1	actin filament−associated protein 1 antisense RNA 1
ASOs	antisense oligonucleotides
BCAR	breast cancer antiestrogen resistance
CAGE	cap analysis gene expression
cDNA	complementary DNA
DANCR	differentiation antagonizing nonprotein-coding RNA
ER	estrogen receptor
H19	H19-imprinted maternally expressed transcript
HER2	human epidermal growth factor receptor 2
HIF1A-AS2	hypoxia-inducible factor 1A-antisense RNA 2
HOTAIR	HOX transcript antisense RNA
LINK-A	long intergenic noncoding RNA for kinase activation

LNA	locked nucleic acid
LncRNAs	long noncoding RNAs
MALAT1	metastasis-associated lung adenocarcinoma transcript 1
MDR	multidrug resistance protein
MIR503HG	miR-503 host gene
miRNAs	microRNAs
ncRNAs	noncoding RNAs
NEAT1	noncoding nuclear-enriched abundant transcript 2
NGS	next-generation sequencing
OS	overall survival
PR	progesterone receptor
RNA seq	RNA sequencing
RT-qPCR	reverse transcription–quantitative polymerase chain reaction
SAGE	serial analysis of gene expression
siRNAs	small interfering RNAs
SPRY4-IT1	sprouty4-intron 1
TNBC	triple-negative breast cancer
UCA1	urothelial cancer-associated 1
XIST	X-inactive-specific transcript

8.1 Introduction

Breast cancer is a common and heterogeneous solid tumor in females (Ferlay et al., 2015). Based on the status of hormone receptors, breast cancer is divided into four categories: (1) luminal A, (2) luminal B, (3) human epidermal growth factor receptor 2 (HER2), and (4) basal-like type. The luminal subtypes (A and B) are positive for estrogen receptor (ER) or progesterone receptor (PR) or both and negative for HER2. In addition, reduced proliferation is reported in the luminal A subtype. HER2 subtype can be both luminal and nonluminal with overexpression of HER2. The basal-like subtype usually does not express ER, PR, and HER2 but expresses basal markers. Thus basal-like subtypes mostly constitute triple-negative breast cancer (TNBC) (Harbeck & Gnant, 2017). While basal-like subtype is highly aggressive, a favorable prognosis is reported for luminal A subtype. Depending on the breast cancer subtypes, anti-HER2 therapy, endocrine therapy, and chemotherapy have been used as treatment options. These therapies have minimally improved the survival of patients. Yet, the development of resistance over time limits the efficacy of these therapeutic options.

Breast cancer, like other cancer types, is associated with a dysregulation in numerous genes. Once considered to be a "dark matter" or a "transcriptional noise," noncoding RNAs (ncRNAs) are now known to play significant roles in the pathogenesis of several cancer types. microRNAs (miRNAs) and long noncoding RNAs (lncRNAs) constitute the major types of ncRNAs. While miRNAs are relatively well characterized, very little is known about lncRNAs. Unlike miRNAs, lncRNAs are equal to or more than 200 nucleotides in length (Li, Ma, Huang, Chen, & Zhou, 2017). The lncRNAs are now known to play a crucial role during normal and pathological conditions. Advancement in RNA sequencing (RNA-seq) techniques, epigenomics, and computational biology has revealed numerous lncRNAs (Clark et al., 2015; Guttman et al., 2009). Encyclopedia of DNA Elements Project Consortium has unraveled more than 120,000 lncRNAs in the human genome (Jalali, Gandhi, & Scaria, 2016). lncRNAs can modulate gene expression at the level of transcription, epigenetics, and translation (Ezkurdia et al., 2014; Schmitt &

Chang, 2016; Yang et al., 2018). The aberrant expression of lncRNAs is reported to be a major contributor to tumorigenesis (Cipolla et al., 2018; Qi & Du, 2013). Due to deregulation and modulation of multiple cancer-related targets, the lncRNAs could be targeted therapeutically. The functions of lncRNAs in tumorigenesis, and chemoresistance of multiple cancer types, including breast cancer, are thoroughly investigated (Ma et al., 2017). The high level of lncRNAs is often associated with worse clinicopathological outcomes. The diagnostic and prognostic significance of lncRNAs has been reported for breast cancer. In this chapter, we review the potential of lncRNAs in breast cancer patients. More specifically, we discuss the diagnostic, prognostic, and therapeutic potential of lncRNAs. How lncRNAs contribute to breast cancer chemoresistance is also discussed.

8.2 Potential of lncRNAs in the diagnosis of breast cancer

A number of studies have demonstrated the diagnostic potential of lncRNAs. In most cases, upregulation or downregulation in the lncRNAs expression is reported as a tool for breast cancer diagnosis. In one study, an upregulation in the expression of lncRNA RP11−445H22.4 was observed in the serum and tumor tissues of breast cancer patients (Xu, Chen et al., 2015). Similarly, RP11−445H22.4 in the circulating serum was shown to exhibit sensitivity and specificity in the breast cancer diagnosis (Xu, Chen et al., 2015). An integrated expression profile of lncRNA and messenger RNA (mRNA) could also be employed to classify breast cancer types (Liu, Li et al., 2016). The tumorigenicity was associated with H19-imprinted maternally expressed transcript (H19) overexpression in breast cancer (Lottin et al., 2002). The levels of H19 in the plasma of breast cancer patients are significantly correlated with the ER, PR, HER2, and lymph node metastasis. Further, H19 was found to perform better than carbohydrate antigen 153 and carcinoembryonic antigen (Zhang, Luo et al., 2016) in the breast cancer diagnosis. One study conducted in breast cancer cells showed an abnormal expression of 12 lncRNAs (Jiao, Tian, Li, Wang, & Li, 2018). Of these, H19, HOX transcript antisense RNA (HOTAIR), and RP11−445H22.4 were significantly increased compared to the other nine lncRNAs in breast cancer patients (Jiao et al., 2018). In another study, H19 expression was examined in the breast cancer tissues of 52 patients by chromogenic in situ hybridization assay. A varied level of H19 was found in different breast cancer subtypes (Zhang, Weaver et al., 2016). A respective lower to higher H19 level was observed in normal adjacent tissue, ductal carcinoma, and invasive breast carcinoma, respectively (Zhang, Weaver et al., 2016). Thus the level of H19 could be used to classify breast cancer types accurately. One study analyzed the TCGA database for the lncRNAs expression and clinical features in 1097 breast cancer patients (Fan, Ma, & Liu, 2019). From this analysis, FOXCUT, AC091043.1, and AP000924.1 were identified to have strong diagnostic significance for TNBC. The plasma from TNBC patients ($n = 25$), non-TNBC patients ($n = 35$), and normal individuals were analyzed for lncRNAs expression (Liu, Xing, & Liu, 2017). The results revealed an increase in the expression of hypoxia-inducible factor 1A-antisense RNA 2 (HIF1A-AS2), antisense ncRNA in the INK4 locus, and urothelial cancer-associated 1 (UCA1) in TNBC patients. These lncRNAs were proposed to be specific diagnostic biomarkers for TNBC (Liu et al., 2017). BC040587 is significantly reduced in breast cancer tissues. On the other hand, lincRNA-BC2 and lincRNA-BC5 are enhanced in breast cancer (Ding et al., 2014).

8.3 Potential of lncRNAs in the prognosis of breast cancer

The lncRNAs have been demonstrated the potential of prognostic significance. For instance, metastasis-associated lung adenocarcinoma transcript 1 (MALAT1) is overexpressed in numerous cancer types (Goyal et al., 2021). However, in one study, a significant reduction in MALAT1 level was reported in the breast tumor tissues compared to adjacent noncancer tissue (Xu, Sui et al., 2015). Further, patients with a lower level of MALAT1 were reported to have shorter periods of relapse-free survival than those with higher expression (Xu, Sui et al., 2015). Thus the lower level of MALAT1 could be of potential for the prognosis of breast cancer. An analysis from a publicly available database also supports the prognostic values of lncRNAs in breast cancer (Liu, Li et al., 2016). An overexpression of MALAT1 correlates with a poor prognosis. For instance, an elevated level of MALAT1 is correlated with lymph node metastasis and reduced survival of patients (Tian et al., 2018). An increase in MALAT1 is also associated with short-term recurrence-free survival (RFS) of breast cancer patients (Li, Xu et al., 2018).

The potential of HOTAIR in promoting breast cancer pathogenesis has been examined. HOTAIR overexpression correlates with shorter survival in breast cancer patients (Sun, Wu et al., 2019). Patients with low circulating HOTAIR were reported to have a better prognosis than those with high circulating HOTAIR (Lu et al., 2018). HOTAIR is also correlated with the luminal androgen receptor subtype of TNBC (Collina et al., 2019). HOTAIR can independently predict the risk of metastasis, specifically in ER$^+$ breast cancer patients (Gökmen-Polar, Vladislav, Neelamraju, Janga, & Badve, 2015). Additionally, lncRNAs in nonhomologous end joining pathway 1 (Liu, Yang, & Han, 2018), LINC00473 (Bai, Zhao, Li, Ying, & Jiang, 2019), terminal differentiation-induced ncRNA (Dong et al., 2019), and breast cancer antiestrogen resistance 4 (BCAR4) (Meng, Liu, Li, & Fu, 2019; Zhao, Wang, Fang, Li, & Fang, 2019) were overexpressed in breast cancer tissues. The expression of these lncRNAs was further increased during metastasis. Further, these lncRNAs were correlated with poor prognosis of breast cancer patients. An enhanced level of BCAR4 was negatively associated with the overall survival (OS) in breast cancer patients. Thus BCAR4 could be utilized as a negative and unfavorable prognostic marker for breast cancer (Meng et al., 2019; Zhao et al., 2019).

The lncRNAs such as BC016831, immunoglobulin kappa variable, LINC00052, and RP11−434D9.1 were differentially expressed in the TNBC tissues in comparison to non-TNBC tissues (Lv et al., 2016). An abrupt upregulation of differentiation antagonizing nonprotein-coding RNA (DANCR) correlates with worse OS and tumor node metastasis stage in TNBC patients (Sha, Yuan, Liu, Han, & Zhong, 2017). Similarly, lncRNAs such as HIF1A-AS2, lncRNA activated by transforming growth factor-beta (lncRNA-ATB), and nicotinamide phosphoribosyltransferase antisense RNA are negatively correlated with the survival of TNBC patients (Li, Chen et al., 2018; Wang, Zhang, & Han, 2019; Zhang et al., 2019). Some tumor suppressor lncRNAs can also be used as prognostic markers. For example, TNBC patients with a reduced level of miR-503 host gene (MIR503HG) were reported to have a worse prognosis than those with a high level of MIR503HG (Fu et al., 2019).

H19 relates to unfavorable survival of breast cancer patients (Shima et al., 2018). In another study, H19 was correlated with poor prognosis of breast cancer patients (Han, Ma, Liu, & Zhou, 2016). An integrated lncRNA (AK124454, HIF1A-AS2) and mRNA (CHRDL1, FCGR1A, RSAD2) signature could also predict the RFS of TNBC patients (Jiang et al., 2016). The level of sprouty4-intron 1 (SPRY4-IT1) showed a significant increase in breast cancer patients in comparison to normal tissue. An enhanced level of SPRY4-IT1 was correlated with a larger tumor size (Shi et al., 2015). Conversely, the lncRNAs with poor prognosis such as BC040587 (Chi et al., 2014), neuroblastoma-associated transcript 1 (Hu et al., 2015), eosinophil granule ontogeny transcript (Xu, Sui et al., 2015), fibroblast growth factor 14-antisense RNA 2 (Yang et al., 2016), and chemoresistance-associated lncRNA (Li et al. 2017) were downregulated in primary breast cancer samples.

8.4 Potential of lncRNAs in breast cancer therapy

The lncRNAs could also be targeted therapeutically. Several efforts have been taken to correct dysregulation in lncRNAs expression. Depending upon the role of lncRNAs, various lncRNAs-based strategies have been adopted. One strategy is the reintroduction of tumor suppressor lncRNAs to the cells lacking it. The oncogenic lncRNAs can be targeted using antisense oligonucleotides (ASOs), small interfering RNAs (siRNAs), and short/small hairpin RNAs (Matsui & Corey, 2017). ASOs are synthetic single-stranded oligonucleotides with complementarity to target lncRNAs. Because of the ability to form a DNA/RNA heteroduplex, ASOs can undergo cleavage by RNase H. The siRNAs are double-stranded RNAs, the guide strand of which can be loaded into argonaute 2. This results in the formation of RNA-induced silencing complex that can degrade target lncRNAs. While nuclear lncRNAs are susceptible to ASOs, siRNAs predominantly target cytoplasmic lncRNAs (Lennox & Behlke, 2016). The preclinical studies have shown the efficacy of ASOs and siRNAs against breast cancer. For instance, ASO-mediated knockout/knockdown of LINC00673 (Qiao et al., 2019), LINC02273 (Xiu, Chi et al., 2019), and MALAT1 were found to attenuate breast cancer growth and metastasis. Similarly, siRNAs-mediated HOTAIR silencing can reduce the growth and invasion of breast cancer (Li, Li et al., 2017).

The modulation of lncRNAs through RNA nanoparticles has also been utilized to treat multiple cancer types (Lee et al., 2017). The tumor-targeting RGD-PEG-ECO/siDANCR nanoparticles were formulated for the systemic delivery of siDANCR (Vaidya et al., 2019). The nanoparticle-mediated suppression of the oncogenic DANCR was found effective against TNBC. In mice xenograft model for lung cancer, ASO-mediated suppression of MALAT1 was associated with the reduction of tumor metastasis (Gutschner et al., 2013). siRNA-mediated suppression of HOTAIR was associated with a reduction in invasiveness and the reversal of the epithelial–mesenchymal transition process. The proliferation and metastasis of TNBC are enhanced by the lncRNA, TROJAN (Jin et al., 2019). TROJAN correlates with poor patient survival (Jin et al., 2019). Further, ASO-mediated silencing of TROJAN was linked with a suppression in the lung metastasis in a xenograft mouse model (Jin et al., 2019). The silencing of abundant in neuroepithelium area/B-cell translocation gene 3 locus by ASOs was found to suppress the tumorigenic potential of TNBC cells (Xia et al., 2017). ASO-mediated targeting of ncRNA in the aldehyde dehydrogenase 1A was found to reduce tumor growth and stem cell characteristics in TNBC

(Vidovic et al., 2020). The long intergenic ncRNA for kinase activation (LINK-A) is an oncogenic lncRNA known to activate HIF1α in TNBC (Lin et al., 2016). In a recent study, targeting LINK-A by locked nucleic acid (LNA) was found to reduce proliferation in TNBC, but not in non-TNBC cells (Hu et al., 2019). Further, LINK-A LNAs inhibited tumor growth and reduced lung metastasis in MMTV-Tg mice (Hu et al., 2019). An inhibition in the interaction of HOTAIR with polycomb repressive complex 2 or lysine-specific histone demethylase 1 complexes by small molecules is reported to reduce the metastatic potential of breast cancer cells (Tsai, Spitale, & Chang, 2011). BC-819 plasmid carrying diphtheria toxin under the control of the H19 regulatory sequences has been reported to reduce tumor size (Smaldone & Davies, 2010).

8.5 Potential of lncRNAs in predicting breast cancer patient's response to therapeutics

Many lncRNAs are capable of predicting the response of patients to therapeutics. For instance, a high level of HOTAIR and FOXM1 could predict the response of ER$^+$ breast cancer patients to endocrine therapy (Milevskiy et al., 2016). Similarly, X-inactive specific transcript (XIST) could determine the sensitivity of breast cancer patients to histone deacetylase inhibitor abexinostat (Salvador et al., 2013). Fludarabine, an FDA-approved chemotherapeutics, can specifically eliminate breast cancer cells expressing a lower XIST (Xing et al., 2018). The multigene signatures incorporating lncRNAs could predict responsiveness to taxane, and tumor recurrence in TNBC patients (Jiang et al., 2016; Liu, Jiang et al., 2016). MIR2052HG could predict tumor recurrence in the breast cancer patients under aromatase inhibitors (Ingle et al., 2016). Overall, lncRNAs can be used as a predictor of a patient's response to therapeutics. However, these observations should be clinically validated by more studies. This could be helpful for personalized and precision medicine.

8.6 Potential of lncRNAs in predicting chemoresistance and radioresistance in breast cancer patients

The development of resistance by cancer cells over time is a major roadblock for therapy. Recent studies suggest that lncRNAs can contribute to cancer chemoresistance. Thus interrogating the lncRNAs expression can sensitize breast cancer cells to therapeutics. For instance, exosomal UCA1 could enhance resistance to tamoxifen in ER + MCF-7 cells (Xu, Yang, Ren, Wu, & Wang, 2016). Thus UCA1 could be targeted for reversing tamoxifen resistance (Xu et al., 2016). TNBC cells can be sensitized to chemotherapy by silencing noncoding nuclear-enriched abundant transcript 1 (NEAT1) (Shin et al., 2019). Thus NEAT1 could be involved in chemoresistance (Shin et al., 2019). NEAT1 has been reported to promote distant metastasis and 5-FU resistance via miR-129/ZEB2 and miR-211/HMGA2 axes (Li, Wang, Li et al., 2017). The resistance of KPL-4/ADM and MCF-7/ADM cells to vincristine, taxel, and adriamycin can be attenuated by LINC00968 (Sun, Huang et al., 2019). Mechanistically, LINC00968 suppresses the wingless 2/beta-catenin (Wnt2/β-catenin) signaling pathway (Xiu, Liu et al., 2019).

UCA1 also enhances breast cancer cells' sensitivity to tamoxifen via negatively impacting the Wnt/β-catenin pathway. The radiosensitivity of TNBC cells can be enhanced by silencing prostate cancer–associated transcript 6 or tumor protein D52, or by overexpression of miR-185−5p (Shi, Wu, Liu, Chen, & Cong, 2020). The level of H19 is significantly increased in paclitaxel-resistant in comparison to paclitaxel-sensitive TNBC cells (Han et al., 2018). Further, the suppression of H19 can enhance the sensitivity of paclitaxel-resistant cells through the modulation of the AKT/protein kinase B pathway (Han et al., 2018). H19 can also enhance doxorubicin resistance through the involvement of multidrug resistance protein 1 (MDR1) and MDR4 (Zhu et al., 2017). In the tumor tissues of TNBC patients under postoperative radiotherapy, a higher level of actin filament–associated protein 1 antisense RNA1 (AFAP1-AS1) was observed (Bi et al., 2020). The survival of TNBC patients was associated with the level of lncAFAP1-AS1 (Bi et al., 2020). The level of lncAFAP1-AS1 is also correlated with the radioresistance of TNBC patients. The group proposed that the lncAFAP1-AS1 could be used as a predictor of response to radiotherapy. The LNA-mediated targeting of LINK-A (Hu et al., 2019) and BCAR4 (Xing et al., 2014; Zheng et al., 2017) can sensitize breast cancer cells to immune checkpoint inhibitors.

In conclusion, lncRNAs could predict breast cancer cells' response to chemotherapeutic agents. Thus lncRNAs can be targeted for the sensitization of breast cancer cells to chemotherapy. However, more studies should be carried out on patients.

8.7 Experimental methods and tools for analyzing noncoding RNAs in cancer patients

Various strategies have been employed to identify novel and functional ncRNAs in various experimental models (Argaman et al., 2001; Storz, 2002; Washietl, Hofacker, & Stadler, 2005; Wassarman, Repoila, Rosenow, Storz, & Gottesman, 2001). The most commonly used methods include reverse transcription-quantitative polymerase chain reaction (RT-qPCR), next-generation sequencing (NGS), RNA-seq, microarray, TILING array, serial analysis of gene expression (SAGE), and cap analysis gene expression (CAGE).

In RT-qPCR, RNA is first reverse transcribed into complementary DNA (cDNA). Subsequently, cDNA is amplified by qPCR in three repeating steps: denaturation, annealing, and elongation. Both one-step and two-step assays are used for RT-qPCR. Forone-step assay, reverse transcription and PCR are performed in a single tube. In two-step assay, two steps are performed in separate tubes. The reaction buffers, priming strategies, and reaction conditions in two steps are different. The fluorescent labeling of the product in qPCR enables the collection of the data as the reaction progresses. Thus by using RT-qPCR, the product can be measured as the reaction is in progress (exponential phase). This is unlike the standard PCR, which measures the product at the end of the reaction (plateau phase).

The NGS has revolutionized the detection of different types of ncRNAs (Veneziano, Nigita, & Ferro, 2015). RNA-seq involves the detection and quantification of ncRNAs. For this, the construction of RNA-specific cDNA libraries is required (Kukurba & Montgomery, 2015). This is followed by the sequencing of the transcripts of interest. Two types of RNA-seq are reported: small RNA-seq and total RNA-seq. While small RNA-seq is used for the small ncRNAs sequencing, total RNA-seq is suitable for lncRNA sequencing (Podnar, Deiderick, Huerta, & Hunicke-Smith, 2014). RNA-seq provides an elaborated analysis of whole transcriptomes. Using RNA-seq, sequences that differ even in a single nucleotide can be detected (Grillone et al., 2020). The disadvantages of RNA-seq are the complexity in the data analysis and the high deep reads required to detect a low amount of the target. Microarray is used for performing global analysis of the transcriptome in different types of cells, tissues, or pathological conditions. The microarray can be used in the profiling of both miRNAs and lncRNAs (Kukurba & Montgomery, 2015).

TILING array is a microarray-based technique. Tilling arrays require specific sequences as well as contiguous regions for sequencing (Gräf et al., 2007). In recent years, NGS has replaced the TILING array. In SAGE, short-stretches of unbiased cDNAs are prepared through restriction enzymes followed by concatenation, cloning, and sequencing (Velculescu, Zhang, Vogelstein, & Kinzler, 1995). CAGE is a NGS-based technology that allows the generation of a snapshot of the 5′ end of the mRNA. In CAGE, sequencing is followed by cDNAs generation, concatenation, and cloning (de Hoon & Hayashizaki, 2008).

8.8 Summary and conclusion

The lncRNAs are dysregulated in multiple tumor types, including breast cancer. The lncRNAs are often expressed in a disease and cell type—specific manner. Moreover, lncRNAs are generally easy to detect. The lncRNAs have demonstrated potential in the diagnosis and prognosis of breast cancer. The lncRNAs could also be used as a tool in examining the sensitivity of breast cancer cells to chemotherapy. The dysregulation in lncRNAs expression is reported to contribute to cancer chemoresistance. Most of the conclusions on the diagnostic and prognostic significance of lncRNAs are based on the upregulation or downregulation of lncRNAs in the clinical specimens. The dysregulation in the lncRNAs expression could also be due to the conditions other than cancer. Many conclusions on the utility of lncRNAs are based on the observations in the preclinical models. Future studies should focus more on exploring the potential of lncRNAs in the samples from larger cohorts of breast cancer patients.

Acknowledgment

The work was supported in part by the Indian Council of Medical Research, New Delhi in SCG's laboratory (5/13/51/2020/NCD-III). Dr. Challagundla's laboratory is supported in whole or part by NIH/NCI grant CA197074, Buffet Pilot & Pediatric Cancer Research Group Grants at UNMC. Sharma's laboratory is supported by Nitte Research Grant (NU/Dr/NUFR1/NUCSER/2019−20/01).

References

Argaman, L., Hershberg, R., Vogel, J., Bejerano, G., Wagner, E. G. H., Margalit, H., & Altuvia, S. (2001). Novel small RNA-encoding genes in the intergenic regions of Escherichia coli. *Current Biology, 11*, 941−950.

Bai, J., Zhao, W., Li, W., Ying, Z., & Jiang, D. (2019). Long noncoding RNA LINC00473 indicates a poor prognosis of breast cancer and accelerates tumor carcinogenesis by competing endogenous sponging miR-497. *European Review for Medical and Pharmacological Sciences, 23*, 3410−3420.

Bi, Z., Li, Q., Dinglin, X., Xu, Y., You, K., Hong, H., ... Tan, Y. (2020). Nanoparticles (NPs)-meditated LncRNA AFAP1-AS1 silencing to block Wnt/β-catenin signaling pathway for synergistic reversal of radioresistance and effective cancer radiotherapy. *Advanced Science, 7*, 2000915.

Chi, Y., Huang, S., Yuan, L., Liu, M., Huang, N., Zhou, S., ... Wu, J. (2014). Role of BC040587 as a predictor of poor outcome in breast cancer. *Cancer Cell International, 14*, 1–6.

Cipolla, G. A., De Oliveira, J. C., Salviano-Silva, A., Lobo-Alves, S. C., Lemos, D. S., Oliveira, L. C., ... Zambalde, E. P. (2018). Long non-coding RNAs in multifactorial diseases: Another layer of complexity. *Non-coding RNA, 4*, 13.

Clark, M. B., Mercer, T. R., Bussotti, G., Leonardi, T., Haynes, K. R., Crawford, J., ... Chen, W. Y. (2015). Quantitative gene profiling of long noncoding RNAs with targeted RNA sequencing. *Nature Methods, 12*, 339–342.

Collina, F., Aquino, G., Brogna, M., Cipolletta, S., Buonfanti, G., De Laurentiis, M., Di Bonito, M., Cantile, M., & Botti, G. (2019). LncRNA HOTAIR up-regulation is strongly related with lymph nodes metastasis and LAR subtype of triple negative breast cancer. *Journal of Cancer, 10*, 2018.

de Hoon, M., & Hayashizaki, Y. (2008). Deep cap analysis gene expression (CAGE): Genome-wide identification of promoters, quantification of their expression, and network inference. *Biotechniques, 44*, 627–632.

Ding, X., Zhu, L., Ji, T., Zhang, X., Wang, F., Gan, S., ... Yang, H. (2014). Long intergenic non-coding RNAs (lincRNAs) identified by RNA-seq in breast cancer. *PLoS One, 9*, e103270.

Dong, H., Hu, J., Zou, K., Ye, M., Chen, Y., Wu, C., ... Han, M. (2019). Activation of LncRNA TINCR by H3K27 acetylation promotes Trastuzumab resistance and epithelial-mesenchymal transition by targeting MicroRNA-125b in breast cancer. *Molecular Cancer, 18*, 1–18.

Ezkurdia, I., Juan, D., Rodriguez, J. M., Frankish, A., Diekhans, M., Harrow, J., ... Tress, M. L. (2014). Multiple evidence strands suggest that there may be as few as 19 000 human protein-coding genes. *Human Molecular Genetics, 23*, 5866–5878.

Fan, C. N., Ma, L., & Liu, N. (2019). Comprehensive analysis of novel three-long noncoding RNA signatures as a diagnostic and prognostic biomarkers of human triple-negative breast cancer. *Journal of Cellular Biochemistry, 120*, 3185–3196.

Ferlay, J., Soerjomataram, I., Dikshit, R., Eser, S., Mathers, C., Rebelo, M., ... Bray, F. (2015). Cancer incidence and mortality worldwide: Sources, methods and major patterns in GLOBOCAN 2012. *International Journal of Cancer, 136*, E359–E386.

Fu, J., Dong, G., Shi, H., Zhang, J., Ning, Z., Bao, X., ... Xiong, B. (2019). LncRNA MIR503HG inhibits cell migration and invasion via miR-103/OLFM4 axis in triple negative breast cancer. *Journal of Cellular and Molecular Medicine, 23*, 4738–4745.

Gökmen-Polar, Y., Vladislav, I. T., Neelamraju, Y., Janga, S. C., & Badve, S. (2015). Prognostic impact of HOTAIR expression is restricted to ER-negative breast cancers. *Scientific Reports, 5*, 1–6.

Goyal, B., Yadav, S.R.M., Awasthee, N., Gupta, S., Kunnumakkara, A.B., & Gupta, S.C. (2021). *BBA-reviews on cancer, 1875*, 188502.

Gräf, S., Nielsen, F. G., Kurtz, S., Huynen, M. A., Birney, E., Stunnenberg, H., & Flicek, P. (2007). Optimized design and assessment of whole genome tiling arrays. *Bioinformatics (Oxford, England), 23*, i195–i204.

Grillone, K., Riillo, C., Scionti, F., Rocca, R., Tradigo, G., Guzzi, P. H., ... Tassone, P. (2020). Non-coding RNAs in cancer: Platforms and strategies for investigating the genomic "dark matter,". *Journal of Experimental & Clinical Cancer Research, 39*, 1–19.

Gutschner, T., Hämmerle, M., Eißmann, M., Hsu, J., Kim, Y., Hung, G., ... Groß, M. (2013). The noncoding RNA MALAT1 is a critical regulator of the metastasis phenotype of lung cancer cells. *Cancer Research, 73*, 1180–1189.

Guttman, M., Amit, I., Garber, M., French, C., Lin, M. F., Feldser, D., ... Cassady, J. P. (2009). Chromatin signature reveals over a thousand highly conserved large non-coding RNAs in mammals. *Nature, 458*, 223–227.

Han, J., Han, B., Wu, X., Hao, J., Dong, X., Shen, Q., & Pang, H. (2018). Knockdown of lncRNA H19 restores chemo-sensitivity in paclitaxel-resistant triple-negative breast cancer through triggering apoptosis and regulating Akt signaling pathway. *Toxicology and Applied Pharmacology, 359*, 55–61.

Han, L., Ma, P., Liu, S.-M., & Zhou, X. (2016). Circulating long noncoding RNA GAS5 as a potential biomarker in breast cancer for assessing the surgical effects. *Tumor Biology, 37*, 6847–6854.

Harbeck, N., & Gnant, M. (2017). Breast cancer. *Lancet, 389*, 1134–1150.

Hu, P., Chu, J., Wu, Y., Sun, L., Lv, X., Zhu, Y., ... Liu, B. (2015). NBAT1 suppresses breast cancer metastasis by regulating DKK1 via PRC2. *Oncotarget, 6*, 32410.

Hu, Q., Ye, Y., Chan, L.-C., Li, Y., Liang, K., Lin, A., ... Gong, J. (2019). Oncogenic lncRNA downregulates cancer cell antigen presentation and intrinsic tumor suppression. *Nature Immunology, 20*, 835–851.

Ingle, J. N., Xie, F., Ellis, M. J., Goss, P. E., Shepherd, L. E., Chapman, J.-A. W., ... Momozawa, Y. (2016). Genetic polymorphisms in the long noncoding RNA MIR2052HG offer a pharmacogenomic basis for the response of breast cancer patients to aromatase inhibitor therapy. *Cancer Research, 76*, 7012−7023.

Jalali, S., Gandhi, S., & Scaria, V. (2016). Navigating the dynamic landscape of long noncoding RNA and protein-coding gene annotations in GENCODE. *Human Genomics, 10*, 1−12.

Jiang, Y.-Z., Liu, Y.-R., Xu, X.-E., Jin, X., Hu, X., Yu, K.-D., & Shao, Z.-M. (2016). Transcriptome analysis of triple-negative breast cancer reveals an integrated mRNA-lncRNA signature with predictive and prognostic value. *Cancer Research, 76*, 2105−2114.

Jiao, Z., Tian, Q., Li, N., Wang, H., & Li, K. (2018). Plasma long non-coding RNAs (lncRNAs) serve as potential biomarkers for predicting breast cancer. *European Review for Medical and Pharmacological Sciences, 22*, 1994−1999.

Jin, X., Xu, X.-E., Jiang, Y.-Z., Liu, Y.-R., Sun, W., Guo, Y.-J., ... Huang, S.-L. (2019). The endogenous retrovirus-derived long noncoding RNA TROJAN promotes triple-negative breast cancer progression via ZMYND8 degradation. *Science Advances, 5*, eaat9820.

Kukurba, K. R., & Montgomery, S. B. (2015). RNA sequencing and analysis. *Cold Spring Harbor Protocols, 2015*, pdb. top084970.

Lee, T. J., Yoo, J. Y., Shu, D., Li, H., Zhang, J., Yu, J.-G., ... Cui, R. (2017). RNA nanoparticle-based targeted therapy for glioblastoma through inhibition of oncogenic miR-21. *Molecular Therapy, 25*, 1544−1555.

Lennox, K. A., & Behlke, M. A. (2016). Cellular localization of long non-coding RNAs affects silencing by RNAi more than by antisense oligonucleotides. *Nucleic Acids Research, 44*, 863−877.

Li, H., Ma, S. Q., Huang, J., Chen, X. P., & Zhou, H. H. (2017). Roles of long noncoding RNAs in colorectal cancer metastasis. *Oncotarget, 8*, 39859−39876.

Li, M., Li, X., Zhuang, Y., Flemington, E. K., Lin, Z., & Shan, B. (2017). Induction of a novel isoform of the lnc RNA HOTAIR in Claudin-low breast cancer cells attached to extracellular matrix. *Molecular Oncology, 11*, 1698−1710.

Li, R.-H., Chen, M., Liu, J., Shao, C.-C., Guo, C.-P., Wei, X.-L., ... Zhang, G.-J. (2018). Long noncoding RNA ATB promotes the epithelial − mesenchymal transition by upregulating the miR-200c/Twist1 axe and predicts poor prognosis in breast cancer. *Cell Death & Disease, 9*, 1−16.

Li, X., Wang, S., Li, Z., Long, X., Guo, Z., Zhang, G., ... Wen, L. (2017). The lncRNA NEAT1 facilitates cell growth and invasion via the miR-211/HMGA2 axis in breast cancer. *International Journal of Biological Macromolecules, 105*, 346−353.

Li, Y., Wang, B., Lai, H., Li, S., You, Q., Fang, Y., ... Liu, Y. (2017). Long non-coding RNA CRALA is associated with poor response to chemotherapy in primary breast cancer. *Thorac Cancer, 8*, 582−591.

Li, Z., Xu, L., Liu, Y., Fu, S., Tu, J., Hu, Y., & Xiong, Q. (2018). LncRNA MALAT1 promotes relapse of breast cancer patients with postoperative fever. *American Journal of Translational Research, 10*, 3186.

Lin, A., Li, C., Xing, Z., Hu, Q., Liang, K., Han, L., ... Zhang, Y. (2016). The LINK-A lncRNA activates normoxic HIF1α signalling in triple-negative breast cancer. *Nature Cell Biology, 18*, 213−224.

Liu, H., Li, J., Koirala, P., Ding, X., Chen, B., Wang, Y., ... Mo, Y.-Y. (2016). Long non-coding RNAs as prognostic markers in human breast cancer. *Oncotarget, 7*, 20584.

Liu, M., Xing, L.-Q., & Liu, Y.-J. (2017). A three-long noncoding RNA signature as a diagnostic biomarker for differentiating between triple-negative and non-triple-negative breast cancers. *Medicine, 96*.

Liu, X., Yang, B., & Han, J. (2018). Increased long noncoding RNA LINP1 expression and its prognostic significance in human breast cancer. *European Review for Medical and Pharmacological Sciences, 22*, 8749−8754.

Liu, Y.-R., Jiang, Y.-Z., Xu, X.-E., Yu, K.-D., Jin, X., Hu, X., ... Liu, G.-Y. (2016). Comprehensive transcriptome analysis identifies novel molecular subtypes and subtype-specific RNAs of triple-negative breast cancer. *Breast Cancer Research, 18*, 1−10.

Lottin, S., Adriaenssens, E., Dupressoir, T., Berteaux, N., Montpellier, C., Coll, J., ... Curgy, J. J. (2002). Overexpression of an ectopic H19 gene enhances the tumorigenic properties of breast cancer cells. *Carcinogenesis, 23*, 1885−1895.

Lu, R., Zhang, J., Zhang, W., Huang, Y., Wang, N., Zhang, Q., & Qu, S. (2018). Circulating HOTAIR expression predicts the clinical response to neoadjuvant chemotherapy in patients with breast cancer. *Cancer Biomarkers, 22*, 249−256.

Lv, M., Xu, P., Wu, Y., Huang, L., Li, W., Lv, S., ... Jia, X. (2016). LncRNAs as new biomarkers to differentiate triple negative breast cancer from non-triple negative breast cancer. *Oncotarget, 7*, 13047.

Ma, P., Pan, Y., Li, W., Sun, C., Liu, J., Xu, T., & Shu, Y. (2017). Extracellular vesicles-mediated noncoding RNAs transfer in cancer. *Journal of Hematology & Oncology, 10,* 1–11.

Matsui, M., & Corey, D. R. (2017). Non-coding RNAs as drug targets. *Nature Reviews Drug Discovery, 16,* 167–179.

Meng, Y., Liu, Y.-L., Li, K., & Fu, T. (2019). Prognostic value of long non-coding RNA breast cancer anti-estrogen resistance 4 in human cancers: A meta-analysis. *Medicine, 98.*

Milevskiy, M. J., Al-Ejeh, F., Saunus, J. M., Northwood, K. S., Bailey, P. J., Betts, J. A., ... Gee, J. M. (2016). Long-range regulators of the lncRNA HOTAIR enhance its prognostic potential in breast cancer. *Human Molecular Genetics, 25,* 3269–3283.

Podnar, J., Deiderick, H., Huerta, G., & Hunicke-Smith, S. (2014). Next-Generation sequencing RNA-Seq library construction. *Current Protocols in Molecular Biology, 106,* 4.21. 21–24.21. 19.

Qi, P., & Du, X. (2013). The long non-coding RNAs, a new cancer diagnostic and therapeutic gold mine. *Modern Pathology, 26,* 155–165.

Qiao, K., Ning, S., Wan, L., Wu, H., Wang, Q., Zhang, X., ... Pang, D. (2019). LINC00673 is activated by YY1 and promotes the proliferation of breast cancer cells via the miR-515-5p/MARK4/Hippo signaling pathway. *Journal of Experimental & Clinical Cancer Research, 38,* 1–15.

Salvador, M. A., Wicinski, J., Cabaud, O., Toiron, Y., Finetti, P., Josselin, E., ... Bertucci, F. (2013). The histone deacetylase inhibitor abexinostat induces cancer stem cells differentiation in breast cancer with low Xist expression. *Clinical Cancer Research, 19,* 6520–6531.

Schmitt, A. M., & Chang, H. Y. (2016). Long noncoding RNAs in cancer pathways. *Cancer Cell, 29,* 452–463.

Sha, S., Yuan, D., Liu, Y., Han, B., & Zhong, N. (2017). Targeting long non-coding RNA DANCR inhibits triple negative breast cancer progression. *Biology Open, 6,* 1310–1316.

Shi, R., Wu, P., Liu, M., Chen, B., & Cong, L. (2020). Knockdown of lncRNA PCAT6 enhances radiosensitivity in triple-negative breast cancer cells by regulating miR-185-5p/TPD52 axis. *OncoTargets and Therapy, 13,* 3025–3037.

Shi, Y., Li, J., Liu, Y., Ding, J., Fan, Y., Tian, Y., ... Shu, Y. (2015). The long noncoding RNA SPRY4-IT1 increases the proliferation of human breast cancer cells by upregulating ZNF703 expression. *Molecular Cancer, 14,* 1–13.

Shima, H., Kida, K., Adachi, S., Yamada, A., Sugae, S., Narui, K., ... Murata, S. (2018). Lnc RNA H19 is associated with poor prognosis in breast cancer patients and promotes cancer stemness. *Breast Cancer Research and Treatment, 170,* 507–516.

Shin, V. Y., Chen, J., Cheuk, I. W., Siu, M. T., Ho, C. W., Wang, X., ... Kwong, A. (2019). Long non-coding RNA NEAT1 confers oncogenic role in triple-negative breast cancer through modulating chemoresistance and cancer stemness. *Cell Death & Disease, 10,* 270.

Smaldone, M. C., & Davies, B. J. (2010). BC-819, a plasmid comprising the H19 gene regulatory sequences and diphtheria toxin A, for the potential targeted therapy of cancers. *Current Opinion in Molecular Therapeutics, 12,* 607–616.

Storz, G. (2002). An expanding universe of noncoding RNAs. *Science (New York, N.Y.), 296,* 1260–1263.

Sun, M., Wu, D., Zhou, K., Li, H., Gong, X., Wei, Q., ... Zhu, H. (2019). An eight-lncRNA signature predicts survival of breast cancer patients: A comprehensive study based on weighted gene co-expression network analysis and competing endogenous RNA network. *Breast Cancer Research and Treatment, 175,* 59–75.

Sun, X., Huang, T., Zhang, C., Zhang, S., Wang, Y., Zhang, Q., & Liu, Z. (2019). Long non-coding RNA LINC00968 reduces cell proliferation and migration and angiogenesis in breast cancer through up-regulation of PROX1 by reducing hsa-miR-423-5p. *Cell Cycle (Georgetown, Tex.), 18,* 1908–1924.

Tian, T., Wang, M., Lin, S., Guo, Y., Dai, Z., Liu, K., ... Zheng, Y. (2018). The impact of lncRNA dysregulation on clinicopathology and survival of breast cancer: A systematic review and meta-analysis. *Molecular Therapy-Nucleic Acids, 12,* 359–369.

Tsai, M.-C., Spitale, R. C., & Chang, H. Y. (2011). Long intergenic noncoding RNAs: New links in cancer progression. *Cancer Research, 71,* 3–7.

Vaidya, A. M., Sun, Z., Ayat, N., Schilb, A., Liu, X., Jiang, H., ... He, S. (2019). Systemic delivery of tumor-targeting siRNA nanoparticles against an oncogenic LncRNA facilitates effective triple-negative breast cancer therapy. *Bioconjugate Chemistry, 30,* 907–919.

Velculescu, V. E., Zhang, L., Vogelstein, B., & Kinzler, K. W. (1995). Serial analysis of gene expression. *Science (New York, N.Y.), 270,* 484–487.

Veneziano, D., Nigita, G., & Ferro, A. (2015). Computational approaches for the analysis of ncRNA through deep sequencing techniques. *Frontiers in Bioengineering and Biotechnology, 3,* 77.

Vidovic, D., Huynh, T. T., Konda, P., Dean, C., Cruickshank, B. M., Sultan, M., ... Marcato, P. (2020). ALDH1A3-regulated long non-coding RNA NRAD1 is a potential novel target for triple-negative breast tumors and cancer stem cells. *Cell Death & Differentiation, 27*, 363–378.

Wang, Y., Zhang, G., & Han, J. (2019). HIF1A-AS2 predicts poor prognosis and regulates cell migration and invasion in triple-negative breast cancer. *Journal of Cellular Biochemistry, 120*, 10513–10518.

Washietl, S., Hofacker, I. L., & Stadler, P. F. (2005). Fast and reliable prediction of noncoding RNAs. *Proceedings of the National Academy of Sciences, 102*, 2454–2459.

Wassarman, K. M., Repoila, F., Rosenow, C., Storz, G., & Gottesman, S. (2001). Identification of novel small RNAs using comparative genomics and microarrays. *Genes & Development, 15*, 1637–1651.

Xia, Y., Xiao, X., Deng, X., Zhang, F., Zhang, X., Hu, Q., & Sheng, W. (2017). Targeting long non-coding RNA ASBEL with oligonucleotide antagonist for breast cancer therapy. *Biochemical and Biophysical Research Communications, 489*, 386–392.

Xing, F., Liu, Y., Wu, S.-Y., Wu, K., Sharma, S., Mo, Y.-Y., ... Singh, R. (2018). Loss of XIST in breast cancer activates MSN-c-Met and reprograms microglia via exosomal miRNA to promote brain metastasis. *Cancer Research, 78*, 4316–4330.

Xing, Z., Lin, A., Li, C., Liang, K., Wang, S., Liu, Y., ... Hawke, D. H. (2014). lncRNA directs cooperative epigenetic regulation downstream of chemokine signals. *Cell, 159*, 1110–1125.

Xiu, B., Chi, Y., Liu, L., Chi, W., Zhang, Q., Chen, J., ... Wu, J. (2019). LINC02273 drives breast cancer metastasis by epigenetically increasing AGR2 transcription. *Molecular Cancer, 18*, 187.

Xiu, D.-H., Liu, G.-F., Yu, S.-N., Li, L.-Y., Zhao, G.-Q., Liu, L., & Li, X.-F. (2019). Long non-coding RNA LINC00968 attenuates drug resistance of breast cancer cells through inhibiting the Wnt2/β-catenin signaling pathway by regulating WNT2. *Journal of Experimental & Clinical Cancer Research, 38*, 1–18.

Xu, C., Yang, M., Ren, Y., Wu, C., & Wang, L. (2016). Exosomes mediated transfer of lncRNA UCA1 results in increased tamoxifen resistance in breast cancer cells. *European Review for Medical and Pharmacological Sciences, 20*, 4362–4368.

Xu, N., Chen, F., Wang, F., Lu, X., Wang, X., Lv, M., & Lu, C. (2015). Clinical significance of high expression of circulating serum lncRNA RP11-445H22. 4 in breast cancer patients: A Chinese population-based study. *Tumor Biology, 36*, 7659–7665.

Xu, S., Sui, S., Zhang, J., Bai, N., Shi, Q., Zhang, G., ... Liu, F. (2015). Downregulation of long noncoding RNA MALAT1 induces epithelial-to-mesenchymal transition via the PI3K-AKT pathway in breast cancer. *International Journal of Clinical and Experimental Pathology, 8*, 4881.

Yang, F., Liu, Y.-h., Dong, S.-y., Ma, R.-m., Bhandari, A., Zhang, X.-h., & Wang, O.-c. (2016). A novel long noncoding RNA FGF14-AS2 is correlated with progression and prognosis in breast cancer. *Biochemical and Biophysical Research Communications, 470*, 479–483.

Yang, S., Sun, Z., Zhou, Q., Wang, W., Wang, G., Song, J., ... Xia, K. (2018). MicroRNAs, long noncoding RNAs, and circular RNAs: Potential tumor biomarkers and targets for colorectal cancer. *Cancer Management and Research, 10*, 2249.

Zhang, H., Zhang, N., Liu, Y., Su, P., Liang, Y., Li, Y., ... Sang, Y. (2019). Epigenetic regulation of NAMPT by NAMPT-AS drives metastatic progression in triple-negative breast cancer. *Cancer Research, 79*, 3347–3359.

Zhang, K., Luo, Z., Zhang, Y., Zhang, L., Wu, L., Liu, L., ... Liu, J. (2016). Circulating lncRNA H19 in plasma as a novel biomarker for breast cancer. *Cancer Biomarkers, 17*, 187–194.

Zhang, Z., Weaver, D. L., Olsen, D., deKay, J., Peng, Z., Ashikaga, T., & Evans, M. F. (2016). Long non-coding RNA chromogenic in situ hybridisation signal pattern correlation with breast tumour pathology. *Journal of Clinical Pathology, 69*, 76–81.

Zhao, W., Wang, Z., Fang, X., Li, N., & Fang, J. (2019). Long noncoding RNA Breast cancer antiestrogen resistance 4 is associated with cancer progression and its significant prognostic value. *Journal of Cellular Physiology, 234*, 12956–12963.

Zheng, X., Han, H., Liu, G. P., Ma, Y. X., Pan, R. L., Sang, L. J., ... Wang, W. (2017). Lnc RNA wires up Hippo and Hedgehog signaling to reprogramme glucose metabolism. *The EMBO Journal, 36*, 3325–3335.

Zhu, Q.-N., Wang, G., Guo, Y., Peng, Y., Zhang, R., Deng, J.-L., ... Zhu, Y.-S. (2017). LncRNA H19 is a major mediator of doxorubicin chemoresistance in breast cancer cells through a cullin4A-MDR1 pathway. *Oncotarget, 8*, 91990.

9

Noncoding ribonucleic acids in gastric cancer patients

Rachel Sexton, Najeeb Al-Hallak, Bayan Al-Share, Anteneh Tesfaye and Asfar S. Azmi

Department of Oncology, Wayne State University School of Medicine, Detroit, MI, United States

9.1 Introduction

Based on a 2018 GLOBOCAN study, stomach cancer is the fifth most common neoplasm and the third most deadly cancer worldwide (Hu et al., 2012). There are three main subtypes of gastric cancer and the subtypes vary with the demographic and geographic location. The incidence of the well-differentiated subtype, or intestinal gastric cancer, is trending down in recent years due to the discovery of *Helicobacter pylori*, and subsequent eradication, and how these bacteria influence this subtype of gastric cancer (Kang & Friedländer, 2015). Although there is a decrease in the intestinal subtype, there is an influx of poorly differentiated, or diffuse, gastric cancer subtype within the United States and throughout the world (Chen et al., 2019; Kang & Friedländer, 2015). The cause of this cancer is due to genetic predisposition and environmental factors such as diet and smoking (Chen et al., 2019).

Small noncoding RNAs are noncoding RNAs, smaller than 200 nucleotides, that bind to messenger RNA (mRNAs) or other noncoding RNAs and function to control DNA synthesis or protein expression to elicit cellular responses (Quan et al., 2020). There are various classes of small noncoding RNAs, including microRNAs (miRNAs), piwi RNAs, circular RNAs (circRNAs), and long noncoding RNAs (lncRNAs). Since the sequencing of the human genome and the realization that only 20,000 genes were encoded within the genome, the majority of the sequences were considered as "junk DNA." Within recent years the focus has been turned toward this junk DNA and it has been realized that although these sequences do not encode proteins, they in fact have important roles within human cells. Small noncoding RNAs have been found to play a major role in gastric cancer growth, pathogenesis, invasion, migration, and overall influence on the survival of patients. Notably, it was found, as early as 2003, that noncoding RNAs may contribute to gastrointestinal cancer pathogenesis (Martinez, Enfield, Rowbotham, & Lam, 2016).

miRNAs are small noncoding RNA sequences, 18–30 nucleotides in length, highly conserved and complementary to coding genes (Quan et al., 2020). In 1993 two small and complementary RNA transcripts were found to control the *Caenorhabditis elegans lin*-4 gene (Martinez et al., 2016). This discovery led to the birth of a new scientific field and ultimately led to the premise that the human genome is dynamic and contains more than just coding RNA sequences. Since the first human genome sequencing was complete in 2003, various discoveries have been made about small noncoding RNA sequences and their functions in normal and pathogenic conditions (Heather & Chain, 2016; Hüttenhofer & Vogel, 2006; Yang, Qiao, Song, Bao, & Ma, 2020). The biogenesis of noncoding RNAs differ depending on the type, but they all have similarities that they are a by-product of normal mRNA processing and translation.

9.1.1 MicroRNAs

miRNAs are a well-studied noncoding RNA family that functions to control gene expression. Its biogenesis occurs through normal transcription processes via RNA polymerase II/III, and they are either transcribed in clusters or through the processing of introns from premature transcripts. The canonical biogenesis pathway involves a host of proteins, including RNA-binding protein DiGeorge Syndrome Critical Region 8 (DGCR8), and Drosha, a member of the ribonuclease III enzyme family (Derrien et al., 2012). DGCR8 recognizes methylated sequences on the pre-miRNA transcripts, by methyltransferase like 3 (METTL3), like the Guanine Guanine Adenosine Cytosine motif that binds to the RGAC recognition sequence on METTL3 and cleaves the sequence leaving a 3' overhang (Derrien et al., 2012). Nuclear Export protein 5 transports this pre-miRNA sequence into the nucleus where it is further processed by Dicer (Derrien et al., 2012). The noncanonical biogenesis pathway occurs through the splicing of mRNA introns that are transported to the nucleus via nuclear export protein 1 (Derrien et al., 2012). In normal cells, miRNAs are involved with a host of cellular processes, including regulating aging, fat metabolism, development, and cellular differentiation (Dastmalchi, Safaralizadeh, & Banan Khojasteh, 2019). The function of miRNAs within gastric cancer has only been elucidated in recent years and has recently become a major focus of gastric cancer research into pathogenesis and drug resistance as well as and biomarker discovery (Ku & Lin, 2014; Zimmermann, Bilusic, Lorenz, & Schroeder, 2010).

9.1.2 PiwiRNAs

PiRNA interacting RNAs (PiRNAs) are another class of small noncoding RNAs, 26–31 nucleotide segments in length, and are highly conserved and complementary to coding genes (Juzenas et al., 2017). The biogenesis of piRNAs occurs differently than with miRNAs as they are derived from RNA transcripts of transposons and they interact with PIWI proteins to function. PIWI proteins belong to a large family of proteins, formally known as the Argonaute (AGO)/PIWI protein family, which consist of AGO and PIWI proteins. PIWI proteins are expressed in the germline of organisms, whereas the AGO proteins are expressed throughout all cells (Nobili, Lionetti, & Neri, 2016). These two protein families have different structural makeups, such as different binding domains, which were characterized in the *Drosophila* model (Nobili et al., 2016). There are two main types of piRNA biogenesis which include the primary pathway and the ping-pong pathway. The primary pathway of piRNA biogenesis is very complex and involves a host of proteins, including PIWI, Zuc (zucchini), Vrenteno (Vret), and TDRD proteins (Tudor protein family). The intricacies of this biogenesis pathway are beyond the scope of this chapter but include loading of the piRNA precursor onto PIWI to become processed into a complete transcript (Nobili et al., 2016). The ping-pong pathway is a secondary biogenesis cycle occurring in germ cells where the piRNA transcript is processed and shuffled through a host of proteins, mainly Ago3 and Aub, for processing (Nobili et al., 2016). Irrespective of the route of biogenesis, the roles of piRNAs are to silence specific genes and transposons at all levels of transcription (Derrien et al., 2012). In gastric cancer, piRNAs are found within both the tissues and bloodstream and may function to regulate key genes that promote gastric cancer pathogenesis (Chong et al., 2010).

9.1.3 Circular RNAs

circRNAs are noncoding RNAs that are alternatively single-stranded spliced pre-mRNAs that are expressed similar to their cognate mRNA in eukaryotes and mammals. They are formed through back splicing where either a singular or group of exons can ligate together and form a circRNA structure, where the 5′ terminus of an upstream exon ligates to the 3′ terminus of a downstream exon (Lekka & Hall, 2018). CircRNA function is currently being studied but they have been known to regulate mammal development in normal cellular environments (Ivey & Srivastava, 2015). In cancer, they play a role in development and progression as well as acting as sponges for miRNAs. Sponging is a term which means these two small noncoding RNAs, circRNAs and miRNAs, can bind to one another and elicit a cellular response. This has been commonly seen in gastric cancer to control and perpetuate disease (Yu & Kuo, 2019).

9.1.4 Long noncoding RNAs

lncRNAs are another class of small noncoding RNAs, greater than 200 nucleotides in length, that are made through (1) the actions of RNA polymerase II and III and (2) epigenetic modifications and can operate within the cell in a variety of ways, including signaling molecules, decoy molecules, guides, and scaffolds for other noncoding RNAs. Taken from an mRNA transcript, the introns alone, or intergenic RNAs, or introns and exons can fold upon themselves and form hairpins that are then cleaved and process which result lncRNAs that are functionally similar to mRNAs but do not encode proteins (Wang, Lu, Wang, Liu, & Jiao, 2018). LncRNAs regulate DNA histones and epigenetic modifications in human cells (Fang & Fullwood, 2016). In gastric cancer, lncRNAs function to control and maintain disease and can act as sponges to miRNAs, similar to circRNA function (Kietrys, Velema, & Kool, 2017; Zhang, Huang, et al., 2017).

9.2 Experimental methods and tools for analyzing noncoding RNAs in gastric cancer patients

There are various ways to study noncoding RNA expression in gastric cancer patient tissues, but the main categories for noncoding RNA studies include database analysis and experimental methods.

Experimental methods mainly incorporate some type of sequencing platform. The act of sequencing was discovered by the British scientist Frederick Sanger in 1977 which is where the term Sanger sequencing was coined. Sequencing was basic and included running digested nuclear material, either DNA or RNA, on a gel and identified by looking at the terminated 3′ end of the nucleotide chain (Sárközy, Kahán, & Csont, 2018). Improvements have been made on this early sequencing method where fluorescent labeled dideoxynucleotide triphosphate are read by the order that the fluorescent signals are detected using software, based on size separation. Clearly, technologies have advanced since these initial discoveries and there are complex ways to study noncoding RNAs using newer sequencing methods. One way to analyze noncoding RNAs is with 2D fingerprinting (Shen et al., 2020). As with early sequencing methods, the RNA needs to be isolated from patient's tumor samples. This can be collected and isolated for total RNA using established methods (see Samples Used Section for further detail). Once isolated, the total RNA is labeled to have an extra phosphate group (PO4) attached to their 3′ end and are partially or completely digested with various RNAses (Shen et al., 2020). The samples are loaded and run into a denatured polyacrylamide gel onto cellulose acetate paper. This step consists of the first dimension of 2D fingerprinting. The second dimension uses thin-layer chromatography on Diethylethanolamine-cellulose paper to visualize the RNA bands which the bands can be used to identify the sequence of the RNAs of interest (Shen et al., 2020).

More complex sequencing platforms can be performed since the discovery of next-generation sequencing (NGS), which was adapted from the Sanger sequencing method. NGS utilizes cDNA library preparations, which are pools of cDNA that is labeled on their 3′ ends by Cytidine triphosphate and poly(A) polymerase and then made to have a 5′ adapter performed by the T4 RNA ligase (Ma, Shen, Kapesa, & Zeng, 2016). After reverse transcription, with oligo primers, the cDNA library is complete and is assessed using sequencing techniques or cloned for functional studies (Ma et al., 2016). NGS is a deep and/or parallel sequencing platform that utilizes a cDNA library, which is attached to a solid surface and amplified using the polymerase chain reaction (PCR). Sequences are then read and assembled by software and/or mapped back to a reference genome (Ma et al., 2016).

Microarray analyses can be utilized to study the noncoding RNAs. The basis of the microarray is to use an immobilized surface, such as glass slides, quartz, fiber optic slides, or microchips, with anchored short DNA elements (probes) to look at large-scale gene expression within an RNA or DNA sample. For noncoding RNAs, samples of total RNA are labeled with biotin and added to the microchips. The RNA hybridizes with the probe and the fluorescent emission is recorded using a scanner. Downstream data analysis will provide the expression data for the noncoding RNAs, by mapping to a reference genome, and downstream analyses can be performed, including gene ontology and pathway computational analyses (Slack & Chinnaiyan, 2019).

Genomic systematic evolution of ligands with exponential enrichment is another tool that can identify novel noncoding RNAs. This system can be utilized because it identifies low expressing RNA species. This method can screen entire genomes for noncoding RNAs that bind to global proteins and may elicit a function in gastric cancer. Similar to coimmunoprecipitation and protein tagging, a bait protein in its active conformation is needed to draw out noncoding RNA sequences that bind to it (Lee, Feinbaum, Ambros, & The, 1993). Genetic material, or source DNA, is needed as well. This genetic material needs to have a reference genome available, as with all sequencing applications. Primers are also needed, which need to be carefully selected to avoid primer dimer formation, self-complementation and no flanking sequences along with randomized regions within the primer (Lee et al., 1993). The basis of this technique is to utilize genomic aptamers that are functional domains within the noncoding RNA molecules that bind ligands. The original starting material for this process is contained in DNA libraries. These libraries construct the actual RNA library that will be assessed and undergo various rounds of selection and amplification to allow the enrichment of the lower expressing nonprotein-coding RNAs within the sample using radio-labeled nucleotides (Lee et al., 1993). After the addition of the bait protein the specific and nonspecific RNAs are separated using nitrocellulose filtration or other separation methods, such as chromatography, and thus RNA recovery is measured through scintillation counting, or the counting of radioactive material by measuring photon emissions (Michael, O'Connor, Van Holst Pellekaan, Young, & James, 2003). The RNAs within the pool are reverse transcribed and amplified using PCR techniques and sequenced with the Roche 454 sequencing platform, a type of DNA sequencing that can perform longer reads than other sequencing platforms, like Illumina (Weick & Miska, 2014). Once sequenced, algorithms are used to map the sequences back to the genome and clusters are generated to identify these sequences. Further functional studies can be utilized to determine the validity of the finding, including coimmunoprecipitation and chemical probing and/or foot-printing as discussed earlier (Lee et al., 1993).

Computational methods can also be utilized to assess noncoding RNA pools. This method can be utilized after sequencing has been performed and/or can be used retrospectively through public access to sequencing data in the GEO-NCBI data repository or a similar platform. To assess fresh sequencing data, there are two types of analysis that can be performed, either (1) sequence homology or (2) common features of the sequences (Wang & Dong, 2019). Sequence homology is best associated with the BLAST software or FASTA. These software use model predicting and sequence alignment to predict sequence similarity and structure in relation to the human reference genome (Wang & Dong, 2019). Sequence features, such as AT rich or GC rich regions or secondary structure formations, can also be performed with three different software: CRITICA, CST miner, and EST scan. Prediction models are built into these algorithms as well as other complex algorithms to ensure that what is being assessed is indeed noncoding RNAs (Wang & Dong, 2019). Depending on the type of noncoding RNA, there are different software and computational methods that can be applied. For example, with miRNAs there are a variety of publicly available tools, including MiRAlign, which investigates miRNAs in different species to identify novel miRNAs, miR Deep, MiRank, miRanalyzer, and mirTool which have complex algorithms that filter out the known RNA species (like tRNAs and mRNAs) by mapping back to the genome (Bunnik & Le Roch, 2013; Hou, 2018; Quinn & Chang, 2016). GENCODE is a software that uses machine learning to identify lncRNAs (Zhang, Wu, Chen, & Chen, 2019).

It is clear that sequencing is at the heart of all noncoding RNA discoveries, whether they be novel or not, and this technique is a mainstay within noncoding RNA research in gastric cancer. Once the expression of noncoding RNAs is found, other tools exist to preform downstream functional studies. For miRNAs, mimics and inhibitors can be transiently transfected into gastric cancer cell lines to determine how the miRNAs impact growth and cellular survival. Noncoding RNAs can be transfected within a stable vector, along with a DNA sequence, and transfected into a stable cell line to determine downstream noncoding RNA-transcription factor binding or noncoding RNA–gene regulation using the dual luciferase assay technique. Specific small noncoding RNA extraction kits and RT-qPCR probes are commercially available to study the noncoding RNA expression within gastric cancer cell lines or available tissues.

9.3 Expression of noncoding RNAs in gastric cancer patients

Small noncoding RNAs have either oncogenic or tumor-suppressive properties within gastric cancer. There is emerging evidence that miRNAs, lncRNAs, circRNAs, and piwi RNAs can influence gastric cancer cellular proliferation and cellular survival. These small noncoding RNAs can be differentially expressed in gastric cancer patients.

miRNA signatures have been detected by analyzing both tumor and serum samples from gastric cancer patients. Various studies were preformed either with retrospective analysis, such as utilizing publicly available databases, or from fresh tissues. For example, an miRNA microarray analysis found that four upregulated miRNAs, miR-181a-5p, miR-106b-5p, miR-199a-3p, and miR-148a-3p, and four were down regulated: miR-21–5p, miR-222–3p, miR-221–3p, and miR-146a-5p, and contributed to the potential of disease progression and may contribute to peritoneal metastasis observed in gastric cancer patients (Esteller, 2011). Serum analysis from gastric cancer patients showed that circulating miRNAs miR-133, miR-17, and miR-25 have potential diagnostic value in Iranian patients with gastric cancer and may be used as a biomarker (Link & Kupcinskas, 2018). The consensus of studies shows that miR-21–5p, miR-25–3p, miR-93, and miR-181a-3p are important with different disease subtypes of gastric cancer and varying populations (Li et al., 2015; López-Urrutia, Bustamante Montes, Ladrón de Guevara Cervantes, Pérez-Plasencia, & Campos-Parra, 2019; Lu et al., 2019; Huerta & Burke, 2016). Literature searches have put into perspective the vast amount of characterized miRNAs within gastric cancer but due to the high heterogeneity of this disease, these expression patterns should be taken only within the context of the disease subtype or population being studied.

Various lncRNAs have been found to be unregulated in gastric cancer patients and cell lines, including PTPRG-AS1, FOXP4-AS1, BLACAT2, ZXF2, and UCC, that control cell proliferation in gastric cancer and predict overall outcomes in patients (Gao et al., 2020). Studies can utilize both datasets and patient tissues and in recent years there has been more of a focus on discovering lnc-RNA signatures in gastric cancer in hopes of discovering gastric cancer biomarkers. For example, Song et al. have shown that two lncRNAs, RP11–366F6.2 and RP5–881L22.5, were associated with improved gastric cancer overall survival, while ERICH3-ASI was linked to unfavorable survival and prognosis (Han et al., 2020). While looking at small noncoding RNA signatures in gastric cancer, it is safe to assume that these cellular components interact and work with each other to regulate gastric cancer disease and progression. Gao et. al found that a variety of lncRNAs can act as sponges to regulate miRNA behavior, notably ENSG00000214548, which can downregulate miR-21 to induce epithelial-to-mesenchymal transition (EMT) (Zhu et al., 2010). Clearly, the complexity of these interactions is only beginning to be studied and more work is needed to fully elucidate the function of these lncRNAs.

circRNA signatures have been recently studied for their prognostic value in gastric cancer. circRNAs are desirable for biomarker studies due to the fact that their circular conformation protects from endonuclease activity, a process that can degrade other noncoding RNAs within the body. Thus circRNAs can be assessed in noninvasive blood-based assays. For example, it was found that a dual-circRNA signature, circ_0021087 and circ_0005051, could be used as a noninvasive biomarker for gastric cancer (Feng et al., 2018). CircRNAs can also act as miRNA sponges, like lncRNAs, and can regulate transcription and gene expression. CDR1 was found to act as an miRNA sponge to miR-7 and regulate developmental pathways, including the Phosphoinositide-3-kinase (PI3K)/AKT pathway, which stimulates gastric cancer growth (Binang et al., 2020; ZiaSarabi, Sorayayi, Hesari, & Ghasemi, 2019) CircHIAT1, a normally downregulated circRNA, can act as a sponge to miR-21, a notable miRNA found in various gastric cancer analyses, and regulate cellular proliferation and apoptosis (Forchelet et al., 2018a).

Due to their novelty and recent emergence within the field, piRNA signature studies are beginning in gastric cancer. Recently *Martinez* et al. detailed a transcriptome signature for piwiRNA expression in gastric cancer (Song, Wu, & Guan, 2020). The authors found that most piRNAs were located in protein-coding sequences and FR222326 piRNA was associated with overall survival in gastric cancer patients and the piRNA signature of FR290353, FR064000, and FR387750 (FR157678) correlated with regression-free survival of gastric cancer patients (Song et al., 2020). Vinasco-Sandoval et al. (2020) found that the expression of two piRNAs has potential risk for gastric cancer: piR-48966 and piR-31335. Due to the complexity and the novelty of these small RNA species, more investigations are needed to further describe piRNAs in gastric cancer.

Although there is emerging evidence of the prognostic value of small noncoding RNAs in gastric cancer, all of the studies mentioned are preclinical. Further experimental validation is needed to bring forth a clear-cut noncoding RNA biomarker that can be used in either gastric cancer diagnoses or guiding treatment decisions.

9.4 Sample types used for analyzing noncoding RNAs (tumor biopsies, liquid biopsies, etc.)

Various biological material can be used to analyze noncoding RNAs, including tumor biopsies, liquid biopsies, and cell lines. Within a tumor, noncoding RNAs are found in abundance. Tumor tissue can be used, either fresh or fixed (paraffin or formalin), to preform noncoding RNA analysis. Fresh tumors need to be collected, stored in ice, and processed almost immediately to preserve the integrity of the small noncoding RNA expression network. RNA is extracted from the tissue using an organic dissociating agent such as qizol and isolated using a commercially available small noncoding RNA kit according to standard methods (Tapia et al., 2014). Depending on the type of the sample, whether obtained from a tumor biopsy or resection, the whole tumor or portions of the tumor are removed and portioned for a variety of uses, including biopsy, staining, and research. In some cases, normal tissue adjacent to the tumor can be collected and used as a comparison for analysis.

Liquid biopsies also can be used, either blood or extracted serum, as it is known that noncoding RNAs can circulate in the bloodstream and are shed by tumor. In this way a blood collection is less invasive than a surgery and more testing can be performed quickly and economically. Small noncoding RNAs are located within the plasma, the clear portion of the blood remaining after the removal of all cellular components, within different blood components (T-cells, B-cells, natural killer cells, neutrophils, and erythrocytes), exosomes and with whole blood (Rodríguez et al., 2020). Blood separation and individual cell isolation can be performed with commonly used techniques, such as centrifugation or newer technology, which use either hollow fibers and gravity, or capillary-driven microfluidic technology to separate blood at a much smaller scale (Forchelet et al., 2018b; Hornsey, McColl, Drummond, & Prowse, 2005). Exosome isolation is more complex as it requires centrifugation (either differential or density gradient), size exclusion chromatography (column packed pours beads), filtration (ultrafiltration membranes), polymer-based precipitation, immune separation (Enzyme-linked immunosorbent assay- or bead-based), and sieving (filtration or pressure electrophoresis) (Yakimchuk, 2015).

Gastric cancer cell lines can be used as a tool to analyze noncoding RNAs. Various gastric cancer cell lines exist. Some commercially available cell lines include NCI-N87, AGS, KATO-III, SNU-1, SNU-16, and SNU-5. Various laboratories within the United States and throughout the world have characterized and established their own gastric cancer cell lines, including MKN45, GC1415, and GC1436. Normal gastric cell lines, such as GES-1, are few and not commercially available. Gastric cancer cell lines can be used for the extraction of noncoding RNAs with trizol or qizol and such methods have been extensively published. Cell lines serve as an excellent source for the proof of concept studies that can be validated in patient tissues. Not only can gastric cancer cell lines be used in vitro but cell lines also can be implanted into immunocompromised mice, either subcutaneously or orthotopically, and tumor and/or serum can be collected for circulating noncoding RNAs with the previous methods.

9.5 Cell signaling pathways modulated by noncoding RNAs in gastric cancer patients

Various cell signaling pathways are modulated by small noncoding RNAs, including the PI3K/AKT, Ras, Slit/Robo, Jak/STAT, and Wnt/B-catenin pathway. Components of the previous oncogenic cellular pathways, such as PI3K, Ras, Jak, and Wnt, complementary bind to the small noncoding RNAs. The result of this complementary binding induces changes to the pathway, either stimulation of oncogenic genes or inhibition of pathway inhibiting proteins, resulting in cellular perturbations. Signaling pathways influence gastric cancer cellular growth in a variety of ways. All of these signaling pathways mentioned earlier involve binding of a stimuli molecule (such as a growth factor) to a growth receptor located within the extracellular membrane. Once bound, internal changes occur through a phosphorylation cascade of downstream proteins ultimately leads to physiological changes. Next, we will discuss how noncoding RNAs modulate these growth-stimulating pathways within gastric cancer.

PI3K/AKT signaling is an important part of gastric cancer pathogenesis. This pathway controls gastric cancer growth and progression by influencing cell division, metabolism, and metastasis (Matsuoka & Yashiro, 2014). After stimuli is bound to a growth factor receptor (such as epidermal growth factor receptor or insulin), a phosphorylation cascade results in downstream proteins such as PIP3, PDK1, and AKT being phosphorylated and ultimately activated leading to phenotypic changes stated earlier. Various miRNAs are involved with PI3K/AKT pathway regulation, either as activators or suppressors or other inhibitory proteins. Naturally, many are downregulated such as miRNA 141, miRNA-146a, and miRNA-29c to perpetuate processes such as metastases, cell migration, and invasion (Hu, Zhu, Xiong, Xue, & Zhou, 2019). Others noncoding RNAs are upregulated and act in an oncogenic fashion, such as miR-25 and miR-296–5p, which stimulate activity of antiapoptotic proteins and activate cellular growth (Hu et al., 2019). Additional small noncoding RNAs are involved in targeting the PI3K/AKT pathway such as the lncRNA AK023391 which upregulates PI3K/AKT signaling through the targeting and activation of FOXO3a, a member of the forehead transcription factor family that is induced downstream of AKT signaling (Huang et al., 2017; Zhang, Tang, Hadden, & Rishi, 2011).

Mutant Ras is involved with perpetuating a variety of cancers, including gastric cancer. Ras signaling is a highly complex that is activated by stimuli binding to growth factor receptors or tyrosine kinase receptors (TKRs). After phosphorylation of effector molecules GRB2/SOS, Ras is phosphorylated which, in turn, phosphorylates downstream molecules such as Mitogen-activated protein kinase kinase, extracellular signal-regulated kinase and downstream transcription factors such as ELK-1 (Downward & Targeting, 2003). A variety of Ras proteins exist (H-Ras, N-Ras, R-Ras, and K-Ras) and mutations within these proteins have all been found to induce gastric cancer and/or correlate to overall survival. For example, KRAS mutations or amplifications were found in 50%–60% of intestinal-type gastric cancers within a Japanese cohort, whereas 57% of total patients had high intratumor heterogeneity to KRAS status (Hewitt et al., 2019). Noncoding RNAs have been shown to bind to and stimulate this pathway, including various miRNAs such as let-7 and miR-761 (Johnson et al., 2005; Zhang, Sui, & Sui, 2019). Let-7 binds to the 3′ UTR of Ras genes altering and activating its expression (Johnson et al., 2005). MiR-761 is naturally downregulated in gastric cancer cells due to its naive function of negatively regulating the interaction between Ras and Rap interaction 1, a tyrosine kinase activator that regulates EMT and interacts with Rat sarcoma virus (Zhang, Sui, et al., 2019). Gu, Ren, Zheng, Hu, and Hu (2019) preformed a comprehensive analysis of lncRNAs within gastric cancer and found the lncRNA prostate androgen-regulated transcript 1 is downregulated and this downregulation perpetuates signaling favorable to gastric cancer by altering Ras signaling.

The wingless integration site pathway is a highly conserved growth pathway involved in controlling stem cell pluripotency. This complex pathway involves a variety of Wnt proteins (19 types) which bind to receptors and coreceptors (Frizzled, low-density lipoprotein, and receptor TKRs). The resulting binding induces the expression of the transcriptional activator beta-catenin (canonical signaling) or induces expression of cellular components not involved with beta-catenin (noncanonical) (Koushyar, Powell, Vincan, & Phesse, 2020). As this pathway is highly intricate and involves a variety of proteins, effectors, and activator molecules through a phosphorylation cascade, the important points to note are that activation of this pathway induces a variety of cellular changes to gastric cancer and induces gastric cancer growth through cell cycle progression, matrix remodeling, and cell migration. A variety of lncRNAs can stimulate this pathway making gastric cancer cells favorable toward growth through either their upregulation or downregulation predicting more favorable survival. In this direction, LOC400043 was found to be downregulated in gastric cancer and negatively activates Wnt signaling (Jafarzadeh & Soltani, 2020). LINC01503 and LINC00460 positively promotes Wnt signaling to induce gastric cancer proliferation and invasion (Ding, Shi, Xie, & Zhu, 2020; Zhang, Xu, Wang, & Guo, 2019).

There are clearly a large variety of small noncoding RNAs that influence gastric cancer signaling. Further experimental validation is needed to elucidate the roles of all noncoding RNAs in gastric cancer.

9.6 Noncoding RNAs as prognostic and predictive marker for gastric cancer patients

Currently, there are very little biomarkers that are used clinically to monitor patient outcomes but the ones used include carcinogenic embryonic antigen, CA19−19, CA72−4, and CA125. These markers are circulating proteins or mRNA within the blood that monitor patient responses to treatment and guide treatment decisions (Lu et al., 2020).

Novel studies are underway to assess noncoding RNA biomarkers for their predictive potential. Current biomarker studies involving noncoding RNAs are based on blood analyses. Blood-based biomarkers are much more accessible for studies due to the ease of collection, rather than an invasive procedure involving tumor collection. Various publications have focused on the noninvasive route for biomarker studies and have focused on studying exosomes, miRNAs, circRNAs, and lncRNAs. Exosomes are stable lipid enclosed vesicles that circulate within the blood and other tissues and are being investigated as predictive biomarkers for gastric cancer patients as they can house different miRNAs, including miR-221, miR-107, and miR-20a-3p [full panel is detailed in Lu et al. (2020)]. Kim et al. found that circulating miRNAs (like miR-21, miR-146a, and miR-148a) can predict lymph node metastasis in gastric cancer, whereas Sierzega et al. state that miR-21 and miR-331 were found within the serum and are being investigated further for their biomarker potential (Kim et al., 2013; Sierzega et al., 2017). CircRNAs are attractive candidates for biomarkers due to their circular shape, stability, and resistance to degradation by intracellular RNAses. Not only do that act alone to regulate gastric cancer signaling but also can act as sponges to other noncoding RNAs, such as miRNAs. For example, circPVT1 inhibits miR-125 to suppress anticancerous signaling and allow the gastric tumor to continue proliferation (Chen et al., 2017; Yang et al., 2020). circLARP4 has been shown alone to inversely correlate to tumor progression and size while also interacting with miR-424 to inhibit gastric cancer growth and is downregulated in its normal state in gastric cancer (Liu et al., 2018; Yang et al., 2020). Finally, circ_100269 can suppress cell death through p53 by inhibiting miR-630 (Yang et al., 2020; Zhang, Liu, et al., 2017). Long-noncoding

RNAs are also being investigated for their potential as gastric cancer biomarkers. Liu et al. detailed how LINC00941 may be a probably prognostic biomarker as it was found within the cancer genome atlas (TCGA) via a retrospective study (Liu et al., 2019). Tian et al. found a 12 lncRNA signature that has potential as a prognostic biomarker in gastric cancer. They assert that larger panels of biomarkers hold more weight and are a better model for biomarker studies (Tian et al., 2017). Some of the lncRNAs that are probable biomarkers include LINC00340, LOC293174, and LOC100133985 (Tian et al., 2017).

To preform biomarker studies, appropriate statistical investigation is needed to ensure that the intended biomarkers are not only specific but also can be validated in a reproducible fashion. Predictive biomarker studies need to compare baseline values to changes over time of the expression of the noncoding RNA with or without treatment (Matsuoka & Yashiro, 2018). Prognostic biomarkers need to consider the baseline noncoding RNA level, the changes over time with or without treatment, and how the clinical outcome (survival) correlates (Matsuoka & Yashiro, 2018). Either experimental data or retrospective analyses can be utilized to identify the candidate biomarkers. Additionally, validation is needed, including model-based analyses, statistical significance, and clinical relevance in collaboration with a biostatistician.

9.7 Diagnostic value of small noncoding RNAs in gastric cancer

The value of noncoding RNAs is not limited to being only biomarkers for gastric cancer after the disease has been discovered but also can be used as a diagnostic tool. The direction of gastric cancer research is moving toward finding ways to identify this disease in its earlier stages, or precancerous dysplastic stages. Gastric cancer is commonly found in the metastatic stages due to its silent nature by a lack of early symptoms or warning signs. Not only are noncoding RNAs found within the blood and tumor but also have been located in other places such as the gastric juice, urine, and stool. LncRNAs, such as LINC00982, can be found in the gastric juice of patients and may or may not correlate with similar expressions in tumor tissues (Yuan et al., 2020). The lncRNA AA174084 has also been found within the gastric juice and can be correlated with tumor size, stage, and subtype of disease (Yuan et al., 2020).

Obtaining gastric juice may prove to be difficult as it involves an invasive procedure to obtain the juice directly from the stomach. Although this is the case, this may prove to be a novel avenue of research since the gastric juice, or digestive juices within the stomach, remains at a very low pH and the small noncoding RNAs that remain within are relatively stable. Low pH conditions tend to destroy proteins and other cellular components during the digestion process but identifying lncRNAs that are stably expressed within the juice does hold some diagnostic potential. Theoretically, those patients who are at a higher risk of developing gastric cancer can be screened and monitored for gastric cancer development at early time points using gastric juice screenings.

A more attractive diagnostic tool would be with analyzing noncoding RNAs in either urine or fecal samples as it is less invasive. This has been successfully created with the at-home screening colorectal cancer kit Cologuard which tests for changes within the DNA that can indicate the presence of a tumor or polyp. Although this tests for DNA changes, further work is needed to identify stably expressed noncoding RNAs (found in exosomes or other noncoding RNAs such as circRNAs) that can be used for diagnostic purposes. There is a need for more exploration of these alternative diagnostic avenues. Like *H. pylori* has been identified as a causative factor in gastric cancer development while identifying ways to detect and eradicate early to control cancer disease progression, utilizing noncoding RNAs as a diagnostic tool could detect gastric cancer in its earlier states which will lead to more favorable overall survival outcomes (Wroblewski, Peek, & Wilson, 2010).

9.8 Potential of noncoding RNAs in predicting chemotherapy resistance and radiotherapy resistance in gastric cancer patients

As the field of oncology leans more toward precision medicine approaches rather than a blanket treatment approach, noncoding RNAs hold great value. Currently, chemotherapeutic resistance is a major hurdle in the management of gastric cancer patients. As explained earlier, most patients have their disease detected during metastatic stages. At this point the tumor has evolved significantly from its initial state and has acquired mutations with the activation of pathways related to drug efflux, resistance to apoptosis, increases in DNA repair, and changes in signaling related to drug metabolism (Mansoori, Mohammadi, Davudian, Shirjang, & Baradaran, 2017). A variety if chemotherapies are used to treat gastric cancer, including platinum agents, taxols, antimetabolites, targeted therapeutics, and most recently immune-checkpoint inhibitors. Aside from chemotherapy, radiation can be used as a great palliative therapy in gastric cancer. According to treatment guidelines set by the National Comprehensive Cancer Network, radiation is given normally in patients with nonoperable tumors or those that have significant comorbidities that makes them ineligible for surgery (Shinde, Novak, Amini, & Jen Chen, 2019). Depending on the age of the patient and their overall health, radiation can be given alone or in conjunction with other chemotherapeutics such as 5-florouracil (5-FU) (Shinde et al., 2019). Genetic alterations impact tumor sensitivity to chemotherapy and alterations within specific genes, such as drug efflux genes, contribute to chemoresistance. Likewise, various noncoding RNAs can indirectly contribute to radiation resistance and chemoresistance through noncoding RNA—mRNA interactions.

Various miRNAs have been shown to cause resistance to platinum-based compounds, including miR-106a, miR-25, and miR-141 (Wei et al., 2020). Platinum compounds, such as cisplatin, carboplatin, and oxaliplatin, are commonly used chemotherapeutic regimens in gastric cancer. The mechanism of action of platinum-based compounds is to intercalate within the DNA at specific guanine—cytosine regions and through these DNA lesions causes cellular death through disruption of transcription (Hato, Khong, De Vries, & Joost Lesterhuis, 2014). miR-106 has been found to interact with RUNX family transcription factor 3 (RUNX3) to cause multidrug resistance and enhance drug efflux (Zhang, Lu, & Cai, 2013). miR-25 targets and inhibits FOXO3a, an important tumor suppressor involved in cell cycle regulation, and knockdown of FOXO3a was found to decrease sensitivity to cisplatin compounds (He, Qi, Chen, & Cao, 2017; Zhang et al., 2013). MiR-141 was found to target and activate Kelch-like ECH-associated protein 1, a ubiquitin ligase reactive to stress conditions, and when a patient is infected with *H. Pylori* it causes downregulation of this tumor suppressor miRNA (Zhang et al., 2013; Zhou, Su, Zhu, & Zhang, 2014; uniprot.org. UniProtKB-Q13134KEAP1_Human).

Paclitaxel, a microtubule inhibitor that inhibits cell growth, is frequently used in gastric cancer in combination with trastuzumab, a monoclonal HER2 antibody. A number of miRNAs impact paclitaxel sensitivity in gastric cancer, including miR-155—5p (Zhang et al., 2013). MiR-155—5p is located in circulating exosomes and targets and inhibits the Erythroid transcription factor-binding protein 3 which inhibits apoptosis and induces EMT (Zhang et al., 2013; Wang et al., 2019). Trastuzumab targets HER2 amplifications in gastric cancer, a genetic mutation found in 10% of gastric cancer patients. Examples of miRNAs interfering with trastuzumab effectivity include miR-21 and miR-125b (Zhang et al., 2013). MiR-21 upregulation targets and inhibits the action of tumor suppressor PTEN to inhibit apoptosis in HER2+ gastric cancer cells (Eto et al., 2014). Although the mechanism is unknown, miR-125b has been associated with trastuzumab resistance in HER2-positive gastric cancer through a patient cohort analysis (Sui et al., 2017).

Antimetabolites, such as 5-FU, are used in gastric cancer treatment in therapeutic combinations with platinum compounds and/or radiation. The 5-FU inhibits the action of thymidylate synthase, an enzyme that catalyzes the conversion of deoxyuridine monophosphate to deoxythymidine monophosphate, an enzyme important in DNA synthesis. The 5-FU is commonly used in combination with platinum compounds within the FOLFOX (5-FU, oxaliplatin, and folinic acid) and XELOX (rcapecitabine/5-FU prodrug and oxaliplatin) regimens. miRNAs such as miR-147 and miR-17 can promote 5-FU resistance by suppressing the PTEN tumor suppressor and death effector domain containing protein (DEDD), respectively. In both cases the suppression of these proteins inhibits apoptosis (Zhang et al., 2013; uniprot.org, UniProtKB-075618 DEDD_HUMAN). Although radiation resistance with noncoding RNAs has not been studied extensively in gastric cancer, it has been found that in colorectal cancer, the lncRNA PVT1 plays an important role in radiation resistance (Martinez-Barriocanal, Arango, & Dopeso, 2020). PVT1 was found to be highly expressed in HCT116 colon cancer cell lines that were resistance to radiation (Martinez-Barriocanal et al., 2020). The common signaling markers targeted by noncoding RNAs point that besides colon cancer certain noncoding RNAs may have a role in resistance to radiotherapy that requires further investigations. Fig. 9.1 summarizes the implications of noncoding RNAs in gastric cancer, including a repertoire of biological samples that can be studied and some of the pathways that are modulated by these cellular components.

9.9 Summary and conclusion

It is clear that noncoding RNAs play an important role in gastric cancer development, progression and also define overall disease characteristics. Gastric cancer is a heterogeneous disease that involves a variety of genetic aberrations, overactive proteins, and genetic intervention from noncoding RNAs. Herein, we have described how noncoding RNA signatures can be used in diagnostics, in the classification of patient, and can also help in further understanding of their roles as drivers of gastric cancer pathways and resistance to chemotherapeutics. It is clear that more work is needed that is focused on a subset of noncoding RNAs that can be used as a predictive or prognostic biomarker within gastric cancer, and studying noncoding RNAs provides a rich area of potential research opportunities.

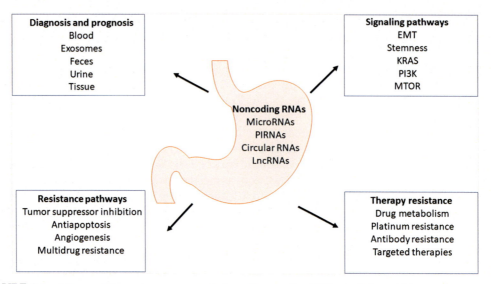

FIGURE 9.1   Scheme of the various processes that small noncoding RNAs modulate within gastric cancer.

References

Binang, H., Wang, Y., Tewara, M., Du, L., Shi, S., Li, N., ... Wang, C. (2020). Expression levels and associations of five long non-coding RNAs in gastric cancer and their clinical significance. *Oncology Letters*, *19*(3). Available from https://doi.org/10.3892/ol.2020.11311.

Bunnik, E. M., & Le Roch, K. G. (2013). An introduction to functional genomics and systems biology. *Advances in Wound Care*, *2*, 490–498. Available from https://doi.org/10.1089/wound.2012.0379.

Chen, J., Li, Y., Zheng, Q., Bao, C., He, J., Chen, B., Lyu, D., Zheng, B., Xu, Y., Long, Z., et al. (2017). Circular RNA profile identifies circPVT1 as a proliferative factor and prognostic marker in gastric cancer. *Cancer Letters*, *388*, 208–219. Available from https://doi.org/10.1016/j.canlet.2016.12.006.

Chen, L., Heikkinen, L., Wang, C., Yang, Y., Sun, H., & Wong, G. (2019). Trends in the development of miRNA bioinformatics tools. *Briefings in Bioinformatics*, *20*(5), 1836–1852. Available from https://doi.org/10.1093/bib/bby054.

Chong, M. M. W., Zhang, G., Cheloufi, S., Neubert, T. A., Hannon, G. J., & Littman, D. R. (2010). Canonical and alternate functions of the microRNA biogenesis machinery. *Genes and Development*, *24*(17), 1951–1960. Available from https://doi.org/10.1101/gad.1953310.

Dastmalchi, N., Safaralizadeh, R., & Banan Khojasteh, S. M. (2019). The correlation between microRNAs and *Helicobacter pylori* in gastric cancer. *Pathogens and Disease*, *77*(4). Available from https://doi.org/10.1093/femspd/ftz039.

Derrien, T., Johnson, R., Bussotti, G., Tanzer, A., Djebali, S., Tilgner, H., ... Guigó, R. (2012). The GENCODE v7 catalog of human long noncoding RNAs: Analysis of their gene structure, evolution, and expression. *Genome Research*, *22*(9), 1775–1789. Available from https://doi.org/10.1101/gr.132159.111.

Ding, J., Shi, F., Xie, G., & Zhu, Y. (2020). Long non-coding RNA LINC01503 promotes gastric cancer cell proliferation and invasion by regulating Wnt signaling. *Digestive Diseases and Sciences*, *66*, 134.

Downward, J., & Targeting, R. A. S. (2003). Signaling pathways in cancer. *Nature Reviews Cancer*, *3*, 11–22.

Esteller, M. (2011). Non-coding RNAs in human disease. *Nature Reviews Genetics*, *12*(12), 861–874. Available from https://doi.org/10.1038/nrg3074.

Eto, K., Iwatsuki, M., Watanabe, M., Ida, S., Ishimoto, T., Iwagami, S., ... Baba, H. (2014). The microRNA-21/PTEN pathway regulates the sensitivity of HER2-positive gastric cancer cells to trastuzumab. *Annals of Surgical Oncology*, *21*(1), 343–350. Available from https://doi.org/10.1245/s10434-013-3325-7.

References

Fang, Y., & Fullwood, M. J. (2016). Roles, functions, and mechanisms of long non-coding RNAs in cancer. *Genomics, Proteomics and Bioinformatics*, 14(1), 42−54. Available from https://doi.org/10.1016/j.gpb.2015.09.006.

Feng, Y., Bai, F., You, Y., Bai, F., Wu, C., Xin, R., ... Nie, Y. (2018). Dysregulated microrna expression profiles in gastric cancer cells with high peritoneal metastatic potential. *Experimental and Therapeutic Medicine*, 16(6), 4602−4608. Available from https://doi.org/10.3892/etm.2018.6783.

Forchelet, D., Beguin, S., Sajic, T., Bararpour, N., Pataky, Z., Frias, M., Grabherr, S., Augsburger, M., Liu, Y., Charnley, M., et al. (2018a). Separation of blood microsamples by exploding sedimentation at the microscale. *Nature*, 8.

Forchelet, D., Beguin, S., Sajic, T., Bararpour, N., Pataky, Z., Frias, M., Grabherr, S., Augsburger, M., Liu, Y., Charnley, M., et al. (2018b). Separation of blood microsamples by exploding sedimentation at the microscale. *Nature*, 8(14101).

Gao, Y., Wang, J. W., Ren, J. Y., Guo, M., Guo, C. W., Ning, S. W., & Yu, S. (2020). Long noncoding RNAs in gastric cancer: From molecular dissection to clinical application. *World Journal of Gastroenterology*, 26(24), 3401−3412. Available from https://doi.org/10.3748/wjg.v26.i24.3401.

Gu, W., Ren, J. H., Zheng, X., Hu, X. Y., & Hu, M. (2019). Comprehensive analysis of expression profiles of long non-coding RNAs with associated ceRNA network involved in gastric cancer progression. *Molecular Medicine Reports*, 20, 2209−2218. Available from https://doi.org/10.3892/mmr.2019.10478.

Han, L., Zhang, X., Wang, A., Ji, Y., Cao, X., Qin, Q., ... Huang, H. (2020). A dual-circular RNA signature as a non-invasive diagnostic biomarker for gastric cancer. *Frontiers in Oncology*, 10. Available from https://doi.org/10.3389/fonc.2020.00184.

Hato, S. V., Khong, A., De Vries, J. M. I., & Joost Lesterhuis, W. (2014). Molecular pathways: The immunogenic effects of platinum-based chemotherapeutics. *Clinical Cancer Research: An Official Journal of the American Association for Cancer Research*, 20. Available from https://doi.org/10.1158/1078-0432.CCR-13-3141.

He, J., Qi, H., Chen, G., & Cao, C. (2017). MicroRNA-25 contributes to cisplatin resistance in gastric cancer cells by inhibiting forehead box 03a. *Oncology Letters*, 14(5), 6097−6102. Available from https://doi.org/10.3892/ol.2017.6982.

Heather, J. M., & Chain, B. (2016). The sequence of sequencers: The history of sequencing DNA. *Genomics*, 107(1), 1−8. Available from https://doi.org/10.1016/j.ygeno.2015.11.003.

Hewitt, L. C., Saito, Y., Wang, T., Matsuda, Y., Oosting, J., Silvia, A. N. S., Slaney, H. L., Hutchins, G., Tan, P., Yoshikawa, T., et al. (2019). KRAS status is related to histological phenotype in gastric cancer: Results from a large multicentre study. *Gastric Cancer: Official Journal of the International Gastric Cancer Association and the Japanese Gastric Cancer Association*, 22, 1193−1203.

Hornsey, V. S., McColl, K., Drummond, O., & Prowse, C. V. (2005). Separation of whole blood into plasma and red blood cells by using a hollow-fibre filtration system. *Vox Sanguinis*, 89(2), 81−85. Available from https://doi.org/10.1111/j.1423-0410.2005.00660.x.

Hou, X. (2018). Liquid scintillation counting for determination of radionuclides in environmental and nuclear application. *Journal of Radioanalytical and Nuclear Chemistry*, 318(3), 1597−1628. Available from https://doi.org/10.1007/s10967-018-6258-6.

Hu, B., Hajj, N. E., Sittler, S., Lammert, N., Barnes, R., & Meloni-Ehrig, A. (2012). Gastric cancer: Classification, histology and application of molecular pathology. *Journal of Gastrointestinal Oncology*, 3(3), 251−261. Available from https://doi.org/10.3978/j.issn.2078-6891.2012.021.

Hu, M., Zhu, S., Xiong, S., Xue, X., & Zhou, X. (2019). MicroRNAs and the PTEN/PI3K/AKT pathway in gastric cancer (Review). *Oncology Letters*. Available from https://doi.org/10.3892/or.2019.6962.

Huang, Y., Zhang, J., Hou, L., Wang, G., Liu, H., Zhang, R., ... Zhu, J. (2017). LncRNA AK023391 promotes tumorigenesis and invasion of gastric cancer through activation of the PI3K/Akt signaling. *BMC*, 36(194).

Huerta, L., & Burke, M. (2016). Functional genomics II Common technologies and data analysis methods. EMBL-EBI Training. Accessed 12.10.20. Available from https://doi.org/10.6019/TOL.FunGenII-c.2016.00001.1EMBL-EBI.uniprot.org, UniProtKB-075618 (DEDD_HUMAN). Database Accessed 12.07.20. EMBL-EBI. uniprot.org UniProtKB-Q13134 (KEAP1_Human). Database Accessed 12.07.20.

Hüttenhofer, A., & Vogel, J. (2006). Experimental approaches to identify non-coding RNAs. *Nucleic Acids Research*, 34(2), 635−646. Available from https://doi.org/10.1093/nar/gkj469.

Ivey, K. N., & Srivastava, D. (2015). microRNAs as developmental regulators. *Cold Spring Harbor Perspectives in Biology*, 7(7), a008144. Available from https://doi.org/10.1101/cshperspect.a008144.

Jafarzadeh, M., & Soltani, B. M. (2020). Long noncoding RNA LOC400043(LINC02381) inhibits gastric cancer progression through regulating Wnt signaling pathway. *Frontiers in Oncology, 10*, 562253. Available from https://doi.org/10.3389/fonc.2020.562253.

Johnson, S. M., Grosshans, H., Shingara, J., Byrom, M., Jarvis, R., Cheng, A., ... Slack, F. J. (2005). Ras is regulated by the let-7 microRNA family. *Cell., 120*(5). Available from https://doi.org/10.1016/j.cell.2005.01.014.

Juzenas, S., Venkatesh, G., Hübenthal, M., Hoeppner, M. P., Du, Z. G., Paulsen, M., ... Hemmrich-Stanisak, G. (2017). A comprehensive, cell specific microRNA catalogue of human peripheral blood. *Nucleic Acids Research, 45*(16), 9290–9301. Available from https://doi.org/10.1093/nar/gkx706.

Kang, W., & Friedländer, M. R. (2015). Computational prediction of miRNA genes from small RNA sequencing data. *Frontiers in Bioengineering and Biotechnology, 3*. Available from https://doi.org/10.3389/fbioe.2015.00007.

Kietrys, A. M., Velema, W. A., & Kool, E. T. (2017). Fingerprints of modified RNA bases from deep sequencing profiles. *Journal of the American Chemical Society, 139*(47), 17074–17081. Available from https://doi.org/10.1021/jacs.7b07914.

Kim, S. Y., Jeon, T. Y., Choi, C. I., Kim, D. H., Kim, D. H., Kim, G. H., ... Kim, H. H. (2013). Validation of circulating miRNA biomarkers for predicting lymph node metastasis in gastric cancer. *The Journal of Molecular Diagnostics, 15*(5), 661–669. Available from https://doi.org/10.1016/j.jmoldx.2013.04.004.

Koushyar, S., Powell, A. G., Vincan, E., & Phesse, T. J. (2020). Targeting Wnt signaling for the treatment of gastric cancer. *International Journal of Molecular Sciences, 21*(11), 3927. Available from https://doi.org/10.3390/ijms21113927.

Ku, H. Y., & Lin, H. (2014). PIWI proteins and their interactors in piRNA biogenesis, germline development and gene expression. *National Science Review, 1*(2), 205–218. Available from https://doi.org/10.1093/nsr/nwu014.

Lee, R. C., Feinbaum, R. L., Ambros, Vs, & The, C. (1993). Elegans heterochronic gene lin-4 encodes small RNAs with antisense complementarity to lin-14. *Cell, 75*(5). Available from https://doi.org/10.1016/0092-8674.

Lekka, E., & Hall, J. (2018). Noncoding RNAs in disease. *FEBS Letters, 592*(17), 2884–2900. Available from https://doi.org/10.1002/1873-3468.13182.

Li, B. S., Zuo, Q. F., Zhao, Y. L., Xiao, B., Zhuang, Y., Mao, X. H., ... Guo, G. (2015). MicroRNA-25 promotes gastric cancer migration, invasion and proliferation by directly targeting transducer of ERBB2, 1 and correlates with poor survival. *Oncogene, 34*(20), 2556–2565. Available from https://doi.org/10.1038/onc.2014.214.

Link, A., & Kupcinskas, J. (2018). MicroRNAs as non-invasive diagnostic biomarkers for gastric cancer: Current insights and future perspectives. *World Journal of Gastroenterology, 24*(30), 3313–3329. Available from https://doi.org/10.3748/wjg.v24.i30.3313.

Liu, H., Liu, Y., Bian, Z., Zhang, J., Zhang, R., Chen, X., ... Zhu, J. (2018). Circular RNA YAP1 inhibits the proliferation and invasion of gastric cancer cells by regulating the miR-367-5p/p27 (Kip1) axis. *Molecular Cancer, 17*, 151. Available from https://doi.org/10.1186/s12943-018-0902-1.

Liu, H., Wu, N., Zhang, Z., Zhong, X. D., Zhang, H., Guo, H., ... Liu, Y. (2019). Long non-coding RNA LINC00941 as a potential biomarker promotes the proliferation and metastasis of gastric cancer. *Frontiers in Genetics, 10*. Available from https://doi.org/10.3389/fgene.2019.00005.

López-Urrutia, E., Bustamante Montes, L. P., Ladrón de Guevara Cervantes, D., Pérez-Plasencia, C., & Campos-Parra, A. D. (2019). Crosstalk between long non-coding RNAs, micro-RNAs and mRNAs: Deciphering molecular mechanisms of master regulators in cancer. *Frontiers in Oncology, 9*. Available from https://doi.org/10.3389/fonc.2019.00669.

Lu, Q., Chen, Y., Sun, D., Wang, S., Ding, K., Liu, M., ... Zhou, F. (2019). MicroRNA-181a functions as an oncogene in gastric cancer by targeting caprin-1. *Frontiers in Pharmacology, 9*. Available from https://doi.org/10.3389/fphar.2018.01565.

Lu, X., Zhang, Y., Xie, G., Ding, Y., Cong, H., & Xuan, S. (2020). Exosomal non-coding RNAs: Novel biomarkers with emerging clinical applications in gastric cancer (Review). *Medicine Reports, 22*, 4091–4100. Available from https://doi.org/10.3892/mmr.2020.11519.

Ma, J., Shen, H., Kapesa, L., & Zeng, S. (2016). Lauren classification and individualized chemotherapy in gastric cancer (Review). *Oncology Letters, 11*(5), 2959–2964. Available from https://doi.org/10.3892/ol.2016.4337.

Mansoori, B., Mohammadi, A., Davudian, S., Shirjang, S., & Baradaran, B. (2017). The different mechanisms of cancer drug resistance: A brief review. *Advanced Pharmaceutical Bulletin, 7*(3), 339–348. Available from https://doi.org/10.15171/apb.2017.041.

Martinez, V. D., Enfield, K. S. S., Rowbotham, D. A., & Lam, W. L. (2016). An atlas of gastric PIWI-interacting RNA transcriptomes and their utility for identifying signatures of gastric cancer recurrence. *Gastric Cancer:*

Official Journal of the International Gastric Cancer Association and the Japanese Gastric Cancer Association, 19(2), 660−665. Available from https://doi.org/10.1007/s10120-015-0487-y.

Martinez-Barriocanal, A., Arango, D., & Dopeso, H. (2020). PVT1 long-noncoding RNA in gastrointestinal cancer. *Frontiers in Oncology, 10*. Available from https://doi.org/10.3389/fonc.2020.00038.

Matsuoka, T., & Yashiro, M. (2018). Biomarkers of gastric cancer: Current topics and future perspective. *World Journal of Gastroenterology: WJG, 24*(26), 2818−2832. Available from https://doi.org/10.3748/wjg.v24.i26.2818.

Matsuoka, T., & Yashiro, M. (2014). The role of PI3K/AKT/mTOR singling in gastric carcinoma. *Cancers (Basel), 6*(3), 1441−1463. Available from https://doi.org/10.3390/cancers6031441.

Michael, M. Z., O'Connor, S. M., Van Holst Pellekaan, N. G., Young, G. P., & James, R. J. (2003). Reduced accumulation of specific microRNAs in colorectal neoplasia. *Molecular Cancer Research, 1*(12), 882−891.

Nobili, L., Lionetti, M., & Neri, A. (2016). Long non-coding RNAs in normal and malignant hematopoiesis. *Oncotarget, 7*(31), 50666−50681. Available from https://doi.org/10.18632/oncotarget.9308.

Quan, J., Dong, D., Lun, Y., Sun, B., Sun, H., Wang, Q., & Yuan, G. (2020). Circular RNA circHIAT1 inhibits proliferation and epithelial-mesenchymal transition of gastric cancer cell lines through downregulation of miR-21. *Journal of Biochemical and Molecular Toxicology, 34*(4). Available from https://doi.org/10.1002/jbt.22458.

Quinn, J. J., & Chang, H. Y. (2016). Unique features of long non-coding RNA biogenesis and function. *Nature Reviews Genetics, 17*(1), 47−62. Available from https://doi.org/10.1038/nrg.2015.10.

Rodríguez, A., Duyvejonck, H., van Belleghem, J. D., Gryp, T., van Simaey, L., Vermeulen, S., . . . Vaneechoutte, M. (2020). Comparison of procedures for RNA-extraction from peripheral blood mononuclear cells. *PLoS One, 15*(2). Available from https://doi.org/10.1371/journal.pone.0229423.

Sárközy, M., Kahán, Z., & Csont, T. (2018). A myriad of roles of miR-25 in health and disease. *Oncotarget, 9*(30), 21580−21612. Available from https://doi.org/10.18632/oncotarget.24662.

Shen, E., Wang, X., Liu, X., Lv, M., Zhang, L., Zhu, G., & Sun, Z. (2020). MicroRNA-93-5p promotes epithelial-mesenchymal transition in gastric cancer by repressing tumor suppressor AHNAK expression. *Cancer Cell International, 20*(1). Available from https://doi.org/10.1186/s12935-019-1092-7.

Shinde, A., Novak, J., Amini, A., & Jen Chen, Y. (2019). The evolving role of radiation therapy for resectable and unresectable gastric cancer. *Translational Gastroenterology and Hepatology, 4*. Available from https://doi.org/10.21037/tgh.2019.08.06.

Sierzega, M., Kaczor, M., Kolodziejczyk, P., Kulig, J., Sanak, M., & Richer, P. (2017). Evaluation of serum microRNA biomarkers for gastric cancer based on blood and tissue pools profiling: The importance of miR-21 and miR-331. *Nature, 117*, 266−273.

Slack, F. J., & Chinnaiyan, A. M. (2019). The role of non-coding RNAs in oncology. *Cell, 179*(5), 1033−1055. Available from https://doi.org/10.1016/j.cell.2019.10.017.

Song, P., Wu, L., & Guan, W. (2020). Genome-wide identification and characterization of DNA methylation and long non-coding RNA expression in gastric cancer. *Frontiers in Genetics, 11*. Available from https://doi.org/10.3389/fgene.2020.00091.

Sui, M., Jiao, A., Zhai, H., Wang, Y., Wang, Y., Sun, D., & Li, P. (2017). Upregulation of miR-125b is associated with poor prognosis and trastuzumab resistance in HER2-positive gastric cancer. *Experimental and Therapeutic Medicine, 14*(1), 657−663. Available from https://doi.org/10.3892/etm.2017.4548.

Tapia, O., Riquelme, I., Leal, P., Sandoval, A., Aedo, S., Weber, H., . . . Roa, J. C. (2014). The PI3K/AKT/mTOR pathway is activated in gastric cancer with potential prognostic and predictive significance. *Virchows Archiv, 465*(1), 25−33. Available from https://doi.org/10.1007/s00428-014-1588-4.

Tian, X., Zhu, X., Yan, T., Yu, C., Shen, C., Hong, J., . . . Fang, J. Y. (2017). Differentially expressed lncRNAs in gastric cancer patients: A potential biomarker for gastric cancer prognosis. *Journal of Cancer, 8*(13), 2575−2586. Available from https://doi.org/10.7150/jca.19980.

uniprot.org, UniProtKB-075618 (DEDD_HUMAN). Accessed 12.07.20.

uniprot.org. UniProtKB-Q13134(KEAP1_Human). Accessed 12.07.20.

Vinasco-Sandoval, T., Moreira, F. C., Vidal, A. F., Pinto, P., Ribeiro-Dos-santos, A. M., Cruz, R. L. S., . . . Santos, S. (2020). Global analyses of expressed piwi-interacting RNAs in gastric cancer. *International Journal of Molecular Sciences, 21*(20), 1−16. Available from https://doi.org/10.3390/ijms21207656.

Wang, K. W., & Dong, M. (2019). Role of circular RNAs in gastric cancer: Recent advances and prospects. *World. Journal of Gastrointestinal Oncology, 11*(6), 459−469. Available from https://doi.org/10.4251/wjgo.v11.i6.459.

Wang, M., Qui, R., Yu, S., Xu, X., Li, G., Gu, R., ... Shen, B. (2019). Paclitaxel-resistance gastric cancer MGC-803 cells promote epithelial-to-mesenchymal transition and chemoresistance in paclitaxel-sensitive cells via exosomal delivery of miR-144-4p. *International Journal of Oncology, 54*(1), 326–338. Available from https://doi.org/10.3892/ijo,2018.4601.

Wang, Y., Lu, T., Wang, Q., Liu, J., & Jiao, W. (2018). Circular RNAs: Crucial regulators in the human body (Review). *Oncology Reports, 40*(6), 3119–3135. Available from https://doi.org/10.3892/or.2018.6733.

Wei, L., Sun, J., Zhang, N., Zheng, Y., Wang, X., Lv, L., ... Yang, M. (2020). Noncoding RNAs in gastric cancer: Implication for drug resistance. *Molecular Cancer, 19*(62).

Weick, E.-M., & Miska, E. A. (2014). piRNAs: From biogenesis to function. *Development (Cambridge, England), 141*(18), 3458–3471. Available from https://doi.org/10.1242/dev.094037.

Wroblewski, L. E., Peek, R. M., Jr, & Wilson, K. T. (2010). *Helicobacter pylori* and gastric cancer: Factors that modulate disease risk. *Clinical Microbiology Reviews, 23*(4), 713–739. Available from https://doi.org/10.1128/CMR.00011-10.

Yakimchuk, K. (2015). Exosomes: Isolation and characterization methods and specific markers. *Labome, 5*. Available from https://doi.org/10.13070/mm.en.5.1450.

Yang, C. M., Qiao, G. L., Song, L. N., Bao, S., & Ma, L. J. (2020). Circular RNAs in gastric cancer: Biomarkers for early diagnosis (Review). *Oncology Letters, 20*(1), 465–473. Available from https://doi.org/10.3892/ol.2020.11623.

Yu, C. Y., & Kuo, H. C. (2019). The emerging roles and functions of circular RNAs and their generation. *Journal of Biomedical Science, 26*(1). Available from https://doi.org/10.1186/s12929-019-0523-z.

Yuan, L., Xu, Z. Y., Ruan, S. M., Mo, S., Qin, J. J., & Cheng, X. D. (2020). Long non-coding RNAs towards precision medicine in gastric cancer: Early diagnosis, treatment and drug resistance. *Molecular Cancer, 19*(96).

Zhang, Y., Liu, H., Li, W., Yu, J., Li, J., Shen, Z., ... Li, G. (2017). CircRNA_100269 is downregulated in gastric cancer and suppresses tumor cell growth by targeting miR-630. *Aging (Albany NY), 9*, 1585–1594.

Zhang, P., Wu, W., Chen, Q., & Chen, M. (2019). Non-Coding RNAs and their integrated networks. *Journal of Integrative Bioinformatics, 16*(3). Available from https://doi.org/10.1515/jib-2019-0027.

Zhang, Q., Sui, Y., & Sui, X. (2019). MicroRNA-761 inhibits the metastasis of gastric cancer by negatively regulating Ras and Rap interactor 1. *Oncology Letters, 18*(3), 3097–3103. Available from https://doi.org/10.3892/ol.2019.10645.

Zhang, S., Xu, J., Wang, H., & Guo, H. (2019). Downregulation of long noncoding RNA LINC00460 expression surpasses tumor growth in vitro and in vivo in gastric cancer. *Cancer Biomarkers, 24*(4), 429–437. Available from https://doi.org/10.3233/CBM-182177.

Zhang, X., Tang, N., Hadden, T. J., & Rishi, A. K. (2011). Biochimica et biophysica acta (BBA). *Molecular Cell Research, 1813*(11), 1978–1986. Available from https://doi.org/10.1016/j.bbamcr.2011.03.010.

Zhang, Y., Huang, H., Zhang, D., Qiu, J., Yang, J., Wang, K., ... Yang, J. (2017). A review on recent computational methods for predicting noncoding RNAs. *BioMed Research International, 2017*. Available from https://doi.org/10.1155/2017/9139504.

Zhang, Y., Lu, Q., & Cai, X. (2013). MicroRNA-106a induces multidrug resistance in gastric cancer by targeting RUNX3. *FEBS Letters, 587*(18), 3069–3075. Available from https://doi.org/10.1016/j.febslet.2013.06.058.

Zhou, X., Su, J., Zhu, L., & Zhang, G. (2014). *Helicobacter pylori* modulates cisplatin sensitivity in gastric cancer by down-regulating miRNA-141 expression. *Helicobacter, 19*(3), 174–181. Available from https://doi.org/10.1111/hel.12120.

Zhu, E., Zhao, F., Xu, G., Hou, H., Zhou, L. L., Li, X., ... Wu, J. (2010). MirTools: MicroRNA profiling and discovery based on high-throughput sequencing. *Nucleic Acids Research, 38*(2), W392–W397. Available from https://doi.org/10.1093/nar/gkq393.

ZiaSarabi, P., Sorayayi, S., Hesari, A. R., & Ghasemi, F. (2019). Circulating microRNA-133, microRNA-17 and microRNA-25 in serum and its potential diagnostic value in gastric cancer. *Journal of Cellular Biochemistry, 120*(8), 12376–12381. Available from https://doi.org/10.1002/jcb.28503.

Zimmermann, B., Bilusic, I., Lorenz, C., & Schroeder, R. (2010). Genomic SELEX: A discovery tool for genomic aptamers. *Methods (San Diego, Calif.), 52*(2), 125–132. Available from https://doi.org/10.1016/j.ymeth.2010.06.004.

Noncoding RNAs in prostate cancer patients

Atiyeh Al-e-Ahmad[1,2,3],*, Nahid Neamati[2,3],*, Emadoddin Moudi[4], Simin Younesi[5] and Hadi Parsian[2,3]

[1]Student Research Committee, Babol University of Medical Sciences, Babol, Iran [2]Cellular and Molecular Biology Research Center, Health Research Institute, Babol University of Medical Sciences, Babol, Iran [3]Department of Clinical Biochemistry, Babol University of Medical Sciences, Babol, Iran [4]Department of Urology, Babol University of Medical Sciences, Babol, Iran [5]School of Health and Biomedical Sciences, RMIT University, Melbourne, VIC, Australia

10.1 Introduction

Prostate cancer (PCa) is one of the most common malignancies in men, worldwide. It caused 1.6 million deaths (out of 8.7 million cancer-related deaths) during years 2005–15 (Fitzmaurice et al., 2017). PCa staging is determined using digital rectal examination (DRE) in four categories (T1–T4; the higher scale the worse malignancy). Moreover, it is graded according to Gleason score with a range of 2–10 score (Gleason & Mellinger, 1974). Patients with score ≥7 are at higher risk for recurrence (D'Amico et al., 1998).

Measuring serum levels of prostate-specific antigen (PSA) as a screening tool helps PCa diagnose. PSA, as a PCa biomarker, is a prostate tissue-specific antigen but is not cancer-specific. Therefore, its use as an ideal tumor marker for PCa due to false-positive results is still controversial (Dall'Era et al., 2008). Nowadays, the current gold standard for diagnosing PCa is needle core biopsy, which usually indicates the result following increased serum PSA levels. However, a false-negative result is also possible (Litwin & Tan, 2017).

A growing body of evidence points the importance of genomic studies in diagnosis of PCa (Bijnsdorp, van Royen, Verhaegh, & Martens-Uzunova, 2017; Sartori & Chan, 2014).

* Jointly first authors.

Today, it is well understood that only a few parts of the human genome transcript into protein-coding genes. However, the majority of that comprises noncoding RNAs (ncRNA) (Mattick & Makunin, 2006). ncRNAs are classified into small and large ncRNAs according to their length. Further subclassification into different groups is based on their function or transcriptional location on DNA (Martens-Uzunova, Olvedy, & Jenster, 2013). There is a lot of evidence introducing ncRNA as a useful cancer-related biomarker for diagnosis, prognosis, and therapeutic monitoring in various cancers such as PCa (Chen et al., 2019; Cho, 2010; Fabris et al., 2016; Heydari et al., 2018; Hua, Chen, & He, 2019; Iyer et al., 2015; Nashtahosseini, Aghamaali, Sadeghi, Heydari, & Parsian, 2021).

In this chapter, we will review the various laboratory methods for the study and evaluation of ncRNA. In clinic, the most ideal method is the one that is laboratory-feasible and cost-effective. General identification of ncRNAs is performed by techniques such as microarray and RNA sequencing (RNA-seq). In addition, further techniques such as in situ hybridization (ISH), quantitative real-time Polymerase Chain Reaction (qRT-PCR), and immunoprecipitation (IP) help us to further investigate and confirm ncRNAs in PCa. Investigation and identification of ncRNAs by practical laboratory methods is relatively time-consuming and costly. Recently, bioinformatics methods have been introduced in research studies. In this chapter, bioinformatics techniques in the study of ncRNAs involved in PCa will also be reviewed and introduced along with some databases.

As mentioned, serum PSA levels with DRE and needle core biopsy are helpful in screening and diagnosing PCa. However, there are limitations in terms of the sensitivity and specificity of each of these methods. Genomic studies especially the determination of ncRNAs status have some advantages. First, they are present in serum and other bodily fluids; sampling of body fluids such as serum, plasma, and urine is simple and less invasive. As a second advantage, ncRNAs are mainly produced by cancer cells and their altered values can be tumor-specific reflection. It should be noted in clinical studies, the observation of standard procedures in sampling, transferring, and storing of biological specimens is very important. In this chapter, various samples that have been tested for the determination of ncRNAs status in PCa are also introduced. In addition, some biospecimen repositories will be introduced to help researchers preparing different biological samples of PCa.

ncRNAs are important molecules in posttranscriptional gene regulation of biological functions in various cellular processes (Hua et al., 2019). For example, microRNAs (miRNAs) (18–22 nucleotide ncRNA) by direct targeting messenger RNAs (mRNAs) can modulate cellular gene expression, cellular proliferation, differentiation, and apoptosis (Fabris et al., 2016). Long ncRNAs are noncoding transcripts (longer than 200 nucleotides) without protein-coding potential, acting as post-transcription regulators in the cellular processes (Martens-Uzunova et al., 2014). To better understanding ncRNAs' function in PCa pathogenesis, cell signaling pathways modulated by ncRNAs in PCa patients are also discussed. As diagnosis and monitoring of treatment strategies are big challenges in the clinical management of PCa, ncRNA has been suggested as a potential diagnostic, prognostic, and therapeutic target in this disease. In this chapter, we will overview the application of ncRNA as a beneficial biomarker for PCa.

10.2 Experimental methods and tools for analyzing ncRNAs in prostate cancer patients

10.2.1 NcRNA profiling in prostate cancer

Given the fact that the production and expression of ncRNAs are specific to their tissue and progenitor cells, ncRNA profiling is considered a suitable tool for identifying ncRNAs as specific biomarkers associated with the development, progression, or metastasis of cancer (Grillone et al., 2020). Two main approaches for screening and initial detection of ncRNA, including Microarray technology and RNA-seq, have been developed for ncRNA profiling.

10.2.2 Microarray

In this technique, an array chip contains a cluster of specific probes. Hybridization of the labeled RNA and probes occurs based on nucleic acid complementary. ncRNAs profiling based on the target RNA hybridization with its complementary synthetic probe has been proposed as a powerful tool for detecting ncRNAs. Jones et al. investigated serum-based miRNA profile in 15 men diagnosed with PCa using Taqman Array Human MicroRNA Pool A Cards v.2 (Thermo Fisher Scientific). Three endogenous controls (RNU6, RNU44, and RNU48) and one nonhuman ath-miR-159a (as negative control) were composed in their array card. Jones et al. (2018) suggested miR-186−5p as an oncogene miRNA in PCa. Their results were confirmed by qRT-PCR.

10.2.3 RNA sequencing

Next-generation sequencing, as well as RNA-seq, is a common tool for identifying the sequence and frequency of RNAs in biological samples (Derrien et al., 2012). The discovery of transcription sequences in various types and classes of ncRNAs is possible using these techniques. In addition, the study of a RNA transcript from small sample sizes is possible by single-cell RNA-seq (Song, 2016). To identify the sequence of a RNA transcript, preparation of complementary DNA (cDNA) library is needed (Chandra Gupta & Nandan Tripathi YJIjoc, 2017). Serial analysis of gene expression (SAGE) and cap analysis gene expression (CAGE) are two main sequencing techniques. SAGE is based on restriction enzyme mediation followed by cloning and sequencing of short cDNA (9-bp SAGE tags) sequences. The produced cDNA-tag has to be sequenced followed by concatenation and cloning (Velculescu, Zhang, Vogelstein, & Kinzler, 1995). In CAGE, 5′ end of RNA sequence produces. So, as an advantage, CAGE is efficient to identify the exact location and analysis of 5′ end of RNA transcript (Song, 2016). Ren et al. using RNA-seq proposed substantial diversities in long noncoding RNAs (lncRNA) expression levels, gene fusion, alternative splicing, and somatic mutation between PCa tumor and their adjusted normal tissues. They also identified some novel gene fusions and different lncRNA expressions in prostate tumors (Ren et al., 2012).

10.2.4 ncRNA validation in prostate cancer

RNA screening by microarray or sequencing provides an overall landscape about ncRNA expression. However, the resulted data have to be confirmed through some laboratory techniques, including qRT-PCR, ISH assays, and northern blot (NB).

10.2.4.1 Reverse transcription quantitative PCR

Traditionally, qRT-PCR is one of the most simple, reliable, and cost-effective techniques used for quantifying gene expression in biological samples. Similar to gene expression assessment, qRT-PCR is considered a useful method for ncRNA quantification after amplifying the cDNA produced by reverse transcriptase and specific primers such as random Hexamer, oligo dT, or specific stem-loop RT primers. Using microarray, Ge et al. analyzed lncRNA expression profiles of PCa patients versus a control group. They identified some lncRNAs up-/downregulated in PCa patients. Their candidate lncRNAs were confirmed by qRT-PCR in five other prostate tumor specimens versus seven specimens as a control group (Ge et al., 2020).

10.2.4.2 Northern blot analysis

In NB technique, existing RNAs in a sample are separated by agarose gel electrophoresis. The RNA has to be transferred onto a nylon membrane, hybridized and tracked with its complementary RNA-probe (Cruz & Houseley, 2020). Determination of specific ncRNA abundance and its different splicing variants is possible by NB (Song, 2016; Streit, Michalski, Erkan, Kleeff, & Friess, 2009). Tang et al. performed a microarray analysis of miRNA profile in different PCa cell lines. As a finding, they reported different expressions of miR-200c and miR-205 between androgen-dependent and androgen-independent PCa cells. Validation of their investigated miRNAs was performed by northern blotting (Tang et al., 2011).

10.2.4.3 In situ hybridization

ncRNAs tracking is possible by hybridization of fluorescent probe which complementary binds to a nucleic acid sequence. ISH and fluorescence ISH (FISH) are two main techniques efficient for determination subcellular distribution of ncRNA (Subramanian, Li, Hara, & Lal, 2015). It means the detection of the exact subcellular location of a nucleic acid sequence, RNA target, or chromosome depends on hybridization with fluorophore-labeled complementary probe (Song, 2016; Subramanian et al., 2015). Tao et al. reported two lncRNAs, including PCAT3 and PCAT9, which are involved in tumor progression. They used computational analysis, qRT-PCR, and RNA-FISH to propose miR-203, a target for PCAT3 and PCAT9, which can be served as a promising factor for PCa diagnostic and therapeutic target (Tao, Tian, & Zhang, 2018). ISH is useful to detect a specific nucleic acid sequence in frozen tissue or formalin-fixed paraffin-embedded without requiring a fluorescent probe (Song, 2016).

10.2.5 Investigation of ncRNA interactions

NcRNA interacts with the other counterparts, including protein, DNA, chromatin, or other RNAs, and their interactions can be detected using experimental or computational methods. Using computational methods, we are able to predict the inter- or intramolecular interaction of ncRNA with the other RNA molecules (Gong, Ju, Shao, & Zhang, 2018).

Studies in the field of ncRNA—protein interaction allow us to have a better understanding of ncRNA function, the pathways, and mechanisms attributed to them within target tissue cells. To this end, various in vitro and in vivo techniques have been developed. RNA immunoprecipitation (RIP) methods are useful methods for the identification of ncRNA−protein interactions using antibody-targeted ncRNA−protein complex (Gagliardi & Matarazzo, 2016). To identify ncRNA−protein interaction researchers can isolate precipitated RNA−protein−antibody complex and couple it with microarray (RIP-chip) or sequencing (RIP-seq) as modified techniques (Atkinson, Marguerat, & Bähler, 2012). IP techniques are useful to precipitate miRNAs bound to Argonat or to be applied as photoactivatable ribonucleoside−enhanced cross-linking and IP and high-throughput sequencing coupled with ultra violet (UV) cross-linking and IP (Subramanian et al., 2015).

10.2.6 Secondary structures of ncRNAs

Understanding the second and tertiary structures of ncRNAs is another new area of research in the field of ncRNAs. The interaction between ncRNAs and their target DNA extremely depends on the complementarity of their nucleic acid sequences. However, their interaction with the protein is entirely influenced by the structure of the ncRNAs. It has even been reported that this interaction largely depends on the presence of some specific domains and motifs that are completely conserved (Hua et al., 2019). There are some well-known softwares for predicting the secondary structure of RNA according to their primary sequences. For example, RNA structure (https://rna.urmc.rochester.edu/RNAstructure.html) (Bellaousov, Reuter, Seetin, & Mathews, 2013), RNAfold (http://rna.tbi.univie.ac.at/cgi-bin/RNAWebSuite/RNAfold.cgi) (Denman, 1993), and MFold (http://rothlab.ucdavis.edu/genhelp/mfold.html) (Zuker, 2003) are software to predict the secondary structure of DNA and RNA.

10.3 Datasets and informatics for analyzing noncoding RNAs

Computational experiments and bioinformatics are useful methods to study ncRNAs and predict their specific functions. These methods have some priorities over experimental assays. For example, experimental techniques are usually expensive and time-consuming. However, computational methods are more cost-effective and time-efficient, so we can better identify new ncRNAs or predict their features.

10.3.1 Homology-based methods

Similarities of structure or sequence between specific ncRNAs may arise from their same origin (Zhang et al., 2017). Generally, for a known RNA sequence, homology methods rely on similarities in sequence, structure, or a combination of both sequence and structure.

There are some databases that help us comparing our query RNA sequence with the already submitted RNA sequences in the database. If our query RNA is statistically similar to a specific existed RNA in the database, we can conclude that they are familiar RNAs. BLAST and BLAT are two well-known databases assisting with homology base sequence investigation. The more high-scoring segment pair meeting the less expected value (*E*-value) means better homology.

Although sequence-based homology is rapid and relatively simple, genomic variations of ncRNAs and their poor conserved sequence are inevitable. Structural-based homology or a combination of both sequence and structure is proposed instead of a single sequence-based homology (Zhang et al., 2017). RNAz helps researchers to compare the secondary structure and thermodynamic features of a query RNA (Gruber, Findeiß, Washietl, Hofacker, & Stadler, 2010). Infernal or Rfam uses a combination of sequence and secondary structure of ncRNA to predict their homology alignment (Kalvari et al., 2021). Some databases available for ncRNA evaluation are listed in Table 10.1.

10.3.2 De novo methods for ncRNA predictions

As mentioned earlier, in homology-based methods, a query RNA has to be compared with an identified RNA sequence. ncRNAs have less conserved sequences or structures. Moreover, probably the query RNA sequence does not exist in database. So, identification of new or less conserved ncRNA sequence may encounter some limitations. In de novo methods, identification of ncRNA is based on their primary sequence, for example, an adenine and thymine (AT)-rich or a specific K-mer sequence within its primary structure, without a requirement for comparison with known RNAs (Panwar, Arora, & Raghava, 2014; Wang, Reyes, Chua, & Gaasterland, 2004; Zhang et al., 2017).

TABLE 10.1 Databases of noncoding RNAs (ncRNAs) homology assessment.

Database	Application	PCa-related studies
BLAST	Useful for recognition similarities between query nucleic acid sequence and existed one in database. It helps identifying members of a gene family (Kuznetsov & Bollin, 2021)	Singh et al. (2020)
BLAT	A web based design for finding similarities (95% or more) of a query sequence (Kent, 2002)	Chen and Wei (2015)
Infernal	Infernal is a database for searching based on similarities on RNA structure and sequence (Nawrocki & Eddy, 2013)	McCown, Wang, Jaeger, and Brown (2019)
Rfam	A collection of RNA families information (sequence alignments, secondary structures, and covariance models) is presented on Rfam database (Kalvari et al., 2018, 2021)	Xu et al. (2010)
RNAz	A robust program, useful to predict the secondary structure of a specific RNA sequence according to their thermodynamically features (Gruber et al., 2010)	Lu et al. (2011)
LNCipedia	A public database for lncRNA sequence (Volders et al., 2019)	Sun et al. (2018)
RNAcentral	A comprehensive ncRNA sequence collection (RNAcentral Consortium, 2017)	Ge et al. (2020)

lncRNA, Long noncoding RNA; *PCa*, prostate cancer.

10.3.3 Special miRNA, circularRNA, and lncRNA databases

Various types of ncRNAs have their own features. For example, miRNAs are too short to check with homology-based assays. lncRNAs are longer but exhibit unique features that are different from mRNA or other long RNAs. Some databases regarding ncRNA features are listed in Table 10.2.

10.4 Sample types used for analyzing ncRNAs (tumor biopsies, liquid biopsies, etc.)

Some important challenges have to be considered to improve research quality management and practical outcomes or clinical experiments. For example, a strict management of

TABLE 10.2 Web based resources for microRNAs (miRNAs), long noncoding RNAs (lncRNAs), and small interfering RNAs (siRNAs).

Database	Applications	PCa-related studies
miRBase	miRNA database represents information of the hairpin structure of miRNA transcript, location, and sequence of the mature miRNA sequence (Kozomara, Birgaoanu, & Griffiths-Jones, 2019)	Avci, Harman, Dodurga, Susluer, and Gunduz (2013), Ye et al. (2013)
oncomiRDB	Database of functional miRNAs on the cellular process (e.g., apoptosis, proliferation, and invasion) (Tong, Ru, & Zhang, 2018; Wang, Gu, Wang, & Ding, 2014)	Liu, Xie et al. (2013)
HMDD (the Human microRNA Disease Database)	A full comprehensive database of the association between diseases and miRNAs expression (Huang et al., 2019)	Li, Li, Nie, You, and Bao (2020), Zhao, Kuang, Feng, Zou, and Wang (2019)
miRTarBase	Consists more than 360,000 identified miRNA–target interactions, validated using western blot, microarray, and next-generation sequencing (Huang et al., 2020)	Feng et al. (2013), Wu et al. (2020)
miRDB	An online database aimed to predict the target and function of miRNAs (Chen & Wang, 2020)	Ostadrahimi et al. (2018), Wu et al. (2020), Zhang et al. (2020)
deepBase	A popular database for prediction function of small and long noncoding RNAs (Xie et al., 2021)	
The Cancer Genome Atlas (TCGA)	A landmark cancer genomics program database of various cancer types	Lehrer and Rheinstein (2020), Wu et al. (2020)
LncRNA Disease	A database of disease-associated lncRNA or siRNA (Bao et al., 2019)	Liu et al. (2015)
Gene Ontology (GO)	A comprehensive database of functional genes (Consortium, 2017)	Wang et al. (2020)
LncBook	LncBook: a curated knowledgebase of human long noncoding RNAs (Ma et al., 2019)	Wang, Cao, Tang, Yang, and Du (2020)
LncExpDB	A comprehensive database for lncRNA expression (Li et al., 2021)	

PCa, Prostate cancer.

collecting, processing, storing, and data registering has to be considered the same for all the experiment biospecimens (Moore et al., 2011; Srigley et al., 2012).

10.4.1 New aspects of RNA-based biomarkers discovery in prostate cancer

Traditionally, diagnosis of PCa is possible by DRE and serum PSA followed by transrectal ultrasound (TRUS)-guided biopsy (Descotes, 2019). Serum PSA level serves as a good screening biomarker, but it is not tumor-specific and is also produced by other prostate cells (Alberts, Schoots, & Roobol, 2015). Recent advances in molecular technology lead to the development of new diagnostic, prognostics, and therapeutic approaches. Martignano et al. (2017) proposed urinary RNA-base biomarker evaluation other than prostate biopsy for diagnosis of PCa. Di Meo, Bartlett, Cheng, Pasic, and Yousef (2017) suggested liquid biopsy as a minimally invasive method for biomarker evaluation in urological malignancies. Urabe et al. (2019) offered circulating miR-1185−2−3p and miR-17−3p with high sensitivity and specificity as blood-based biomarkers in PCa patients.

10.4.2 Prostate cancer biospecimen repositories

There are some biospecimen repositories to help researchers in the genomic and clinical experiment fields (Table 10.3). Different types of biological samples include blood, urine, saliva, tissue, hair, and nail, which are collected with a standard method and are available for experimental researches (Ragin & Park, 2014).

10.5 Cell signaling pathways modulated by ncRNAs in prostate cancer

There is increasing evidence of the pivotal regulating role of ncRNAs in molecular events leading to PCa progression. Here, a review of the most mentioned functions of ncRNAs in regulating of molecular pathways is presented.

TABLE 10.3 Biospecimen repositories for experimental researches.

Biobank	Accessibility
Prostate Cancer Biorepository Network (PCBN)	https://prostatebiorepository.org/specimens
Biospecimen and Pathology Core (BPC)	https://prostatespore.weill.cornell.edu/cores/biospecimen-and-pathology
PLCO Data and Biospecimens	https://cdas.cancer.gov/learn/plco/biospecimens/
Biobank Resource Centre (BRC)	https://biobanking.org/
Canadian Tissue Repository Network (Ctrnet)	https://www.ctrnet.ca/en/home/
International Society for Biological and Environmental Repositories (ISBER)	https://www.isber.org/page/BPR

10.5.1 Phosphatase and tensin homolog/phosphoinositide 3-kinase/AkT/mammalian target of rapamycin pathway

Phosphoinositide 3-kinase (PI3K)/Akt/mammalian target of rapamycin (mTOR) pathway is an intracellular signaling pathway involved in regulating protein synthesis, glucose metabolism, growth, proliferation, survival, differentiation, stem cell–like properties, migration, and angiogenesis. The PI3K pathway activation leads to an increase in the nutrients' transporters, glucose uptake, amino acid, and other energy sources. It is believed that PI3K pathway is mediated by phosphatase and tensin homolog (PTEN), a tumor suppressor that negatively regulates Akt expression. PTEN inactivation due to somatic mutations as well as deletions results in overactivation of the PI3K/AKT signaling pathway and the downstream mTOR activity, glucose and lipid metabolism alteration, stimulation of cell cycle progression, and cell proliferation and invasion (Endzeliņš et al., 2016; Kasomva et al., 2018).

In addition to genetic inactivation, Loss of PTEN can also occur via miRNA downregulation. For instance, MiR-21 can inhibit PTEN expression to promote PCa cell proliferation and invasion (Fig. 10.1: Pathway 1) (Yang, Guo, & Shao, 2017). Moreover, MiR-21 is reported to suppress radiation-induced Akt activation in malignant glioma cells. Besides, evidence showed that miR-21 antagonizing leads to ionizing radiation sensitization along with suppressing PI3K/Akt/mTOR pathway in malignant glioma and also esophageal cancer cell lines, suggesting it worthy of considering in PCa development, as well (Gwak et al., 2012; Kasomva et al., 2018; Wu, Qi, Deng, Luo, & Yang, 2016; Yang et al., 2017). PTEN can be further suppressed by miR-4534. PCa cell lines have shown an overexpression in miR-4534 with oncogenic effects by targeting tumor suppressor PTEN gene leading to its downregulation. MiR-4534 knockdown resulted in decreased downstream pAkt and induced p21, p53, and p73 that are involved in impaired cell proliferation, migration/invasion, and induced G0/G1 cell cycle arrest and apoptosis in PCa. The antitumorigenic effects of depletion of miR-4534 were also confirmed in a PCa xenograft nude mouse model, suggesting its oncogenic potential both in vitro and in vivo (Nip et al., 2016).

MiR-106b-93–25 cluster is also overexpressed in PCa cells and can downregulate PTEN, potentiating cellular transformation. MiR-106b-93–25 cluster overexpression in a transgenic mouse model reduced PTEN expression and initiated prostate tumorigenesis, as well. This cluster also targets a pseudogene-derived RNA highly homologous to PTEN (PTENP1) which its expression seems highly correlated with PTEN. The mRNA transcript of PTENP1 carries the same miRNA-binding sites as PTEN which appears to compete with PTEN mRNA for miR-19b/20a. It indicates that PTENP1 might function as a competitive endogenous RNA (ceRNA) and decoy for miRNAs targeting PTEN, promoting tumor development (Poliseno, Salmena, Riccardi et al., 2010; Poliseno, Salmena, Zhang et al., 2010; Tay et al., 2011).

The PI3K/AKT pathway is also targeted by a lncRNA called Linc00963 in PCa cell lines. Knockdown of Linc00963 reduced cell proliferation, motility, invasion, the expression of epidermal growth factor receptor (EGFR) and phosphorylation levels of AKT, and facilitated cell apoptosis. Linc00963 was found to stimulate EGFR expression and promote AKT signaling, thus promoting the transition from androgen-dependent to androgen-independent PCa with enhanced cell proliferation, migration, and invasion (Wang, Han

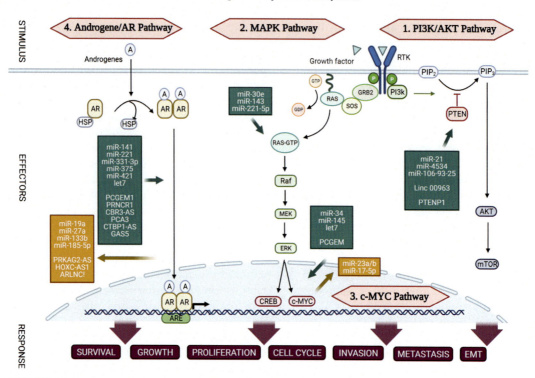

FIGURE 10.1 Noncoding RNAs involved in PCa cell signaling pathways:
Pathway 1. PTEN/PI3K/AKT/mTOR pathway: miR-21, miR-4534, miR-106–93-25, Linc 00963, and PTENP1 pseudogene;
Pathway 2. MAPK pathway: miR-30e, miR-143, and miR-221–5p;
Pathway 3. c-Myc pathway: miR-34, miR-145, let7, and PCGEM are affecting on and miR-23a/b and miR-17–5p affected by c-Myc pathway;
Pathway 4. AR signaling pathway: miR-141, miR-221, miR-331–3p, miR-375, miR-421, let7, PCGEM1, PRNCR1, CBR3-AS, PCA3, CTBP1-AS, and GAS5 are affecting on and miR-19a, miR-27a, miR-133b, miR-185–5p, PRKAG2-AS, HOXC-AS1, ARLNC1 are affected by AR signaling pathway. *PCa*, Prostate cancer.

et al., 2014). A more recent study revealed that Linc00963 acts in a Linc00963/miR-655/TRIM24 ceRNA network to promote cell proliferation and tumor growth of castration-resistant prostate cancer (CRPC) cells in vitro and in vivo and suggests that Linc00963 could be considered a novel therapeutic target for CRPC (Bai et al., 2021).

10.5.2 Mitogen-activated protein kinase pathway

The mitogen-activated protein kinase (MAPK) pathway mediates extracellular growth signals via transmembrane receptors, including receptor tyrosine kinases and G protein−coupled receptors to modulate the expression of several genes that eventually regulate cell proliferation, differentiation, survival, senescence, migration, and apoptosis.

Conformational changes in receptors trigger the RAS/RAF/MEK/ERK cascade to regulate MAPK target genes (Shorning, Dass, Smalley, & Pearson, 2020). The MAPK pathway can also be targeted by ncRNA mediators (Fig. 10.1: Pathway 2).

It has been reported that miR-143 is overexpressed in PCa cells and inhibits their proliferation, migration, and chemo-sensitivity via suppressing Kirsten rat sarcoma viral oncogene homolog expression, by targeting the EGFR/RAS/ MAPK pathway (Jackson, Grabowska, & Ratan, 2014; Xu et al., 2011). MAPK/ERK signaling pathway is also affected by miR-221−5p in PCa cells. MiR-221−5p targets the suppressors of cytokine signaling protein 1, a tumor suppressor gene, which mediates the proliferation and migration via MAPK/ERK signaling pathway and epithelial-to-mesenchymal transition (EMT) features of PCa cells in vivo (Shao, Ma, Zhang, & Zhu, 2018). MiR-30e seems to be another mediator of MAPK signaling pathway via suppressing its specific target, that is, M3 muscarinic acetylcholine receptor and inhibit the adhesion, migration, invasion, and progression of the cell cycle in PCa cells. Therefore miR-30e was proposed to be considered a candidate target in PCa treatment (Zheng, Zhang, Liu, Chen, & Zhang, 2019).

10.5.3 c-Myc pathway

The MYC gene is producing Myc protein that is a transcriptional factor known to be involved in G1/S phase of cell cycle transition via activating cyclin D2 and CDK2, promoting cell proliferation. Overexpression of MYC in PCa is reported by several studies (Benassi et al., 2012; Fleming et al., 1986). This could happen as a consequence of amplifying MYC location on chromosome 8q24 as well as alternative posttranscriptional mechanisms. MYC is known to be targeted by several miRNAs including miR-34, miR-145, and let-7c, in PCa initiation or progression (Nadiminty et al., 2012; Sachdeva et al., 2009; Yamamura et al., 2012). MYC is also known to modulate the transcription of several miRNAs commonly deregulated in PCa, including the miR-23a/b and miR-17−5p cluster (Fig. 10.1: Pathway 3) (Gao et al., 2009; O'Donnell, Wentzel, Zeller, Dang, & Mendell, 2005).

There are reports on the regulatory role of lncRNAs on MYC, as well. PCa gene expression marker 1 (PCGEM1) as an androgen-induced prostate-specific lncRNA appears to activate c-Myc along with androgen receptor (AR) regulating the central metabolic pathways in PCa cells (Hung et al., 2014).

10.5.4 AR signaling pathway

The AR is a transcription factor regulating the expression of its target genes such as PSA or NK3 homeobox 1 gene (NKX3.1) in response to androgen ligands in target cells via AR signaling pathway. AR singling with a pivotal role in tumor initiation, maintenance, and progression is known to be affected in many PCa cases. Interestingly, there are numerous reports on ncRNAs both affecting on and modulated by AR signaling pathway (Fig. 10.1: Pathway 4) (Kumar & Lupold, 2016; Sharma & Baruah, 2019; Weiss, Plass, & Gerhauser, 2014).

In the first category lies a long list of ncRNAs. There are several lncRNAs, including PCGEM1, PCa ncRNA 1, CBR3-antisense, PCA3, and C-terminal-binding protein 1-antisense upregulated in PCa, suggested to play an AR coactivating role in PCa progression (Chung et al., 2011; Cui et al., 2013; Ferreira et al., 2012; Petrovics et al., 2004; Takayama et al., 2013).

Nevertheless, reports on lncRNA growth arrest–specific 5 (GAS5) expression in PCa seem rather controversial. While some studies report the downregulation of GAS5 (Mourtada-Maarabouni, Pickard, Hedge, Farzaneh, & Williams, 2009) in PCa, other researchers showed that high GAS5 expression promoted apoptosis and inhibited cell proliferation in PCa cell lines, whereas the knockdown of GAS5 showed an opposite effect. Besides, UV-C irradiation and chemotherapeutic drugs induced cancer cell death via GAS5 overexpression, while downregulation of GAS5 represented the opposite effects (Pickard, Mourtada-Maarabouni, & Williams, 2013; Romanuik et al., 2010).

There are also several studies reporting the modulating role of miRNAs on AR signaling in PCa, including miR-141, miR-221, miR-222, miR-331–3p, miR-375, mir-421, and miR-let7c (Chu et al., 2014; Epis, Giles, Barker, Kendrick, & Leedman, 2009; Meng et al., 2016; Nadiminty et al., 2012; Sun et al., 2009; Xiao et al., 2012). Moreover, Östling et al. identified 71 unique miRNAs that influenced the level of AR in human PCa cells by miRNA screening in a panel of human PCa cell lines and quantifying the changes in AR protein content using protein lysate microarrays. It revealed that most of the miRNAs target the 3′UTR region of AR gene, suggesting the AR pathway as a promising therapeutic target in PCa (Östling et al., 2011).

As in the second category, there are various ncRNAs regulated by activation of AR signaling. Among lncRNAs, PRKAG2 antisense RNA1 showed a significant upregulation in AR-positive PCa cells after androgen deprivation. In contrast, HOXC cluster antisense RNA 1, which is positively regulated by AR, did not show a similar upregulation, but the silencing of these two lncRNAs promotes tumor growth, showing repression of AR and AR variant expression (Takayama, Fujimura, Suzuki, & Inoue, 2020). This suggests that the different regulatory pathways by AR signaling will have different impacts on the expression level of the lncRNAs.

Another interesting lncRNA is AR-regulated long ncRNA 1 (ARLNC1) which is strongly associated with AR signaling in a two-sided way. Not only ARLNC1 was induced by AR protein, but also it stabilized the AR transcript via RNA–RNA interaction. Besides, ARLNC1 knockdown suppresses the AR signaling by AR downregulation in vitro and in vivo, implying a positive feedback loop that potentiates AR signaling during PCa progression (Zhang et al., 2018).

Certain miRNAs are suggested to be regulated by AR signaling, as well. AR binds to androgen response elements within *cis*-regulatory regions that regulate the expression of miRNAs in PCa. In other words, AR induces oncogenic miRNAs, which act as downstream effectors of AR signaling, and represses tumor-suppressor miRNAs (Tay et al., 2011). Among all miRNAs, miR-19a, miR-27a, miR-133b, and miR-185–5p are most reported to be upregulated, whereas miR-221/222 and miR-421 are downregulated by AR (Shorning et al., 2020; Wang, Han et al., 2014; Wu et al., 2016).

10.6 NcRNAs as biomarkers for prostate cancer

The main issue in PCa clinical management appears to be diagnosis and follow-up monitoring after therapies. Regarding the vast involvement of ncRNAs in PCa tumorigenesis, aggressiveness, and response to therapy via different pathways previously discussed, the diagnostic, prognostic, and predictive potential of ncRNAs in PCa are summarized in the following sections.

10.6.1 Diagnostic value

The DRE, PSA testing, TRUS, prostate MRI, and TRUS-guided transrectal prostate biopsy are the current PCa screening and diagnostic methods used frequently by physicians. Initial staging, risk stratification, active surveillance, and focal therapy result in diagnostic accuracy that is the major challenge. Considering the low sensitivity of DRE as well as the low specificity of PSA testing, the combination of these two is suggested to improve the diagnostic accuracy (Takayama et al., 2020; Xu, Deng, Wang, & Zhu, 2019). From a far minimally invasive point of view, ncRNAs are extremely stable and can easily be isolated from a small amount of sample from various biological fluids such as serum, plasma, urine, saliva, breast milk, tears, and semen (Kanwal, Plaga, Liu, Shukla, & Gupta, 2017).

MiRNA profiling in plasma or serum of patients with different types of PCa has been performed by several studies to identify miRNA signatures with remarkably high diagnostic values. There have been several studies reporting the possible diagnostic utility of a miRNA signature composed of miR-141 and miR-375. This combination has also shown a strong correlation with tumor staging and prognosis in PCa patients (Agaoglu et al., 2011; Brase et al., 2011; Bryant et al., 2012; Haldrup et al., 2014; Kachakova et al., 2015). Another set of combination comprising miR-141 along with miR-151−3p and miR-16 was found to be a good differentiator of localized and metastatic PCa (Watahiki et al., 2013). Furthermore, miR-21 has shown a great potential for stage-specific classification of PCa due to the fact that its expression level was much higher in androgen-independent PCa cells than in androgen-dependent ones (Li, Li, Sha, Sun, & Huang, 2009). MiR-21 is also altered in urine and in combination with miR-19a/19b has shown a greater diagnostic power than PSA (Stuopelytė, Daniūnaitė, Jankevičius, & Jarmalaitė, 2016).

It has been a fair debate on the diagnostic potential of lncRNAs in the same manner. An FDA-approved lncRNA PCA3 recently showed more specificity and sensitivity than PSA (Groskopf et al., 2006; Leyten et al., 2014). However, given its low detection rate and intraindividual variability, PCA3 failed to efficiently differentiate between lower- and higher grade PCa (Calonge et al., 2014; Wei et al., 2014). Another lncRNA with a diagnostic value is Metastasis Associated Lung Adenocarcinoma Transcript 1 (MALAT1) which its expression is in coordination with PSA levels and Gleason scores as well as with tumor stage and castration-resistant PCa (Ren et al., 2013). The urinary MALAT1 is more accurate than PSA and %fPSA [free-to-total ratio (f/t) PSA] as an independent predictor of PCa before biopsy (Wang, Ren et al., 2014).

10.6.2 Prognostic value

Active surveillance, radiation therapy (RT), and radical prostatectomy (RP) are current options in early-stage PCa, and androgen deprivation is added on in the case of metastatic or high-risk localized disease. Routine PCa prognosis is performed based on the tumour, node and metastasis (TNM) staging with the addition of Gleason score and PSA levels. PSA is known and measured as an indicator of relapse following RP or RT. Therefore, detectable PSA levels after treatment are considered biochemical recurrence (BCR) due to the presence of a residual tumor. BCR has the potential value for predicting the sign of clinical progression or metastasis (Fang & Gao, 2014; Matin, Jeet, Clements, Yousef, & Batra, 2016).

On the other hand, several studies have been carried out to establish a miRNA expression profile as a promising approach for improving PCa prognosis. A miRNA expression signature, including 16 miRNAs, has been proposed to distinguish the relapse within 10 years after RP (Tong et al., 2009). A subsequent study reported the elevated expression of 25 miRNAs in PCa tissue among which miR-31, miR-96, miR-125b, miR-205, and miR-222 were correlated with Gleason score (Schaefer et al., 2010). An independent study suggested a miRNA index quote called miQ consisted of four miRNA—two upregulated (miR-96–5p and miR-183–5p) and two downregulated (miR-145–5p and miR-221–5p)—as a prognostic tool in PCa to forecast tumor aggressiveness, metastasis, and overall survival with high accuracy (Larne et al., 2013). The miQ was further investigated in other cohorts with different sizes, methodologies, and experimental designs and the result suggested it as superior to PSA in predicting diagnosis. Thus, miQ was proposed to be used as a clinical tool for PCa diagnosis and as a prognostic marker of PCa progression (Fabris et al., 2016). MicroRNA-96 alone was also confirmed to correlate with tumor aggressiveness, postoperative outcome, and recurrence (Haflidadóttir et al., 2013).

Another well-studied miRNA in PCa recurrence is miR-21 which has shown a strong association with PCa recurrence. MiR-21 as a known onco-miR appeared to be overexpressed in recurrent PCa suggesting its potential to indicate recurrence-free survival (Leite et al., 2015; Li et al., 2012; Liu, Gao et al., 2013; Melbø-Jørgensen et al., 2014; Selcuklu, Donoghue, & Spillane, 2009). In addition, several studies suggested the prognostic potential of other miRNAs such as miR-15/16, miR-21, miR-34b, miR-126, miR-143, miR-205, miR-224, miR-135b, miR-194, miR-222, and miR-125b in PCa (Aghdam et al., 2019; Ni et al., 2017).

Tumor progression despite androgen deprivation therapy is referred to as CRPC with a wide range of clinical manifestations from elevated PSA to new metastasis. Overexpression of miR-21 has been reported in both tissue and plasma of CRPC patients (Ribas et al., 2009; Shen et al., 2012). Beside, miR-21, there are other miRNAs associated with the onset of CRPC as well. Among all, miR-221/-222 were found to be strongly upregulated, while miR-23b/-27b were downregulated in the prostatic tissue and bone marrow of CRPC patients (Goto et al., 2014; Ishteiwy, Ward, Dykxhoorn, & Burnstein, 2012).

Several lncRNAs such as TMPO-AS1, FALEC, SChLAP1, BDNF-AS, TINCR, PCAT1, LINC00844 have been studied for diagnostic accuracy and prognostic potential of PCa (Dong, Ding, Li, Xue, & Liu, 2018; Huang et al., 2018; Li et al., 2018; Lingadahalli et al., 2018; Mehra et al., 2016; Rezatabar et al., 2020; Zhao et al., 2017). A recent review recommended that SChLAP1 overexpression in PCa tissue is highly associated with BCR, clinical progression, and PCa-specific mortality (Xu et al., 2019).

10.6.3 Predictive value

Another aspect of miRNAs' function is their response to various treatment options. RT is introduced as the gold standard for localized high-risk PCa patients who cannot undergo RP (Fabris et al., 2016). Therefore suggesting a miRNA signature to predict the RT response could improve PCa treatment. The first report on miRNA signature in response to RT comprised six miRNAs (miR-133b, miR-143, miR-145b, miR-196a, miR-218, and miR-512) downregulated on either androgen-dependent or androgen-resistant PCa cells. Among them, miR-521 showed the most significant changes and was further investigated. Overexpression of miR-521 with miR-521 mimics sensitized PCa cells to radiation treatment, while its knockdown resulted in radiation resistance. These findings suggested the miR-521 as a conceivable target to improve the radiation treatment on PCa cells (Josson, Sung, Lao, Chung, & Johnstone, 2008). Several other miRNAs are reported to be altered in PC cells exposed to different doses of radiation. Expressions of six miRNAs (miR-9, miR-22, miR-25, miR-30a, miR-550a, and miR-548h) were enhanced and 16 other miRNAs (let-7c/d/e, miR-15a, miR-17, miR-30d, miR-92a, miR-125a, miR-197, miR-221, miR-320b, miR-342, miR-361, miR-374a, miR-501, and miR-671) were downregulated after radiation (Leung et al., 2014).

The alternative choice of treatments after RP and also when androgen-deprivation therapies fail include taxane-based chemotherapy, drugs targeting the AR [enzalutamide (Enz)], or androgen synthesis (abiraterone acetate). Unfortunately, most of the investigations on the role of miRNAs as predictors of chemotherapy response are based on preliminary in vitro studies and there are only few at the clinical level (Fabris et al., 2016). After miR profiling on docetaxel (DTX)-resistant and sensitive cell lines to identify candidate circulating miRNA biomarkers, a signature comprising six miRNAs (miR-20a, miR-146a, miR-200c, miR-200b, miR-222, and miR-301b) has been introduced to be associated with DTX chemotherapy outcome in CRPC patients and suggested to have a potential to distinguish responders from nonresponders (Lin et al., 2014). MiR-21 has also been found to have a fundamental role in controlling the response to taxanes on a human level. Serum miR-21 levels have shown higher levels in patients with hormone-refractory PCa who were resistant to DTX-based chemotherapy compared to those sensitive to chemotherapy. Therefore miR-21 was suggested as a marker to indicate the transformation to hormone-refractory disease, and a potential predictor for the efficacy of DTX-based chemotherapy (Zhang et al., 2011).

Another miR profiling in patients with metastatic CRPC (mCRPC) resulted in identifying a miRNA signature of five components (miR-141, miR-200a, miR-200c, miR-210, and miR-375) correlated with treatment response as assessed by change in PSA. MiR-210 was the most significantly elevated amongst mCRPC patients undergoing therapy. Induction of hypoxia in cultured PCa cells led to increased miR-210 levels as a target of the hypoxia-responsive HIF-1α signaling pathway. Therefore miR-210 was proposed as a predictive biomarker to identify a biologically distinct, subset of mCRPC patients with tumor-associated hypoxia for whom the development of alternative therapeutic approaches could be considered (Cheng et al., 2013).

Among lncRNAs, PCAT1 seems to have the potential to predict the response to Poly ADP-ribose polymerase 1 (PARP1) inhibitor therapy in PCa. PCAT-1 mediates the double-stranded DNA breaks repair and inhibited homologous recombination via binding to the 3′UTR of BRACA2 tumor suppressor. Enhanced expression of PCAT-1 could predict the sensitivity of PCa xenografts to PARP1 inhibitor called olaparib (Prensner et al., 2014). There is similar evidence on GAS5 and NEAT1 in response to mTOR inhibitors androgen or AR antagonists, consequently (Chakravarty et al., 2014; Yacqub-Usman, Pickard, & Williams, 2015). LncRNAs are also expected to predict BCR. For instance, PCa patients with lower expression of PCAT29 have shown a higher risk of biochemical recurrence (Malik et al., 2014).

10.7 Therapeutic potential of ncRNAs in prostate cancer patients

Several preliminary studies were attempting to show the use of ncRNAs as an effective choice of treatment for PCa; most of which are needed to be verified on humans. Regarding miRNAs, this approach includes a wide range of applications such as small molecule inhibitors, miRNA vectors, miRNA mimics, miRNA sponges, the clustered regularly interspaced short palindromic repeat-Cas9 technology, antisense oligonucleotides (ASOs), and miR-mask oligonucleotides (Matin et al., 2016). Delivering miRNAs to target cells can be achieved by whether viral-based vectors, including retroviruses, lentiviruses, adenoviruses, and adeno-associated viruses, or nonviral-based vectors such as physical and chemical approaches. The latter benefits from less toxicity and immunogenicity along with no limitation on transferred miRNA size. Nevertheless, the risk of lower transfection efficiency, shorter half-live, and less biodegradability should be taken into consideration (Nayerossadat, Maedeh, & Ali, 2012; Wang et al., 2015; Yang, 2015).

The report from the first-in-human phase I clinical trial of miRNA cancer therapy showed that the treatment of the patients with refractory advanced solid tumors with MRX34—a liposomal miR-34a mimic—accompanied by dexamethasone premedication was associated with acceptable safety and showed evidence of antitumor activity (Beg et al., 2017). The therapeutic role of miR-34a on PCa in vitro and in vivo has been previously established (Liu et al., 2011).

Interatomur injection of miR-199–3p mimics inhibited an oncogene named the Aurora kinase A in PCa xenograft mouse model. Downregulation of miR-199–3p showed a strong inverse correlation with tumor stage and Gleason score (Qu et al., 2014). Using miR-29b mimic in a more recent study showed the same results in a similar model as well. The elevated expression of miR-29b showed a strong association with the expression of proapoptotic gene BCL2L11 (Bim) in PC3 cells which was further verified in miR-29b-treated xenograft tumors (Sur, Steele, Shi, & Ray, 2019).

MiR-15 and miR-16 as putative tumor suppressors are reported to be downregulated in PCa. It has been shown that loss of miR-15 and miR-16 and gain of miR-21 in PCa cells led to the activation of Transforming growth factor beta (TGF-β) and Hedgehog signaling, mediating the local invasion, distant bone marrow colonization, and osteolysis (Bonci et al., 2016). Transient transfection of synthetic miR-16 inhibited PCa cell proliferation. Delivery of synthetic miR-16 into PCa xenograft model could be promoted by atelocollagen. Tail vein injection of miR-16 with atelocollagen in a bone metastasis model reduced the prostate tumors growth in bone (Takeshita et al., 2010). Injection of antagomir of miR-15a and miR-16 in normal prostate tissue promoted significant hyperplasia in mice. Knockdown of miR-15a and miR-16 enhanced the survival, proliferation, and invasiveness of untransformed prostate cells, which later became tumorigenic in immune-deficient mice. In contrast, reconstitution of miR-15a and miR-16–1 expression results in growth arrest, apoptosis, and marked regression of prostate tumor xenografts. These findings suggest that both miR-15a and miR-16 act as tumor suppressors and may be a promising effort to the future treatment of PCa (Bonci et al., 2008). There are also studies reporting the therapeutic potential of miR-34, MiR-124, mir-200b, miR-203, miR-890, and let-7a/g in PCa progression and metastasis (Dong et al., 2010; Esposito et al., 2014; Hatano et al., 2015; Liu et al., 2011; Shi et al., 2015; Viticchiè et al., 2011; Williams, Veliceasa, Vinokour, & Volpert, 2013).

On the other hand, lncRNAs seem to be a possible candidate for a therapeutic application targeted by RNA interference technology using miRNAs, small interfering RNAs (siRNAs), small hairpin RNAs (shRNAs), and ASOs (Xu et al., 2019). The siRNA silencing of MALAT1 inhibited cell growth, invasion, and migration and induced the G0/G1 phase cell cycle arrest in CRPC cells. Besides, intratumoral delivery of the specific siRNA in PCa xenografts in castrated male nude mice confirmed the reduction in growth and metastasis along with improved survival (Ren et al., 2013). Moreover, the in vivo knockdown of a lncRNA called second chromosome locus associated with prostate-1 with shRNA proved a decrease in metastasis (Prensner et al., 2014). Another study reported that MEG3 could act as a miRNA sponge and suppress the PCa development by targeting the miR-9-5p/QKI-5 axis (Wu et al., 2019).

10.8 Potential of ncRNAs in predicting chemo-resistance and radioresistance in prostate cancer patients

Altered expression of miRNAs is reported to be in correlation with resistance to radiotherapy, hormone therapy, and chemotherapy (ChunJiao, Huan, ChaoYang, & GuoMei, 2014; Halimi et al., 2013; Halimi et al., 2014; Korpela, Vesprini, & Liu, 2015; Sun et al., 2015). MiRNAs are known to mediate the development of resistant PCa via several mechanisms, including cell growth and proliferation, autophagy, androgen signaling, apoptosis avoidance, invasion and metastasis, multiple drug resistance transporters, EMT, and cancer stem cells (CSCs) (Li & Mahato, 2014; Razdan, de Souza, & Roberts, 2018).

10.8.1 Chemo-resistance

DTX is the current chemotherapeutic choice for CRPC patients who develop metastasis, particularly to the bones. However, many of the patients receiving DTX develop resistance, eventually. There are several miRNAs implicated in chemo-resistance, including miRNAs-15a/miR-16, miRNA-31, miRNA-34a, miRNA-128, miR-143, and miR-200c. One of the well-examined miRNA families in resistant PCa-related studies is miR-34. Mir-34a acts as a tumor suppressor by inhibiting cell proliferation, migration, invasion, and EMT and inducing apoptosis in vitro, in addition to suppressing tumor growth and metastasis, in vivo (Razdan et al., 2018). MiR-34a is also associated with chemo-resistance via androgen signaling, apoptosis avoidance, and CSCs (Fujita et al., 2008; Kashat et al., 2012; Kojima, Fujita, Nozawa, Deguchi, & Ito, 2010). MiR-34b, on the other hand, seems to be more involved in suppressing EMT via reducing mesenchymal markers, including vimentin, ZO1, N-cadherin, and Snail while increasing epithelial markers such as E-cadherin. MiR-34b was found to inhibit PCa by demethylation, active chromatin modification, and AKT pathways suggesting (Majid et al., 2013).

MiR-200 family is another group controlling the epithelial characteristics, and their loss contributes to EMT which is tightly associated with metastasis and chemo-resistance. In particular, miRNA-200c showed a significant downregulation in PCa DTX-resistant cells which resulted in the inhibition of proliferation, migration, and invasion via suppressing EMT DTX-resistant cells and also showed significant increase in stemness and invasiveness. Besides, DTX therapy also caused a downregulation in E-cadherin levels and upregulation of mesenchymal markers in

PCa patients (Puhr et al., 2012; Shi et al., 2014). Both miR34a and miR200c are reported to be downregulated in a MiR profiling among PCa cells. Treatment with a combination of paclitaxel and cyclopamine seems to restore the expression of miR-34a and miR-200c, confirming their role in modulating chemo-resistance (Singh et al., 2012).

There is also growing evidence on the role of lncRNAs in chemotherapy resistance. MALAT1 expression is reported to be elevated in DTX-resistant PCa samples and cells; upregulated MALAT1, boosted PCa cell proliferation, migration, and invasion, but reduced apoptosis in spite of DTX treatment. MALAT1 was found that can be act as a sponge for miR-145–5p and subsequently increase the A-kinase anchor protein 12 (AKAP12) levels which resulted in DTX-chemo-resistance of PCa in vitro and in vivo. Therefore the MALAT1/miR-145–5p/AKAP12 axis was introduced to play a major role in DTX-resistance (Xue, Lu, Xu, Zhou, & He, 2018). MALAT1 could also participate in resistance against Enz as an antiandrogen drug that has been used for mCRPC. MALAT1 promoted the expression of the AR splice variant 7 (AR-V7) in Enz-resistant PCa cell lines and PCa patients who received Enz treatment. Targeting MALAT1/AR-V7 axis with specific siRNA or AR-V7 degradation enhancer ASC-J9 in Enz-resistant PCa cell lines and mouse models suppressed the Enz-resistant PCa progression (Wang et al., 2017). Another recent study demonstrated that NEAT1 lncRNA and miR-34a are consequently up- and downregulated, respectively, in PCa tissues, parental PCa cells, and DTX-resistant PCa cells. The knockdown of NEAT1 leads to the improvement of sensitivity to DTX in DTX-resistant PCa cells in vitro and in vivo. On the contrary, inhibition of miR-34a partially overturned NETA1-knockdown-induced sensitivity to DTX. Thus NEAT1 is believed to act as a ceRNA for miR-34a in DTX-resistant PCa cells to enhance the expression of rearranged during transfection (RET) implying that targeting the NEAT1/miR-34a/RET axis must be taken into consideration for PCa chemotherapy improvement (Tian, Zhang, Zhao, Li, & Zhu, 2017).

10.8.2 Radioresistance

RT is often used as a primary treatment or adjuvant therapy after RP or as palliative therapy in patients who develop metastasis to relieve bone pain. The response rate for those with localized tumors is up to 60%, while for those having recurrent PCa within 5 years it seems to reduce to 45%. Since a considerable amount of the patients may show resistance to RT, a long list of indices has come to help. miRNAs have also collected interest as a potential indicator of radioresistance. Among all, miR-95, miR-106b, and miR-521 are shortlisted miRNAs having direct implications in resistance to RT (Razdan et al., 2018). Expression of miR-95 in radioresistant PC3 cells was elevated which itself led to increasing invasiveness, growth, and radioresistance, eventually. In addition, upregulation of miR-95 also increased tumor growth and resistance to radiation in tumor xenografts, suggesting it as a critical determinant of radioresistance (Huang et al., 2013).

10.9 Conclusion

Evidence from the literature suggests a wide-ranging role of ncRNAs in PCa via cell cycle progression, migration, invasion, apoptosis, and metastasis. Expression of ncRNAs is altered in PCa cancer cells by both genetic and epigenetic mechanisms, potentially affecting both oncogene and tumor suppressor gene regulations. The noninvasive sampling of circular RNAs in human biofluids facilitates the detection and prognosis of PCa. Recently, much research has focused on utilizing the therapeutic potential of ncRNAs in PCa treatment. Besides, there are several ncRNA targets to predict and overcome resistance to therapeutic options. With consideration of the multifarious role of ncRNAs, further research is necessary to improve new strategies for PCa detection and the efficacy of disease therapeutics utilizing all types of ncRNAs especially miRNAs and lncRNAs. At last, using these novel molecules can improve the quality of human life and it would not be far for those physicians to use these interesting molecules in all aspects of PCa, including diagnosis, treatment, prognosis, and treatment monitoring, in the near future.

References

Agaoglu, F. Y., Kovancilar, M., Dizdar, Y., Darendeliler, E., Holdenrieder, S., Dalay, N., et al. (2011). Investigation of miR-21, miR-141, and miR-221 in blood circulation of patients with prostate cancer. *Tumor Biology, 32*(3), 583–588.

Aghdam, A. M., Amiri, A., Salarinia, R., Masoudifar, A., Ghasemi, F., & Mirzaei, H. (2019). MicroRNAs as diagnostic, prognostic, and therapeutic biomarkers in prostate cancer. *Critical Reviews in Eukaryotic Gene Expression, 29*(2).

Alberts, A. R., Schoots, I. G., & Roobol, M. J. (2015). Prostate-specific antigen-based prostate cancer screening: Past and future. *International Journal of Urology, 22*(6), 524–532.

Exploring long non-coding RNAs through sequencing. In S. R. Atkinson, S. Marguerat, & J. Bähler (Eds.), *Seminars in cell & developmental biology;*. Elsevier.

Avci, C. B., Harman, E., Dodurga, Y., Susluer, S. Y., & Gunduz, C. (2013). Therapeutic potential of an anti-diabetic drug, metformin: Alteration of miRNA expression in prostate cancer cells. *Asian Pacific Journal of Cancer Prevention, 14*(2), 765–768.

Bai, M., He, C., Shi, S., Wang, M., Ma, J., Yang, P., et al. (2021). Linc00963 promote cell proliferation and tumor growth in castration-resistant prostate cancer by modulating miR-655/TRIM24 axis. *Frontiers in Oncology, 11*.

Bao, Z., Yang, Z., Huang, Z., Zhou, Y., Cui, Q., & Dong, D. (2019). LncRNADisease 2.0: An updated database of long non-coding RNA-associated diseases. *Nucleic acids Research, 47*(D1), D1034–D1037.

Beg, M. S., Brenner, A. J., Sachdev, J., Borad, M., Kang, Y.-K., Stoudemire, J., et al. (2017). Phase I study of MRX34, a liposomal miR-34a mimic, administered twice weekly in patients with advanced solid tumors. *Investigational New Drugs, 35*(2), 180–188.

Bellaousov, S., Reuter, J. S., Seetin, M. G., & Mathews, D. H. (2013). RNAstructure: Web servers for RNA secondary structure prediction and analysis. *Nucleic Acids Research, 41*(W1), W471–W474.

Benassi, B., Flavin, R., Marchionni, L., Zanata, S., Pan, Y., Chowdhury, D., et al. (2012). MYC is activated by USP2a-mediated modulation of microRNAs in prostate cancer. *Cancer Discovery, 2*(3), 236–247.

Bijnsdorp, I. V., van Royen, M. E., Verhaegh, G. W., & Martens-Uzunova, E. S. (2017). The non-coding transcriptome of prostate cancer: Implications for clinical practice. *Molecular Diagnosis & Therapy, 21*(4), 385–400.

Bonci, D., Coppola, V., Musumeci, M., Addario, A., Giuffrida, R., Memeo, L., et al. (2008). The miR-15a–miR-16-1 cluster controls prostate cancer by targeting multiple oncogenic activities. *Nature Medicine, 14*(11), 1271–1277.

Bonci, D., Coppola, V., Patrizii, M., Addario, A., Cannistraci, A., Francescangeli, F., et al. (2016). A microRNA code for prostate cancer metastasis. *Oncogene, 35*(9), 1180–1192.

Brase, J. C., Johannes, M., Schlomm, T., Fälth, M., Haese, A., Steuber, T., et al. (2011). Circulating miRNAs are correlated with tumor progression in prostate cancer. *International Journal of Cancer, 128*(3), 608–616.

Bryant, R., Pawlowski, T., Catto, J., Marsden, G., Vessella, R., Rhees, B., et al. (2012). Changes in circulating microRNA levels associated with prostate cancer. *British Journal of Cancer, 106*(4), 768–774.

Calonge, N., Klein, R. D., Berg, J. S., Campos-Outcalt, D., Djulbegovic, B., Ganiats, T., et al. (2014). Recommendations from the EGAPP Working Group: Does PCA3 testing for the diagnosis and management of prostate cancer improve patient health outcomes? *Genetics in Medicine, 16*(4), 338.

Chakravarty, D., Sboner, A., Nair, S. S., Giannopoulou, E., Li, R., Hennig, S., et al. (2014). The oestrogen receptor alpha-regulated lncRNA NEAT1 is a critical modulator of prostate cancer. *Nature Communications, 5*(1), 1–16.

Chandra Gupta, S., & Nandan Tripathi, Y. (2017). Potential of long non-coding RNAs in cancer patients: From biomarkers to therapeutic targets. *International Journal of Cancer, 140*(9), 1955–1967.

Chen, S., Huang, V., Xu, X., Livingstone, J., Soares, F., Jeon, J., et al. (2019). Widespread and functional RNA circularization in localized prostate cancer. *Cell, 176*(4), 831–843, e22.

Chen, Y., & Wang, X. (2020). miRDB: An online database for prediction of functional microRNA targets. *Nucleic Acids Research, 48*(D1), D127–D131.

Chen, Y., & Wei, J. (2015). Identification of pathogen signatures in prostate cancer using RNA-seq. *PLoS One, 10*(6), e0128955.

Cheng, H. H., Mitchell, P. S., Kroh, E. M., Dowell, A. E., Chéry, L., Siddiqui, J., et al. (2013). Circulating microRNA profiling identifies a subset of metastatic prostate cancer patients with evidence of cancer-associated hypoxia. *PLoS One, 8*(7), e69239.

Cho, W. C. (2010). MicroRNAs: Potential biomarkers for cancer diagnosis, prognosis and targets for therapy. *The International Journal of Biochemistry & Cell Biology, 42*(8), 1273–1281.

Chu, M., Chang, Y., Li, P., Guo, Y., Zhang, K., & Gao, W. (2014). Androgen receptor is negatively correlated with the methylation-mediated transcriptional repression of miR-375 in human prostate cancer cells. *Oncology Reports, 31*(1), 34–40.

Chung, S., Nakagawa, H., Uemura, M., Piao, L., Ashikawa, K., Hosono, N., et al. (2011). Association of a novel long non-coding RNA in 8q24 with prostate cancer susceptibility. *Cancer Science, 102*(1), 245–252.

ChunJiao, S., Huan, C., ChaoYang, X., & GuoMei, R. (2014). Uncovering the roles of miRNAs and their relationship with androgen receptor in prostate cancer. *IUBMB Life, 66*(6), 379–386.

Consortium, G. O. (2017). Expansion of the gene ontology knowledgebase and resources. *Nucleic Acids Research, 45*(D1), D331–D338.

Cruz, C., & Houseley, J. (2020). Protocols for northern analysis of exosome substrates and other noncoding RNAs. *The Eukaryotic RNA Exosome:* (pp. 83–103). Springer.

The prostate cancer-up-regulated long noncoding RNA PlncRNA-1 modulates apoptosis and proliferation through reciprocal regulation of androgen receptor. In Z. Cui, S. Ren, J. Lu, F. Wang, W. Xu, Y. Sun, et al. (Eds.), *Urologic oncology: Seminars and original investigations;*. Elsevier.

Dall'Era, M. A., Cooperberg, M. R., Chan, J. M., Davies, B. J., Albertsen, P. C., Klotz, L. H., et al. (2008). Active surveillance for early-stage prostate cancer: Review of the current literature. *Cancer: Interdisciplinary International Journal of the American Cancer Society, 112*(8), 1650–1659.

D'Amico, A. V., Whittington, R., Malkowicz, S. B., Schultz, D., Blank, K., Broderick, G. A., et al. (1998). Biochemical outcome after radical prostatectomy, external beam radiation therapy, or interstitial radiation therapy for clinically localized prostate cancer. *Journal of the American Medical Association, 280*(11), 969–974.

Denman, R. B. (1993). Using RNAFOLD to predict the activity of small catalytic RNAs. *Biotechniques, 15*(6), 1090–1095.

Derrien, T., Johnson, R., Bussotti, G., Tanzer, A., Djebali, S., Tilgner, H., et al. (2012). The GENCODE v7 catalog of human long noncoding RNAs: Analysis of their gene structure, evolution, and expression. *Genome Research, 22*(9), 1775–1789.

Descotes, J.-L. (2019). Diagnosis of prostate cancer. *Asian Journal of Urology, 6*(2), 129–136.

Di Meo, A., Bartlett, J., Cheng, Y., Pasic, M. D., & Yousef, G. M. (2017). Liquid biopsy: A step forward towards precision medicine in urologic malignancies. *Molecular Cancer, 16*(1), 1–14.

Dong, L., Ding, H., Li, Y., Xue, D., & Liu, Y. (2018). LncRNA TINCR is associated with clinical progression and serves as tumor suppressive role in prostate cancer. *Cancer Management and Research, 10*, 2799.

Dong, Q., Meng, P., Wang, T., Qin, W., Qin, W., Wang, F., et al. (2010). MicroRNA let-7a inhibits proliferation of human prostate cancer cells in vitro and in vivo by targeting E2F2 and CCND2. *PLoS One, 5*(4), e10147.

Endzeliņš, E., Melne, V., Kalniņa, Z., Lietuvietis, V., Riekstiņa, U., Llorente, A., et al. (2016). Diagnostic, prognostic and predictive value of cell-free miRNAs in prostate cancer: A systematic review. *Molecular Cancer, 15*(1), 1–13.

Epis, M. R., Giles, K. M., Barker, A., Kendrick, T. S., & Leedman, P. J. (2009). miR-331-3p regulates ERBB-2 expression and androgen receptor signaling in prostate cancer. *Journal of Biological Chemistry, 284*(37), 24696–24704.

Esposito, C. L., Cerchia, L., Catuogno, S., De Vita, G., Dassie, J. P., Santamaria, G., et al. (2014). Multifunctional aptamer-miRNA conjugates for targeted cancer therapy. *Molecular Therapy, 22*(6), 1151–1163.

Fabris, L., Ceder, Y., Chinnaiyan, A. M., Jenster, G. W., Sorensen, K. D., Tomlins, S., et al. (2016). The potential of microRNAs as prostate cancer biomarkers. *European Urology*, *70*(2), 312–322.

Fang, Y., & Gao, W. (2014). Roles of microRNAs during prostatic tumorigenesis and tumor progression. *Oncogene*, *33*(2), 135–147.

Feng, J., Huang, C., Diao, X., Fan, M., Wang, P., Xiao, Y., et al. (2013). Screening biomarkers of prostate cancer by integrating microRNA and mRNA microarrays. *Genetic Testing and Molecular Biomarkers*, *17*(11), 807–813.

Ferreira, L. B., Palumbo, A., de Mello, K. D., Sternberg, C., Caetano, M. S., de Oliveira, F. L., et al. (2012). PCA3 noncoding RNA is involved in the control of prostate-cancer cell survival and modulates androgen receptor signaling. *BMC Cancer*, *12*(1), 1–15.

Fitzmaurice, C., Allen, C., Barber, R. M., Barregard, L., Bhutta, Z. A., Brenner, H., et al. (2017). Global, regional, and national cancer incidence, mortality, years of life lost, years lived with disability, and disability-adjusted life-years for 32 cancer groups, 1990 to 2015: A systematic analysis for the global burden of disease study. *JAMA Oncology*, *3*(4), 524–548.

Fleming, W. H., Hamel, A., MacDonald, R., Ramsey, E., Pettigrew, N. M., Johnston, B., et al. (1986). Expression of the c-myc protooncogene in human prostatic carcinoma and benign prostatic hyperplasia. *Cancer Research*, *46*(3), 1535–1538.

Fujita, Y., Kojima, K., Hamada, N., Ohhashi, R., Akao, Y., Nozawa, Y., et al. (2008). Effects of miR-34a on cell growth and chemoresistance in prostate cancer PC3 cells. *Biochemical and Biophysical Research Communications*, *377*(1), 114–119.

Gagliardi, M., & Matarazzo, M. R. (2016). *RIP: RNA Immunoprecipitation. Polycomb group proteins* (pp. 73–86). Springer.

Gao, P., Tchernyshyov, I., Chang, T.-C., Lee, Y.-S., Kita, K., Ochi, T., et al. (2009). c-Myc suppression of miR-23a/b enhances mitochondrial glutaminase expression and glutamine metabolism. *Nature*, *458*(7239), 762–765.

Ge, S., Mi, Y., Zhao, X., Hu, Q., Guo, Y., Zhong, F., et al. (2020). Characterization and validation of long noncoding RNAs as new candidates in prostate cancer. *Urologic Oncology*, *20*(1), 1–15.

Gleason, D. F., & Mellinger, G. T. (1974). Prediction of prognosis for prostatic adenocarcinoma by combined histological grading and clinical staging. *The Journal of Urology*, *111*(1), 58–64.

Gong, J., Ju, Y., Shao, D., & Zhang, Q. C. (2018). Advances and challenges towards the study of RNA-RNA interactions in a transcriptome-wide scale. *Quantitative Biology*, *6*(3), 239–252.

Goto, Y., Kojima, S., Nishikawa, R., Enokida, H., Chiyomaru, T., Kinoshita, T., et al. (2014). The microRNA-23b/27b/24-1 cluster is a disease progression marker and tumor suppressor in prostate cancer. *Oncotarget*, *5*(17), 7748.

Grillone, K., Riillo, C., Scionti, F., Rocca, R., Tradigo, G., Guzzi, P. H., et al. (2020). Non-coding RNAs in cancer: Platforms and strategies for investigating the genomic "dark matter.". *Journal of Experimental & Clinical Cancer Research*, *39*(1), 1–19.

Groskopf, J., Aubin, S. M., Deras, I. L., Blase, A., Bodrug, S., Clark, C., et al. (2006). APTIMA PCA3 molecular urine test: Development of a method to aid in the diagnosis of prostate cancer. *Clinical Chemistry*, *52*(6), 1089–1095.

Gruber, A. R., Findeiß, S., Washietl, S., Hofacker, I. L., & Stadler, P. F. (2010). *RNAz 2.0: Improved noncoding RNA detection. Biocomputing 2010* (pp. 69–79). World Scientific;.

Gwak, H.-S., Kim, T. H., Jo, G. H., Kim, Y.-J., Kwak, H.-J., Kim, J. H., et al. (2012). *Silencing of microRNA-21 confers radio-sensitivity through inhibition of the PI3K/AKT pathway and enhancing autophagy in malignant glioma cell lines*.

Haflidadóttir, B. S., Larne, O., Martin, M., Persson, M., Edsjö, A., Bjartell, A., et al. (2013). Upregulation of miR-96 enhances cellular proliferation of prostate cancer cells through FOXO1. *PLoS One*, *8*(8), e72400.

Haldrup, C., Kosaka, N., Ochiya, T., Borre, M., Høyer, S., Orntoft, T. F., et al. (2014). Profiling of circulating microRNAs for prostate cancer biomarker discovery. *Drug Delivery and Translational Research*, *4*(1), 19–30.

Halimi, M., Parsian, H., Asghari, S. M., Sariri, R., Moslemi, D., & Yeganeh, F. (2013). MicroRNAs: Are they indicators for prediction of response to radiotherapy in breast cancer? *Journal of Medical Hypotheses and Ideas*, *7*(2), 59–64.

Halimi, M., Parsian, H., Asghari, S. M., Sariri, R., Moslemi, D., Yeganeh, F., et al. (2014). Clinical translation of human microRNA 21 as a potential biomarker for exposure to ionizing radiation. *Translational Research*, *163*(6), 578–584.

Hatano, K., Kumar, B., Zhang, Y., Coulter, J. B., Hedayati, M., Mears, B., et al. (2015). A functional screen identifies miRNAs that inhibit DNA repair and sensitize prostate cancer cells to ionizing radiation. *Nucleic Acids Research*, *43*(8), 4075–4086.

Heydari, N., Nikbakhsh, N., Sadeghi, F., Farnoush, N., Khafri, S., Bastami, M., et al. (2018). Overexpression of serum MicroRNA-140-3p in premenopausal women with newly diagnosed breast cancer. *Gene, 655*, 25–29.

Hua, J. T., Chen, S., & He, H. H. (2019). Landscape of noncoding RNA in prostate cancer. *Trends in Genetics, 35*(11), 840–851.

Huang, H.-Y., Lin, Y.-C.-D., Li, J., Huang, K.-Y., Shrestha, S., Hong, H.-C., et al. (2020). miRTarBase 2020: Updates to the experimentally validated microRNA–target interaction database. *Nucleic Acids Research, 48*(D1), D148–D154.

Huang, W., Su, X., Yan, W., Kong, Z., Wang, D., Huang, Y., et al. (2018). Overexpression of AR-regulated lncRNA TMPO-AS1 correlates with tumor progression and poor prognosis in prostate cancer. *The Prostate, 78*(16), 1248–1261.

Huang, X., Taeb, S., Jahangiri, S., Emmenegger, U., Tran, E., Bruce, J., et al. (2013). miRNA-95 mediates radioresistance in tumors by targeting the sphingolipid phosphatase SGPP1. *Cancer Research, 73*(23), 6972–6986.

Huang, Z., Shi, J., Gao, Y., Cui, C., Zhang, S., Li, J., et al. (2019). HMDD v3. 0: A database for experimentally supported human microRNA–disease associations. *Nucleic Acids Research, 47*(D1), D1013–D1017.

Hung, C.-L., Wang, L.-Y., Yu, Y.-L., Chen, H.-W., Srivastava, S., Petrovics, G., et al. (2014). A long noncoding RNA connects c-Myc to tumor metabolism. *Proceedings of the National Academy of Sciences, 111*(52), 18697–18702.

Ishteiwy, R. A., Ward, T. M., Dykxhoorn, D. M., & Burnstein, K. L. (2012). The microRNA-23b/-27b cluster suppresses the metastatic phenotype of castration-resistant prostate cancer cells. *PLoS One, 7*(12), e52106.

Iyer, M. K., Niknafs, Y. S., Malik, R., Singhal, U., Sahu, A., Hosono, Y., et al. (2015). The landscape of long noncoding RNAs in the human transcriptome. *Nature Genetics, 47*(3), 199–208.

Jackson, B. L., Grabowska, A., & Ratan, H. L. (2014). MicroRNA in prostate cancer: Functional importance and potential as circulating biomarkers. *BMC Cancer, 14*(1), 1–10.

Jones, D. Z., Schmidt, M. L., Suman, S., Hobbing, K. R., Barve, S. S., Gobejishvili, L., et al. (2018). Micro-RNA-186-5p inhibition attenuates proliferation, anchorage independent growth and invasion in metastatic prostate cancer cells. *BioMed Central Cancer, 18*(1), 421.

Josson, S., Sung, S. Y., Lao, K., Chung, L. W., & Johnstone, P. A. (2008). Radiation modulation of microRNA in prostate cancer cell lines. *The Prostate, 68*(15), 1599–1606.

Kachakova, D., Mitkova, A., Popov, E., Popov, I., Vlahova, A., Dikov, T., et al. (2015). Combinations of serum prostate-specific antigen and plasma expression levels of let-7c, miR-30c, miR-141, and miR-375 as potential better diagnostic biomarkers for prostate cancer. *DNA and Cell Biology, 34*(3), 189–200.

Kalvari, I., Nawrocki, E. P., Argasinska, J., Quinones-Olvera, N., Finn, R. D., Bateman, A., et al. (2018). Non-coding RNA analysis using the Rfam database. *Current Protocols in Bioinformatics, 62*(1), e51.

Kalvari, I., Nawrocki, E. P., Ontiveros-Palacios, N., Argasinska, J., Lamkiewicz, K., Marz, M., et al. (2021). Rfam 14: Expanded coverage of metagenomic, viral and microRNA families. *Nucleic Acids Research, 49*(D1), D192–D200.

Kanwal, R., Plaga, A. R., Liu, X., Shukla, G. C., & Gupta, S. (2017). MicroRNAs in prostate cancer: Functional role as biomarkers. *Cancer Letters, 407*, 9–20.

Kashat, M., Azzouz, L., Sarkar, S. H., Kong, D., Li, Y., & Sarkar, F. H. (2012). Inactivation of AR and Notch-1 signaling by miR-34a attenuates prostate cancer aggressiveness. *American Journal of Translational Research, 4*(4), 432.

Kasomva, K., Sen, A., Paulraj, M. G., Sailo, S., Raphael, V., Puro, K.-U., et al. (2018). Roles of microRNA in prostate cancer cell metabolism. *The International Journal of Biochemistry & Cell Biology, 102*, 109–116.

Kent, W. J. (2002). BLAT—the BLAST-like alignment tool. *Genome Research, 12*(4), 656–664.

Kojima, K., Fujita, Y., Nozawa, Y., Deguchi, T., & Ito, M. (2010). MiR-34a attenuates paclitaxel-resistance of hormone-refractory prostate cancer PC3 cells through direct and indirect mechanisms. *The Prostate, 70*(14), 1501–1512.

Korpela, E., Vesprini, D., & Liu, S. (2015). MicroRNA in radiotherapy: MiRage or miRador? *British Journal of Cancer, 112*(5), 777–782.

Kozomara, A., Birgaoanu, M., & Griffiths-Jones, S. (2019). miRBase: From microRNA sequences to function. *Nucleic Acids Research, 47*(D1), D155–D162.

Kumar, B., & Lupold, S. E. (2016). MicroRNA expression and function in prostate cancer: A review of current knowledge and opportunities for discovery. *Asian Journal of Andrology, 18*(4), 559.

Kuznetsov, A., & Bollin, C. J. (2021). NCBI genome workbench: Desktop software for comparative genomics, visualization, and GenBank data submission. *Multiple sequence alignment* (pp. 261–295). Springer.

Larne, O., Martens-Uzunova, E., Hagman, Z., Edsjö, A., Lippolis, G., den Berg, M. S. V. V., et al. (2013). miQ—A novel microRNA based diagnostic and prognostic tool for prostate cancer. *International Journal of Cancer, 132* (12), 2867–2875.

Lehrer, S., & Rheinstein, P. H. (2020). Co-occurrent alterations of Alzheimer's genes and prostate cancer genes in prostate cancer. *Cancer Genomics-Proteomics, 17*(3), 271–275.

Leite, K. R., Reis, S. T., Viana, N., Morais, D. R., Moura, C. M., Silva, I. A., et al. (2015). Controlling RECK miR21 promotes tumor cell invasion and is related to biochemical recurrence in prostate cancer. *Journal of Cancer, 6* (3), 292.

Leung, C.-M., Li, S.-C., Chen, T.-W., Ho, M.-R., Hu, L.-Y., Liu, W. S., et al. (2014). Comprehensive microRNA profiling of prostate cancer cells after ionizing radiation treatment. *Oncology Reports, 31*(3), 1067–1078.

Leyten, G. H., Hessels, D., Jannink, S. A., Smit, F. P., de Jong, H., Cornel, E. B., et al. (2014). Prospective multicentre evaluation of PCA3 and TMPRSS2-ERG gene fusions as diagnostic and prognostic urinary biomarkers for prostate cancer. *European Urology, 65*(3), 534–542.

Li, F., & Mahato, R. I. (2014). MicroRNAs and drug resistance in prostate cancers. *Molecular Pharmaceutics, 11*(8), 2539–2552.

Li, J., Li, Z., Nie, R., You, Z., & Bao, W. (2020). FCGCNMDA: Predicting miRNA-disease associations by applying fully connected graph convolutional networks. *Molecular Genetics and Genomics, 295*, 1197–1209.

Li, T., Li, D., Sha, J., Sun, P., & Huang, Y. (2009). MicroRNA-21 directly targets MARCKS and promotes apoptosis resistance and invasion in prostate cancer cells. *Biochemical and Biophysical Research Communications, 383*(3), 280–285.

Li, T., Li, R.-S., Li, Y.-H., Zhong, S., Chen, Y.-Y., Zhang, C.-M., et al. (2012). miR-21 as an independent biochemical recurrence predictor and potential therapeutic target for prostate cancer. *The Journal of Urology, 187*(4), 1466–1472.

Li, W., Dou, Z., We, S., Zhu, Z., Pan, D., Jia, Z., et al. (2018). Long noncoding RNA BDNF-AS is associated with clinical outcomes and has functional role in human prostate cancer. *Biomedicine & Pharmacotherapy, 102*, 1105–1110.

Li, Z., Liu, L., Jiang, S., Li, Q., Feng, C., Du, Q., et al. (2021). LncExpDB: An expression database of human long non-coding RNAs. *Nucleic Acids Research, 49*(D1), D962–D968.

Lin, H.-M., Castillo, L., Mahon, K., Chiam, K., Lee, B. Y., Nguyen, Q., et al. (2014). Circulating microRNAs are associated with docetaxel chemotherapy outcome in castration-resistant prostate cancer. *British Journal of Cancer, 110*(10), 2462–2471.

Lingadahalli, S., Jadhao, S., Sung, Y. Y., Chen, M., Hu, L., Chen, X., et al. (2018). Novel lncRNA LINC00844 regulates prostate cancer cell migration and invasion through AR signaling. *Molecular Cancer Research, 16*(12), 1865–1878.

Litwin, M. S., & Tan, H.-J. (2017). The diagnosis and treatment of prostate cancer: A review. *Journal of the American Medical Association, 317*(24), 2532–2542.

Liu, C., Kelnar, K., Liu, B., Chen, X., Calhoun-Davis, T., Li, H., et al. (2011). The microRNA miR-34a inhibits prostate cancer stem cells and metastasis by directly repressing. *Nature Medicine, 17*(2), 211–CD45.

Liu, H. P., Gao, Z. H., Cui, S. X., Wang, Y., Li, B. Y., Lou, H. X., et al. (2013). Chemoprevention of intestinal adenomatous polyposis by acetyl-11-keto-beta-boswellic acid in APCMin/ + mice. *International Journal of Cancer, 132* (11), 2667–2681.

Liu, Y., Xie, Q. R., Wang, B., Shao, J., Zhang, T., Liu, T., et al. (2013). Inhibition of SIRT6 in prostate cancer reduces cell viability and increases sensitivity to chemotherapeutics. *Protein & Cell, 4*(9), 702–710.

Liu, Y., Zhang, R., Qiu, F., Li, K., Zhou, Y., Shang, D., et al. (2015). Construction of a lncRNA–PCG bipartite network and identification of cancer-related lncRNAs: A case study in prostate cancer. *Molecular BioSystems, 11* (2), 384–393.

Lu, Y., Zhang, Z., Yu, H., Zheng, S. L., Isaacs, W. B., Xu, J., et al. (2011). Functional annotation of risk loci identified through genome-wide association studies for prostate cancer. *The Prostate, 71*(9), 955–963.

Ma, L., Cao, J., Liu, L., Du, Q., Li, Z., Zou, D., et al. (2019). LncBook: A curated knowledgebase of human long non-coding RNAs. *Nucleic Acids Research, 47*(D1), D128–D134.

Majid, S., Dar, A. A., Saini, S., Shahryari, V., Arora, S., Zaman, M. S., et al. (2013). miRNA-34b inhibits prostate cancer through demethylation, active chromatin modifications, and AKT pathways. *Clinical Cancer Research, 19* (1), 73–84.

Malik, R., Patel, L., Prensner, J. R., Shi, Y., Iyer, M. K., Subramaniyan, S., et al. (2014). The lncRNA PCAT29 inhibits oncogenic phenotypes in prostate cancer. *Molecular Cancer Research, 12*(8), 1081–1087.

Martens-Uzunova, E. S., Böttcher, R., Croce, C. M., Jenster, G., Visakorpi, T., & Calin, G. A. (2014). Long noncoding RNA in prostate, bladder, and kidney cancer. *European Urology, 65*(6), 1140–1151.

Martens-Uzunova, E. S., Olvedy, M., & Jenster, G. (2013). Beyond microRNA—novel RNAs derived from small non-coding RNA and their implication in cancer. *Cancer Letters, 340*(2), 201–211.

Martignano, F., Rossi, L., Maugeri, A., Gallà, V., Conteduca, V., De Giorgi, U., et al. (2017). Urinary RNA-based biomarkers for prostate cancer detection. *Clinica Chimica Acta, 473*, 96–105.

Matin, F., Jeet, V., Clements, J. A., Yousef, G. M., & Batra, J. (2016). MicroRNA theranostics in prostate cancer precision medicine. *Clinical Chemistry, 62*(10), 1318–1333.

Mattick, J. S., & Makunin, I. V. (2006). Non-coding RNA. *Human Molecular Genetics, 15*(Suppl. 1), R17–R29.

McCown, P. J., Wang, M. C., Jaeger, L., & Brown, J. A. (2019). Secondary structural model of human MALAT1 reveals multiple structure–function relationships. *International Journal of Molecular Sciences, 20*(22), 5610.

Mehra, R., Udager, A. M., Ahearn, T. U., Cao, X., Feng, F. Y., Loda, M., et al. (2016). Overexpression of the long non-coding RNA SChLAP1 independently predicts lethal prostate cancer. *European Urology, 70*(4), 549–552.

Melbø-Jørgensen, C., Ness, N., Andersen, S., Valkov, A., Dønnem, T., Al-Saad, S., et al. (2014). Stromal expression of MiR-21 predicts biochemical failure in prostate cancer patients with Gleason score 6. *PLoS One, 9*(11), e113039.

Meng, D., Yang, S., Wan, X., Zhang, Y., Huang, W., Zhao, P., et al. (2016). A transcriptional target of androgen receptor, miR-421 regulates proliferation and metabolism of prostate cancer cells. *The International Journal of Biochemistry & Cell Biology, 73*, 30–40.

Moore, H. M., Kelly, A. B., Jewell, S. D., McShane, L. M., Clark, D. P., Greenspan, R., et al. (2011). Biospecimen reporting for improved study quality (BRISQ). *Cancer Cytopathology, 119*(2), 92–102.

Mourtada-Maarabouni, M., Pickard, M., Hedge, V., Farzaneh, F., & Williams, G. (2009). GA.5, a non-protein-coding RNA, controls apoptosis and is downregulated in breast cancer. *Oncogene, 28*(2), 195–208.

Nadiminty, N., Tummala, R., Lou, W., Zhu, Y., Zhang, J., Chen, X., et al. (2012). MicroRNA let-7c suppresses androgen receptor expression and activity via regulation of Myc expression in prostate cancer cells. *Journal of Biological Chemistry, 287*(2), 1527–1537.

Nashtahosseini, Z., Aghamaali, M. R., Sadeghi, F., Heydari, N., & Parsian, H. (2021). Circulating Status of microRNAs 660-5p and 210-3p in Breast Cancer Patients. *The Journal of Gene Medicine*, e3320.

Nawrocki, E. P., & Eddy, S. R. (2013). Infernal 1.1: 100-fold faster RNA homology searches. *Bioinformatics, 29*(22), 2933–2935.

Nayerossadat, N., Maedeh, T., & Ali, P. A. (2012). Viral and nonviral delivery systems for gene delivery. *Advanced Biomedical Research*, 1.

Ni, J., Bucci, J., Chang, L., Malouf, D., Graham, P., & Li, Y. (2017). Targeting microRNAs in prostate cancer radiotherapy. *Theranostics, 7*(13), 3243.

Nip, H., Dar, A. A., Saini, S., Colden, M., Varahram, S., Chowdhary, H., et al. (2016). Oncogenic microRNA-4534 regulates PTEN pathway in prostate cancer. *Oncotarget, 7*(42), 68371.

O'Donnell, K. A., Wentzel, E. A., Zeller, K. I., Dang, C. V., & Mendell, J. T. (2005). c-Myc-regulated microRNAs modulate E2F1 expression. *Nature, 435*(7043), 839–843.

Ostadrahimi, S., Fayaz, S., Parvizhamidi, M., Abedi-Valugerdi, M., Hassan, M., Kadivar, M., et al. (2018). Downregulation of miR-1266-5P, miR-185-5P and miR-30c-2 in prostatic cancer tissue and cell lines. *Oncology Letters, 15*(5), 8157–8164.

Östling, P., Leivonen, S.-K., Aakula, A., Kohonen, P., Mäkelä, R., Hagman, Z., et al. (2011). Systematic analysis of microRNAs targeting the androgen receptor in prostate cancer cells. *Cancer Research, 71*(5), 1956–1967.

Panwar, B., Arora, A., & Raghava, G. P. (2014). Prediction and classification of ncRNAs using structural information. *BMC Genomics, 15*(1), 1–13.

Petrovics, G., Zhang, W., Makarem, M., Street, J. P., Connelly, R., Sun, L., et al. (2004). Elevated expression of PCGEM1, a prostate-specific gene with cell growth-promoting function, is associated with high-risk prostate cancer patients. *Oncogene, 23*(2), 605–611.

Pickard, M., Mourtada-Maarabouni, M., & Williams, G. (2013). Long non-coding RNA GAS5 regulates apoptosis in prostate cancer cell lines. *Biochimica et Biophysica Acta (BBA)-Molecular Basis of Disease, 1832*(10), 1613−1623.

Poliseno, L., Salmena, L., Riccardi, L., Fornari, A., Song, M. S., Hobbs, R. M., et al. (2010). Identification of the miR-106b ~25 microRNA cluster as a proto-oncogenic PTEN-targeting intron that cooperates with its host gene MCM7 in transformation. *Science Signaling, 3*(117), ra29-ra.

Poliseno, L., Salmena, L., Zhang, J., Carver, B., Haveman, W. J., & Pandolfi, P. P. (2010). A coding-independent function of gene and pseudogene mRNAs regulates tumour biology. *Nature, 465*(7301), 1033−1038.

Prensner, J. R., Zhao, S., Erho, N., Schipper, M., Iyer, M. K., Dhanasekaran, S. M., et al. (2014). Nomination and validation of the long noncoding RNA SChLAP1 as a risk factor for metastatic prostate cancer progression: A multi-institutional high-throughput analysis. *The Lancet Oncology, 15*(13), 1469.

Puhr, M., Hoefer, J., Schäfer, G., Erb, H. H., Oh, S. J., Klocker, H., et al. (2012). Epithelial-to-mesenchymal transition leads to docetaxel resistance in prostate cancer and is mediated by reduced expression of miR-200c and miR-205. *The American Journal of Pathology, 181*(6), 2188−2201.

Qu, Y., Huang, X., Li, Z., Liu, J., Wu, J., Chen, D., et al. (2014). miR-199a-3p inhibits aurora kinase A and attenuates prostate cancer growth: New avenue for prostate cancer treatment. *The American Journal of Pathology, 184* (5), 1541−1549.

Ragin, C., & Park, J. Y. (2014). Biospecimens, biobanking and global cancer research collaborations. *Ecancermedicalscience, 8*.

Razdan, A., de Souza, P., & Roberts, T. L. (2018). Role of MicroRNAs in treatment response in prostate cancer. *Current Cancer Drug Targets, 18*(10), 929−944.

Ren, S., Liu, Y., Xu, W., Sun, Y., Lu, J., Wang, F., et al. (2013). Long noncoding RNA MALAT-1 is a new potential therapeutic target for castration resistant prostate cancer. *The Journal of Urology, 190*(6), 2278−2287.

Ren, S., Peng, Z., Mao, J.-H., Yu, Y., Yin, C., Gao, X., et al. (2012). RNA-seq analysis of prostate cancer in the Chinese population identifies recurrent gene fusions, cancer-associated long noncoding RNAs and aberrant alternative splicings. *Cell Research, 22*(5), 806−821.

Rezatabar, S., Moudi, E., Sadeghi, F., Khafri, S., Kopi, T. A., & Parsian, H. (2020). Evaluation of the plasma level of long non-coding RNA PCAT1 in prostatic hyperplasia and newly diagnosed prostate cancer patients. *The Journal of Gene Medicine, 22*(10), e3239.

Ribas, J., Ni, X., Haffner, M., Wentzel, E. A., Salmasi, A. H., Chowdhury, W. H., et al. (2009). miR-21: An androgen receptor−regulated microRNA that promotes hormone-dependent and hormone-independent prostate cancer growth. *Cancer Research, 69*(18), 7165−7169.

RNAcentral Consortium. (2017). RNAcentral: A comprehensive database of non-coding RNA sequences. *Nucleic Acids Research, 45*(D1), D128−D134.

Romanuik, T. L., Wang, G., Morozova, O., Delaney, A., Marra, M. A., & Sadar, M. D. (2010). LNCaP Atlas: Gene expression associated with in vivo progression to castration-recurrent prostate cancer. *BMC Medical Genomics, 3*(1), 1−19.

Sachdeva, M., Zhu, S., Wu, F., Wu, H., Walia, V., Kumar, S., et al. (2009). p53 represses c-Myc through induction of the tumor suppressor miR-145. *Proceedings of the National Academy of Sciences, 106*(9), 3207−3212.

Sartori, D. A., & Chan, D. W. (2014). Biomarkers in prostate cancer: What's new? *Current Opinion in Oncology, 26* (3), 259.

Schaefer, A., Jung, M., Mollenkopf, H. J., Wagner, I., Stephan, C., Jentzmik, F., et al. (2010). Diagnostic and prognostic implications of microRNA profiling in prostate carcinoma. *International Journal of Cancer, 126*(5), 1166−1176.

Selcuklu, S. D., Donoghue, M. T., & Spillane, C. (2009). miR-21 as a key regulator of oncogenic processes. *Biochemical Society Transactions, 37*(4), 918−925.

Shao, N., Ma, G., Zhang, J., & Zhu, W. (2018). miR-221-5p enhances cell proliferation and metastasis through post-transcriptional regulation of SOCS1 in human prostate cancer. *BMC Urology, 18*(1), 1−9.

Sharma, N., & Baruah, M. M. (2019). The microRNA signatures: Aberrantly expressed miRNAs in prostate cancer. *Clinical and Translational Oncology, 21*(2), 126−144.

Shen, J., Hruby, G. W., McKiernan, J. M., Gurvich, I., Lipsky, M. J., Benson, M. C., et al. (2012). Dysregulation of circulating microRNAs and prediction of aggressive prostate cancer. *Prostate, 72*(13), 1469−1477.

Shi, R., Xiao, H., Yang, T., Chang, L., Tian, Y., Wu, B., et al. (2014). Effects of miR-200c on the migration and invasion abilities of human prostate cancer Du145 cells and the corresponding mechanism. *Frontiers of Medicine, 8* (4), 456−463.

Shi, X. B., Ma, A. H., Xue, L., Li, M., Nguyen, H. G., Yang, J. C., et al. (2015). miR-124 and androgen receptor signaling inhibitors repress prostate cancer growth by downregulating androgen receptor splice variants, EZH2, and Src. *Cancer Research, 75*(24), 5309−5317.

Shorning, B. Y., Dass, M. S., Smalley, M. J., & Pearson, H. B. (2020). The PI3K-AKT-mTOR pathway and prostate cancer: At the crossroads of AR, MAPK, and WNT signaling. *International Journal of Molecular Sciences, 21*(12), 4507.

Singh, J. P., Dagar, M., Dagar, G., Kumar, S., Rawal, S., Sharma, R. D., et al. (2020). Activation of GPR56, a novel adhesion GPCR, is necessary for nuclear androgen receptor signaling in prostate cells. *PLoS One, 15*(9), e0226056.

Singh, S., Chitkara, D., Mehrazin, R., Behrman, S. W., Wake, R. W., & Mahato, R. I. (2012). Chemoresistance in prostate cancer cells is regulated by miRNAs and Hedgehog pathway. *PLoS One, 7*(6), e40021.

Song, E. (2016). *The long and short non-coding RNAs in cancer biology:*. Springer.

Srigley, J. R., Humphrey, P. A., Amin, M. B., Chang, S. S., Egevad, L., Epstein, J. I., et al. (2012). Protocol for the examination of specimens from patients with carcinoma of the prostate gland. *Prostate, 3*, 0.

Streit, S., Michalski, C. W., Erkan, M., Kleeff, J., & Friess, H. (2009). Northern blot analysis for detection and quantification of RNA in pancreatic cancer cells and tissues. *Nature Protocols, 4*(1), 37.

Stuopelytė, K., Daniūnaitė, K., Jankevičius, F., & Jarmalaitė, S. (2016). Detection of miRNAs in urine of prostate cancer patients. *Medicina, 52*(2), 116−124.

Subramanian, M., Li, X. L., Hara, T., & Lal, A. (2015). A biochemical approach to identify direct microRNA targets. *Regulatory non-coding RNAs:* (pp. 29−37). Springer.

Sun, T., Wang, Q., Balk, S., Brown, M., Lee, G.-S. M., & Kantoff, P. (2009). The role of microRNA-221 and microRNA-222 in androgen-independent prostate cancer cell lines. *Cancer Research, 69*(8), 3356−3363.

Sun, X., Li, Y., Yu, J., Pei, H., Luo, P., & Zhang, J. (2015). miR-128 modulates chemosensitivity and invasion of prostate cancer cells through targeting ZEB1. *Japanese Journal of Clinical Oncology, 45*(5), 474−482.

Sun, Y., Liu, J., Chu, L., Yang, W., Liu, H., Li, C., et al. (2018). Long noncoding RNA SNHG12 facilitates the tumorigenesis of glioma through miR-101-3p/FOXP1 axis. *Gene, 676*, 315−321.

Sur, S., Steele, R., Shi, X., & Ray, R. B. (2019). miRNA-29b inhibits prostate tumor growth and induces apoptosis by increasing bim expression. *Cells, 8*(11), 1455.

Takayama, K., Horie-Inoue, K., Katayama, S., Suzuki, T., Tsutsumi, S., Ikeda, K., et al. (2013). Androgen-responsive long noncoding RNA CTBP1-AS promotes prostate cancer. *The EMBO Journal, 32*(12), 1665−1680.

Takayama, K.-I., Fujimura, T., Suzuki, Y., & Inoue, S. (2020). Identification of long non-coding RNAs in advanced prostate cancer associated with androgen receptor splicing factors. *Communications Biology, 3*(1), 1−14.

Takeshita, F., Patrawala, L., Osaki, M., Takahashi, R.-U., Yamamoto, Y., Kosaka, N., et al. (2010). Systemic delivery of synthetic microRNA-16 inhibits the growth of metastatic prostate tumors via downregulation of multiple cell-cycle genes. *Molecular Therapy, 18*(1), 181−187.

Tang, X., Tang, X., Gal, J., Kyprianou, N., Zhu, H., & Tang, G. (2011). Detection of microRNAs in prostate cancer cells by microRNA array. *MicroRNAs in development* (pp. 69−88). Springer.

Tao, F., Tian, X., & Zhang, Z. (2018). The PCAT3/PCAT9-miR-203-SNAI2 axis functions as a key mediator for prostate tumor growth and progression. *Oncotarget, 9*(15), 12212.

Tay, Y., Kats, L., Salmena, L., Weiss, D., Tan, S. M., Ala, U., et al. (2011). Coding-independent regulation of the tumor suppressor PTEN by competing endogenous mRNAs. *Cell, 147*(2), 344−357.

Tian, X., Zhang, G., Zhao, H., Li, Y., & Zhu, C. (2017). Long non-coding RNA NEAT1 contributes to docetaxel resistance of prostate cancer through inducing RET expression by sponging miR-34a. *RSC Advances, 7*(68), 42986−42996.

Tong, A., Fulgham, P., Jay, C., Chen, P., Khalil, I., Liu, S., et al. (2009). MicroRNA profile analysis of human prostate cancers. *Cancer Gene Therapy, 16*(3), 206−216.

Tong, Y., Ru, B., & Zhang, J. (2018). miRNACancerMAP: An integrative web server inferring miRNA regulation network for cancer. *Bioinformatics, 34*(18), 3211−3213.

Urabe, F., Matsuzaki, J., Yamamoto, Y., Kimura, T., Hara, T., Ichikawa, M., et al. (2019). Large-scale circulating microRNA profiling for the liquid biopsy of prostate cancer. *Clinical Cancer Research, 25*(10), 3016−3025.

Velculescu, V. E., Zhang, L., Vogelstein, B., & Kinzler, K. W. J. S. (1995). Serial analysis of gene expression. *Science, 270*(5235), 484−487.

Viticchiè, G., Lena, A. M., Latina, A., Formosa, A., Gregersen, L. H., Lund, A. H., et al. (2011). MiR-203 controls proliferation, migration and invasive potential of prostate cancer cell lines. *Cell Cycle, 10*(7), 1121−1131.

Volders, P.-J., Anckaert, J., Verheggen, K., Nuytens, J., Martens, L., Mestdagh, P., et al. (2019). LNCipedia 5: Towards a reference set of human long non-coding RNAs. *Nucleic Acids Research, 47*(D1), D135−D139.

Wang, D., Gu, J., Wang, T., & Ding, Z. (2014). OncomiRDB: A database for the experimentally verified oncogenic and tumor-suppressive microRNAs. *Bioinformatics, 30*(15), 2237−2238.

Wang, F., Ren, S., Chen, R., Lu, J., Shi, X., Zhu, Y., et al. (2014). Development and prospective multicenter evaluation of the long noncoding RNA MALAT-1 as a diagnostic urinary biomarker for prostate cancer. *Oncotarget, 5*(22), 11091.

Wang, H., Jiang, Y., Peng, H., Chen, Y., Zhu, P., & Huang, Y. (2015). Recent progress in microRNA delivery for cancer therapy by non-viral synthetic vectors. *Advanced Drug Delivery Reviews, 81*, 142−160.

Wang, L., Han, S., Jin, G., Zhou, X., Li, M., Ying, X., et al. (2014). Linc00963: A novel, long non-coding RNA involved in the transition of prostate cancer from androgen-dependence to androgen-independence. *International Journal of Oncology, 44*(6), 2041−2049.

Wang, R., Sun, Y., Li, L., Niu, Y., Lin, W., Lin, C., et al. (2017). Preclinical study using Malat1 small interfering RNA or androgen receptor splicing variant 7 degradation enhancer ASC-J9® to suppress enzalutamide-resistant prostate cancer progression. *European Urology, 72*(5), 835−844.

Wang, S., Cao, Y., Tang, X., Yang, Y., and Du, P. (2020). Identification of androgen receptor variant 7-related RNAs affecting abiraterone efficacy in castration-resistant prostate cancer treatment by RNA-sequencing.

Wang, S., Su, W., Zhong, C., Yang, T., Chen, W., Chen, G., et al. (2020). An eight-circRNA assessment model for predicting biochemical recurrence in prostate cancer. *Frontiers in Cell and Developmental Biology, 8*.

Wang, X.-J., Reyes, J. L., Chua, N.-H., & Gaasterland, T. (2004). Prediction and identification of *Arabidopsis thaliana* microRNAs and their mRNA targets. *Genome Biology, 5*(9), 1−15.

Watahiki, A., Macfarlane, R. J., Gleave, M. E., Crea, F., Wang, Y., Helgason, C. D., et al. (2013). Plasma miRNAs as biomarkers to identify patients with castration-resistant metastatic prostate cancer. *International Journal of Molecular Sciences, 14*(4), 7757−7770.

Wei, J. T., Feng, Z., Partin, A. W., Brown, E., Thompson, I., Sokoll, L., et al. (2014). Can urinary PCA3 supplement PSA in the early detection of prostate cancer? *Journal of Clinical Oncology, 32*(36), 4066.

Weiss, M., Plass, C., & Gerhauser, C. (2014). Role of lncRNAs in prostate cancer development and progression. *Biological Chemistry, 395*(11), 1275−1290.

Williams, L. V., Veliceasa, D., Vinokour, E., & Volpert, O. V. (2013). miR-200b inhibits prostate cancer EMT, growth and metastasis. *PLoS One, 8*(12), e83991.

Wu, M., Huang, Y., Chen, T., Wang, W., Yang, S., Ye, Z., et al. (2019). LncRNA MEG3 inhibits the progression of prostate cancer by modulating miR-9-5p/QKI-5 axis. *Journal of Cellular and Molecular Medicine, 23*(1), 29−38.

Wu, Y.-P., Lin, X.-D., Chen, S.-H., Ke, Z.-B., Lin, F., Chen, D.-N., et al. (2020). Identification of prostate cancer-related circular RNA through bioinformatics analysis. *Frontiers in Genetics, 11*.

Wu, Y.-R., Qi, H.-J., Deng, D.-F., Luo, Y.-Y., & Yang, S.-L. (2016). MicroRNA-21 promotes cell proliferation, migration, and resistance to apoptosis through PTEN/PI3K/AKT signaling pathway in esophageal cancer. *Tumor Biology, 37*(9), 12061−12070.

Xiao, J., Gong, A. Y., Eischeid, A. N., Chen, D., Deng, C., Young, C. Y., et al. (2012). miR-141 modulates androgen receptor transcriptional activity in human prostate cancer cells through targeting the small heterodimer partner protein. *The Prostate, 72*(14), 1514−1522.

Xie, F., Liu, S., Wang, J., Xuan, J., Zhang, X., Qu, L., et al. (2021). deepBase v3. 0: Expression atlas and interactive analysis of ncRNAs from thousands of deep-sequencing data. *Nucleic Acids Research, 49*(D1), D877−D883.

Xu, B., Niu, X., Zhang, X., Tao, J., Wu, D., Wang, Z., et al. (2011). miR-143 decreases prostate cancer cells proliferation and migration and enhances their sensitivity to docetaxel through suppression of KRAS. *Molecular and Cellular Biochemistry, 350*(1), 207−213.

Xu, G., Wu, J., Zhou, L., Chen, B., Sun, Z., Zhao, F., et al. (2010). Characterization of the small RNA transcriptomes of androgen dependent and independent prostate cancer cell line by deep sequencing. *PLoS One, 5*(11), e15519.

Xu, Y.-H., Deng, J.-L., Wang, G., & Zhu, Y.-S. (2019). Long non-coding RNAs in prostate cancer: Functional roles and clinical implications. *Cancer Letters, 464*, 37−55.

Xue, D., Lu, H., Xu, H. Y., Zhou, C. X., & He, X. Z. (2018). Long noncoding RNA MALAT 1 enhances the docetaxel resistance of prostate cancer cells via miR-145-5p-mediated regulation of AKAP 12. *Journal of Cellular and Molecular Medicine, 22*(6), 3223–3237.

Yacqub-Usman, K., Pickard, M. R., & Williams, G. T. (2015). Reciprocal regulation of GAS5 lncRNA levels and mTOR inhibitor action in prostate cancer cells. *The Prostate, 75*(7), 693–705.

Yamamura, S., Saini, S., Majid, S., Hirata, H., Ueno, K., Deng, G., et al. (2012). MicroRNA-34a modulates c-Myc transcriptional complexes to suppress malignancy in human prostate cancer cells. *PLoS One, 7*(1), e29722.

Yang, N. (2015). An overview of viral and nonviral delivery systems for microRNA. *International Journal of Pharmaceutical Investigation, 5*(4), 179.

Yang, Y., Guo, J.-X., & Shao, Z.-Q. (2017). miR-21 targets and inhibits tumor suppressor gene PTEN to promote prostate cancer cell proliferation and invasion: An experimental study. *Asian Pacific Journal of Tropical Medicine, 10*(1), 87–91.

Ye, L., Li, S., Ye, D., Yang, D., Yue, F., Guo, Y., et al. (2013). Livin expression may be regulated by miR-198 in human prostate cancer cell lines. *European Journal of Cancer, 49*(3), 734–740.

Zhang, H. L., Yang, L. F., Zhu, Y., Yao, X. D., Zhang, S. L., Dai, B., et al. (2011). Serum miRNA-21: Elevated levels in patients with metastatic hormone-refractory prostate cancer and potential predictive factor for the efficacy of docetaxel-based chemotherapy. *The Prostate, 71*(3), 326–331.

Zhang, S., Yu, J., Sun, B.-F., Hou, G.-Z., Yu, Z.-J., & Luo, H. (2020). MicroRNA-92a Targets SERTAD3 and regulates the growth, invasion, and migration of prostate cancer cells via the p53 pathway. *OncoTargets and therapy, 13*, 5495.

Zhang, Y., Huang, H., Zhang, D., Qiu, J., Yang, J., Wang, K., et al. (2017). A review on recent computational methods for predicting noncoding RNAs. *BioMed Research International, 2017*.

Zhang, Y., Pitchiaya, S., Cieslik, M., Niknafs, Y. S., Tien, J. C., Hosono, Y., et al. (2018). Analysis of the androgen receptor-regulated lncRNA landscape identifies a role for ARLNC1 in prostate cancer progression. *Nature Genetics, 50*(6), 814–824.

Zhao, H., Kuang, L., Feng, X., Zou, Q., & Wang, L. (2019). A novel approach based on a weighted interactive network to predict associations of MiRNAs and diseases. *International Journal of Molecular Sciences, 20*(1), 110.

Zhao, R., Sun, F., Bei, X., Wang, X., Zhu, Y., Jiang, C., et al. (2017). Upregulation of the long non-coding RNA FALEC promotes proliferation and migration of prostate cancer cell lines and predicts prognosis of PCa patients. *The Prostate, 77*(10), 1107–1117.

Zheng, X.-M., Zhang, P., Liu, M.-H., Chen, P., & Zhang, W.-B. (2019). MicroRNA-30e inhibits adhesion, migration, invasion and cell cycle progression of prostate cancer cells via inhibition of the activation of the MAPK signaling pathway by downregulating CHRM3. *International Journal of Oncology, 54*(2), 443–454.

Zuker, M. (2003). Mfold web server for nucleic acid folding and hybridization prediction. *Nucleic Acids Research, 31*(13), 3406–3415.

Noncoding RNAs in liver cancer patients

Julie Sanceau and Angélique Gougelet
Centre de Recherche des Cordeliers, Sorbonne Université, Inserm, Université de Paris, Team "Oncogenic Functions of Beta-Catenin Signaling in the Liver", Paris, France

Abbreviations

AAV	adeno-associated virus
ABCA1	ATP-binding cassette A1
AFP	α-fetoprotein
AMPKα	AMP-activated protein kinase
AMO	anti-miRNA oligonucleotide
ASO	antisense oligonucleotide
ATB	Activated by TGF-β
AXIN2	axis inhibition protein 2
Bmf	Bcl2 modifying factor
CAGE	cap analysis gene expression
CHC	chronic HCV with cirrhosis
circRNA	circular RNA
COX2	cyclooxygenase 2
CPT-1	Carnitine palmitoyltransferase 1
DANCR	differentiation antagonizing nonprotein-coding RNA
DDIT4	DNA damage inducible transcript 4
DEANR1	definitive endoderm-associated lncRNA1
DIGIT	divergent to goosecoid induced by TGF-β family signaling
EGFR	epidermal growth factor receptor
EMT	epithelial-to-mesenchymal transition
EPCAM	epithelial cell adhesion molecule
EZH2	enhancer of zeste homolog 2
FAS	Fatty acid synthetase
FISH	fluorescent in situ hybridization
GSC	goosecoid
HB	hepatoblastoma
HBV	hepatitis B virus

HCC	hepatocellular carcinoma
HCV	hepatitis C virus
HEIH	highly expressed in HCC
HEXIM2	Hexamethylene Bis-Acetamide Inducible 2
HOTAIR	homeobox transcript antisense RNA
HSC	hepatic stellate cells
HULC	highly upregulated in liver cancer
iCCA	intrahepatic cholangiocarcinoma
IL	interleukin
KC	Kupffer cell
lincRNA	long intergenic noncoding RNA
LNA	locked nucleic acid
lncRNA	long noncoding RNA
LSEC	liver cancer stem cells
MAFLD	metabolic-associated fatty Liver disease
MALAT-1	metastasis-associated lung adenocarcinoma transcript
Mcl-1	myeloid leukemia cell differentiation protein 1
MEG3	maternally expressed 3
miRNA	microRNA
MVB	multivesicular bodies
NAFLD	Non-Alcoholic Fatty Liver Disease
NASH	Non-alcoholic steatohepatitis
ncRNA	noncoding RNA
NGS	next-generation sequencing
NLK	Nemo-like kinase
NO	nitric oxide
OS	overall survival
P-bodies	processing bodies
PD-L1	programmed cell death ligand 1
PDX	patient-derived xenograft
piRNA	Piwi-interacting RNA
PGE2	prostaglandin E2
PRC2	polycomb repressive complex 2
pre-miRNA	microRNA precursor
pri-miRNA	primary transcript of miRNA
PTEN	phosphatase and tensin homolog
PTP1B	Polypyrimidine tract-binding protein 1
PVT1	plasmacytoma variant translocation 1
RFS	recurrence-free survival
RISC	RNA-induced silencing complex
ROCK2	Rho associated coiled-coil containing protein kinase 2
qRT-PCR	quantitative reverse transcription PCR
RXRα	retinoid X receptor
SAGE	Serial analysis of gene expression
SAMMSON	Survival associated mitochondrial melanoma specific oncogenic noncoding RNA
siRNA	small-interfering RNA
SCD-1	Stearoyl-coA desaturase 1
snRNA	small nuclear RNA
snoRNA	small nucleolar RNA
SREBP-1c	steroid response−binding protein 1c
SRSF1	serine and arginine rich splicing factor 1
TGF-β	transforming growth factor-β
TIMP3	Tissue inhibitor of metalloproteinase-3

TNF-α	tumor necrosis factor α
TRBP	transactivation response element RNA-binding protein
TUG-1	taurine upregulated gene 1
UCA1	urothelial cancer—associated 1
UTR	untranslated transcribed region

11.1 Introduction

11.1.1 Liver functions

By its central position and the vital functions that it serves, the liver plays essential roles in metabolisms and detoxification to maintain the systemic homeostasis. The liver is a highly differentiated organ divided into lobules organized in hepatocyte plates surrounded by sinusoids, which drain blood from the portal vein to the centrilobular vein. Liver functions are ensured by close cooperations between cells dispersed in specialized zones, including hepatocytes, the most prevalent cells, cholangiocytes, sinusoidal endothelial cells, and immune cells such as Kupffer cells (KC), the liver macrophages. The hepatocyte population represents 60% of liver cells and assumes the main metabolic functions accordingly to their localization around the portal vein or around the centrilobular vein, a phenomenon defined as the liver metabolic zonation (Jungermann & Kietzmann, 1996). Hepatocyte functions are regulated by surrounding nonparenchymal cells which also play a crucial protective role against liver injuries induced by gut-derived antigens. KC constitute 80% of the whole-body macrophages, and around 30% of the nonparenchymal liver cells (Jenne & Kubes, 2013). They reside within the sinusoids, in close proximity with innate immune cells such as natural killer, natural killer-T cells, and dendritic cells. They favor a tolerogenic response to these incoming immunoreactive products. Due to constant exposure to pathogens and their derivative materials, the peculiar immune system spread in the liver, and particularly KCs, orchestrate liver tolerance by the secretion of several factors creating an inflammatory equilibrium (Crispe, 2009). They can not only produce anti-inflammatory as well as immunosuppressive factors, such as interleukin (IL)-10, nitric oxide (NO), transforming growth factor-β (TGF-β), or the arachidonic acid metabolite prostaglandin E2, but also proinflammatory cytokines, especially TNF-α (tumor necrosis factor α), IL-1β and IL-6 (Horst, Neumann, Diehl, & Tiegs, 2016).

However, in the case of repeated deleterious insults, the resulting metabolic dysfunctions and KC function dysregulations contribute to chronic inflammation and could favor the emergence of chronic liver diseases. Liver diseases range from simple steatosis, also known as nonalcoholic fatty liver disease (NAFLD), recently renamed metabolic-associated fatty liver disease (MAFLD) to critical steatohepatitis [nonalcoholic steatohepatitis (NASH/MASH)] characterized by steatosis together with inflammation, fibrosis, and cycles of hepatocyte necrosis/proliferation. Liver disease could finally evolve to cirrhosis and to cancer — cancer development rarely occurring in a healthy liver (around 10% of cases). Progression from NAFLD to NASH results from several disorders such as gut dysbiosis, lipotoxicity causing oxidative and endoplasmic reticulum stress, and aberrant secretion of different patterns of cytokines and adipokines (TNF-α, IL-6, leptin, resistin, etc.) (Tilg & Moschen, 2010). Recent advances have shown that injured hepatocytes transmit proinflammatory signals to KCs, which release proinflammatory and profibrogenic cytokines to activate hepatic stellate cells (HSC) toward fibrotic phenotype. In this context, cancer development is facilitated by proinflammatory molecules such as IL-1, IL-6, TNF-α, lymphotoxin-β together with immune escape and release of reactive oxygen species promoting DNA damages (Sircana, Paschetta, Saba, Molinaro, & Musso, 2019).

11.1.2 Liver cancers

11.1.2.1 Liver cancer in adults

Primary liver cancer is the sixth most common cancer in the world and the fourth leading cause of cancer mortality worldwide, accounting for more than 800,000 newly diagnosed people and more than 700,000 deaths each year (Bray et al., 2018). Hepatocellular carcinoma (HCC), arising from the malignant transformation of hepatocytes, is the most common primary liver cancer, which constitutes 75% of primary tumors. The second most frequent primary cancer is intrahepatic cholangiocarcinoma (iCCA), developing from the small intrahepatic bile duct epithelium, which comprises around 15% of primary liver malignancies. These two types of tumors share risk factors, including chronic viral hepatitis B and C (HBV and HCV, respectively), toxin exposure, alcohol, cirrhosis, and NASH. In addition, HCC and iCCA are estimated from two to three times more common in men than in women, partly due to gender discrepancy in risk factors. For both tumors, there is also a geographic variability in incidence, with a higher incidence in East Asia than in western countries (Banales et al., 2020; Llovet et al., 2016). Both malignancies also sadly share poor prognosis features with an overall survival (OS) around 15% at 5 years.

Important efforts using integrative genomics approaches have been deployed to stratify primary liver tumors in adult. Regarding the numbers of mutations, HCC and CCA tumors display a median amount in the mutational spectrum of cancers with more than 30 genomic alterations in CCA (Nakamura et al., 2015) and more than 40 in HCC (Vogelstein et al., 2013) (from 40 to 80 mutations according to studies). Common cell signaling pathways are mutationally affected in primary liver tumors, numbers of them being frequently mutated in cancer such as DNA damage and genomic instability (*TP53, CDKN2A, CCND1, ATM*), Wnt/β-catenin pathway (*APC*), kinase signaling (*ERBB2, BRAF, PIK3CA, PTEN, FGFR*), and immune response (*JAK–STAT3*). Frequent mutations in epigenetic regulators and chromatin remodelers are also encountered, including NADPH metabolism (*IDH1* and *IDH2*), SWI–SNF (SWItch/Sucrose Non-Fermentable) complex (*ARID1A, ARID2, SMARCA2*), and histone modifications (*MLL2, MML3, KMT2C*) (Jusakul et al., 2017; Schulze et al., 2015). The integration of these transcriptomic data together with pathological analyses such as proliferation or differentiation status, immune infiltrates, allows clinicians to classify liver tumors in different subgroups defined by their molecular alterations that are finally tightly associated with their phenotypes.

HCCs are classified in two large groups and six subclasses (G1–G6): one proliferative subclass associated with mutations in *TP53* or *AXIN1* and high chromosomal instability, displaying steatosis and undifferentiated features (G1–G3), while the second one contains tumors with activating mutations in *CTNNB1*, encoding β-catenin with higher chromosome stability, lower proliferation rate, and with phenotypical traits of differentiated hepatocytes and frequent cholestasis (G5–G6) (Boyault et al., 2007; Calderaro et al., 2017). More recently, this classification has been enriched with immune aspects. A quarter of

HCC exhibits markers of inflammatory response and thus considered the immune class, while other HCCs belong to an exhausted immune response group with poorly infiltrated tumors in particular for T lymphocytes and resistant to immunotherapies (Sia et al., 2017). This latter group is enriched in tumors with *CTNNB1* activation (Harding et al., 2019).

For iCCA, there is no consensus for an international stratification regarding their origins, histological features, or molecular alterations. iCCA tumors are heterogeneous dependently on their localization and the associated risk factors, which render its clinical management highly challenging. However, recent reports have aimed at connecting morphological and molecular characteristics to stratify iCCAs. Two groups of iCCAs either proliferative or inflammatory with various mutations and variations in copy numbers were identified. The first proliferative group includes less differentiated tumors associated with poor survival and high recurrence, while the second inflammatory group contains differentiated tumors with good survival and low recurrence (Sia et al., 2013). Another multianalysis compiling genetic and epigenetic data defined four groups of iCCAs of different prognosis according to genomic alterations, DNA methylation status, and modulated signaling pathways, either oncogenic (Wnt, growth factors) or immune pathways such as PD-L1 (programmed cell death ligand 1) (Jusakul et al., 2017). Last year, PD-L1 positivity has been confirmed as a factor of poor clinical outcome in iCCA, in favor of clinical tests with immunotherapies (Kriegsmann et al., 2019).

11.1.2.2 Liver cancer in children

Primary liver cancer is also encountered in children. Hepatoblastoma (HB), the most common primary liver tumor in children, accounts for around 1% of pediatric cancers. Inversely to liver cancer in adult, OS of HB patients is relatively high, around 75%, since the majority of patients respond to neoadjuvant chemotherapy and subsequent resection. Nevertheless, new optimized therapies are needed for relapsed patients or those with metastases, even if the survival of children with metastatic HB has been largely improved these three last decades thanks to international efforts deployed to standardize tumor stratification and therapeutic options. The origin of HB remains quite misunderstood but this embryonal tumor exhibits histologic heterogeneity ranging from epithelial tumors emerging from fetal or embryonic hepatocytes to mixed epithelial/mesenchymal tumors. HB exhibits a limited number of mutations (less than 4 per tumor). Aberrant activation of Wnt/β-catenin pathway is a key driver event in HB following mutations in *CTNNB1* (Wei et al., 2000). By whole genome sequencing, mutations in other components of this pathway such as *APC* or *AXIN* have also been identified, as well as members of antioxidant response (*NFE2L2*) or epigenetic remodelers (*MLL2*, *ARID1A*, or *INI1* recently shown as absent in the aggressive rhabdoid tumors) (Calvisi & Solinas, 2020; Eichenmüller et al., 2014). Last year, a study of Carrillo-Reixach et al. (2020) integrates genomic and transcriptomic data with epigenomic analyses and defined two epigenomics clusters regarding the level of CpG methylation and defined the DLK1/DIO3 locus from 14q32 region as an interesting biomarker. Nevertheless, even if molecular and immunohistochemical characterization of HB have pointed out new biomarkers, they are mostly not used in clinical practice and neither therapeutically targetable.

In sum, despite the numerous efforts conducted to improve the comprehension of liver cancer pathogenesis, it appears crucial to identify new molecular actors, which could be useful for its diagnosis and its treatment. In consideration with the pleiotropic roles of noncoding RNAs (ncRNAs) in physiopathology and their deregulations in several cancers, in tumor cells as well as in their microenvironment as a way of material exchange, these tiny regulators appear as promising candidates. In this chapter, we will review the main data concerning the impact of microRNAs (miRNAs) and long noncoding RNAs (lncRNAs) in primary liver cancers with a major focus on HCC regarding their benefit as diagnosis, prognosis, and therapeutic tools.

11.2 NcRNA roles in liver development and functions

NcRNAs are arbitrarily divided into two categories, small [<200 nucleotides (nt)] and lncRNAs (>200 nt) (Kapranov et al., 2007). Small ncRNAs mainly include (1) piwi-interacting RNAs, mainly identified in germ cells to silence genetic elements such as retrotransposons (Ozata, Gainetdinov, Zoch, O'Carroll, & Zamore, 2019), (2) small nucleolar RNAs (snoRNAs) that guide chemical modifications of other RNAs (Dupuis-Sandoval, Poirier, & Scott, 2015), (3) small nuclear RNAs (snRNAs), found within the spliceosome (Matera & Wang, 2014), and (4) the most widely characterized miRNAs, involved in post-transcriptional gene silencing (He & Hannon, 2004). lncRNAs are generally 1000–10,000 residues in length and refer to natural antisense transcripts, other regulatory lncRNAs, and circular RNAs produced from thousands of genes with covalent linked ends (Mercer, Dinger, & Mattick, 2009).

11.2.1 MicroRNAs

Since their discovery at the end of the 1990s, miRNAs are the best characterized small ncRNAs mainly ranging from 21 to 25 nucleotides in size. Primary transcripts of miRNAs are transcribed by the RNA polymerase II, as polycistronic transcripts in about 15% of cases (Fig. 11.1). They are processed by Drosha/Pasha to generate miRNA precursors and then by Dicer/transactivation response element RNA-binding protein to become mature. After integration in RNA-induced silencing complex containing Argonaute protein, their matching with complementary sequences mainly located in the 3'UTR (untranslated transcribed region) of messenger RNAs (mRNAs) results in most cases in translation inhibition and accelerated mRNA degradation in subcellular compartments called processing bodies (Hutvágner & Zamore, 2002; Lee, Jeon, Lee, Kim, & Kim, 2002).

MiRNA expression levels are tissue-specific. They finely regulate precise patterns of genes during development, including in the liver. Their central roles in liver functions have been supported by works performed on conditional knockout mice depleted from Dicer or Drosha, their constitutive depletion being embryonic lethal. In particular, Dicer deletion results in a loss of liver metabolic zonation and favors tumor emergence (Sekine, Ogawa, Ito, et al., 2009; Sekine, Ogawa, Mcmanus, Kanai, & Hebrok, 2009). Some miRNAs are highly expressed in embryonic livers such as miR-92, miR-106b, miR-124, miR-125a-5p, miR-125b, miR-140-3p, miR-205, and miR-206 (Tzur et al., 2009), while the adult liver exhibits high expression of miRNAs required for their metabolic and detoxification functions

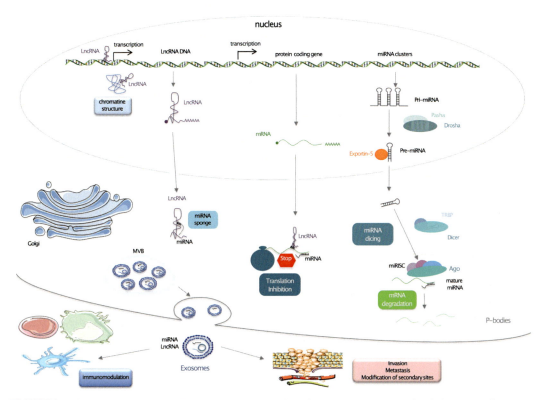

FIGURE 11.1 MiRNA and lncRNA syntheses and modes of action. NcRNAs are divided into small ncRNAs (<200 nt), including miRNAs, and long noncoding RNAs (>200 nt). Pre-miRNAs processed from pri-miRNAs by Drosha and Pasha complexes are translocated into the cytoplasm by exportin-5, diced by Dicer/TRBP complexes and then integrated in RISC complex to impair translation and accelerate mRNA degradation. LncRNAs not only control transcription, either positively or negatively, but also mRNA stability or translation and chromatin architecture. They also could act as miRNA sponges. Disequilibrium in their expression could favor liver disease and promote liver tumorigenesis. NcRNAs could also be secreted from tumor and immune cells into exosomes to modulate cell fate in their neighborhood or at a distance to favor metastasis. *LncRNA*, Long noncoding RNA; *miRNA*, microRNA; *mRNA*, messenger RNA; *MVB*, multivesicular bodies; *NcRNA*, noncoding RNAs; *P-bodies*, processing bodies; *pri-miRNA*, primary transcript of miRNA; *RISC*, RNA-induced silencing complex; *TRBP*, trans-activation response element RNA-binding protein.

such as miR-122 (Hsu et al., 2012), miR-192 (Liu et al., 2017), miR-199a/b (Lee et al., 2012), let-7a (Frost & Olson, 2011), miR-21 (Calo et al., 2016), miR-29 (Hung et al., 2019), and miR-33 (Rayner et al., 2010).

During liver development a critical step is the formation and migration of hepatoblasts from the definitive endoderm, which needs to be tightly regulated. This step notably involves TGF-β signaling, which is regulated by miR-302 and miR-20a, by a process reminiscent of an epithelial-to-mesenchymal transition (EMT) (Wei et al., 2013). MiR-495 and miR-218 are also pivotal regulators of this step through the regulation of HNF-6 (Onecut-1), a central transcription factor regulating liver bud expansion (Simion et al., 2010).

The TGF-β signaling is also a driving force for cell fate decision during the differentiation of hepatocytes and biliary cells from the same progenitor cells, the hepatoblasts (Clotman et al., 2005). At this step the miR-23b/27b/24−1 cluster controls the terminal differentiation into hepatocytes or cholangiocytes through the negative regulation of TGF-β expression (Rogler et al., 2009). For a normal biliary morphogenesis, miR-30a and miR-30c are required. Their targets are, in particular, epidermal growth factor receptor (EGFR) and activin A, involved in biliary cell proliferation and differentiation (Hand et al., 2009). The final step of hepatocyte maturation and proper differentiation is governed by miR-122, the most abundant miRNA accounting for 70% of total liver miRNAs. This miRNA is, in particular, a target of the transcription factor HNF-4α, a master regulator of hepatocyte differentiation (Laudadio et al., 2012). MiR-122 is used by HCV for its replication through an unusual binding in the 5′UTR region of the viral RNA — the first therapeutic strategy based on miRNA inhibition in liver being an anti-miR-122. MiR-122 also plays pleiotropic roles in metabolism (e.g., iron and lipid metabolisms, cholesterol synthesis) (Vaulont, 2011; Wen & Friedman, 2012). MiR-33a/b also play central regulatory roles in liver metabolic function. It regulates lipid metabolism and cholesterol trafficking through the targeting of sterol transporters ABCA1 and ABCG1 (ATP-binding cassette A1 and G1) (Marquart, Allen, Ory, & Baldan, 2010), carnitine palmitoyltransferase 1 (Gerin et al., 2010) and AMP-activated protein kinase (Davalos et al., 2011). Finally, miR-27a is able to impair lipid accumulation by retinoid X receptor (Ji et al., 2009), fatty acid synthetase, and stearoyl-CoA desaturase targeting (Zhang, Yang, Trottier, Barbier, & Wang, 2017).

11.2.2 Long noncoding RNAs

LncRNAs are the most common ncRNAs accounting for more than 80% of ncRNAs and which currently know the biggest attention from the scientific community. LncRNAs are transcribed by the RNA polymerase II from intergenic regions (long intergenic noncoding RNAs), introns or enhancers in sense, antisense, or bidirectional manner (Derrien et al., 2012). They are frequently polyadenylated. Their sequences are poorly conserved among species, challenging the comprehension of their underlying mechanisms. As for other ncRNAs, the number of lncRNAs is directly proportional to organism complexity. They exhibit a great diversity of sequences and structures. They also display a tissue-specific pattern of expression and are finely regulated in time and space across the stages of development. In contrast to other ncRNAs, lncRNAs are predominantly nuclear and interact with ribonucleoproteins to control (1) transcription, either positively (ribosome guide, scaffold for transcriptional activators or chromatin remodelers, signal for transcription) or negatively (decoy, recruitment of transcriptional repressors), (2) mRNA stability and translation (miRNA sponge), and (3) chromatin architecture (chromosome loops and interchromosomal interactions) (Fig. 11.1) (Marchese, Raimondi, & Huarte, 2017). Panels of lncRNAs have been shown to play essential roles during embryogenesis (Ulitsky, Shkumatava, Jan, Sive, & Bartel, 2011) and tumorigenesis (Huarte, 2015), including in the liver.

During liver development, lncRNA signatures have been identified overtime. Two lncRNAs are particularly involved in endoderm differentiation: the lncRNA definitive endoderm-associated lncRNA1, which promotes activin-mediated endoderm differentiation (Jiang, Liu, Liu, Zhang, & Zhang, 2015) and divergent to goosecoid (GSC) induced by

TGF-β family signaling, which regulates GSC and hexamethylene bis-acetamide inducible 2 expression to promote endoderm differentiation (Daneshvar et al., 2020).

In the adult liver, highly expressed lncRNAs are required for proper metabolic functions, particularly for lipid metabolism. MALAT-1 (metastasis-associated lung adenocarcinoma transcript) regulates the stability of SREBP-1c (steroid response—binding protein 1c) on the promoter regions of genes involved in lipid metabolism and improves insulin sensitivity and glucose tolerance (Yan, Chen, & Chen, 2016). SREBP-1c is also regulated by a human-specific lncRNA, LncHR1 repressing triglyceride accumulation (Li, Piontek, et al., 2017). Another key player in glucose and lipid metabolism in the liver is polypyrimidine tract-binding protein 1, an RNA-binding protein that regulates mRNA stability and pre-mRNA splicing. Several studies have reported PTBP1 interactions with a number of lncRNAs such as H19 to regulate SREBP-1c expression and function (Liu et al., 2018), MEG3 (maternally expressed 3) to promote cholestasis (Zhang, Sun, Zhou, & Tang, 2017) or hLMR1 to favor cholesterol synthesis (Ruan et al., 2021). Cholesterol accumulation is also regulated by Lnc-HC which destabilizes *CYP7A1* and *ABCA1* mRNAs (Lan et al., 2016). Finally, another way for lncRNAs to control liver lipid content is impacting lipid trafficking through the disturbance of apolipoproteins by APOA1-AS (Halley et al., 2014) and APOA4-AS (Qin et al., 2016).

Because of their pleiotropic activities and their abilities to modulate development, cell proliferation and differentiation, as well as liver metabolic functions, ncRNAs notably miRNAs and lncRNAs, emerged as key contributors to liver disease establishment and tumor progression. In particular, deregulated expressions of embryonic ncRNAs are found in liver diseases and tumors suggesting that the reactivation of an embryonic ncRNA program may contribute to liver cancer.

11.3 Noncoding RNA detection

11.3.1 Tissular versus circulating ncRNAs

Genomic studies in cancers have revealed that the great majority of cancer-associated polymorphisms occur in noncoding regions of the genome (>80%), including lncRNA and miRNA loci. In particular, more than half of genes encoding miRNAs are located in cancer-associated regions or fragile sites (Calin et al., 2004). NcRNA abnormal expressions contribute to tumor progression, in which they play an ambivalent role of tumor suppressors or oncogenes. NcRNA expression is submitted to a great variety of genetic regulations, similarly to coding genes, - such as gene amplification or deletion, promoter hypermethylation, translocation, or single nucleotide polymorphism. A global analysis of ncRNA expression in different types of tumors highlighted specific ncRNA signatures according to their cellular and tissular origins for miRNAs (Lu et al., 2005; Volinia et al., 2006) and lncRNAs (Iyer et al., 2015).

In addition to their detections in tissues and tumors, ncRNA are also present in several body fluids such as blood, urine, and saliva, supporting that circulating ncRNAs are relatively resistant to RNAse activities present in body fluids. Some of them are also resistant to extreme pH and storage conditions opening the way toward noninvasive diagnosis tools and clinical monitoring. Circulating ncRNAs could be of several origins either complexed to Argonaute or high-density lipoproteins, or encapsulated into extracellular membranous vesicles of different size like apoptotic bodies (>500 nm), microvesicles (100–1000 nm), and exosomes (<100 nm) (Fritz et al., 2016). Exosomes are double-membrane cell-derived microvesicles, containing thousands of lipids, proteins, and nucleic acids from coding or noncoding regions (Keerthikumar et al., 2016). Most cells, and particularly immune cells, are physiologically secreting exosomes originated from multivesicular bodies, fusion of which with the plasma membrane is orchestrated by Rab proteins (Fig. 11.1) (Théry, Zitvogel, & Amigorena, 2002). Importantly, exosomes and their parental cells exhibit different patterns of ncRNAs supporting the existence of specific packaging (Guduric-Fuchs et al., 2012). Once in the proximity of recipient cells, exosomes act through direct stimulation via ligands exposed at their surfaces or could transfer their content into the target cells to modify their transcriptional and epigenetic programs following different ways of internalization (Valadi et al., 2007). In physiological conditions, exosomes are involved in essential biological processes such as angiogenesis, coagulation, inflammation, and immune response (antigen presentation, immune activation or suppression, and immunotolerance) (Robbins & Morelli, 2014). However, during tumorigenesis, tumor cells are important conveyers of exosomes to communicate not only with their associated stroma particularly immune cells to impair antitumor immune response but also with more distant cells to favor metastasis (Kalluri, 2016). Therefore ncRNAs have the potential to be used as diagnostic, prognostic, and predictive biomarkers in cancer, including for liver tumors, as we will detail thereafter.

11.3.2 Methods of ncRNA analyses

11.3.2.1 Bioinformatics tools

Regarding the growing numbers of ncRNAs newly identified, databases compiling genetic information (i.e., their chromosome localization, sequence), their molecular and cellular functions, as well as their targets constitute a key prerequisite for optimal research and clinical application. In the last two decades, several databases for miRNAs as well as lncRNAs appeared (Fig. 11.2). They either offer repositories compiling genetic information for different organisms (human, mouse, rat, etc.) [miRbase (Griffiths-Jones, 2004), LNCipedia (Volders et al., 2013), LncRNAdb (Amaral, Clark, Gascoigne, Dinger, & Mattick, 2011), and LncRNome (Bhartiya et al., 2013)]. Others have listed functional interactions already published [i.e., miRNet (Fan et al., 2016), Diana tools for both miRNAs (Vergoulis et al., 2012) and lncRNAs (Paraskevopoulou et al., 2013)] or bioinformatics tools to predict ncRNA targetome and network [i.e., miRDB (Chen and Wang, 2020), miRcode (Jeggari, Marks, & Larsson, 2012) using algorithms of prediction like TargetScan or miRanda for miRNAs, RegRNA (Chang et al., 2013) or IntaRNA2 for lncRNAs (Mann, Wright, & Backofen, 2017)]. Some laboratories also provide databases concerning ncRNA regulations in specific diseases such as LncTarD (Zhao et al., 2020), LncRNADisease (Chen et al., 2013), Lnc2Cancer (Ning et al., 2016), or HMDD (Human microRNA Disease Database) (Huang, Shi, et al., 2019). Concerning circulating ncRNAs, some repositories have also been developed such as Vesiclepedia (Kalra et al., 2012) or Exocarta (Mathivanan, Fahner, Reid, & Simpson, 2012).

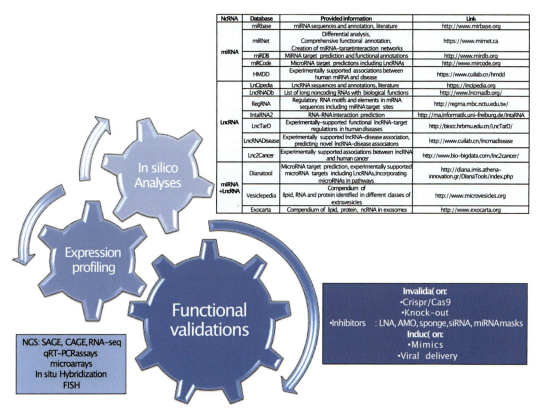

FIGURE 11.2 Tools and methods for ncRNA analysis. These recent decades, a number of databases have been developed to compile annotations and sequences of ncRNAs, references and algorithms of prediction for their potent targets and networks. Their detection in tissue and body fluids has been facilitated by NGS techniques such as SAGE, CAGE, and RNA-seq. They are also detected with current methods used for nucleic acids such as quantitative RT-PCR, microarrays, in situ hybridization, and FISH. After expression profiling, functional validations are required to confirm their associations with cancer initiation and progression. Multiple strategies for ncRNA targeting have been developed, either for inhibition with LNA, AMO, siRNA, sponge, masks, Crispr/Cas9 or knocking out strategies, or inversely restoration using mimics or viral delivery. *AMO*, Anti-miRNA oligonucleotide; *CAGE*, cap analysis gene expression; *FISH*, fluorescent in situ hybridization; *LNA*, locked nucleic acid; *ncRNA*, noncoding RNA; *NGS*, next-generation sequencing; *RT-PCR*, reverse transcription PCR; *SAGE*, serial analysis of gene expression; *siRNA*, small-interfering RNA.

11.3.2.2 Techniques of detection

Regarding the low abundance of miRNAs and lncRNAs in body fluids and tumors (<0.1% total RNAs) and their differential expression patterns in physiopathology, another key requirement for their clinical significance is the robustness of isolation methods and detection platforms. Isolation of ncRNAs could be performed on total RNAs or enriched fractions from tissue or liquid biopsies with several ready-to-use kits utilizing chemicals, columns, or beads, efficiency of which is strongly dependent on the amount of input material (El-Khoury, Pierson, Kaoma, Bernardin, & Berchem, 2016). For ncRNA-loaded vesicles, different exosome purification strategies are available from the gold standard method using ultracentrifugation, which is hardly compatible with clinical use, to size-exclusion chromatography, ultrafiltration, polyethylene glycol–based precipitation, or even enrichment with exosome-specific biomarkers (CD9, CD63, CD81, and epithelial cell adhesion molecule). All these methods of isolation exhibit advantages and disadvantages for the yield of exosomes obtained and the degree of contamination of extracted RNA (An, Wu, Zhu, & Lubman, 2018).

For their detection, next-generation sequencing (NGS)-based technologies (serial analysis of gene expression, cap analysis gene expression, RNA-seq), quantitative reverse transcription PCR (qRT-PCR) assays, and microarrays are the most frequent techniques used for ncRNA quantification with all their limitations (Grillone et al., 2020) (Fig. 11.2). Quantitative RT-PCR is quick and highly sensitive from small amounts of materials but only allows analyzing a small number of RNAs, while microarrays and NGS overcome these limitations with additional readout systems needed. LncRNAs resemble mRNAs regarding their splicing and the presence of 5′-cap and poly(A) tail and, thus, can be profiled together with mRNAs, while miRNAs, devoid of tail, must be specifically detected. NcRNAs could be either detected by Taqman or SYBR green technology. In TaqMan assay, miRNAs are reverse transcribed using a specific stem loop RT primer, while addition of a poly-A tail to miRNA is the approach for SYBR green. For lncRNA, classical RT is performed using random primers, the main difficulties being the design of detection probes since they are expressed from GC-rich intergenic regions with frequent antisense transcription.

A major problem for miRNA relative quantification is the way of normalization. Several endogenous small ncRNAs, either miRNAs or snRNAs, are used to normalize miRNA expression data. The normalization method should be chosen with cautions since (1) snRNAs and miRNAs are from different origins and (2) even if snRNAs are ubiquitously expressed, their expression is not fully stable as shown for RNU6, the most frequently normalization used, which displays significant variability in healthy individuals and in patients according to pathologies and treatments (Lou et al., 2015). This is particularly the case for circulating miRNAs in liver disease and could potentially lead to misinterpretation (Benz et al., 2013). Current trends preferentially adopt the most stable endogenous miRNA standards in the disease of interest and normalization on at least two ncRNAs rather than a single ncRNA. Algorithms to select better candidates have been created such as geNorm (Schlotter et al., 2009) and NormFinder (Andersen, Jensen, & Ørntoft, 2004). Finally, the use of external standards, such as cel-miR-39 from *Caenorhabditis elegans*, enables to reconcile differences in miRNA recovery and cDNA synthesis between all samples but it does not remove the bias inherent in starting materials.

Finally, the expression of ncRNAs could also be visualized in situ at the single cell level on tissue- and tumor-slides thanks to advances in in situ hybridization techniques, notably in innovative probes based on locked nucleic acid (LNA) technology, with strong and specific binding to complementary RNA targets. They are also of particular interest to explore which cells express ncRNA in tumors, either tumor cells or stroma cells. Another interesting technique is fluorescent in situ hybridization, using long single-stranded oligoprobes of 20 nt in size labeled with fluorophore. These techniques are particularly interesting for lncRNA, cytoplasmic or nuclear sublocalization of which could inform on their biological functions. This approach is also quantitative because the count of individual probe signals is related to the number of lncRNA molecules in each cell. Interestingly, as a clinical promise, SChLAP1 RNA-ISH has been recently demonstrated to be a novel tissue-based biomarker predicting lethal prostate cancers (Mehra et al., 2016).

11.4 Expression of ncRNAs in liver cancers

As previously mentioned, ncRNAs are key regulators of liver functions and their deregulations may be a source of liver diseases and consequential cancers. Besides the explosion of studies and discrepancies according to cohorts and techniques of isolation, detection, and normalization (see earlier), global analyses of ncRNA expressions unraveled common patterns of aberrantly expressed ncRNAs in liver tumors. All these ncRNAs regulate tumor growth through the modulation of key cellular actors of cell cycle, apoptosis, EMT, and immune response as we will detail in the next paragraph (Fig. 11.3).

11.4.1 MicroRNAs

Dozens of miRNAs have been described aberrantly expressed in HCC. Here, we will only list data obtained from more than one individual study.

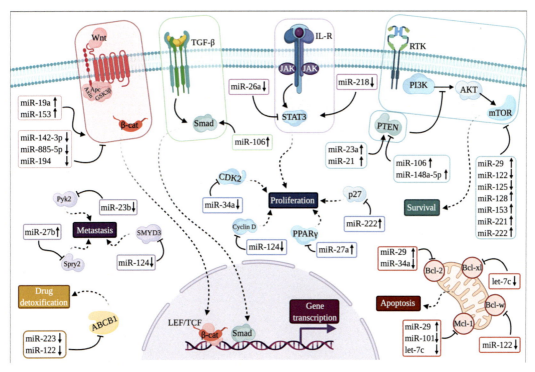

FIGURE 11.3 Cell signaling pathways modulated by miRNAs in liver cancer patients. Up and down arrow: respectively, upregulation and downregulation of miRNAs expression in HCC. *HCC*, Hepatocellular carcinoma; *miRNA*, microRNA. Source: *Made with biorender.com.*

11.4.1.1 MicroRNA signatures in liver tumors

In HCC, several miRNAs have been found upregulated such as miR-155 (Yan et al., 2013), the cluster miR-17−92 (Yang et al., 2010), miR-21 (Meng et al., 2007), miR-221/miR-222 (Pineau et al., 2010), miR-224 (Wang et al., 2008), and miR-23 (Kutay et al., 2006). Inversely, a loss of let-7 (Lan et al., 2011), miR-101 (Su et al., 2009), miR-122 (Ladeiro et al., 2008), miR-124 (Furuta et al., 2009), miR-125b (Li et al., 2008), miR-199 (Hou et al., 2011), miR-26a (Yang et al., 2013), and miR-29 (Xiong et al., 2010) has been described (Fig. 11.4). In addition to common miRNA signatures, HCCs also exhibit distinct miRNA patterns accordingly to associated risk factors, either genetic mutations or etiologies, and sex. Tumors mutated for *CTNNB1* are characterized by low level of miR-375, while a loss of miR-126* is characteristic of alcohol-related HCCs. Inversely, miR-96 overexpression is related to HBV-HCCs (Zucman-Rossi, 2010) and miR-18a with HCC in women (Liu et al., 2009). A signature of 20 miRNAs has been also found

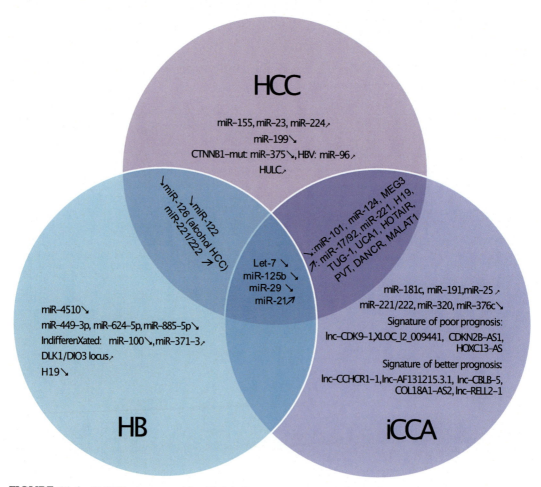

FIGURE 11.4 NcRNA signatures identified in liver cancers. *HB*, Hepatoblastoma; *HCC*, hepatocellular carcinoma; *iCCA*, intrahepatic cholangiocarcinoma.

associated with metastasis, including miR-122, miR-124, and miR-34a (Budhu et al., 2008). Other EMT-associated miRNAs have been unveiled in HCC (e.g., miR-130a, miR-148a, miR-181a, miR-199, or miR-34a) as regulators of TGF-β signaling (Brockhausen et al., 2015; Zhou et al., 2017) and c-MET/Snail axis (Li, Fu, Tie et al., 2009; Liu, Tu, & Liu, 2014; Zhang et al., 2014). MiR-26a (Yang, Zhang, et al., 2014) and miR-29b (Fang et al., 2011) are important regulators of HCC angiogenesis, another crucial component of metastasis.

In HB, a study have identified an upregulation of miR-129-5p, miR-181b, miR-184, miR-19a, miR-220a, miR-221, miR-222, miR-320a, and miR-330-3p and a downregulation of miR-106b, miR-122, miR-126*, miR-148a/b, miR-29b-1*, miR-148a, miR-301a, and miR-9/9* (Cairo et al., 2010) (Fig. 11.4). Two clusters of miRNAs, miR-100/let-7a-2/miR-125b-1 and miR-371−3, discriminate HB according to their differentiation status and their aggressive features. The four miRNAs, miR-100, miR-371, miR-373, and let-7a, are also discriminant for a subtype of HCCs characterized by their poor differentiation and higher invasiveness, which supports the existence of an miRNA signature common to HB and HCC. More recently, several miRNAs have been found decreased in HB and all related to the Wnt/β-catenin pathway: miR-4510 (Cartier et al., 2017), let-7i-3p, miR-449b-3p, miR-624-5p, and miR-885-5p (Indersie et al., 2017). Last year, an integrative study also showed that the expression profile of the biggest cluster of miRNAs called the DLK1/DIO3 locus distinguishes two groups of HB (Carrillo-Reixach et al., 2020). This cluster contains coding RNAs such as DLK1, a fetal biomarker highly expressed in HB, 3 lncRNAs, multiple tandemly repeated C/D snoRNAs, and 54 miRNAs. Tumors with high expression of the DLK1/DIO3 locus are mostly mutated in *CTNNB1* and harbor progenitor features and clinical aggressiveness.

In CCA, some miRNA deregulations are common to HCC with the induction of miR-17−92 cluster (Zhu, Han, Lu, & Wu, 2014), miR-21 (Selaru et al., 2009), and miR-26a (Zhang, Han, & Wu, 2012), and decrease in miR-125b (Lin et al., 2016) and miR-29 (Mott, Kobayashi, Bronk, & Gores, 2007) (Fig. 11.4). Other miRNAs play oncogenic roles in iCCAs [e.g., miR-181c (Wang et al., 2016), miR-191 (Li, Cheng, et al., 2017), and miR-25 (Razumilava et al., 2012)] or inversely tumor-suppressive functions [e.g., miR-320 (Chen et al., 2009) and miR-376c (Nakaoka, Saito, & Saito, 2017)]. MiRNA profiles differ mostly according to tumor grade and differentiation (Plieskatt et al., 2014). A 23-miRNA signature, including miR-200c related to EMT, discriminates two phenotypes of iCCA with either stem cell−like features or with mature hepatocyte traits (Oishi et al., 2012). Other pro- and antimetastatic miRNAs have been identified in iCCA such as miR-214 regulating Twist (Li et al., 2012), miR-204 targeting Slug (Qiu et al., 2013), and miR-34a on TGF-β signaling (Qiao et al., 2015).

11.4.1.2 Examples of deregulated miRNAs and signaling pathways in HCC

MiR-21 is one prototypical oncogenic miRNA induced in a plethora of cancers and also in liver cancers. It plays a crucial role in the pathogenesis of liver disease as well as in tumorigenesis by regulations of multiple biological processes (metabolic and proliferation signaling pathways, immune response and invasion) (Zhang, Yang, Kusumanchi, Han, & Liangpunsakul, 2020). MiR-21 is induced in HCC and CCA cells (Karakatsanis et al., 2013; Zhu et al., 2012), as well as in liver cancer stem cells (LCSC) (Li et al., 2010), inflammatory macrophages (Thulasingam et al., 2011), and HSC (Wei, Feng, Li, Xu, & Fan, 2013). Interestingly, miR-21 can also be detected in the serum from HCC (Zhou et al., 2011; Xu et al., 2011) and CCA patients (Wang et al., 2015). It constitutes a potent biomarker for diagnosis and prognosis, a part that we will detail in the next sections.

Another well-studied miRNA in cancers, including in liver tumors, is miR-221. This miRNA is a key player during fibrosis and chronic liver disease (Ogawa et al., 2012) but also in tumorigenesis. A number of studies showed its induction in HCC samples (Fornari et al., 2008; Garofalo et al., 2009; Gramantieri et al., 2009; Le Sage et al., 2007; Pineau et al., 2010; Wong et al., 2010). MiR-221 is also induced in extrahepatic CCA (Li et al., 2015), while it is inversely reduced in iCCA (Karakatsanis et al., 2013). MiR-221 induction has been observed not only in tumor cells but also in HSCs and in KCs. In HCC, miR-221 upregulation is in part dependent on c-MET oncogene and the c-Jun transcription factor (Garofalo et al., 2009). MiR-221 regulates multiple targets involved in proliferation [PTEN (phosphatase and TENsin homolog)] (Garofalo et al., 2009), p27/p57 (Fornari et al., 2008), and DNA damage inducible transcript 4 (Pineau et al., 2010), but also apoptosis regulators Bcl-2-modifying factor (Gramantieri et al., 2009) or genes critical for invasion such as axis inhibition protein 2 (Dong et al., 2019) and tissue inhibitor of metalloproteinase-3 (Garofalo et al., 2009). MiR-221 is also a potential candidate for noninvasive cancer diagnosis with higher expression in serum in the case of liver cancer and associated risk factors (see earlier). It also constitutes an miRNA of interest for miRNA inhibition−based therapies with multiple approaches tested (sponge, LNA, or decoy) (see Section 11.6).

Inversely, miR-101 is described as an miRNA with tumor-suppressive functions in HCC (Su et al., 2009) and in CCA (Zhang, Han, Zhu, Song, & Wu, 2013). Its expression is particularly regulated by epigenetic events such as c-MYC-dependent silencing by EZH2 (enhancer of zeste homolog 2) (Wang, Zhang, et al., 2014). MiR-101 impairs tumor progression by a feedback loop on EZH2 (Xu et al., 2014), cyclooxygenase 2 and rho-associated coiled-coil containing protein kinase 2 (Zheng et al., 2015), Mcl-1 (myeloid leukemia cell differentiation protein 1) (Su et al., 2009), Girdin (Cao et al., 2016), or nemo-like kinase (Shen et al., 2014). It also regulates invasion capacities of cancer cells through the regulation of the oncogene FOS (Li et al., 2009) and of the proangiogenic factor vascular endothelial growth factor-C (Liu, Wang, Mao, Zou, & Fan, 2016). In addition, this miRNA plays roles in cancer-associated fibroblasts (Yang, Lu, et al., 2016), KC (Wei et al., 2015) or HSC (Meroni et al., 2019). This miRNA is also of clinical significance not only as a key player in drug response and a therapeutic option (Xu, Liao, et al., 2017) but also a diagnosis factor (Moshiri et al., 2018).

11.4.2 Long noncoding RNAs

11.4.2.1 LncRNA signatures in liver cancer

Research on lncRNAs is more recent than those on miRNAs, and lncRNA signatures associated with liver cancer subclasses are for that moment lowly defined. Nevertheless, numbers of lncRNAs have been found aberrantly deregulated in liver cancers—the first identified being HULC (highly upregulated in liver cancer) (Panzitt et al., 2007). Several other lncRNAs are also upregulated in HCC and CCA such as H19 (Xu et al., 2017), TUG-1 (taurine-upregulated gene 1) (Xu et al., 2017; Huang, Shi, et al., 2019), HOTAIR (homeobox transcript antisense RNA) (Qin et al., 2018; Yang, Zhou, et al., 2011), DANCR (differentiation antagonizing nonprotein-coding RNA) (Wang, Zhang, et al., 2019; Yuan et al., 2016), PVT1 (plasmacytoma variant translocation 1)

(Yu et al., 2018), and UCA1 (urothelial cancer—associated 1) (Wang, Ying, et al., 2015; Xu, Yao, et al., 2017). Inversely, decrease in MEG3 expression is found in HCC (Anwar et al., 2012) and CCA (Li et al., 2019). A nine-lncRNA signature has been recently associated with the survival of CCA with four lncRNAs found highly expressed in iCCA with poor prognosis (lnc-CDK9−1, XLOC_l2_009441, CDKN2B-AS1, and HOXC13-AS) and five highly expressed in iCCA of better prognosis (lnc-CCHCR1−1, lnc-AF131215.3.1, lnc-CBLB-5, COL18A1-AS2, and lnc-RELL2−1) (Angenard, Merdrignac, Louis, Edeline, & Coulouarn, 2019). Interestingly, in this study, the lncRNA ANRIL has been identified as a poor prognosis biomarker in iCCA and in HCC. The DLK1/DIO3 locus containing three lncRNAs MEG3, MEG8, and MEG9 could discriminate two groups of HB as previously mentioned (Carrillo-Reixach et al., 2020). HBs are also characterized by the deregulation of another imprinted locus, the IGF2/H19 locus leading to the downregulation of H19 (Ross, Radloff, & Davies, 2000).

Concerning HCC, specific patterns of up- and downregulated lncRNAs have been identified in each subclass of HCC from S1 to S3 (Yang et al., 2017). In particular the lncRNA RP11−166D19.1 is specifically downregulated in S2 subclass. LncRNAs mostly affect tumor progression such as MEG3 (Braconi et al., 2011), H19 (Matouk et al., 2007), MALAT-1 (Lai et al., 2012), and TUG-1 (Huang et al., 2015) or tumor invasion such as DANCR (Yuan et al., 2016), hPVT1 (Wang, Yuan, et al., 2014), lncTCF7 (Wang, He, et al., 2015), Linc-GALH (Xu et al., 2019), or ATB (activated by TGF-β) (Jang et al., 2017; Yuan et al., 2014). Importantly, several lncRNAs promote HCC through the specific recruitment of epigenetic regulators on genes involved in liver carcinogenesis. Some of them recruit PRC2 (polycomb repressive complex 2) to silence genes by H3K27 trimethylation such as TUG-1 (Huang et al., 2015), lncRNA-HEIH (highly expressed in HCC) (Yang et al., 2011), or lnc-β-Catm (Zhu et al., 2016). Others inversely promote HCC-related genes by interaction with epigenetic activators and SWI/SNF remodelers such as lncTCF7 (Wang, He, et al., 2015), H19 (Zhang et al., 2013), or MALAT-1 (Huang, Wang, Hu, & Cao, 2019). Finally, others like ID2-AS1 impair HDAC8 recruitment on its adjacent gene ID2 and suppress metastasis (Zhou et al., 2020).

11.4.2.2 Examples of deregulated lncRNAs and signaling pathways in HCC

The first identified lncRNA, HULC, is upregulated in tumors and plasma from HCC patients and associated with HBV infection (Panzitt et al., 2007). Indeed, it has been shown that HBx regulates HULC expression, which, in turn, impairs the expression of its neighbored gene p18, leading to cell proliferation (Du et al., 2012). HULC also modulates the activity of multiple miRNAs involved in HCC metastasis such as miR-107 (Lu et al., 2016) and miR-200 (Li et al., 2016), and in proliferation such as miR-186 (Wang et al., 2017) and miR-372 (Lin et al., 2018). Similarly, under oxidative stress, HULC sponges miR-372−373 and upregulates the chemokine receptor of CXCR4, involved in CCA inflammation, migration, and invasion (Wang, Ye, et al., 2016).

Another lncRNA overexpressed in HCC and in several other cancers is MALAT-1. It predicts tumor recurrence after liver transplantation (Lai et al., 2012). MALAT-1 is able to transform normal hepatocytes and favors HCC cell proliferation [upregulation of MYC, Cyclin D1, and serine and arginine rich splicing factor 1 (SRSF1) and activator of mTORC1 pathway] (Malakar et al., 2017). SRSF1 is able to induce the degradation of MALAT-1 and to disrupt YAP/TCF/β-catenin activating complexes bound on its promoter region (Wang, Wang, et al., 2014). MALAT-1 sponges several miRNAs such as miR-125a (Liu, Qiu, et al., 2019), miR-146a (Peng et al., 2020), miR-30a (Pan et al., 2018), and miR-204 (Zhouhua et al., 2017) to favor HCC cell growth and invasion. A similar interaction between MALAT-1 and miR-204 has been also associated with progression and invasion of hilar CCA through downstream regulation of CXCR4 (Tan, Huang, & Li, 2017). The oncogenic role of MALAT-1 is also central under hypoxia through the regulation of miR-200 (Zhao et al., 2019), HIF-1α (Luo, Sun, et al., 2016), and HIF-2α (Luo, Liu, et al., 2016).

HOTAIR is also reported as upregulated in HCC and related to lymph node metastasis, larger tumor size, recurrence, and poor survival following liver transplantation (Ishibashi et al., 2013; Yang, Zhou, et al., 2011). HOTAIR is highly expressed both in CCA and also correlated to tumor size, TNM stage and postoperative recurrence (Qin et al., 2018). In particular, HOTAIR is overexpressed in HBV-related HCC through HBx-mediated proteasomal degradation of its two negative regulators SUZ12 and ZNF198 (Zhang, Diab, et al., 2015). Interactions of HOTAIR with PRC2 complex also play a significant role in HSC during fibrogenesis by controlling H3K27me3 level at the promoter region of MEG3 (Bian et al., 2017). HOTAIR is also able to repress miR-122 expression via DNA methylation (Cheng et al., 2018). It reduces SETD2 expression to favor LCSC growth by inhibiting P300 and POLII recruitment on SETD2 promoter (Li, An, et al., 2015).

In summary, while ncRNA genes were for years considered a junk part the genome, they are now considered tiny regulators of cell fate and behavior through their pleiotropic targets and mechanisms of action. In the following sections, we will address how these regulators constitute tumor and circulating biomarkers for liver cancer diagnosis, prognosis, and treatment.

11.5 Noncoding RNA relevance in liver cancer diagnosis and prognosis

11.5.1 Noncoding RNA as potential diagnostic tools

As previously mentioned, liver cancer is highly lethal and often detected when it reached an advanced stage. Detection of liver tumors is done by ultrasonography, which allows the detection of nodules smaller than 1 cm and by the measurement of the plasmatic level of different liver enzymes such as α-fetoprotein (AFP), or alanine transaminase and aspartate transaminase associated with liver dysfunction. Since risk patients are well defined, the necessity to develop new precise diagnostic tool is relevant. The detection of exosomal ncRNAs in body fluids is, thus, a valuable noninvasive diagnosis tool for cancer, including HCC. This has been extensively described in the literature for miRNAs (Gougelet & Colnot, 2016), such as miR-221 (Li, Wang, Yu, Chen, & Luo, 2011), miR-224 (Lin, Lu, Yu, Liu, & Zhou, 2016), and miR-21 (Tomimaru et al., 2012). Serum miR-16, let-7f, and miR-21 are potential indicators of tumor size and HCC recurrence (Ge et al., 2014). Plasma miR-21 is also significantly higher in hepatitis-associated HCC patients than in chronic hepatitis patients and healthy volunteers (Tomimaru et al., 2012). Interestingly, circulating miR-21 level is decreased after surgery. A seven-lncRNA signature has been identified in HCC and a combination PVT1/uc002mbe.2 showed potential as a diagnostic marker for HCC to distinguish HCC patients from healthy volunteers (Yu et al., 2016). An overexpression of HULC also significantly correlates with tumor size and capsular invasion (Li et al., 2015).

Since each risk factor is associated with specific mutations and metabolic features, it may be necessary to identify specific biomarkers depending of patient status. Here we tried to create a nonlimited list of circulating ncRNAs that could be used to diagnose and predict liver cancer development. However, it must be taken into consideration that most articles do not converge toward the same list of ncRNAs as risk factors because they have selected few miRNAs in a list of 100 differentially expressed ones with variable criteria.

11.5.1.1 In HBV-related HCCs

Compared to healthy donors, HCC patients show a lower level of miR-122, miR-194, miR-223, miR-23b, miR-26a, and miR-27a together with a higher level of miR-192, miR-21, and miR-801 that can provide a high-diagnostic accuracy for HCC (Zhou et al., 2011). In another multicenter study the panel miR-133a, miR-143, miR-145, miR-192, miR-29a, miR-29c, and miR-505 is found differentially expressed and could detect early-stage HCCs with better accuracy than AFP (Lin et al., 2015). In a cohort of more than 100 patients, 5 circulating miRNAs (miR-100-5p, miR-122-5p, miR-125b-5p, miR-885-5p, and miR-148a-3p) are significantly upregulated in HCC (Jin et al., 2019). MiR-34a-5p is a biomarker for cirrhosis and disease progression because it can differentiate cirrhotic and HCC patients from healthy donors and HBV patients. MiR-885-5p and miR-100-5p also allow to distinguish cirrhotic patient but with less sensitivity. In this study, miR-1972, miR193a-5p, miR-214-3p, and miR-365a-3p can be used to discern HCC patients from non-HCC subjects (either cirrhotic, HBV, or healthy) with a highest performance for miR-1972. MiR-122 and let-7b are also upregulated in HBV patient serum and able to differentiate early HCC (Hung et al., 2016). In a small cohort of Asiatic patients, five miRNAs are also found overexpressed (miR-3200, miR-339, miR-342, miR-3615, miR-4508, and miR-935) and two downregulated (miR-182 and miR-4485) during chronic HBV-associated cirrhosis (Riazalhosseini, Mohamed, Apalasamy, Langmia, & Mohamed, 2017).

Several other works have identified circulating miRNA signatures able to discriminate HBV-related HCC patients from HBV tumor-free patients. For example, Mjelle et al. show that HBV-related HCC patients exhibit an upregulation of miR-194-5p and miR-122-3p (Mjelle et al., 2019), while Wen et al. identify an upregulation of miR-20a-5p, miR-25-3p, miR-30a-5p, miR-92a-3p, miR-132-3p, miR-185-5p, miR-320a, miR-324-3p, miR-192-5p, miR-21-5p, and miR-375 in HBV-related HCCs (Wen et al., 2015). The panel of miR-20a-5p, miR-320a, miR-324-3p, and miR-375 in combination with AFP was tested on two separate cohorts to diagnose HCC and showed pretty accurate results. Downregulation of circulating miR-223-3p is also associated with HCC development compared to HBV patients and healthy donors. Its association with AFP is highly sensitive for the detection of early HCC (85%) and intermediate and advanced HCC (100%) (Pratedrat et al., 2020). MiR-223-3p together with miR-92a-3p, miR-122-5p, and miR-125b-5p also differentiate HBV-HCCs from healthy donors (Giray et al., 2014). Plasma DANCR levels were also higher in patients with HBV-related HCC than in chronic HBV cirrhosis and healthy groups, with an AUC (Area under rhe curve) superior to AFP (Ma et al., 2016).

11.5.1.2 In HCV-related HCCs

Studies concerning serum miRNA signatures in HCV-associated HCCs are mainly performed in Egyptian cohorts. Early HCV-related HCC could be differentiated from

HCV tumor-free patients with several serum miRNAs, particularly on the basis of miR-122 (Amr, Elmawgoud Atia, Elazeem Elbnhawy, & Ezzat, 2017) and miR-125a expression (Oura et al., 2019). A panel of miR-146a, miR-19a, miR-192, and miR-195 can differentiate HCC patients not only from healthy donors but also from HCV patients with chronic liver disease (Motawi, Shaker, El-Maraghy, & Senousy, 2015). Interestingly, miR-19a can grade the level of liver disease from early fibrosis stage to cirrhosis and finally to HCC (Mourad et al., 2018). HCV-infected patients developing HCC show a lower serum expression of miR-125a, miR-139, miR-145, and miR-199a, and a higher expression of miR-224 in comparison with patients with a chronic cirrhosis due to HCV without HCC [chronic HCV with cirrhosis (CHC)] or not (Mourad et al., 2018). Those miRNAs upregulated in CHC patients, with or without cirrhosis, are decreased when HCC arises. Depending on HCV variants, the expression of these serum miRNAs may vary when compared to healthy donors. Interestingly, the present study also showed that miR-199a and miR-224 are differently expressed compared to healthy donors. Another study from Asia highlighted different miRNA panels to discriminate HCV-related HCC from healthy donors (miR-122, miR-29b, and miR-885-5p with AFP), from cirrhosis (miR-122, miR-22, miR-221, and miR-885-5p with AFP) and from CHC (miR-22, miR-199a-3p with AFP) (Zekri et al., 2016). In HCV patients, serum UCA1 and WRAP53 were found significantly higher in HCC than in chronic HCV infection or in healthy volunteers and associated with advanced HCCs (Kamel et al., 2016). Plasma MALAT-1 levels also increased progressively with liver disease and hepatitis-related HCCs and found higher during HCV than HBV (Konishi et al., 2016).

11.5.1.3 In HCC related to metabolic disorders

Deregulation of circulating miRNA levels has been extensively reported during liver disease and particularly during NAFLD. We could cite miR-122, miR-192, miR-21, miR-34a, and miR-375 (Cermelli, Ruggieri, Marrero, Ioannou, & Beretta, 2011; Pirola et al., 2015). However, little is known about miRNA signatures in NAFLD/NASH-related HCC. A recent work showed a significant overexpression of miR-34a and miR-221, and a decreased expression of miR-16, miR-23-3p, miR-122-5p, miR-198, and miR-199a-3p in diabetic patients developing HCC as compared to diabetics with cirrhosis only (Elemeery et al., 2019).

11.5.1.4 In CCA

As in HCC, cholangiocarcinoma cells hold the ability to secrete ncRNAs into the bloodstream, which also gives a diagnostic ability to miRNAs. Several teams work to highlight ncRNA signatures and the following ncRNAs show some promising features. Plasma miR-21 and miR-221 are highly expressed in iCCA patients compared with healthy subjects (Correa-Gallego et al., 2016). MiR-26a (Wang, He, Sui, et al., 2015) and miR-191 are also more highly expressed in the serum of CCA patients—in vitro analysis showing that miR-191 promotes cell proliferation, infiltration, and metastasis (Li, Zhou, et al., 2017). Bernuzzi et al. profiled 90 serum samples and found that 33 miRNAs are differentially expressed in CCA as compared to healthy volunteers, with an induction of both miR-194 and miR-483-5p confirmed in the validation cohort of CCA patients as compared to controls (Bernuzzi et al., 2016).

11.5.1.5 In HB

Plasma miR-21 was found significantly higher in patients with HB compared with control group and could serve as a diagnostic biomarker (Liu, Chen, & Liu, 2016). Plasmatic miR-21 is equivalent to AFP in diagnosing HB, while exosomal miR-21 was significantly more accurate as compared to AFP levels.

11.5.2 Prognostic potential of noncoding RNAs in liver cancer patients

11.5.2.1 Risk-scoring systems

Since patient 5-year survival rate is only of 15%, it is necessary to develop new robust prognostic methods to guide patient treatment. Recently, several teams worked on creating different risk-scoring system to predict HCC patient prognosis based on tumor lncRNA expression. A five-lncRNA signature (RP11-325L7.2, DKFZP434L187, RP11-100L22.4, DLX2-AS1, and RP11-104L21.3) is associated with prognosis (Sun et al., 2019). Similarly, (Su et al., 2020) create a risk factor model based on miRNA and mRNA expression that predict the prognosis outcome. Ye et al. (2020) have worked differently and compared cirrhotic HCC with noncirrhotic ones to create a risk-scoring system based on lncRNA expression to predict cirrhotic HCC patient OS. Higher expression of LINC02086 and LINC00880 is associated with poor OS, while higher expression of AC13647.3 and LINC01549 is associated with better OS. Higher expression of SH3RF3-AS1, AC104117.3, AC136475.3, LINC00239, and MRPL23-AS1 is also correlated to a better RFS (recurrence-free survival). In alcohol-related HCCs with worst survival rate in comparison with nondrinkers, an lncRNA risk-scoring system correlates an increase in AC012640.1, AC01345.2 with worse OS and increase in AC062004.1 and LINC02334 with better OS (Bucci et al., 2016). Additionally, an increase in ERVH48–1, LINC02043, and LINC01605 predicts a worse RFS, whereas AC062004.1 and AL139385 increase predicts a better RFS (Luo et al., 2020).

11.5.2.2 Prognostic miRNA signatures in HCC

Sathipati et al. have identified an miRNA signature associated with HCC stage and OS: a tumor signature of let-7i, miR-424, miR-512, miR-518, miR-550a, miR-574, and miR-549 is significantly associated with patient OS (Yerukala Sathipati & Ho, 2020). In addition, miR-135a, an miRNA upregulated in HCC, has also been described as a promising prognostic factor (Huang et al., 2017), since its expression is associated with recurrence after tumor resection (von Felden et al., 2017). Some serum miRNAs related to HCC could also constitute potent prognostic factors. In HBV patients an increase in circulating miR-128, miR-139-5p, miR-382-5p, and miR-410 and an inverse decrease in miR-424-5p and miR-101-3p are linked to the worst prognosis group (Jin et al., 2019). MiR-424-5p and miR-101-3p lower expressions are correlated to tumor size and invasion, while high miR-128 expression is inversely correlated. MiR-139-5p is found associated with nodule multiplicity and tumor invasion, while miR-382-5p expression correlates with larger tumors and higher level of AFP. A higher expression of miR-21 in HBV patients was also correlated to a shorter time to recurrence and worse OS (Wang, Zhang, Wang, et al., 2014), while a negative correlation between miR-223-3p and OS was recently found (Pratedrat et al., 2020). In HCV, serum miR-16, miR-34a, and miR-221 also display prognostic values (Mourad et al., 2018).

11.5.2.3 Prognostic lncRNA signatures in HCC

An interesting work from (Zheng et al., 2018) identifies several lncRNAs that are deregulated in HCC in alcohol drinkers as compared to nondrinkers and adjacent nontumor tissues (lnc-AFM-2:1, lnc-HSP3−2:3, lnc-TBC1D9B-1:6 and 8; lnc-M6PR-3:1, lnc-AC022098.1−1:3 and 4; and lnc-EXO1−8:2, lnc-CALM2−2:4 and 5). In particular, increase in lnc-HSP3−2:3 expression in *CTNNB1*-mutated HCC and in CALM2−2:5 is associated with poor OS. Overexpression of HEIH is also significantly associated with tumor recurrence and reduced OS in HBV-related HCCs (Yang, Zhang, et al., 2011). Inversely, an upregulation of HULC in tumor tissues is positively associated with HCC survival, while H19 overexpression is associated with poor outcomes and HCC aggressiveness (Yang, Lu, et al., 2015). Finally, CASC9, another lncRNA upregulated in HCCs and in serum, is associated with lower recurrence after surgery (Gramantieri et al., 2018).

11.5.2.4 Prognostic miRNA and lncRNA signatures in CCA

Concerning the use of ncRNAs as prognostic biomarkers in CCA, several miRNAs in tumor and in serum show an interesting efficacy to predict CCA behavior. MiR-29a constitutes a good prognostic biomarker for CCA patient, associated with differentiation, clinical stage, and metastatic status of CCA (Deng & Chen, 2017), as well as miR-200a (Chen, Yang, Wang, & Wang, 2015). MiR-122 is also a promising biomarker since it is markedly decreased in CCA and associated with bad prognosis (Liu, Jiang, et al., 2015). MiR-203 is associated with tumor size and stage, differentiation status, and inversely with OS (Li et al., 2015). MiR-590-3p is downregulated in both CCA serum and tumors and specially linked to tumors with metastatic potential (Zu et al., 2017). Concerning serum, miR-21 increase is linked to a more metastatic and invasive CCA (Huang et al., 2013), while miR-191 increase is found associated with advanced tumor stage and poor OS (Li, Zhou, et al., 2017) as well as miR-26a (Wang, Zhang, Zhang, et al., 2015). Concerning lncRNA, overexpression of CCAT1 is associated with aggressive malignant behavior and poor OS (Jiang et al., 2017). MALAT-1 is also associated with tumor stage, larger tumor size and metastasis, and poor prognosis in CCA patients (Tan et al., 2017).

11.5.3 Noncoding RNAs as predictive markers for liver cancer patients

Numbers of ncRNAs associated with liver cancer prognosis are also associated with recurrence after treatment. As previously mentioned, HOTAIR is reported to be associated with recurrence in both HCC (Yang et al., 2011) and CCA (Qin et al., 2018). A higher expression of the lncRNAs HEIH (Yang, Zhang, et al., 2011) and CASC9 (Gramantieri et al., 2018) is also associated with lower recurrence after HCC surgery. When miR-151-3p is upregulated and miR-126 is downregulated after CCA resection, patients have a better prognosis (McNally et al., 2013). Ge et al. (2014) also discover that miR-16 downregulation was correlated to tumors smaller than 5 cm in diameters and let-7f upregulation with tumors bigger than 5 cm with higher risk of early recurrence. A higher expression of miR-21 in HBV patients was also correlated to a shorter time to recurrence (Wang, Zhang, Wang, et al., 2014). Interestingly, a plasmatic signature of nine miRNAs (miR-15b, miR-107, miR-122, miR-125b, miR-200a, miR-30a, miR-320, miR-374b, and miR-645) is significantly associated with OS during treatment with regorafenib, a tyrosine kinase receptor inhibitor (Teufel et al., 2019).

A recent study interestingly established a link between circulating levels of miR193a-5p and early HCC outcomes after surgery (Loosen et al., 2020). In patient serum collected before tumor resection and after liver transplant, a preoperative miR193a-5p level above a cutoff value 3.57 significantly reduces OS about 451 days instead of 1158 days for the group below this cutoff. This new finding could be a novel indicator to perform or not surgery on early-stage HCC patients.

11.5.4 Roles of microRNA in drug resistance

As we have previously mentioned, liver cancers are currently resistant to conventional anticancer treatments such as chemotherapy and radiotherapy in first line or in second line, after treatments with sorafenib for HCC and chemotherapy for CCA. With their pleiotropic roles, ncRNAs regulate several physiological pathways, including drug efflux and subsequent resistance to anticancer agents. There are now several in vitro studies showing their implication in drug resistance. Here, we will resume data focused on sorafenib [for other drugs such as doxorubicin, please see Manna and Sarkar (2020)].

LncRNA implication in sorafenib resistance is still in its infancy. The LncRNA H19 leads to sorafenib resistance through the regulation of miR-675 expression (Xu et al., 2020). Recently, MALAT-1 (Fan et al., 2020) and HOTAIR (Tang et al., 2020) have been found involved in sorafenib resistance in HCC cells. Growing literature reports miRNA deregulations associated with drug resistance: miR-122 downregulation is a major event in liver cancer and this could participate to resistance to several drugs, including sorafenib (Shu, Fan, Long, Zhou, & Wang, 2016; Wu et al., 2015; Xu et al., 2011).

Autophagy plays a central role in resistance to targeted therapy in many cancers, including in the liver. In Huh-7 cells an upregulation of miR-21 has been observed in sorafenib-resistant cells (He et al., 2015). MiR-21 notably targets PTEN leading to AKT activation—an anti-miR-21 being able to overcome sorafenib resistance through the modulation of autophagy. AKT-associated autophagy is a frequent event participating to drug resistance. Therefore some ncRNAs have been associated with AKT regulation and sorafenib response in HCC: miR-153, miR-494, and miR-10a-5p (Tang et al., 2016), miR-19a-3p (Jiang et al., 2018), miR-622 (Dietrich et al., 2018), and the lncRNAs SNHG3 (Zhang et al., 2019) and HEIH (Shen et al., 2020). Another important actor of the PI3K/AKT pathway, as previously mentioned, is PTEN, which is a target of miRNAs modulating sorafenib response such as miR-216a/miR-217 targeting the PTEN/Smad7 axis (Xia, Ooi, & Hui, 2013), miR-222 (Liu et al., 2014), and miR-494 (Liu, Liu, et al., 2015). Other miRNAs have been implicated in autophagy regulation and sorafenib response such as miR-142-3p (Zhang et al., 2018) and miR-375 (Zhao et al., 2018).

Apoptosis is another key cellular process modulated by miRNAs in sorafenib response. MiR-221 was upregulated in sorafenib-resistant HCC nodules and higher serum miR-221 levels are associated with increased disease progression under sorafenib treatment.

In addition, sorafenib resistance is associated with miR-221-targeting of caspase-3 (Fornari et al., 2017). MiR-221 is also a regulator of sensitivity of CCA cells to gemcitabine, a cytidine analog actually used in CCA, together with miR-205 and miR-29b (Okamoto, Miyoshi, & Murawaki, 2013). A potentiation of sorafenib-induced apoptosis is also dependent on several miRNAs: miR-193b targeting Mcl-1 (Mao et al., 2014), miR-34a regulating Bcl-2 (Yang et al., 2014) or the let-7 family of miRNAs targeting Bcl-xL (Shimizu et al., 2010), miR-181a regulating RASSF1 (Azumi, Tsubota, Sakabe, & Shiota, 2016), and miR-27b targeting p53/CYP1B1 (Mu et al., 2015). MiR-338-3p targeting HIF-1α modulates hypoxia, a key process involved in sorafenib resistance (Xu, Zhao, et al., 2014).

Importantly, the cluster miR-17/92, by the plethora of its targets, is a key regulator of drug resistance through the regulation of IL-6 and EGFR pathways regulating PTEN/AKT/mTOR/HIF1-α, Ras/MAP kinase and JAK/STAT signaling downstream (Awan et al., 2017). Therapies against this cluster to overcome drug resistance and impair progression appear of particular relevance.

11.5.5 Therapeutic potential of noncoding RNAs in liver cancer patients

Regarding the central roles of ncRNAs in liver cancer initiation, progression, and metastasis and the multiple signaling cascades they regulate, ncRNAs appear as attractive therapeutic opportunities for personalized strategies. MiRNA-based therapies have been particularly successful for liver diseases and cancers since this type of molecules are preferentially delivered into the liver (Roberts et al., 2006).

Two reciprocal approaches are currently developed, particularly for miRNAs, which have been extensively studied these recent decades (Fig. 11.5).

The first approach consists to increase the expression of tumor-suppressive miRNAs by miRNA delivery with vectors or mimics encapsulated into nanoparticles and liposomes. For oncogenic miRNAs, reduction of their expression has been obtained by several technologies based on complementary sequences such as anti-miRNA oligonucleotides (AMOs), antagomirs, or LNA. Unlike single-strand AMOs, which demonstrate poor stability after intravenous delivery, LNAs chemically modified by the addition of a methylene bridge in the ribose present higher affinity for their complementary RNAs and become resistant to nuclease degradation—giving more promising results in clinical tests. To improve AMO/antagomir affinity and stability in vivo, some chemical modifications have been added, for example, addition of 2'-O-methyl or 2'-O-methyoxyethyl groups or synthesis into phosphorothioate backbones (Lennox & Behlke, 2011). Other approaches have been tested to target multiple miRNAs by miRNA sponges, competitive inhibitors containing binding sites for several miRNAs, or multiple-target anti-miRNA antisense oligodeoxyribonucleotides—a single sequence being able to silence several miRNAs or miRNAs with common seeds. Another strategy is the use of miRNA masks, complementary to the binding sites located in mRNAs targeted by the miRNA of therapeutic interest.

The first miRNA-based therapy to enter into clinical trials in 2008 was an anti-miR-122 used against HCV and based on the LNA technology—the virus utilizing the hepatocytic miR-122 for its replication. Miravirsen succeeds in the treatment of chronic HCV in primates (Lanford et al., 2010) and in patients with chronic HCV infection (Janssen et al., 2013).

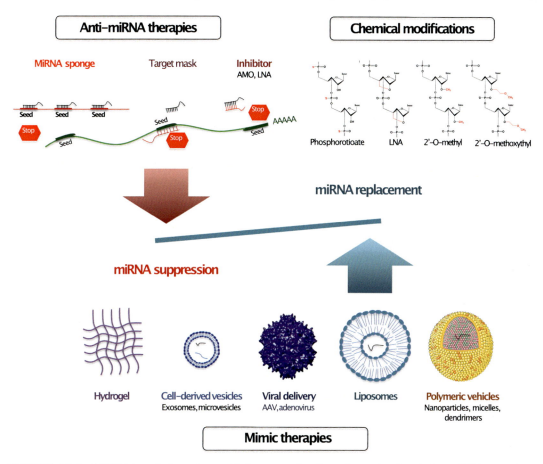

FIGURE 11.5 MiRNA-based therapies multiple approaches have been developed for miRNA targeting. Several strategies have been tested to suppress oncogenic miRNA such as miRNA sponge, inhibitors, or seed masks. Inversely, miRNA expression could be restored not only by using mimics encapsulated into exosomes, nanoparticles, liposomes, or hydrogel but also by viral delivery. Some of these strategies have been successfully tested in clinical trials such as LNA with miravirsen against miR-122, or mimics encapsulated into liposomes with mrx34. *AAV*, Adenovirus, associated virus; *AMO*, anti-miRNA oligonucleotide; *LNA*, locked nucleic acid; *miRNA*, microRNA.

Miravirsen dose dependently decreases plasma miR-122 and HCV RNA levels with no observable viral resistance and adverse events. Importantly, none of the patients have developed HCC, regarding the association between reduced levels of miR-122 and HCC. Thereafter, miR-122 inhibition was also tested with tiny-LNAs, seed-targeting 8-mer LNAs, which showed better distribution in tissue in relation with their shorter sequence and better efficiency in vitro and in vivo as compared to LNAs (Obad et al., 2011).

Some antagomirs have also been tested in vivo against liver cancer, in particular against miR-221, which is frequently upregulated in HCC and associated with aggressive HCC features. An anti-miR-221 oligonucleotide chemically modified with cholesterol- and

2'-O-methyl phosphorothioate successfully delays HCC development in an orthotopic HCC mouse model through reduction in cell proliferation, increase in apoptosis, and arrest of cell cycle (Park et al., 2011). A 2-O-metoxyethyl-antagomiR against miR-191 also gave significant results in an orthotopic mouse model of HCC. This antagomiR halves tumor mass in association with the reduction of tumor cell proliferation and increase in apoptosis (Elyakim et al., 2010). Last year, Su, Wu, Xia, Wu, and Liu (2020) have successfully silenced miR-21 in HCC orthotopic xenografts by using DNA tetrahedral nanostructures, a drug delivery system with promising clinical potentials exhibiting high biocompatibility, nuclease stability, and transmembrane passage ability (Hu et al., 2017).

In CCA, ncRNA-based therapy research is in its infancy. In 2018 a study demonstrates successful antitumor activity of nanoparticles encapsulating CXCR4 antagonists and a hairpin anti-miR-201 in a xenograft model of Mz-ChA-1 cells (Xie et al., 2018).

The converse approach for miRNA-based therapies consists to restore miRNA expression via a double-strand RNA mimicking the activity of the endogenous miRNA, and, so-called mimic-based therapies. In this case, mimics need to be protected from RNAse degradation through loading into nanoparticles or liposomes. The first mimic-based therapies, which entered into clinical trials, targeted miR-34a, an miRNA frequently decreased in cancers (MRX34). Despite interesting results obtained with MRX34 in mouse models and in patients, including in HBV-related HCC patients, the clinical trial was stopped due to detrimental effects associated with off-targets impact of miR-34a mimics and toxicity of liposomes leading to severe immune dysfunction and patient deaths (Beg et al., 2017). MiR-122 reexpression has also been tested. The cationic lipid nanoparticle LNP-DP1 with miR-122 mimic suppressed HCC development and angiogenesis without systemic toxicity (Hsu et al., 2013). Ultrasound-guided nanodroplets with miR-122 mimic also impaired HCC xenograft development (Guo et al., 2020), as well as gold nanocages coloaded with doxorubicin chemotherapy (Huang et al., 2018). Delivery of miRNA sequences using viral vectors is also an alternative to increase miRNA expression. Delivery of miR-26 mimic with an adeno-associated virus (AAV) vehicle in a mouse model of HCC mutated for MYC impaired tumor cell proliferation. Tumor reduction results from a loss of cyclin E2 and D2 in tumors and apoptosis induction without detrimental effects on nontumor tissue (Kota et al., 2009). An adenoviral delivery of miR-199a also reduced tumor growth in Hep3B cell xenografts and in HCC induced by diethylnitrosamine (Callegari et al., 2013). The use of viral vectors raised the question of safety, since AAV vectors could integrate into DNA. Additionally, at the end of 2020, a patient treated with viral vector against hemophilia developed HCC (http://uniqure.com/patients/clinical-hold-hemophilia-b.php). Even if the relationship between tumor development and AAV5 treatment has not been proved for that moment, cautions have to be kept.

For lncRNA-based therapies the same approaches than those successfully tested for miRNAs could be used, with additional strategies related to their pleiotropic mode of action and their longer size. For example, small-interfering RNA (siRNA) could be useful for lncRNA knocking down as well as molecules to sterically inhibit lncRNA-binding function on proteins. However, for that moment, no studies have been published for HCC but encouraging results were obtained with gapmer, chimeric RNA–DNA–RNA antisense oligonucleotides (ASOs), where RNA residues contain chemically modified backbone. MALAT-1 inhibition gave interesting results either with ASO in models of mammary tumors (Arun et al., 2016) or with LNA in multiple myeloma (Amodio et al., 2018). LNA-Gapmer technology was also promising in PDX (patient-derived xenograft) from melanoma patients with ASO against survival-associated mitochondrial melanoma-specific oncogenic noncoding RNA SAMMSON (Leucci et al., 2016), another oncogenic lncRNA, including in HCC (Yang, Cai, Hu, & Tu, 2019). More recently, nanoparticles delivering a TUG-1-ASO displayed antitumor activity in brain and pancreatic tumor PDX models (Watanabe et al., 2019). As a sponge method, a peptide nucleic acid complementary to the single-stranded domain of interaction between HOTAIR and EZH2 efficiently suppressed HOTAIR-dependent EZH2 activity and increased chemotherapy sensitivity of ovarian and breast cancers (Özeş et al., 2017) Thus common oncogenic lncRNAs appear as potent druggable ncRNAs, opening potential new avenues in molecular medicine based on lncRNAs.

In conclusion, a number of challenges remains to improve the efficiency of miRNA-based therapies and reduce the current limitations concerning their stability, administration, specificity, distribution, and secondary effects. The use of exosomes, which are endogenous vesicles with multiple biological functions, could be an attractive modality for personalized cancer treatment (Gougelet, 2018). Multiple clinical trials are ongoing with exosomes in several pathologies and cancers such as malignant pleural effusion, diabetes, acute ischemia stroke, pancreatic, or colon cancers. They are used as a delivery system for chemotherapeutic agents (Tang et al., 2012), drugs (metformin, NCT03109873), siRNA (KrasG12D, NCT03608631), ASO (IGF-1R NCT02507583), and ncRNA (miR-124, NCT03384433). In liver cancers, encouraging works have been obtained in HCC and CCA models following the injection of HSC-derived exosomes either loaded with miR-335-5p (Wang, Li, Piontek, Sakaguchi, & Selaru, 2018) and miR-195 (Li, Cheng, et al., 2017), respectively. The intratumor administration of exosomes derived from LCSC containing miR-451 and miR-31 also exerts antitumor activity (Fonsato et al., 2012).

Nevertheless, the clinical use of patient-derived exosomes appears expensive, time-consuming, and complex regarding the need of rigorous methods for their preparation and loading. Another major problem is the reproducibility in preparations since experiments showed that only 20% of content were common in three independent preparations of exosomes from mesenchymal stem cells (Chai et al., 2012). An attractive solution is the synthesis of bioengineered exosome mimetics, which permits reproducible and precisely characterized preparations of exosomes (Kim et al., 2020). This could help to obtain a gold standard method for their clinical application, particularly for liver cancers, a paradigm of inflammation-related cancers with highly specialized immune system and refractory to conventional treatments.

11.6 Summary and conclusion

Since their discovery during the 1990s, the study of ncRNAs is an expanding domain that allows the identification of an ncRNA signature discriminant for tumors, including for liver cancers. The enhanced understanding of ncRNA impacts in tumor cells and in their microenvironment unravels the great potential of these regulatory molecules as tools for diagnosis, prognosis, prediction of response to treatment and therapeutic strategies in liver cancers, which are tumors of bad prognosis with few therapeutic options and diagnosed in advanced stages. Interestingly, regarding their tumor expression and circulating panels, ncRNAs integrate all liver cancer features not only related to tumor proliferation, migration, invasion, apoptosis but also inflammation, metabolism, hypoxia, and immune escape. NcRNA research, particularly for lncRNA, is in its infancy and only a small proportion of deregulated ncRNAs in liver tumors have been characterized. Although a great number of studies help to decipher the different targets of one miRNA, the main challenge remains to determine for each type of tumor the preferential targets of one miRNA and how an mRNA targeted by multiple deregulated ncRNAs is expressed in that context. Additional studies in relevant animal models of liver cancers are, thus, required to determine precisely each function of ncRNA in liver pathophysiology and the molecular mechanisms engaged in response to ncRNA deregulations. This is required to avoid nonexpected effects if ncRNA-based therapies are considered.

From a clinical point of view, tissue and tumor specificity of ncRNA signatures render them attractive for therapies in cancers. Despite the growing body of knowledge, only few ncRNAs have entered into phases I and II trials and no ncRNA-based therapies have entered into clinical use. If recent advances have been made on nucleic acid therapeutics, improvements are still needed regarding their safety and specificity (delivery, off-target effects, competition with endogenous ncRNAs, etc.). With their preferential delivery into the liver, many mimics and inhibitors of ncRNAs have been tested in preclinical studies for liver disease and cancers and would hopefully be translated into clinical practice in the near future. Additionally, with their release into body fluids via exosome secretion, many ncRNAs and especially miRNAs can serve as potential diagnostic, prognostic, and predictive biomarkers for liver cancers. Several preclinical and clinical trials are ongoing on circulating ncRNAs such as NCT02448056 in HCC and NCT03102268 in CCA and could open new avenues for a noninvasive diagnosis of liver cancers, obtained at early stage and with better sensitivity and specificity.

Regarding all these results and ongoing projects, it is expected that ncRNAs will play more prominent roles in the clinical management of liver cancer patients in the coming future.

Acknowledgments

AG was supported by the Ligue Nationale Contre le Cancer and the Agence Nationale de la Recherche (DLK1-EPILIV 2018–2021). JS was supported by the Université de Paris.

References

Amaral, P. P., Clark, M. B., Gascoigne, D. K., Dinger, M. E., & Mattick, J. S. (2011). LncRNAdb: A reference database for long noncoding RNAs. *Nucleic Acids Research*, 39(1), D146–D151. Available from https://doi.org/10.1093/nar/gkq1138.

Amodio, N., Raimondi, L., Juli, G., Stamato, M. A., Caracciolo, D., Tagliaferri, P., & Tassone, P. (2018). MALAT1: A druggable long non-coding RNA for targeted anti-cancer approaches. *Journal of Hematology and Oncology*, 11(1). Available from https://doi.org/10.1186/s13045-018-0606-4.

Amr, K. S., Elmawgoud Atia, H. A., Elazeem Elbnhawy, R. A., & Ezzat, W. M. (2017). Early diagnostic evaluation of miR-122 and miR-224 as biomarkers for hepatocellular carcinoma. *Genes and Diseases*, 4(4), 215–221. Available from https://doi.org/10.1016/j.gendis.2017.10.003.

References

An, M., Wu, J., Zhu, J., & Lubman, D. M. (2018). Comparison of an optimized ultracentrifugation method vs size-exclusion chromatography for isolation of exosomes from human serum. *Journal of Proteome Research, 17*(10), 3599–3605. Available from https://doi.org/10.1021/acs.jproteome.8b00479.

Andersen, C. L., Jensen, J. L., & Ørntoft, T. F. (2004). Normalization of real-time quantitative reverse transcription-PCR data: A model-based variance estimation approach to identify genes suited for normalization, applied to bladder and colon cancer data sets. *Cancer Research, 64*(15), 5245–5250. Available from https://doi.org/10.1158/0008-5472.CAN-04-0496.

Angenard, G., Merdrignac, A., Louis, C., Edeline, J., & Coulouarn, C. (2019). Expression of long non-coding RNA ANRIL predicts a poor prognosis in intrahepatic cholangiocarcinoma. *Digestive and Liver Disease, 51*(9), 1337–1343. Available from https://doi.org/10.1016/j.dld.2019.03.019.

Anwar, S. L., Krech, T., Hasemeier, B., Schipper, E., Schweitzer, N., Vogel, A., ... Lehmann, U. (2012). Loss of imprinting and allelic switching at the DLK1-MEG3 locus in human hepatocellular carcinoma. *PLoS ONE, 7*(11). Available from https://doi.org/10.1371/journal.pone.0049462.

Arun, G., Diermeier, S., Akerman, M., Chang, K. C., Wilkinson, J. E., Hearn, S., ... Spector, D. L. (2016). Differentiation of mammary tumors and reduction in metastasis upon Malat1 lncRNA loss. *Genes and Development, 30*(1), 34–51. Available from https://doi.org/10.1101/gad.270959.115.

Awan, F. M., Naz, A., Obaid, A., Ikram, A., Ali, A., Ahmad, J., ... Janjua, H. A. (2017). MicroRNA pharmacogenomics based integrated model of miR-17–92 cluster in sorafenib resistant HCC cells reveals a strategy to forestall drug resistance. *Scientific Reports, 7*(1). Available from https://doi.org/10.1038/s41598-017-11943-1.

Azumi, J., Tsubota, T., Sakabe, T., & Shiota, G. (2016). miR-181a induces sorafenib resistance of hepatocellular carcinoma cells through downregulation of RASSF1 expression. *Cancer Science, 107*(9), 1256–1262. Available from https://doi.org/10.1111/cas.13006.

Banales, J. M., Marin, J. J. G., Lamarca, A., Rodrigues, P. M., Khan, S. A., Roberts, L. R., ... Gores, G. J. (2020). Cholangiocarcinoma 2020: The next horizon in mechanisms and management. *Nature Reviews Gastroenterology and Hepatology, 17*(9), 557–588. Available from https://doi.org/10.1038/s41575-020-0310-z.

Beg, M. S., Brenner, A. J., Sachdev, J., Borad, M., Kang, Y. K., Stoudemire, J., ... Hong, D. S. (2017). Phase I study of MRX34, a liposomal miR-34a mimic, administered twice weekly in patients with advanced solid tumors. *Investigational New Drugs, 35*(2), 180–188. Available from https://doi.org/10.1007/s10637-016-0407-y.

Benz, F., Roderburg, C., Cardenas, D. V., Vucur, M., Gautheron, J., Koch, A., ... Luedde, T. (2013). U6 is unsuitable for normalization of serum miRNA levels in patients with sepsis or liver fibrosis. *Experimental and Molecular Medicine, 45*(9). Available from https://doi.org/10.1038/emm.2013.81.

Bernuzzi, F., Marabita, F., Lleo, A., Carbone, M., Mirolo, M., Marzioni, M., et al. (2016). *Clin Exp Immunol, 185*, 61–71. Available from https://doi.org/10.1111/cei.12776, In press.

Bhartiya, D., Pal, K., Ghosh, S., Kapoor, S., Jalali, S., Panwar, B., ... Scaria, V. (2013). LncRNome: A comprehensive knowledgebase of human long noncoding RNAs. *Database, 2013*. Available from https://doi.org/10.1093/database/bat034.

Bian, E. B., Wang, Y. Y., Yang, Y., Wu, B. M., Xu, T., Meng, X. M., ... Li, J. (2017). Hotair facilitates hepatic stellate cells activation and fibrogenesis in the liver. *Biochimica et Biophysica Acta — Molecular Basis of Disease, 1863*(3), 674–686. Available from https://doi.org/10.1016/j.bbadis.2016.12.009.

Boyault, S., Rickman, D. S., De Reyniès, A., Balabaud, C., Rebouissou, S., Jeannot, E., ... Zucman-Rossi, J. (2007). Transcriptome classification of HCC is related to gene alterations and to new therapeutic targets. *Hepatology, 45*(1), 42–52. Available from https://doi.org/10.1002/hep.21467.

Braconi, C., Kogure, T., Valeri, N., Huang, N., Nuovo, G., Costinean, S., ... Patel, T. (2011). MicroRNA-29 can regulate expression of the long non-coding RNA gene MEG3 in hepatocellular cancer. *Oncogene, 30*(47), 4750–4756. Available from https://doi.org/10.1038/onc.2011.193.

Bray, F., Ferlay, J., Soerjomataram, I., Siegel, R. L., Torre, L. A., & Jemal, A. (2018). Global cancer statistics 2018: GLOBOCAN estimates of incidence and mortality worldwide for 36 cancers in 185 countries. *CA Cancer Journal for Clinicians, 68*(6), 394–424. Available from https://doi.org/10.3322/caac.21492.

Brockhausen, J., Tay, S. S., Grzelak, C. A., Bertolino, P., Bowen, D. G., d'Avigdor, W. M., ... Mccaughan, G. W. (2015). miR-181a mediates TGF-β-induced hepatocyte EMT and is dysregulated in cirrhosis and hepatocellular cancer. *Liver International, 35*(1), 240–253. Available from https://doi.org/10.1111/liv.12517.

Bucci, L., Garuti, F., Camelli, V., Lenzi, B., Farinati, F., Giannini, E. G., ... Gemini, S. (2016). Comparison between alcohol- and hepatitis C virus-related hepatocellular carcinoma: Clinical presentation, treatment and outcome. *Alimentary Pharmacology and Therapeutics, 43*(3), 385–399. Available from https://doi.org/10.1111/apt.13485.

Budhu, A., Jia, H. L., Forgues, M., Liu, C. G., Goldstein, D., Lam, A., ... Xin, W. W. (2008). Identification of metastasis-related microRNAs in hepatocellular carcinoma. *Hepatology*, 47(3), 897–907. Available from https://doi.org/10.1002/hep.22160.

Cairo, S., Wang, Y., De Reyniès, A., Duroure, K., Dahan, J., Redon, M. J., ... Buendia, M. A. (2010). Stem cell-like micro-RNA signature driven by Myc in aggressive liver cancer. *Proceedings of the National Academy of Sciences of the United States of America*, 107(47), 20471–20476. Available from https://doi.org/10.1073/pnas.1009009107.

Calderaro, J., Couchy, G., Imbeaud, S., Amaddeo, G., Letouzé, E., Blanc, J. F., ... Zucman-Rossi, J. (2017). Histological subtypes of hepatocellular carcinoma are related to gene mutations and molecular tumour classification. *Journal of Hepatology*, 67(4), 727–738. Available from https://doi.org/10.1016/j.jhep.2017.05.014.

Calin, G. A., Sevignani, C., Dumitru, C. D., Hyslop, T., Noch, E., Yendamuri, S., ... Croce, C. M. (2004). Human microRNA genes are frequently located at fragile sites and genomic regions involved in cancers. *Proceedings of the National Academy of Sciences of the United States of America*, 101(9), 2999–3004. Available from https://doi.org/10.1073/pnas.0307323101.

Callegari, E., Elamin, B. K., D'Abundo, L., Falzoni, S., Donvito, G., Moshiri, F., ... Sabbioni, S. (2013). Anti-tumor activity of a miR-199-dependent oncolytic adenovirus. *PLoS ONE*, 8(9). Available from https://doi.org/10.1371/journal.pone.0073964.

Calo, N., Ramadori, P., Sobolewski, C., Romero, Y., Maeder, C., Fournier, M., ... Foti, M. (2016). Stress-activated miR-21/miR-21* in hepatocytes promotes lipid and glucose metabolic disorders associated with high-fat diet consumption. *Gut*, 65(11), 1871–1881. Available from https://doi.org/10.1136/gutjnl-2015-310822.

Calvisi, D. F., & Solinas, A. (2020). Hepatoblastoma: Current knowledge and promises from preclinical studies. *Translational Gastroenterology and Hepatology*, 5. Available from https://doi.org/10.21037/tgh.2019.12.03.

Cao, K., Li, J., Zhao, Y., Wang, Q., Zeng, Q., He, S., ... Cao, P. (2016). miR-101 inhibiting cell proliferation, migration and invasion in hepatocellular carcinoma through downregulating girdin. *Molecules and Cells*, 39(2), 96–102. Available from https://doi.org/10.14348/molcells.2016.2161.

Carrillo-Reixach, J., Torrens, L., Simon-Coma, M., Royo, L., Domingo-Sàbat, M., Abril-Fornaguera, J., ... Armengol, C. (2020). Epigenetic footprint enables molecular risk stratification of hepatoblastoma with clinical implications. *Journal of Hepatology*, 73(2), 328–341. Available from https://doi.org/10.1016/j.jhep.2020.03.025.

Cartier, F., Indersie, E., Lesjean, S., Charpentier, J., Hooks, K. B., Ghousein, A., ... Grosset, C. F. (2017). New tumor suppressor microRNAs target glypican-3 in human liver cancer. *Oncotarget*, 8(25), 41211–41226. Available from https://doi.org/10.18632/oncotarget.17162.

Cermelli, S., Ruggieri, A., Marrero, J. A., Ioannou, G. N., & Beretta, L. (2011). Circulating microRNAs in patients with chronic hepatitis C and non-alcoholic fatty liver disease. *PLoS ONE*, 6(8). Available from https://doi.org/10.1371/journal.pone.0023937.

Chai, L. R., Sim, T. S., Ju, T. B., Kwan, S. S., Fatih, A., de K., D. P., ... Kiang, L. S. (2012). Proteolytic potential of the MSC exosome proteome: Implications for an exosome-mediated delivery of therapeutic proteasome. *International Journal of Proteomics*, 2012, 1–14. Available from https://doi.org/10.1155/2012/971907.

Chang, T. H., Huang, H. Y., Hsu, J. B. K., Weng, S. L., Horng, J. T., & Huang, H. D. (2013). An enhanced computational platform for investigating the roles of regulatory RNA and for identifying functional RNA motifs. *BMC Bioinformatics*, 14. Available from https://doi.org/10.1186/1471-2105-14-S2-S4.

Chen, Y., & Wang, X. (2020). MiRDB: An online database for prediction of functional microRNA targets. *Nucleic Acids Research*, 48(1), D127–D131. Available from https://doi.org/10.1093/nar/gkz757.

Chen, L., Yan, H. X., Yang, W., Hu, L., Yu, L. X., Liu, Q., ... Wang, H. Y. (2009). The role of microRNA expression pattern in human intrahepatic cholangiocarcinoma. *Journal of Hepatology*, 50(2), 358–369. Available from https://doi.org/10.1016/j.jhep.2008.09.015.

Chen, G., Wang, Z., Wang, D., Qiu, C., Liu, M., Chen, X., ... Cui, Q. (2013). LncRNADisease: A database for long-non-coding RNA-associated diseases. *Nucleic Acids Research*, 41(1), D983–D986. Available from https://doi.org/10.1093/nar/gks1099.

Chen, C., Yang, D., Wang, Q., & Wang, X. (2015). Expression and clinical pathological significance of miR-200a in concurrent cholangiocarcinoma associated with hepatolithiasis. *Medical Science Monitor*, 21, 3585–3590. Available from https://doi.org/10.12659/MSM.895013.

Cheng, D., Deng, J., Zhang, B., He, X., Meng, Z., Li, G., ... Zhou, J. (2018). LncRNA HOTAIR epigenetically suppresses miR-122 expression in hepatocellular carcinoma via DNA methylation. *EBioMedicine*, 36, 159–170. Available from https://doi.org/10.1016/j.ebiom.2018.08.055.

Clotman, F., Jacquemin, P., Plumb-Rudewiez, N., Pierreux, C. E., Van Der Smissen, P., Dietz, H. C., ... Lemaigre, F. P. (2005). Control of liver cell fate decision by a gradient of TGFβ signaling modulated by Onecut transcription factors. *Genes and Development*, 19(16), 1849–1854. Available from https://doi.org/10.1101/gad.340305.

Correa-Gallego, C., Maddalo, D., Doussot, A., Kemeny, N., Kingham, T. P., Allen, P. J., ... Ventura, A. (2016). Circulating plasma levels of MicroRNA-21 and MicroRNA-221 are potential diagnostic markers for primary intrahepatic cholangiocarcinoma. *PLoS ONE*, 11(9). Available from https://doi.org/10.1371/journal.pone.0163699.

Crispe, I. N. (2009). The liver as a lymphoid organ. *Annual Review of Immunology*, 27, 147–163. Available from https://doi.org/10.1146/annurev.immunol.021908.132629.

Daneshvar, K., Ardehali, M. B., Klein, I. A., Hsieh, F. K., Kratkiewicz, A. J., Mahpour, A., ... Mullen, A. C. (2020). lncRNA DIGIT and BRD3 protein form phase-separated condensates to regulate endoderm differentiation. *Nature Cell Biology*, 22(10), 1211–1222. Available from https://doi.org/10.1038/s41556-020-0572-2.

Davalos, A., Goedeke, L., Smibert, P., Ramirez, C. M., Warrier, N. P., Andreo, U., ... Fernandez-Hernando, C. (2011). miR-33a/b contribute to the regulation of fatty acid metabolism and insulin signaling. *Proceedings of the National Academy of Sciences*, 9232–9237. Available from https://doi.org/10.1073/pnas.1102281108.

Deng, Y., & Chen, Y. (2017). Increased expression of miR-29a and its prognostic significance in patients with cholangiocarcinoma. *Oncology Research and Treatment*, 40(3), 128–132. Available from https://doi.org/10.1159/000455869.

Derrien, T., Johnson, R., Bussotti, G., Tanzer, A., Djebali, S., Tilgner, H., ... Guigó, R. (2012). The GENCODE v7 catalog of human long noncoding RNAs: Analysis of their gene structure, evolution, and expression. *Genome Research*, 22(9), 1775–1789. Available from https://doi.org/10.1101/gr.132159.111.

Dietrich, P., Koch, A., Fritz, V., Hartmann, A., Bosserhoff, A. K., & Hellerbrand, C. (2018). Wild type Kirsten rat sarcoma is a novel microRNA- 622-regulated therapeutic target for hepatocellular carcinoma and contributes to sorafenib resistance. *Gut*, 67(7), 1328–1341. Available from https://doi.org/10.1136/gutjnl-2017-315402.

Dong, Y., Zhang, N., Zhao, S., Chen, X., Li, F., & Tao, X. (2019). MiR-221-3p and miR-15b-5p promote cell proliferation and invasion by targeting Axin2 in liver cancer. *Oncology Letters*, 18(6), 6491–6500. Available from https://doi.org/10.3892/ol.2019.11056.

Du, Y., Kong, G., You, X., Zhang, S., Zhang, T., Gao, Y., ... Zhang, X. (2012). Elevation of highly up-regulated in liver cancer (HULC) by hepatitis B virus X protein promotes hepatoma cell proliferation via down-regulating p18. *Journal of Biological Chemistry*, 287(31), 26302–26311. Available from https://doi.org/10.1074/jbc.M112.342113.

Dupuis-Sandoval, F., Poirier, M., & Scott, M. S. (2015). The emerging landscape of small nucleolar RNAs in cell biology. *Wiley Interdisciplinary Reviews: RNA*, 6(4), 381–397. Available from https://doi.org/10.1002/wrna.1284.

Eichenmüller, M., Trippel, F., Kreuder, M., Beck, A., Schwarzmayr, T., Häberle, B., ... Kappler, R. (2014). The genomic landscape of hepatoblastoma and their progenies with HCC-like features. *Journal of Hepatology*, 61(6), 1312–1320. Available from https://doi.org/10.1016/j.jhep.2014.08.009.

Elemeery, M. N., Mohamed, M. A., Madkour, M. A., Shamseya, M. M., Issa, N. M., Badr, A. N., ... Pan, C. H. (2019). MicroRNA signature in patients with hepatocellular carcinoma associated with type 2 diabetes. *World Journal of Gastroenterology*, 25(42), 6322–6341. Available from https://doi.org/10.3748/wjg.v25.i42.6322.

El-Khoury, V., Pierson, S., Kaoma, T., Bernardin, F., & Berchem, G. (2016). Assessing cellular and circulating miRNA recovery: The impact of the RNA isolation method and the quantity of input material. *Scientific Reports*, 6. Available from https://doi.org/10.1038/srep19529.

Elyakim, E., Sitbon, E., Faerman, A., Tabak, S., Montia, E., Belanis, L., ... Yerushalmi, N. (2010). hsa-miR-191 is a candidate oncogene target for hepatocellular carcinoma therapy. *Cancer Research*, 70(20), 8077–8087. Available from https://doi.org/10.1158/0008-5472.CAN-10-1313.

Fan, Y., Siklenka, K., Arora, S. K., Ribeiro, P., Kimmins, S., & Xia, J. (2016). miRNet – dissecting miRNA-target interactions and functional associations through network-based visual analysis. *Nucleic Acids Research*, 44(1), W135–W141. Available from https://doi.org/10.1093/nar/gkw288.

Fan, L., Huang, X., Chen, J., Zhang, K., Gu, Y. H., Sun, J., & Cui, S. Y. (2020). Long noncoding RNA MALAT1 contributes to sorafenib resistance by targeting miR-140-5p/aurora-a signaling in hepatocellular carcinoma. *Molecular Cancer Therapeutics*, 19(5), 1197–1209. Available from https://doi.org/10.1158/1535-7163.MCT-19-0203.

Fang, J. H., Zhou, H. C., Zeng, C., Yang, J., Liu, Y., Huang, X., ... Zhuang, S. M. (2011). MicroRNA-29b suppresses tumor angiogenesis, invasion, and metastasis by regulating matrix metalloproteinase 2 expression. *Hepatology*, 54(5), 1729–1740. Available from https://doi.org/10.1002/hep.24577.

Fonsato, V., Collino, F., Herrera, M. B., Cavallari, C., Deregibus, M. C., Cisterna, B., ... Camussi, G. (2012). Human liver stem cell-derived microvesicles inhibit hepatoma growth in SCID mice by delivering antitumor microRNAs. *Stem Cells, 30*(9), 1985–1998. Available from https://doi.org/10.1002/stem.1161.

Fornari, F., Gramantieri, L., Ferracin, M., Veronese, A., Sabbioni, S., Calin, G. A., ... Negrini, M. (2008). MiR-221 controls CDKN1C/p57 and CDKN1B/p27 expression in human hepatocellular carcinoma. *Oncogene, 27*(43), 5651–5661. Available from https://doi.org/10.1038/onc.2008.178.

Fornari, F., Pollutri, D., Patrizi, C., La Bella, T., Marinelli, S., Casadei Gardini, A., ... Gramantieri, L. (2017). In hepatocellular carcinoma miR-221 modulates sorafenib resistance through inhibition of caspase-3–mediated apoptosis. *Clinical Cancer Research, 23*(14), 3953–3965. Available from https://doi.org/10.1158/1078-0432.CCR-16-1464.

Fritz, J. V., Heintz-Buschart, A., Ghosal, A., Wampach, L., Etheridge, A., Galas, D., & Wilmes, P. (2016). Sources and functions of extracellular small RNAs in human circulation. *Annual Review of Nutrition, 36*, 301–336. Available from https://doi.org/10.1146/annurev-nutr-071715-050711.

Frost, R. J. A., & Olson, E. N. (2011). Control of glucose homeostasis and insulin sensitivity by the Let-7 family of microRNAs. *Proceedings of the National Academy of Sciences*, 21075–21080. Available from https://doi.org/10.1073/pnas.1118922109.

Furuta, M., Kozaki, K. I., Tanaka, S., Arii, S., Imoto, I., & Inazawa, J. (2009). miR-124 and miR-203 are epigenetically silenced tumor-suppressive microRNAs in hepatocellular carcinoma. *Carcinogenesis, 31*(5), 766–776. Available from https://doi.org/10.1093/carcin/bgp250.

Garofalo, M., Di Leva, G., Romano, G., Nuovo, G., Suh, S. S., Ngankeu, A., ... Croce, C. M. (2009). miR-221&222 regulate TRAIL resistance and enhance tumorigenicity through PTEN and TIMP3 downregulation. *Cancer Cell, 16*(6), 498–509. Available from https://doi.org/10.1016/j.ccr.2009.10.014.

Ge, W., Yu, D. C., Li, Q. G., Chen, X., Zhang, C. Y., & Ding, Y. T. (2014). Expression of serum miR-16, let-7f, and miR-21 in patients with hepatocellular carcinoma and their clinical significances. *Clinical Laboratory, 60*(3), 427–434. Available from https://doi.org/10.7754/Clin.Lab.2013.130133.

Gerin, I., Clerbaux, L. A., Haumont, O., Lanthier, N., Das, A. K., Burant, C. F., ... Bommer, G. T. (2010). Expression of miR-33 from an SREBP2 intron inhibits cholesterol export and fatty acid oxidation. *Journal of Biological Chemistry, 285*(44), 33652–33661. Available from https://doi.org/10.1074/jbc.M110.152090.

Giray, B. G., Emekdas, G., Tezcan, S., Ulger, M., Serin, M. S., Sezgin, O., ... Tiftik, E. N. (2014). Profiles of serum microRNAs; miR-125b-5p and miR223-3p serve as novel biomarkers for HBV-positive hepatocellular carcinoma. *Molecular Biology Reports, 41*(7), 4513–4519. Available from https://doi.org/10.1007/s11033-014-3322-3.

Gougelet, A. (2018). Exosomal microRNAs as a potential therapeutic strategy in hepatocellular carcinoma. *World Journal of Hepatology, 10*(11), 785–789. Available from https://doi.org/10.4254/wjh.v10.i11.785.

Gougelet, A., & Colnot, S. (2016). Hepatocellular carcinoma diagnosis: Circulating microRNAs emerge as robust biomarkers. *Clinics and Research in Hepatology and Gastroenterology, 40*(4), 367–369. Available from https://doi.org/10.1016/j.clinre.2015.12.010.

Gramantieri, L., Fornari, F., Ferracin, M., Veronese, A., Sabbioni, S., Calin, G. A., ... Negrini, M. (2009). MicroRNA-221 targets Bmf in hepatocellular carcinoma and correlates with tumor multifocality. *Clinical Cancer Research, 15*(16), 5073–5081. Available from https://doi.org/10.1158/1078-0432.CCR-09-0092.

Gramantieri, L., Baglioni, M., Fornari, F., Laginestra, M. A., Ferracin, M., Indio, V., ... Giovannini, C. (2018). LncRNAs as novel players in hepatocellular carcinoma recurrence. *Oncotarget, 9*(80), 35085–35099. Available from https://doi.org/10.18632/oncotarget.26202.

Griffiths-Jones, S. (2004). The microRNA Registry. *Nucleic Acids Research, 32*(Database issue), D109–D111. Available from https://doi.org/10.1093/nar/gkh023.

Grillone, K., Riillo, C., Riillo, C., Scionti, F., Rocca, R., Rocca, R., ... Tassone, P. (2020). Non-coding RNAs in cancer: Platforms and strategies for investigating the genomic \dark matter\. *Journal of Experimental and Clinical Cancer Research, 39*(1). Available from https://doi.org/10.1186/s13046-020-01622-x.

Guduric-Fuchs, J., O'Connor, A., Camp, B., O'Neill, C. L., Medina, R. J., & Simpson, D. A. (2012). Selective extracellular vesicle-mediated export of an overlapping set of microRNAs from multiple cell types. *BMC Genomics, 13*(1). Available from https://doi.org/10.1186/1471-2164-13-357.

Guo, H., Xu, M., Cao, Z., Li, W., Chen, L., Xie, X., ... Liu, J. (2020). Ultrasound-assisted miR-122-loaded polymeric nanodroplets for hepatocellular carcinoma gene therapy. *Molecular Pharmaceutics, 17*(2), 541–553. Available from https://doi.org/10.1021/acs.molpharmaceut.9b00983.

Halley, P., Kadakkuzha, B. M., Faghihi, M. A., Magistri, M., Zeier, Z., Khorkova, O., ... Wahlestedt, C. (2014). Regulation of the apolipoprotein gene cluster by a long noncoding RNA. *Cell Reports*, *6*(1), 222−230. Available from https://doi.org/10.1016/j.celrep.2013.12.015.

Hand, N. J., Master, Z. R., Eauclaire, S. F., Weinblatt, D. E., Matthews, R. P., & Friedman, J. R. (2009). The microRNA-30 family is required for vertebrate hepatobiliary development. *Gastroenterology*, *136*(3), 1081−1090. Available from https://doi.org/10.1053/j.gastro.2008.12.006.

Harding, J. J., Nandakumar, S., Armenia, J., Khalil, D. N., Albano, M., Ly, M., ... Abou-Alfa, G. K. (2019). Prospective genotyping of hepatocellular carcinoma: Clinical implications of next-generation sequencing for matching patients to targeted and immune therapies. *Clinical Cancer Research*, *25*(7), 2116−2126. Available from https://doi.org/10.1158/1078-0432.CCR-18-2293.

He, L., & Hannon, G. J. (2004). MicroRNAs: Small RNAs with a big role in gene regulation. *Nature Reviews Genetics*, *5*(7), 522−531. Available from https://doi.org/10.1038/nrg1379.

He, C., Dong, X., Zhai, B., Jiang, X., Dong, D., Li, B., ... Sun, X. (2015). MiR-21 mediates sorafenib resistance of hepatocellular carcinoma cells by inhibiting autophagy via the PTEN/Akt pathway. *Oncotarget*, *6*(30), 28867−28881. Available from https://doi.org/10.18632/oncotarget.4814.

Horst, A. K., Neumann, K., Diehl, L., & Tiegs, G. (2016). Modulation of liver tolerance by conventional and non-conventional antigen-presenting cells and regulatory immune cells. *Cellular and Molecular Immunology*, *13*(3), 277−292. Available from https://doi.org/10.1038/cmi.2015.112.

Hou, J., Lin, L., Zhou, W., Wang, Z., Ding, G., Dong, Q., ... Cao, X. (2011). Identification of miRNomes in human liver and hepatocellular carcinoma reveals miR-199a/b-3p as therapeutic target for hepatocellular carcinoma. *Cancer Cell*, *19*(2), 232−243. Available from https://doi.org/10.1016/j.ccr.2011.01.001.

Hsu, S. H., Wang, B., Kota, J., Yu, J., Costinean, S., Kutay, H., ... Ghoshal, K. (2012). Essential metabolic, anti-inflammatory, and anti-tumorigenic functions of miR-122 in liver. *Journal of Clinical Investigation*, *122*(8), 2871−2883. Available from https://doi.org/10.1172/JCI63539.

Hsu, S. H., Yu, B., Wang, X., Lu, Y., Schmidt, C. R., Lee, R. J., ... Ghoshal, K. (2013). Cationic lipid nanoparticles for therapeutic delivery of siRNA and miRNA to murine liver tumor. *Nanomedicine: Nanotechnology, Biology, and Medicine*, *9*(8), 1169−1180. Available from https://doi.org/10.1016/j.nano.2013.05.007.

Hu, Y., Chen, Z., Zhang, H., Li, M., Hou, Z., Luo, X. X., & Xue, X. (2017). Development of DNA tetrahedron-based drug delivery system. *Drug Delivery*, *24*(1), 1295−1301. Available from https://doi.org/10.1080/10717544.2017.1373166.

Huang, Q., Liu, L., Liu, C. H., You, H., Shao, F., Xie, F., ... Zhang, C. H. (2013). MicroRNA-21 regulates the invasion and metastasis in cholangiocarcinoma and may be a potential biomarker for cancer prognosis. *Asian Pacific Journal of Cancer Prevention*, *14*(2), 829−834. Available from https://doi.org/10.7314/APJCP.2013.14.2.829.

Huang, M. D., Chen, W. M., Qi, F. Z., Sun, M., Xu, T. P., Ma, P., & Shu, Y. Q. (2015). Long non-coding RNA TUG1 is up-regulated in hepatocellular carcinoma and promotes cell growth and apoptosis by epigenetically silencing of KLF2. *Molecular Cancer*, *14*(1). Available from https://doi.org/10.1186/s12943-015-0431-0.

Huang, K. T., Kuo, I. Y., Tsai, M. C., Wu, C. H., Hsu, L. W., Chen, L. Y., ... Chen, K. D. (2017). Factor VII-induced MicroRNA-135a inhibits autophagy and is associated with poor prognosis in hepatocellular carcinoma. *Molecular Therapy − Nucleic Acids*, *9*, 274−283. Available from https://doi.org/10.1016/j.omtn.2017.10.002.

Huang, S., Liu, Y., Xu, X., Ji, M., Li, Y., Song, C., ... Hu, Y. (2018). Triple therapy of hepatocellular carcinoma with microRNA-122 and doxorubicin co-loaded functionalized gold nanocages. *Journal of Materials Chemistry B*, *6*(15), 2217−2229. Available from https://doi.org/10.1039/c8tb00224j.

Huang, M., Wang, H., Hu, X., & Cao, X. (2019). lncRNA MALAT1 binds chromatin remodeling subunit BRG1 to epigenetically promote inflammation-related hepatocellular carcinoma progression. *OncoImmunology*, *8*(1). Available from https://doi.org/10.1080/2162402X.2018.1518628.

Huang, Z., Shi, J., Gao, Y., Cui, C., Zhang, S., Li, J., ... Cui, Q. (2019). HMDD v3.0: A database for experimentally supported human microRNA-disease associations. *Nucleic Acids Research*, *47*(1), D1013−D1017. Available from https://doi.org/10.1093/nar/gky1010.

Huarte, M. (2015). The emerging role of lncRNAs in cancer. *Nature Medicine*, *21*(11), 1253−1261. Available from https://doi.org/10.1038/nm.3981.

Hung, C. H., Hu, T. H., Lu, S. N., Kuo, F. Y., Chen, C. H., Wang, J. H., ... Chiu, Y. C. (2016). Circulating microRNAs as biomarkers for diagnosis of early hepatocellular carcinoma associated with hepatitis B virus. *International Journal of Cancer*, *138*(3), 714−720. Available from https://doi.org/10.1002/ijc.29802.

Hung, Y. H., Kanke, M., Kurtz, C. L., Cubitt, R. L., Bunaciu, R. P., Zhou, L., ... Sethupathy, P. (2019). MiR-29 regulates de novo lipogenesis in the liver and circulating triglyceride levels in a Sirt1-dependent manner. *Frontiers in Physiology, 10*. Available from https://doi.org/10.3389/fphys.2019.01367.

Hutvágner, G., & Zamore, P. D. (2002). A microRNA in a multiple-turnover RNAi enzyme complex. *Science, 297*(5589), 2056–2060. Available from https://doi.org/10.1126/science.1073827.

Indersie, E., Lesjean, S., Hooks, K. B., Sagliocco, F., Ernault, T., Cairo, S., ... Grosset, C. F. (2017). MicroRNA therapy inhibits hepatoblastoma growth in vivo by targeting β-catenin and Wnt signaling. *Hepatology Communications*, 168–183. Available from https://doi.org/10.1002/hep4.1029.

Ishibashi, M., Kogo, R., Shibata, K., Sawada, G., Takahashi, Y., Kurashige, J., ... Mori, M. (2013). Clinical significance of the expression of long non-coding RNA HOTAIR in primary hepatocellular carcinoma. *Oncology Reports, 29*(3), 946–950. Available from https://doi.org/10.3892/or.2012.2219.

Iyer, M. K., Niknafs, Y. S., Malik, R., Singhal, U., Sahu, A., Hosono, Y., ... Chinnaiyan, A. M. (2015). The landscape of long noncoding RNAs in the human transcriptome. *Nature Genetics, 47*(3), 199–208. Available from https://doi.org/10.1038/ng.3192.

Jang, S. Y., Kim, G., Park, S. Y., Lee, Y. R., Kwon, S. H., Kim, H. S., ... Hur, K. (2017). Clinical significance of lncRNA-ATB expression in human hepatocellular carcinoma. *Oncotarget, 8*(45), 78588–78597. Available from https://doi.org/10.18632/oncotarget.21094.

Janssen, H. L. A., Reesink, H. W., Lawitz, E. J., Zeuzem, S., Rodriguez-Torres, M., Patel, K., ... Hodges, M. R. (2013). Treatment of HCV infection by targeting microRNA. *New England Journal of Medicine, 368*(18), 1685–1694. Available from https://doi.org/10.1056/NEJMoa1209026.

Jeggari, A., Marks, D. S., & Larsson, E. (2012). miRcode: A map of putative microrna target sites in the long noncoding transcriptome. *Bioinformatics, 28*(15), 2062–2063. Available from https://doi.org/10.1093/bioinformatics/bts344.

Jenne, C. N., & Kubes, P. (2013). Immune surveillance by the liver. *Nature Immunology, 14*(10), 996–1006. Available from https://doi.org/10.1038/ni.2691.

Ji, J., Zhang, J., Huang, G., Qian, J., Wang, X., & Mei, S. (2009). Over-expressed microRNA-27a and 27b influence fat accumulation and cell proliferation during rat hepatic stellate cell activation. *FEBS Letters, 583*(4), 759–766. Available from https://doi.org/10.1016/j.febslet.2009.01.034.

Jiang, W., Liu, Y., Liu, R., Zhang, K., & Zhang, Y. (2015). The lncRNA DEANR1 facilitates human endoderm differentiation by activating FOXA2 expression. *Cell Reports, 11*(1), 137–148. Available from https://doi.org/10.1016/j.celrep.2015.03.008.

Jiang, X. M., Li, Z. L., Li, J. L., Zheng, W. Y., Li, X. H., Cui, Y. F., & Sun, D. J. (2017). LncRNA CCAT1 as the unfavorable prognostic biomarker for cholangiocarcinoma. *European Review for Medical and Pharmacological Sciences, 21*(6), 1242–1247.

Jiang, X. M., Yu, X. N., Liu, T. T., Zhu, H. R., Shi, X., Bilegsaikhan, E., ... Zhu, J. M. (2018). microRNA-19a-3p promotes tumor metastasis and chemoresistance through the PTEN/Akt pathway in hepatocellular carcinoma. *Biomedicine and Pharmacotherapy, 105*, 1147–1154. Available from https://doi.org/10.1016/j.biopha.2018.06.097.

Jin, Y., Wong, Y. S., Goh, B. K. P., Chan, C. Y., Cheow, P. C., Chow, P. K. H., ... Lee, C. G. L. (2019). Circulating microRNAs as potential diagnostic and prognostic biomarkers in hepatocellular carcinoma. *Scientific Reports, 9*(1). Available from https://doi.org/10.1038/s41598-019-46872-8.

Jungermann, K., & Kietzmann, T. (1996). Zonation of parenchymal and nonparenchymal metabolism in liver. *Annual Review of Nutrition, 16*, 179–203. Available from https://doi.org/10.1146/annurev.nu.16.070196.001143.

Jusakul, A., Cutcutache, I., Yong, C. H., Lim, J. Q., Huang, M. N., Padmanabhan, N., ... Tan, P. (2017). Whole-genome and epigenomic landscapes of etiologically distinct subtypes of cholangiocarcinoma. *Cancer Discovery, 7*(10), 1116–1135. Available from https://doi.org/10.1158/2159-8290.CD-17-0368.

Kalluri, R. (2016). The biology and function of exosomes in cancer. *Journal of Clinical Investigation, 126*(4), 1208–1215. Available from https://doi.org/10.1172/JCI81135.

Kalra, H., Simpson, R., Ji, H., Aikawa, E., Altevogt, P., Askenase, P., ... Mathivanan, S. (2012). Vesiclepedia: A compendium for extracellular vesicles with continuous community annotation. *PLoS Biology, 10*, e1001450. Available from https://doi.org/10.1371/journal.pbio.1001450.

Kamel, M. M., Matboli, M., Sallam, M., Montasser, I. F., Saad, A. S., & El-Tawdi, A. H. F. (2016). Investigation of long noncoding RNAs expression profile as potential serum biomarkers in patients with hepatocellular carcinoma. *Translational Research, 168*, 134–145. Available from https://doi.org/10.1016/j.trsl.2015.10.002.

Kapranov, P., Cheng, J., Dike, S., Nix, D. A., Duttagupta, R., Willingham, A. T., ... Gingeras, T. R. (2007). RNA maps reveal new RNA classes and a possible function for pervasive transcription. *Science*, *316*(5830), 1484–1488. Available from https://doi.org/10.1126/science.1138341.

Karakatsanis, A., Papaconstantinou, I., Gazouli, M., Lyberopoulou, A., Polymeneas, G., & Voros, D. (2013). Expression of microRNAs, miR-21, miR-31, miR-122, miR-145, miR-146a, miR-200c, miR-221, miR-222, and miR-223 in patients with hepatocellular carcinoma or intrahepatic cholangiocarcinoma and its prognostic significance. *Molecular Carcinogenesis*, *52*, 297–303. Available from https://doi.org/10.1002/mc.21864.

Keerthikumar, S., Chisanga, D., Ariyaratne, D., Al Saffar, H., Anand, S., Zhao, K., ... Mathivanan, S. (2016). ExoCarta: A web-based compendium of exosomal cargo. *Journal of Molecular Biology*, *428*(4), 688–692. Available from https://doi.org/10.1016/j.jmb.2015.09.019.

Kim, H., Kim, D., Nam, H., Moon, S., Kwon, Y. J., & Lee, J. B. (2020). Engineered extracellular vesicles and their mimetics for clinical translation. *Methods*, *177*, 80–94. Available from https://doi.org/10.1016/j.ymeth.2019.10.005.

Konishi, H., Ichikawa, D., Yamamoto, Y., Arita, T., Shoda, K., Hiramoto, H., ... Otsuji, E. (2016). Plasma level of metastasis-associated lung adenocarcinoma transcript 1 is associated with liver damage and predicts development of hepatocellular carcinoma. *Cancer Science*, *107*(2), 149–154. Available from https://doi.org/10.1111/cas.12854.

Kota, J., Chivukula, R. R., O'Donnell, K. A., Wentzel, E. A., Montgomery, C. L., Hwang, H. W., ... Mendell, J. T. (2009). Therapeutic microRNA delivery suppresses tumorigenesis in a murine liver cancer model. *Cell*, *137*(6), 1005–1017. Available from https://doi.org/10.1016/j.cell.2009.04.021.

Kriegsmann, M., Roessler, S., Kriegsmann, K., Renner, M., Longuespée, R., Albrecht, T., ... Goeppert, B. (2019). Programmed cell death ligand 1 (PD-L1, CD274) in cholangiocarcinoma — Correlation with clinicopathological data and comparison of antibodies. *BMC Cancer*, *19*(1). Available from https://doi.org/10.1186/s12885-018-5254-0.

Kutay, H., Bai, S., Datta, J., Motiwala, T., Pogribny, I., Frankel, W., ... Ghoshal, K. (2006). Downregulation of miR-122 in the rodent and human hepatocellular carcinomas. *Journal of Cellular Biochemistry*, *99*(3), 671–678. Available from https://doi.org/10.1002/jcb.20982.

Ladeiro, Y., Couchy, G., Balabaud, C., Bioulac-Sage, P., Pelletier, L., Rebouissou, S., & Zucman-Rossi, J. (2008). MicroRNA profiling in hepatocellular tumors is associated with clinical features and oncogene/tumor suppressor gene mutations. *Hepatology*, *47*(6), 1955–1963. Available from https://doi.org/10.1002/hep.22256.

Lai, M. C., Yang, Z., Zhou, L., Zhu, Q. Q., Xie, H. Y., Zhang, F., ... Zheng, S. S. (2012). Long non-coding RNA MALAT-1 overexpression predicts tumor recurrence of hepatocellular carcinoma after liver transplantation. *Medical Oncology*, *29*(3), 1810–1816. Available from https://doi.org/10.1007/s12032-011-0004-z.

Lan, F. F., Wang, H., Chen, Y. C., Chan, C. Y., Ng, S. S., Li, K., ... Kung, H. F. (2011). Hsa-let-7g inhibits proliferation of hepatocellular carcinoma cells by downregulation of c-Myc and upregulation of p16INK4A. *International Journal of Cancer*, *128*(2), 319–331. Available from https://doi.org/10.1002/ijc.25336.

Lan, X., Yan, J., Ren, J., Zhong, B., Li, J., Li, Y., ... Lu, S. (2016). A novel long noncoding RNA Lnc-HC binds hnRNPA2B1 to regulate expressions of Cyp7a1 and Abca1 in hepatocytic cholesterol metabolism. *Hepatology*, *64*(1), 58–72. Available from https://doi.org/10.1002/hep.28391.

Lanford, R. E., Hildebrandt-Eriksen, E. S., Petri, A., Persson, R., Lindow, M., Munk, M. E., ... Rum, H. (2010). Therapeutic silencing of microRNA-122 in primates with chronic hepatitis C virus infection. *Science*, *327*(5962), 198–201. Available from https://doi.org/10.1126/science.1178178.

Laudadio, I., Manfroid, I., Achouri, Y., Schmidt, D., Wilson, M. D., Cordi, S., ... Lemaigre, F. P. (2012). A feedback loop between the liver-enriched transcription factor network and miR-122 controls hepatocyte differentiation. *Gastroenterology*, *142*(1), 119–129. Available from https://doi.org/10.1053/j.gastro.2011.09.001.

Le Sage, C., Nagel, R., Egan, D. A., Schrier, M., Mesman, E., Mangiola, A., ... Agami, R. (2007). Regulation of the p27Kip1 tumor suppressor by miR-221 and miR-222 promotes cancer cell proliferation. *EMBO Journal*, *26*(15), 3699–3708. Available from https://doi.org/10.1038/sj.emboj.7601790.

Lee, Y., Jeon, K., Lee, J. T., Kim, S., & Kim, V. N. (2002). MicroRNA maturation: Stepwise processing and subcellular localization. *EMBO Journal*, *21*(17), 4663–4670. Available from https://doi.org/10.1093/emboj/cdf476.

Lee, C. G., Kim, Y. W., Kim, E. H., Meng, Z., Huang, W., Hwang, S. J., & Kim, S. G. (2012). Farnesoid X receptor protects hepatocytes from injury by repressing miR-199a-3p, which increases levels of LKB1. *Gastroenterology*, *142*(5), 1206–e7. Available from https://doi.org/10.1053/j.gastro.2012.01.007.

Lennox, K. A., & Behlke, M. A. (2011). Chemical modification and design of anti-miRNA oligonucleotides. *Gene Therapy*, *18*(12), 1111−1120. Available from https://doi.org/10.1038/gt.2011.100.

Leucci, E., Vendramin, R., Spinazzi, M., Laurette, P., Fiers, M., Wouters, J., ... Marine, J. C. (2016). Melanoma addiction to the long non-coding RNA SAMMSON. *Nature*, *531*(7595), 518−522. Available from https://doi.org/10.1038/nature17161.

Li, W., Xie, L., He, X., Li, J., Tu, K., Wei, L., ... Gu, J. (2008). Diagnostic and prognostic implications of microRNAs in human hepatocellular carcinoma. *International Journal of Cancer*, *123*(7), 1616−1622. Available from https://doi.org/10.1002/ijc.23693.

Li, N., Fu, H., Tie, Y., Hu, Z., Kong, W., Wu, Y., & Zheng, X. (2009). miR-34a inhibits migration and invasion by down-regulation of c-Met expression in human hepatocellular carcinoma cells. *Cancer Letters*, *275*(1), 44−53. Available from https://doi.org/10.1016/j.canlet.2008.09.035.

Li, S., Fu, H., Wang, Y., Tie, Y., Xing, R., Zhu, J., ... Zheng, X. (2009). MicroRNA-101 regulates expression of the v-fos FBJ murine osteosarcoma viral oncogene homolog (FOS) oncogene in human hepatocellular carcinoma. *Hepatology*, *49*(4), 1194−1202. Available from https://doi.org/10.1002/hep.22757.

Li, R., Qian, N., Tao, K., You, N., Wang, X., & Dou, K. (2010). MicroRNAs involved in neoplastic transformation of liver cancer stem cells. *Journal of Experimental and Clinical Cancer Research*, *29*(1). Available from https://doi.org/10.1186/1756-9966-29-169.

Li, J., Wang, Y., Yu, W., Chen, J., & Luo, J. (2011). Expression of serum miR-221 in human hepatocellular carcinoma and its prognostic significance. *Biochemical and Biophysical Research Communications*, *406*(1), 70−73. Available from https://doi.org/10.1016/j.bbrc.2011.01.111.

Li, B., Han, Q., Zhu, Y., Yu, Y., Wang, J., & Jiang, X. (2012). Down-regulation of miR-214 contributes to intrahepatic cholangiocarcinoma metastasis by targeting Twist. *FEBS Journal*, *279*(13), 2393−2398. Available from https://doi.org/10.1111/j.1742-4658.2012.08618.x.

Li, H., An, J., Wu, M., Zheng, Q., Gui, X., Li, T., ... Lu, D. (2015). LncRNA HOTAIR promotes human liver cancer stem cell malignant growth through downregulation of SETD2. *Oncotarget*, *6*(29), 27847−27864. Available from https://doi.org/10.18632/oncotarget.4443.

Li, J., Gao, B., Huang, Z., Duan, T., Li, D., Zhang, S., ... Cheng, K. (2015). Prognostic significance of microRNA-203 in cholangiocarcinoma. *International Journal of Clinical and Experimental Pathology*, *8*(8), 9512−9516. Available from http://www.ijcep.com/files/ijcep0010965.pdf.

Li, J., Wang, X., Tang, J., Jiang, R., Zhang, W., Ji, J., & Sun, B. (2015). HULC and Linc00152 act as novel biomarkers in predicting diagnosis of hepatocellular carcinoma. *Cellular Physiology and Biochemistry*, *37*(2), 687−696. Available from https://doi.org/10.1159/000430387.

Li, J., Yao, L., Li, G., Ma, D., Sun, C., Gao, S., ... Gao, F. (2015). MiR-221 promotes epithelial-mesenchymal transition through targeting PTEN and forms a positive feedback loop with β-catenin/c-Jun signaling pathway in extrahepatic cholangiocarcinoma. *PLoS ONE*, *10*(10). Available from https://doi.org/10.1371/journal.pone.0141168.

Li, S. P., Xu, H. X., Yu, Y., He, J. D., Wang, Z., Xu, Y. J., ... Shen, Z. Y. (2016). LncRNA HULC enhances epithelial-mesenchymal transition to promote tumorigenesis and metastasis of hepatocellular carcinoma via the miR-200a-3p/ZEB1 signaling pathway. *Oncotarget*, *7*(27), 42431−42446. Available from https://doi.org/10.18632/oncotarget.9883.

Li, D., Cheng, M., Niu, Y., Chi, X., Liu, X., Fan, J., ... Yang, W. (2017). Identification of a novel human long noncoding RNA that regulates hepatic lipid metabolism by inhibiting SREBP-1c. *International Journal of Biological Sciences*, *13*(3), 349−357. Available from https://doi.org/10.7150/ijbs.16635.

Li, H., Zhou, Z. Q., Yang, Z. R., Tong, D. N., Guan, J., Shi, B. J., ... Zhang, Z. Y. (2017). MicroRNA-191 acts as a tumor promoter by modulating the TET1–p53 pathway in intrahepatic cholangiocarcinoma. *Hepatology*, *66*(1), 136−151. Available from https://doi.org/10.1002/hep.29116.

Li, L., Piontek, K., Ishida, M., Fausther, M., Dranoff, J. A., Fu, R., ... Selaru, F. M. (2017). Extracellular vesicles carry microRNA-195 to intrahepatic cholangiocarcinoma and improve survival in a rat model. *Hepatology*, *65*(2), 501−514. Available from https://doi.org/10.1002/hep.28735.

Li, J., Jiang, X., Li, C., Liu, Y., Kang, P., Zhong, X., & Cui, Y. (2019). LncRNA-MEG3 inhibits cell proliferation and invasion by modulating Bmi1/RNF2 in cholangiocarcinoma. *Journal of Cellular Physiology*, *234*(12), 22947−22959. Available from https://doi.org/10.1002/jcp.28856.

Lin, X. J., Chong, Y., Guo, Z. W., Xie, C., Yang, X. J., Zhang, Q., ... Zhuang, S. M. (2015). A serum microRNA classifier for early detection of hepatocellular carcinoma: A multicentre, retrospective, longitudinal biomarker identification study with a nested case-control study. *The Lancet Oncology*, 16(7), 804–815. Available from https://doi.org/10.1016/S1470-2045(15)00048-0.

Lin, K. Y., Ye, H., Han, B. W., Wang, W. T., Wei, P. P., He, B., ... Chen, Y. Q. (2016). Genome-wide screen identified let-7c/miR-99a/miR-125b regulating tumor progression and stem-like properties in cholangiocarcinoma. *Oncogene*, 35(26), 3376–3386. Available from https://doi.org/10.1038/onc.2015.396.

Lin, L., Lu, B., Yu, J., Liu, W., & Zhou, A. (2016). Serum miR-224 as a biomarker for detection of hepatocellular carcinoma at early stage. *Clinics and Research in Hepatology and Gastroenterology*, 40(4), 397–404. Available from https://doi.org/10.1016/j.clinre.2015.11.005.

Lin, Z., Lu, Y., Meng, Q., Wang, C., Li, X., Yang, Y., ... Lu, D. (2018). miR372 promotes progression of liver cancer cells by upregulating erbB-2 through enhancement of YB-1. *Molecular Therapy – Nucleic Acids*, 11, 494–507. Available from https://doi.org/10.1016/j.omtn.2018.04.001.

Liu, W. H., Yeh, S. H., Lu, C. C., Yu, S. L., Chen, H. Y., Lin, C. Y., ... Chen, P. J. (2009). MicroRNA-18a prevents estrogen receptor-α expression, promoting proliferation of hepatocellular carcinoma cells. *Gastroenterology*, 136(2), 683–693. Available from https://doi.org/10.1053/j.gastro.2008.10.029.

Liu, Z., Tu, K., & Liu, Q. (2014). Effects of microRNA-30a on migration, invasion and prognosis of hepatocellular carcinoma. *FEBS Letters*, 588(17), 3089–3097. Available from https://doi.org/10.1016/j.febslet.2014.06.037.

Liu, K., Liu, S., Zhang, W., Ji, B., Wang, Y., & Liu, Y. (2014). miR-222 regulates sorafenib resistance and enhance tumorigenicity in hepatocellular carcinoma. *International Journal of Oncology*, 45(4), 1537–1546. Available from https://doi.org/10.3892/ijo.2014.2577.

Liu, K., Liu, S., Zhang, W., Jia, B., Tan, L., Jin, Z., & Liu, Y. (2015). miR-494 promotes cell proliferation, migration and invasion, and increased sorafenib resistance in hepatocellular carcinoma by targeting PTEN. *Oncology Reports*, 34(2), 1003–1010. Available from https://doi.org/10.3892/or.2015.4030.

Liu, N., Jiang, F., He, T. L., Zhang, J. K., Zhao, J., Wang, C., ... Cui, Y. F. (2015). The roles of microRNA-122 overexpression in inhibiting proliferation and invasion and stimulating apoptosis of human cholangiocarcinoma cells. *Scientific Reports*, 5. Available from https://doi.org/10.1038/srep16566.

Liu, W., Chen, S., & Liu, B. (2016). Diagnostic and prognostic values of serum exosomal microRNA-21 in children with hepatoblastoma: A Chinese population-based study. *Pediatric Surgery International*, 32(11), 1059–1065. Available from https://doi.org/10.1007/s00383-016-3960-8.

Liu, Z., Wang, J., Mao, Y., Zou, B., & Fan, X. (2016). MicroRNA-101 suppresses migration and invasion via targeting vascular endothelial growth factor-C in hepatocellular carcinoma cells. *Oncology Letters*, 11(1), 433–438. Available from https://doi.org/10.3892/ol.2015.3832.

Liu, X. L., Cao, H. X., Wang, B. C., Xin, F. Z., Zhang, R. N., Zhou, D., ... Fan, J. G. (2017). MiR-192-5p regulates lipid synthesis in non-Alcoholic fatty liver disease through SCD-1. *World Journal of Gastroenterology*, 23(46), 8140–8151. Available from https://doi.org/10.3748/wjg.v23.i46.8140.

Liu, C., Yang, Z., Wu, J., Zhang, L., Lee, S., Shin, D. J., ... Wang, L. (2018). Long noncoding RNA H19 interacts with polypyrimidine tract-binding protein 1 to reprogram hepatic lipid homeostasis. *Hepatology*, 67(5), 1768–1783. Available from https://doi.org/10.1002/hep.29654.

Liu, S., Qiu, J., He, G., Liang, Y., Wang, L., Liu, C., & Pan, H. (2019). LncRNA MALAT1 acts as a miR-125a-3p sponge to regulate FOXM1 expression and promote hepatocellular carcinoma progression. *Journal of Cancer*, 10(26), 6649–6659. Available from https://doi.org/10.7150/jca.29213.

Llovet, J. M., Zucman-Rossi, J., Pikarsky, E., Sangro, B., Schwartz, M., Sherman, M., & Gores, G. (2016). Hepatocellular carcinoma. *Nature Reviews Disease Primers*, 2. Available from https://doi.org/10.1038/nrdp.2016.18.

Loosen, S. H., Wirtz, T. H., Roy, S., Vucur, M., Castoldi, M., Schneider, A. T., ... Luedde, T. (2020). Circulating levels of microRNA193a-5p predict outcome in early stage hepatocellular carcinoma. *PLoS ONE*, 15(9). Available from https://doi.org/10.1371/journal.pone.0239386.

Lou, G., Ma, N., Xu, Y., Jiang, L., Yang, J., Wang, C., ... Gao, X. (2015). Differential distribution of U6 (RNU6-1) expression in human carcinoma tissues demonstrates the requirement for caution in the internal control gene selection for microRNA quantification. *International Journal of Molecular Medicine*, 36(5), 1400–1408. Available from https://doi.org/10.3892/ijmm.2015.2338.

Lu, J., Getz, G., Miska, E. A., Alvarez-Saavedra, E., Lamb, J., Peck, D., ... Golub, T. R. (2005). MicroRNA expression profiles classify human cancers. *Nature*, 435(7043), 834–838. Available from https://doi.org/10.1038/nature03702.

Lu, Z., Xiao, Z., Liu, F., Cui, M., Li, W., Yang, Z., ... Zhang, X. (2016). Long non-coding RNA HULC promotes tumor angiogenesis in liver cancer by up-regulating sphingosine kinase 1 (SPHK1). *Oncotarget, 7*(1), 241–254. Available from https://doi.org/10.18632/ONCOTARGET.6280.

Luo, F., Liu, X., Ling, M., Lu, L., Shi, L., Lu, X., ... Liu, Q. (2016). The lncRNA MALAT1, acting through HIF-1α stabilization, enhances arsenite-induced glycolysis in human hepatic L-02 cells. *Biochimica et Biophysica Acta — Molecular Basis of Disease, 1862*(9), 1685–1695. Available from https://doi.org/10.1016/j.bbadis.2016.06.004.

Luo, F., Sun, B., Li, H., Xu, Y., Liu, Y., Liu, X., ... Zhang, A. (2016). A MALAT1/HIF-2α feedback loop contributes to arsenite carcinogenesis. *Oncotarget, 7*(5), 5769–5787. Available from https://doi.org/10.18632/oncotarget.6806.

Luo, Y., Ye, J., Wei, J., Zhang, J., & Li, Y. (2020). Long non-coding RNA-based risk scoring system predicts prognosis of alcohol-related hepatocellular carcinoma. *Molecular Medicine Reports, 22*(2), 997–1007. Available from https://doi.org/10.3892/mmr.2020.11179.

Ma, X., Wang, X., Yang, C., Wang, Z., Han, B., Wu, L., & Zhuang, L. (2016). DANCR acts as a diagnostic biomarker and promotes tumor growth and metastasis in hepatocellular carcinoma. *Anticancer Research, 36*(12), 6389–6398. Available from https://doi.org/10.21873/anticanres.11236.

Malakar, P., Shilo, A., Mogilevsky, A., Stein, I., Pikarsky, E., Nevo, Y., ... Karni, R. (2017). Long noncoding RNA MALAT1 promotes hepatocellular carcinoma development by SRSF1 upregulation and mTOR activation. *Cancer Research, 77*(5), 1155–1167. Available from https://doi.org/10.1158/0008-5472.CAN-16-1508.

Mann, M., Wright, P. R., & Backofen, R. (2017). IntaRNA 2.0: Enhanced and customizable prediction of RNA-RNA interactions. *Nucleic Acids Research, 45*(1), W435–W439. Available from https://doi.org/10.1093/nar/gkx279.

Manna, D., & Sarkar, D. (2020). Non-coding RNAs: Regulating disease progression and therapy resistance in hepatocellular carcinoma. *Cancers, 12*(5). Available from https://doi.org/10.3390/cancers12051243.

Mao, K., Zhang, J., He, C., Xu, K., Liu, J., Sun, J., ... Xiao, Z. (2014). Restoration of miR-193b sensitizes hepatitis B virus-associated hepatocellular carcinoma to sorafenib. *Cancer Letters, 352*(2), 245–252. Available from https://doi.org/10.1016/j.canlet.2014.07.004.

Marchese, F. P., Raimondi, I., & Huarte, M. (2017). The multidimensional mechanisms of long noncoding RNA function. *Genome Biology, 18*(1). Available from https://doi.org/10.1186/s13059-017-1348-2.

Marquart, T. J., Allen, R. M., Ory, D. S., & Baldan, A. (2010). miR-33 links SREBP-2 induction to repression of sterol transporters. *Proceedings of the National Academy of Sciences*, 12228–12232. Available from https://doi.org/10.1073/pnas.1005191107.

Matera, A. G., & Wang, Z. (2014). A day in the life of the spliceosome. *Nature Reviews Molecular Cell Biology, 15*(2), 108–121. Available from https://doi.org/10.1038/nrm3742.

Mathivanan, S., Fahner, C. J., Reid, G. E., & Simpson, R. J. (2012). ExoCarta 2012: Database of exosomal proteins, RNA and lipids. *Nucleic Acids Research, 40*(1), D1241–D1244. Available from https://doi.org/10.1093/nar/gkr828.

Matouk, I. J., DeGroot, N., Mezan, S., Ayesh, S., Abu-Lail, R., Hochberg, A., & Galun, E. (2007). The H19 non-coding RNA is essential for human tumor growth. *PLoS ONE, 2*(9). Available from https://doi.org/10.1371/journal.pone.0000845.

McNally, M. E., Collins, A., Wojcik, S. E., Liu, J., Henry, J. C., Jiang, J., ... Bloomston, M. (2013). Concomitant dysregulation of microRNAs miR-151-3p and miR-126 correlates with improved survival in resected cholangiocarcinoma. *HPB, 15*(4), 260–264. Available from https://doi.org/10.1111/j.1477-2574.2012.00523.x.

Mehra, R., Udager, A. M., Ahearn, T. U., Cao, X., Feng, F. Y., Loda, M., ... Chinnaiyan, A. M. (2016). Overexpression of the long non-coding RNA SChLAP1 independently predicts lethal prostate cancer. *European Urology, 70*(4), 549–552. Available from https://doi.org/10.1016/j.eururo.2015.12.003.

Meng, F., Henson, R., Wehbe-Janek, H., Ghoshal, K., Jacob, S. T., & Patel, T. (2007). MicroRNA-21 regulates expression of the PTEN tumor suppressor gene in human hepatocellular cancer. *Gastroenterology, 133*(2), 647–658. Available from https://doi.org/10.1053/j.gastro.2007.05.022.

Mercer, T. R., Dinger, M. E., & Mattick, J. S. (2009). Long non-coding RNAs: Insights into functions. *Nature Reviews Genetics, 10*(3), 155–159. Available from https://doi.org/10.1038/nrg2521.

Meroni, M., Longo, M., Erconi, V., Valenti, L., Gatti, S., Fracanzani, A. L., & Dongiovanni, P. (2019). Mir-101-3p downregulation promotes fibrogenesis by facilitating hepatic stellate cell transdifferentiation during insulin resistance. *Nutrients, 11*(11). Available from https://doi.org/10.3390/nu11112597.

Mjelle, R., Dima, S. O., Bacalbasa, N., Chawla, K., Sorop, A., Cucu, D., ... Popescu, I. (2019). Comprehensive transcriptomic analyses of tissue, serum, and serum exosomes from hepatocellular carcinoma patients. *BMC Cancer*, *19*(1). Available from https://doi.org/10.1186/s12885-019-6249-1.

Moshiri, F., Salvi, A., Gramantieri, L., Sangiovanni, A., Guerriero, P., De Petro, G., ... Negrini, M. (2018). Circulating miR-106b-3p, miR-101-3p and miR-1246 as diagnostic biomarkers of hepatocellular carcinoma. *Oncotarget*, *9*(20), 15350−15364. Available from https://doi.org/10.18632/oncotarget.24601.

Motawi, T. K., Shaker, O. G., El-Maraghy, S. A., & Senousy, M. A. (2015). Serum MicroRNAs as potential biomarkers for early diagnosis of hepatitis C virus-related hepatocellular carcinoma in egyptian patients. *PLoS ONE*, *10*(9). Available from https://doi.org/10.1371/journal.pone.0137706.

Mott, J. L., Kobayashi, S., Bronk, S. F., & Gores, G. J. (2007). mir-29 regulates Mcl-1 protein expression and apoptosis. *Oncogene*, *26*(42), 6133−6140. Available from https://doi.org/10.1038/sj.onc.1210436.

Mourad, L., El-Ahwany, E., Zoheiry, M., Abu-Taleb, H., Hassan, M., Ouf, A., ... Zada, S. (2018). Expression analysis of liver-specific circulating microRNAs in HCV-induced hepatocellular carcinoma in Egyptian patients. *Cancer Biology and Therapy*, *19*(5), 400−406. Available from https://doi.org/10.1080/15384047.2018.1423922.

Mu, W., Hu, C., Zhang, H., Qu, Z., Cen, J., Qiu, Z., ... Hui, L. (2015). MiR-27b synergizes with anticancer drugs via p53 activation and CYP1B1 suppression. *Cell Research*, *25*(4), 477−495. Available from https://doi.org/10.1038/cr.2015.23.

Nakamura, H., Arai, Y., Totoki, Y., Shirota, T., Elzawahry, A., Kato, M., ... Shibata, T. (2015). Genomic spectra of biliary tract cancer. *Nature Genetics*, *47*(9), 1003−1010. Available from https://doi.org/10.1038/ng.3375.

Nakaoka, T., Saito, Y., & Saito, H. (2017). Aberrant DNA methylation as a biomarker and a therapeutic target of cholangiocarcinoma. *International Journal of Molecular Sciences*, *18*(6). Available from https://doi.org/10.3390/ijms18061111.

Ning, S., Zhang, J., Wang, P., Zhi, H., Wang, J., Liu, Y., ... Li, X. (2016). Lnc2Cancer: A manually curated database of experimentally supported lncRNAs associated with various human cancers. *Nucleic Acids Research*, *44*(1), D980−D985. Available from https://doi.org/10.1093/nar/gkv1094.

Obad, S., Dos Santos, C. O., Petri, A., Heidenblad, M., Broom, O., Ruse, C., ... Kauppinen, S. (2011). Silencing of microRNA families by seed-targeting tiny LNAs. *Nature Genetics*, *43*(4), 371−380. Available from https://doi.org/10.1038/ng.786.

Ogawa, T., Enomoto, M., Fujii, H., Sekiya, Y., Yoshizato, K., Ikeda, K., & Kawada, N. (2012). MicroRNA-221/222 upregulation indicates the activation of stellate cells and the progression of liver fibrosis. *Gut*, *61*(11), 1600−1609. Available from https://doi.org/10.1136/gutjnl-2011-300717.

Oishi, N., Kumar, M. R., Roessler, S., Ji, J., Forgues, M., Budhu, A., ... Wang, X. W. (2012). Transcriptomic profiling reveals hepatic stem-like gene signatures and interplay of miR-200c and epithelial-mesenchymal transition in intrahepatic cholangiocarcinoma. *Hepatology*, *56*(5), 1792−1803. Available from https://doi.org/10.1002/hep.25890.

Okamoto, K., Miyoshi, K., & Murawaki, Y. (2013). miR-29b, miR-205 and miR-221 enhance chemosensitivity to gemcitabine in HuH28 human cholangiocarcinoma cells. *PLoS ONE*, *8*(10). Available from https://doi.org/10.1371/journal.pone.0077623.

Oura, K., Fujita, K., Morishita, A., Iwama, H., Nakahara, M. A. I., Tadokoro, T., ... Masaki, T. (2019). Serum microRNA-125a-5p as a potential biomarker of HCV-associated hepatocellular carcinoma. *Oncology Letters*, *18*(1), 882−890. Available from https://doi.org/10.3892/ol.2019.10385.

Ozata, D. M., Gainetdinov, I., Zoch, A., O'Carroll, D., & Zamore, P. D. (2019). PIWI-interacting RNAs: Small RNAs with big functions. *Nature Reviews Genetics*, *20*(2), 89−108. Available from https://doi.org/10.1038/s41576-018-0073-3.

Pan, Y., Tong, S., Cui, R., Fan, J., Liu, C., Lin, Y., ... Yu, X. (2018). Long non-coding MALAT1 functions as a competing endogenous RNA to regulate vimentin expression by sponging miR-30a-5p in hepatocellular carcinoma. *Cellular Physiology and Biochemistry*, *50*(1), 121−135. Available from https://doi.org/10.1159/000493962.

Panzitt, K., Tschernatsch, M. M. O., Guelly, C., Moustafa, T., Stradner, M., Strohmaier, H. M., ... Zatloukal, K. (2007). Characterization of HULC, a novel gene with striking up-regulation in hepatocellular carcinoma, as noncoding RNA. *Gastroenterology*, *132*(1), 330−342. Available from https://doi.org/10.1053/j.gastro.2006.08.026.

Paraskevopoulou, M. D., Georgakilas, G., Kostoulas, N., Reczko, M., Maragkakis, M., Dalamagas, T. M., & Hatzigeorgiou, A. G. (2013). DIANA-LncBase: Experimentally verified and computationally predicted microRNA targets on long non-coding RNAs. *Nucleic Acids Research, 41*(1), D239−D245. Available from https://doi.org/10.1093/nar/gks1246.

Park, J. K., Kogure, T., Nuovo, G. J., Jiang, J., He, L., Kim, J. H., ... Schmittgen, T. D. (2011). miR-221 silencing blocks hepatocellular carcinoma and promotes survival. *Cancer Research, 71*(24), 7608−7616. Available from https://doi.org/10.1158/0008-5472.CAN-11-1144.

Peng, N., He, J., Li, J., Huang, H., Huang, W., Liao, Y., & Zhu, S. (2020). Long noncoding RNA MALAT1 inhibits the apoptosis and autophagy of hepatocellular carcinoma cell by targeting the microRNA-146a/PI3K/Akt/mTOR axis. *Cancer Cell International, 20*(1). Available from https://doi.org/10.1186/s12935-020-01231-w.

Pineau, P., Volinia, S., McJunkin, K., Marchio, A., Battiston, C., Terris, B., ... Dejean, A. (2010). miR-221 overexpression contributes to liver tumorigenesis. *Proceedings of the National Academy of Sciences of the United States of America, 107*(1), 264−269. Available from https://doi.org/10.1073/pnas.0907904107.

Pirola, C. J., Gianotti, T. F., Castaño, G. O., Mallardi, P., Martino, J. S., Ledesma, M. M. G. L., ... Sookoian, S. (2015). Circulating microRNA signature in non-alcoholic fatty liver disease: From serum non-coding RNAs to liver histology and disease pathogenesis. *Gut, 64*(5), 800−812. Available from https://doi.org/10.1136/gutjnl-2014-306996.

Plieskatt, J. L., Rinaldi, G., Feng, Y., Peng, J., Yonglitthipagon, P., Easley, S., ... Bethony, J. M. (2014). Distinct miRNA signatures associate with subtypes of cholangiocarcinoma from infection with the tumourigenic liver fluke *Opisthorchis viverrini*. *Journal of Hepatology, 61*(4), 850−858. Available from https://doi.org/10.1016/j.jhep.2014.05.035.

Pratedrat, P., Chuaypen, N., Nimsamer, P., Payungporn, S., Pinjaroen, N., Sirichindakul, B., & Tangkijvanich, P. (2020). Diagnostic and prognostic roles of circulating miRNA-223-3p in hepatitis B virus−related hepatocellular carcinoma. *PLoS ONE, 15*(4). Available from https://doi.org/10.1371/journal.pone.0232211.

Qiao, P., Li, G., Bi, W., Yang, L., Yao, L., & Wu, D. (2015). microRNA-34a inhibits epithelial mesenchymal transition in human cholangiocarcinoma by targeting Smad4 through transforming growth factor-beta/Smad pathway. *BMC Cancer, 15*(1). Available from https://doi.org/10.1186/s12885-015-1359-x.

Qin, W., Li, X., Xie, L., Li, S., Liu, J., Jia, L., ... Chen, Z. (2016). A long non-coding RNA, APOA4-AS, regulates APOA4 expression depending on HuR in mice. *Nucleic Acids Research, 44*(13), 6423−6433. Available from https://doi.org/10.1093/nar/gkw341.

Qin, W., Kang, P., Xu, Y., Leng, K., Li, Z., Huang, L., ... Zhong, X. (2018). Long non-coding RNA HOTAIR promotes tumorigenesis and forecasts a poor prognosis in cholangiocarcinoma. *Scientific Reports, 8*(1). Available from https://doi.org/10.1038/s41598-018-29737-4.

Qiu, Y. H., Wei, Y. P., Shen, N. J., Wang, Z. C., Kan, T., Yu, W. L., ... Zhang, Y. J. (2013). MiR-204 inhibits epithelial to mesenchymal transition by targeting slug in intrahepatic cholangiocarcinoma cells. *Cellular Physiology and Biochemistry, 32*(5), 1331−1341. Available from https://doi.org/10.1159/000354531.

Rayner, K. J., Suárez, Y., Dávalos, A., Parathath, S., Fitzgerald, M. L., Tamehiro, N., ... Fernández-Hernando, C. (2010). MiR-33 contributes to the regulation of cholesterol homeostasis. *Science, 328*(5985), 1570−1573. Available from https://doi.org/10.1126/science.1189862.

Razumilava, N., Bronk, S. F., Smoot, R. L., Fingas, C. D., Werneburg, N. W., Roberts, L. R., & Mott, J. L. (2012). miR-25 targets TNF-related apoptosis inducing ligand (TRAIL) death receptor-4 and promotes apoptosis resistance in cholangiocarcinoma. *Hepatology, 55*(2), 465−475. Available from https://doi.org/10.1002/hep.24698.

Riazalhosseini, B., Mohamed, R., Apalasamy, Y. D., Langmia, I. M., & Mohamed, Z. (2017). Circulating microRNA as a marker for predicting liver disease progression in patients with chronic hepatitis B. *Revista Da Sociedade Brasileira de Medicina Tropical, 50*(2), 161−166. Available from https://doi.org/10.1590/0037-8682-0416-2016.

Robbins, P. D., & Morelli, A. E. (2014). Regulation of immune responses by extracellular vesicles. *Nature Reviews Immunology, 14*(3), 195−208. Available from https://doi.org/10.1038/nri3622.

Roberts, J., Palma, E., Sazani, P., Ørum, H., Cho, M., & Kole, R. (2006). Efficient and persistent splice switching by systemically delivered LNA oligonucleotides in mice. *Molecular Therapy, 14*(4), 471−475. Available from https://doi.org/10.1016/j.ymthe.2006.05.017.

Rogler, C. E., LeVoci, L., Ader, T., Massimi, A., Tchaikovskaya, T., Norel, R., & Rogler, L. E. (2009). MicroRNA-23b cluster microRNAs regulate transforming growth factor-beta/bone morphogenetic protein signaling and liver stem cell differentiation by targeting Smads. *Hepatology, 50*(2), 575−584. Available from https://doi.org/10.1002/hep.22982.

Ross, J. A., Radloff, G. A., & Davies, S. M. (2000). H19 and IGF-2 allele-specific expression in hepatoblastoma. *British Journal of Cancer, 82*(4), 753−756. Available from https://doi.org/10.1054/bjoc.1999.0992.

Ruan, X., Li, P., Ma, Y., Jiang, C. F., Chen, Y., Shi, Y., ... Cao, H. (2021). Identification of human long noncoding RNAs associated with nonalcoholic fatty liver disease and metabolic homeostasis. *Journal of Clinical Investigation*, 131(1). Available from https://doi.org/10.1172/JCI136336.

Schlotter, Y. M., Veenhof, E. Z., Brinkhof, B., Rutten, V. P. M. G., Spee, B., Willemse, T., & Penning, L. C. (2009). A GeNorm algorithm-based selection of reference genes for quantitative real-time PCR in skin biopsies of healthy dogs and dogs with atopic dermatitis. *Veterinary Immunology and Immunopathology*, 129(1−2), 115−118. Available from https://doi.org/10.1016/j.vetimm.2008.12.004.

Schulze, K., Imbeaud, S., Letouzé, E., Alexandrov, L. B., Calderaro, J., Rebouissou, S., ... Zucman-Rossi, J. (2015). Exome sequencing of hepatocellular carcinomas identifies new mutational signatures and potential therapeutic targets. *Nature Genetics*, 47(5), 505−511. Available from https://doi.org/10.1038/ng.3252.

Sekine, S., Ogawa, R., Ito, R., Hiraoka, N., McManus, M. T., Kanai, Y., & Hebrok, M. (2009). Disruption of Dicer1 induces dysregulated fetal gene expression and promotes hepatocarcinogenesis. *Gastroenterology*, 136(7), 2304−e4. Available from https://doi.org/10.1053/j.gastro.2009.02.067.

Sekine, S., Ogawa, R., Mcmanus, M. T., Kanai, Y., & Hebrok, M. (2009). Dicer is required for proper liver zonation. *Journal of Pathology*, 219(3), 365−372. Available from https://doi.org/10.1002/path.2606.

Selaru, F. M., Olaru, A. V., Kan, T., David, S., Cheng, Y., Mori, Y., ... Meltzer, S. J. (2009). MicroRNA-21 is overexpressed in human cholangiocarcinoma and regulates programmed cell death 4 and tissue inhibitor of metalloproteinase 3. *Hepatology*, 49(5), 1595−1601. Available from https://doi.org/10.1002/hep.22838.

Shen, Q., Bae, H. J., Eun, J. W., Kim, H. S., Park, S. J., Shin, W. C., ... Nam, S. W. (2014). MiR-101 functions as a tumor suppressor by directly targeting nemo-like kinase in liver cancer. *Cancer Letters*, 344(2), 204−211. Available from https://doi.org/10.1016/j.canlet.2013.10.030.

Shen, Q., Jiang, S., Wu, M., Zhang, L., Su, X., & Zhao, D. (2020). Lncrna HEIH confers cell sorafenib resistance in hepatocellular carcinoma by regulating mir-98-5p/pi3k/akt pathway. *Cancer Management and Research*, 12, 6585−6595. Available from https://doi.org/10.2147/CMAR.S241383.

Shimizu, S., Takehara, T., Hikita, H., Kodama, T., Miyagi, T., Hosui, A., ... Hayashi, N. (2010). The let-7 family of microRNAs inhibits Bcl-xL expression and potentiates sorafenib-induced apoptosis in human hepatocellular carcinoma. *Journal of Hepatology*, 52(5), 698−704. Available from https://doi.org/10.1016/j.jhep.2009.12.024.

Shu, X. L., Fan, C. B., Long, B., Zhou, X., & Wang, Y. (2016). The anti-cancer effects of cisplatin on hepatic cancer are associated with modulation of miRNA-21 and miRNA-122 expression. *European Review for Medical and Pharmacological Sciences*, 20(21), 4459−4465.

Sia, D., Hoshida, Y., Villanueva, A., Roayaie, S., Ferrer, J., Tabak, B., ... Llovet, J. M. (2013). Integrative molecular analysis of intrahepatic cholangiocarcinoma reveals 2 classes that have different outcomes. *Gastroenterology*, 144(4), 829−840. Available from https://doi.org/10.1053/j.gastro.2013.01.001.

Sia, D., Jiao, Y., Martinez-Quetglas, I., Kuchuk, O., Villacorta-Martin, C., Castro de Moura, M., ... Llovet, J. M. (2017). Identification of an immune-specific class of hepatocellular carcinoma, based on molecular features. *Gastroenterology*, 153(3), 812−826. Available from https://doi.org/10.1053/j.gastro.2017.06.007.

Simion, A., Laudadio, I., Prévot, P. P., Raynaud, P., Lemaigre, F. P., & Jacquemin, P. (2010). MiR-495 and miR-218 regulate the expression of the Onecut transcription factors HNF-6 and OC-2. *Biochemical and Biophysical Research Communications*, 391(1), 293−298. Available from https://doi.org/10.1016/j.bbrc.2009.11.052.

Sircana, A., Paschetta, E., Saba, F., Molinaro, F., & Musso, G. (2019). Recent insight into the role of fibrosis in nonalcoholic steatohepatitis-related hepatocellular carcinoma. *International Journal of Molecular Sciences*, 20(7). Available from https://doi.org/10.3390/ijms20071745.

Su, H., Yang, J. R., Xu, T., Huang, J., Xu, L., Yuan, Y., & Zhuang, S. M. (2009). MicroRNA-101, down-regulated in hepatocellular carcinoma, promotes apoptosis and suppresses tumorigenicity. *Cancer Research*, 69(3), 1135−1142. Available from https://doi.org/10.1158/0008-5472.CAN-08-2886.

Su, J., Wu, F., Xia, H., Wu, Y., & Liu, S. (2020). Accurate cancer cell identification and microRNA silencing induced therapy using tailored DNA tetrahedron nanostructures. *Chemical Science*, 11(1), 80−86. Available from https://doi.org/10.1039/c9sc04823e.

Su, Z. J., Lin, C. C., Pan, J. H., Zhang, J. H., Han, T., & Pan, Q. (2020). Prediction of poor prognosis of HCC by early warning model for co-expression of miRNA and mRNA based on bioinformatics analysis. *Technology in Cancer Research & Treatment*, 19. Available from https://doi.org/10.1177/1533033820959353, 1533033820959353.

Sun, Y., Zhang, F., Wang, L., Song, X., Jing, J., Zhang, F., . . . Liu, H. (2019). A five lncRNA signature for prognosis prediction in hepatocellular carcinoma. *Molecular Medicine Reports, 19*(6), 5237−5250. Available from https://doi.org/10.3892/mmr.2019.10203.

Tan, X., Huang, Z., & Li, X. (2017). Long non-coding RNA MALAT1 interacts with miR-204 to modulate human hilar cholangiocarcinoma proliferation, migration, and invasion by targeting CXCR4. *Journal of Cellular Biochemistry, 118*(11), 3643−3653. Available from https://doi.org/10.1002/jcb.25862.

Tang, K., Zhang, Y., Zhang, H., Xu, P., Liu, J., Ma, J., . . . Huang, B. (2012). Delivery of chemotherapeutic drugs in tumour cell-derived microparticles. *Nature Communications, 3*. Available from https://doi.org/10.1038/ncomms2282.

Tang, S., Tan, G., Jiang, X., Han, P., Zhai, B., Dong, X., . . . Sun, X. (2016). An artificial lncRNA targeting multiple miRNAs overcomes sorafenib resistance in hepatocellular carcinoma cells. *Oncotarget, 7*(45), 73257−73269. Available from https://doi.org/10.18632/oncotarget.12304.

Tang, X., Zhang, W., Ye, Y., Li, H., Cheng, L., Zhang, M., . . . Yu, J. (2020). LncRNA hotair contributes to sorafenib resistance through suppressing miR-217 in hepatic carcinoma. *BioMed Research International, 2020*. Available from https://doi.org/10.1155/2020/9515071.

Teufel, M., Seidel, H., Köchert, K., Meinhardt, G., Finn, R. S., Llovet, J. M., & Bruix, J. (2019). Biomarkers associated with response to regorafenib in patients with hepatocellular carcinoma. *Gastroenterology, 156*(6), 1731−1741. Available from https://doi.org/10.1053/j.gastro.2019.01.261.

Théry, C., Zitvogel, L., & Amigorena, S. (2002). Exosomes: Composition, biogenesis and function. *Nature Reviews Immunology, 2*(8), 569−579. Available from https://doi.org/10.1038/nri855.

Thulasingam, S., Massilamany, C., Gangaplara, A., Dai, H., Yarbaeva, S., Subramaniam, S., . . . Reddy, J. (2011). miR-27b*, an oxidative stress-responsive microRNA modulates nuclear factor-kB pathway in RAW 264.7 cells. *Molecular and Cellular Biochemistry, 352*(1−2), 181−188. Available from https://doi.org/10.1007/s11010-011-0752-2.

Tilg, H., & Moschen, A. R. (2010). Evolution of inflammation in nonalcoholic fatty liver disease: The multiple parallel hits hypothesis. *Hepatology, 52*(5), 1836−1846. Available from https://doi.org/10.1002/hep.24001.

Tomimaru, Y., Eguchi, H., Nagano, H., Wada, H., Kobayashi, S., Marubashi, S., . . . Mori, M. (2012). Circulating microRNA-21 as a novel biomarker for hepatocellular carcinoma. *Journal of Hepatology, 56*(1), 167−175. Available from https://doi.org/10.1016/j.jhep.2011.04.026.

Tzur, G., Israel, A., Levy, A., Benjamin, H., Meiri, E., Shufaro, Y., . . . Galun, E. (2009). Comprehensive gene and microRNA expression profiling reveals a role for microRNAs in human liver development. *PLoS ONE, 4*(10). Available from https://doi.org/10.1371/journal.pone.0007511.

Ulitsky, I., Shkumatava, A., Jan, C. H., Sive, H., & Bartel, D. P. (2011). Conserved function of lincRNAs in vertebrate embryonic development despite rapid sequence evolution. *Cell, 147*(7), 1537−1550. Available from https://doi.org/10.1016/j.cell.2011.11.055.

Valadi, H., Ekström, K., Bossios, A., Sjöstrand, M., Lee, J. J., & Lötvall, J. O. (2007). Exosome-mediated transfer of mRNAs and microRNAs is a novel mechanism of genetic exchange between cells. *Nature Cell Biology, 9*(6), 654−659. Available from https://doi.org/10.1038/ncb1596.

Vaulont, S. (2011). miR-122, un microARN "à tout fer". *Medecine/Sciences, 27*(8−9), 704−706. Available from https://doi.org/10.1051/medsci/2011278010.

Vergoulis, T., Vlachos, I. S., Alexiou, P., Georgakilas, G., Maragkakis, M., Reczko, M., . . . Hatzigeorgiou, A. G. (2012). TarBase 6.0: Capturing the exponential growth of miRNA targets with experimental support. *Nucleic Acids Research, 40*(1), D222−D229. Available from https://doi.org/10.1093/nar/gkr1161.

Vogelstein, B., Papadopoulos, N., Velculescu, V. E., Zhou, S., Diaz, L. A., & Kinzler, K. W. (2013). Cancer genome landscapes. *Science, 340*(6127), 1546−1558. Available from https://doi.org/10.1126/science.1235122.

Volders, P. J., Helsens, K., Wang, X., Menten, B., Martens, L., Gevaert, K., . . . Mestdagh, P. (2013). LNCipedia: A database for annotated human lncRNA transcript sequences and structures. *Nucleic Acids Research, 41*(1), D246−D251. Available from https://doi.org/10.1093/nar/gks915.

Volinia, S., Calin, G. A., Liu, C. G., Ambs, S., Cimmino, A., Petrocca, F., . . . Croce, C. M. (2006). A microRNA expression signature of human solid tumors defines cancer gene targets. *Proceedings of the National Academy of Sciences of the United States of America, 103*(7), 2257−2261. Available from https://doi.org/10.1073/pnas.0510565103.

von Felden, J., Heim, D., Schulze, K., Krech, T., Ewald, F., Nashan, B., . . . Wege, H. (2017). High expression of micro RNA-135A in hepatocellular carcinoma is associated with recurrence within 12 months after resection. *BMC Cancer, 17*(1). Available from https://doi.org/10.1186/s12885-017-3053-7.

Wang, F., Li, L., Piontek, K., Sakaguchi, M., & Selaru, F. M. (2018). Exosome miR-335 as a novel therapeutic strategy in hepatocellular carcinoma. *Hepatology*, 67(3), 940–954. Available from https://doi.org/10.1002/hep.29586.

Wang, Y., Lee, A. T. C., Ma, J. Z. I., Wang, J., Ren, J., Yang, Y., ... Lee, C. G. L. (2008). Profiling microRNA expression in hepatocellular carcinoma reveals microRNA-224 up-regulation and apoptosis inhibitor-5 as a microRNA-224-specific target. *Journal of Biological Chemistry*, 283(19), 13205–13215. Available from https://doi.org/10.1074/jbc.M707629200.

Wang, F., Yuan, J. H., Wang, S. B., Yang, F., Yuan, S. X., Ye, C., ... Sun, S. H. (2014). Oncofetal long noncoding RNA PVT1 promotes proliferation and stem cell-like property of hepatocellular carcinoma cells by stabilizing NOP2. *Hepatology*, 60(4), 1278–1290. Available from https://doi.org/10.1002/hep.27239.

Wang, J., Wang, H., Zhang, Y., Zhen, N., Zhang, L., Qiao, Y., ... Sun, F. (2014). Mutual inhibition between YAP and SRSF1 maintains long non-coding RNA, Malat1-induced tumourigenesis in liver cancer. *Cellular Signalling*, 26(5), 1048–1059. Available from https://doi.org/10.1016/j.cellsig.2014.01.022.

Wang, L., Zhang, X., Jia, L. T., Hu, S. J., Zhao, J., Yang, J. D., ... Yang, A. G. (2014). C-Myc-mediated epigenetic silencing of MicroRNA-101 contributes to dysregulation of multiple pathways in hepatocellular carcinoma. *Hepatology*, 59(5), 1850–1863. Available from https://doi.org/10.1002/hep.26720.

Wang, W. Y., Zhang, H. F., Wang, L., Ma, Y. P., Gao, F., Zhang, S. J., & Wang, L. C. (2014). MiR-21 expression predicts prognosis in hepatocellular carcinoma. *Clinics and Research in Hepatology and Gastroenterology*, 38(6), 715–719. Available from https://doi.org/10.1016/j.clinre.2014.07.001.

Wang, F., Ying, H. Q., He, B. S., Pan, Y. Q., Deng, Q. W., Sun, H. L., ... Wang, S. K. (2015). Upregulated lncRNA-UCA1 contributes to progression of hepatocellular carcinoma through inhibition of miR-216b and activation of FGFR1/ERK signaling pathway. *Oncotarget*, 6(10), 7899–7917. Available from https://doi.org/10.18632/oncotarget.3219.

Wang, L. J., He, C. C., Sui, X., Cai, M. J., Zhou, C. Y., Ma, J. L., ... Zhu, Q. (2015). MiR-21 promotes intrahepatic cholangiocarcinoma proliferation and growth in vitro and in vivo by targeting PTPN14 and PTEN. *Oncotarget*, 6(8), 5932–5946. Available from https://doi.org/10.18632/oncotarget.3465.

Wang, L. J., Zhang, K. L., Zhang, N., Ma, X. W., Yan, S. W., Cao, D. H., & Shi, S. J. (2015). Serum miR-26a as a diagnostic and prognostic biomarker in cholangiocarcinoma. *Oncotarget*, 6(21), 18631–18640. Available from https://doi.org/10.18632/oncotarget.4072.

Wang, Y., He, L., Du, Y., Zhu, P., Huang, G., Luo, J., ... Fan, Z. (2015). The long noncoding RNA lncTCF7 promotes self-renewal of human liver cancer stem cells through activation of Wnt signaling. *Cell Stem Cell*, 16(4), 413–425. Available from https://doi.org/10.1016/j.stem.2015.03.003.

Wang, W. T., Ye, H., Wei, P. P., Han, B. W., He, B., Chen, Z. H., & Chen, Y. Q. (2016). LncRNAs H19 and HULC, activated by oxidative stress, promote cell migration and invasion in cholangiocarcinoma through a ceRNA manner. *Journal of Hematology and Oncology*, 9(1), 1–12. Available from https://doi.org/10.1186/s13045-016-0348-0.

Wang, J., Xie, C., Pan, S., Liang, Y., Han, J., Lan, Y., ... Liu, L. (2016). N-myc downstream-regulated gene 2 inhibits human cholangiocarcinoma progression and is regulated by leukemia inhibitory factor/MicroRNA-181c negative feedback pathway. *Hepatology*, 64(5), 1606–1622. Available from https://doi.org/10.1002/hep.28781.

Wang, Y., Chen, F., Zhao, M., Yang, Z., Li, J., Zhang, S., ... Zhang, X. (2017). The long noncoding RNA HULC promotes liver cancer by increasing the expression of the HMGA2 oncogene via sequestration of the microRNA-186. *Journal of Biological Chemistry*, 292(37), 15395–15407. Available from https://doi.org/10.1074/jbc.M117.783738.

Wang, N., Zhang, C., Wang, W., Liu, J., Yu, Y., Li, Y., ... Miao, L. (2019). Long noncoding RNA DANCR regulates proliferation and migration by epigenetically silencing FBP1 in tumorigenesis of cholangiocarcinoma. *Cell Death and Disease*, 10(8). Available from https://doi.org/10.1038/s41419-019-1810-z.

Watanabe, S., Hayashi, K., Toh, K., Kim, H. J., Liu, X., Chaya, H., ... Kataoka, K. (2019). In vivo rendezvous of small nucleic acid drugs with charge-matched block catiomers to target cancers. *Nature Communications*, 10(1). Available from https://doi.org/10.1038/s41467-019-09856-w.

Wei, Y., Fabre, M., Branchereau, S., Gauthier, F., Perilongo, G., & Buendia, M. A. (2000). Activation of β-catenin in epithelial and mesenchymal hepatoblastomas. *Oncogene*, 19(4), 498–504. Available from https://doi.org/10.1038/sj.onc.1203356.

Wei, J., Feng, L., Li, Z., Xu, G., & Fan, X. (2013). MicroRNA-21 activates hepatic stellate cells via PTEN/Akt signaling. *Biomedicine and Pharmacotherapy*, 67(5), 387–392. Available from https://doi.org/10.1016/j.biopha.2013.03.014.

Wei, W., Hou, J., Alder, O., Ye, X., Lee, S., Cullum, R., ... Hoodless, P. A. (2013). Genome-wide microRNA and messenger RNA profiling in rodent liver development implicates mir302b and mir20a in repressing transforming growth factor-beta signaling. *Hepatology*, 57(6), 2491–2501. Available from https://doi.org/10.1002/hep.26252.

Wei, X., Tang, C., Lu, X., Liu, R., Zhou, M., He, D., ... Wu, Z. (2015). MiR-101 targets DUSP1 to regulate the TGF-ß secretion in sorafenib inhibits macrophage-induced growth of hepatocarcinoma. *Oncotarget*, 6(21), 18389–18405. Available from https://doi.org/10.18632/oncotarget.4089.

Wen, J., & Friedman, J. R. (2012). miR-122 regulates hepatic lipid metabolism and tumor suppression. *Journal of Clinical Investigation*, 122(8), 2773–2776. Available from https://doi.org/10.1172/JCI63966.

Wen, Y., Han, J., Chen, J., Dong, J., Xia, Y., Liu, J., ... Hu, Z. (2015). Plasma miRNAs as early biomarkers for detecting hepatocellular carcinoma. *International Journal of Cancer*, 137(7), 1679–1690. Available from https://doi.org/10.1002/ijc.29544.

Wong, Q. W. L., Ching, A. K. K., Chan, A. W. H., Choy, K. W., To, K. F., Lai, P. B. S., & Wong, N. (2010). MiR-222 overexpression confers cell migratory advantages in hepatocellular carcinoma through enhancing AKT signaling. *Clinical Cancer Research*, 16(3), 867–875. Available from https://doi.org/10.1158/1078-0432.CCR-09-1840.

Wu, Q., Liu, H. O., Liu, Y. D., Liu, W. S., Pan, D., Zhang, W. J., ... Gu, J. X. (2015). Decreased expression of hepatocyte nuclear factor 4α (Hnf4α)/Microrna-122 (miR-122) axis in hepatitis B virus-associated hepatocellular carcinoma enhances potential oncogenic GALNT10 protein activity. *Journal of Biological Chemistry*, 290(2), 1170–1185. Available from https://doi.org/10.1074/jbc.M114.601203.

Xia, H., Ooi, L. L. P. J., & Hui, K. M. (2013). MicroRNA-216a/217-induced epithelial-mesenchymal transition targets PTEN and SMAD7 to promote drug resistance and recurrence of liver cancer. *Hepatology*, 58(2), 629–641. Available from https://doi.org/10.1002/hep.26369.

Xie, Y., Wang, Y., Li, J., Hang, Y., Jaramillo, L., Wehrkamp, C. J., ... Oupický, D. (2018). Cholangiocarcinoma therapy with nanoparticles that combine downregulation of microRNA-210 with inhibition of cancer cell invasiveness. *Theranostics*, 8(16), 4305–4320. Available from https://doi.org/10.7150/thno.26506.

Xiong, Y., Fang, J. H., Yun, J. P., Yang, J., Zhang, Y., Jia, W. H., & Zhuang, S. M. (2010). Effects of microrna-29 on apoptosis, tumorigenicity, and prognosis of hepatocellular carcinoma. *Hepatology*, 51(3), 836–845. Available from https://doi.org/10.1002/hep.23380.

Xu, J., Wu, C., Che, X., Wang, L., Yu, D., Zhang, T., ... Lin, D. (2011). Circulating MicroRNAs, miR-21, miR-122, and miR-223, in patients with hepatocellular carcinoma or chronic hepatitis. *Molecular Carcinogenesis*, 50(2), 136–142. Available from https://doi.org/10.1002/mc.20712.

Xu, Y., Xia, F., Ma, L., Shan, J., Shen, J., Yang, Z., ... Qian, C. (2011). MicroRNA-122 sensitizes HCC cancer cells to adriamycin and vincristine through modulating expression of MDR and inducing cell cycle arrest. *Cancer Letters*, 310(2), 160–169. Available from https://doi.org/10.1016/j.canlet.2011.06.027.

Xu, H., Zhao, L., Fang, Q., Sun, J., Zhang, S., Zhan, C., ... Jiang, B.-H. (2014). MiR-338-3p inhibits hepatocarcinoma cells and sensitizes these cells to sorafenib by targeting hypoxia-induced factor 1α. *PLoS ONE*, 9, e115565. Available from https://doi.org/10.1371/journal.pone.0115565.

Xu, L., Beckebaum, S., Iacob, S., Wu, G., Kaiser, G. M., Radtke, A., ... Cicinnati, V. R. (2014). MicroRNA-101 inhibits human hepatocellular carcinoma progression through EZH2 downregulation and increased cytostatic drug sensitivity. *Journal of Hepatology*, 60(3), 590–598. Available from https://doi.org/10.1016/j.jhep.2013.10.028.

Xu, F., Liao, J. Z., Xiang, G. Y., Zhao, P. X., Ye, F., Zhao, Q., & He, X. X. (2017). MiR-101 and doxorubicin codelivered by liposomes suppressing malignant properties of hepatocellular carcinoma. *Cancer Medicine*, 6(3), 651–661. Available from https://doi.org/10.1002/cam4.1016.

Xu, Y., Ge, Z., Zhang, E., Zuo, Q., Huang, S., Yang, N., ... Sun, L. (2017). The lncRNA TUG1 modulates proliferation in trophoblast cells via epigenetic suppression of RND3. *Cell Death and Disease*, 8(10). Available from https://doi.org/10.1038/cddis.2017.503.

Xu, Y., Yao, Y., Leng, K., Li, Z., Qin, W., Zhong, X., ... Cui, Y. (2017). Long non-coding RNA UCA1 indicates an unfavorable prognosis and promotes tumorigenesis via regulating AKT/GSK-3β signaling pathway in cholangiocarcinoma. *Oncotarget*, 8(56), 96203–96214. Available from https://doi.org/10.18632/oncotarget.21884.

Xu, Y., Wang, Z., Jiang, X., & Cui, Y. (2017). Overexpression of long noncoding RNA H19 indicates a poor prognosis for cholangiocarcinoma and promotes cell migration and invasion by affecting epithelial-mesenchymal transition. *Biomedicine and Pharmacotherapy*, 92, 17–23. Available from https://doi.org/10.1016/j.biopha.2017.05.061.

Xu, X., Lou, Y., Tang, J., Teng, Y., Zhang, Z., Yin, Y., ... Tan, Z. (2019). The long non-coding RNA Linc-GALH promotes hepatocellular carcinoma metastasis via epigenetically regulating Gankyrin. *Cell Death and Disease*, *10*(2). Available from https://doi.org/10.1038/s41419-019-1348-0.

Xu, Y., Liu, Y., Li, Z., Li, H., Li, X., Yan, L., ... Xue, F. (2020). Long non-coding RNA H19 is involved in sorafenib resistance in hepatocellular carcinoma by upregulating miR-675. *Oncology Reports*, *44*(1), 165–173. Available from https://doi.org/10.3892/or.2020.7608.

Yan, X. L., Jia, Y. L., Chen, L., Zeng, Q., Zhou, J. N., Fu, C. J., ... Pei, X. T. (2013). Hepatocellular carcinoma-associated mesenchymal stem cells promote hepatocarcinoma progression: Role of the S100A4-miR155-SOCS1-MMP9 axis. *Hepatology*, *57*(6), 2274–2286. Available from https://doi.org/10.1002/hep.26257.

Yan, C., Chen, J., & Chen, N. (2016). Long noncoding RNA MALAT1 promotes hepatic steatosis and insulin resistance by increasing nuclear SREBP-1c protein stability. *Scientific Reports*, *6*. Available from https://doi.org/10.1038/srep22640.

Yang, F., Yin, Y., Wang, F., Wang, Y., Zhang, L., Tang, Y., & Sun, S. (2010). miR-17-5p promotes migration of human hepatocellular carcinoma cells through the p38 mitogen-activated protein kinase-heat shock protein 27 pathway. *Hepatology*, *51*(5), 1614–1623. Available from https://doi.org/10.1002/hep.23566.

Yang, F., Zhang, L., Huo, X. S., Yuan, J. H., Xu, D., Yuan, S. X., ... Sun, S. H. (2011). Long noncoding RNA high expression in hepatocellular carcinoma facilitates tumor growth through enhancer of zeste homolog 2 in humans. *Hepatology*, *54*(5), 1679–1689. Available from https://doi.org/10.1002/hep.24563.

Yang, Z., Zhou, L., Wu, L. M., Lai, M. C., Xie, H. Y., Zhang, F., & Zheng, S. S. (2011). Overexpression of long noncoding RNA HOTAIR predicts tumor recurrence in hepatocellular carcinoma patients following liver transplantation. *Annals of Surgical Oncology*, *18*(5), 1243–1250. Available from https://doi.org/10.1245/s10434-011-1581-y.

Yang, X., Liang, L., Zhang, X. F., Jia, H. L., Qin, Y., Zhu, X. C., ... Qin, L. X. (2013). MicroRNA-26a suppresses tumor growth and metastasis of human hepatocellular carcinoma by targeting interleukin-6-Stat3 pathway. *Hepatology*, *58*(1), 158–170. Available from https://doi.org/10.1002/hep.26305.

Yang, X., Zhang, X. F., Lu, X., Jia, H. L., Liang, L., Dong, Q. Z., ... Qin, L. X. (2014). MicroRNA-26a suppresses angiogenesis in human hepatocellular carcinoma by targeting hepatocyte growth factor-cMet pathway. *Hepatology*, *59*(5), 1874–1885. Available from https://doi.org/10.1002/hep.26941.

Yang, F., Li, Q. J., Gong, Z. B., Zhou, L., You, N., Wang, S., ... Dou, K. F. (2014). MicroRNA-34a targets Bcl-2 and sensitizes human hepatocellular carcinoma cells to sorafenib treatment. *Technology in Cancer Research and Treatment*, *13*(1), 77–86. Available from https://doi.org/10.7785/tcrt.2012.500364.

Yang, Z., Lu, Y., Xu, Q., Tang, B., Park, C. K., & Chen, X. (2015). HULC and H19 played different roles in overall and disease-free survival from hepatocellular carcinoma after curative hepatectomy: A preliminary analysis from gene expression omnibus. *Disease Markers*, *2015*. Available from https://doi.org/10.1155/2015/191029.

Yang, J., Lu, Y., Lin, Y. Y., Zheng, Z. Y., Fang, J. H., He, S., & Zhuang, S. M. (2016). Vascular mimicry formation is promoted by paracrine TGF-β and SDF1 of cancer-associated fibroblasts and inhibited by miR-101 in hepatocellular carcinoma. *Cancer Letters*, *383*(1), 18–27. Available from https://doi.org/10.1016/j.canlet.2016.09.012.

Yang, Y., Chen, L., Gu, J., Zhang, H., Yuan, J., Lian, Q., ... Lu, Z. J. (2017). Recurrently deregulated lncRNAs in hepatocellular carcinoma. *Nature Communications*, *8*. Available from https://doi.org/10.1038/ncomms14421.

Yang, S., Cai, H., Hu, B., & Tu, J. (2019). LncRNA SAMMSON negatively regulates miR-9-3p in hepatocellular carcinoma cells and has prognostic values. *Bioscience Reports*, *39*(7). Available from https://doi.org/10.1042/BSR20190615.

Ye, J., Li, H., Wei, J., Luo, Y., Liu, H., Zhang, J., & Luo, X. (2020). Risk scoring system based on lncRNA expression for predicting survival in hepatocellular carcinoma with cirrhosis. *Asian Pacific Journal of Cancer Prevention*, *21*(6), 1787–1795. Available from https://doi.org/10.31557/APJCP.2020.21.6.1787.

Yerukala Sathipati, S., & Ho, S. Y. (2020). Novel miRNA signature for predicting the stage of hepatocellular carcinoma. *Scientific Reports*, *10*(1). Available from https://doi.org/10.1038/s41598-020-71324-z.

Yu, J., Han, J., Zhang, J., Li, G., Liu, H., Cui, X., ... Wang, C. (2016). The long noncoding RNAs PVT1 and uc002mbe.2 in sera provide a new supplementary method for hepatocellular carcinoma diagnosis. *Medicine (United States)*, *95*(31). Available from https://doi.org/10.1097/MD.0000000000004436.

Yu, Y., Zhang, M., Liu, J., Xu, B., Yang, J., Wang, N., ... Miao, L. (2018). Long non-coding RNA PVT1 promotes cell proliferation and migration by silencing ANGPTL4 expression in cholangiocarcinoma. *Molecular Therapy — Nucleic Acids*, *13*, 503–513. Available from https://doi.org/10.1016/j.omtn.2018.10.001.

Yuan, J. H., Yang, F., Wang, F., Ma, J. Z., Guo, Y. J., Tao, Q. F., ... Sun, S. H. (2014). A long noncoding RNA activated by TGF-β promotes the invasion-metastasis cascade in hepatocellular carcinoma. *Cancer Cell, 25*(5), 666−681. Available from https://doi.org/10.1016/j.ccr.2014.03.010.

Yuan, Sx, Wang, J., Yang, F., Tao, Qf, Zhang, J., Wang, Ll, ... Zhou, W. P. (2016). Long noncoding RNA DANCR increases stemness features of hepatocellular carcinoma by derepression of CTNNB1. *Hepatology, 63*(2), 499−511. Available from https://doi.org/10.1002/hep.27893.

Özeş, A. R., Wang, Y., Zong, X., Fang, F., Pilrose, J., & Nephew, K. P. (2017). Therapeutic targeting using tumor specific peptides inhibits long non-coding RNA HOTAIR activity in ovarian and breast cancer. *Scientific Reports, 7*(1). Available from https://doi.org/10.1038/s41598-017-00966-3.

Zekri, A. R. N., Youssef, A. S. E. D., El-Desouky, E. D., Ahmed, O. S., Lotfy, M. M., Nassar, A. A. M., & Bahnassey, A. A. (2016). Serum microRNA panels as potential biomarkers for early detection of hepatocellular carcinoma on top of HCV infection. *Tumor Biology, 37*(9), 12273−12286. Available from https://doi.org/10.1007/s13277-016-5097-8.

Zhang, J., Han, C., & Wu, T. (2012). MicroRNA-26a promotes cholangiocarcinoma growth by activating β-catenin. *Gastroenterology, 143*(1), 246−e8. Available from https://doi.org/10.1053/j.gastro.2012.03.045.

Zhang, J., Han, C., Zhu, H., Song, K., & Wu, T. (2013). MiR-101 inhibits cholangiocarcinoma angiogenesis through targeting vascular endothelial growth factor (VEGF). *American Journal of Pathology, 182*(5), 1629−1639. Available from https://doi.org/10.1016/j.ajpath.2013.01.045.

Zhang, L., Yang, F., Yuan, J. H., Yuan, S. X., Zhou, W. P., Huo, X. S., ... Sun, S. H. (2013). Epigenetic activation of the MiR-200 family contributes to H19-mediated metastasis suppression in hepatocellular carcinoma. *Carcinogenesis, 34*(3), 577−586. Available from https://doi.org/10.1093/carcin/bgs381.

Zhang, J. P., Zeng, C., Xu, L., Gong, J., Fang, J. H., & Zhuang, S. M. (2014). MicroRNA-148a suppresses the epithelial-mesenchymal transition and metastasis of hepatoma cells by targeting Met/Snail signaling. *Oncogene, 33*(31), 4069−4076. Available from https://doi.org/10.1038/onc.2013.369.

Zhang, H., Diab, A., Fan, H., Mani, S. K. K., Hullinger, R., Merle, P., & Andrisani, O. (2015). PLK1 and HOTAIR accelerate proteasomal degradation of SUZ12 and ZNF198 during hepatitis B virus-induced liver carcinogenesis. *Cancer Research, 75*(11), 2363−2374. Available from https://doi.org/10.1158/0008-5472.CAN-14-2928.

Zhang, L., Yang, Z., Trottier, J., Barbier, O., & Wang, L. (2017). Long noncoding RNA MEG3 induces cholestatic liver injury by interaction with PTBP1 to facilitate shp mRNA decay. *Hepatology, 65*(2), 604−615. Available from https://doi.org/10.1002/hep.28882.

Zhang, M., Sun, W., Zhou, M., & Tang, Y. (2017). MicroRNA-27a regulates hepatic lipid metabolism and alleviates NAFLD via repressing FAS and SCD1. *Scientific Reports, 7*(1). Available from https://doi.org/10.1038/s41598-017-15141-x.

Zhang, K., Chen, J., Zhou, H., Chen, Y., Zhi, Y., Zhang, B., ... Zhang, C. (2018). PU.1/microRNA-142-3p targets ATG5/ATG16L1 to inactivate autophagy and sensitize hepatocellular carcinoma cells to sorafenib. *Cell Death and Disease, 9*(3). Available from https://doi.org/10.1038/s41419-018-0344-0.

Zhang, P. F., Wang, F., Wu, J., Wu, Y., Huang, W., Liu, D., ... Ke, A. W. (2019). LncRNA SNHG3 induces EMT and sorafenib resistance by modulating the miR-128/CD151 pathway in hepatocellular carcinoma. *Journal of Cellular Physiology, 234*(3), 2788−2794. Available from https://doi.org/10.1002/jcp.27095.

Zhang, T., Yang, Z., Kusumanchi, P., Han, S., & Liangpunsakul, S. (2020). Critical role of microRNA-21 in the pathogenesis of liver diseases. *Frontiers in Medicine, 7*. Available from https://doi.org/10.3389/fmed.2020.00007.

Zhao, P., Li, M., Wang, Y., Chen, Y., He, C., Zhang, X., ... Xiang, G. (2018). Enhancing anti-tumor efficiency in hepatocellular carcinoma through the autophagy inhibition by miR-375/sorafenib in lipid-coated calcium carbonate nanoparticles. *Acta Biomaterialia, 72*, 248−255. Available from https://doi.org/10.1016/j.actbio.2018.03.022.

Zhao, Z. B., Chen, F., & Bai, X. F. (2019). Long noncoding RNA MALAT1 regulates hepatocellular carcinoma growth under hypoxia via sponging microRNA-200a. *Yonsei Medical Journal, 60*(8), 727−734. Available from https://doi.org/10.3349/ymj.2019.60.8.727.

Zhao, H., Shi, J., Zhang, Y., Xie, A., Yu, L., Zhang, C., ... Li, X. (2020). LncTarD: A manually-curated database of experimentally-supported functional lncRNA-target regulations in human diseases. *Nucleic Acids Research, 48*(1), D118−D126. Available from https://doi.org/10.1093/nar/gkz985.

Zheng, F., Liao, Y. J., Cai, M. Y., Liu, T. H., Chen, S. P., Wu, P. H., ... Xie, D. (2015). Systemic delivery of microRNA-101 potently inhibits hepatocellular carcinoma in vivo by repressing multiple targets. *PLoS Genetics, 11*(2), 1−21. Available from https://doi.org/10.1371/journal.pgen.1004873.

Zheng, H., Li, P., Kwok, J. G., Korrapati, A., Li, W. T., Qu, Y., ... Ongkeko, W. M. (2018). Alcohol and hepatitis virus-dysregulated lncRNAs as potential biomarkers for hepatocellular carcinoma. *Oncotarget, 9*(1), 224−235. Available from https://doi.org/10.18632/oncotarget.22921.

Zhou, J., Yu, L., Gao, X., Hu, J., Wang, J., Dai, Z., ... Fan, J. (2011). Plasma microRNA panel to diagnose hepatitis B virus-related hepatocellular carcinoma. *Journal of Clinical Oncology, 29*(36), 4781−4788. Available from https://doi.org/10.1200/JCO.2011.38.2697.

Zhou, S. J., Liu, F. Y., Zhang, A. H., Liang, H. F., Wang, Y., Ma, R., ... Sun, N. F. (2017). MicroRNA-199b-5p attenuates TGF-β1-induced epithelial-mesenchymal transition in hepatocellular carcinoma. *British Journal of Cancer, 117*(2), 233−244. Available from https://doi.org/10.1038/bjc.2017.164.

Zhou, Y., Huan, L., Wu, Y., Bao, C., Chen, B., Wang, L., ... He, X. (2020). LncRNA ID2-AS1 suppresses tumor metastasis by activating the HDAC8/ID2 pathway in hepatocellular carcinoma. *Cancer Letters, 469*, 399−409. Available from https://doi.org/10.1016/j.canlet.2019.11.007.

Zhouhua, H., Xuwen, X., Ledu, Z., Xiaoyu, F., Shuhui, T., Jiebin, Z., ... Shuiping, L. (2017). The long non-coding RNA MALAT1 promotes the migration and invasion of hepatocellular carcinoma by sponging miR-204 and releasing SIRT1. *Tumor Biology, 39*(7). Available from https://doi.org/10.1177/1010428317718135, 101042831771813.

Zhu, Q., Wang, Z., Hu, Y., Li, J., Li, X., Zhou, L., & Huang, Y. (2012). miR-21 promotes migration and invasion by the miR-21-PDCD4-AP-1 feedback loop in human hepatocellular carcinoma. *Oncology Reports, 27*(5), 1660−1668. Available from https://doi.org/10.3892/or.2012.1682.

Zhu, H., Han, C., Lu, D., & Wu, T. (2014). MiR-17-92 cluster promotes cholangiocarcinoma growth evidence for PTEN as downstream target and IL-6/Stat3 as upstream activator. *American Journal of Pathology, 184*(10), 2828−2839. Available from https://doi.org/10.1016/j.ajpath.2014.06.024.

Zhu, P., Wang, Y., Huang, G., Ye, B., Liu, B., Wu, J., ... Fan, Z. (2016). Lnc-β-Catm elicits EZH2-dependent β-catenin stabilization and sustains liver CSC self-renewal. *Nature Structural and Molecular Biology, 23*(7), 631−639. Available from https://doi.org/10.1038/nsmb.3235.

Zu, C., Liu, S., Cao, W., Liu, Z., Qiang, H., Li, Y., ... Li, J. (2017). MiR-590-3p suppresses epithelial-mesenchymal transition in intrahepatic cholangiocarcinoma by inhibiting SIP1 expression. *Oncotarget, 8*(21), 34698−34708. Available from https://doi.org/10.18632/oncotarget.16150.

Zucman-Rossi, J. (2010). Molecular classification of hepatocellular carcinoma. *Digestive and Liver Disease,* S235−S241. Available from https://doi.org/10.1016/s1590-8658(10)60511-7.

12

Noncoding ribonucleic acids in gallbladder cancer patients

Bela Goyal[1], Tarunima Gupta[1], Sweety Gupta[2] and Amit Gupta[3]

[1]Department of Biochemistry, All India Institute of Medical Sciences, Rishikesh, India
[2]Department of Radiation Oncology, All India Institute of Medical Sciences, Rishikesh, India
[3]Department of General Surgery, All India Institute of Medical Sciences, Rishikesh, India

12.1 Introduction

Gallbladder carcinoma (GBC) is usually diagnosed at an advanced stage due to the absence of specific symptoms. The 5-year survival rate in the range of 19%–62% depending on whether an early diagnosis was made or the patient presented with late symptoms (Siegel, Miller, & Jemal, 2020). GBC presents with a wide range of genetic heterogeneity. The current biomarkers used for diagnosis and management (CA 19–9, CEA, CA 125, CA 242, CA 15–3) may not be helpful for early diagnosis in most patients (Montalvo-Jave et al., 2019). There is an essential need to discover novel biomarkers for early diagnosis and prognosis. Furthermore, to date surgery remains the only curative option available for GBC which can only be done at an early stage. Adjuvant therapy with chemotherapy and radiotherapy provides the only palliation and has failed to demonstrate much improvement in overall survival (Yang, Zhang, Li, & Wang, 2016). Therefore identifying a diagnostic biomarker that can detect GBC in its initial stage and exploring potential therapeutic targets is the major interest of the scientific community.

In the past decades, noncoding Ribonucleic acids (ncRNA) have emerged as an important tool as molecular biomarkers and as therapeutic targets in GBC. ncRNAs are a large and heterogeneous class of RNAs. Based on their size, ncRNAs are classified into small ncRNAs like microRNAs (miRNAs), small interfering RNAs, PIWI-interacting RNAs (piRNA), transfer RNA-derived small RNAs, and small nucleolar RNAs. On the other end of the spectrum are the large-sized (>200 ntds) long-noncoding RNAs (lncRNAs), circular

RNAs (circRNAs), and pseudogenes. Out of these miRNAs and lnRNAs are most extensively studied with respect to GBC (Djebali et al., 2012). This is also illustrated in Fig. 12.1.

Surprisingly, only 10% of the total RNAs encode for proteins, whereas 90% is noncoding. Earlier considered unnecessary, ncRNAs have emerged as a key regulator of the genome at transcriptomic and epigenetic levels. NcRNAs are implicated in the regulation of normal cellular processes as well as diseased conditions notably, cancers. Cancer is characterized by hallmarks of unregulated cell proliferation, metastasis, invasion. Dysregulated expression of ncRNAs has been implicated in the regulation of tumorigenesis of cancer cells via a multitude of mechanisms involving molecular signaling pathways (Slack and Chinnaiyan, 2019).

The stability of some of the small ncRNAs in blood makes them an interesting candidate as a diagnostic biomarker for liquid biopsy. Moreover, they are amenable to be targeted for developing novel targeted therapy (Djebali et al., 2012; Hombach and Kretz, 2016; Slack and Chinnaiyan, 2019). A surge in this field has been observed with the recent FDA approval of RNA targeting drugs in 2018 based on RNA interference (RNAi) or antisense oligonucleotide approach in neurodegenerative disorders (Adams et al., 2018; Wurster et al., 2019). Moreover, ongoing trials targeting miRNAs are also afoot in the field of oncology (Beg et al., 2017; Seto et al., 2018).

Out of a huge repertoire of lncRNAs, miRNAs and lncRNAs are most extensively studied with respect to GBC. Biogenesis, biological functions with respect to cancer hallmarks, utility as potential diagnostic or prognostic biomarker, and therapeutic target amenability of miRNAs and lncRNAs will be discussed in subsequent sections.

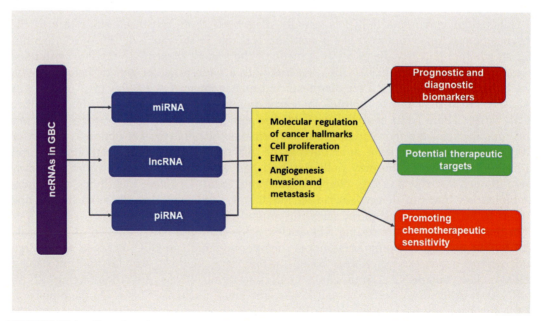

FIGURE 12.1 Clinical applications of noncoding RNAs in gallbladder cancer.

12.2 MiRNAs in gallbladder carcinoma

MiRNAs are single-stranded nonprotein-coding RNAs approximately 20–24 nucleotides in length. Mature miRNA sequences are highly conservative in nature. miRNAs act as posttranslational inhibitors by binding to the complementary sequences in 3' untranslated regions (UTR) of messenger RNA (mRNA), thus stimulating target mRNAs translational repression and degradation (Jansson and Lund, 2012). Recent studies have shifted our understanding of how miRNAs interact with their targets, which include not only mRNAs but also long noncoding RNAs (lncRNAs), pseudogenes, and circRNAs. These may act as competing endogenous RNA (ceRNA) to regulate other RNA transcripts by competing for shared miRNAs (Salmena, Poliseno, Tay, Kats, & Pandolfi, 2011; Tay, Rinn, & Pandolfi, 2014). Many studies have indicated that miRNAs play important role in modulating cell proliferation, migration, invasion, apoptosis, radio- and chemosensitivity, and cancer stem cell phenotype. Circulating miRNAs could be promising potential biomarkers for GBC because detectable miRNAs in the bodily fluids are stable and can be measured using noninvasive methods (Wu, Yang, & Li, 2009). miRNAs although still in the experimental phase are being proven to have specific regulatory functions in gallbladder cancer pathways, emerging evidence suggests that miRNAs might prove helpful to map and/or develop signatures for early diagnosis of GBC.

12.2.1 Biogenesis and biological functions of miRNA

Initially synthesized from longer precursors in the nucleus, a hairpin structure referred to as a pri-miRNA, miRNAs are processed through a series of events into mature, double-stranded miRNAs. Initially, miRNA genes are transcribed by RNA polymerase II. Then pri-miRNAs are cleaved in the nucleus by Drosha, to form a precursor-miRNA (premiRNA), shorter sequence of approximately 70 nucleotides in length. Pre-miRNAs are then translocated to the cytoplasm by Exportin 5, where Dicer (an miRNA-specific nuclease) cleaves pre-miRNAs resulting in double-stranded miRNA of approximately 22 nucleotides in length. Subsequently, the mature miRNA is incorporated into the RNA-inducing silencing complex (RISC) by associating with Argonaute proteins. It is this complex with the miRNA that will target the 3' UTR of mRNA that is responsible for the decrease in target gene expression, via either degradation of the mRNA or inhibition of protein synthesis (Felekkis, Touvana, Stefanou, & Deltas, 2010; Ha and Kim, 2014; O'Brien, Hayder, Zayed, & Peng, 2018). With the ability to interact with multiple target genes, miRNAs have been proven to influence many important biological processes such as cell growth, tissue differentiation, cell proliferation, embryonic development, and apoptosis. Dysregulated miRNA plays a critical role in the progression of various cancers (Sayed and Abdellatif, 2011).

12.2.2 MicroRNAs in pathogenesis and as a therapeutic target in gallbladder carcinoma

The cancer hallmarks constitute the basic principle of complexities of neoplastic disease and thus comprise the biological capabilities acquired during the stepwise process of developing human tumors. They include sustaining proliferative signaling, evading growth suppressors, resisting cell death, enabling replicative immortality, inducing angiogenesis, activating invasion and metastasis, reprogramming energy metabolism, and evading immune destruction (Montalvo-Jave et al., 2019).

miRNAs have been shown to affect target genes that are involved in the regulation of cell proliferation, apoptosis, metastasis, epithelial—mesenchymal transition (EMT), and poor survival. Ongoing research of miRNA-binding site has indicated several potential gene targets of cancer-associated miRNA that are important to control cell adhesion, neovascularization, and tissue invasion (Alanazi, Hoffmann, & Adelson, 2015; Bueno and Malumbres, 2011; Gallach, Calabuig-Fariñas, Jantus-Lewintre, & Camps, 2014; Lima et al., 2011; Peng and Croce, 2016). Various miRNAs have been demonstrated to influence these hallmarks and function as either oncogenic or tumor suppressive under certain conditions. Most of the studies have, however, shown miRNAs to be primarily tumor suppressors in GBC. Few studies have depicted oncogenic effect as well as depicted in Table 12.1.

Experimental studies have shown that oncogenic miRNAs like miR-20a and miR-182 promote proliferation and invasion in GBC cell lines in response to Transforming Growth Factor beta (TGF-β) (Chang et al., 2013; Qiu et al., 2014). miR-20a affected invasion via modulating miR-20a/Smad7/β-catenin axis (Chang et al., 2013), whereas upregulated miR-182 suppressed cell adhesion molecule 1 resulting in EMT thereby promoting invasion and metastasis of GBC (Qiu et al., 2014).

Most of the miRNAs have been studied to be oncosuppressor by silencing tumor-promoting genes regulating molecular pathways involved in tumorigenesis. Receptor tyrosine kinase pathways are known to be the most commonly modulated pathway responsible for tumor proliferation and angiogenesis. miR-143—3p modulated GBC tumor growth and angiogenesis via PI3K/AKT pathway by downregulating placental growth factor (Jin, Hu, et al., 2018). Similarly, transcription factor mitogen-activated protein kinase (MAPK) has been demonstrated to be downregulated by miR 136 and miRNA-29c-5p (Niu et al., 2019; Shu et al., 2017). Upregulation of miR 136 has been associated with reduced proliferation and angiogenesis in experimental studies on GBC by inhibiting MAP2K4 via the Jun N-terminal kinase (JNK) signaling pathway (Niu et al., 2019). Additionally, other cell proliferating pathways like Janus kinase - Signal Transducers and Activators of Transcription pathways (JAK-STAT pathways) have been found to be regulated by miR-145a-5p in GBC (Goeppert et al., 2019). In a study by Goeppert et al., profiling of 40 GBC and 8 normal gallbladder tissues was performed and miR-145a5p was found to be the most downregulated miRNA in GBC relative to normal tissue. On experimental studies in GBC cell lines, miR-145a-5p mimic promoted cell proliferation via the downregulation of JAK-STAT pathways (Goeppert et al., 2019).

Angiogenic factor vascular endothelial growth factor (VEGF) primarily responsible for invasion and metastasis of cancer cells has also been shown to be modulated by miR-1 in GBC (Letelier et al., 2014). Similarly, epidermal growth factor receptor is downregulated by miRNA-146b-5p (Cai et al., 2015).

Telomere lengthening has been a known factor responsible for tumorigenesis. Mir-34a has been found to affect telomere lengthening and tumor proliferation in GBC (Jin et al., 2014).

Not only in cancer hallmarks, but miRNAs have also been implicated in drug sensitivity in GBC. GBC response to gemcitabine was seen to be accentuated by miR-218—5p by the suppression of multidrug resistance 1 expression (Wang et al., 2017). Similarly, miR-223 has been shown to improve the sensitivity of GBC cells to docetaxel by modulating STMN1 gene expression (Lu et al., 2016).

TABLE 12.1 Expression of microRNAs (miRNAs) in gallbladder cancer.

miRNA	Expression levels	Model used	Target gene/pathway	Clinical outcome	Reference
miR-20a	Upregulation	Cell lines/clinical samples	SMAD7	Prognostic factor (overall survival), potential therapeutic target	Chang et al. (2013)
miR-182	Upregulation	Cell lines/mice/clinical samples	CADM1	Prognostic marker	Qiu et al. (2014)
miR-143–3p	Downregulation	Cell lines/clinical samples	PI3K/AKT pathway	Prognostic factor (overall survival), potential therapeutic target	Jin, Hu, et al. (2018)
miR-136	Downregulation	Cell lines/mice/clinical samples	MAP2K4	Potential therapeutic target	Niu, Li, and Li (2019)
miR-29c-5p	Downregulation	Cell lines/clinical samples	MAPK pathway/CPEB4 gene	Prognostic factor (overall survival)	Shu et al. (2017)
miR-145a-5p	Downregulation	Cell lines/clinical samples	STAT1	Prognostic factor (overall survival), potential therapeutic target	Goeppert et al. (2019)
miR-1	Downregulation	Cell lines/clinical samples	VEGF-A	Prognostic marker	Letelier et al. (2014)
miR-146b-5p	Downregulation	Cell lines/mice/clinical samples	EGFR	Prognostic factor, potential therapeutic target	Cai et al. (2015)
miR-34a-5p	Downregulation	Cell lines/mice/clinical samples	PNUTS	Poor prognostic factor, potential therapeutic target	Jin et al. (2014)
mir-218–5p	Downregulation	Cell lines/clinical samples	MDR1 expression	Prognostic marker, promotes gemcitabine sensitivity	Wang et al. (2017)
miR-223	Downregulation	Cell lines/clinical samples	STMN1	Prognostic marker, promotes docetaxel sensitivity	Lu et al. (2016)
miR-372	Downregulation	Cell lines/clinical samples	Chloride intracellular channel 1 (CLIC1)	Prognostic factor (overall survival)	Zhou, Cheng, Peng, Liu, and Jiang (2017)
miR-124	Downregulation	Cell lines/clinical samples	HIPK3	Potential therapeutic target	Kai et al. (2018)
miR-26a	Downregulation	Cell lines/clinical samples	HMGA2	Prognostic marker	Zhou et al. (2014)

(Continued)

TABLE 12.1 (Continued)

miRNA	Expression levels	Model used	Target gene/pathway	Clinical outcome	Reference
miR-30d-5p	Downregulation	Cell lines/mice/clinical samples	Lactate dehydrogenase-A (LDHA)	Poor prognostic factor (overall survival), potential therapeutic target	He et al. (2018)
miR-1–5p	Downregulation	Cell lines/mice/clinical samples	NOTCH-2	Prognostic marker	Hua, Song, Ma, and Li (2019)
miR-139–5p	Downregulation	Cell lines/mice/clinical samples	Pyruvate kinase M2 (PKM2)	Poor prognostic factor (overall survival), potential therapeutic target	Chen et al. (2018)
miR-19a	Upregulation	Cell lines/mice/clinical samples	PLGF/cMyc	Poor prognostic factor (overall survival)	Li et al. (2018)

Similarly, a multitude of miRNAs has been implicated in the pathogenesis of GBC by targeting mRNAs of various transcription factors thereby regulating cancer hallmarks (Chen et al., 2018; He et al., 2018; Hua et al., 2019; Kai et al., 2018; Li et al., 2018; Lu et al., 2016; Zhou et al., 2014; Zhou et al., 2017). These miRNAs henceforth are amenable for therapeutic targeting. Further clinical studies may be designed for the validation of miRNA-based strategies to be put into clinical utility.

12.2.3 MicroRNAs as biomarkers in gallbladder carcinoma

miRNAs are the most abundant type of cell-free RNA found in the blood. They can also be contained in exosomes, apoptotic bodies, protein miRNA complexes. Their stability in body fluids makes miRNAs suitable candidates as tumor biomarkers. They are resistant to extreme pH and temperatures, as well as against degradation by RNAses. They can withstand repeated freeze/thaw cycles. miRNAs in various body fluids, such as cerebrospinal fluid, pleural effusion, urine, breast milk, saliva, and blood, have been shown to be useful tumor biomarkers. The abundance of miRNAs in the blood correlates with that of their tumors of origin (Wang, Chen, & Sen, 2016).

Many methods are available to identify and quantify miRNA expression. Current methodologies used for detecting miRNAs include quantitative Polymerase Chain Reaction (PCR), in situ hybridization, microarrays, and RNA sequencing. Microarrays and RNA sequencing helps in profiling GBC for differentially expressed miRNAs as compared to normal tissue and also in identifying novel miRNAs that may be implicated in the pathogenesis and may be used as potential diagnostic and prognostic biomarkers. Letelier et al. profiled GBC tissue and normal gallbladder tissue samples. Five of the most deregulated miRNAs were also validated by Real Time Quantitative Reverse Transcription PCR (qRT-PCR) in GBC tissue samples. The most downregulated miRNAs (miR-1, miR-133, miR-143, and miR-145) as depicted by pathway enrichment studies targeted various cell signaling pathways like VEGF, ErbB3, TGF-β, and Wingless related integration site (WNT) (Letelier et al., 2014). Similarly, Goeppert et al. (2019) profiled 40 GBC and 8 normal gallbladder tissues and 24 miRNAs were found to be significantly deregulated in GBC tissues. Furthermore, downregulation of these miRNAs was found to be associated with poor overall survival in patients.

Most of the miRNAs have been explored for their prognostic utility wherein dysregulation of miRNAs in GBC tissues has been correlated with poor overall survival in GBC patients.

Expression of miR-155 was observed to be significantly higher in GBC tissue relative to normal tissue samples. Moreover, elevated expression of miR-155 was found to be associated with significantly advanced clinicopathological status in terms of Tumor Node Metastasis (TNM) stage and differentiation of the tumor. Additionally, upregulation of miR-155 was found to be associated with lower median overall survival in GBC patients (Kono et al., 2013; Zhang, Chen, & Qin, 2015). Similarly, other miRNAs such as miR-372, miR-124, miR-26a, miR-30d-5p, and miR-1–5p have been studied in GBC tissues and normal tissues by qRT-PCR (He et al., 2018; Hua et al., 2019; Kai et al., 2018; Zhou et al., 2014; Zhou et al., 2017). Reduced expression of these miRNAs has been found to be associated with significantly lower overall survival in patients.

Besides GBC tissues, circulatory miRNA levels in body fluids like blood and bile have also been explored in GBC. One of the studies performed microarray in peripheral blood samples of GBC patients and healthy controls. They identified 11 miRNAs to be significantly dysregulated in GBC patients relative to healthy controls. Out of which let-7a, miR-21, miR-187, miR-143, miR-202, and miR-335 were in corroboration with qRT-PCR assay as well. Additionally, miR-187, miR-143, and miR-202 miRNAs showed a significant correlation with clinic-pathological features of GBC patients (Li and Pu, 2015). Shigehara et al. concluded that miRNAs are present in bile samples of biliary tract cancers (BTC), including GBC. They observed differential expression of miR-9, miR-302c, miR-199a-3p, and miR-222 in malignant BTC patients as compared to benign cholelithiasis patients. Furthermore, these miRNAs showed high diagnostic sensitivity and specificity for BTC (Shigehara et al., 2011). Liquid biopsy being minimally invasive is the way forward for cancer diagnosis. Utilizing miRNAs owing to their stability and ease of detection seems to be a very suitable candidate as a tumor marker. With the advent of new technology, it is imperative that more and more studies need to be undertaken which are well planned and standardized till miRNAs are introduced in clinical practice as a biomarker and as a therapeutic target in GBC.

12.3 LncRNAs in gallbladder carcinoma

LncRNA may be defined as those ncRNA transcripts that have a length of >200 nucleotides. They have been shown to play a vital role in transcriptional and posttranscriptional regulation of gene expression responsible for cell growth, proliferation, DNA damage repair, etc., (Quinn and Chang, 2016). The altered expression of lncRNA has been implicated in the regulation of a number of physiological and pathological processes like cell growth, differentiation and disease like cancers, neurodegenerative illnesses, cardiovascular diseases, etc. (Batista and Chang, 2013).

12.3.1 Biogenesis and biological functions of LncRNA

LncRNAs similar to mRNA are transcribed by RNA polymerase II (RNA Pol II). They are mainly transcribed from an exonic or intergenic portion of the genome. After being transcribed, similar to mRNA, they are 5′ capped and 3′ polyadenylated. LncRNAs are very commonly subjected to alternative splicing via interaction with splicing factors, chromatin remodeling, and epigenetic regulation resulting in heterogeneity. Based on their structural and functional aspects, lncRNAs have been classified into five main categories: sense lncRNAs, antisense lncRNAs, intronic lncRNAs, Exonic lncRNAs, and bidirectional lncRNAs (Dhanoa, Sethi, Verma, Arora, & Mukhopadhyay, 2018). Even though lncRNAs are noncoding, they play a key role in the regulation of multiple cellular processes. They exert their effect through a variety of mechanisms based on their location. Nuclear lncRNAs primarily regulate transcription via interacting with transcription factor or chromatin remodeling factors or acting as a scaffold for splicing factors or regulatory molecules. Cytoplasmic lncRNAs on the other hand mediate their action through the regulation of mRNA stability and translation via interaction with miRNA or regulatory proteins, thus modulating cellular signaling processes and pathways in GBC (Schmitt and Chang, 2016).

12.3.2 LncRNAs in pathogenesis and as a therapeutic target in gallbladder carcinoma

Dysregulated expression of lncRNAs has most commonly been associated with cancer progression. Similar to miRNAs, lncRNAs also show oncogenic as well as tumor-suppressive characteristics in a number of cancers like lung cancer, breast cancer, hematological cancers, hepatobiliary cancers, and colorectal cancers (Zhang, Xia, Lu, Zhang, & Zhu, 2016). Elevated expression of lncRNA metastasis—associated lung adenocarcinoma transcript 1 (MALAT1), LINC00152, ANRIL, PAGBC, MYC-regulated long ncRNAs (MINCR), etc. and reduced expression of lncRNAs like maternally expressed gene 3 (MEG3), Low expression in Tumor (LET), and gallbladder cancer-associated supressor of pyruvate carboxylase (GCASPC) have been observed in gallbladder cancer as depicted in Table 12.2.

Experimental studies have demonstrated altered expression of these lncRNAs to be associated with various cancer hallmarks like cell proliferation, evasion of growth suppressors, inhibition of apoptosis, induction of angiogenesis, activation of invasion and metastasis, reprogramming energy metabolism, and evasion of immune destruction.

This is mediated via the regulation of various cell proliferation signaling pathways like PI3K/AKT pathway, JAK-STAT pathway, ERK/MAPK pathway, or via acting as miRNA sponge to regulate epithelial—mesenchymal progression, invasion, and metastasis.

lncRNA MALAT1 has been the most commonly studied lncRNA in GBC. Elevated expression of MALAT1 has been associated with GBC proliferation and metastasis in various studies via modulating ERK/MAPK pathway or via acting as ceRNA for miR-206, miR-363—3p (Goyal et al., 2021; Wang, Zhang, Wu, Zhang, Weng, et al., 2016; Wang, Zhang, Wu, Weng, Zhang, et al., 2016; Wu et al., 2014). Additionally, these studies also demonstrated that knockdown of MALAT1 resulted in reduced proliferation of GBC cell lines and reduction of GBC volume in vivo model indicating its oncogenic role.

TABLE 12.2 Expression of long noncoding RNAs in gallbladder cancer.

LncRNA	Expression levels	Model used	Target gene/pathway	Clinical outcome	Reference
MALAT1	Upregulation	Cell lines/clinical samples	ERK/MAPK pathway	Potential therapeutic target	Wu et al. (2014)
MALAT1	Upregulation	Cell lines/mice/clinical samples	miR-206	Poor prognostic factor (overall survival), potential therapeutic target	Wang, Zhang, Wu, Zhang (2016)
MALAT1	Upregulation	Cell lines/clinical samples	Myeloid cell leukemia-1 (MCL-1), miR-363–3p	Potential therapeutic target	Wang, Zhang, Wu, Weng, Zhang, et al. (2016)
MALAT1	Upregulation	Cell lines/clinical samples	–	Poor prognostic factor (overall survival), potential therapeutic target	Goyal et al. (2021)
LINC00152	Upregulation	Cell lines/clinical samples	SP1/LINC00152/PI3K/AKT pathway	Poor prognostic factor, potential therapeutic target	Cai et al. (2016)
LINC00152	Upregulation	Cell lines/clinical samples	LINC00152/miR-138/HIF-1α pathway	Poor prognostic factor (overall survival), potential therapeutic target	Cai et al. (2017)
H 19	Upregulation	Cell lines/clinical samples	H19/miR-194–5p/AKT2 axis	Poor prognostic factor, potential therapeutic target	Wang, Wu, et al. (2016)
H 19	Upregulation	Clinical samples	–	Poor prognostic factor (overall survival)	Wang et al. (2015)
H 19	Upregulation	Cell lines/clinical samples	H19/miR-342–3p/FOXM1	Potential therapeutic target	Wang, Ma, et al. (2016)
PAGBC	Upregulation	Clinical samples	ceRNA	Poor prognostic factor (overall survival)	Wu, Wang, Li, et al. (2017)
MINCR	Upregulation	Cell lines/mice/clinical samples	MINCR/miR-26a-5p/EZH2 axis	Poor prognostic factor, potential therapeutic target	Wang, Yang, et al. (2016)
PVT1	Upregulation	Cell lines/mice/clinical samples	PVT1/miR-143/HK2 axis	Potential therapeutic target	Chen et al. (2019)
HOTAIR	Upregulation	Cell lines/clinical samples	c-myc/HOTAIR/miRNA-130a axis	Potential therapeutic target	Ma et al. (2014)
ANRIL	Upregulation	Cell lines/mice/clinical samples	–	Poor prognostic factor, potential therapeutic target	Liu et al. (2016)

(Continued)

TABLE 12.2 (Continued)

LncRNA	Expression levels	Model used	Target gene/pathway	Clinical outcome	Reference
AFAP1-AS1	Upregulation	Cell lines/clinical samples	–	Poor prognostic factor (overall survival), potential therapeutic target	Ma, Wang, et al. (2016)
ROR	Upregulation	Cell lines/clinical samples	–	Poor prognostic factor, potential therapeutic target	Wang, Zhang, Wu, Weng, Zhou, et al. (2016)
HEGBC	Upregulation	Cell lines/mice/clinical samples	HEGBC/IL-11/STAT3 axis	Poor prognostic factor, potential therapeutic target	Yang, Gao, Wu, Feng, and Xu (2018)
Loc344887	Upregulation	Cell lines/clinical samples	Nrf2/Loc34487	Potential therapeutic target	Wu, Wang, Ou, et al. (2017)
TPTE2	Upregulation	Cell lines	TPTE2P1 pseudogene	Potential therapeutic target	Lv et al. (2015)
MEG3	Downregulation	Cell lines/mice/clinical samples	–	Potential therapeutic target	Liu et al. (2016)
MEG3	Downregulation	Cell lines/mice/clinical samples	MEG3/EZH2/LATS2 axis	Prognostic factor, potential therapeutic target	Jin, Cai, et al. (2018)
LET	Downregulation	Cell lines/clinical samples	–	Good prognostic factor, potential therapeutic target	Ma et al. (2015)
GCSAPC	Downregulation	Cell lines/mice/clinical samples	GCAPSC/miR-17–3p/pyruvate carboxylase	Potential therapeutic target	Ma, Zhang, et al. (2016)

Upregulation of LncRNA LINC00152 has been shown to be associated with increased cell proliferation, invasion, reduced apoptosis and tumorigenesis in vivo and in vitro studies in GBC (Cai et al., 2016; Cai et al., 2017). LINC00152 was demonstrated to be upregulated by a transcription factor specificity protein 1 (SP1) which, in turn, activated PI3K/AKT pathway to promote tumorigenesis (Cai et al., 2016). Another study showed that LINC00152 promoted EMT and metastasis in GBC by upregulating hypoxia inducible factor-1α. This is mediated by sponging oncosuppressor miRNA miR-138 (Cai et al., 2017). Similarly, lncRNA H 19 has also been found to be upregulated in GBC and promote its proliferation and EMT via H19/miR-194–5p/AKT2 axis or Twist expression and H19/miR-342–3p/FOXM1, respectively (Wang et al., 2015; Wang, Wu, et al., 2016; Wang, Ma, et al., 2016). MiRNA sponging of oncosuppressive miRNAs has been explained as a major mechanism by which lncRNAs exert their tumor promoting effects in GBC. LncRNAs like prognosis-associated gallbladder cancer (PAGBC) promote tumor growth and metastasis by activating AKT/mTOR pathway via sponging of miR-133b, miR-511 (Wu, Wang, Li, et al., 2017). Similarly, MINCR upregulation promoted EMT and invasion of GBC in the experimental model via its ceRNA activity on miR-26a-5p thereby regulating enhancer of zeste homolog 2 (Wang, Yang, et al., 2016). Likewise, lncRNA PVT1 upregulated the expression of HK2 via sponging miR-143 (Chen et al., 2019). LncRNA HOX transcript antisense RNA has also been studied to downregulate miR-130a via ceRNA network pathway under the influence of c Myc gene in GBC (Ma et al., 2014).

Additional studied lncRNAs that are oncogenic in GBC include antisense ncRNA at the INK4 locus (ANRIL), actin filament—associated protein 1 antisense RNA 1 (AFAP1-AS1), lncRNA ROR, HEGBC, and UCA1 (Cai et al., 2017; Liu et al., 2016; Ma, Wang, et al., 2016; Wang, Zhang, Wu, Weng, Zhou, et al., 2016; Yang et al., 2018).

Some of the lncRNAs are known to originate from pseudogenes and exert their action by acting as ceRNA and thereby competing for shared miRNAs on mRNAs resulting in altered expression of associated genes (Lou, Ding, & Fu, 2020). Two such lncRNAs studied in GBC are NmrA-like family domain containing 1 pseudogene (Loc344887) (Wu, Wang, Ou, et al., 2017) and TPTE2P1 (Lv et al., 2015). Their knockdown has been associated with reduced tumor proliferation, EMT, and metastasis in GBC.

Although most of the studied lncRNAs have shown tumor promoting oncogenic activity in GBC, few of the lncRNAs like MEG3, LET, and GCASPC via the GCASPC/miR-17−3p/pyruvate carboxylase axis have demonstrated tumor suppressive activity (Jin, Cai, et al., 2018; Liu et al., 2016; Ma et al., 2015; Ma, Zhang, et al., 2016).

All these studies elucidate the role of lncRNAs in pathogenesis of GBC and thereby make them a potential therapeutic target in GBC.

12.3.3 LncRNA as a biomarker

LncRNAs have the potential to be an advantageous candidate as a diagnostic and prognostic biomarker for cancer owing to their noninvasiveness and high stability in body fluids like blood, bile, exosomes, urine, etc. (Bolha, Ravnik-Glavač, and Glavač, 2017). Furthermore, this is preferable for cancer like GBC where obtaining biopsy is not feasible in most cases. In GBC, lncRNAs are most commonly studied as prognostic markers where Kaplan—Meir curves have been explored to determine the potential utility of lncRNA as a suitable prognostic biomarker. The utility of lncRNAs as a diagnostic marker for early detection of GBC has also been explored by some studies. Differential expression of lncRNAs in cancer tissue relative to normal tissue and in different stages of cancers and their secretion in body fluids forms the basis of them being utilized as a biomarker. Techniques like qRT-PCR, microarray, and sequencing have been employed for determining the differential expression of lncRNA in cancers.

Elevated expression of LncRNA MALAT1 and LINC00152 in GBC tissues has been positively correlated with tumor stage, lymph node metastasis, TNM stage, and negatively with the overall survival of patients (Cai et al., 2017; Goyal et al., 2021; Wang, Zhang, Wu, Zhang, et al., 2016). MALAT1 has also been studied for its role in identifying metastasis and recurrence in GBC patients (Seto et al., 2018). Similarly, significantly elevated expression of lncRNA H19 has been observed in GBC as compared to adjacent normal tissue that positively correlated with larger tumor size, lymph node metastasis and showed association with poor overall survival (Wang et al., 2015). Similarly, upregulated lncRNAs PAGBC, MINCR, AFAP1-AS1, etc. in GBC tissues as observed by qRT-PCR assay were found to be associated with poor overall survival in GBC (Ma, Wang, et al., 2016; Wang, Yang, et al., 2016; Wu, Wang, Li, et al., 2017).

One of the studies explored microarray to identify differentially expressed lncRNA in GBC tissue relative to normal tissue. Among 17 identified lncRNAs, on gene coexpression analysis with expressed proteins, PABGBC, was identified as the most coexpressed lncRNAs. Furthermore, PABGBC was found to be an independent prognostic marker for determining overall survival in GBC patients (Wu, Wang, Li, et al., 2017).

In a similar study, lncRNA HEGBC was identified to be greater than fivefolds upregulated by studying publicly available National Center for Biotechnology Information dataset Gene Expression Omnibus (GEO). On validation in patient samples by qRT-PCR, a significant overexpression of HEGBC was observed in GBC tissues relative to normal tissues and was found to be associated with poor overall survival in patients (Yang et al., 2018).

Another study by Zhang, Geng, Meng, Meng, and Wang (2018) analyzed ncRNA profile of GBC samples from GEO database and also performed RNA-binding protein analysis using starbase software. lncRNA forkhead box P2 (FOXP2) and FOXF1 adjacent noncoding developmental regulatory RNA (FENDRR) were found to be most commonly coexpressed lncRNAs in their study.

Besides acting as poor prognostic markers, some of the lncRNAs have also been identified as a good prognostic marker where elevated expression of lncRNAs like MEG3, LET, and GCASPC have been associated with improved overall survival in GBC patients (Jin, Cai, et al., 2018; Ma et al., 2015; Ma, Zhang, et al., 2016).

LncRNA looks very promising as a biomarker as well as a therapeutic target. However, there is still paucity of data in GBC. Extensive and well-planned research studies and clinical trials are required before lncRNAs are introduced into clinical practice for early diagnosis and a therapeutic modality. Newer technology like NGS needs to be explored more for the identification of novel lncRNAs in the case of GBC.

12.4 PiRNAs in gallbladder carcinoma

PiRNAs are classified in category of small ncRNAs of 24–31 nucleotides. They have been identified to regulate mRNA expression by interacting with piwi protein family. PiRNAs are known to be transcribed by RNA Pol II from piRNA clusters in genome. After being transcribed, they are 5' capped and 3' polyadenylated and transported to cytoplasm to Yb body. In the cytoplasm, piRNAs interact with piwi protein family and recruit H3K9me3: histone 3 lysine 9 trimethylation silencing machinery components. This results in transcriptional silencing of genes by methylation. PiRNAs also mediate posttranscriptional silencing via piRNA-induced silencing complexes (pi-RISCs) (Liu et al., 2019).

PiRNAs are still at a very nascent stage in GBC. In a study by Gu et al., authors studied exosomal small RNAs by sequencing in plasma of four GBC patients. Interestingly, their study showed the upregulation of 323 piRNAs and downregulation of 191 piRNAs. On validation in blood samples of 50 GBC patients, a significant upregulation of piR-10506469 was observed in exosomes of these patients which subsequently got reduced after surgery (Gu et al., 2020).

12.5 Limitations of clinical utility of ncRNAs in gallbladder carcinoma

Research on ncRNAs as diagnostic and therapeutic biomarkers is in its initial stages in GBC, and there are some challenges in their clinical application. First, the origin of circulating ncRNAs has not been fully elucidated. Circulating ncRNAs in serum/plasma have a heterogeneous origin, including from blood cells, endothelia cells, and other high blood flow organs, which demonstrates that the expression of tumor-specific ncRNA signatures may be masked by circulating ncRNAs from other origins (Pritchard et al., 2012; Williams et al., 2013). Another issue is that most ncRNA studies use small sample sizes, a limited number of screened ncRNAs in cell lines and animal models. Despite the potential of these biomarkers, they are yet to be validated in independent cohorts and lack large-scale prospective studies. Therefore they have not yet passed the discovery phase and are not ready for use in clinical practice. Clinical implementation depends mainly on the added clinical value that the validation phases provide, but the lack of samples with adequate clinical follow-up, robust screening tests, and financial resources make it difficult to move forward in solving this problem. Moreover, detection of ncRNAs is expensive and more complex, thus necessitating exploration of an improved and well-standardized technique for analysis. Additionally, it is very important to use appropriate controls that are well matched in age, race, gender, etiology, and severity of underlying gallbladder disease when evaluating the diagnostic potential of ncRNAs in GBC. Finally, despite promising results, the efficacy of ncRNA as biomarkers should be carefully evaluated because the response depends on the epigenetic and genetic profile of each individual (Grasedieck et al., 2013; Pritchard et al., 2012; Williams et al., 2013). These approaches are still at an early stage, but with the development of new technologies, especially improving the specific delivery of tumor-suppressor ncRNAs into damaged tissues, this strategy may become an important tool in the diagnosis and treatment of GBC.

12.6 Conclusion

ncRNAs regulate the progression and pathogenesis of various malignancies. Increasing research studies suggest that ncRNAs play a crucial role in various pathways that regulate tumor suppression and oncogenic functions in gallbladder cancers. The overall survival of GBC patients depends on early detection and radical resection of tumor growth via surgery and early chemotherapy. Unfortunately, a major portion of patients are present to the medical/surgical oncology Out Patient Department (OPDs) with late symptoms. Hence, tumor biomarkers are required for early detection of GBC in high-risk patients. ncRNAs are emerging as promising prospects for tumor detection biomarkers as diagnostic and prognostic markers as well as therapeutic targets. miRNAs are stable in blood and other body fluids (such as bile) and, thus can be measured using noninvasive methods. ncRNAs in GBC are still in their initial stages of research, that is, the ongoing research in cancer cell lines and animal models with pilot studies being conducted on GBC patients to detect the role of a particular ncRNA involved in functional or regulatory changes leading to the development of GBC. Further, large-scale studies and clinical trials are required for validating these miRNA signatures so as to increase the rate of early detection in GBC patients and to develop targeted therapies to increase the survival rates. The increasing online literature suggests emerging pieces of evidence of ncRNAs and their gene expression patterns in GBC might prove helpful to map or develop panels of biomarkers of GBC. Various recent technological advances like next-generation sequencing, whole-genome sequencing, or RNA sequencing are able to detect various proteins and molecular targets that are useful for characterization and identification of cellular actions of ncRNAs and their response to newer drug therapies. ncRNAs as biomarkers can be advantageous to detect the prognosis, tumor type, response to treatment, and progression of GBC which might prove to be helpful for developing precision medicine for the GBC subgroup of patients. ncRNA-based diagnostic and therapeutic protocols might develop changes in the current treatment strategies of GBC.

References

Adams, D., Gonzalez-Duarte, A., O'Riordan, W. D., Yang, C. C., Ueda, M., Kristen, A. V., ... Suhr, O. B. (2018). Patisiran, an RNAi therapeutic, for hereditary transthyretin amyloidosis. *The New England Journal of Medicine*, 379(1), 11–21. Available from https://doi.org/10.1056/NEJMoa1716153.

Alanazi, I., Hoffmann, P., & Adelson, D. L. (2015). MicroRNAs are part of the regulatory network that controls EGF induced apoptosis, including elements of the JAK/STAT pathway, in A431 cells. *PLoS One*, 10(3), e0120337. Available from https://doi.org/10.1371/journal.pone.0120337.

Batista, P. J., & Chang, H. Y. (2013). Long noncoding RNAs: Cellular address codes in development and disease. *Cell*, 152(6), 1298–1307. Available from https://doi.org/10.1016/j.cell.2013.02.012.

Beg, M. S., Brenner, A. J., Sachdev, J., Borad, M., Kang, Y. K., Stoudemire, J., ... Hong, D. S. (2017). Phase I study of MRX34, a liposomal miR-34a mimic, administered twice weekly in patients with advanced solid tumors. *Investigational New Drugs*, 35(2), 180–188. Available from https://doi.org/10.1007/s10637-016-0407-y.

Bolha, L., Ravnik-Glavač, M., & Glavač, D. (2017). Long noncoding RNAs as biomarkers in cancer. *Disease Narkers*, 2017, 7243968. Available from https://doi.org/10.1155/2017/7243968.

Bueno, M. J., & Malumbres, M. (2011). MicroRNAs and the cell cycle. *Biochimica et Biophysica Acta*, 1812(5), 592–601. Available from https://doi.org/10.1016/j.bbadis.2011.02.002.

Cai, J., Xu, L., Cai, Z., Wang, J., Zhou, B., & Hu, H. (2015). MicroRNA-146b-5p inhibits the growth of gallbladder carcinoma by targeting epidermal growth factor receptor. *Molecular Medicine Reports*, 12(1), 1549–1555. Available from https://doi.org/10.3892/mmr.2015.3461.

Cai, Q., Jin, L., Wang, S., Zhou, D., Wang, J., Tang, Z., & Quan, Z. (2017). Long non-coding RNA UCA1 promotes gallbladder cancer progression by epigenetically repressing p21 and E-cadherin expression. *Oncotarget*, 8(29), 47957–47968. Available from https://doi.org/10.18632/oncotarget.18204.

Cai, Q., Wang, Z., Wang, S., Weng, M., Zhou, D., Li, C., ... Quan, Z. (2017). Long non-coding RNA LINC00152 promotes gallbladder cancer metastasis and epithelial-mesenchymal transition by regulating HIF-1α via miR-138. *Open Biology*, 7(1), 160247. Available from https://doi.org/10.1098/rsob.160247.

Cai, Q., Wang, Z. Q., Wang, S. H., Li, C., Zhu, Z. G., Quan, Z. W., & Zhang, W. J. (2016). Upregulation of long non-coding RNA LINC00152 by SP1 contributes to gallbladder cancer cell growth and tumor metastasis via PI3K/AKT pathway. *American Journal of Translational Research*, 8(10), 4068–4081.

Chang, Y., Liu, C., Yang, J., Liu, G., Feng, F., Tang, J., ... Wang, H. (2013). MiR-20a triggers metastasis of gallbladder carcinoma. *Journal of Hepatology*, 59(3), 518–527. Available from https://doi.org/10.1016/j.jhep.2013.04.034.

Chen, J., Yu, Y., Chen, X., He, Y., Hu, Q., Li, H., ... Sun, R. (2018). MiR-139-5p is associated with poor prognosis and regulates glycolysis by repressing PKM2 in gallbladder carcinoma. *Cell Proliferation*, 51(6), e12510. Available from https://doi.org/10.1111/cpr.12510.

Chen, J., Yu, Y., Li, H., Hu, Q., Chen, X., He, Y., ... Sun, R. (2019). Long non-coding RNA PVT1 promotes tumor progression by regulating the miR-143/HK2 axis in gallbladder cancer. *Molecular Cancer*, 18(1), 33. Available from https://doi.org/10.1186/s12943-019-0947-9.

Dhanoa, J. K., Sethi, R. S., Verma, R., Arora, J. S., & Mukhopadhyay, C. S. (2018). Long non-coding RNA: Its evolutionary relics and biological implications in mammals: A review. *Journal of Animal Science and Technology, 60*, 25. Available from https://doi.org/10.1186/s40781-018-0183-7.

Djebali, S., Davis, C. A., Merkel, A., Dobin, A., Lassmann, T., Mortazavi, A., ... Gingeras, T. R. (2012). Landscape of transcription in human cells. *Nature, 489*(7414), 101−108. Available from https://doi.org/10.1038/nature11233.

Felekkis, K., Touvana, E., Stefanou, C. H., & Deltas, C. (2010). microRNAs: A newly described class of encoded molecules that play a role in health and disease. *Hippokratia, 14*(4), 236−240.

Gallach, S., Calabuig-Fariñas, S., Jantus-Lewintre, E., & Camps, C. (2014). MicroRNAs: Promising new antiangiogenic targets in cancer. *BioMed Research International, 2014*, 878450. Available from https://doi.org/10.1155/2014/878450.

Goeppert, B., Truckenmueller, F., Ori, A., Fritz, V., Albrecht, T., Fraas, A., ... Roessler, S. (2019). Profiling of gallbladder carcinoma reveals distinct miRNA profiles and activation of STAT1 by the tumor suppressive miRNA-145-5p. *Scientific Reports, 9*(1), 4796. Available from https://doi.org/10.1038/s41598-019-40857-3.

Goyal, B., Yadav, S., Awasthee, N., Gupta, S., Kunnumakkara, A. B., & Gupta, S. C. (2021). Diagnostic, prognostic, and therapeutic significance of long non-coding RNA MALAT1 in cancer. *Biochimica et Biophysica Acta. Reviews on Cancer, 1875*(2), 188502, Advance online publication. Available from https://doi.org/10.1016/j.bbcan.2021.188502.

Grasedieck, S., Sorrentino, A., Langer, C., Buske, C., Döhner, H., Mertens, D., & Kuchenbauer, F. (2013). Circulating microRNAs in hematological diseases: Principles, challenges, and perspectives. *Blood, 121*(25), 4977−4984. Available from https://doi.org/10.1182/blood-2013-01-480079.

Gu, X., Wang, C., Deng, H., Qing, C., Liu, R., Liu, S., & Xue, X. (2020). Exosomal piRNA profiling revealed unique circulating piRNA signatures of cholangiocarcinoma and gallbladder carcinoma. *Acta Biochimica et Biophysica Sinica, 52*(5), 475−484. Available from https://doi.org/10.1093/abbs/gmaa028.

Ha, M., & Kim, V. N. (2014). Regulation of microRNA biogenesis. *Nature Reviews. Molecular Cell Biology, 15*(8), 509−524. Available from https://doi.org/10.1038/nrm3838.

He, Y., Chen, X., Yu, Y., Li, J., Hu, Q., Xue, C., ... Cui, G. (2018). LDHA is a direct target of miR-30d-5p and contributes to aggressive progression of gallbladder carcinoma. *Molecular Carcinogenesis, 57*(6), 772−783. Available from https://doi.org/10.1002/mc.22799.

Hombach, S., & Kretz, M. (2016). Non-coding RNAs: Classification, biology and functioning. *Advances in Experimental Medicine and Biology, 937*, 3−17. Available from https://doi.org/10.1007/978-3-319-42059-2_1.

Hua, C. B., Song, S. B., Ma, H. L., & Li, X. Z. (2019). MiR-1−5p is down-regulated in gallbladder carcinoma and suppresses cell proliferation, migration and invasion by targeting Notch2. *Pathology, Research and Practice, 215*(1), 200−208. Available from https://doi.org/10.1016/j.prp.2018.10.013.

Jansson, M. D., & Lund, A. H. (2012). MicroRNA and cancer. *Molecular Oncology, 6*(6), 590−610. Available from https://doi.org/10.1016/j.molonc.2012.09.006.

Jin, K., Xiang, Y., Tang, J., Wu, G., Li, J., Xiao, H., ... Zhao, J. (2014). miR-34 is associated with poor prognosis of patients with gallbladder cancer through regulating telomere length in tumor stem cells. *Tumour Biology: The Journal of the International Society for Oncodevelopmental Biology and Medicine, 35*(2), 1503−1510. Available from https://doi.org/10.1007/s13277-013-1207-z.

Jin, L., Cai, Q., Wang, S., Wang, S., Mondal, T., Wang, J., & Quan, Z. (2018). Long noncoding RNA MEG3 regulates LATS2 by promoting the ubiquitination of EZH2 and inhibits proliferation and invasion in gallbladder cancer. *Cell Death & Disease, 9*(10), 1017. Available from https://doi.org/10.1038/s41419-018-1064-1.

Jin, Y. P., Hu, Y. P., Wu, X. S., Wu, Y. S., Ye, Y. Y., Li, H. F., ... Liu, Y. B. (2018). miR-143-3p targeting of ITGA6 suppresses tumour growth and angiogenesis by downregulating PLGF expression via the PI3K/AKT pathway in gallbladder carcinoma. *Cell Death & Disease, 9*(2), 182. Available from https://doi.org/10.1038/s41419-017-0258-2.

Kai, D., Yannian, L., Yitian, C., Dinghao, G., Xin, Z., & Wu, J. (2018). Circular RNA HIPK3 promotes gallbladder cancer cell growth by sponging microRNA-124. *Biochemical and Biophysical Research Communications, 503*(2), 863−869. Available from https://doi.org/10.1016/j.bbrc.2018.06.088.

Kono, H., Nakamura, M., Ohtsuka, T., Nagayoshi, Y., Mori, Y., Takahata, S., ... Tanaka, M. (2013). High expression of microRNA-155 is associated with the aggressive malignant behavior of gallbladder carcinoma. *Oncology Reports, 30*(1), 17−24. Available from https://doi.org/10.3892/or.2013.2443.

Letelier, P., García, P., Leal, P., Álvarez, H., Ili, C., López, J., ... Roa, J. C. (2014). miR-1 and miR-145 act as tumor suppressor microRNAs in gallbladder cancer. *International Journal of Clinical and Experimental Pathology, 7*(5), 1849–1867.

Li, G., & Pu, Y. (2015). MicroRNA signatures in total peripheral blood of gallbladder cancer patients. *Tumour Biology: The journal of the International Society for Oncodevelopmental Biology and Medicine, 36*(9), 6985–6990. Available from https://doi.org/10.1007/s13277-015-3412-4.

Li, H., Jin, Y., Hu, Y., Jiang, L., Liu, F., Zhang, Y., ... Liu, Y. (2018). The PLGF/c-MYC/miR-19a axis promotes metastasis and stemness in gallbladder cancer. *Cancer Science, 109*(5), 1532–1544. Available from https://doi.org/10.1111/cas.13585.

Lima, R. T., Busacca, S., Almeida, G. M., Gaudino, G., Fennell, D. A., & Vasconcelos, M. H. (2011). MicroRNA regulation of core apoptosis pathways in cancer. *European Journal of Cancer (Oxford, England: 1990), 47*(2), 163–174. Available from https://doi.org/10.1016/j.ejca.2010.11.005.

Liu, B., Shen, E. D., Liao, M. M., Hu, Y. B., Wu, K., Yang, P., ... Chen, W. D. (2016). Expression and mechanisms of long non-coding RNA genes MEG3 and ANRIL in gallbladder cancer. *Tumour Biology: The Journal of the International Society for Oncodevelopmental Biology and Medicine, 37*(7), 9875–9886. Available from https://doi.org/10.1007/s13277-016-4863-y.

Liu, Y., Dou, M., Song, X., Dong, Y., Liu, S., Liu, H., ... Xu, W. (2019). The emerging role of the piRNA/piwi complex in cancer. *Molecular Cancer, 18*(1), 123. Available from https://doi.org/10.1186/s12943-019-1052-9.

Lou, W., Ding, B., & Fu, P. (2020). Pseudogene-derived lncRNAs and their miRNA sponging mechanism in human cancer. *Frontiers in Cell and Developmental Biology, 8*, 85. Available from https://doi.org/10.3389/fcell.2020.00085.

Lu, W., Hu, Y., Ma, Q., Zhou, L., Jiang, L., Li, Z., ... Liu, Y. (2016). miR-223 increases gallbladder cancer cell sensitivity to docetaxel by downregulating STMN1. *Oncotarget, 7*(38), 62364–62376. Available from https://doi.org/10.18632/oncotarget.11634.

Lv, W., Wang, L., Lu, J., Mu, J., Liu, Y., & Dong, P. (2015). Downregulation of TPTE2P1 inhibits migration and invasion of gallbladder cancer cells. *Chemical Biology & Drug Design, 86*(4), 656–662. Available from https://doi.org/10.1111/cbdd.12533.

Ma, F., Wang, S. H., Cai, Q., Zhang, M. D., Yang, Y., & Ding, J. (2016). Overexpression of LncRNA AFAP1-AS1 predicts poor prognosis and promotes cells proliferation and invasion in gallbladder cancer. *Biomedicine & Pharmacotherapy = Biomedecine & Pharmacotherapie, 84*, 1249–1255. Available from https://doi.org/10.1016/j.biopha.2016.10.064.

Ma, M. Z., Kong, X., Weng, M. Z., Zhang, M. D., Qin, Y. Y., Gong, W., ... Quan, Z. W. (2015). Long non-coding RNA-LET is a positive prognostic factor and exhibits tumor-suppressive activity in gallbladder cancer. *Molecular Carcinogenesis, 54*(11), 1397–1406. Available from https://doi.org/10.1002/mc.22215.

Ma, M. Z., Li, C. X., Zhang, Y., Weng, M. Z., Zhang, M. D., Qin, Y. Y., ... Quan, Z. W. (2014). Long non-coding RNA HOTAIR, a c-Myc activated driver of malignancy, negatively regulates miRNA-130a in gallbladder cancer. *Molecular Cancer, 13*, 156. Available from https://doi.org/10.1186/1476-4598-13-156.

Ma, M. Z., Zhang, Y., Weng, M. Z., Wang, S. H., Hu, Y., Hou, Z. Y., ... Quan, Z. W. (2016). Long noncoding RNA GCASPC, a target of miR-17-3p, negatively regulates pyruvate carboxylase-dependent cell proliferation in gallbladder cancer. *Cancer Research, 76*(18), 5361–5371. Available from https://doi.org/10.1158/0008-5472.CAN-15-3047.

Montalvo-Jave, E. E., Rahnemai-Azar, A. A., Papaconstantinou, D., Deloiza, M. E., Tsilimigras, D. I., Moris, D., ... Pawlik, T. M. (2019). Molecular pathways and potential biomarkers in gallbladder cancer: A comprehensive review. *Surgical Oncology, 31*, 83–89.

Niu, J., Li, Z., & Li, F. (2019). Overexpressed microRNA-136 works as a cancer suppressor in gallbladder cancer through suppression of JNK signaling pathway via inhibition of MAP2K4. *American Journal of Physiology. Gastrointestinal and Liver Physiology, 317*(5), G670–G681. Available from https://doi.org/10.1152/ajpgi.00055.2019.

O'Brien, J., Hayder, H., Zayed, Y., & Peng, C. (2018). Overview of microRNA biogenesis, mechanisms of actions, and circulation. *Frontiers in Endocrinology, 9*, 402. Available from https://doi.org/10.3389/fendo.2018.00402.

Peng, Y., & Croce, C. M. (2016). The role of microRNAs in human cancer. *Signal Transduction and Targeted Therapy, 1*, 15004. Available from https://doi.org/10.1038/sigtrans.2015.4.

Pritchard, C. C., Kroh, E., Wood, B., Arroyo, J. D., Dougherty, K. J., Miyaji, M. M., ... Tewari, M. (2012). Blood cell origin of circulating microRNAs: A cautionary note for cancer biomarker studies. *Cancer Prevention Research (Philadelphia, PA)*, 5(3), 492−497. Available from https://doi.org/10.1158/1940-6207.CAPR-11-0370.

Qiu, Y., Luo, X., Kan, T., Zhang, Y., Yu, W., Wei, Y., ... Jiang, X. (2014). TGF-β upregulates miR-182 expression to promote gallbladder cancer metastasis by targeting CADM1. *Molecular BioSystems*, 10(3), 679−685. Available from https://doi.org/10.1039/c3mb70479c.

Quinn, J. J., & Chang, H. Y. (2016). Unique features of long non-coding RNA biogenesis and function. *Nature Reviews Genetics*, 17(1), 47−62. Available from https://doi.org/10.1038/nrg.2015.10.

Salmena, L., Poliseno, L., Tay, Y., Kats, L., & Pandolfi, P. P. (2011). A ceRNA hypothesis: The Rosetta Stone of a hidden RNA language? *Cell*, 146(3), 353−358. Available from https://doi.org/10.1016/j.cell.2011.07.014.

Sayed, D., & Abdellatif, M. (2011). MicroRNAs in development and disease. *Physiological Reviews*, 91(3), 827−887. Available from https://doi.org/10.1152/physrev.00006.2010.

Schmitt, A. M., & Chang, H. Y. (2016). Long noncoding RNAs in cancer pathways. *Cancer Cell*, 29(4), 452−463. Available from https://doi.org/10.1016/j.ccell.2016.03.010.

Seto, A. G., Beatty, X., Lynch, J. M., Hermreck, M., Tetzlaff, M., Duvic, M., & Jackson, A. L. (2018). Cobomarsen, an oligonucleotide inhibitor of miR-155, co-ordinately regulates multiple survival pathways to reduce cellular proliferation and survival in cutaneous T-cell lymphoma. *British Journal of Haematology*, 183(3), 428−444. Available from https://doi.org/10.1111/bjh.15547.

Shigehara, K., Yokomuro, S., Ishibashi, O., Mizuguchi, Y., Arima, Y., Kawahigashi, Y., ... Uchida, E. (2011). Real-time PCR-based analysis of the human bile microRNAome identifies miR-9 as a potential diagnostic biomarker for biliary tract cancer. *PLoS One*, 6(8), e23584. Available from https://doi.org/10.1371/journal.pone.0023584.

Shu, Y. J., Bao, R. F., Jiang, L., Wang, Z., Wang, X. A., Zhang, F., ... Liu, Y. B. (2017). MicroRNA-29c-5p suppresses gallbladder carcinoma progression by directly targeting CPEB4 and inhibiting the MAPK pathway. *Cell Death and Differentiation*, 24(3), 445−457. Available from https://doi.org/10.1038/cdd.2016.146.

Siegel, R. L., Miller, K. D., & Jemal, A. (2020). Cancer statistics, 2020. *CA A Cancer J Clin*, 70, 7−30.

Slack, F. J., & Chinnaiyan, A. M. (2019). The role of non-coding RNAs in oncology. *Cell*, 179(5), 1033−1055. Available from https://doi.org/10.1016/j.cell.2019.10.017.

Tay, Y., Rinn, J., & Pandolfi, P. P. (2014). The multilayered complexity of ceRNA crosstalk and competition. *Nature*, 505(7483), 344−352. Available from https://doi.org/10.1038/nature12986.

Wang, H., Zhan, M., Xu, S. W., Chen, W., Long, M. M., Shi, Y. H., ... Wang, J. (2017). miR-218-5p restores sensitivity to gemcitabine through PRKCE/MDR1 axis in gallbladder cancer. *Cell Death & Disease*, 8(5), e2770. Available from https://doi.org/10.1038/cddis.2017.178.

Wang, J., Chen, J., & Sen, S. (2016). MicroRNA as biomarkers and diagnostics. *Journal of Cellular Physiology*, 231(1), 25−30. Available from https://doi.org/10.1002/jcp.25056.

Wang, S. H., Ma, F., Tang, Z. H., Wu, X. C., Cai, Q., Zhang, M. D., ... Quan, Z. W. (2016). Long non-coding RNA H19 regulates FOXM1 expression by competitively binding endogenous miR-342-3p in gallbladder cancer. *Journal of Experimental & Clinical Cancer Research: CR*, 35(1), 160. Available from https://doi.org/10.1186/s13046-016-0436-6.

Wang, S. H., Wu, X. C., Zhang, M. D., Weng, M. Z., Zhou, D., & Quan, Z. W. (2016). Long noncoding RNA H19 contributes to gallbladder cancer cell proliferation by modulated miR-194-5p targeting AKT2. *Tumour Biology: The Journal of the International Society for Oncodevelopmental Biology and Medicine*, 37(7), 9721−9730. Available from https://doi.org/10.1007/s13277-016-4852-1.

Wang, S. H., Wu, X. C., Zhang, M. D., Weng, M. Z., Zhou, D., & Quan, Z. W. (2015). Upregulation of H19 indicates a poor prognosis in gallbladder carcinoma and promotes epithelial-mesenchymal transition. *American Journal of Cancer Research*, 6(1), 15−26.

Wang, S. H., Yang, Y., Wu, X. C., Zhang, M. D., Weng, M. Z., Zhou, D., ... Quan, Z. W. (2016). Long non-coding RNA MINCR promotes gallbladder cancer progression through stimulating EZH2 expression. *Cancer Letters*, 380(1), 122−133. Available from https://doi.org/10.1016/j.canlet.2016.06.019.

Wang, S. H., Zhang, M. D., Wu, X. C., Weng, M. Z., Zhou, D., & Quan, Z. W. (2016). Overexpression of LncRNA-ROR predicts a poor outcome in gallbladder cancer patients and promotes the tumor cells proliferation, migration, and invasion. *Tumour Biology: The Journal of the International Society for Oncodevelopmental Biology and Medicine*, 37(9), 12867−12875. Available from https://doi.org/10.1007/s13277-016-5210-z.

Wang, S. H., Zhang, W. J., Wu, X. C., Weng, M. Z., Zhang, M. D., Cai, Q., ... Quan, Z. W. (2016). The lncRNA MALAT1 functions as a competing endogenous RNA to regulate MCL-1 expression by sponging miR-363-3p in gallbladder cancer. *Journal of Cellular and Molecular Medicine, 20*(12), 2299–2308. Available from https://doi.org/10.1111/jcmm.12920.

Wang, S. H., Zhang, W. J., Wu, X. C., Zhang, M. D., Weng, M. Z., Zhou, D., ... Quan, Z. W. (2016). Long noncoding RNA Malat1 promotes gallbladder cancer development by acting as a molecular sponge to regulate miR-206. *Oncotarget, 7*(25), 37857–37867. Available from https://doi.org/10.18632/oncotarget.9347.

Williams, Z., Ben-Dov, I. Z., Elias, R., Mihailovic, A., Brown, M., Rosenwaks, Z., & Tuschl, T. (2013). Comprehensive profiling of circulating microRNA via small RNA sequencing of cDNA libraries reveals biomarker potential and limitations. *Proceedings of the National Academy of Sciences of the United States of America, 110*(11), 4255–4260. Available from https://doi.org/10.1073/pnas.1214046110.

Wu, F., Yang, Z., & Li, G. (2009). Role of specific microRNAs for endothelial function and angiogenesis. *Biochemical and Biophysical Research Communications, 386*(4), 549–553. Available from https://doi.org/10.1016/j.bbrc.2009.06.075.

Wu, X. C., Wang, S. H., Ou, H. H., Zhu, B., Zhu, Y., Zhang, Q., ... Li, H. (2017). The NmrA-like family domain containing 1 pseudogene Loc344887 is amplified in gallbladder cancer and promotes epithelial-mesenchymal transition. *Chemical Biology & Drug Design, 90*(3), 456–463. Available from https://doi.org/10.1111/cbdd.12967.

Wu, X. S., Wang, F., Li, H. F., Hu, Y. P., Jiang, L., Zhang, F., ... Liu, Y. B. (2017). LncRNA-PAGBC acts as a microRNA sponge and promotes gallbladder tumorigenesis. *EMBO Reports, 18*(10), 1837–1853. Available from https://doi.org/10.15252/embr.201744147.

Wu, X. S., Wang, X. A., Wu, W. G., Hu, Y. P., Li, M. L., Ding, Q., ... Liu, Y. B. (2014). MALAT1 promotes the proliferation and metastasis of gallbladder cancer cells by activating the ERK/MAPK pathway. *Cancer Biology & Therapy, 15*(6), 806–814. Available from https://doi.org/10.4161/cbt.28584.

Wurster, C. D., Winter, B., Wollinsky, K., Ludolph, A. C., Uzelac, Z., Witzel, S., ... Kocak, T. (2019). Intrathecal administration of nusinersen in adolescent and adult SMA type 2 and 3 patients. *Journal of Neurology, 266*(1), 183–194. Available from https://doi.org/10.1007/s00415-018-9124-0.

Yang, G., Zhang, L., Li, R., & Wang, L. (2016). The role of microRNAs in gallbladder cancer. *Molecular and Clinical Oncology, 5*(1), 7–13. Available from https://doi.org/10.3892/mco.2016.905.

Yang, L., Gao, Q., Wu, X., Feng, F., & Xu, K. (2018). Long noncoding RNA HEGBC promotes tumorigenesis and metastasis of gallbladder cancer via forming a positive feedback loop with IL-11/STAT3 signaling pathway. *Journal of Experimental & Clinical Cancer Research: CR, 37*(1), 186. Available from https://doi.org/10.1186/s13046-018-0847-7.

Zhang, L., Geng, Z., Meng, X., Meng, F., & Wang, L. (2018). Screening for key lncRNAs in the progression of gallbladder cancer using bioinformatics analyses. *Molecular Medicine Reports, 17*(5), 6449–6455. Available from https://doi.org/10.3892/mmr.2018.8655.

Zhang, R., Xia, L. Q., Lu, W. W., Zhang, J., & Zhu, J. (2016). LncRNAs and cancer (Review). *Oncology Letters, 12*, 1233–1239. Available from https://doi.org/10.3892/ol.2016.4770.

Zhang, X. L., Chen, J. H., & Qin, C. K. (2015). MicroRNA-155 expression as a prognostic factor in patients with gallbladder carcinoma after surgical resection. *International Journal of Clinical and Experimental Medicine, 8*(11), 21241–21246.

Zhou, H., Guo, W., Zhao, Y., Wang, Y., Zha, R., Ding, J., ... Ma, B. (2014). MicroRNA-26a acts as a tumor suppressor inhibiting gallbladder cancer cell proliferation by directly targeting HMGA2. *International Journal of Oncology, 44*(6), 2050–2058. Available from https://doi.org/10.3892/ijo.2014.2360.

Zhou, N., Cheng, W., Peng, C., Liu, Y., & Jiang, B. (2017). Decreased expression of hsa-miR-372 predicts poor prognosis in patients with gallbladder cancer by affecting chloride intracellular channel 1. *Molecular Medicine Reports, 16*, 7848–7854. Available from https://doi.org/10.3892/mmr.2017.7520.

13

Clinical implications of noncoding RNAs in neuroblastoma patients

Anup S. Pathania[1], Oghenetejiri V. Smith[1], Philip Prathipati[2], Subash C. Gupta[3] and Kishore B. Challagundla[1,4]

[1]Department of Biochemistry and Molecular Biology & The Fred and Pamela Buffett Cancer Center, University of Nebraska Medical Center, Omaha, NE, United States [2]Laboratory of Bioinformatics, National Institutes of Biomedical Innovation, Health and Nutrition, Osaka, Japan [3]Department of Biochemistry, Institute of Science, Banaras Hindu University, Varanasi, India [4]The Child Health Research Institute, University of Nebraska Medical Center, Omaha, NE, United States

13.1 Introduction

Neuroblastoma (NB) is a tumor of immature nerve cells called neuroblasts that commonly develops from adrenal glands located on top of the kidneys. Normal neuroblasts mature into nerve cells or adrenal medulla cells situated in the center of the adrenal gland, whereas immature neuroblasts do not develop normally and become cancerous. NB is a childhood cancer mostly diagnosed in children under 5 and accounts for about 15% of all cancer-related deaths in children (Colon and Chung, 2011; Maris, 2010; Park, Eggert, & Caron, 2010). Although NB is a rare type of tumor, many children have a good chance of surviving after treatment unless highly aggressive and insensitive to current therapies (Colon and Chung, 2011; Zhou et al., 2015). Therefore there is an urgent need for better treatment options to improve the poor prognosis rate of high-risk NB. Our continual improvement in understanding NB biology is the only way to identify new factors that promote NB development and change the clinical outcomes of patients. The discovery of the role of noncoding RNAs in regulating a vast array of genes and in cancer development opens up new horizons in target-based therapies. The noncoding RNAs do not code for protein but regulate many proteins' functions and participate in almost every cellular process (Bushati and Cohen, 2007; Iwasaki,

Siomi, & Siomi, 2015; Yao, Wang, & Chen, 2019). In recent years, many studies have shown that noncoding RNAs are involved in NB development and play an imperative role in patients' response to NB therapy. These studies emphasize the potential of noncoding RNAs as therapeutic targets and predictable biomarkers in the diagnosis of NB. This chapter discusses the role of different types of noncoding RNAs in NB development, therapy resistance, and potential as predictive/prognostic markers NB. Further, strategies that can be used to target noncoding RNAs for NB treatment are also discussed.

13.2 Types of noncoding RNAs

13.2.1 MicroRNAs

MicroRNAs (miRNAs) are 18–24 nucleotides in length, single-stranded RNA molecules that bind to the three prime untranslated regions (3'-UTR) of target messenger RNA (mRNAs) and induce their degradation (Kim, Han, & Siomi, 2009; Lee et al., 2010). Due to their potential to bind with a vast array of mRNAs, they appear as critical negative regulators of gene expression and functions. miRNA deregulation and its contribution to tumor growth and development are well known in cancer (Hayes, Peruzzi, & Lawler, 2014; Peng and Croce, 2016). Based on their target gene, miRNAs can function as tumor promoters, known as oncomiRs or tumor suppressors. For instance, if miRNA target genes are involved in suppressing cell growth or promoting cell differentiation, their downregulations can stimulate uncontrolled cell proliferation (La et al., 2018; Le et al., 2009). On the contrary, if target genes are oncogenes, miRNAs can prevent tumor development by negatively regulating their expression (Buechner et al., 2011). MiRNA biogenesis begins with the transcription of a pri-miRNA (primary transcript of mature miRNA) sequence by RNA polymerase II followed by its cleavage by microprocessor complex comprising nuclear RNAse III-type protein, Drosha, and its cofactor, the DiGeorge syndrome critical region eight (Han et al., 2004; Han et al., 2006; Lee et al., 2004). The cleaved pri-miRNA has a 60–70 nucleotide hair-spin structure known as precursor-miRNA (pre-miRNA), which is subsequently exported from the nucleus via GTP-dependent Exportin-5 (Yi, Qin, Macara, & Cullen, 2003). In the cytoplasm, the pre-miRNA is further cleaved into mature double-stranded miRNA by the RNase enzyme Dicer and its cofactor transactivation-responsive RNA-binding protein (Kim and Kim, 2007; Wilson et al., 2015). After this, the miRNA duplex binds to an enzymatic complex known as the RNA-induced silencing complex (RISC) consists of Argonaute (AGO) proteins, small RNAs, and double-stranded RNA-binding proteins (Gregory, Chendrimada, Cooch, & Shiekhattar, 2005; Kobayashi and Tomari, 2016). The miRNA-RISC complex binds to target or complementary mRNA sequences and induces mRNA degradation or repress mRNA translation.

13.2.2 Long noncoding RNAs

Long noncoding RNAs (LncRNAs) are any noncoding RNA sequences that are longer than 200 nucleotides (Pathania and Challagundla, 2021). This arbitrary cutoff permits many lncRNAs, which can be further categorized by genomic location, structure, or regulatory action (Ma, Bajic, & Zhang, 2013; Pathania and Challagundla, 2021; Wilusz, Sunwoo, & Spector, 2009). While they require less processing than miRNA, lncRNAs are transcribed by RNA polymerase II and perform functions in the nucleus and cytoplasm (Yan and Bu, 2021). Based on functions, lncRNAs can be classified into *cis*-acting, which regulates the genes present near the site of their transcription, or trans-acting that leaves the site of transcription and acts on distant genes (Cho et al., 2018; Pavlaki et al., 2018; Tripathi et al., 2010; Zhang, Chao et al., 2019). The lncRNAs can regulate gene functions by recruiting gene regulatory factors or regulating DNA elements such as enhancers or silencers near the transcription site or by promoting the assembly of transcription or spliceosome machinery in the vicinity of a gene to affect its transcription. In addition, LncRNAs can facilitate protein—complex interactions or sponge miRNAs to block its mRNA silencing activity in the cytosol (Zhang, Zhou et al., 2019). LncRNAs can regulate various cell functions, including cell proliferation (Li, Jia, Cheng, Liu, & Song, 2018), migration (Peng et al., 2018), invasion (Sun et al., 2014), angiogenesis (Zhao et al., 2018), and metastasis (Xu et al., 2020) and are involved in the development of multiple cancers (Huarte, 2015).

13.2.3 P-element-induced wimpy testis (Piwi)-interacting RNAs

P-element-induced wimpy testis (Piwi)-interacting RNAs (piRNAs) are small, single-stranded, 23—36 nucleotide in length, noncoding RNAs that form RNA—protein complexes with PIWI proteins. PIWI belongs to one of the two subfamilies of a Piwi/Argonaute/Zwille (PAZ)/PIWI domain protein family involved in the regulation of RNA silencing (Clyde, 2019; Thomson and Lin, 2009). The other subfamily consists of AGO proteins that bind to miRNAs and small interfering RNAs (siRNAs) and are present in most cell types (Meister, 2013). In contrast, PIWI proteins are germline-specific; therefore, their interacting piRNAs are primarily present in germ cells (Clyde, 2019). The majority of piRNAs are produced from genetic regions called piRNA clusters that yield single-stranded piRNAs precursors (Yamanaka, Siomi, & Siomi, 2014). They are transported out of the nucleus and processed into 5'monophosphate containing pre-piRNAs first by the RNA helicase Armitage and then by endonuclease zucchini. Subsequently, the pre-piRNAs are loaded onto PIWI protein, which facilitates piRNAs trimming by the exonuclease Nibbler followed by methylation at its 2'oxygen by the small RNA 2'-*O*-methyltransferase, Hen1 to enhance piRNA stability (Horwich et al., 2007; Houwing et al., 2007; Ishizu, Kinoshita, Hirakata, Komatsuzaki, & Siomi, 2019; Kawaoka, Izumi, Katsuma, & Tomari, 2011; Kirino and Mourelatos, 2007; Saito et al., 2007). This generates functional piRNAs that are further amplified by Ago3 and Aubergine (Aub) proteins (Czech and Hannon, 2016; Nagao et al., 2010). Once processed, piRNAs can regulate many functions, including transposon silencing (Halic and Moazed, 2009), mRNA silencing (Barckmann et al., 2015; Gou et al., 2015), chromatin assembly (Radion et al., 2018), and cell growth (Mai et al., 2018).

13.2.4 Circular RNAs

Circular RNAs (CircRNAs) are a type of long noncoding RNAs that are physically circular via covalent attachment, typically due to an event termed back splicing

(Kristensen et al., 2019; Sanger, Klotz, Riesner, Gross, & Kleinschmidt, 1976). In this noncanonical splicing event, a looping between upstream and downstream splice sites of pre-mRNA occurs because of base pairing between the two regions (Ivanov et al., 2015; Kristensen et al., 2019). Due to their circular structure, they are protected from exonuclease degradation and are, therefore, more stable than most linear noncoding RNAs (Liu, Li et al., 2019). CircRNAs can regulate cell functions through various mechanisms, including sponging miRNAs or RNA-binding proteins (Zhang, Wang et al., 2019), inhibiting mRNA translation because the exons of corresponding genes are incorporated into circRNAs (Ashwal-Fluss et al., 2014; Ragan, Goodall, Shirokikh, & Preiss, 2019) and translate into small peptides that can regulate protein functions (Zhang et al., 2018).

13.3 Role of noncoding RNAs in neuroblastoma growth and development

Noncoding RNAs play an essential role in the differentiation of neural cells (Salvatori, Biscarini, & Morlando, 2020; Zhao, Liu, Zhang, & Zhang, 2020). Dysregulation of noncoding RNAs can profoundly affect neural differentiation, leading to NB development from the neural crest (Chen and Stallings, 2007; Prajapati et al., 2019; Zhao, Li et al., 2020).

13.3.1 MicroRNAs

A study by Chen and Stallings (2007) shows the differential expression of miRNAs during NB development, which correlates with the disease prognosis. The miRNAs are differentially expressed in different genetic subtypes of NB tumors. The treatment of MYCN proto-oncogene (MYCN)-amplified NB tumors with retinoic acid, a drug that induces neural differentiation, increases miR-184 expression. The overexpressed miR-184 induces cell cycle arrest and apoptosis in NB cells suggesting that miR-184 could be associated with neural differentiation and growth (Chen and Stallings, 2007). Another miRNA, let-7, regulates *MYCN* and is negatively associated with *MYCN* amplification and poor outcome in NB patients. The let-7 belongs to the tumor-suppressive miRNA family and is suppressed by various mechanisms in NB; the overexpression of *LIN28B*, an inhibitor of let-7 biogenesis, let-7 sponging by *MYCN* mRNA, and let-7 chromosomal loss contributes to let-7 downregulation, which can lead to NB development (Powers et al., 2016). Many miRNAs regulate *MYCN* during NB progression, and an imbalance in this regulation could lead to *MYCN* amplification (Buechner et al., 2011; Cheng et al., 2020; He et al., 2017; Ooi et al., 2018; Zhao, Shelton et al., 2020). The majority of miRNAs targeting *MYCN* show downregulation during *MYCN*-driven NB tumor development (Beckers et al., 2015). *MYCN* can also regulate miRNAs' expression, which, in turn, favors *MYCN* overexpression and transcriptional activation of *MYCN*-target genes (Schulte et al., 2008); for example, miR204 that directly binds to the coding region of *MYCN* mRNA and causes its degradation and is regulated by *MYCN*. The binding of MYCN to the miR-204 promoter represses its transcription, which promotes NB cell proliferation and tumor development (Ooi et al., 2018). In another mechanism, MYCN directly binds and promotes the transcription of phospholipase C gamma 2 (PLAGL2), a potent activator of *MYCN* transcription. PLAGL2 is downregulated in NB cells by miR506−3p, which binds to its 3′-UTRs. MYCN confers its upregulation by prompting PLAGL2 transcription to overcome miR506−3p inhibition (Zhao, Shelton et al., 2020). Some miRNAs enhance MYCN in NB cells by targeting its negative regulators. miR-221 promotes MYCN expression by binding to its negative regulator, serine/threonine-*protein* kinase Nemo-like kinase (*NLK*), which, in turn, decreases NLK-mediated lymphoid enhancer-binding factor 1 (LEF1) phosphorylation. LEF1 hypophosphorylation increases MYCN levels and enhances NB cell proliferation (Ota et al., 2012). In NB patients, miR-221 shows a positive correlation with *MYCN* status and poor survival (Ota et al., 2012). Furthermore, miRNAs may regulate many other genes as their targets to control NB cell growth. miR-124 negatively regulates transcription factor, E74-like factor (ELF) 4, and its target genes to promote NB cell differentiation. ELF4 induces cell proliferation by inhibiting genes involved in neural differentiation, including doublecortin, NTRK1, NTNG1/-2, NRP1, and ROBO2, thus maintaining cells undifferentiated state (Kosti et al., 2020). Moreover, miR-149 inhibits NB cell proliferation and increases doxorubicin sensitivity by directly repressing the mRNA of proliferative proteins, including cell division cycle 42 and B-cell lymphoma 2 (Mao, Zhang, Cheng, & Xu, 2019). miR-542−3p downregulates cell survival protein survivin, inhibiting cell proliferation and inducing apoptosis in NB xenografts (Althoff et al., 2015).

13.3.2 Long noncoding RNAs

LncRNAs regulate cell functions at transcription and translational levels and control regulation more than miRNAs. Deregulation of LncRNAs expression, associated with altered lncRNA functions, has been reported in NB (Liu, Tee et al., 2019; Rombaut et al., 2019; Sahu, Ho, Juan, & Huang, 2018). Comparing RNA seq data between primary NB tumors and neuroblasts reveals differential expression of long intergenic noncoding (linc) RNAs, suggesting their dysregulation in NB (Rombaut et al., 2019). Many oncogenic or tumor suppressor lincRNAs are present up and downstream of key NB driver genes, including *MYCN*, *ALK*, and *PHOX2B*. Some lincRNAs like myocardial infarction−associated transcript (MIAT) and maternally expressed 3 (MEG3) are abundantly expressed in NB and are associated with NB driver genes *MYCN* and *PHOX2B* activity (Rombaut et al., 2019). MIAT silencing decreases *MYCN* and *c-Myc* mRNA and protein levels and inhibits proliferation in NB cells (Bountali, Tonge, & Mourtada-Maarabouni, 2019; Feriancikova et al., 2021). MEG3 is a 1.6-kb transcript of the MEG3 gene that has tumor-suppressive functions in different cancers. The high MEG3 expression is associated with higher 5-year event-free survival and overall survival in NB patients. On the other hand, MEG3 overexpression inhibits cell proliferation, migration, and invasion in NB cells (Ye et al., 2020). A case study by Novak et al. (2020) in an infant with aggressive NB, which is insensitive to different chemotherapy combinations, found aberrant MEG3 and MEG8 methylation. The degree of methylation is associated with the aggressiveness of NB in the patient. Compared with the low-risk group, MEG3 and *MEG8* expression are significantly lower in both primary and relapse tumor samples of the patient (Novak et al., 2020).

Genome-wide association studies, which are used to detect the association of specific genetic variations with the disease, have found several loci in NB that show a strong association with increased NB risk and aggressiveness (Diskin et al., 2012; Wang et al., 2011). Mondal et al. (2018) have found two sense—antisense lncRNA pairs, CASC15, and NBAT1, which transcribe antisense to CASC15 at one of such loci 6p22.3. These lncRNAs are suppressed in high-risk NB tumors, and their deletion promotes tumor phenotype. Silencing CASC15 and NBAT1 in NB cells promotes nuclear localization of deubiquitinase USP36, stabilizing chromodomain helicase DNA-binding protein 7 (CHD7) by preventing its ubiquitination. CHD7 binds to the promoter of the protumor gene SRY-Box Transcription Factor 9 and positively regulates its transcription, which inhibits neural differentiation (Mondal et al., 2018). NBAT1 is also known to regulate p53 shuttling between nucleus and cytosol by destabilizing nuclear exporter protein CRM1 or exportin. Knockdown of NBAT1 in NB cells stabilizes CRM1 through USP36, which prevents its ubiquitination-mediated degradation. High CRM1 expression promotes p53 cytosol accumulation and thus, decreases the expression of p53-regulated genes p21, MDM2, and GADD45A and reduces drug sensitivity (Mitra et al., 2021).

Furthermore, many oncogenic lncRNAs have been reported overexpressed in NB tissues and cells and associated with poor prognosis in NB patients (Liu, Tee et al., 2019). For example, lncRNA NB1 is upregulated in MYCN-amplified NB cell lines, and its overexpression increases NB cell proliferation and survival. NB1 promotes the phosphorylation of Extracellular signal-regulated kinase (ERK) and MYCN by upregulating GTPase-activating protein and an oncogene, Disheveled, EGL-10, and pleckstrin domain-containing 1B or DEPDC1B expression. Mechanistic details reveal that NB1 directly binds to ribosomal protein L35 (RPL35), a component of a larger subunit of ribosomes, the organelles that catalyze protein synthesis. NB1-RPL35 binding increases the translation of eukaryotic translation elongation factor 2 or E2F1 gene. E2F1 binds to the DEPDC1B gene promoter and enhances its transcription, resulting in ERK phosphorylation and consequent MYCN phosphorylation and stabilization (Liu, Tee et al., 2019). Another lncRNA, pancEts-1, is overexpressed in NB tissues and positively correlates with cancer progression, metastasis, and death in NB patients (Li et al., 2018). pancEts-1 promotes NB cell proliferation through its direct binding with heterogeneous nuclear ribonucleoprotein K (hnRNPK), a pre-mRNA-binding protein that regulates many genes, including β-catenin. pancEts-1 facilitates the interaction between hnRNPK and β-catenin, which prevents β-catenin proteasomal degradation and stabilizes its protein levels. This increases β-catenin nuclear translocation and the transcription of its target genes (Li et al., 2018). The WNT (wingless-type MMTV integration site family)/ β-catenin signaling plays an important role in NB progression, relapse, and chemoresistance development (Becker and Wilting, 2019; Flahaut et al., 2009). Moreover, β-catenin is also regulated by intergenic NB highly expressed 1 (NHEG1) lncRNA, which is located at chromosome 6q23.3 and found upregulated in NB (Zhao, Li et al., 2020). NHEG1 is under transcriptional control of T-cell factor (TCF) and LEF family transcriptional factors, including LEF1/TCF7L2. NHEG1 directly interacts with ATP-dependent RNA processing protein, DEAD-Box Helicase 5 (DDX5), and prevents its degradation. DDX5 stabilization facilitates its binding to β-catenin, which increases the transcription of β-catenin-regulated genes and promotes NB cell proliferation (Zhao, Li et al., 2020).

13.3.3 P-element-induced wimpy testis (Piwi)-interacting RNAs

Deregulation of piRNAs and their interacting partner PIWI proteins have been observed in several human cancers (Liu, Dou et al., 2019). piRNAs and PIWI proteins are involved in cell growth, migration, and metastasis (Ma et al., 2020; Mai et al., 2018; Shi, Yang, Liu, Yang, & Lin, 2020; Sohn, Jo, & Park, 2019). piRNAs show tumor-promoting and tumor-suppressing properties in different cancer types and can regulate cancer cells sensitivity to genotoxic drugs (Liu, Dou et al., 2019; Weng, Li, & Goel, 2019). The study of piRNA transcriptome in two NB cell lines reveals multiple piRNAs with higher numbers reside within chromosomes 1, 2, 6, 11, and 17 (Roy and Mallick, 2018). Most piRNAs are of retrotransposon origin, and a majority of them originate from transposable element family, short interspersed nuclear elements repeat of class I (Roy and Mallick, 2018). One example is piRNA 39980, which can target 387 differentially expressed transcripts in NB. The overexpression of piRNA 39980 in NB cells promotes growth, migration, and invasion, suggesting that it could play a role in NB cell metastasis (Roy, Das, Jain, & Mallick, 2020). Mechanistically, piRNA 39980 directly binds to the 3′-UTR of Janus Kinase 3 (JAK3) and inhibits its expression. This is associated with decreased JAK3-regulated genes, including p16, p21, pRB, and p53, and reversal of cellular senescence phenotypes in NB cells (Roy et al., 2020). Altogether, there are very few piRNAs, the role of which is reported in NB growth and progression, suggesting that we are still at an early stage in our understanding of these piRNAs. Since the piRNAs can regulate a wide variety of functions, unraveling their role can increase our knowledge of NB.

13.3.4 Circular RNAs

Due to their broad expression in a wide variety of mammalian and human cancer cells, it is not surprising that circRNAs are gaining considerable interest in cancer research latterly. The advancements in RNA sequence technology and data analysis algorithms enable us to detect diverse RNA species and their expression in different cell types. In recent years, hundreds of circRNAs have been detected that show deregulation in various human cancers. One of the mechanisms by which circRNAs can regulate gene functions is through miRNAs sponging, which affects miRNA-regulated mRNA expression. For example, circRNA 7 located on the antisense strand of gene cerebellar degeneration−related protein 1 sponges tumor-suppressor miR-7 and makes it less available to its target oncogenes. This increases later expression, which can promote cancer cell proliferation and growth (Hansen, Kjems, & Damgaard, 2013). Another circRNA, circAGO2, was found upregulated in NB, inhibits AGO2-miRNA binding, and represses AGO2/miRNA-mediated gene silencing in cancer cells (Chen et al., 2019). AGO2 forms RISC with miRNAs and guides their binding to target mRNA for RNA silencing. circAGO2 interacts with human antigen R (HuR) protein and facilitates translocating from the nucleus to the cytoplasm. In the cytosol, circAGO2 promotes HuR enrichment on the 3′-UTR of target genes, suppressing HuR binding with these genes and AGO2/miRNA-mediated gene silencing (Chen et al., 2019). Furthermore, the overexpression of circ DGKB (diacylglycerol kinase beta) in NB cells increases cell proliferation, migration, and invasion, whereas downregulation reduces basal apoptosis and tumor growth, suggesting its protumor functions

(Yang, Yu et al., 2020). CircDGKB overexpression decreases miR-873 binding to target GLI1 mRNA, which increases its expression. GLI1 upregulation increases cell proliferation, migration, and invasion and decreases basal apoptosis in NB cells (Yang, Yu et al., 2020).

Moreover, intron-containing circular RNA, circ-CUX1 (cut-like homeobox 1) generated from the CUX1 gene regulates CUX1 functions in NB. circ-CUX1 directly interacts with Ewing sarcoma (EWS) RNA-Binding Protein 1 (EWSR1) and promotes its interaction with transcription factor MYC-associated zinc finger protein (MAZ), resulting in its activation. The activated MAZ increases the transcription of its regulated genes, including CUX1. CUX1 upregulation enhances glucose uptake, lactate production, and ATP levels in NB cells associated with increased tumor growth, proliferation, and invasion (Li et al., 2019). Circular RNA circPDE5A regulates NB growth by sponging miR-362–5p functions through direct binding. miR-362–5p overexpression decreases the expression of its target protein, nucleolar protein 4 like, and inhibits NB growth and proliferation, suggesting the oncogenic nature of circPDE5A (Chen, Lin, Hu, Li, & Wu, 2021). Moreover, knockdown of circSLC45A4, which is highly conserved and one of the highest expressed circRNAs in the human embryonic frontal cortex, induces spontaneous neuronal differentiation in NB cell line SH-SY5Y (Suenkel, Cavalli, Massalini, Calegari, & Rajewsky, 2020). In conclusion, emerging studies suggesting the pivotal role of circRNAs in NB development; therefore, circRNAs can be a promising target for NB therapy.

13.4 Clinical significance of noncoding RNAs in neuroblastoma

The changes in the expression of noncoding RNAs are frequently detected in NB, suggesting that they could be used as determinants of diagnosis and prognosis in NB patients. This session will discuss some examples of noncoding RNAs highlighting their potential as a predictive, prognostic, and diagnostic biomarker in NB.

13.4.1 MicroRNAs

A study by Ramraj et al. in a clinically translatable mice model harboring high-risk metastatic NB identified 25 miRNAs that show either high or low serum levels, specifically in animals with high-risk metastatic disease (Saito et al., 2007). Among them, four miRNAs, miR-1308, miR-1908, miR-639, and miR-628, were deregulated in the sera of patients with other cancers; for example, miR1308 in melanoma (Jones et al., 2012), miR1908 and miR-639 in breast cancer (Ahmed Ismaila, Mansoura, & Fawzyb, 2018; Zhu et al., 2020), and miR-628 in prostate cancer (Srivastava et al., 2014). Furthermore, Zeka et al. (2018) identified some circulating miRNAs in NB patients' sera that show strong association with NB stage, MYCN status, and overall survival. The serum levels of identified miRNAs, including miR-10b-3p, miR-323a-3p, miR-129–5p, miR-218–5p, miR-375, miR-149–5p, miR-490–5p, miR-873–3p, and miR-124–3p, increase proportionally from stage 1 to stage 4 NB and positively correlates with tumor burden. Consistently, studies in immunodeficient mice engrafted with SH-SY5Y NB cells show a proportional increase of these miRNAs during the progression of the disease. In addition, NB patients who responded well to the therapy and have reduced tumors show less serum concertation of these miRNAs than nonresponders suggesting that miRNAs have the potential to be utilized as biomarkers for NB progression and burden (Zeka et al., 2018). Another miRNA, mir-199a-3p, present in exosomes, expresses higher in the plasma of NB patients compared to normal controls. Exosomal mir-199a-3p was found twofold higher in the NB patients with unfavorable histology compared to favorable histology (Ma et al., 2019). A review by Ikegaki and Shimada (2019) provides a piece of detailed information about histopathologic analyses of NB recommended by International Neuroblastoma Pathology Classification. Furthermore, expression profiling of exosomal miRNAs in high-risk NB patients before and after chemotherapy treatment reveals differential expression of many miRNAs suggesting their association with the disease. Three miRNAs, let-7b, miR-29c, and miR-342, which are tumor-suppressive, have been downregulated in patients with inadequate therapy responses. In contrast, their levels remain unchanged in those who responded well to the therapy (Morini et al., 2019). Lin et al. found global downregulation of miRNAs' expression in patients with advanced NB. The group discovered 33 miRNAs that showed clean differential expression between low- or intermediate-risk and aggressive stage 4 MYCN-amplified NB tumors (Lin et al., 2010). A similar study of a cohort of 69 patients diagnosed with NB reveals distinctive miRNA expression profiles of MYCN single-copy number and MYCN-amplified tumors (Schulte et al., 2010). Low expression of miR-487b and miR-410 located at the chromosome 14q32.31 locus is strongly associated with higher NB stage and MYCN amplification. In addition, the overall and disease-free survival in low-risk NB patients expressing high levels of miR-487b and miR-410 are better than those with lower expression, which suggested that they may have a role in the prognosis of the disease (Gattolliat et al., 2011). Moreover, tumor-suppressor miR-149 is significantly downregulated in NB cells, and high-risk NB tumors and found higher in patients with better overall survival (Xu et al., 2017). In conclusion, the previously described examples highlight the potential of miRNAs as a candidate marker to assess NB prognosis, staging, and treatment response.

13.4.2 Long noncoding RNAs

LncRNAs have a distinct expression profile in NB that correlates with their prognosis and response to therapy (Pandey et al., 2014; Yarmishyn et al., 2014). Therefore LncRNAs can be used as an indicator of the disease status and therapeutic response. Studies that proposed lncRNA as biomarkers for cancer diagnosis have soared in recent years, but the rate of approval of lncRNAs as a cancer biomarker is still minimal (de Kok et al., 2002). One of the main factors contributing to this is tumor heterogeneity, which poses a significant challenge to cancer biomarker research. This section will discuss the clinical applications of some lncRNAs as a predictive or prognostic biomarker in highly heterogeneous NB tumors. The first example is lncRNA HOXD cluster antisense RNA 1 (HOXD-AS1), expressed on the HOXD locus located on chromosome 2q31.2, and has oncogenic properties. HOXD-AS1 shows distinct expression between different NB stages and positively correlates with the progression of the disease (Yarmishyn et al., 2014). Another lncRNA, NB-associated transcript-1 (NBAT-1), shows promising results in predicting the clinical outcome of NB patients (Pandey et al., 2014). NBAT-1 is located at chromosome 6p22.3, and its knockdown impairs neuronal differentiation and increases cell viability in NB cells. NBAT-1 is lower in high-risk NB tumors, and patients with high NBAT-1 expression have better 5-year overall and event-free survival than low NBAT-1 expression. Lower NBAT-1 in high-risk patients is due to hypermethylation of CpGs flanking the NBAT-1, which results in the inactivation of the NBAT-1 promoter (Pandey et al., 2014). Furthermore, differential expression analysis between stage 4 and stage 4S NB patients from two NB cohorts reveals four lncRNAs, LOC283177, LOC1019281001, LNC00839, and FIRRE that shows distinct expression in patients with good clinical outcome (Meng, Fang, Zhao, & Feng, 2020). LOC283177 and LOC1019281001 are higher in the low-risk stage 4S group, MYCN-nonamplified group, and age <18 months group and significantly associated with good overall survival. In contrast, LNC00839 and FIRRE are associated with poor overall survival and upregulates in patients in the high-risk groups, including stage 4, MYCN amplification, and age <18 months (Meng et al., 2020).

In a similar study, Sahu et al. (2018) identify 16-lncRNA-based prognostic signatures associated with overall and event-free survival in NB patients and can discriminate patients into risk groups. Further, lncRNAs forkhead box D3 antisense RNA 1 (FOXD3-AS1) and LINC01268 are associated with a favorable outcome in NB patients. FOXD3-AS1 is lower in NB cases with death, progression, advanced INSS stages or MYCN amplification and negatively correlates with the disease's aggressiveness. This implies that FOXD3-AS1 can act as an independent prognostic marker for a favorable outcome of NB patients. Moreover, tumor suppressor lncRNA, cancer susceptibility 15 (CASC15-S) located at 6p.22 locus shows significant association with NB susceptibility. Depletion of CASC15-S increases NB cell proliferation and enhances its invasive properties. CASC15-S expression shows a robust negative correlation with the clinical factors associated with NB progression, such as age, MYCN amplification, and 1p/11q deletion, and positively correlates with the overall survival of patients. Further, lncRNA noncoding RNA activated by DNA damage (NORAD) expresses higher in NB patients with high event-free survival and overall survival (Yu et al., 2020). The examples mentioned earlier indicate that lncRNA signatures in NB can have the potential to assess NB risk. Therefore investigating their expression changes at the genomic levels or in body fluids could provide us better diagnostic markers in the future.

13.4.3 P-element-induced wimpy testis (Piwi)-interacting RNAs

piRNAs express lower in somatic cells, but their deregulation in different cancers makes them a potential candidate in biomarker research. There are reports projecting piRNAs as a potential biomarker candidate in some diseases, but in cancer, especially NB, our understanding is still at an early stage. Some examples of piRNAs associated with the tumor staging and overall patient survival include piR-823 in colorectal cancer (Sabbah et al., 2021), gastric cancer (Cui et al., 2011), renal cell carcinoma (Iliev et al., 2016), and piR-651 in classical Hodgkin lymphoma (Cordeiro et al., 2016); piR-5937 and piR-28876 in colon cancer (Vychytilova-Faltejskova et al., 2018); and piR-54265 in colorectal cancer (Mai et al., 2020). In addition, since piRNAs can regulate a wide range of cellular processes from repressing retrotransposon activity to tissue-specific protein expression and crucial in neural development, they may play a key role in NB development and diagnosis (Czech et al., 2018; Ozata, Gainetdinov, Zoch, O'Carroll, & Zamore, 2019; Rojas-Rios and Simonelig, 2018).

13.4.4 Circular RNAs

Studies have explored the circRNA expression correlation with cancer progression and diagnosis in the last few years. Many circRNA found upregulated or downregulated in the plasma of cancer patients shows association with disease prognosis and therapeutic response (Chen et al., 2021; Fang, Yao, Mao, Zhong, & Xu, 2021; Yang, Yu et al., 2020; Yang et al., 2021; Zhang, Zhang et al., 2020). The deregulation of circRNAs in NB cells, tissues, and patients and their role in NB development suggests that they may have the prognostic and predictive potential for NB. For instance, circ-CUX1, which is upregulated in NB tissues, shows a strong positive association with lower survival, advanced tumor (T), node (N), and metastasis (M) (TNM) stage, and positive lymph node metastasis in NB patients (Zhang, Zhang et al., 2020). Likewise, circDGKB is upregulated in the blood samples of NB patients compared to healthy subjects (Yang, Yu et al., 2020). The other circRNAs found upregulated in NB tissues compared to matched normal tissues include circAGO2 (Chen et al., 2019), circ_0132817 derived from CUX1 gene (Fang et al., 2021), circRNA phosphodiesterase 5A (Chen et al., 2021), and circRNA kinesin family member 2A (Yang et al., 2021). Furthermore, circRNA expression data of 39 NB cell lines reveal the upregulation of 7 circRNAs located within MYCN-amplified regions compared to normal cells. Some target genes of these circRNAs show a strong correlation with overall survival and event-free survival in low-risk and high-risk NB patients (Zhang, Zhou et al., 2020). This suggests that circRNAs may indirectly involve regulating proteins that correlate with cancer prognosis and improved outcome in patients. Overall, our understanding of circRNAs regulation in NB is inadequate as there are limited studies to solidify the notion that circRNAs are critical for NB growth. Therefore to be explored for their biomarker potential, further research is needed on their mechanistic insight and role in NB.

13.5 Therapeutic implications and targeting strategies for noncoding RNAs in neuroblastoma

Emerging studies have suggested the role of noncoding RNAs in therapy resistance and immune escape mechanisms. The up- or downregulation of noncoding RNAs provides considerable protection to cancer cells during cancer treatment, suggesting that they are essential players to consider for developing new therapies for cancer treatment.

13.5.1 Therapeutic potential of noncoding RNAs in neuroblastoma

Anticancer therapies can alter noncoding RNAs expression, associated with drug response and chemoresistance. One example is monoallelic deletion at 13q14.3 locus harboring p53 inducible-tumor suppressor miRNAs, miR-15, and miR-16 during etoposide treatment to NB cells (Marengo et al., 2018). Etoposide-resistant cells have reduced levels of miR-15 and miR-16 compared to parental cells, which correlates with the upregulation of a polycomb complex protein BMI-1 (Marengo et al., 2018). BMI-1 is an MYCN target protein that is highly expressed in NB and essential for its development (Cui et al., 2007). miR-15 and miR-16 also facilitate tumor-suppressing effects of CXCR4 inhibitor BL-8040 in NB cells. CXCR4 is overexpressed in NB cells and promotes tumor growth (Russell, Hicks, Okcu, & Nuchtern, 2004). BL-8040 sensitizes CXCR4 overexpressing cells by upregulating miR-15 and miR-16, suppressing NB cell growth and survival (Klein et al., 2018). Likewise, miR-137 can sensitizes doxorubicin-resistant NB cells to doxorubicin by downregulating constitutive androstane receptor (CAR), a nuclear receptor protein involved in drug metabolism. CAR inhibition reduces the expression of its target protein, multidrug resistance 1, which increases doxorubicin accumulation in cells and sensitizes the resistant NB cells to doxorubicin (Takwi et al., 2014). Furthermore, oncogenic lncRNA, NORAD promotes doxorubicin resistance through sponging tumor suppressor miR-144–3p. The suppression of miR-144–3p increases the expression of its target gene HDAC8, which accelerates the growth of NB cells and promotes doxorubicin resistance (Wang et al., 2020). LncRNA small nucleolar RNA host gene 16 (SNHG16) knockdown sensitizes cisplatin resistance NB cells to the drug and induces apoptosis. Mechanistically, SNHG16 overexpression decreases miR-338-3p and upregulates its target gene PLK4, associated with cisplatin resistance and high malignancy phenotype in NB (Xu, Sun, Wang, Sun, & Liu, 2020). Moreover, targeted inhibition of oncogenic circular RNA circ-CUX1 with inhibitory peptide EIP-22, blocks circ-CUX1 interaction with EWSR1, reduces tumor growth, and decreases ATP levels in mice xenograft tumors. In addition, EIP-22 synergizes the tumor-suppressing effects of glycolysis inhibitors, 2-deoxy-glucose, and 3-bromopyruvate in NB cells. This suggests that circ-CUX1 inhibition could be a promising therapy for NB treatment. Some other examples of noncoding RNAs as a therapeutic target in NB are described in the Table 13.1.

13.5.2 Targeting strategies for noncoding RNAs

The deregulation of noncoding RNAs in human cancers makes them a promising target for developing novel therapeutic approaches for cancer treatment. However, different types of noncoding RNAs require different targeting strategies to restore RNA functions in the cells. For example, miRNA-based therapeutics involve synthetic oligonucleotides as a substitute for downregulated tumor-suppressor miRNAs or as an anti-miRNA oligonucleotide to neutralize oncogenic miRNAs. These oligonucleotides are stable and resistant to nuclear degradation due to chemical modification; thus, they are more efficient than the native miRNAs in delivering functions (Bader, Brown, Stoudemire, & Lammers, 2011; Hutvagner, Simard, Mello, & Zamore, 2004; Lima, Cerqueira, Figueiredo, Oliveira, & Azevedo, 2018). Another example includes locked nucleic acid (LNA), a class of RNA analogues. There is an extra bond between 2′ oxygen and 4′ carbon of the ribose ring of RNA nucleotide, which locks the ring and provides very stable conformation to the oligonucleotide. LNA-modified oligonucleotides are thermally stable and show a strong affinity to the complementary sequence, thus an efficient molecule for RNA targeting in vivo

TABLE 13.1 Noncoding RNAs, their association with neuroblastoma patient's clinical features, and role in drug response.

Noncoding RNAs	Correlation with patient's clinical features	Target gene(s)	Functions in drug response
miRNAs			
miR-124–3p (Nolan et al., 2020)	Downregulated and associates with poor OS	ACTN4, MYH9, VIM, and PLEC	Sensitizes resistant cells to cisplatin and etoposide treatment
miR-497 (Soriano et al., 2016)	Low expression correlates with worse progression-free survival	WEE1, CHEK1, BCL2, CDC25A, and VEGFA	Decreases proliferation of NB cells resistant to cisplatin, etoposide, and melphalan
miR-155 (Challagundla et al., 2015)	Upregulated in patients with high CD163 + TAM infiltration in NB tumors	TERF1	Induces cisplatin resistance by targeting telomerase inhibitor TERF1, which increases telomerase activity, and telomere length in NB cells and promote their growth
miR-129 (Wang, Li, Xu, Zheng, & Li, 2018; Yang et al., 2021)	Downregulated in NB compared to normal tissues	MYO10	Sensitizes NB cells to cyclophosphamide through MYO10 suppression
miR-204 (Ryan et al., 2012)	Downregulated in high-risk NB patients	BCL2 and NTRK2	Sensitizes NB cells to cisplatin and etoposide treatment
LncRNAs			
SNHG7 (Jia et al., 2020; Wang, Wang, & Zhang, 2020)	SNHG7 high in NB tissues. Low SNHG7 patients have longer OS than high SNHG7 patients	miR-329–3p	SNHG7 sponges miR-329–3p, which increases miR target gene MYO10 and decreases NB cell sensitivity to cisplatin
NEAT1 (Yang, Ye, Wang, & Xia, 2020)	Overexpressed in NB cells	miR-326	NEAT1 overexpression decreases NB sensitivity to cisplatin. NEAT1 sponges miR-326, which increases JAK1/STAT3 signaling and promotes NB growth
XIST (Yang, Zhang et al., 2020)	Overexpressed in NB cells and tissues	miR-375	XIST silencing sensitizes NB cells to radiation. XIST sponges miR-375 leading to an increase in its target gene L1CAM, which is associated with increased NB growth and less radio sensitivity
USMycN (Liu et al., 2014)	High USMycN associated with poor prognosis and OS	DNA- and RNA-binding protein, NONO	Antisense oligonucleotides targeting USMycN reduces tumor growth. USMycN binds to NONO, which increases NMyc transcription

ACTN4, Actinin alpha 4; L1CAM, L1 cell adhesion molecule protein; MYH9, myosin heavy chain 9; MYO10, myosin X; NEAT1, nuclear-enriched abundant transcript 1; NTRK2, neurotrophic receptor tyrosine kinase 2; PLEC, plectin; SNHG7, small nucleolar RNA host gene 7; TERF1, telomeric repeat binding factor 1; USMycN, upstream of MycN; VIM, vimentin; XIST, X-inactive specific transcript; OS, Overall survival; NONO, non-POU domain-containing octamer-binding protein; POU, Pit-Oct-Unc

(Grunweller and Hartmann, 2007). Moreover, some chemical compounds also have the potential to modulate miRNAs expression in cells. For example, small molecule aza-flavanones can target miRNAs and inhibit their tumorigenic potential. Aza-flavanones can specifically target oncogenic miR-4644 without affecting the general miRNA biogenesis pathway and inhibits its functions (Chandrasekhar et al., 2012). The other example includes aminoglycosides that can hinder the processing of miR-27a, which results in the formation of less mature miR-27a, an oncogenic miRNA involved in metabolic programming and chemoresistance (Barisciano et al., 2020; Bose, Jayaraj, Kumar, & Maiti, 2013).

Similar strategies can be applied to target oncogenic lncRNAs, like the use of stable, synthetic antisense oligonucleotides that can bind to target lncRNAs and promote their degradation through RNase H-dependent cleavage mechanism. The presence of RNase H in both cytoplasm and the nucleus provides efficient cleavage of nuclear and cytosolic lncRNAs, thus enhancing these oligonucleotides effectiveness in lncRNAs silencing. The other approach includes the use of clustered, regularly interspaced, short palindromic repeats/Cas9 technology where nuclease-inactive dCas9 fused with Kruppel-associated box repressor is directed by guide RNAs to the regulatory regions of lncRNAs for transcriptional silencing (Thakore et al., 2015). LncRNA functions can also be inhibited by blocking lncRNA–protein interactions, resulting in the loss of function of target lncRNAs. For example, alkaloid ellipticine can inhibit brain-derived neurotrophic factor antisense (*BDNF*-AS) lncRNA interaction with enhancer of zeste homolog 2, which upregulates lncRNA target protein *BDNF* transcription (Pedram Fatemi et al., 2015). Macrocyclic peptide NP-C86 can inhibit the binding of lncRNA growth arrest-specific 5 (GAS5) with RNA helicase UPF1, which results in the upregulation of GAS5 (Shi et al., 2019). GAS5 is tumor suppressor lncRNA involved in suppressing genes regulating cell growth, apoptosis, and drug resistance (Ji, Dai, Yeung, & He, 2019). Furthermore, synthetic circular RNAs can be used to sponge oncogenic miRNAs aberrantly overexpressed in tumors. For example, Liu et al. developed a synthetic circRNA miR sponge that can inhibit oncomiR miR-21 functions. This circRNA has multiple miR-21-binding sites and has more resistance to nuclease digestion than native circRNAs. The synthetic circRNA effectively sponges miR-21, which increases the expression of miR-21 downstream proteins and inhibits cell proliferation. In summary, it is evident that noncoding RNA deregulation is a common phenomenon in cancer. Therefore targeting these RNAs with different approaches could be a promising therapeutic option for cancer treatment.

13.6 Conclusion

Studies from the past three decades have shown that the so-called junk or noncoding DNA harbors the genomic regions transcribed into noncoding RNAs. The noncoding RNAs are the master regulator of cells that regulates almost every cellular function. The deregulation of these RNAs is a common phenomenon in cancer, including NB, which plays a critical role in tumorigenesis (Fig. 13.1). The up- or downregulation of noncoding RNAs during tumor development and their association with patient survival and therapeutic response suggests that they may be exploitable targets for cancer treatment and prognosis. Tremendous progress has been made in this direction, and many

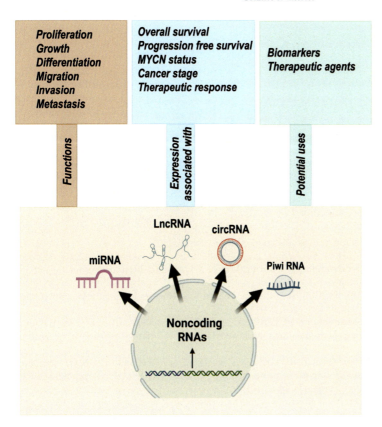

FIGURE 13.1 Noncoding RNAs and their role in NB. *NB*, Neuroblastoma.

pharmaceutical companies have already been working on noncoding RNAs-based therapeutics to treat cancer and other diseases. Considering the progress made in noncoding RNA research in both academia and the pharmaceutical sector during the past decade, we are not far away from the development and commercialization of new RNA-based therapeutics in the treatment of cancer and NB.

Acknowledgments

Dr. Challagundla's laboratory is supported in whole or part by NIH/NCI grant CA197074, Buffet Pilot & Pediatric Cancer Research Group Grants at UNMC, and the Department of Biochemistry & Molecular Biology start-up grants. Figures were prepared using BioRender.com.

Conflict of interest

The authors declare no conflict of interest.

References

Ahmed Ismaila, A. I. A., Mansoura, O. A., & Fawzyb, A. (2018). Diagnostic significance of miR-639 and miR-10b in βreast cancer patients. *Meta Gene, 19*, 155–159. (February 2019).

Althoff, K., Lindner, S., Odersky, A., Mestdagh, P., Beckers, A., Karczewski, S., ... Schulte, J. H. (2015). miR-542-3p exerts tumor suppressive functions in neuroblastoma by downregulating survivin. *International Journal of Cancer. Journal International du Cancer, 136*(6), 1308–1320.

Ashwal-Fluss, R., Meyer, M., Pamudurti, N. R., Ivanov, A., Bartok, O., Hanan, M., ... Kadener, S. (2014). circRNA biogenesis competes with pre-mRNA splicing. *Molecular Cell, 56*(1), 55–66.

Bader, A. G., Brown, D., Stoudemire, J., & Lammers, P. (2011). Developing therapeutic microRNAs for cancer. *Gene Therapy, 18*(12), 1121–1126.

Barckmann, B., Pierson, S., Dufourt, J., Papin, C., Armenise, C., Port, F., ... Simonelig, M. (2015). Aubergine iCLIP reveals piRNA-dependent decay of mRNAs involved in germ cell development in the early embryo. *Cell Reports, 12*(7), 1205–1216.

Barisciano, G., Colangelo, T., Rosato, V., Muccillo, L., Taddei, M. L., Ippolito, L., ... Sabatino, L. (2020). miR-27a is a master regulator of metabolic reprogramming and chemoresistance in colorectal cancer. *British Journal of Cancer, 122*(9), 1354–1366.

Becker, J., & Wilting, J. (2019). WNT signaling in neuroblastoma. *Cancers (Basel), 11*(7).

Beckers, A., Van Peer, G., Carter, D. R., Mets, E., Althoff, K., Cheung, B. B., ... Speleman, F. (2015). MYCN-targeting miRNAs are predominantly downregulated during MYCN driven neuroblastoma tumor formation. *Oncotarget, 6*(7), 5204–5216.

Bose, D., Jayaraj, G. G., Kumar, S., & Maiti, S. (2013). A molecular-beacon-based screen for small molecule inhibitors of miRNA maturation. *ACS Chemical Biology, 8*(5), 930–938.

Bountali, A., Tonge, D. P., & Mourtada-Maarabouni, M. (2019). RNA sequencing reveals a key role for the long non-coding RNA MIAT in regulating neuroblastoma and glioblastoma cell fate. *International Journal of Biological Macromolecules, 130*, 878–891.

Buechner, J., Tomte, E., Haug, B. H., Henriksen, J. R., Lokke, C., Flaegstad, T., & Einvik, C. (2011). Tumour-suppressor microRNAs let-7 and mir-101 target the proto-oncogene MYCN and inhibit cell proliferation in MYCN-amplified neuroblastoma. *British Journal of Cancer, 105*(2), 296–303.

Bushati, N., & Cohen, S. M. (2007). microRNA functions. *Annual Review of Cell and Developmental Biology, 23*, 175–205.

Challagundla, K. B., Wise, P. M., Neviani, P., Chava, H., Murtadha, M., Xu, T., ... Fabbri, M. (2015). Exosome-mediated transfer of microRNAs within the tumor microenvironment and neuroblastoma resistance to chemotherapy. *Journal of the National Cancer Institute, 107*(7).

Chandrasekhar, S., Pushpavalli, S. N., Chatla, S., Mukhopadhyay, D., Ganganna, B., Vijeender, K., ... Bhadra, U. (2012). aza-Flavanones as potent cross-species microRNA inhibitors that arrest cell cycle. *Bioorganic & Medicinal Chemistry Letters, 22*(1), 645–648.

Chen, Y., Lin, L., Hu, X., Li, Q., & Wu, M. (2021). Silencing of circular RNA circPDE5A suppresses neuroblastoma progression by targeting the miR-362-5p/NOL4L axis. *The International Journal of Neuroscience*, 1–11.

Chen, Y., & Stallings, R. L. (2007). Differential patterns of microRNA expression in neuroblastoma are correlated with prognosis, differentiation, and apoptosis. *Cancer Research, 67*(3), 976–983.

Chen, Y., Yang, F., Fang, E., Xiao, W., Mei, H., Li, H., ... Tong, Q. (2019). Circular RNA circAGO2 drives cancer progression through facilitating HuR-repressed functions of AGO2-miRNA complexes. *Cell Death and Differentiation, 26*(7), 1346–1364.

Cheng, J., Xu, L., Deng, L., Xue, L., Meng, Q., Wei, F., & Wang, J. (2020). RNA N(6)-methyladenosine modification is required for miR-98/MYCN axis-mediated inhibition of neuroblastoma progression. *Scientific Reports, 10*(1), 13624.

Cho, S. W., Xu, J., Sun, R., Mumbach, M. R., Carter, A. C., Chen, Y. G., ... Chang, H. Y. (2018). Promoter of lncRNA gene PVT1 is a tumor-suppressor DNA boundary element. *Cell, 173*(6), 1398–1412, e22.

Clyde, D. (2019). piRNAs make sense of retroviral invaders. *Nature Reviews Genetics, 20*(12), 704.

Colon, N. C., & Chung, D. H. (2011). Neuroblastoma. *Advances in Pediatrics, 58*(1), 297–311.

Cordeiro, A., Navarro, A., Gaya, A., Diaz-Beya, M., Gonzalez-Farre, B., Castellano, J. J., ... Monzo, M. (2016). PiwiRNA-651 as marker of treatment response and survival in classical Hodgkin lymphoma. *Oncotarget, 7*(29), 46002–46013.

Cui, H., Hu, B., Li, T., Ma, J., Alam, G., Gunning, W. T., & Ding, H. F. (2007). Bmi-1 is essential for the tumorigenicity of neuroblastoma cells. *The American Journal of Pathology, 170*(4), 1370−1378.

Cui, L., Lou, Y., Zhang, X., Zhou, H., Deng, H., Song, H., ... Guo, J. (2011). Detection of circulating tumor cells in peripheral blood from patients with gastric cancer using piRNAs as markers. *Clinical Biochemistry, 44*(13), 1050−1057.

Czech, B., & Hannon, G. J. (2016). One loop to rule them all: The ping-pong cycle and piRNA-guided silencing. *Trends in Biochemical Sciences, 41*(4), 324−337.

Czech, B., Munafo, M., Ciabrelli, F., Eastwood, E. L., Fabry, M. H., Kneuss, E., & Hannon, G. J. (2018). piRNA-guided genome defense: From biogenesis to silencing. *Annual Review of Genetics, 52*, 131−157.

de Kok, J. B., Verhaegh, G. W., Roelofs, R. W., Hessels, D., Kiemeney, L. A., Aalders, T. W., ... Schalken, J. A. (2002). DD3(PCA3), a very sensitive and specific marker to detect prostate tumors. *Cancer Research, 62*(9), 2695−2698.

Diskin, S. J., Capasso, M., Schnepp, R. W., Cole, K. A., Attiyeh, E. F., Hou, C., ... Maris, J. M. (2012). Common variation at 6q16 within HACE1 and LIN28B influences susceptibility to neuroblastoma. *Nature Genetics, 44*(10), 1126−1130.

Fang, Y., Yao, Y., Mao, K., Zhong, Y., & Xu, Y. (2021). Circ_0132817 facilitates cell proliferation, migration, invasion and glycolysis by regulating the miR-432-5p/NOL4L axis in neuroblastoma. *Experimental Brain Research, 239*.

Feriancikova, B., Feglarova, T., Krskova, L., Eckschlager, T., Vicha, A., & Hrabeta, J. (2021). MIAT is an upstream regulator of NMYC and the disruption of the MIAT/NMYC axis induces cell death in NMYC amplified neuroblastoma cell lines. *International Journal of Molecular Sciences, 22*(7).

Flahaut, M., Meier, R., Coulon, A., Nardou, K. A., Niggli, F. K., Martinet, D., ... Gross, N. (2009). The Wnt receptor FZD1 mediates chemoresistance in neuroblastoma through activation of the Wnt/beta-catenin pathway. *Oncogene, 28*(23), 2245−2256.

Gattolliat, C. H., Thomas, L., Ciafre, S. A., Meurice, G., Le Teuff, G., Job, B., ... Benard, J. (2011). Expression of miR-487b and miR-410 encoded by 14q32.31 locus is a prognostic marker in neuroblastoma. *British Journal of Cancer, 105*(9), 1352−1361.

Gou, L. T., Dai, P., Yang, J. H., Xue, Y., Hu, Y. P., Zhou, Y., ... Liu, M. F. (2015). Pachytene piRNAs instruct massive mRNA elimination during late spermiogenesis. *Cell Research, 25*(2), 266.

Gregory, R. I., Chendrimada, T. P., Cooch, N., & Shiekhattar, R. (2005). Human RISC couples microRNA biogenesis and posttranscriptional gene silencing. *Cell, 123*(4), 631−640.

Grunweller, A., & Hartmann, R. K. (2007). Locked nucleic acid oligonucleotides: The next generation of antisense agents? *BioDrugs: Clinical Immunotherapeutics, Biopharmaceuticals and Gene Therapy, 21*(4), 235−243.

Halic, M., & Moazed, D. (2009). Transposon silencing by piRNAs. *Cell, 138*(6), 1058−1060.

Han, J., Lee, Y., Yeom, K. H., Kim, Y. K., Jin, H., & Kim, V. N. (2004). The Drosha-DGCR8 complex in primary microRNA processing. *Genes & Development, 18*(24), 3016−3027.

Han, J., Lee, Y., Yeom, K. H., Nam, J. W., Heo, I., Rhee, J. K., ... Kim, V. N. (2006). Molecular basis for the recognition of primary microRNAs by the Drosha-DGCR8 complex. *Cell, 125*(5), 887−901.

Hansen, T. B., Kjems, J., & Damgaard, C. K. (2013). Circular RNA and miR-7 in cancer. *Cancer Research, 73*(18), 5609−5612.

Hayes, J., Peruzzi, P. P., & Lawler, S. (2014). MicroRNAs in cancer: Biomarkers, functions and therapy. *Trends in Molecular Medicine, 20*(8), 460−469.

He, X. Y., Tan, Z. L., Mou, Q., Liu, F. J., Liu, S., Yu, C. W., ... Zou, L. (2017). microRNA-221 enhances MYCN via targeting nemo-like kinase and functions as an oncogene related to poor prognosis in neuroblastoma. *Clinical Cancer Research: An Official Journal of the American Association for Cancer Research, 23*(11), 2905−2918.

Horwich, M. D., Li, C., Matranga, C., Vagin, V., Farley, G., Wang, P., & Zamore, P. D. (2007). The Drosophila RNA methyltransferase, DmHen1, modifies germline piRNAs and single-stranded siRNAs in RISC. *Current biology: CB, 17*(14), 1265−1272.

Houwing, S., Kamminga, L. M., Berezikov, E., Cronembold, D., Girard, A., van den Elst, H., ... Ketting, R. F. (2007). A role for Piwi and piRNAs in germ cell maintenance and transposon silencing in Zebrafish. *Cell, 129*(1), 69−82.

Huarte, M. (2015). The emerging role of lncRNAs in cancer. *Nature Medicine, 21*(11), 1253−1261.

Hutvagner, G., Simard, M. J., Mello, C. C., & Zamore, P. D. (2004). Sequence-specific inhibition of small RNA function. *PLoS Biology, 2*(4), E98.

Ikegaki, N., & Shimada, H. (2019). Subgrouping of unfavorable histology neuroblastomas with immunohistochemistry toward precision prognosis and therapy stratification, C. International Neuroblastoma Pathology *JCO Precision Oncology, 3*.

Iliev, R., Fedorko, M., Machackova, T., Mlcochova, H., Svoboda, M., Pacik, D., ... Slaby, O. (2016). Expression levels of PIWI-interacting RNA, piR-823, are deregulated in tumor tissue, blood serum and urine of patients with renal cell carcinoma. *Anticancer Research, 36*(12), 6419−6423.

Ishizu, H., Kinoshita, T., Hirakata, S., Komatsuzaki, C., & Siomi, M. C. (2019). Distinct and collaborative functions of Yb and armitage in transposon-targeting piRNA biogenesis. *Cell Reports, 27*(6), 1822−1835, e8.

Ivanov, A., Memczak, S., Wyler, E., Torti, F., Porath, H. T., Orejuela, M. R., ... Rajewsky, N. (2015). Analysis of intron sequences reveals hallmarks of circular RNA biogenesis in animals. *Cell Reports, 10*(2), 170−177.

Iwasaki, Y. W., Siomi, M. C., & Siomi, H. (2015). PIWI-interacting RNA: Its biogenesis and functions. *Annual Review of Biochemistry, 84*, 405−433.

Ji, J., Dai, X., Yeung, S. J., & He, X. (2019). The role of long non-coding RNA GAS5 in cancers. *Cancer Management and Research, 11*, 2729−2737.

Jia, J., Zhang, D., Zhang, J., Yang, L., Zhao, G., Yang, H., & Wang, J. (2020). Long non-coding RNA SNHG7 promotes neuroblastoma progression through sponging miR-323a-5p and miR-342-5p. *Biomedicine & Pharmacotherapy = Biomedecine & Pharmacotherapie, 128*, 110293.

Jones, C. I., Zabolotskaya, M. V., King, A. J., Stewart, H. J., Horne, G. A., Chevassut, T. J., & Newbury, S. F. (2012). Identification of circulating microRNAs as diagnostic biomarkers for use in multiple myeloma. *British Journal of Cancer, 107*(12), 1987−1996.

Kawaoka, S., Izumi, N., Katsuma, S., & Tomari, Y. (2011). 3′ end formation of PIWI-interacting RNAs in vitro. *Molecular Cell, 43*(6), 1015−1022.

Kim, V. N., Han, J., & Siomi, M. C. (2009). Biogenesis of small RNAs in animals. *Nature Reviews Molecular Cell Biology, 10*(2), 126−139.

Kim, Y. K., & Kim, V. N. (2007). Processing of intronic microRNAs. *The EMBO Journal, 26*(3), 775−783.

Kirino, Y., & Mourelatos, Z. (2007). The mouse homolog of HEN1 is a potential methylase for Piwi-interacting RNAs. *RNA (New York, N.Y.), 13*(9), 1397−1401.

Klein, S., Abraham, M., Bulvik, B., Dery, E., Weiss, I. D., Barashi, N., ... Peled, A. (2018). CXCR4 promotes neuroblastoma growth and therapeutic resistance through miR-15a/16-1-mediated ERK and BCL2/cyclin D1 pathways. *Cancer Research, 78*(6), 1471−1483.

Kobayashi, H., & Tomari, Y. (2016). RISC assembly: Coordination between small RNAs and Argonaute proteins. *Biochimica et Biophysica Acta, 1859*(1), 71−81.

Kosti, A., Du, L., Shivram, H., Qiao, M., Burns, S., Garcia, J. G., ... Penalva, L. O. F. (2020). ELF4 is a target of miR-124 and promotes neuroblastoma proliferation and undifferentiated state. *Molecular Cancer Research: MCR, 18*(1), 68−78.

Kristensen, L. S., Andersen, M. S., Stagsted, L. V. W., Ebbesen, K. K., Hansen, T. B., & Kjems, J. (2019). The biogenesis, biology and characterization of circular RNAs. *Nature Reviews Genetics, 20*(11), 675−691.

La, T., Liu, G. Z., Farrelly, M., Cole, N., Feng, Y. C., Zhang, Y. Y., ... Zhang, X. D. (2018). A p53-responsive miRNA network promotes cancer cell quiescence. *Cancer Research, 78*(23), 6666−6679.

Le, M. T., Teh, C., Shyh-Chang, N., Xie, H., Zhou, B., Korzh, V., ... Lim, B. (2009). MicroRNA-125b is a novel negative regulator of p53. *Genes & Development, 23*(7), 862−876.

Lee, L. W., Zhang, S., Etheridge, A., Ma, L., Martin, D., Galas, D., & Wang, K. (2010). Complexity of the microRNA repertoire revealed by next-generation sequencing. *RNA (New York, N.Y.), 16*(11), 2170−2180.

Lee, Y., Kim, M., Han, J., Yeom, K. H., Lee, S., Baek, S. H., & Kim, V. N. (2004). MicroRNA genes are transcribed by RNA polymerase II. *The EMBO Journal, 23*(20), 4051−4060.

Li, D., Wang, X., Mei, H., Fang, E., Ye, L., Song, H., ... Tong, Q. (2018). Long noncoding RNA pancEts-1 promotes neuroblastoma progression through hnRNPK-mediated beta-catenin stabilization. *Cancer Research, 78*(5), 1169−1183.

Li, H., Jia, Y., Cheng, J., Liu, G., & Song, F. (2018). LncRNA NCK1-AS1 promotes proliferation and induces cell cycle progression by crosstalk NCK1-AS1/miR-6857/CDK1 pathway. *Cell Death & Disease, 9*(2), 198.

Li, H., Yang, F., Hu, A., Wang, X., Fang, E., Chen, Y., ... Tong, Q. (2019). Therapeutic targeting of circ-CUX1/EWSR1/MAZ axis inhibits glycolysis and neuroblastoma progression. *EMBO Molecular Medicine, 11*(12), e10835.

Lima, J. F., Cerqueira, L., Figueiredo, C., Oliveira, C., & Azevedo, N. F. (2018). Anti-miRNA oligonucleotides: A comprehensive guide for design. *RNA Biology, 15*(3), 338−352.

Lin, R. J., Lin, Y. C., Chen, J., Kuo, H. H., Chen, Y. Y., Diccianni, M. B., ... Yu, A. L. (2010). microRNA signature and expression of Dicer and Drosha can predict prognosis and delineate risk groups in neuroblastoma. *Cancer Research, 70*(20), 7841−7850.

Liu, C. X., Li, X., Nan, F., Jiang, S., Gao, X., Guo, S. K., ... Chen, L. L. (2019). Structure and degradation of circular RNAs regulate PKR activation in innate immunity. *Cell, 177*(4), 865−880, e21.

Liu, P. Y., Erriquez, D., Marshall, G. M., Tee, A. E., Polly, P., Wong, M., ... Liu, T. (2014). Effects of a novel long noncoding RNA, lncUSMycN, on N-Myc expression and neuroblastoma progression. *Journal of the National Cancer Institute, 106*(7).

Liu, P. Y., Tee, A. E., Milazzo, G., Hannan, K. M., Maag, J., Mondal, S., ... Liu, T. (2019). The long noncoding RNA lncNB1 promotes tumorigenesis by interacting with ribosomal protein RPL35. *Nature Communications, 10*(1), 5026.

Liu, Y., Dou, M., Song, X., Dong, Y., Liu, S., Liu, H., ... Xu, W. (2019). The emerging role of the piRNA/piwi complex in cancer. *Molecular Cancer, 18*(1), 123.

Ma, C., Zhang, L., Wang, X., He, S., Bai, J., Li, Q., ... Zhu, D. (2020). piRNA-63076 contributes to pulmonary arterial smooth muscle cell proliferation through acyl-CoA dehydrogenase. *Journal of Cellular and Molecular Medicine, 24*(9), 5260−5273.

Ma, J., Xu, M., Yin, M., Hong, J., Chen, H., Gao, Y., ... Mo, X. (2019). Exosomal hsa-miR199a-3p promotes proliferation and migration in neuroblastoma. *Frontiers in Oncology, 9*, 459.

Ma, L., Bajic, V. B., & Zhang, Z. (2013). On the classification of long non-coding RNAs. *RNA Biology, 10*(6), 925−933.

Mai, D., Ding, P., Tan, L., Zhang, J., Pan, Z., Bai, R., ... Lin, D. (2018). PIWI-interacting RNA-54265 is oncogenic and a potential therapeutic target in colorectal adenocarcinoma. *Theranostics, 8*(19), 5213−5230.

Mai, D., Zheng, Y., Guo, H., Ding, P., Bai, R., Li, M., ... Lin, D. (2020). Serum piRNA-54265 is a New Biomarker for early detection and clinical surveillance of human colorectal cancer. *Theranostics, 10*(19), 8468−8478.

Mao, F., Zhang, J., Cheng, X., & Xu, Q. (2019). miR-149 inhibits cell proliferation and enhances chemosensitivity by targeting CDC42 and BCL2 in neuroblastoma. *Cancer Cell International, 19*, 357.

Marengo, B., Monti, P., Miele, M., Menichini, P., Ottaggio, L., Foggetti, G., ... Domenicotti, C. (2018). Etoposide-resistance in a neuroblastoma model cell line is associated with 13q14.3 mono-allelic deletion and miRNA-15a/16-1 down-regulation. *Scientific Reports, 8*(1), 13762.

Maris, J. M. (2010). Recent advances in neuroblastoma. *The New England Journal of Medicine, 362*(23), 2202−2211.

Meister, G. (2013). Argonaute proteins: Functional insights and emerging roles. *Nature Reviews Genetics, 14*(7), 447−459.

Meng, X., Fang, E., Zhao, X., & Feng, J. (2020). Identification of prognostic long noncoding RNAs associated with spontaneous regression of neuroblastoma. *Cancer Medicine, 9*(11), 3800−3815.

Mitra, S., Muralidharan, S. V., Marco, M. D., Juvvuna, P. K., Kosalai, S. T., Reischl, S., ... Kanduri, C. (2021). Subcellular distribution of p53 by the p53-responsive lncRNA NBAT1 determines chemotherapeutic response in neuroblastoma. *Cancer Research, 81*(6), 1457−1471.

Mondal, T., Juvvuna, P. K., Kirkeby, A., Mitra, S., Kosalai, S. T., Traxler, L., ... Kanduri, C. (2018). Sense-antisense lncRNA pair encoded by locus 6p22.3 determines neuroblastoma susceptibility via the USP36-CHD7-SOX9 regulatory axis. *Cancer Cell, 33*(3), 417−434, e7.

Morini, M., Cangelosi, D., Segalerba, D., Marimpietri, D., Raggi, F., Castellano, A., ... Varesio, L. (2019). Exosomal microRNAs from longitudinal liquid biopsies for the prediction of response to induction chemotherapy in high-risk neuroblastoma patients: A proof of concept SIOPEN study. *Cancers (Basel), 11*(10).

Nagao, A., Mituyama, T., Huang, H., Chen, D., Siomi, M. C., & Siomi, H. (2010). Biogenesis pathways of piRNAs loaded onto AGO3 in the drosophila testis. *RNA (New York, N.Y.), 16*(12), 2503−2515.

Nolan, J. C., Salvucci, M., Carberry, S., Barat, A., Segura, M. F., Fenn, J., ... Piskareva, O. (2020). A context-dependent role for MiR-124-3p on cell phenotype, viability and chemosensitivity in neuroblastoma in vitro. *Frontiers in Cell and Developmental Biology, 8*, 559553.

Novak, E. M., Gimenez, T. M., Neves, N. H., Vince, C. C., Krepischi, A. C. V., Lapa, R. M., ... Filho, V. O. (2020). MEG3 and MEG8 aberrant methylation in an infant with neuroblastoma. *Pediatric Blood & Cancer, 67*(9), e28328.

Ooi, C. Y., Carter, D. R., Liu, B., Mayoh, C., Beckers, A., Lalwani, A., ... Marshall, G. M. (2018). Network modeling of microRNA-mRNA interactions in neuroblastoma tumorigenesis identifies miR-204 as a direct inhibitor of MYCN. *Cancer Research*, 78(12), 3122–3134.

Ota, S., Ishitani, S., Shimizu, N., Matsumoto, K., Itoh, M., & Ishitani, T. (2012). NLK positively regulates Wnt/beta-catenin signalling by phosphorylating LEF1 in neural progenitor cells. *The EMBO Journal*, 31(8), 1904–1915.

Ozata, D. M., Gainetdinov, I., Zoch, A., O'Carroll, D., & Zamore, P. D. (2019). PIWI-interacting RNAs: Small RNAs with big functions. *Nature Reviews Genetics*, 20(2), 89–108.

Pandey, G. K., Mitra, S., Subhash, S., Hertwig, F., Kanduri, M., Mishra, K., ... Kanduri, C. (2014). The risk-associated long noncoding RNA NBAT-1 controls neuroblastoma progression by regulating cell proliferation and neuronal differentiation. *Cancer Cell*, 26(5), 722–737.

Park, J. R., Eggert, A., & Caron, H. (2010). Neuroblastoma: Biology, prognosis, and treatment. *Hematology/Oncology Clinics of North America*, 24(1), 65–86.

Pathania, A. S., & Challagundla, K. B. (2021). Exosomal long non-coding RNAs: Emerging players in the tumor microenvironment. *Molecular Therapy – Nucleic Acids*, 23, 1371–1383.

Pavlaki, I., Alammari, F., Sun, B., Clark, N., Sirey, T., Lee, S., ... Vance, K. W. (2018). The long non-coding RNA Paupar promotes KAP1-dependent chromatin changes and regulates olfactory bulb neurogenesis. *The EMBO Journal*, 37(10).

Pedram Fatemi, R., Salah-Uddin, S., Modarresi, F., Khoury, N., Wahlestedt, C., & Faghihi, M. A. (2015). Screening for small-molecule modulators of long noncoding RNA-protein interactions using AlphaScreen. *Journal of Biomolecular Screening: The Official Journal of the Society for Biomolecular Screening*, 20(9), 1132–1141.

Peng, Y., & Croce, C. M. (2016). The role of microRNAs in human cancer. *Signal Transduction and Targeted Therapy*, 1, 15004.

Peng, Z., Wang, J., Shan, B., Li, B., Peng, W., Dong, Y., ... Duan, C. (2018). The long noncoding RNA LINC00312 induces lung adenocarcinoma migration and vasculogenic mimicry through directly binding YBX1. *Molecular Cancer*, 17(1), 167.

Powers, J. T., Tsanov, K. M., Pearson, D. S., Roels, F., Spina, C. S., Ebright, R., ... Daley, G. Q. (2016). Multiple mechanisms disrupt the let-7 microRNA family in neuroblastoma. *Nature*, 535(7611), 246–251.

Prajapati, B., Fatma, M., Fatima, M., Khan, M. T., Sinha, S., & Seth, P. K. (2019). Identification of lncRNAs associated with neuroblastoma in cross-sectional databases: Potential biomarkers. *Frontiers in Molecular Neuroscience*, 12, 293.

Radion, E., Morgunova, V., Ryazansky, S., Akulenko, N., Lavrov, S., Abramov, Y., ... Kalmykova, A. (2018). Key role of piRNAs in telomeric chromatin maintenance and telomere nuclear positioning in Drosophila germline. *Epigenetics Chromatin*, 11(1), 40.

Ragan, C., Goodall, G. J., Shirokikh, N. E., & Preiss, T. (2019). Insights into the biogenesis and potential functions of exonic circular RNA. *Scientific Reports*, 9(1), 2048.

Rojas-Rios, P., & Simonelig, M. (2018). piRNAs and PIWI proteins: Regulators of gene expression in development and stem cells. *Development (Cambridge, England)*, 145(17).

Rombaut, D., Chiu, H. S., Decaesteker, B., Everaert, C., Yigit, N., Peltier, A., ... Mestdagh, P. (2019). Integrative analysis identifies lincRNAs up- and downstream of neuroblastoma driver genes. *Scientific Reports*, 9(1), 5685.

Roy, J., Das, B., Jain, N., & Mallick, B. (2020). PIWI-interacting RNA 39980 promotes tumor progression and reduces drug sensitivity in neuroblastoma cells. *Journal of Cellular Physiology*, 235(3), 2286–2299.

Roy, J., & Mallick, B. (2018). Investigating piwi-interacting RNA regulome in human neuroblastoma. *Genes, Chromosomes & Cancer*, 57(7), 339–349.

Russell, H. V., Hicks, J., Okcu, M. F., & Nuchtern, J. G. (2004). CXCR4 expression in neuroblastoma primary tumors is associated with clinical presentation of bone and bone marrow metastases. *Journal of Pediatric Surgery*, 39(10), 1506–1511.

Ryan, J., Tivnan, A., Fay, J., Bryan, K., Meehan, M., Creevey, L., ... Stallings, R. L. (2012). MicroRNA-204 increases sensitivity of neuroblastoma cells to cisplatin and is associated with a favourable clinical outcome. *British Journal of Cancer*, 107(6), 967–976.

Sabbah, N. A., Abdalla, W. M., Mawla, W. A., AbdAlMonem, N., Gharib, A. F., Abdul-Saboor, A., ... Raafat, N. (2021). piRNA-823 is a unique potential diagnostic non-invasive biomarker in colorectal cancer patients. *Genes (Basel)*, 12(4).

Sahu, D., Ho, S. Y., Juan, H. F., & Huang, H. C. (2018). High-risk, expression-based prognostic long noncoding RNA signature in neuroblastoma. *JNCI Cancer Spectrum, 2*(2), pky015.

Saito, K., Sakaguchi, Y., Suzuki, T., Suzuki, T., Siomi, H., & Siomi, M. C. (2007). Pimet, the Drosophila homolog of HEN1, mediates 2′-O-methylation of Piwi- interacting RNAs at their 3′ ends. *Genes & Development, 21*(13), 1603−1608.

Salvatori, B., Biscarini, S., & Morlando, M. (2020). Non-coding RNAs in nervous system development and disease. *Frontiers in Cell and Developmental Biology, 8*, 273.

Sanger, H. L., Klotz, G., Riesner, D., Gross, H. J., & Kleinschmidt, A. K. (1976). Viroids are single-stranded covalently closed circular RNA molecules existing as highly base-paired rod-like structures. *Proceedings of the National Academy of Sciences of the United States of America, 73*(11), 3852−3856.

Schulte, J. H., Schowe, B., Mestdagh, P., Kaderali, L., Kalaghatgi, P., Schlierf, S., ... Schramm, A. (2010). Accurate prediction of neuroblastoma outcome based on miRNA expression profiles. *International Journal of Cancer. Journal International du Cancer, 127*(10), 2374−2385.

Schulte, J. H., Horn, S., Otto, T., Samans, B., Heukamp, L. C., Eilers, U. C., ... Berwanger, B. (2008). MYCN regulates oncogenic microRNAs in neuroblastoma. *International Journal of Cancer. Journal International du Cancer, 122*(3), 699−704.

Shi, S., Yang, Z. Z., Liu, S., Yang, F., & Lin, H. (2020). PIWIL1 promotes gastric cancer via a piRNA-independent mechanism. *Proceedings of the National Academy of Sciences of the United States of America, 117*(36), 22390−22401.

Shi, Y., Parag, S., Patel, R., Lui, A., Murr, M., Cai, J., & Patel, N. A. (2019). Stabilization of lncRNA GAS5 by a small molecule and its implications in diabetic adipocytes. *Cell Chemical Biology, 26*(3), 319−330, e6.

Sohn, E. J., Jo, Y. R., & Park, H. T. (2019). Downregulation MIWI-piRNA regulates the migration of Schwann cells in peripheral nerve injury. *Biochemical and Biophysical Research Communications, 519*(3), 605−612.

Soriano, A., Paris-Coderch, L., Jubierre, L., Martinez, A., Zhou, X., Piskareva, O., ... Segura, M. F. (2016). MicroRNA-497 impairs the growth of chemoresistant neuroblastoma cells by targeting cell cycle, survival and vascular permeability genes. *Oncotarget, 7*(8), 9271−9287.

Srivastava, A., Goldberger, H., Dimtchev, A., Marian, C., Soldin, O., Li, X., ... Kumar, D. (2014). Circulatory miR-628-5p is downregulated in prostate cancer patients. *Tumour Biology: The Journal of the International Society for Oncodevelopmental Biology and Medicine, 35*(5), 4867−4873.

Suenkel, C., Cavalli, D., Massalini, S., Calegari, F., & Rajewsky, N. (2020). A highly conserved circular RNA is required to keep neural cells in a progenitor state in the mammalian brain. *Cell Reports, 30*(7), 2170−2179, e5.

Sun, N. X., Ye, C., Zhao, Q., Zhang, Q., Xu, C., Wang, S. B., ... Li, W. (2014). Long noncoding RNA-EBIC promotes tumor cell invasion by binding to EZH2 and repressing E-cadherin in cervical cancer. *PLoS One, 9*(7), e100340.

Takwi, A. A., Wang, Y. M., Wu, J., Michaelis, M., Cinatl, J., & Chen, T. (2014). miR-137 regulates the constitutive androstane receptor and modulates doxorubicin sensitivity in parental and doxorubicin-resistant neuroblastoma cells. *Oncogene, 33*(28), 3717−3729.

Thakore, P. I., D'Ippolito, A. M., Song, L., Safi, A., Shivakumar, N. K., Kabadi, A. M., ... Gersbach, C. A. (2015). Highly specific epigenome editing by CRISPR-Cas9 repressors for silencing of distal regulatory elements. *Nature Methods, 12*(12), 1143−1149.

Thomson, T., & Lin, H. (2009). The biogenesis and function of PIWI proteins and piRNAs: Progress and prospect. *Annual Review of Cell and Developmental Biology, 25*, 355−376.

Tripathi, V., Ellis, J. D., Shen, Z., Song, D. Y., Pan, Q., Watt, A. T., ... Prasanth, K. V. (2010). The nuclear-retained noncoding RNA MALAT1 regulates alternative splicing by modulating SR splicing factor phosphorylation. *Molecular Cell, 39*(6), 925−938.

Vychytilova-Faltejskova, P., Stitkovcova, K., Radova, L., Sachlova, M., Kosarova, Z., Slaba, K., ... Slaby, O. (2018). Circulating PIWI-interacting RNAs piR-5937 and piR-28876 are promising diagnostic biomarkers of colon cancer. *Cancer Epidemiology, Biomarkers & Prevention: A Publication of the American Association for Cancer Research, Cosponsored by the American Society of Preventive Oncology, 27*(9), 1019−1028.

Wang, B., Xu, L., Zhang, J., Cheng, X., Xu, Q., Wang, J., & Mao, F. (2020). LncRNA NORAD accelerates the progression and doxorubicin resistance of neuroblastoma through up-regulating HDAC8 via sponging miR-144-3p. *Biomedicine & Pharmacotherapy = Biomedecine & Pharmacotherapie, 129*, 110268.

Wang, K., Diskin, S. J., Zhang, H., Attiyeh, E. F., Winter, C., Hou, C., ... Maris, J. M. (2011). Integrative genomics identifies LMO1 as a neuroblastoma oncogene. *Nature, 469*(7329), 216−220.

Wang, S. Y., Wang, X., & Zhang, C. Y. (2020). LncRNA SNHG7 enhances chemoresistance in neuroblastoma through cisplatin-induced autophagy by regulating miR-329-3p/MYO10 axis. *European Review for Medical and Pharmacological Sciences, 24*(7), 3805–3817.

Wang, X., Li, J., Xu, X., Zheng, J., & Li, Q. (2018). miR-129 inhibits tumor growth and potentiates chemosensitivity of neuroblastoma by targeting MYO10. *Biomedicine & Pharmacotherapy = Biomedecine & Pharmacotherapie, 103*, 1312–1318.

Weng, W., Li, H., & Goel, A. (2019). Piwi-interacting RNAs (piRNAs) and cancer: Emerging biological concepts and potential clinical implications. *Biochimica et Biophysica Acta — Reviews on Cancer, 1871*(1), 160–169.

Wilson, R. C., Tambe, A., Kidwell, M. A., Noland, C. L., Schneider, C. P., & Doudna, J. A. (2015). Dicer-TRBP complex formation ensures accurate mammalian microRNA biogenesis. *Molecular Cell, 57*(3), 397–407.

Wilusz, J. E., Sunwoo, H., & Spector, D. L. (2009). Long noncoding RNAs: Functional surprises from the RNA world. *Genes & Development, 23*(13), 1494–1504.

Xu, L., Huan, L., Guo, T., Wu, Y., Liu, Y., Wang, Q., ... He, X. (2020). LncRNA SNHG11 facilitates tumor metastasis by interacting with and stabilizing HIF-1alpha. *Oncogene, 39*(46), 7005–7018.

Xu, Y., Chen, X., Lin, L., Chen, H., Yu, S., & Li, D. (2017). MicroRNA-149 is associated with clinical outcome in human neuroblastoma and modulates cancer cell proliferation through Rap1 independent of MYCN amplification. *Biochimie, 139*, 1–8.

Xu, Z., Sun, Y., Wang, D., Sun, H., & Liu, X. (2020). SNHG16 promotes tumorigenesis and cisplatin resistance by regulating miR-338-3p/PLK4 pathway in neuroblastoma cells. *Cancer Cell International, 20*, 236.

Yamanaka, S., Siomi, M. C., & Siomi, H. (2014). piRNA clusters and open chromatin structure. *Mobile DNA, 5*, 22.

Yan, H., & Bu, P. (2021). Non-coding RNA in cancer. *Essays in Biochemistry*, EBC20200032.

Yang, B., Ye, X., Wang, J., & Xia, S. (2020). Long noncoding RNA nuclear-enriched abundant transcript 1 regulates proliferation and apoptosis of neuroblastoma cells treated by cisplatin by targeting miR-326 through Janus kinase/signal transducer and activator of transcription 3 pathway. *NeuroReport, 31*(17), 1189–1198.

Yang, H., Zhang, X., Zhao, Y., Sun, G., Zhang, J., Gao, Y., ... Zhu, H. (2020). Downregulation of lncRNA XIST represses tumor growth and boosts radiosensitivity of neuroblastoma via modulation of the miR-375/L1CAM axis. *Neurochemical Research, 45*(11), 2679–2690.

Yang, J., Yu, L., Yan, J., Xiao, Y., Li, W., Xiao, J., ... Yu, X. (2020). Circular RNA DGKB promotes the progression of neuroblastoma by targeting miR-873/GLI1 axis. *Frontiers in Oncology, 10*, 1104.

Yang, Y., Pan, H., Chen, J., Zhang, Z., Liang, M., & Feng, X. (2021). CircKIF2A contributes to cell proliferation, migration, invasion and glycolysis in human neuroblastoma by regulating miR-129-5p/PLK4 axis. *Molecular and Cellular Biochemistry*.

Yao, R. W., Wang, Y., & Chen, L. L. (2019). Cellular functions of long noncoding RNAs. *Nature Cell Biology, 21*(5), 542–551.

Yarmishyn, A. A., Batagov, A. O., Tan, J. Z., Sundaram, G. M., Sampath, P., Kuznetsov, V. A., & Kurochkin, I. V. (2014). HOXD-AS1 is a novel lncRNA encoded in HOXD cluster and a marker of neuroblastoma progression revealed via integrative analysis of noncoding transcriptome. *BMC Genomics, 15*(Suppl 9), S7.

Ye, M., Lu, H., Tang, W., Jing, T., Chen, S., Wei, M., ... Dong, K. (2020). Downregulation of MEG3 promotes neuroblastoma development through FOXO1-mediated autophagy and mTOR-mediated epithelial-mesenchymal transition. *International Journal of Biological Sciences, 16*(15), 3050–3061.

Yi, R., Qin, Y., Macara, I. G., & Cullen, B. R. (2003). Exportin-5 mediates the nuclear export of pre-microRNAs and short hairpin RNAs. *Genes & Development, 17*(24), 3011–3016.

Yu, Y., Chen, F., Jin, Y., Yang, Y., Wang, S., Zhang, J., ... Ni, X. (2020). Downregulated NORAD in neuroblastoma promotes cell proliferation via chromosomal instability and predicts poor prognosis. *Acta Biochimica Polonica, 67*(4), 595–603.

Zeka, F., Decock, A., Van Goethem, A., Vanderheyden, K., Demuynck, F., Lammens, T., ... Vandesompele, J. (2018). Circulating microRNA biomarkers for metastatic disease in neuroblastoma patients. *JCI Insight, 3*(23).

Zhang, L., Zhou, H., Li, J., Wang, X., Zhang, X., Shi, T., & Feng, G. (2020). Comprehensive characterization of circular RNAs in neuroblastoma cell lines. *Technology in Cancer Research & Treatment, 19*, 1533033820957622.

Zhang, M., Zhao, K., Xu, X., Yang, Y., Yan, S., Wei, P., ... Zhang, N. (2018). A peptide encoded by circular form of LINC-PINT suppresses oncogenic transcriptional elongation in glioblastoma. *Nature Communications, 9*(1), 4475.

Zhang, Q., Chao, T. C., Patil, V. S., Qin, Y., Tiwari, S. K., Chiou, J., ... Rana, T. M. (2019). The long noncoding RNA ROCKI regulates inflammatory gene expression. *The EMBO Journal, 38*(8).

Zhang, X., Wang, S., Wang, H., Cao, J., Huang, X., Chen, Z., ... Xu, Z. (2019). Circular RNA circNRIP1 acts as a microRNA-149-5p sponge to promote gastric cancer progression via the AKT1/mTOR pathway. *Molecular Cancer*, *18*(1), 20.

Zhang, X., Zhang, J., Liu, Q., Zhao, Y., Zhang, W., & Yang, H. (2020). Circ-CUX1 accelerates the progression of neuroblastoma via miR-16-5p/DMRT2 axis. *Neurochemical Research*, *45*(12), 2840−2855.

Zhang, X., Zhou, Y., Chen, S., Li, W., Chen, W., & Gu, W. (2019). LncRNA MACC1-AS1 sponges multiple miRNAs and RNA-binding protein PTBP1. *Oncogenesis*, *8*(12), 73.

Zhao, J., Du, P., Cui, P., Qin, Y., Hu, C., Wu, J., ... Huang, G. (2018). LncRNA PVT1 promotes angiogenesis via activating the STAT3/VEGFA axis in gastric cancer. *Oncogene*, *37*(30), 4094−4109.

Zhao, X., Li, D., Yang, F., Lian, H., Wang, J., Wang, X., ... Tong, Q. (2020). Long noncoding RNA NHEG1 drives beta-catenin transactivation and neuroblastoma progression through interacting with DDX5. *Molecular Therapy: The Journal of the American Society of Gene Therapy*, *28*(3), 946−962.

Zhao, Y., Liu, H., Zhang, Q., & Zhang, Y. (2020). The functions of long non-coding RNAs in neural stem cell proliferation and differentiation. *Cell & Bioscience*, *10*, 74.

Zhao, Z., Shelton, S. D., Oviedo, A., Baker, A. L., Bryant, C. P., Omidvarnia, S., & Du, L. (2020). The PLAGL2/MYCN/miR-506-3p interplay regulates neuroblastoma cell fate and associates with neuroblastoma progression. *Journal of Experimental & Clinical Cancer Research: CR*, *39*(1), 41.

Zhou, M. J., Doral, M. Y., DuBois, S. G., Villablanca, J. G., Yanik, G. A., & Matthay, K. K. (2015). Different outcomes for relapsed vs refractory neuroblastoma after therapy with (131)I-metaiodobenzylguanidine ((131)I-MIBG). *European Journal of Cancer*, *51*(16), 2465−2472.

Zhu, Y., Wang, Q., Xia, Y., Xiong, X., Weng, S., Ni, H., ... Lin, Y. (2020). Evaluation of MiR-1908-3p as a novel serum biomarker for breast cancer and analysis its oncogenic function and target genes. *BMC Cancer*, *20*(1), 644.

14

Potential clinical application of lncRNAs in pediatric cancer

Ravindresh Chhabra[1], Priyasha Neyol[1], Sonali Bazala[2], Ipsa Singh[2], Masang Murmu[2], Uttam Sharma[2], Tushar Singh Barwal[2] and Aklank Jain[2]

[1]Department of Biochemistry, Central University of Punjab, Bathinda, India
[2]Department of Zoology, Central University of Punjab, Bathinda, India

14.1 Introduction

14.1.1 Pediatric cancer

Pediatric cancer, also referred to as childhood cancer, is often used to describe the cancers occurring in individuals below 15 years of age. In many cancer registries across the world, this age threshold has been kept at 19 years. Although pediatric cancers are rare, they are globally the leading cause of death by disease in children. A population-based registry study on the incidence of childhood cancer supported by International Agency for Research on Cancer revealed that around 200,000 children between 0 and 15 years were diagnosed with cancer per year (Steliarova-Foucher et al., 2017). The common childhood cancers include leukemia, lymphoma, neuroblastoma (NB), osteosarcoma, retinoblastoma, rhabdomyosarcoma, and Wilms tumor out of which NB, retinoblastoma, rhabdomyosarcoma, and Wilms tumor occur exclusively in children (Fig. 14.1). Even, the other childhood cancers are different from the adult cancers as these are usually not caused by environmental factors or lifestyle. Also, these cancers do not arise because of hereditary genetic mutations. There are, however, few reports which suggest that certain genetic disorders like Down syndrome may increase the risk of childhood cancer (Goldsby et al., 2018; Mateos, Barbaric, Byatt, Sutton, & Marshall, 2015; Xavier, Ge, & Taub, 2009).

The young growing children often respond well to the intense chemotherapy and radiotherapy considering that at a young age they are usually not afflicted with other age-related pathological conditions. But the major concern is that they usually have more years

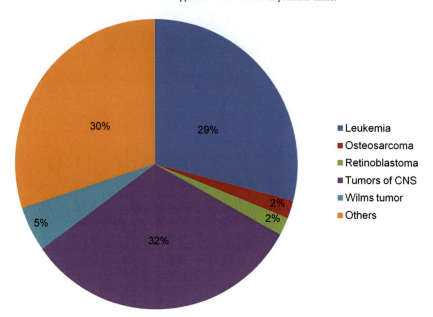

FIGURE 14.1 The incidence of cancers in the children below 15 years of age. The percentage corresponds to their respective proportion among total childhood cancers. The tumors of the CNS include neuroblastoma that alone accounts for 6% of the total pediatric cancers. The other categories include lymphomas, rhabdomyosarcoma, germ cell tumors, and liver tumors. The data have been adapted from the webpage of American Society of Clinical Oncology (https://www.cancer.net/cancer-types/childhood-cancer/introduction). *CNS*, Central nervous system.

than the adults to develop the long-term side effects of the therapeutic regimen. Some of these side effects include infertility, stunted growth, cardiovascular disorders, and psychological issues. There is also the concern that the kids who have received chemotherapy and radiotherapy are more likely to have a cancer relapse and, therefore require continuous follow-up throughout their life.

The treatment methods of pediatric cancers include surgery, radiotherapy, chemotherapy, immunotherapy, and stem cell therapy. Immunotherapy is a relatively new technique that targets cancerous cells and involves the use of monoclonal antibodies, cancer vaccines, and chimeric antigen receptor T-cell therapy (CAR-T therapy). The first CAR-T therapy approved for pediatric cancer by US FDA was Kymriah for the treatment of the acute lymphoblastic leukemia in 2017 (Prasad, 2018). In this treatment the patient's T cells are genetically modified ex vivo which are then infused back to kill cancer cells expressing CD19. For the patients who do not respond to the CD19 CAR-T therapy, CD22 CAR-T therapy was approved by the FDA in 2019 (Fry et al., 2018).

14.1.2 Long noncoding RNA

The existence of the noncoding RNAs in human systems had been well established but it was mostly considered the junk. It was only in the late 1990s when scientists discovered the importance of the noncoding RNAs in the regulation of gene expression. Since then, there has been a tremendous growth in the scientific research involving noncoding RNAs. The noncoding RNAs have been further classified into small noncoding RNAs (<200 bases) and long noncoding RNAs (lncRNAs) (>200 bases). The size of lncRNAs can vary from a few hundred bases to tens of kilobases. MicroRNAs (miRNAs), small interfering RNAs (siRNAs), piwi-interacting RNA (piRNAs), and small nucleolar RNAs (snoRNAs) comprise the class of small noncoding RNAs. The miRNAs, siRNAs, and piRNAs regulate the gene expression by targeting messenger RNAs (mRNAs), whereas the snoRNAs are involved in the posttranscriptional modification of different cellular RNAs. lncRNAs, on the other hand, have diverse functional attributes (Sharma et al., 2021; Sharma, Barwal, Acharya, Singh et al., 2020; Sharma, Barwal, Acharya, Tamang et al., 2020; Sharma, Barwal, Malhotra et al., 2020; Tamang et al., 2019; Barwal, Sharma, Vasquez, Prakash, & Jain, 2020). The lncRNAs have also been classified on the basis of their genomic loci and their function (Khandelwal, Bacolla, Vasquez, & Jain, 2015). The broad functional classes of lncRNAs include the *cis*-acting lncRNAs which regulates the expression of genes in its immediate vicinity and the *trans*-acting lncRNAs that regulate the expression of unrelated genes. The more complicated functional classification of lncRNAs has been previously reviewed (Kirk et al., 2018; Kopp & Mendell, 2018). The categorization of lncRNAs on the basis of their genomic loci include intronic lncRNAs, intergenic lncRNAs, sense lncRNAs, antisense lncRNAs, promoter lncRNAs, pseudogene lncRNAs, and enhancer lncRNAs (Chhabra, 2017; Khandelwal et al., 2015).

LncRNAs are known to be present in different cellular compartments, which also influences their function (Flippot et al., 2016). They can be localized within the nucleus where they may effect histone modifications or DNA methylations [e.g., Metastasis Associated Lung Adenocarcinoma Transcript 1 (*MALAT1*), Nuclear Paraspeckle Assembly Transcript 1 (*NEAT1*)] or they can be cytoplasmic where they may regulate the gene expression by binding to mRNAs, miRNAs, or proteins [e.g., *H19*, Small Nucleolar RNA Host Gene 1 (*SNHG1*)] (Khandelwal et al., 2015). In the case of pediatric cancers, quite often, the lncRNAs are shown to act as ceRNAs (competing endogenous RNAs) which binds a specific miRNA, thereby blocking the miRNA binding to its target mRNA, resulting in the upregulation of target mRNA (Fig. 14.2). In some instances the lncRNAs can be present in the nucleus as well as in the cytoplasm [e.g., HOX Transcript Antisense RNA (*HOTAIR*)]. Like the protein coding genes, the lncRNAs are majorly transcribed by Pol II complex. The total number of lncRNAs transcripts reported so far are 127,802 (Volders et al., 2019) but hardly 1% of them have been functionally characterized. The diverse function of lncRNAs obviates their involvement in disease pathologies.

14.2 Experimental and bioinformatics tools for studying lncRNAs

The huge number of lncRNAs in the human genome and the excessive use of deep sequencing for lncRNAs warrant the use of computational analysis to understand the functional relevance of lncRNAs. In this section, we highlight the key databases and bioinformatics tools that are presently used for the study of lncRNAs.

The discovery of novel lncRNA sequences with the use of high throughput sequencing technologies is often accompanied by the challenges of characterizing it. One of the first

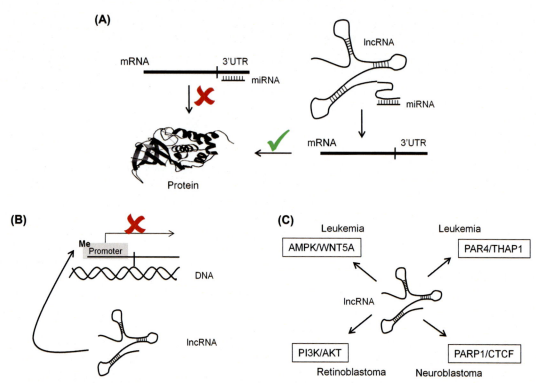

FIGURE 14.2 The varied functions of lncRNAs in pediatric cancers: (A) majority of the lncRNAs implicated in childhood cancers competitively bind the miRNA. The bound miRNA is unable to attach to the 3' UTR of mRNA, thus enabling the protein expression. (B) lncRNAs can also regulate gene transcription by either directly binding to the promoter region or by altering the methylation status of the promoter for instance lncRNA 91H in osteosarcoma. (C) lncRNAs in pediatric cancer have also been demonstrated to alter signaling pathways related to cell cycle, migration, invasion, and apoptosis. A few of these altered gene axes have been represented in the figure. *lncRNA*, Long noncoding RNA; *miRNA*, microRNAs.

objectives in characterizing a novel lncRNA is to confirm if its sequence is noncoding. Experimentally, it would involve the cloning of lncRNA sequence in an expression vector followed by in vitro transcription and translation reaction to detect the formation of any peptide sequence. Several computation tools have also been developed which can predict whether the sequence codes for any peptide or not. Some of these tools include Coding–Noncoding Index (Sun et al., 2013; Zhou et al., 2016), Coding-Potential Assessment Tool (Wang et al., 2013), and coding potential calculator (Kang et al., 2017). The function of an lncRNA is mostly dependent on its subcellular localization. This may be accomplished experimentally by RNA Fluorescence in situ hybridization (FISH) or by subcellular fractionation followed by Real-Time Quantitative Reverse Transcription PCR (qRT-PCR). Several researchers opine that the sequence of lncRNA could also play a role in determining the cellular localization of the lncRNA and based on that, computational tools such as lncLocator and iLoc-lncRNA have been developed which predict the nuclear or cytoplasmic presence of a lncRNA (Cao, Pan, Yang, Huang, & Shen, 2018; Su et al.,

2018). The outcome of research on previously characterized lncRNAs has been compiled in multiple databases but the most frequently cited are LNCipedia (Volders et al., 2013; Volders et al., 2019) and NONCODE (Zhao, Yuan, & Chen, 2016). LNCipedia is a comprehensive database for human lncRNA sequence and annotation and contains 127,802 transcripts and 56,946 genes (Volders et al., 2019). NONCODE, on the other hand, contains information on lncRNAs from multiple species. It provides advanced annotations, including predicted function, disease association, and tissue-specific expression profiles from 16 human tissues and 8 cell lines. The experimentally proven lncRNA-disease associations and lncRNA interactions with different nucleic acids and proteins in humans have been compiled in LncRNADisease database (Chen et al., 2013). It also provides tools to predict the plausible association of novel lncRNAs with diseases. All these databases are web-based with an interactive user interface.

The discovery of lncRNAs for their biomarker and therapeutic potential usually begins with high throughput sequencing of patient samples. The differential expression of lncRNAs helps narrow down the search for discovering lncRNAs with biomarker or therapeutic potential. Since lncRNAs are found to be relatively stable in circulating body fluids, including blood, plasma, serum, urine, and even saliva, their exclusive presence in circulation could serve as diagnostic and prognostic marker (Barwal et al., 2020). The presence of lncRNA PCA3 in urine for prostate cancer was the first lncRNA-based biomarker test approved by FDA in 2012 (Cui et al., 2016). Interestingly, quite a few lncRNAs have been identified in pediatric cancer patient samples that could serve as both diagnostic and prognostic markers.

The development of an lncRNA-based therapeutic would not only require its functional characterization but also the elucidation of its interacting partners. The overexpression and knockdown studies using cloning and siRNAs or CRISPR-Cas9, respectively, can help determine the functions of lncRNA but it is generally not that straightforward. The overexpression studies need to consider the localization of the overexpressed lncRNA and the knockdown studies may not work for *cis*-acting lncRNAs as the siRNAs might silence their regulated gene too. It is rather interesting to note that lncRNAs can have mRNA, miRNA, DNA, chromatin, or proteins as their interacting partners. RNA immunoprecipitation and chromatin isolation by RNA purification assay followed by DNA isolation or proteomic analysis are two experimental techniques that can help delineate the interacting partners of lncRNAs.

14.3 LncRNAs in pediatric cancer

In the recent years lncRNAs have become a focus of intense research in pathophysiological conditions. The following section talks about functions and applications of lncRNAs in different pediatric cancers, some of which have also been summarized in Table 14.1.

14.3.1 Leukemia

Leukemia is by far the most common childhood cancer. There are two distinct types of leukemia—acute lymphocytic leukemia (ALL) and acute myelocytic leukemia (AML). As

TABLE 14.1 Role of long noncoding RNAs (lncRNAs) in different pediatric cancers.

LncRNA	Expression in cancer	Molecular mechanism	Function	Reference
Leukemia				
HOTAIR	Upregulated	Modulates c-kit expression by competitively binding miR-193a	HOTAIR knockdown inhibits cell growth and induces apoptosis	Xing et al. (2015)
HOTTIP	Upregulated	HOTTIP could bind with miR-608 and upregulates expression of DDA1	HOTTIP promotes cell cycle progression and proliferative ability of AML cells	Zhuang, Li, and Ma (2019)
LINC00909	Upregulated	LONC00909 acts as a sponge for miR-625	Knockdown of LINC00909 suppresses cell viability, migration, and invasion and promotes apoptosis	Ma, Wang, and Jiang (2020)
LINP1	Upregulated	LINP1 promotes the HNF4a-AMPK/WNT5A signaling pathway	Knockdown of LINP1 expression suppresses glucose uptake and AML cell maintenance	Shi et al. (2019)
MEG3	Downregulated	MEG3 blocks the function of miR-184	Overexpression of MEG3 inhibits cell proliferation and invasion in vitro	Li, Zi, Wang, and Li (2018)
MIAT	Upregulated	MIAT acts as a sponge to miR-495 (tumor suppressor) in AML cells	Knockdown of MIAT inhibits colony formation, proliferation and increases apoptosis of AML cells	Wang et al. (2019)
SBF2-AS1	Upregulated	SBF2-AS1 upregulates the expression of ZFP91 by sponging miR-188–5p	SBF2-AS1 inhibition reduces cell proliferative ability, induces apoptosis and arrest cells in G0/G1 phase	Tian et al. (2019)
SNHG3	Upregulated	SNHG3 modulates SRGN expression by competitively binding to miR-758–3p	Knockdown of SRGN or SNHG3 mediates cell apoptosis and inhibits proliferation	Peng, Zhang, and Xin (2020)
T-ALL-R-LncR1	Upregulated	T-ALL-R-LncR1 prevents the activation of caspase 3 by inhibiting Par-4/THAP1 protein complex	Knockdown of T-ALL-Lnc-R1 induces cells to undergo PAR4 mediated apoptosis	Zhang, Xu, and Lu (2014)
Neuroblastoma				
DUXAP8	Upregulated	DUXAP8 sponges miR-29, and NOL4L is regulated by DUXAP8/miR-29 axis in NB cells	DUXAP8 levels are positively correlated with NB tumor stages and negatively relate to patient survival. Knockdown of DUXAP8 inhibits proliferation, colony formation, cycle, and motility of NB cells	Nie, Li, Zhao, Wang, and Liu (2020)
FOXD3-AS1	Downregulated	FOXD3-AS1 interacts with PARP1 to inhibit the poly ADP-ribosyl ation and activation of CTCF, which results in derepressed expression of downstream tumor suppressor genes	It inhibits NB progression	Zhao et al. (2018)
HCP5	Upregulated	HCP5 competitively binds to miR186–5p, which targets MAP3K2	Downregulation of HCP5 expression inhibits NB cell proliferation	Zhu et al. (2020)
NBAT1	Downregulated	NBAT1 causes epigenetic silencing of tumor inducing genes	Loss of NBAT1 increases cellular proliferation and invasion	Pandey et al. (2014)
SNHG1	Upregulated	SNHG1 represses miR-338p which is known to target PLK4	Knockdown of SNHG1 reduces tumor volume and size	Zhang, Liu, Ma, Zeng, and Zhang (2019)

(Continued)

TABLE 14.1 (Continued)

LncRNA	Expression in cancer	Molecular mechanism	Function	Reference
SNHG4	Upregulated	SNHG4 acts as a ceRNA and sponges miR-377–3p	SNHG4 is negatively correlated to survival of NB patients. Knockdown of SNHG4 represses cell proliferation and induces apoptosis	Yang et al. (2020)
SNHG7	Upregulated	SNHG7 targets miR-323a-5p and miR-342–5p, which are known to inhibit the expression of CCND1	SNHG7 knockdown results in the repression of cell migration, invasion, and glycolysis	Jia et al. (2020)
XIST	Upregulated	XIST modulates L1CAM expression by competitively binding to miR-375	Its downregulation represses tumor growth and sensitizes the tumor to radiotherapy	Yang et al. (2020)
Osteosarcoma				
91H	Upregulated	Methylation of CDK4 promoter, in U2OS cells by 91H siRNA	Knockdown of 91H suppresses the cell proliferation, inhibits migration and invasion, and induces apoptosis	-
EBLNP3P	Upregulated	EBLNP acts as a ceRNA and modulates RAB 10 expression by sponging miR-224 5p	Knockdown of EBLN3P reduces cell proliferation and invasion	-
EPIC1	Downregulated	EPIC1 targets MEFD2 protein and inhibits its expression	EPIC1 prevents cancer progression	-
EWSAT1	Upregulated	EWSAT1 positively regulates another lncRNA MEG3 at the transcriptional level	Knockdown of EWSAT1 inhibits cell proliferation, invasion	-
FER1L4	Downregulated	Functions as ceRNA to miR-18a-5p and modulates the expression of PTEN	Upregulation of FER1L4 inhibits cell proliferation, colony formation, migration, and invasion	Fei et al. (2018)
MIR31HG	Upregulated	Derepresses target genes of MIR-361, including VEGF, FOXM1, and Twist, which leads to upregulated BCL2, CCND1	MIR361HG acts as an oncogene and promotes EMT	-
NR-136400	Downregulated	NR-136400 acts as ceRNA of miRNA-8081 and regulates the expression of TUSC5	It promotes EMT by inhibiting E-cadherin, elevating ZEB1, snail, and fibronectin	-
PVT1	Upregulated	PVT1 acts as molecular sponge to miR-95	Silencing of PVT1 inhibits migration and invasion, proliferation, induces cell cycle arrest, and promotes apoptosis	Zhou et al. (2016)
PVT1	Upregulated	PVT1 acts as molecular sponge to repress miRNA-497, which is responsible for HK2 enzyme function	Its upregulation indicates poor prognosis. Increases glucose uptake, lactate production, and HK2 expression	-
XIST	Upregulated	Regulate YAP expression (hippo signaling in cancer) by functioning as ceRNA and sponging miR-195–5p which is known to target YAP	XIST enhances cancer cell proliferation	-

(Continued)

TABLE 14.1 (Continued)

LncRNA	Expression in cancer	Molecular mechanism	Function	Reference
Retinoblastoma				
AFAP1-AS1	Upregulated	AFAP1-AS1 promotes cell cycle progression	Expression of AFAP1-AS1 is correlated with large tumor size, choroidal invasion, and optic nerve invasion	Hao, Mou, Zhang, Wang, and Yang (2018)
CANT1	Downregulated	CANT1 occupies the promoter of PI3Kγ and prevents binding with hSET1, thereby reducing PI3K/Akt signaling	CANT1 overexpression suppresses RB progression	Ni et al. (2020)
HOTAIR	Upregulated	Targets the cell cycle proteins phospho-RB1, RB1, and CCNE and causes loss of G1/S checkpoints	Knockdown of HOTAIR inhibits proliferation, invasion, causes G0/G1 arrest, and attenuation of notch signaling pathway	Dong et al. (2016)
LNC00202	Upregulated	LNC00202 acts by sponging miR-3619–5p which increases the expression of oncogene RIN1	Depletion of LNC00202 reduces cell proliferation, invasion, and migration	Yan, Su, Ma, Yu, and Chen (2019)
NEAT1	Upregulated	NEAT1 acts as ceRNA for miR204 and regulates CXCR4 expression	Downregulation of NEAT1 decreases proliferation, migration, and increases apoptosis	Zhong, Yang, Li, Li, and Li (2019)
Wilms tumor				
LINC00473	Upregulated	LINC00473 targets miR-195 which further regulates expression of IKKα	Knockdown of LINC00473 inhibits cell viability, induces Bcl-2-dependent apoptosis and G1/S arrest	Zhu et al. (2018)

per the World Health Organization, ALL could further be classified into B-cell ALL and T-cell ALL. An 80% of childhood leukemias are ALL and rest of them are AML. lncRNAs have been demonstrated to have important functions in childhood leukemias. The high throughput transcriptomic profiling of T-cell acute lymphoblastic leukemia cells detected the expression of 1200 lncRNAs. One of these, T-ALL-R-LncR1 was shown to prevent apoptosis by regulating Par-4/THAP1 protein complex in T-cell acute lymphoblastic leukemia. The silencing of T-ALL-R-LncR1 activates the proapoptotic caspase-3 and enhances the Smac protein in T-ALL cells. The expression of T-ALL-R-Lnc-R was also seen in 11 children out of 21 with T-ALL (Zhang et al., 2014), thus implying that T-ALL-R-Lnc-R could be a potential therapeutic target in T-ALL. In addition, NOTCH1 Associated LncRNA In T Cell Acute Lymphoblastic Leukemia 1 (*NALT*), *NEAT1*, and *MALAT1* are also found over expressed in ALL. *NALT* is responsible for promoting cell proliferation by activating notch signaling pathway and *NEAT1* and *MALAT1* act as a sponge for miR-335-p and, in turn, induce the expression of ABC transporter protein, ABCA3, which reduces the bioavailability of drugs and spawns drug resistance (Pouyanrad, Rahgozar, & Ghodousi, 2019).

In AML, multiple lncRNAs have been shown to be upregulated, including *HOTAIR*, HOXA Distal Transcript Antisense RNA (*HOTTIP*), ZNF407 antisense RNA 1 (*LINC00909*), LncRNA In Non-Homologous End Joining Pathway 1 (*LINP1*), *MALAT1*, Myocardial Infarction Associated Transcript (*MIAT*), Microvascular Invasion in HCC

(*MVIH*), SBF2 Antisense RNA 1 (*SBF2-AS1*), and *SNHG3*. Most of the upregulated lncRNAs in AML have independently been shown to control cellular proliferation, cell cycle progression, and apoptosis. *HOTAIR* acts as endogenous competitive RNA binds with miR-193a and modulates c-kit expression, as miR-193a inhibits cell proliferation and induces apoptosis by targeting c-kit in AML cells (Xing et al., 2015). *HOTTIP* also acts as a sponge to miR-608 which targets 1,5-dihydroxy-4,8-dinitro anthraquinone (DDA), inducer of proliferation (Zhuang et al., 2019). *LINC00909* controls Wnt/β-catenin signaling and cell cycle proteins like cyclins (Ma et al., 2020). *MIAT* functions as a sponge to miR-495, a tumor suppressor in AML cells (Tian et al., 2019). LINP1 inhibits glucose metabolism and AML cell maintenance by suppressing HNF4α expression (Shi et al., 2019). *SNHG3* modulates SRGN expression by competitively binding with miR-758–3p and promotes growth in AML (Peng et al., 2020). *SBF2-AS1* upregulates the expression level of ZFP91 by sponging miR-188–5p. The inhibition of *SBF2-AS1* is demonstrated to reduce the tumor growth in vivo.

Among the aforementioned lncRNAs, *MVIH* and *MALAT1* are both associated with the severity of AML (Huang et al., 2017; Jiang, Yu, & Luo, 2020). LINP1 can also serve as prognostic marker as it shows elevated expression at the time of AML diagnosis and shows reduced expression after complete remission (Shi et al., 2019).

In contrast to the other lncRNAs implicated in AML which promote the tumor growth, the expression of Maternally Expressed 3 (*MEG3*) on the other hand is reduced in AML (Li et al., 2018). In normal physiological conditions, *MEG3* acts as a tumor suppressor lncRNA by promoting the binding of p53 to its target. Its downregulation promotes tumor growth.

14.3.2 Neuroblastoma

NB is one of the most common pediatric malignancy diagnosed in the first year of life, and median age of patient at diagnosis is around 18 months. It is a neuroendocrine tumor that arises in developing sympathetic nervous system. As NB is a complicated disease exhibiting genetic and epigenetic predisposition, it is crucial to identify novel molecular mechanisms for effective prognosis and therapeutics. In recent years, lncRNA have helped provide a deeper insight in understanding NB.

The members of SNHG lncRNA family, *SNHG1, SNHG4*, and *SNHG7*, are all highly expressed in NBs. *SNHG1* is associated with poor patient survival and is upregulated by MYCN Proto-Oncogene, BHLH Transcription Factor (*MYCN*) amplification (Sahu et al., 2016). It was shown to target PLK4 by binding to miR-338–3p, and its knockdown results in reducing the tumor volume and size (Zhang et al., 2014). *SNHG4* acts as a ceRNA to miR-377–3p, which is often found downregulated in NB cells. The knockdown of *SNHG4* decreases the proliferation and stimulates apoptosis in in vivo and in vitro conditions (Yang et al., 2020). *SNHG7* is directly related with poor prognosis of NB patients and was shown to regulate the NB cell migration, invasion, and glycolysis. It also acts as a ceRNA for miR-323a-5p and miR-342–5p, which, in turn, targets the CCND1 (Jia et al., 2020). The other upregulated lncRNAs in NB include DLX6 Antisense RNA 1 (*DLX6-AS1*), Double Homeobox A Pseudogene 8 (*DUXAP8*), HLA Complex P5 (*HCP5*), Double Homeobox A Pseudogene 9 (*LINC01296*), and X Inactive Specific Transcript (*XIST*). Most of the upregulated lncRNAs in NB share the functional attribute of acting as a ceRNA to miRNAs. *DUXAP8* sponges the miR-29 and prevents it from targeting NOL4L (Nie et al., 2020),

HCP5 binds miR-186—5p and enhances the expression of MAP3K2 (Zhu et al., 2020), and *XIST* acts as ceRNA to miR-375 and modulates the expression of its target, L1CAM (Yang, Guo, Zhang, & Li, 2020). Interestingly, three independent studies have shown that *DLX6-AS1* can act as ceRNA to multiple miRNAs, including miR-497—5p, miR-107, and miR-506—3p regulating expression of YAP1 (Li et al., 2018), BDNF (Zhang et al., 2019), and STAT2, respectively (Hu, Sun, Hu, & Zhang, 2020). With such diverse roles in NB, it is not surprising that *DLX6-AS1* was implicated in promoting the proliferation, migration, invasion ability, and Epithelial-to-Mesenchymal Transition (EMT) of NB cells. The elevated levels of *LINC01296* were shown to be associated with poor prognosis, advanced disease stage, large tumor size, and poor overall survival (Wang et al., 2013).

The downregulated lncRNAs in NB tissues and cell lines included *FOXD3-AS1* and *NBAT1*. *FOXD3-AS1* is reported to derepress the expression of tumor suppressor genes by interacting with poly(ADP-ribose) polymerase 1 leading to CCCTC-Binding Factor (CTCF) activation (Zhao et al., 2016). The downregulation of *FOXD3-AS1* promotes tumor progression in NB. NBAT1 expression can also be an independent prognostic marker in NB. Interestingly, its expression varies among various subtypes of the disease for instance; it is expressed at lower level in high-risk NB which implies that its expression can serve as a risk factor assessment among NB patients. NBAT1 controls tumor progression by targeting SOX9, Oncostatin M Receptor (OSMR), and Versican (VCAN) which are associated with loss of proliferative and invasive properties (Pandey et al., 2014).

14.3.3 Osteosarcoma

Osteosarcoma, also referred to as osteogenic carcinoma, is the cancer of the bone which commonly occurs in the growing bones of the leg and sometimes in the arms of children. In osteosarcoma, multiple lncRNAs have been shown to be differentially expressed. While H19 Imprinted Maternally Expressed Transcript (*91H*), CDKN2B Antisense RNA 1 (*CDKN2B-AS1*), Endogenous Bornavirus Like Nucleoprotein 3, Pseudogene (*EBLN3P*), Hepatocellular Carcinoma Up-Regulated Long Noncoding RNA (*HULC*), *MALAT1*, MIR31 Host Gene (*MIR31HG*), Pvt1 oncogene (*PVT1*), and *XIST* are upregulated, *EPIC1*, Fer-1 Like Family Member 4 (*FER1L4*), and *NR-136400* are downregulated (Table 14.1). Most of them, including, *CDKN2B-AS1*, *EBLN3P*, *FER1L4*, *MIR31HG*, *NR-136400*, *PVT1*, and *XIST*, act as ceRNA and regulate different miRNA/mRNA axes.

HULC, *MALAT1*, and *FER1L4* have all independently been shown to have correlation with the poor prognosis of the disease (Sun et al., 2013; Zhou et al., 2016). The enhanced expression of *MALAT1* (Liu et al., 2019) and *HULC* (Sun, Yang, Geng, Wang, & Zhang, 2015) and the reduced levels of *FER1L4* (Fei et al., 2018) were found associated with worse outcome in osteosarcoma patients. Knockdown of *HULC* in osteosarcoma cells caused a significant decrease in proliferation, invasion, and migration of cells (Sun et al., 2015). *FER1L4* acts as a ceRNA for miR18-a-5p and the low expression of *FER1L4* could be detrimental for the expression of the miRNA target Phosphatase And Tensin Homolog (PTEN), a tumor suppressor gene. The functions and molecular mechanism of most of the lncRNAs involved in osteosarcoma have been listed in Table 14.1.

14.3.4 Retinoblastoma

Retinoblastoma is the most common cancer of the eye in the childhood. It is often attributed to the mutations occurring in the RB Transcriptional Corepressor 1 (RB1) gene, which can be congenital or can occur sporadically. Retinoblastomas can lead to glaucoma which can result in the loss of vision. If left untreated, the cancer can metastasize to other parts of the body, including brain. The lncRNAs *AFAP1-AS1, HOTAIR, LINC00202,* and *NEAT1* are frequently found to be upregulated in retinoblastoma tissues when compared with paracancerous tissues (Dong et al., 2016; Hao et al., 2018; Wu, Zhou, Mi, & Shen, 2020; Zhong et al., 2019). All these lncRNAs are implicated in the proliferation and invasion of retinoblastoma. While the role of *HOTAIR* is attributed to its regulation of the Notch signaling pathway, the lncRNA *NEAT1*, and *LINC00202* promote the growth of human retinoblastoma by blocking the function of specific miRNAs. *NEAT1* binds miR-204 and upregulates its mRNA target, CXCR4, which is involved in cellular proliferation and metastasis. *LINC00202* has been shown to bind with miR-3619–5p and miR-204–5p in independent studies and upregulates the expression of RIN1 and 3-Hydroxy-3-Methylglutaryl-CoA Reductase (HMGCR), respectively (Wu et al., 2020; Yan et al., 2019). The knockdown of the aforementioned lncRNAs significantly inhibited the tumor progression in vitro and in vivo. Additionally, both lncRNA *LINC00202* and *AFAP1-AS1* are independently associated with poor prognosis in RB. In contrast the expression of lncRNA CASC15-New-Transcript 1 (*CANT1*) is significantly repressed in RB. In the normal physiological state, *CANT1* inhibits the expression of PI3Kγ by competitively binding to its promoter region and preventing the histone H3K4 trimethylation of the PI3Kγ promoter by histone methyltransferase hSET1 (Ni et al., 2020).

14.3.5 Wilms tumor

Wilms tumor, also referred to as nephroblastoma, is the most common cancer of kidneys in children and often occurs before the age of 5. It remains unclear to this day as to what causes Wilms tumor. The lncRNA *LINC00473* is often found elevated in Wilms tumor patient samples. Moreover, the higher expression of *LINC00473* correlates to more aggressive tumors. It was observed that the knockdown of LINC00473 induces the cell cycle arrest and apoptosis by enhancing the miR-195 levels which inhibits the expression of IkappaB Kinase (IKKα) an important protein for nuclear factor-kappa B (NFkB) survival pathway (Zhu et al., 2020). A comprehensive study to identify the lncRNAs acting as competing endogenous RNAs in Wilms tumor patient samples revealed 3 lncRNAs, 3 miRNAs, and 17 mRNAs which had significant effect on the survival rate of patients (Zhang et al., 2020). Out of these three lncRNAs, *MYCNOS* and *DELU2* were highly expressed in the late stages, whereas C8orf31 was highly expressed in the early stages. MYCNOS was further shown to be an independent prognostic marker for Wilms tumor with high statistical significance (Zhang et al., 2020). Another interesting lncRNA implicated in Wilms tumor is *WT1-AS* which originates from the intronic region of WT1 gene, a highly mutated gene in Wilms tumor. *WT1-AS* is observed to regulate the protein expression of WTI by binding to it. But there are contradictory reports on whether *WT1-AS* is a positive or negative regulator of WT1. It is also important to note that both WT1 and the

TABLE 14.2 List of long noncoding RNAs (lncRNAs) that show biomarker potential in pediatric cancers.

Cancer type	LncRNA	Expression in cancer	Reference
Acute myeloid leukemia	ANRIL	Upregulated	-
Acute myeloid leukemia	LINC00899	Upregulated	-
Acute myeloid leukemia	LINP1	Upregulated	Shi et al. (2019)
Acute myeloid leukemia	MVIH	Upregulated	Jiang et al. (2020)
Neuroblastoma	LINC01296	Upregulated	Wang et al. (2019)
Neuroblastoma	NBAT1	Downregulated	Pandey et al. (2014)
Osteosarcoma	HULC	Upregulated	-
Osteosarcoma	MALAT1	Upregulated	Liu et al. (2019)
Retinoblastoma	AFAP1-AS1	Upregulated	Hao et al. (2018)
Retinoblastoma	LINC00202	Upregulated	Yan et al. (2019)

lncRNA, *WT1-AS*, are associated with multiple cancer malignancies so the regulation of WT1 by *WT1-AS* may vary according to the cancer type (Zhang, Fan, Zhang, Jiang, & Qi, 2020).

14.4 Conclusion and perspectives

lncRNAs are a large heterogeneous group of noncoding RNAs with a significantly altered expression profile in different pathological conditions, including the different forms of childhood cancer. The recent studies on lncRNAs highlight their immense potential as diagnostic and prognostic markers in pediatric cancer as some of them exhibit an altered expression in the early stages of cancer that returns to normal physiological levels once the cancer goes into remission. The expression of quite a few lncRNAs (*LINP1, HULC, FER1L4, LINC01296,* and *SNHG1*) also shows a strong correlation with the patient survival outcome (Table 14.2). The presence of lncRNAs in circulatory bodily fluids, including blood, plasma, and urine, further emphasize their biomarker potential. However, most of the research has focused on the role of a single lncRNA and its target. The detailed integrative network map of deregulated lncRNAs and its molecular targets is missing in most of the childhood cancers. This information is imperative if the intent is to see the use of lncRNAs as therapeutic targets in the clinical setting.

Acknowledgment

We would like to acknowledge Central University of Punjab for providing Research Seed Money Grant (GP-25) to R.C and DST for the DST-FIST grant to the Department of Biochemistry, Central University of Punjab. A.J is thankful to DST, India for the Indo-Russia grant (INT/RUS/RFBR/P-311). U.S is thankful to DST for the DST-INSPIRE fellowship (IF180680).

References

Barwal, T. S., Sharma, U., Vasquez, K. M., Prakash, H., & Jain, A. (2020). A panel of circulating long non-coding RNAs as liquid biopsy biomarkers for breast and cervical cancers. *Biochimie, 176*, 62−70. Available from https://doi.org/10.1016/j.biochi.2020.06.012.

Cao, Z., Pan, X., Yang, Y., Huang, Y., & Shen, H.B. (2018). The lncLocator: A subcellular localization predictor for long non-coding RNAs based on a stacked ensemble classifier. In *Bioinformatics* (34, Issue 13, pp. 2185−2194). Oxford University Press. Available from https://doi.org/10.1093/bioinformatics/bty085.

Chen, G., Wang, Z., Wang, D., Qiu, C., Liu, M., Chen, X., ... Cui, Q. (2013). LncRNADisease: A database for long-non-coding RNA-associated diseases. *Nucleic Acids Research, 41*(1), D983−D986. Available from https://doi.org/10.1093/nar/gks1099.

Chhabra, R. (2017). The epigenetics of noncoding RNA. In *Handbook of epigenetics: The new molecular and medical genetics* (pp. 47−59). Elsevier. Available from https://doi.org/10.1016/B978-0-12-805388-1.00004-3.

Cui, Y., Cao, W., Li, Q., Shen, H., Liu, C., Deng, J., ... Shao, Q. (2016). Evaluation of prostate cancer antigen 3 for detecting prostate cancer: A systematic review and meta-analysis. *Scientific Reports, 6*. Available from https://doi.org/10.1038/srep25776.

Dong, C., Liu, S., Lv, Y., Zhang, C., Gao, H., Tan, L., & Wang, H. (2016). Long non-coding RNA HOTAIR regulates proliferation and invasion via activating Notch signalling pathway in retinoblastoma. *Journal of Biosciences, 41*(4), 677−687. Available from https://doi.org/10.1007/s12038-016-9636-7.

Fei, D., Zhang, X., Liu, J., Tan, L., Xing, J., Zhao, D., & Zhang, Y. (2018). Long noncoding RNA FER1L4 suppresses tumorigenesis by regulating the expression of PTEN Targeting miR-18a-5p in Osteosarcoma. *Cellular Physiology and Biochemistry, 51*(3), 1364−1375. Available from https://doi.org/10.1159/000495554.

Flippot, R., Malouf, G. G., Su, X., Mouawad, R., Spano, J. P., & Khayat, D. (2016). Cancer subtypes classification using long non-coding RNA. *Oncotarget, 7*(33), 54082−54093. Available from https://doi.org/10.18632/oncotarget.10213.

Fry, T. J., Shah, N. N., Orentas, R. J., Stetler-Stevenson, M., Yuan, C. M., Ramakrishna, S., ... Mackall, C. L. (2018). CD22-targeted CAR T cells induce remission in B-ALL that is naive or resistant to CD19-targeted CAR immunotherapy. *Nature Medicine, 24*(1), 20−28. Available from https://doi.org/10.1038/nm.4441.

Goldsby, R. E., Stratton, K. L., Raber, S., Ablin, A., Strong, L. C., Oeffinger, K., ... Leisenring, W. M. (2018). Long-term sequelae in survivors of childhood leukemia with Down syndrome: A childhood cancer survivor study report. *Cancer, 124*(3), 617−625. Available from https://doi.org/10.1002/cncr.31065.

Hao, F., Mou, Y., Zhang, L., Wang, S., & Yang, Y. (2018). LncRNA AFAP1-AS1 is a prognostic biomarker and serves as oncogenic role in retinoblastoma. *Bioscience Reports, 38*(3). Available from https://doi.org/10.1042/BSR20180384.

Hu, Y., Sun, H., Hu, J., & Zhang, X. (2020). Lncrna dlx6-as1 promotes the progression of neuroblastoma by activating stat2 via targeting mir-506-3p. *Cancer Management and Research, 12*, 7451−7463. Available from https://doi.org/10.2147/CMAR.S252521.

Huang, J. L., Liu, W., Tian, L. H., Chai, T. T., Liu, Y., Feng, Z., ... Shen, J. Z. (2017). Upregulation of long non-coding RNA MALAT-1 confers poor prognosis and influences cell proliferation and apoptosis in acute monocytic leukemia. *Oncology Reports, 38*(3), 1353−1362. Available from https://doi.org/10.3892/or.2017.5802.

Jia, J., Zhang, D., Zhang, J., Yang, L., Zhao, G., Yang, H., & Wang, J. (2020). Long non-coding RNA SNHG7 promotes neuroblastoma progression through sponging miR-323a-5p and miR-342-5p. *Biomedicine and Pharmacotherapy, 128*. Available from https://doi.org/10.1016/j.biopha.2020.110293.

Jiang, Z., Yu, Q., & Luo, X. (2020). Identification of long non-coding RNA MVIH as a prognostic marker and therapeutic target in acute myeloid leukemia. *Journal of Clinical Laboratory Analysis, 34*(4). Available from https://doi.org/10.1002/jcla.23113.

Kang, Y. J., Yang, D. C., Kong, L., Hou, M., Meng, Y. Q., Wei, L., & Gao, G. (2017). CPC2: A fast and accurate coding potential calculator based on sequence intrinsic features. *Nucleic Acids Research, 45*(1), W12−W16. Available from https://doi.org/10.1093/nar/gkx428.

Khandelwal, A., Bacolla, A., Vasquez, K. M., & Jain, A. (2015). Long non-coding RNA: A new paradigm for lung cancer. *Molecular Carcinogenesis, 54*(11), 1235−1251. Available from https://doi.org/10.1002/mc.22362.

Kirk, J. M., Kim, S. O., Inoue, K., Smola, M. J., Lee, D. M., Schertzer, M. D., ... Calabrese, J. M. (2018). Functional classification of long non-coding RNAs by k-mer content. *Nature Genetics, 50*(10), 1474−1482. Available from https://doi.org/10.1038/s41588-018-0207-8.

Kopp, F., & Mendell, J. T. (2018). Functional classification and experimental dissection of long noncoding RNAs. *Cell*, 172(3), 393–407. Available from https://doi.org/10.1016/j.cell.2018.01.011.

Li, J., Zi, Y., Wang, W., & Li, Y. (2018). Long noncoding RNA MEG3 inhibits cell proliferation and metastasis in chronic myeloid leukemia via targeting miR-184. *Oncology Research*, 26(2), 297–305. Available from https://doi.org/10.3727/096504017X14980882803151.

Liu, M., Yang, P., Mao, G., Deng, J., Peng, G., Ning, X., ... Sun, H. (2019). Long non-coding RNA MALAT1 as a valuable biomarker for prognosis in osteosarcoma: A systematic review and *meta*-analysis. *International Journal of Surgery*, 72, 206–213. Available from https://doi.org/10.1016/j.ijsu.2019.11.004.

Ma, L., Wang, Y. Y., & Jiang, P. (2020). LncRNA LINC00909 promotes cell proliferation and metastasis in pediatric acute myeloid leukemia via miR-625-mediated modulation of Wnt/β-catenin signaling. *Biochemical and Biophysical Research Communications*, 527(3), 654–661. Available from https://doi.org/10.1016/j.bbrc.2020.05.001.

Mateos, M. K., Barbaric, D., Byatt, S.-A., Sutton, R., & Marshall, G. M. (2015). Down syndrome and leukemia: Insights into leukemogenesis and translational targets. *Translational Pediatrics*, 4(2).

Ni, H., Chai, P., Yu, J., Xing, Y., Wang, S., Fan, J., ... Fan, X. (2020). LncRNA CANT1 suppresses retinoblastoma progression by repellinghistone methyltransferase in PI3Kγ promoter. *Cell Death and Disease*, 11(5). Available from https://doi.org/10.1038/s41419-020-2524-y.

Nie, L., Li, C., Zhao, T., Wang, Y., & Liu, J. (2020). LncRNA double homeobox A pseudogene 8 (DUXAP8) facilitates the progression of neuroblastoma and activates Wnt/β-catenin pathway via microRNA-29/nucleolar protein 4 like (NOL4L) axis. *Brain Research*, 1746. Available from https://doi.org/10.1016/j.brainres.2020.146947.

Pandey, G. K., Mitra, S., Subhash, S., Hertwig, F., Kanduri, M., Mishra, K., ... Kanduri, C. (2014). The risk-associated long noncoding RNA NBAT-1 controls neuroblastoma progression by regulating cell proliferation and neuronal differentiation. *Cancer Cell*, 26(5), 722–737. Available from https://doi.org/10.1016/j.ccell.2014.09.014.

Peng, L., Zhang, Y., & Xin, H. (2020). lncRNA SNHG3 facilitates acute myeloid leukemia cell growth via the regulation of miR-758-3p/SRGN axis. *Journal of Cellular Biochemistry*, 121(2), 1023–1031. Available from https://doi.org/10.1002/jcb.29336.

Pouyanrad, S., Rahgozar, S., & Ghodousi, E. S. (2019). Dysregulation of miR-335-3p, targeted by NEAT1 and MALAT1 long non-coding RNAs, is associated with poor prognosis in childhood acute lymphoblastic leukemia. *Gene*, 692, 35–43. Available from https://doi.org/10.1016/j.gene.2019.01.003.

Prasad, V. (2018). Tisagenlecleucel – The first approved CAR-T-cell therapy: Implications for payers and policy makers. *Nature Reviews Clinical Oncology*, 15(1), 11–12. Available from https://doi.org/10.1038/nrclinonc.2017.156.

Sahu, D., Hsu, C. L., Lin, C. C., Yang, T. W., Hsu, W. M., Ho, S. Y., ... Huang, H. C. (2016). Co-expression analysis identifies long noncoding RNA SNHG1 as a novel predictor for event-free survival in neuroblastoma. *Oncotarget*, 7(36), 58022–58037. Available from https://doi.org/10.18632/oncotarget.11158.

Sharma, U., Barwal, T. S., Acharya, V., Singh, K., Rana, M. K., Singh, S. K., ... Jain, A. (2020). Long non-coding RNAs as strategic molecules to augment the radiation therapy in esophageal squamous cell carcinoma. *International Journal of Molecular Sciences*, 21(18), 1–18. Available from https://doi.org/10.3390/ijms21186787.

Sharma, U., Barwal, T. S., Acharya, V., Tamang, S., Vasquez, K. M., & Jain, A. (2020). Cancer susceptibility candidate 9 (CASC9): A novel targetable long noncoding RNA in cancer treatment. *Translational Oncology*, 13(8). Available from https://doi.org/10.1016/j.tranon.2020.100774.

Sharma, U., Barwal, T. S., Khandelwal, A., Malhotra, A., Rana, M. K., Singh Rana, A. P., ... Jain, A. (2021). LncRNA ZFAS1 inhibits triple-negative breast cancer by targeting STAT3. *Biochimie*, 182, 99–107. Available from https://doi.org/10.1016/j.biochi.2020.12.026.

Sharma, U., Barwal, T. S., Malhotra, A., Pant, N., Vivek., Dey, D., ... Jain, A. (2020). Long non-coding RNA TINCR as potential biomarker and therapeutic target for cancer. *Life Sciences*, 257. Available from https://doi.org/10.1016/j.lfs.2020.118035.

Shi, J., Dai, R., Chen, Y., Guo, H., Han, Y., & Zhang, Y. (2019). LncRNA LINP1 regulates acute myeloid leukemia progression via HNF4α/AMPK/WNT5A signaling pathway. *Hematological Oncology*, 37(4), 474–482. Available from https://doi.org/10.1002/hon.2651.

Steliarova-Foucher, E., Colombet, M., Ries, L.A.G., Hesseling, P., Moreno, F., Shin, H.Y., & Stiller, C.A. (Eds.) (2017). International incidence of childhood cancer, Volume III (electronic version). Lyon, France: International Agency for Research on Cancer. Available from: http://iicc.iarc.fr/results/ Accessed 19.09.21.

Su, Z. D., Huang, Y., Zhang, Z. Y., Zhao, Y. W., Wang, D., Chen, W., ... Lin, H. (2018). ILoc-lncRNA: Predict the subcellular location of lncRNAs by incorporating octamer composition into general PseKNC. *Bioinformatics (Oxford, England)*, 34(24), 4196−4204. Available from https://doi.org/10.1093/bioinformatics/bty508.

Sun, L., Luo, H., Bu, D., Zhao, G., Yu, K., Zhang, C., ... Zhao, Y. (2013). Utilizing sequence intrinsic composition to classify protein-coding and long non-coding transcripts. *Nucleic Acids Research*, 41(17), e166. Available from https://doi.org/10.1093/nar/gkt646.

Sun, X. H., Yang, L. B., Geng, X. L., Wang, R., & Zhang, Z. C. (2015). Increased expression of lncRNA HULC indicates a poor prognosis and promotes cell metastasis in osteosarcoma. *International Journal of Clinical and Experimental Pathology*, 8(3), 2994−3000. <http://www.ijcep.com/files/ijcep0005023.pdf>.

Tamang, S., Acharya, V., Roy, D., Sharma, R., Aryaa, A., Sharma, U., ... Jain, A. (2019). SNHG12: An lncRNA as a potential therapeutic target and biomarker for human cancer. *Frontiers in Oncology*, 9. Available from https://doi.org/10.3389/fonc.2019.00901.

Tian, Y. J., Wang, Y. H., Xiao, A. J., Li, P. L., Guo, J., Wang, T. J., & Zhao, D. J. (2019). Long noncoding RNA SBF2-AS1 act as a ceRNA to modulate cell proliferation via binding with miR-188-5p in acute myeloid leukemia. *Artificial Cells, Nanomedicine and Biotechnology*, 47(1), 1730−1737. Available from https://doi.org/10.1080/21691401.2019.1608221.

Volders, P. J., Anckaert, J., Verheggen, K., Nuytens, J., Martens, L., Mestdagh, P., & Vandesompele, J. (2019). Lncipedia 5: Towards a reference set of human long non-coding RNAs. *Nucleic Acids Research*, 47(1), D135−D139. Available from https://doi.org/10.1093/nar/gky1031.

Volders, P. J., Helsens, K., Wang, X., Menten, B., Martens, L., Gevaert, K., ... Mestdagh, P. (2013). LNCipedia: A database for annotated human lncRNA transcript sequences and structures. *Nucleic Acids Research*, 41(1), D246−D251. Available from https://doi.org/10.1093/nar/gks915.

Wang, J., Wang, Z., Yao, W., Dong, K., Zheng, S., & Li, K. (2019). The association between lncRNA LINC01296 and the clinical characteristics in neuroblastoma. *Journal of Pediatric Surgery*, 54(12), 2589−2594. Available from https://doi.org/10.1016/j.jpedsurg.2019.08.032.

Wang, L., Park, H. J., Dasari, S., Wang, S., Kocher, J. P., & Li, W. (2013). CPAT: Coding-potential assessment tool using an alignment-free logistic regression model. *Nucleic Acids Research*, 41(6), e74. Available from https://doi.org/10.1093/nar/gkt006.

Wu, A., Zhou, X., Mi, L., & Shen, J. (2020). LINC00202 promotes retinoblastoma progression by regulating cell proliferation, apoptosis, and aerobic glycolysis through miR-204-5p/HMGCR axis. *Open Life Sciences*, 15(1), 437−448. Available from https://doi.org/10.1515/biol-2020-0047.

Xavier, A. C., Ge, Y., & Taub, J. W. (2009). Down syndrome and malignancies: A unique clinical relationship: a paper from the 2008 William Beaumont hospital symposium on molecular pathology. *The Journal of Molecular Diagnostics*, 11, 371−380. Available from https://doi.org/10.2353/jmoldx.2009.080132.

Xing, C. y, Hu, X. q, Xie, F. y, Yu, Z. j, Li, H. y, Bin-Zhou, Wu, J. b, ... Gao, S. m (2015). Long non-coding RNA HOTAIR modulates c-KIT expression through sponging miR-193a in acute myeloid leukemia. *FEBS Letters*, 589(15), 1981−1987. Available from https://doi.org/10.1016/j.febslet.2015.04.061.

Yan, G., Su, Y., Ma, Z., Yu, L., & Chen, N. (2019). Long noncoding RNA LINC00202 promotes tumor progression by sponging miR-3619-5p in retinoblastoma. *Cell Structure and Function*, 44(1), 51−60. Available from https://doi.org/10.1247/csf.18033.

Yang, H., Guo, J. F., Zhang, M. L., & Li, A. M. (2020). Lncrna snhg4 promotes neuroblastoma proliferation, migration, and invasion by sponging mir-377-3p. *Neoplasma*, 67(5), 1054−1062. Available from https://doi.org/10.4149/neo_2020_191023N1081.

Yang, H., Zhang, X., Zhao, Y., Sun, G. L., Zhang, J., Gao, Y., ... Zhu, H. (2020). Downregulation of lncRNA XIST represses tumor growth and boosts radiosensitivity of neuroblastoma via modulation of the miR-375/L1CAM axis. *Neurochemical Research*, 45(11), 2679−2690. Available from https://doi.org/10.1007/s11064-020-03117-9.

Zhang, F., Zeng, L., Cai, Q., Xu, Z., Liu, R., Zhong, H., ... Xin, H. (2020). Comprehensive analysis of a long non-coding RNA-associated competing endogenous RNA network in Wilms tumor. *Cancer Control: Journal of the Moffitt Cancer Center*, 27(2). Available from https://doi.org/10.1177/1073274820936991.

Zhang, L., Xu, H. G., & Lu, C. (2014). A novel long non-coding RNA T-ALL-R-LncR1 knockdown and Par-4 cooperate to induce cellular apoptosis in T-cell acute lymphoblastic leukemia cells. *Leukemia and Lymphoma*, 55(6), 1373−1382. Available from https://doi.org/10.3109/10428194.2013.829574.

Zhang, N., Liu, F. L., Ma, T. S., Zeng, Z. D., & Zhang, J. J. (2019). LncRNA SNHG1 contributes to tumorigenesis and mechanism by targeting MIR-338-3p to regulate PLK4 in human neuroblastoma. *European Review for Medical and Pharmacological Sciences, 23*(20), 8971–8983. Available from https://doi.org/10.26355/eurrev_201910_19296.

Zhang, Y., Fan, L. J., Zhang, Y., Jiang, J., & Qi, X. W. (2020). Long non-coding Wilms tumor 1 antisense RNA in the development and progression of malignant tumors. *Frontiers in Oncology, 10*. Available from https://doi.org/10.3389/fonc.2020.00035.

Zhao, X., Li, D., Huang, D., Song, H., Mei, H., Fang, E., ... Tong, Q. (2018). Risk-associated long noncoding RNA FOXD3-AS1 inhibits neuroblastoma progression by repressing PARP1-mediated activation of CTCF. *Molecular Therapy, 26*(3), 755–773. Available from https://doi.org/10.1016/j.ymthe.2017.12.017.

Zhao, Y., Yuan, J., & Chen, R. (2016). NONCODEv4: Annotation of noncoding RNAs with emphasis on long noncoding RNAs. In *Methods in molecular biology* (1402, pp. 243–254). Humana Press Inc. Available from https://doi.org/10.1007/978-1-4939-3378-5_19.

Zhong, W., Yang, J., Li, M., Li, L., & Li, A. (2019). Long noncoding RNA NEAT1 promotes the growth of human retinoblastoma cells via regulation of miR-204/CXCR4 axis. *Journal of Cellular Physiology, 234*(7), 11567–11576. Available from https://doi.org/10.1002/jcp.27812.

Zhou, Q., Chen, F., Zhao, J., Li, B., Liang, Y., Pan, W., ... Zheng, D. (2016). Long non-coding RNA PVT1 promotes osteosarcoma development by acting as a molecular sponge to regulate miR-195. *Oncotarget, 7*(50), 82620–82633. Available from https://doi.org/10.18632/oncotarget.13012.

Zhu, K., Wang, L., Zhang, X., Sun, H., Chen, T., Sun, C., ... Su, Y. (2020). LncRNA HCP5 promotes neuroblastoma proliferation by regulating miR-186-5p/MAP3K2 signal axis. *Journal of Pediatric Surgery, 56*. Available from https://doi.org/10.1016/j.jpedsurg.2020.10.011.

Zhu, S., Fu, W., Zhang, L., Fu, K., Hu, J., Jia, W., & Liu, G. (2018). LINC00473 antagonizes the tumour suppressor miR-195 to mediate the pathogenesis of Wilms tumour via IKKα. *Cell Proliferation, 51*(1). Available from https://doi.org/10.1111/cpr.12416.

Zhuang, M. F., Li, L. J., & Ma, J. B. (2019). LncRNA HOTTIP promotes proliferation and cell cycle progression of acute myeloid leukemia cells. *European Review for Medical and Pharmacological Sciences, 23*(7), 2908–2915. Available from https://doi.org/10.26355/eurrev_201904_17569.

Index

Note: Page numbers followed by "*f*" and "*t*" refer to figures and tables, respectively.

A

A-kinase anchor protein 12 (AKAP12), 332
ACA11, 113
Actin filament–associated protein 1 antisense RNA 1 (AFAP1-AS1), 291, 401–402
Activated in RCC with sunitinib resistance (ARSR), 254–255
Acute lymphocytic leukemia (ALL), 437–440
Acute myelocytic leukemia (AML), 437–441
Adeno-associated virus (AAV), 330, 368
α-fetoprotein (AFP), 360
Androgen receptor (AR), 242–243, 325
  signaling pathway, 325–326
  stability, 242–243
Anti-miRNA oligonucleotides (AMOs), 366
Antimetabolites, 308–309
Antisense ncRNA at INK4 locus (ANRIL), 146, 157, 159–160, 179, 401
Antisense oligonucleotides (ASOs), 45–46, 289, 330, 368–369
Apoptosis, 365–366
AR-regulated long ncRNA 1 (ARLNC1), 326
Argonaute (AGO), 298–299, 351–352, 410
Associated protein 1 antisense RNA1 (AFAP1-AS1), 180*t*, 291, 399*t*, 443
ATP-binding cassette A1 and G1 (ABCG1), 350
Atypical teratoid rhabdoid tumor (ATRT), 27, 41–42
Aubergine (Aub), 411
Autophagy, 114, 365
  of RB cells, 190
Autophagy-related 12 protein (ATG12), 39

B

196b, 4
Background adjustment, 70
β-catenin, 414
Biliary tract cancers (BTC), 397
Biochemical recurrence (BCR), 327–329
Bioinformatics
  for ncRNA, 3
    in CRC, 70–72
  tools, 352
Blood samples, 35, 148–151, 158–159, 402, 419
Blood–brain barrier (BBB), 29
Bone marrow endothelial cells (BMEC), 124
Bone marrow niche, ncRNAs interactions with, 124–125
Bone marrow plasma cells (BMPC), 99
Bone marrow stromal cells (BMSC), 109
BRAF (B-Raf proto-oncogene)-activated noncoding RNA (BANCR), 159–160, 179
Brain cancer
  circular RNA, 20–21
  ncRNAs
    aberrant expression profiles of, 30*f*
    advantages and disadvantages of, 33*t*
    data sets for analysis, 21–24
    diagnostic potential of, 35–37
    expression in, 24–30
    ncRNA interactions, 30–31
    in predicting chemoresistance and radioresistance, 40–45
    as predictive marker for, 35–39
    prognostic potential of, 37–39
    therapeutic potential and targeting of, 45–47
    types of, 19–20
    validations of RNA-seq, 31–34
  types of, 17–18
    GBM, 18
    MB, 18
Brain-derived neurotrophic factor antisense (BDNF-AS), 422
Breast cancer, 286
  experimental methods and tools for ncRNAs in, 291–292
  lncRNAs
    in breast cancer therapy, 289–290
    in diagnosis, 287
    in predicting breast cancer patient's response to therapeutics, 290
    in predicting chemoresistance and radioresistance, 290–291
    in prognosis of breast cancer, 288–289
Breast cancer antiestrogen resistance 4 (BCAR4), 288
Bronchoalveolar lavage fluid (BALF), 150–151

C

c-Myc pathway, 325
Cadherin 2 (CDH2), 252–253
Cancer stem cells (CSCs), 24, 81, 331
  ncRNA expression profiles in, 26–30
Cancer susceptibility 15 (CASC15-S), 418
Cancer susceptibility candidate 11 (*CASC11*), 74
Cancer-specific survival (CSS), 246
Cap analysis gene expression (CAGE), 68–69, 291, 317
Carcinoembryonic antigen (CEA), 76–77
CASC11.. See Cancer susceptibility candidate 11 (*CASC11*)
CASC15-New-Transcript 1 (CANT1), 443
Caspase cleavage cascade, 82
Castration-resistant prostate cancer (CRPC), 323–324
CCA
  cells, 357–358
  NCT03102268 in, 370
  prognostic miRNA and lncRNA signatures in, 364
  survival, 358–359
  tumors, 346
Cell signaling pathways
  modulated by ncRNAs in CRC, 77–81
    JAK/STAT signaling pathway, 79
    NF-κB (Nuclear Factor Kappa B) signaling pathway, 81
    Notch signaling pathway, 81
    p53 signaling pathway, 80–81
    PI3K/PTEN/AKT/mTOR signaling pathway, 79–80
    Ras/MAPK-signaling pathway, 80
    Wnt/β-catenin signaling pathway, 77–79
  modulated by ncRNAs in lung cancer, 153–155
  modulated by noncoding RNAs in MM, 119–124
    IL6/JAK/STAT signaling and, 120–121
    MAPK, 123–124
    NFκB signaling, 123–124
    p53 pathway, 121–123
    PI3K/AKT/mTOR, 123–124
  noncoding RNAs in gastric cancer, 304–306
Cell-free miRNA (cf-miR), 76–77
Cellular deficiencies, 97–98
Cerebrospinal fluid (CSF), 24
Chemokine signaling, 82
Chemoresistance. *See also* Radioresistance
  lncRNAs in breast cancer, 290–291
  in lung cancer, 162–165
  ncRNAs in, 40–45
    PCa, 331–332
  noncoding RNAs in renal cancer, 256–257
Chemotherapy resistance, 82–83
  in gastric cancer patients, 308–309
Childhood cancer. *See* Pediatric cancer
Chimeric antigen receptor T-cell therapy (CAR-T therapy), 434
Cholesterol accumulation, 351
Chromatin immunoprecipitation (ChIP), 69
Chromodomain helicase DNA-binding protein 7 (CHD7), 414
Chronic HCV with cirrhosis (CHC), 361–362
Circ-ABCB10, 253
circ-cut-like homeobox 1 (circ-CUX1), 416
CircNFIX, 29, 42–44
Circular intronic RNAs (ciRNAs), 212
Circular RNAs (circRNAs), 2–3, 19–21, 29, 76–77, 100, 108, 142, 177–178, 212, 297–298, 391, 411–412
  in brain cancer, 22t
  brain tumor chemo-and radioresistance, 43t
  diagnostic circular RNA markers of PDAC, 9
  diagnostic potential in brain cancers, 36t
  in gastric cancer, 299
  in lung cancer, 146–147
  in NB
    clinical significance, 419
    growth and development, 415–416
  in pancreatic cancer, 6–7, 8t
  in RB, 178–184
  in RCC
    circRNAs in predicting chemoresistance and radioresistance, 257
    prognostic potential of circRNAs, 249t, 252–253
    therapeutic potential of circRNAs, 255
  in renal cancer, 214–228
    cell signaling pathways modulated by circRNAs, 243
    datasets and informatics for analyzing circRNAs, 214–228
    expression of circRNAs, 233–240
    web-based resources and databases for circRNAs, 226t
Circulating free DNA (cfDNA), 76–77
Circulating tumor cells (CTCs), 76–77
Circulating tumor DNA (ctDNA), 76–77
Clear cell RCC (ccRCC), 211
CLIP-seq to AGO (AGO-CLIP), 31
Colon cancer–associated transcript 1 (*CCAT1*), 73
Colon cancer–associated transcript 2 (*CCAT2*), 73, 158–159
Colorectal cancer (CRC), 66
  cap analysis gene expression, 68–69
  caspase cleavage cascade, 82
  chemokine signaling, 82
  interleukin pathway, 82
  microarray, 67–68
  ncRNAs
    cell signaling pathways modulated by, 77–81

Index 451

clinical applications of, 82–83
dataset and bioinformatics for analyzing, 70–72
diagnostic potential of, 83–84
experimental methods and tools for analyzing, 66–67
expression of, 72–76
as predictive markers, 82–83
prognostic potential of, 84–85
therapeutic potential of, 85
RNA-seq, 69–70
SAGE, 68
Colorectal neoplasia differentially expressed (CRNDE), 74, 80, 118
Competing endogenous RNAs (ceRNAs), 21, 74, 119, 179, 225, 323, 393
changes in levels of, 108–109
databases, 229t
Complementary DNA (cDNA), 291
Computed tomography (CT), 243
Constitutive androstane receptor (CAR), 419–420
Cross-linking immunoprecipitation (CLIP), 31
Cross-linking immunoprecipitation sequencing (CLIP-seq), 214–225

D

Dark matter, 286–287
De novo methods for ncRNA predictions, 320
DEAD-Box Helicase 5 (DDX5), 414
Delta-like protein 3 (DLL3), 154–155
Design-free probe technique, 69
Diacylglycerol kinase beta (DGKB), 415–416
Diagnostic markers, 360
Diagnostic potential of ncRNAs in lung cancer, 158–159
Diagnostic value, 327
Differentiation antagonizing nonprotein-coding RNA (DANCR), 288, 358–359
DiGeorge syndrome, 410
DiGeorge Syndrome Critical Region 8 (DGCR8), 298
Digital droplet PCR (ddPCR), 32
Digital rectal examination (DRE), 315
Disease-free survival (DFS), 38, 76–77, 179, 246
Docetaxel (DTX), 329, 331
Drosophila model, 298–299

E

E74-like factor (ELF), 412–413
Endobronchial ultrasonography-guided transbronchial needle aspiration (EBUS-TBNA), 152
Endoscopic ultrasound (EUS), 2–3
Engineered circRNAs, 255
Enhancer of zeste homolog 2 (EZH2), 184, 358
Enhancer RNAs (eRNAs), 65–66

Enzalutamide (Enz), 329
Epidermal growth factor receptor (EGFR), 145, 323–324, 350
Epigenetics, 177
Epithelial cell adhesion molecule (EpCAM), 188
Epithelial–mesenchymal transition (EMT), 77–79, 145, 179, 225, 302–303, 325, 349, 394
Estrogen receptor (ER), 286
Eukaryotic translation initiation factor 4A3 (EIF4A3), 42
Ewing sarcoma (EWS), 416
EWS RNA-Binding Protein 1 (EWSR1), 416
Expected value (*E*-value), 319–320
Extracellular vesicles (EVs), 29–30

F

Fascin actin–bundling protein 1 (*FSCN1*), 85
Fine needle aspiration (FNA), 2–3
5-florouracil (5-FU), 308
Fluorescence in situ hybridization (FISH), 31–34, 318
Forkhead box P2 (FOXP2), 402
FOXF1 adjacent noncoding developmental regulatory RNA (FENDRR), 402
Free-to-total ratio (f/t), 327

G

Gallbladder cancer
  CCAT1 in, 73
  long noncoding RNAs in, 399t
  microRNAs in, 395t
  noncoding RNAs in, 392f
Gallbladder carcinoma (GBC), 391
  clinical utility of ncRNAs in, 403
  lncRNAs, 397–402
    biogenesis of, 398
    biological functions of, 398
    as biomarker, 401–402
    expression of, 399t
    in pathogenesis and as therapeutic target, 398–401
  miRNAs, 393–397
    biogenesis of, 393
    biological functions of, 393
    as biomarkers, 396–397
    expression of, 395t
    in pathogenesis, 393–396
  piRNAs in, 402
Gastric cancer, 297–298
  circular RNAs, 299
  long noncoding RNAs, 299
  microRNAs, 298
  noncoding RNAs
    cell signaling pathways modulated by, 304–306

Gastric cancer (*Continued*)
  experimental methods and tools for analyzing, 300−302
  expression of, 302−303
  in predicting chemotherapy resistance and radiotherapy resistance, 308−309
  as prognostic and predictive marker, 306−307
  sample types used for analyzing, 303−304
  piwiRNAs, 298−299
  small noncoding RNAs in, 307−308
GBM stem cells (GSCs), 25
Gemcitabine (GEM), 255
GENCODE software, 301
Gene Expression Omnibus database (GEO database), 68, 402
Genetically engineered mouse model (GEMM), 25
Genome-wide association studies, 414
Genomic Data Commons, 24
Genomic studies, 316, 351
Glioblastoma multiforme (GBM), 17−18
GLOBOCAN study, 297
Glucocorticoid receptor (GR), 75−76
Goosecoid (GSC), 350−351
Growth arrest-specific transcript 5 (*GAS5*), 4−6, 75−76, 158−159, 325−326, 419−420, 422
  downregulation, 325−326
  lncRNA, 81, 159−160

H

Hazard ratio (HR), 247
Heat shock protein 27 (HSP27), 82−83
Helicobacter pylori, 297
Hepatic stellate cells (HSC), 345−346
Hepatitis B virus (HBV), 346
  HBV-related HCCs, 361
Hepatitis C virus (HCV), 346
  HCV-related HCCs, 361−362
Hepatoblastoma (HB), 347
Hepatocellular carcinoma (HCC), 346
  HBV-related, 361
  HCV-related, 361−362
  prognostic lncRNA signatures in, 364
  prognostic miRNA signatures in, 363−364
  related to metabolic disorders, 362
  signaling pathways
    in long noncoding RNAs, 359−360
    in microRNAs, 357−358
Hepatocyte, 345
  differentiation, 350
  functions, 345
  necrosis/proliferation, 345−346
Heterogeneous nuclear ribonucleoprotein K (hnRNPK), 414

High-grade gliomas (HGG), 35
High-throughput RNA-sequencing technique, 142
High-throughput sequencing of RNA isolated by CLIP (HITS-CLIP), 31
High-throughput technologies, 65−66
Highly upregulated in liver cancer (HULC), 358−359
Homology-based methods, 319−320
HOX transcript antisense intergenic RNA (*HOTAIR*), 72, 146, 225, 288, 358−360
HOXA transcript at distal tip (HOTTIP), 225
HOXD cluster antisense RNA 1 (HOXD-AS1), 417−418
hsa_-circ_001653, 9
hsa_circRNA_101237, 118
Human antigen R (HuR), 415−416
Human epidermal growth factor receptor 2 (HER2), 286
Human urothelial carcinoma-associated 1 (*UCA1*), 73
Hypoxia-inducible factor (HIF), 190
Hypoxia-inducible factor 1 subunit alpha (HIF-1α), 225

I

Immunoglobulin heavy chain (IgH), 98
Immunoprecipitation (IP), 316
  ncRNA−protein interactions by, 30−31
  RNA, 253
Immunotherapy, 254, 434
In situ hybridization (ISH), 316, 318
Individual-nucleotide resolution CLIP (iCLIP), 31
Insulin-like growth factor-1 receptor, 184
Insulin-like growth factor-2 (IGF-2), 225
Interferon regulatory factor 4 (IRF4), 112−113
Interleukins (ILs), 82
  IL6/JAK/STAT signaling, 120−121
  IL-10, 345
  pathway, 82
International Intraocular Retinoblastoma Classification (IIRC), 178−179
International Staging Scheme (ISS), 126
Intraductal papillary mucinous neoplasm (IPMN), 2−3
Intrahepatic cholangiocarcinoma (iCCA), 346−347
Intraocular tumor patients
  RB, 178−190
  UM, 190−200

J

JAK (Janus Kinase)/STAT (Signal Transducer and Activator of Transcription) signaling pathway, 79
Janus Kinase 3 (JAK3), 415
Junk DNA, 1−2

K

Karnofsky Performance Status (KPS), 38
Kupffer cells (KC), 345

L

Lethal-7 (let-7), 161–162
Leukemia, 437–441
LINC00473, 443–444
Linc00963, 323–324
Liquid biopsies, 304
Liver cancer, 346–348
  in adults, 346–347
  in children, 347–348
  expression of ncRNAs in, 355–360
    long noncoding RNAs, 358–360
    microRNAs, 355–358
  microRNA in drug resistance, 365–366
  ncRNA
    as potential diagnostic tools, 360–363
    as predictive markers, 364–365
  ncRNA detection, 351–355
    bioinformatics tools, 352
    methods of ncRNA analyses, 352–355
    techniques of detection, 353–355
    tissular vs. circulating ncRNAs, 351–352
  ncRNA in liver development and functions, 348–351
    lncRNAs, 350–351
    microRNAs, 348–350
  prognostic potential of noncoding RNAs in, 363–364
  therapeutic potential of noncoding RNAs in, 366–369
Liver cancer stem cells (LCSC), 357–358
Liver functions, 345–346
Lnc-SCYL1–1. See Metastasis-associated lung adenocarcinoma transcript 1 (MALAT1)
LNCipedia, 435–437
LncRNA activated by TGF-beta (lncRNA-ATB), 288
LOC285194, 159–160
Locked nucleic acid (LNA), 289–290, 354–355, 420–422
Long intergenic ncRNA for kinase activation (LINK-A), 289–290
Long intergenic noncoding (linc), 413
Long intergenic noncoding RNAs (lincRNA), 107–108
Long interspersed nuclear element-1 (LINE-1), 82–83
Long noncoding RNAs (lncRNAs), 2–3, 19–20, 65–66, 99–100, 142, 177–178, 212, 286–287, 297–298, 317, 347–348, 391, 393, 410–411, 434–435
  bioinformatics resources, 71t
  in brain cancer, 22t
  brain tumor chemo-and radioresistance, 43t
  in breast cancer therapy, 289–290
  datasets profiling expression in MM, 105t, 107–108
  in diagnosis of breast cancer, 287
  diagnostic lncRNA markers of PDAC, 7–9
  as diagnostic markers, 158–159
  diagnostic potential in brain cancers, 36t
  expression in liver cancer, 358–360
    deregulated lncRNAs, 359–360
    signaling pathways in HCC, 359–360
    signatures in, 358–359
  in gastric cancer, 299
  in GBC, 397–402
  in liver development and functions, 350–351
  in lung cancer, 146
  in MM, 113–118, 115t
    CRNDE, 118
    MALAT1, 114
    MEG3, 117
    NEAT1, 114–117
    PVT1, 117
  in NB
    clinical significance of, 417–418
    growth and development, 413–414
  in pancreatic cancer, 4–6
  in pediatric cancer, 434–435, 438t
    experimental and bioinformatics tools for, 435–437
    leukemia, 437–441
    NB, 441–442
    osteosarcoma, 442
    retinoblastoma, 443
    Wilms tumor, 443–444
  in predicting breast cancer patient's response to therapeutics, 290
  in predicting chemoresistance and radioresistance in breast cancer patients, 290–291
  in prognosis of breast cancer, 288–289
  as prognostic biomarkers, 159–160
  in RB, 178–184, 180t, 183f
  in RCC
    in chemoresistance and radioresistance, 257
    prognostic potential of, 248–252, 251t
  in renal cancer, 214, 215f
    cell signaling pathways modulated by lncRNAs, 242–243
    datasets and informatics for analyzing lncRNAs, 214
    expression of lncRNAs, 228–233
    web-based databases for lncRNAs, 219t
  signatures in multiple myeloma, 127t
  therapeutic potential of, 161
  in UM, 191–193, 192t
Low-grade gliomas (LGG), 35
Luciferase assays, 30–31
Lung adenocarcinoma (LUAD), 143–144
Lung cancer, 142
  circular RNA, 146–147
  long noncoding RNA in, 146
  micro-RNA in, 145

Lung cancer (*Continued*)
  natural antisense transcripts, 147
  ncRNAs
    cell signaling pathways modulated by, 153–155
    datasets and informatics for analyzing, 143–144
    diagnostic potential of, 158–159
    experimental methods and tools for analyzing, 143
    expression of, 144–148
    in predicting chemoresistance and radioresistance, 162–165
    as predictive markers, 155–157
    prognostic potential of, 159–160
    therapeutic potential of, 161–162
  PIWI-interacting RNA, 147
  sample types used for analyzing ncRNAs, 148–153
    BALF, 150–151
    blood, 148–150
    sputum, 151–152
    tumor biopsies, 152–153
  small nucleolar RNA, 147
  T-UCR, 148
  tRNA-derived small RNA, 146
Lymphoid enhancer-binding factor 1 (LEF1), 412–413

M

Magnetic resonance imaging (MRI), 243
Mammalian target of rapamycin pathway (mTOR pathway), 323–324
MAPK phosphatase-1 (MKP-1), 80
Mapping-first algorithms, 69
Maternally expressed gene 3 (MEG3), 117, 157, 351, 413
Matrix metallopeptidases (MMPs), 197
Matrix metalloproteinases 2 (MMP2), 38
Matrix metalloproteinases 9 (MMP9), 38
MB stem cells (MBSCs), 25
MB with extensive nodularity (MBEN), 18
Medulloblastoma (MB), 17–18
Mesenchymal stem cells (MSC), 117
Messenger RNAs (mRNAs), 2–3, 19, 66, 100–107, 142, 287, 297–298, 316, 348, 393, 410, 434–435
Metaanalysis, 39, 148–149, 248
Metabolic disorders, HCC related to, 362
Metastasis-associated lung adenocarcinoma transcript 1 (*MALAT1*), 71–72, 114, 146, 157, 191, 288, 398
Metastatic CRPC (mCRPC), 329
Metastatic RCC-associated transcript 1 (MRCCAT1), 228
Methyltransferase like 3 (METTL3), 298
MGMT. *See* O-methylguanine-DNA methyltransferase (MGMT)
Microarray, 67–68, 300, 317

MicroRNAs (miRNAs), 2–3, 19, 65–66, 99–100, 142, 177–178, 212, 286–287, 297–298, 316, 347–348, 391, 410, 434–435
  biogenesis and mechanism of action, 213f
  in brain cancer, 22t
  in chemoresistance, 164
  as diagnostic markers, 159
  diagnostic microRNA markers of PDAC, 7
  diagnostic potential in brain cancers, 36t
  in drug resistance, 365–366
  in etiology of MM, 108–113
  expression in liver cancer, 355–358
    deregulated miRNAs, 357–358
    signaling pathways in HCC, 357–358
    signatures in, 356–357
  in gastric cancer, 298
  in GBC, 393–397
  in liver development and functions, 348–350
  in lung cancer, 145
  in NB
    clinical significance, 416–417
    growth and development, 412–413
  in pancreatic cancer, 4, 5t
  as prognostic biomarkers, 160
  in radioresistance, 165
  in RB, 184–189
    carcinogenic action, 187t
    dysregulated expression, 184–188
    miRNAs as potential biomarkers, 188–189
    tumor suppressor action, 185t
  in renal cancer, 213
    cell signaling pathways modulated by microRNAs, 241–242
    datasets and informatics for analyzing microRNAs, 213
    expression of microRNAs, 228
    gene targets databases, 216t
  sponges, 108–109, 225
  therapeutic potential of, 161–162
  in UM, 194–199
    carcinogenic action, 196t
    dysregulated expression of, 194–198
    miRNAs as potential biomarkers, 198–199
    tumor suppressor action, 194t
miR-15a/16 cluster, 113
miR-17~92 cluster, 112
miR-21, 4, 82–83, 109
miR-29a, 41
miR-34a, 197, 331
miR-34a-5p, 361
miR-101, 40, 358
miR-106b-93–25 cluster, 323

miR-106b~25 cluster, 109–111
miR-122, 364–365
miR-125 family, 112–113
miR-129, 82–83
miR-141, 244, 308
miR-155, 4, 397
miR-155–5p, 309
miR-181a/b, 109–111
miR-183, 154–155
miR196s, 4
miR-200 family, 331–332
miR-200c, 244
miR-210, 241, 246
miR217, 7
miR-221, 358
miR-301a, 248
miR-302c, 40–41
miR-320, 41
miR-432, 156–157
MiR-590–3p, 364
miR-622, 85–86
Mitogen-activated protein kinase pathway (MAPK pathway), 39, 123–124, 324–325, 394
Monoclonal gammopathy of undetermined significance (MGUS), 97
Monosomy 3 (M3), 191
mRNA response elements (MRE), 19
Multidrug resistance protein 1 (MDR1), 40–41, 291
Multiple myeloma (MM), 97
  datasets analyzing, 100–108
    of circular noncoding RNA in, 105t
    datasets profiling expression of long noncoding RNAs, 105t, 107–108
    datasets profiling expression of short noncoding RNAs, 100–107, 101t
  long noncoding RNAs in, 113–118
  ncRNAs
    affecting interactions with bone marrow niche, 124–125
    cell signaling pathways modulated by, 119–124
    as diagnostic and prognostic biomarkers, 125–130
    implicated in etiology of multiple myeloma, 108–119
    interactions between, 119
    samples and experimental methods for analysis of, 99–100
    therapeutic potential of, 130–131
MYC-associated zinc finger protein (MAZ), 416
Myeloid leukemia cell differentiation protein 1 (Mcl-1), 358
Myocardial infarction–associated transcript (MIAT), 413

N

N-cadherin. See Cadherin 2 (CDH2)
National Cancer Institute (NCI), 24
Natural antisense transcripts (NATs), 142
Natural antisense transcripts, 147
NB highly expressed 1 (NHEG1), 414
NB-associated transcript-1 (NBAT-1), 417–418
Nemo-like kinase (*NLK*), 412–413
Nephroblastoma. See Wilms tumor
Neuroblastoma (NB), 409–410, 433, 441–442
  clinical significance of noncoding RNAs in, 416–419
    circular RNAs, 419
    lncRNAs, 417–418
    microRNAs, 416–417
    piwi-interacting RNAs, 418–419
  noncoding RNAs in growth and development, 412–416
    circular RNAs, 415–416
    LncRNAs, 413–414
    MicroRNAs, 412–413
    Piwi-interacting RNAs, 415
  noncoding RNAs
    targeting strategies for, 420–422
    therapeutic potential of, 419–420
Neuroblastoma Ras (NRAS), 29
Neuroblasts, 409–410
Next-generation sequencing (NGS), 2–3, 24, 66–67, 291, 300, 354
NF-κB (Nuclear Factor Kappa B) signaling pathway, 81, 123–124
Nitric oxide (NO), 345
NK3 homeobox 1 gene (NKX3.1), 325
Nonalcoholic fatty liver disease (NAFLD), 345–346
Nonalcoholic steatohepatitis (NASH), 345–346
NONCODE, 435–437
Noncoding RNA activated by DNA damage (NORAD), 418
Noncoding RNAs (ncRNAs), 1–4, 19, 65–66, 98–99, 142, 177, 212, 286–287, 315–316, 347–348, 391. See also Long noncoding RNAs (lncRNAs)
  in brain cancer, 19–20
    data sets for analysis, 21–24
    expression of ncRNAs, 24–30
  classification of, 4f
  CRC in
    cell signaling pathways modulated by ncRNAs, 77–81
    clinical applications, 82–83
    dataset and bioinformatics for analyzing ncRNAs, 70–72
    experimental methods and tools for analyzing ncRNAs, 66–67
    expression of ncRNAs, 72–76

Noncoding RNAs (ncRNAs) (*Continued*)
    ncRNAs as predictive markers, 82–83
    diagnostic circular RNA markers of PDAC, 9
    diagnostic lncRNA markers of PDAC, 7–9
    diagnostic microRNA markers of PDAC, 7
    experimental methods and tools for breast cancer, 291–292
    in gastric cancer
        cell signaling pathways modulated by ncRNAs, 304–306
        experimental methods and tools for analyzing ncRNAs, 300–302
        expression of ncRNAs, 302–303
        ncRNAs as prognostic and predictive marker for, 306–307
        ncRNAs in predicting chemotherapy resistance and radiotherapy resistance, 308–309
        sample types used for analyzing ncRNAs, 303–304
    implicated in etiology of multiple myeloma, 108–119
    interactions with bone marrow niche, 124–125
    in liver development and functions, 348–351
    in lung cancer, 142
        cell signaling pathways modulated by ncRNAs, 153–155
        datasets and informatics for analyzing ncRNAs, 143–144
        diagnostic potential of ncRNAs, 158–159
        experimental methods and tools for analyzing ncRNAs, 143
        expression of ncRNAs, 144–148
        ncRNAs as predictive markers, 155–157
        ncRNAs in predicting chemoresistance and radioresistance, 162–165
        prognostic potential of ncRNAs, 159–160
        therapeutic potential of ncRNAs, 161–162
    in NB growth and development, 412–416
    PCa in
        cell signaling pathways modulated by ncRNAs, 322–326
        datasets and informatics for analyzing ncRNAs, 319–321
        experimental methods and tools for analyzing ncRNAs, 317–319
        investigation of ncRNA interactions, 318–319
        ncRNAs as biomarkers, 326–329
        ncRNAs in predicting chemo-resistance and radioresistance, 331–332
        sample types used for analyzing ncRNAs, 321–322
        secondary structures of ncRNAs, 319
        therapeutic potential of ncRNAs, 330–331
    in PCs patients, 2–3
    bioinformatics for, 3
    circular RNAs, 6–7
    lncRNAs, 4–6
    microRNAs, 4
    in renal cancer
        cell signaling pathways modulated by ncRNAs, 240–243
        datasets and informatics for analyzing ncRNAs, 213–228
        diagnostic potential of ncRNAs, 243–245
        expression of ncRNAs, 228–240
        prognostic potential of ncRNAs, 246–253
    sample types used for analyzing, 76–77
    samples and experimental methods for analysis in MM, 99–100
    types of, 410–412
Non–small cell lung carcinoma (NSCLC), 142
Normalization, 70
Northern blot (NB), 317–318
Notch signaling pathway, 81
Nuclear Export protein 5, 298
Nuclear paraspeckle assembly transcript 1, 114–117
Nuclear-enriched abundant transcript 1 (NEAT1), 290
Nucleotides (nt), 19, 65–66, 348

O

O-methylguanine-DNA methyltransferase (MGMT), 18
"Off-target" effect, 46
Olaparib, 329
Oncogenic circRNAs. *See* Tumor promoters
OncomiR Cancer Database (OMCD), 144
OncomiRs, 410
Organoids, 26
Osteogenic carcinoma. *See* Osteosarcoma
Osteosarcoma, 442
Overall survival (OS), 18, 74, 98, 178–179, 225, 288, 346

P

P-element-induced wimpy testis-interacting RNAs (Piwi-interacting RNAs), 142, 298–299, 391, 411, 434–435
    biogenesis, 298–299
    in gastric cancer, 298–299
    in GBC, 402
    in lung cancer, 147
    in NB
        clinical significance, 418–419
        growth and development, 415
p300-CBP-associated factor (PCAF), 109
p53 signaling pathway, 80–81, 121–123
Paclitaxel, 309
Pancreatic cancers (PCs), 1–2
    circular RNAs in, 6–7

lncRNAs in, 4–6
microRNAs in, 4
ncRNA in, 2–3
  bioinformatics for analyzing, 3
Pancreatic ductal adenocarcinoma (PDAC), 1–2
  diagnostic circular RNA markers of, 9
  diagnostic lncRNA markers of, 7–9
  diagnostic microRNA markers of, 7
Papillary RCC (pRCC), 211
Patient-derived xenograft (PDX), 368–369
PCa gene expression marker 1 (PCGEM1), 325
Pediatric cancer, 433–434
  in long noncoding RNA, 434–435, 438t
    experimental and bioinformatics tools for studying lncRNAs, 435–437
    leukemia, 437–441
    neuroblastoma, 441–442
    osteosarcoma, 442
    retinoblastoma, 443
    Wilms tumor, 443–444
Percutaneous biopsy, 152
Phosphatase and tensin homolog (PTEN), 109, 188, 323–324, 358
Phosphoinositide-3-kinase (PI3K), 303
  PI3K/AKT signaling pathway, 305, 323–324
  PI3K/AKT/mTOR, 123–124
  PI3K/PTEN/AKT/mTOR signaling pathway, 79–80
Phospholipase C gamma 2 (PLAGL2), 412–413
Photoactivable–ribonucleoside-enhanced-CLIP (PAR-CLIP), 31
Ping-pong pathway of piRNA biogenesis, 298–299
piRNA. See P-element-induced wimpy testis-interacting RNAs (Piwi-interacting RNAs)
piRNA-induced silencing complexes (pi-RISCs), 402
Plasmacytoma variant translocation 1 (PVT1), 117, 146, 179, 358–359
Platinum compounds, 308
Poly ADP-ribose polymerase 1 (PARP1), 329
Polycomb repressive complex 2 (PRC2), 72
Polymerase chain reaction (PCR), 300
Precursor-miRNA (pre-miRNA), 393, 410
Predictive value, 328–329
Primary liver cancer, 346
Primary pathway of piRNA biogenesis, 298–299
Primary transcript of mature miRNA (pri-miRNA), 410
Processing bodies, 348
Progesterone receptor (PR), 286
Prognosis-associated gallbladder cancer (PAGBC), 398–401
Prognostic potential of ncRNAs in lung cancer, 159–160
Prognostic value, 327–328
Progression-free survival (PFS), 38, 246–247

Prostate cancer (PCa), ncRNAs in, 315–316
  cell signaling pathways modulated by ncRNAs, 322–326
  datasets and informatics for analyzing ncRNAs, 319–321
  experimental methods and tools for analyzing ncRNAs, 317–319
  investigation of ncRNA interactions, 318–319
  ncRNAs as biomarkers, 326–329
  in predicting chemo-resistance and radioresistance, 331–332
  sample types used for analyzing ncRNAs, 321–322
  secondary structures of ncRNAs, 319
  therapeutic potential of ncRNAs, 330–331
Prostate cancer-associated transcript 1 (PCAT1), 74, 119
Prostate-specific antigen (PSA), 315
Proteasome inhibitors (PIs), 98
Pseudopodium-enriched atypical kinase 1 (PEAK1), 154–155

Q

Quantitative reverse transcriptase–polymerase chain reaction (qRT-PCR), 245, 316, 354
Quartz-Seq, 69

R

Radiation therapy (RT), 327–328
Radical prostatectomy (RP), 327–328
Radioresistance. See also Chemoresistance
  lncRNAs in breast cancer, 290–291
  in lung cancer, 162–165
  ncRNAs in, 40–45
    PCa, 331–332
  noncoding RNAs in renal cancer, 256–257
Radiotherapy resistance, noncoding RNAs in predicting, 308–309
Ras signaling, 305
Ras/MAPK-signaling pathway, 80
Real-time quantitative polymerase chain reaction (RTqPCR), 143
Rearranged during transfection (RET), 332
Receiver operating characteristic curve (ROC curve), 245
Recurrence-free survival (RFS), 247, 288
Renal cancer
  cell signaling pathways modulated by ncRNAs in, 240–243
    circular RNAs, 243
    long noncoding RNAs, 242–243
    microRNAs, 241–242
  ncRNAs in predicting chemoresistance and radioresistance in, 256–257
  datasets and informatics

Renal cancer (*Continued*)
    for analyzing circular RNAs, 214–228
    for analyzing lncRNAs, 214
    for analyzing microRNAs, 213
  diagnostic potential of ncRNAs in, 243–245
    circular RNAs, 245
    long noncoding RNAs, 245
    microRNAs, 243–244, 244*t*
  expression of ncRNAs, 228–240
    circular RNAs, 233–240
    long noncoding RNAs, 228–233
    microRNAs, 228
  prognostic potential of ncRNAs in, 246–253
    circular RNAs, 252–253
    long noncoding RNAs, 248–252
    miRNAs, 246–248
  therapeutic potential of ncRNAs in, 254–255
    circular RNAs, 255
    long noncoding RNAs, 254–255
    microRNAs, 254
Renal cell carcinoma (RCC), 211
  downregulated lncRNAs in, 240*t*
  downregulated microRNAs in, 234*t*
  upregulated lncRNAs in, 239*t*
  upregulated microRNAs in, 230*t*
Retinoblastoma (RB), 75, 177–190, 443
  circRNA in, 178–184
  lncRNAs in, 178–184
  microRNAs in, 184–189
  therapeutic potential of ncRNAs in, 189–190
Reverse transcription by qPCR (RT-qPCR), 31–32, 291, 318
Ribosomal protein L35 (RPL35), 414
Risk-scoring systems, 363
RNA immunoprecipitation (RIP), 318–319
RNA interference (RNAi), 45–46
RNA polymerase II (RNA Pol II), 398
RNA sequencing (RNA-seq), 68–70, 286–287, 316–317
  deposits, 24
  validations of, 31–34
RNA-based biomarkers discovery in prostate cancer, 322
RNA-binding proteins (RBPs), 20–21, 143, 255
RNA-inducing silencing complex (RISC), 393, 410
RUNX family transcription factor 3 (RUNX3), 308

S
SCARNA22. *See* ACA11
Secondary plasma cell leukemia (sPCL), 100–107
Sequence homology, 301
Sequencing, 300
Serial analysis of gene expression (SAGE), 68, 291, 317
Serine and arginine rich splicing factor 1 (SRSF1), 359–360

Short interfering RNA (siRNA), 27–28
Short noncoding RNAs (sncRNAs), 100
  datasets profiling expression in MM, 100–107, 101*t*
  in MM, 110*t*
Small Cajal body-specific RNAs (scaRNAs), 100–107
Small hairpin RNAs (shRNAs), 143, 331
Small interfering RNAs (siRNAs), 65–66, 212, 289, 331, 434–435
Small ncRNAs, 65–66
  in gastric cancer, 307–308
Small nuclear host gene 1 (*SNHG1*), 71–72
Small nuclear RNAs (snRNAs), 348
Small nucleolar RNA host gene 16 (SNHG16), 4–6, 419–420
Small nucleolar RNAs (snoRNAs), 100–107, 142, 348, 434–435
  in lung cancer, 147
Small-cell lung carcinoma (SCLC), 142
Small-interfering RNA (siRNA), 368–369
Smoldering multiple myeloma (SMM), 97–98
SNORA42, 147
Sonic hedgehog (SHH), 18
Sorafenib resistance-associated lncRNA in RCC (SRLR), 254–255
SOX2OT, 4–6
Specificity protein 1 (SP1), 146, 398–401
Sponge RNAs. *See* Competitive endogenous RNAs (ceRNAs)
Sprouty4-intron 1 (SPRY4-IT1), 288–289
Sputum, 151–152
Squamous-cell carcinoma (SCC), 143–144
Stem-cell biology, 25
Steroid response–binding protein 1c (SREBP-1c), 351
Structural-based homology, 320
Summarization, 70

T
T-cell factor (TCF), 414
Taurine-upregulated gene 1 (TUG1), 125–126
TDRD proteins, 298–299
The Cancer Genome Atlas (TCGA), 24, 70, 143–144
Therapeutic potential of ncRNAs in lung cancer, 161–162
6-thioguanosine (6SG), 31
4-thiouridine (4SU), 31
TILING array, 292
Toxicity, 46
Transcribed ultraconserved region (T-UCR), 142, 148
Transcriptional noise, 286–287
Transforming growth factor-β (TGF-β), 345
Transrectal ultrasound (TRUS), 322
Transthoracic needle aspiration (TTNA).
  See Percutaneous biopsy

Trastuzumab, 309
Triple-negative breast cancer (TNBC), 286
tRNA-derived small RNA (tsRNA), 142, 146
Tumor biopsies, 152–153
Tumor necrosis factor receptor-associated factor-4 (TRAF4), 38
Tumor necrosis factor α (TNF-α), 345
Tumor promoters, 21
Tumor suppressors, 21
Tumor suppressors. *See* OncomiRs
Tumor-associated dendritic cells (TADCs), 82
Tumorigenicity, 287
Tyrosine kinase receptors (TKRs), 305

U

Ultraconserved element 338 (uc.338), 148
3′ untranslated regions (UTR), 393
Urothelial cancer–associated 1 (UCA1), 228, 287, 358–359
Uveal melanoma (UM), 177–178, 190–200
　long noncoding RNAs in, 191–193, 192t
　microRNAs in, 194–199

V

Vascular endothelial growth factor (VEGF), 124–125, 394
Von Hippel–Lindau gene (VHL gene), 228–233, 241
Vrenteno (Vret), 298–299

W

Wilms tumor, 443–444
Wingless (WNT), 18
　integration site pathway, 305–306
　Wnt/β-catenin signaling pathway, 77–79
World Health Organization (WHO), 17

X

X-inactive specific transcript (XIST), 290

Y

Yes-associated protein 1 (YAP1), 85

Z

Zucchini (Zuc), 298–299

Printed in the United States
by Baker & Taylor Publisher Services